NATO ASI Series

Advanced Science Institutes Series

A series presenting the results of activities sponsored by the NATO Science Committee, which aims at the dissemination of advanced scientific and technological knowledge, with a view to strengthening links between scientific communities.

The Series is published by an international board of publishers in conjunction with the NATO Scientific Affairs Division

A **Life Sciences** B **Physics**	Plenum Publishing Corporation London and New York
C **Mathematical and Physical Sciences** D **Behavioural and Social Sciences** E **Applied Sciences**	Kluwer Academic Publishers Dordrecht, Boston and London
F **Computer and Systems Sciences** G **Ecological Sciences** H **Cell Biology** I **Global Environmental Change**	Springer-Verlag Berlin Heidelberg New York London Paris Tokyo Hong Kong Barcelona Budapest

PARTNERSHIP SUB-SERIES

1. **Disarmament Technologies**	Kluwer Academic Publishers
2. **Environment**	Springer-Verlag/Kluwer Acad. Publishers
3. **High Technology**	Kluwer Academic Publishers
4. **Science and Technology Policy**	Kluwer Academic Publishers
5. **Computer Networking**	Kluwer Academic Publishers

The Partnership Sub-Series incorporates activities undertaken in collaboration with NATO's Cooperation Partners, the countries of the CIS and Central and Eastern Europe, in Priority Areas of concern to those countries.

NATO-PCO DATABASE

The electronic index to the NATO ASI Series provides full bibliographical references (with keywords and/or abstracts) to about 50000 contributions from international scientists published in all sections of the NATO ASI Series. Access to the NATO-PCO DATABASE compiled by the NATO Publication Coordination Office is possible in two ways:

- via online FILE 128 (NATO-PCO DATABASE) hosted by ESRIN, Via Galileo Galilei, I-00044 Frascati, Italy.
- via CD-ROM "NATO Science & Technology Disk" with user-friendly retrieval software in English, French and German (© WTV GmbH and DATAWARE Technologies Inc. 1992).

The CD-ROM can be ordered through any member of the Board of Publishers or through NATO-PCO, Overijse, Belgium.

Series I: Global Environmental Change, Vol. 43

Springer
Berlin
Heidelberg
New York
Barcelona
Budapest
Hong Kong
London
Milan
Paris
Santa Clara
Singapore
Tokyo

Chemical Exchange Between the Atmosphere and Polar Snow

Edited by

Eric W. Wolff
British Antarctic Survey
High Cross, Madingley Road
Cambridge CB3 0ET, UK

Roger C. Bales
Department of Hydrology and Water Resources
University of Arizona
Tucson, Arizona 85721, USA

Springer

Published in cooperation with NATO Scientific Affairs Division

Proceedings of the NATO Advanced Research Workshop "Processes of Chemical Exchange Between the Atmosphere and Polar Snow" held at Il Ciocco, Italy, March 19–23, 1995

Library of Congress Cataloging-in-Publication Data

Chemical exchange between the atmosphere and polar snow / edited by Eric W. Wolff, Roger C. Bales.
p. cm. -- (NATO ASI series. Series I, Global environmental change ; vol. 43)
"Proceedings of the NATO Advanced Research Workshop "Processes of Chemical Exchange Between the Atmosphere and Polar Snow," held at Il Ciocco, Italy, March 19-23, 1995"--T.p. verso.
"Published in cooperation with NATO Scientific Affairs Division."
Includes bibliographical references and index.

ISBN-13:978-3-642-64730-7 e-ISBN-13:978-3-642-61171-1
DOI: 10.1007/978-3-642-61171-1

1. Atmospheric chemistry--Polar regions--Congresses. 2. Snow--Polar regions--Congresses. I. Wolff, Eric W., 1957- . II. Bales, Roger C., 1951- . III. North Atlantic Treaty Organization. Scientific Affairs Division. IV. NATO Advanced Research Workshop "Processes of Chemical Exchange Between the Atmosphere and Polar Snow" (1995 : Il Ciocco, Italy) V. Series.
QC879.6.C45 1996
551.5'11'0911--dc20 96-21625
CIP

ISBN-13:978-3-642-64730-7 Springer-Verlag Berlin Heidelberg New York

Softcover reprint of the hardcover 1st edition 1996

Typesetting: Camera ready by the editors
Printed on acid-free paper
SPIN: 10475265 31/3137 - 5 4 3 2 1 0

PREFACE

The NATO Advanced Research Workshop (ARW) was held at Il Ciocco, Tuscany, March 19-23, 1995. The main goal of the workshop was to assess, improve, and summarise understanding of the processes that control the transfer of species between the atmosphere and the record that is finally seen in ice cores. This preface outlines the rationale for holding such a meeting now, and provides an introduction to this proceedings volume.

A previous NATO ARW, held in 1990, concentrated on processes occurring in seasonal snowpacks (Davies et al., 1991), while another, in 1993, looked at the biogeochemical information in polar ice cores (Delmas, 1995). The latter meeting in particular highlighted the problems of ice core interpretation that have been inadequately studied until now. With major deep ice drilling programmes in Greenland recently completed, and proposals in the pipeline for even grander projects in Antarctica, it is critical that the maximum information be obtained from the ice. At present, large amounts of chemical data cannot be understood properly due to a lack of understanding of the air to snow transfer, and of post-depositional modification.

Two international initiatives have started to address these problems. The International Commission for Snow and Ice convened a Snow Atmosphere Chemical Exchange Working Group, which agreed to organise a meeting on the topic. Simultaneously, the International Global Atmospheric Chemistry project of IGBP set up a Polar Atmospheric and Snow Chemistry group to co-ordinate research on polar atmospheric chemistry, and on air-snow transfer. The ARW was the result of efforts by these two groups.

The meeting came at a period when a number of important international experiments had been completed, including the ATM experiments at the deep drilling sites at Summit, Greenland, and other experiments in Antarctica. The meeting aimed to provide a forum where these experiments could be discussed, and the results placed in a process-oriented framework. One output from the meeting would be a statement to the ice core community of our present state of knowledge (Bales and Wolff, 1995). Ultimately, the hope is that a set of quantitative transfer functions can be provided that would allow ice concentrations to be translated to atmospheric concentrations under a range of climatic conditions.

The organisation of this book follows the organisation of the meeting. The first ten chapters form an introduction for those unfamiliar with the background to polar ice and air chemistry. The first three papers discuss ice core records, what they have shown, and some of the problems that have been recognised. For a fuller treatment, the reader is referred to the proceedings of a previous ARW (Delmas, 1995). The next five papers describe what is known about the atmospheric transport and chemical composition of the Arctic and Antarctic atmosphere. Knowledge of the present-day composition of the atmosphere in the polar regions is obviously a prerequisite for deciding what new species are worth measuring in ice cores, and for interpreting the records. Understanding of atmospheric transport is required in order to assess the source regions to the ice sheet, and therefore to elucidate the spatial representativeness of ice core data. These themes were extended by the next two papers, one describing the chemical reactions occurring in the polar troposphere, the other setting polar data in the context of global models of the atmosphere.

The next section of the book covers mainly the recent field data. These were roughly divided between species that are irreversibly-deposited (mainly aerosol species), and those that seem to be reversible (e.g. H_2O_2, HCHO, acidic gases). These data chapters are accompanied by two chapters giving a theoretical underpinning to the main processes at work. Further field data can be found in the extended abstracts printed at the end of this volume. These were chosen from a much larger number of poster papers presented at the workshop. In each case, they add data on further aspects of air-snow exchange that are not covered in detail in the invited chapters.

The theme of looking at individual processes continues in the remaining chapters of the book. Processes of uptake at ice surfaces have been well studied with stratospheric applications in mind, but less well studied in polar tropospheric conditions, although they are probably crucial for the reversible species. This topic is covered by papers covering laboratory and thermodynamic work. Very little is known about photochemistry in ice, although it may be of some importance in the top few metres of ice. Metamorphism is clearly very important in determining processes that occur near the surface of the ice. Both snow ventilation and blowing snow may significantly modulate chemical concentrations, as the two relevant chapters indicate, but far more work is needed to substantiate their significance. The meteorology on a small scale near the surface determines the input of material to the surface. For gases that are eventually enclosed in air bubbles, a different set of problems arise, and these are addressed in one chapter. Finally, one chapter highlights the fact that concentrations may alter even below the snow surface, and a number of processes can be involved.

Beyond the invited papers, a day was devoted to workshop sessions, where the problems of different types of species were extensively discussed and summarised. The three working group reports can be considered a summary of the state-of-the-art for this topic in 1995, and include recommendations for what future work is required. A brief conclusion for the ice core community follows these reports. These last four contributions should give a flavour of the meeting, and indicate the directions this subject is likely to take in the future.

The editors of this volume would like to thank all the participants at the ARW, who helped to make it a very fruitful and exciting meeting. Many delegates were probably surprised to find just how large and important the topic of this meeting had become, and it is clear that many problems must be solved urgently, to match the pace of ice core retrieval.

We thank all those who peer-reviewed the papers in this volume, and our colleagues at the British Antarctic Survey and the University of Arizona, who assisted with editing and preparation of this volume.

Finally, we acknowledge the financial support of NATO, which made the ARW possible, and the additional funds provided by NSF to allow further delegates to attend.

Eric W. Wolff
Roger C. Bales

References

Bales RC, Wolff EW (1995) Interpreting natural climate signals in ice cores. Eos 76: 477-483

Davies TD, Tranter M, Jones HG (eds) (1991) Seasonal snowpacks: Processes of compositional change. NATO ASI Series G28. Springer Berlin etc.

Delmas R (ed) (1995) Ice core studies of global biogeochemical cycles. NATO ASI Series I. Springer Berlin etc.

TABLE OF CONTENTS

THE RECORD OF AEROSOL DEPOSITED SPECIES IN ICE CORES, AND PROBLEMS OF INTERPRETATION

Eric W. Wolff
British Antarctic Survey
Natural Environment Research Council
High Cross
Madingley Road
Cambridge CB3 0ET
UK

1. Introduction

Ice cores have now become established as one of the primary archives of paleoclimatic information, covering timescales from seasonal up to 100,000 years or more. In the polar ice sheets, where there is little or no melting, snow layers build up year by year. Included in them are samples of the atmosphere: trace gases in air bubbles, particles and adsorbed gases, and the water molecules themselves. By drilling into the ice at suitable places, we can collect cores that give profiles of chemical content and physical properties of the ice. These are then used to infer the state of the atmosphere in the past.

For some species, this inversion to a relevant atmospheric variable is either fairly simple, or has been well-calibrated. For instance, for a non-reactive trace gas such as methane, the concentration in the air extracted from gas bubbles is assumed to be identical to the concentration in the atmosphere at the time the bubble was sealed (Raynaud et al. 1993). The oxygen isotopic content of water molecules in ice has long been assumed to represent the temperature at the time of snowfall; recent studies (Cuffey et al. 1995; Johnsen et al. 1995) have used borehole temperatures to calibrate this signal.

For particulate and reactive gas species, it has also until recently been assumed that there was a linear relationship between changes seen in the ice and changes in the atmosphere at the time of snowfall. This has offered the attractive proposition of learning about atmospheric changes representing some of the forcing factors for climate. For instance, there are papers that use ice core data to try to derive past volcanism (using sulphate or acidity measurements), sea ice position (marine ions), atmospheric circulation patterns (dust), marine biological productivity (methanesulphonic acid), stratospheric temperatures (nitrate), pollution (e.g. lead), oxidative

NATO ASI Series, Vol. I 43
Chemical Exchange Between the Atmosphere and Polar Snow
Edited by Eric W. Wolff and Roger C. Bales

capacity of the atmosphere (hydrogen peroxide), and many other parameters. However, there has been a growing realisation that depositional and post-depositional processes confuse the interpretation for many species, and this workshop is the culmination of that awareness.

This chapter, along with two others, aims to show the potential of ice cores to deliver information on paleoatmospheric chemistry. In each case, the chapters will give a flavour of the data that exist, and will explain some of the difficulties involved in interpreting the ice core profiles in terms of atmospheric changes. The processes causing the difficulties will be explored in later chapters. The present chapter concentrates on the chemical species that are delivered to the snow as aerosol (that is, as solid or liquid particles). The following chapters look at acidic gases, including nitric and organic acids (Legrand et al. 1996) and at other gases such as hydrogen peroxide and formaldehyde (Neftel 1996).

2. Aerosol sources, transport and deposition

We can distinguish a number of types of aerosol reaching the polar regions, derived from various sources.

a) Terrestrial material, derived from soil and desert dusts. In ice cores, this is investigated by looking at particle numbers or volume, or by using marker elements such as Al, Si, or Ca.

b) Marine aerosol, from sea spray. The two marker species are sodium and chloride. Both present some problems: sodium can have a significant terrestrial source in some cores, while chloride can be fractionated by reaction with acids, producing HCl which can be volatilised from the particle, so that chloride can be in excess or deficit compared to other marine ions.

c) Sulphur species. Methanesulphonic acid (MSA) comes from oxidation of dimethyl sulphide (DMS), produced by marine algae. Sulphate is also produced from this source, though it has additional sources from volcanism (particularly sporadic spikes after major eruptive volcanoes), sea salt aerosol, and in recent times pollution.

d) Pollution. Apart from sulphate, there have been major studies involving a range of heavy metals, such as lead (Pb). Work has also been carried out on organochlorine compounds, though these may not be deposited exclusively in the aerosol phase.

e) Exotic material. The possibility must exist to study sources such as pollen, diatoms, and various biological molecules. Most of these species are in relatively small concentrations, and there is only a little work in the literature. Cosmic debris is another potential, but probably minor, source to polar ice.

We can identify three factors that influence ice core concentrations for aerosol species. The first one is the source strength (which is often what we hope to derive from an ice core). The second factor is the efficiency of transport from the source to the atmosphere above the ice core site, and the final one is the efficiency of deposition from the free troposphere to the snow surface. All these factors can potentially have a large effect on the concentrations seen in ice cores. In general, post-depositional changes are less important for aerosol species than for volatile gases, although there is some evidence for limited movement of aerosol species within firn or deeper ice (Wolff 1996), and sublimation of water vapour can affect snow concentrations at sites with low snow accumulation rates.

In the absence of changes in transport or deposition efficiency, it may be possible to interpret changes in ice cores linearly in terms of changes in the atmosphere (although the quantitative derivation of actual atmospheric concentrations will still require a knowledge of deposition processes). This condition might apply in comparisons over timescales of decades to centuries, where it could be a reasonable assumption that climatic factors have been fairly constant. It is unlikely to be the case for comparisons between seasons, or for comparisons between periods with strongly changing climate such as between glacials and interglacials. For these cases, changes in transport and deposition efficiency must also be considered.

Because aerosol particles have relatively short lifetimes in the atmosphere (generally days to weeks), they are not well mixed in the atmosphere. In order to understand the geographical significance of results from single ice cores, it is crucial to understand their transport to the polar regions. An outline of the species present in the polar atmospheres, and some basics of transport to Greenland and Antarctica, are given in other chapters. Although transport has often been thought of mainly in terms of changing wind speeds, of more importance is the overall circulation pattern (Mayewski et al. 1994), and the transport pathways (Davidson et al. 1993) to the ice core site. The geographical positions of Greenland and Antarctica (Fig. 1) are very different, with Greenland influenced by nearby continents, while Antarctica is shielded from continental sources by a wide ocean barrier.

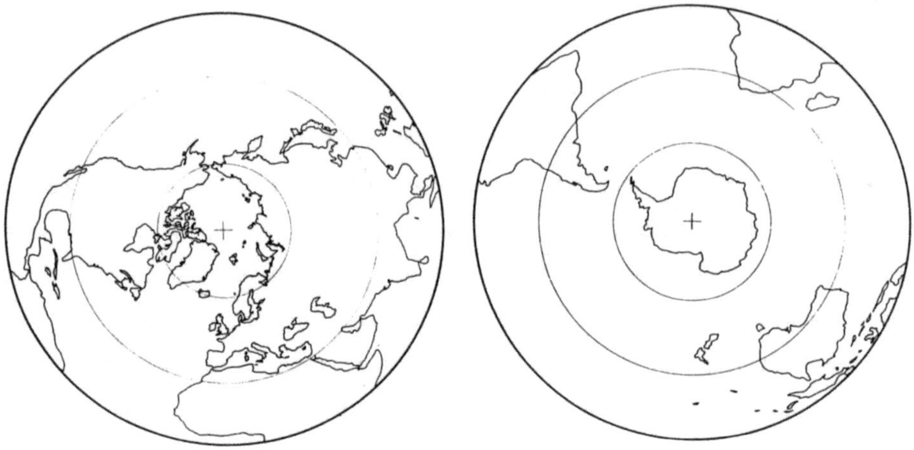

Fig. 1. Maps showing the geographical locations of Greenland and Antarctica, where most ice cores have been extracted. Reprinted with permission from Wolff 1990, *Antarct. Sci.*, 2, 191.

Two simple examples can be considered to illustrate the possible effects of transport for Greenland. At the present time, back trajectories show that air arriving in Greenland can originate from North America, the Atlantic, or the Arctic basin - any change in the proportion of storm tracks coming from each direction could significantly alter the delivery of pollutants to Greenland, even if the source strengths remained constant. And the delivery of terrestrial dust to Greenland would depend strongly on whether air masses arriving there could have passed over desert areas, even if the area of desert and the entrainment had stayed constant - such a change is likely one of the main causes of the large increase in dust seen in Greenland ice in the last glacial.

Included in the efficiency of transport of aerosol to a site is the effect of removal of aerosol during transport. It has been argued (Hansson 1994) that changes in the ratio of atmospheric lifetime (regulated by deposition en route to the ice core site) to transport time could play a major role in glacial/interglacial concentration changes.

Finally, even if the source strength and transport efficiency are unchanged, deposition efficiency will also affect ice core concentrations, and several papers in this volume address this issue. Deposition takes place by wet, dry and fog deposition (Bergin et al. 1995), and a change in efficiency of any of these mechanisms may alter snow concentrations. Two simple ways in which such changes may occur are (a) a change in snow accumulation rate (such as the 2-3 fold change between the late glacial maximum (LGM) and present in Greenland) would affect the balance between wet and dry deposition; (b) a change in particle size would affect the deposition efficiency even if atmospheric concentrations remained identical.

3. Examples of glaciochemical time series

Noting all the difficulties discussed above, I will next present a few examples of ice core records that have been obtained, and discuss the possible interpretations placed upon them. This is not intended to be a comprehensive list, and the reader is referred to previous review works for more detailed information (Delmas and Legrand 1989; Delmas 1995).

3.1. Recent centuries - the pollution record

Records covering the period of industrialisation can be obtained in relatively shallow cores, or even deep hand-dug pits. Over this time period, there is a reasonable probability that the climatic factors affecting transport pathways and deposition efficiencies have remained relatively constant, so that changes seen in the snow should represent changes in the atmosphere, and can be interpreted in terms of changing source strengths with some confidence. There is still the possibility that anthropogenic particles may have a different size spectrum to their naturally-produced counterparts for a given species, and that this could affect deposition efficiencies.

The clearest example presented of concentrations in snow and ice that have increased due to pollution has been for lead (Pb). In Greenland, increases of more than 100-fold in concentrations since pre-industrial times were first reported over 20 years ago (Murozumi et al. 1969). Most of the increase seems to have been due to the increase in emissions from motor vehicles, and this study played a significant role in the debate about reductions in the lead content of fuels. Subsequent work has confirmed the earlier findings (Fig. 2) and shown that abatement measures have led to a significant reduction in lead content of Greenland snow (Candelone et al. 1995). Detailed interpretation of this result requires a consideration of the likely sources of lead to Greenland (North America and Eurasia), and this has been aided by measurements of the lead isotopic content (Rosman et al. 1994). It has also been shown that lead concentrations were somewhat elevated above natural levels even in Greek and Roman times (Hong et al. 1994) - evidence that even these earlier civilisations had the ability to affect the atmosphere on a hemispheric scale.

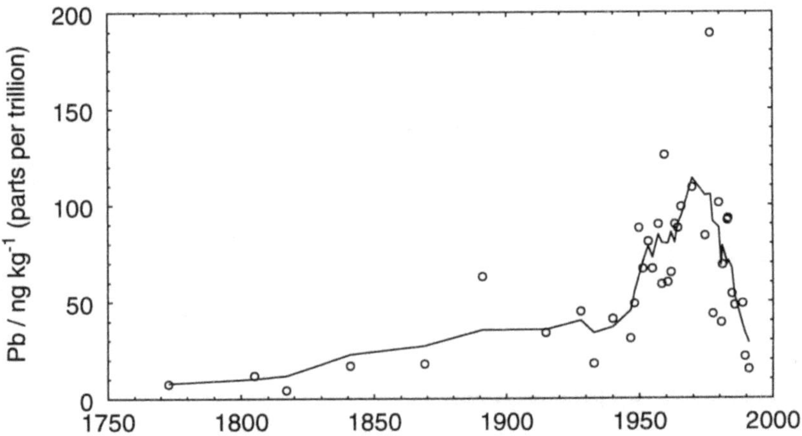

Fig. 2. The record of lead concentrations in Greenland snow over the last two centuries (data from cores and pits at Summit, central Greenland) (Boutron et al. 1991; Savarino et al. 1994; Candelone et al. 1995). Where there was more than one data point in a year, these have been combined. The line is a 5 point running mean.

A profile of lead (Fig. 3) for the last 70 years for Antarctica has also been produced (Wolff and Suttie 1994). Concentrations are much lower than in Greenland, as would be expected since Antarctica is much more remote from the main sources of industrial lead. Nonetheless, an increase (and a recent decrease) are seen, and this was attributed to a combination of increased smelting and vehicle emissions in the southern hemisphere.

The effect of pollution reaching Greenland has also been seen for sulphate, with a three-fold increase during the last century (Mayewski et al. 1990) at a site near Dye 3, south Greenland. No increases in sulphate due to pollution are apparent for Antarctic snow to date.

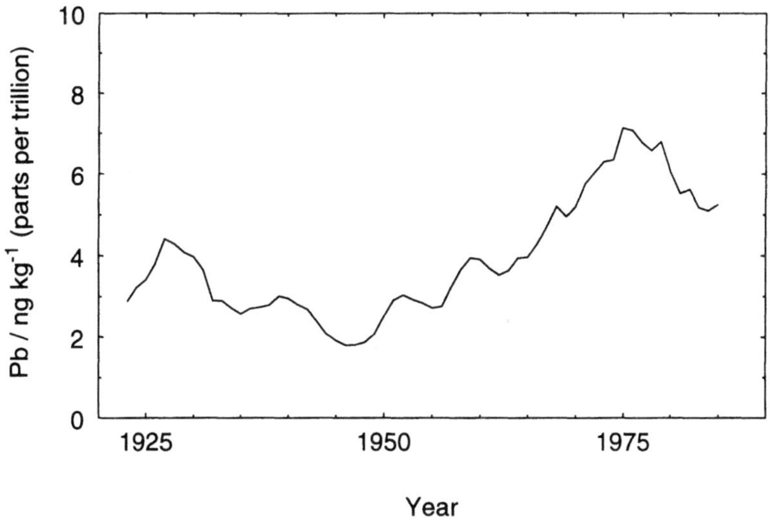

Fig. 3. The record of lead concentrations in Antarctic snow over the last 70 years (Wolff and Suttie 1994). Data are from an 8 m deep snow pit dug in Coats Land.

3.2. S species in Antarctica: the last century and the longer perspective

In a study of a South Pole firn core and snow pit (Legrand and Feniet-Saigne 1991), the level of non-sea salt sulphate was fairly stable over the last 70 years, apart from perturbations due to the volcanic eruptions of Agung in 1963 and El Chichon in 1982 (Fig. 4). However, the MSA profile for the same period showed several short-term increases which appeared to be correlated with major El Niño Southern Oscillation (ENSO) events. The authors suggested that these peaks were due to enhanced emissions of DMS at high latitudes (high latitude emissions give higher ratios of MSA to non-sea salt sulphate after oxidation). This could be related to higher wind speeds (leading to greater sea-air exchange of DMS), or perhaps to changing extents of sea ice. In any case, it would be very valuable if this finding could be used to extend our knowledge of El Niño climatology to earlier periods. Some caution, and further confirmation of the result, is needed however, since no relationship with ENSO events is apparent in a coastal Antarctic ice core (Pasteur et al. 1995). It will be necessary to define the geographical area over which the ENSO link applies before it can be interpreted more confidently.

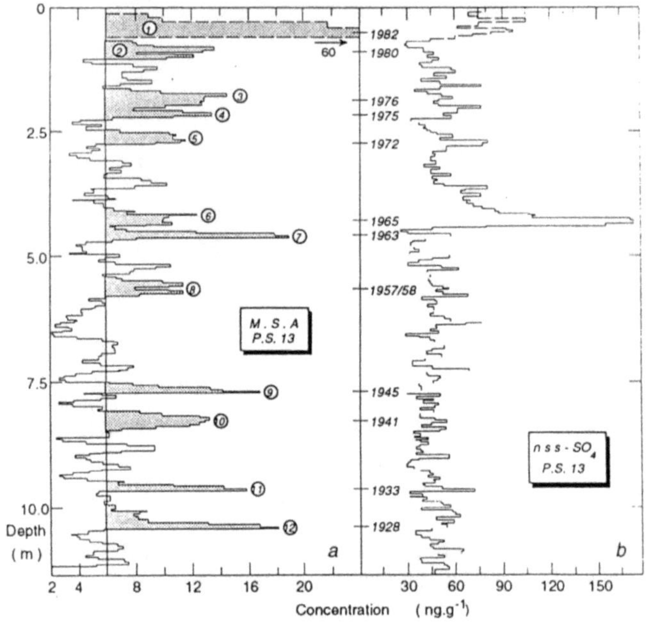

Fig. 4. Non-sea salt sulphate and MSA over the last 70 years at South Pole, Antarctica. Reprinted from Legrand and Feniet-Saigne (1991).

Attempts have been made to extend consideration of S species records back through the last glacial cycle. At Vostok, central Antarctica, non-sea salt sulphate increased by a factor two in the coldest stages of the last glacial compared to the present, while MSA concentrations increased by a factor 5-6 (Legrand et al. 1991). The increases are associated with a large increase in the Ca content of the ice (discussed later, and presumably derived from terrestrial dust sources), and with a decrease in snow accumulation rate. It is thus hard at this stage to separate possible effects on the sulphate concentration: a change in source strength from biogenic emissions, a change in transport efficiency due to a possible change of speciation and of transport strength, or a change in deposition efficiency (due for instance to a change in speciation and snow accumulation rate). The increased ratio of MSA to non-sea salt sulphate during cold periods could be used to help isolate the marine biogenic source strength from other factors. It has been suggested that it could be a result of the decreased temperature, causing a change in the ratio of MSA to sulphate produced from DMS, or to an increase in the amount of high latitude (high ratio) emissions reaching the Antarctic plateau (Legrand 1995). However, while the present-day factors influencing this ratio remain poorly understood, it is not possible to decipher this information yet.

In Greenland, sulphate increased strongly in cold periods of the last glacial cycle (along with Ca), while MSA concentrations decreased (Hansson and Saltzman 1993). One explanation is that the increased sea ice cover in the North Atlantic closed off the source of DMS, the precursor to MSA. However, again, a greater understanding of production, transport and deposition of MSA is required before these results can be understood and interpreted with confidence.

3.3. The Holocene record of volcanism

Large volcanic eruptions deposit sulphuric acid (and other components) onto the polar ice sheets, and these show up as sporadic increases in both sulphate and acidity (commonly

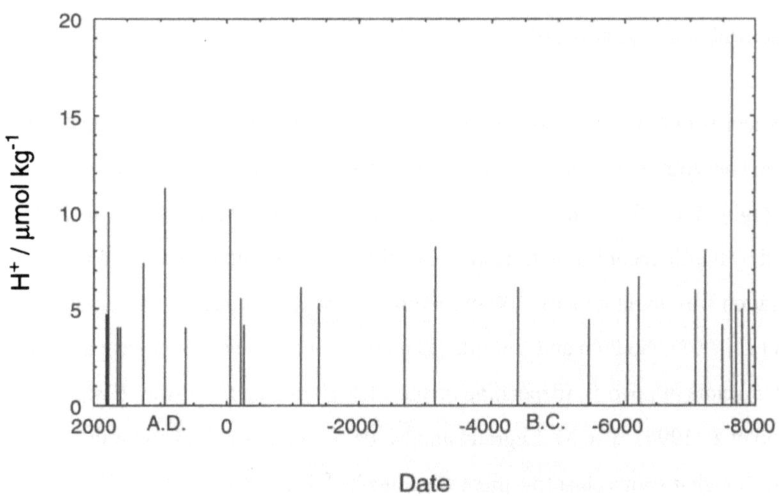

Figure 5. The record of major volcanic eruptions (>4 μmol kg^{-1}) recorded in the acidity record from Greenland ice (data from Crete and Camp Century) (Hammer et al. 1980). No data were available for the section from AD 553 to AD 43. The two most recent volcanoes shown are Tambora (1815) and Laki (1783). Reprinted with permission from Hammer et al. 1980, *Nature* 288, 233. Copyright (1980) Macmillan Magazines Ltd.

measured by electrical methods). Records covering the entire Holocene period (Fig. 5) have been published (Hammer et al. 1980; Zielinski et al. 1994).

A distinction must be made between relatively local volcanoes (such as some Icelandic volcanoes which show up strongly in Greenland records), and those that emit sulphate into the stratosphere, where it may be globally dispersed and in some cases is found in all records in both hemispheres. In both cases, but particularly the former, the location of the eruption and the location of the core site (Clausen and Hammer 1988) exert a very strong influence on the amount of sulphate found in a particular core. This does not pose a problem where the aim of obtaining a volcanic record is mainly to provide time markers that can be used to fix timescales between different cores, and between core chronologies and known historical eruptions. However, often the aim is to try to estimate the amount of material injected into the stratosphere, and to make comparisons with temperature to assess the volcano-climate linkage. In this case, the use of several ice core sites together (Robock and Free 1995) may help to reduce local noise, and allow an assessment of global impact.

3.4. Terrestrial and marine aerosol records of the last 100 kyr - Greenland

Records of either dust content or calcium (as a marker of terrestrial dust) show a huge increase in concentration at sites all around Greenland during the cold periods of the last glacial cycle (Cragin et al. 1977; Fuhrer et al. 1993; Hansson 1994; Mayewski et al. 1994). This is illustrated by the Ca record (Fig. 6) from the GRIP core (Summit, central Greenland), where the concentration was about a factor 100 higher in the last glacial maximum than in the Holocene (Fuhrer et al. 1993). Sodium and chloride (as markers of marine aerosol) were also raised, albeit by smaller factors of up to 10 (depending on the site) (Herron and Langway 1985; Hansson 1994; Mayewski et al. 1994) (also M. Legrand and M. de Angelis, personal communication).

Although it seems clear that there were much higher concentrations of both terrestrial and marine aerosol over Greenland in the last glacial period (with possible effects on optical depth causing climate feedbacks (Harvey 1988)), it remains difficult to quantify the effects, and particularly the geographic extent of them, because of the confusing effect of changes in transport pathways, removal efficiencies en route to Greenland, and deposition efficiencies.

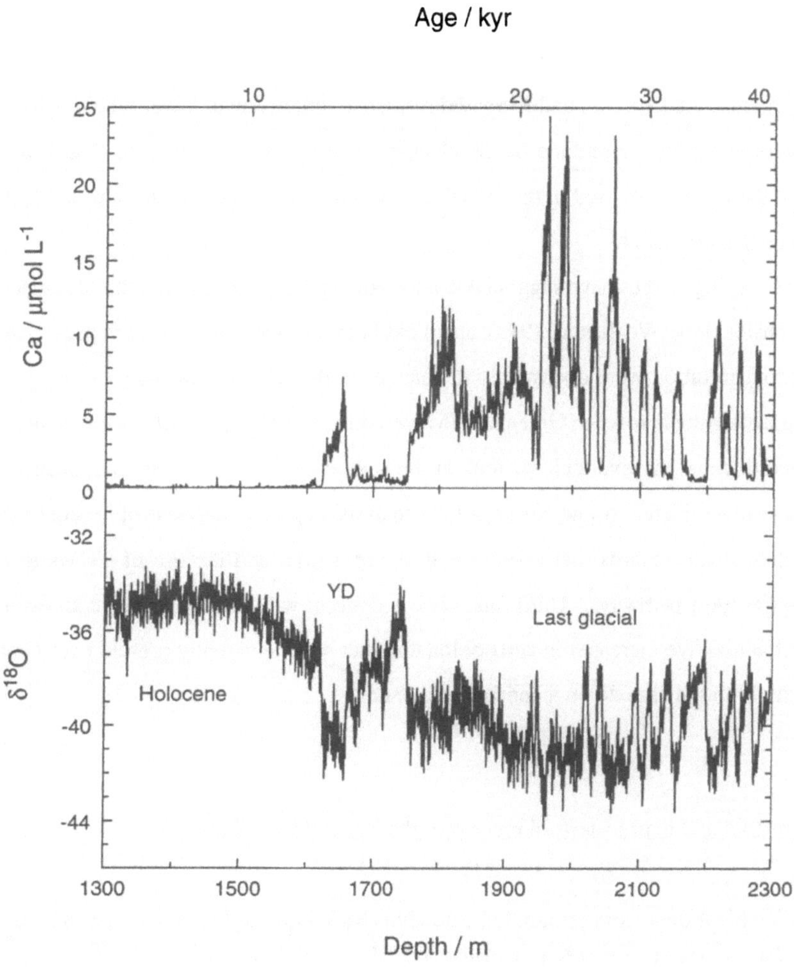

Fig. 6. Profile of calcium concentrations (top) and oxygen isotope values in the GRIP ice core, Summit, Greenland, over the last 40 kyr (Fuhrer et al. 1993). Adapted from Fuhrer et al. 1993, *Atmos. Envir.*, 27A, 1875. Copyright 1993 Elsevier Science Ltd., Oxford, UK. The dating is that used in recent GRIP papers (S.J. Johnsen, personal communication).

The snow accumulation rate was up to a factor 3 lower in the last glacial period, and this could in itself have led to an increased concentration as the relative influence of dry deposition would have increased. However, consideration of the way in which concentrations varied from year to year depending on the snow accumulation rate (derived from detection of individual annual layers) has suggested that this is not a very significant factor (Alley et al. 1995), at least at the Summit site, and that atmospheric concentrations of both types of aerosol were genuinely elevated by very large factors. A simple model suggested that decreased transport times (due to possibly higher wind speeds) and increased residence times (due to lower precipitation) might alone account for the magnitude of the changes seen in the Renland core (Hansson 1994). However, the parameters used in the model are speculative, and further work would be needed to advance this argument.

Modelling studies have suggested that the atmospheric circulation in the last glacial was altered considerably (Webb et al. 1993), and it has been suggested that the increases in dust and sea salt concentration were due mainly to changes in the size and intensity of the circulation system transporting aerosol to Greenland (Mayewski et al. 1994). Further steps in interpreting these records must involve a closer look at the source areas for the dust components (this is underway using mineralogy and Nd/Sr isotope analysis), a proper synthesis of ice core data with other paleoclimate records that document dust deposition, and the use of GCMs to look at possible transport pathways. Until this work is done, it will be hard to make an estimate of whether the massive increases in aerosol loading that were apparently present over Greenland could have affected climate on a hemispheric scale.

3.5. Terrestrial and marine aerosol records of the last 100 kyr - Antarctica

Chemical data from Antarctica extending back into the last glacial period exist for a number of sites (Legrand and Delmas 1988; Legrand et al. 1988), though most of the available data refer to sites with very low snow accumulation rate on the East Antarctic plateau. Fig. 7 shows the aluminium (as a terrestrial marker) and sodium (marine aerosol marker) profiles for the Vostok core. The main feature, as in Greenland, is large increases during glacial periods in both components. Al is about a factor 30 higher in the last glacial maximum (LGM) than the Holocene, with sodium about a factor 5 higher. Somewhat smaller increases are seen at sites that are at lower altitude and somewhat nearer the coast (Legrand and Delmas 1988).

Again, it seems that there was a large increase in aerosol content of the atmosphere over Antarctica, but as in Greenland, it is hard to quantify this or to assign a cause at present. Because of the very low accumulation rates at the central Antarctic sites, dry deposition is probably important, and the halving of accumulation rate in the LGM would therefore be expected to

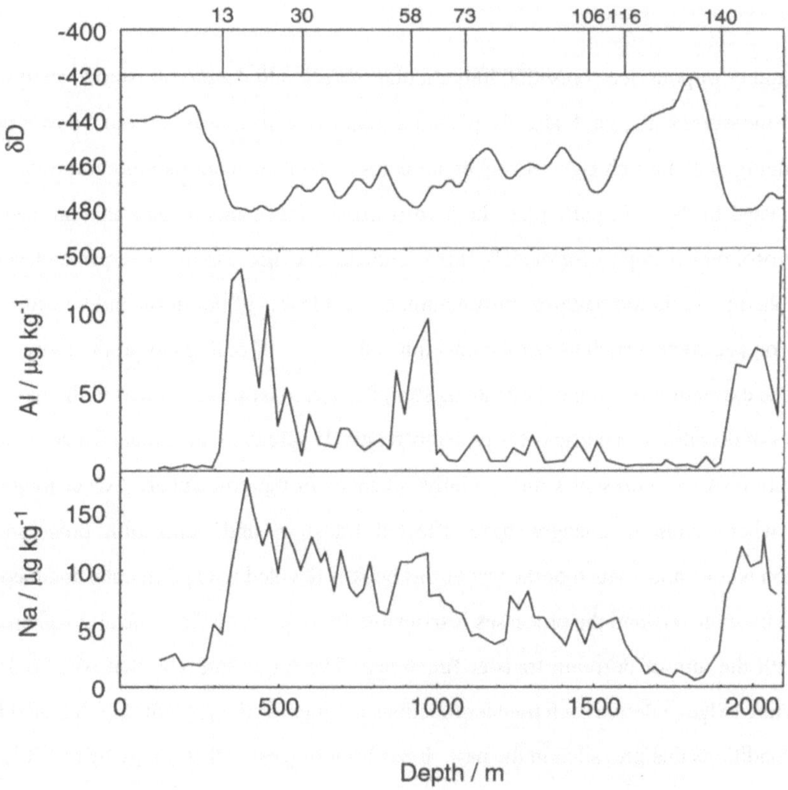

Fig. 7. Sodium and aluminium concentrations, along with the smoothed deuterium profile, for the Vostok core, Antarctica over the last 160,000 years (Legrand et al. 1988). The numbers along the top are ages (in kyr BP) according to the 1985 Vostok timescale. Adapted from Legrand et al. 1988, *Atmos. Envir.*, 22, 319-320. Copyright 1988 Elsevier Science Ltd., Oxford, UK.

significantly affect snow concentrations. Factors such as a change in the seasonality of snowfall might also play a role - the input of sea salt to the central Antarctic plateau in present times is highly sporadic. Remaining changes have been ascribed to increased wind speed, and increased size of arid areas in the southern continents (Legrand et al. 1988). The next steps in deciphering this data must include gaining a better understanding of factors that control deposition at low accumulation sites, a better knowledge of transport to the Antarctic plateau and the present-day seasonality of it, and extraction of further cores to understand the spatial pattern of the increase.

4. Discussion and conclusions

Studies of polar ice cores, and in particular oxygen and deuterium isotope analyses and trace gas measurements, have already played a major role in improving our knowledge and understanding of climate change. However, ice cores certainly contain far more information than has been used to date. In particular, the aerosol component contains data that are needed to derive aerosol optical depths for models of past climate, data that can improve parameterisations of atmospheric circulation patterns, information on past levels of pollution and volcanism, and probably on parameters such as sea-ice position and aspects of biological productivity.

The extraction of such data is not as straightforward as was assumed in the past. Over timescales of decades to centuries, it is probably relatively safe to assume that changes in aerosol concentration in ice represent similar relative changes in the atmosphere. Over longer time periods, where climatic changes have affected transport and deposition processes, this assumption is not valid. Much of the rest of this book is devoted to experimental and theoretical studies to try to understand the processes responsible for deposition from the atmosphere to the snow, with the aim of deriving transfer functions. The requirement is that we should have enough knowledge to derive such transfer functions not just for the present-day, but also for the climate conditions that prevailed in the past. It has been suggested that it may be possible under some conditions to use procedures involving calculation of fluxes in individual annual layers to bypass the detailed process studies (Alley et al. 1995). This is a promising approach that deserves a wider test, but will not be easily applied to most of the low accumulation rate Antarctic sites that will yield deep ice cores in the next few years. A considerable effort in studies of deposition, transport and other issues alluded to in this paper will be needed if full value is to be obtained from the new cores.

References

Alley RB, Finkel RC, Nishiizumi K, Anandakrishnan S, Shuman CA, Mershon GR, Zielinski GA, Mayewski PA (1995) Changes in continental and sea-salt atmospheric loadings in central Greenland during the most recent deglaciation: model-based estimates. J Glaciol 41:503-514

Bergin MH, Jaffrezo J-L, Davidson CI, Dibb JE, Pandis SN, Hillamo R, Maenhaut W, Kuhns HD, Makela T (1995) The contributions of snow, fog, and dry deposition to the summer flux of anions and cations at Summit, Greenland. J geophys Res 100:16275-16288

Boutron CF, Görlach U, Candelone J-P, Bolshov MA, Delmas RJ (1991) Decrease in anthropogenic lead, cadmium and zinc in Greenland snows since the late 1960s. Nature 353:153-156

Candelone J-P, Hong S, Pellone C, Boutron CF (1995) Post-Industrial Revolution changes in large-scale atmospheric pollution of the northern hemisphere by heavy metals as documented in central Greenland snow and ice. J geophys Res 100:16605-16616

Clausen HB, Hammer CU (1988) The Laki and Tambora eruptions as revealed in Greenland ice cores from 11 locations. Ann Glaciol 10:16-22

Cragin JH, Herron MM, Langway CC Jr, Klouda G (1977) Interhemispheric comparison of changes in the composition of atmospheric precipitation during the late Cenozoic era. In: Dunbar MJ (ed) Polar Oceans. Arctic Institute of North America, Calgary

Cuffey KM, Clow GD, Alley RB, Stuiver M, Waddington ED, Saltus RW (1995) Large Arctic temperature change at the Wisconsin-Holocene glacial transition. Science 270:455-458

Davidson CI, Jaffrezo J-L, Small MJ, Summers PW, Olson MP, Borys RD (1993) Trajectory analysis of source regions influencing the south Greenland ice sheet during the Dye 3 gas and aerosol sampling program. Atmos Environ 27A:2739-2749

Delmas RJ (ed) (1995) NATO ASI Series I, 30: Ice core studies of global biogeochemical cycles. Springer, Berlin etc

Delmas RJ, Legrand M (1989) Long-term changes in the concentrations of major chemical compounds (soluble and insoluble) along deep ice cores. In: Oeschger H, Langway CC Jr (eds) Dahlem Konferenzen: The environmental record in glaciers and ice sheets. John Wiley, Chichester

Fuhrer K, Neftel A, Anklin M, Maggi V (1993) Continuous measurements of hydrogen peroxide, formaldehyde, calcium and ammonium concentrations along the new GRIP core from Summit, central Greenland. Atmos Environ 27A:1873-1880

Hammer CU, Clausen HB, Dansgaard W (1980) Greenland ice sheet evidence of post-glacial volcanism and its climatic impact. Nature 288:230-235

Hansson ME (1994) The Renland ice core: A Northern Hemisphere record of aerosol composition over 120,000 years. Tellus 46B:390-418

Hansson ME, Saltzman ES (1993) The first Greenland ice core record of methanesulfonate and sulfate over a full glacial cycle. Geophys Res Lett 20:1163-1166

Harvey LDD (1988) Climatic impact of ice-age aerosols. Nature 334:333-335

Herron MM, Langway CC Jr (1985) Chloride, nitrate, and sulfate in the Dye 3 and Camp Century,Greenland ice cores. In: Langway CC Jr, Oeschger H, Dansgaard W (eds) Geophysical Monograph, 33: Greenland ice core: geophysics, geochemistry, and the environment. American Geophysical Union, Washington, DC

Hong S, Candelone J-P, Patterson CC, Boutron CF (1994) Greenland ice evidence of hemispheric lead pollution two millenia ago by Greek and Roman civilizations. Science 265:1841-1843

Johnsen SJ, Dahl-Jensen D, Dansgaard W, Gundestrup N (1995) Greenland palaeotemperatures derived from GRIP bore hole temperature and ice core isotope profiles. Tellus 47B:624-629

Legrand M (1995) Sulphur-devrived species in polar ice: A review. In: Delmas RJ (ed) NATO ASI Series I, 30: Ice core studies of global biogeochemical cycles. Springer, Berlin etc.

Legrand MR, Delmas RJ (1988) Soluble impurities in four Antarctic ice cores over the last 30 000 years. Ann Glaciol 10:116-120

Legrand M, Feniet-Saigne C (1991) Methanesulfonic acid in South Polar snow layers: a record of strong El Nino? Geophys Res Lett 18:187-190

Legrand MR, Lorius C, Barkov NI, Petrov VN (1988) Vostok (Antarctica) ice core: atmospheric chemistry changes over the last climatic cycle (160,000 years). Atmos Environ 22:317-331

Legrand M, Feniet-Saigne C, Saltzman ES, Germain C, Barkov NI, Petrov VN (1991) Ice-core record of oceanic emissions of dimethylsulphide during the last climate cycle. Nature 350:144-146

Legrand M, Leopold A, Domine F (1996) Acidic gases (HCl, HF, HNO_3, HCOOH, and CH_3COOH): a review of ice core data and some preliminary discussions on their air-snow relationships. In: Wolff EW, Bales RC (eds) NATO ASI Series I: Chemical Exchange between the Atmosphere and Polar Snow. Springer Verlag, Berlin etc (this volume)

Mayewski PA, Lyons WB, Spencer MJ, Twickler MS, Buck CF, Whithlow S (1990) An ice-core record of atmospheric response to anthropogenic sulphate and nitrate. Nature 346:554-556

Mayewski PA, Meeker LD, Whitlow S, Twickler MS, Morrison MC, Bloomfield P, Bond GC, Alley RB, Gow AJ, Grootes PM, Meese DA, Ram M, Taylor KC, Wumkes W (1994) Changes in atmospheric circulation and ocean ice cover over the North Atlantic during the last 41,000 years. Science 263:1747-1751

Murozumi M, Chow TJ, Patterson C (1969) Chemical concentrations of pollutant lead aerosols, terrestrial dusts and sea salts in greenland and antarctic snow strata. Geochim Cosmochim acta 33:1247-1294

Neftel A (1996) The record of gases and reactive species in ice cores, and problems of interpretation. In: Wolff EW, Bales RC (eds) NATO ASI Series I: Chemical Exchange between the Atmosphere and Polar Snow. Springer Verlag, Berlin etc (this volume)

Pasteur EC, Mulvaney R, Peel DA, Saltzman ES, Whung P-Y (1995) A 340-year record of biogenic sulphur from the Weddell Sea area, Antarctica. Ann Glaciol 21:169-174

Raynaud D, Jouzel J, Barnola JM, Chappellaz J, Delmas RJ, Lorius C (1993) The ice record of greenhouse gases. Science 259:926-934

Robock A, Free MP (1995) Ice cores as an index of global volcanism from 1850 to the present. J geophys Res 100:11549-11567

Rosman KJR, Chisholm W, Boutron CF, Candelone JP, Hong S (1994) Isotopic evidence to account for changes in the concentration of lead in greenland snow between 1960 and 1988. Geochim Cosmochim acta 58:3265-3269

Savarino J, Boutron CF, Jaffrezo J-L (1994) Short-term variations of Pb, Cd, Zn and Cu in recent Greenland snow. Atmos Environ 28:1731-1737

Webb T III, Ruddiman WF, Street-Perrott FA, Markgraf V, Kutzbach JE, Bartlein PJ, Wright HE Jr, Prell WL (1993) Climatic changes during the past 18,000 years: regional syntheses, mechanisms, and causes. In: Wright HEJ, Kutzbach JE, Webb T III, Ruddiman WF, Street-Perrott FA, Bartlein PJ (eds) Global climates since the last glacial maximum. University of Minnesota Press, Minneapolis

Wolff EW (1996) Location, movement and reactions of impurities in polar ice. In: Wolff EW, Bales RC (eds) NATO ARW: Chemical Exchange between the Atmosphere and Polar Snow. Springer Verlag, Berlin etc (this volume)

Wolff EW, Suttie ED (1994) Antarctic snow record of southern hemisphere lead pollution. Geophys Res Lett 21:781-784

Zielinski GA, Mayewski PA, Meeker LD, Whitlow S, Twickler MS, Morrison M, Meese DA, Gow AJ, Alley RB (1994) Record of volcanism since 7000 B.C. from the GISP2 Greenland ice core and implications for the volcano-climate system. Science 264:948-952

ACIDIC GASES (HCl, HF, HNO_3, HCOOH, AND CH_3COOH): A REVIEW OF ICE CORE DATA AND SOME PRELIMINARY DISCUSSIONS ON THEIR AIR-SNOW RELATIONSHIPS

Michel Legrand, Anne Léopold, and Florent Dominé
Laboratoire de Glaciologie et Géophysique de l'Environnement du CNRS
B.P. 96
38402 St Martin d'Hères Cedex
France

1. INTRODUCTION

In spite of their relative remoteness, the atmosphere of the polar regions is already disturbed by human activities. The development of the so called "ozone hole" over Antarctica in spring and the Arctic haze in winter represent good examples of the fragility of these remote atmospheres. One interesting characteristic of these high latitude regions lies in solid precipitation which has been accumulating over polar ice caps for the last several hundred thousands years. Assuming a sufficient knowledge of the relation linking the composition of the snow and that of the atmosphere at the time of deposition, ice cores extracted from Greenland and Antarctica offer a unique possibility to reconstruct the chemical composition of our pre-industrial atmosphere. They can also help to investigate the variability of our natural atmosphere over time periods as long as several thousands of years.

Since the 70s glaciochemical studies have dealt with an increasing number of various species trapped in snow. The first ice core study was achieved by Murozumi et al. [1969] who showed a 200 fold increase of lead level over recent decades in Greenland ice due to automobile emissions. Some years after this pioneering work, the studies of deep ice cores first extracted from Greenland (Camp Century) and Antarctica (Byrd and Dome C) revealed that higher inputs of sea salt and dust characterized the ice age [Cragin, 1977; Petit et al., 1981]. Stimulated by the potential effect of aerosols in forcing the climate, numerous works have then dealt with the glaciochemistry of marine and continental aerosols [see Legrand, 1995a; Wolff, 1996a, for reviews]. In addition to such studies of primary aerosols, investigations of sulphur derived species (sulphate and later on methanesulfonate) permitted the reconstruction of volcanism in the past following the initial study of Hammer [1977] and were useful in studying the role of sulphur

NATO ASI Series, Vol. I 43
Chemical Exchange Between the Atmosphere and Polar Snow
Edited by Eric W. Wolff and Roger C. Bales

emissions from the marine biota on the global sulphur cycle [see Legrand, 1995b, for a review]. With the exception of methanesulfonate [Legrand et al., 1992a; Mulvaney et al., 1992] and perhaps magnesium [Wolff, 1996b] for which a high mobility in ice is suspected, sea salt, dust and sulphur derived species are non reactive, non volatile and are therefore irreversibility trapped in snow layers. One of the major difficulties in providing valid interpretation of aerosol ice records in terms of changes of the aerosol load of the past atmosphere is related to the various incorporation modes of aerosols in snow which include both wet (snowflakes, diamond dusts, riming) and dry (direct impaction onto the surface, wind pumping effect, see Waddington et al. [1996]) deposition. Thus, changes in meteorological conditions may have modulated trace compound concentrations in ice in the past even if atmospheric concentrations remained unchanged.

Since the mid 1980s, glaciochemical investigations have also been directed towards species like H_2O_2 and HCHO which can help to reconstruct the past oxidizing capacity of the atmosphere [Neftel, 1996]. In addition, numerous Greenland and Antarctic ice records of water soluble gaseous species derived from nitrogen, halogen, and hydrocarbon cycles spanning time periods from the last few decades back to the entire glacial-interglacial cycle (100,000 years) are already available. Although they contain information of great interest for atmospheric chemists on changes of biogenic cycles in the past, the ice records of these acidic gaseous species remain sometimes difficult to translate in terms of atmospheric information. Indeed, the incorporation of these gaseous species in snow involves several complex physical processes including dissolution in supercooled water cloud droplets or in the so called quasi-liquid layer [Conklin and Bales, 1993], adsorption or condensation onto ice surfaces and diffusion. Our present poor knowledge of these low temperature processes renders the interpretation of ice records more complicated, in particular those spanning time periods over which changes in climatic conditions had occurred. Furthermore, an additional complexity is introduced for acidic species because their incorporation in snow is likely dependent on the acidity of the atmosphere which is not a conservative property over the past. Finally, as for H_2O_2 [Sigg and Neftel, 1988] and HCHO [Staffelbach et al., 1991], possible redistribution in the firn after deposition of these volatile species takes place under certain conditions [Silvente and Legrand, 1995; De Angelis and Legrand, 1995].

This paper focusses on the records of mineral (HCl, HF, and HNO_3) and carboxylic (formic: HCOOH and acetic: CH_3COOH) acids from Greenland and Antarctica firn and ice cores and their atmospheric significance. We first review available ice records and examine in what extent reliable information can be extracted in terms of chemical composition of past atmospheres. Indeed, it must be stressed that some ice records of these species do not necessary reflect atmospheric signals. We first start with firn records which span the recent time period from the pre-industrial era to present-days and represent a potential way to investigate the impact of human activities on the natural budget of these acidic species. However, in a few cases, recent trends exhibited by firn cores certainly are related to post depositional phenomena which can very significantly disturb initial atmospheric signals of these volatile species. Then, we discuss changes in acidic species levels seen in deep ice cores spanning the last climatic cycle and examine to what extent some of these changes can be attributed to changes in atmospheric concentrations due to modification of source strength or/and transport following past climate fluctuations. Based on a few recent measurements of these acidic species in the summer atmosphere at Summit (central Greenland), and on preliminary laboratory experiments dealing with adsorption, diffusion, and solubility in ice, the second part of this paper is dedicated to preliminary investigations of which physical and chemical processes may be involved in incorporating strong (HCl and HNO_3) and weak (HCOOH and CH_3COOH) acids in Greenland snow.

2. ICE CORE RECORDS

Given the complexity and the number of processes involved, in particular for acidic gaseous species, the reliable reconstruction of past polar atmospheric concentrations from ice records is impossible at present. What can be done, at best, is to examine relative changes depicted by ice core depth profiles. With the aim of extracting valid atmospheric information from ice core records of volatile acidic species, several points need to be kept in mind. Although being not understood so far, we can expect that mechanisms of incorporation of these gases into solid precipitation to be dependent on meteorological conditions. The temperature experienced during the incorporation is likely to be a key parameter on which depends the amount of supercooled water versus ice present in clouds for instance. Also, adsorption, diffusion and solubility are temperature dependent. Mechanisms of incorporation are also dependent on the acidity of the atmosphere, particularly for weak acids. In addition to temperature, other meteorological variables may be of importance when considering post depositional effect (wind speed at the surface responsible for the ventilation, metamorphism of fresh snowflakes). All these parameters have likely changed in the past depending on climatic conditions experienced by these regions. Therefore it is not always possible at present to determine whether a measured change in concentration in ice is due to an actual change in atmospheric concentrations or to a change in other parameters affecting incorporation in ice.

2.1. Calculation of acidic nitrate, chloride, fluoride, formate and acetate

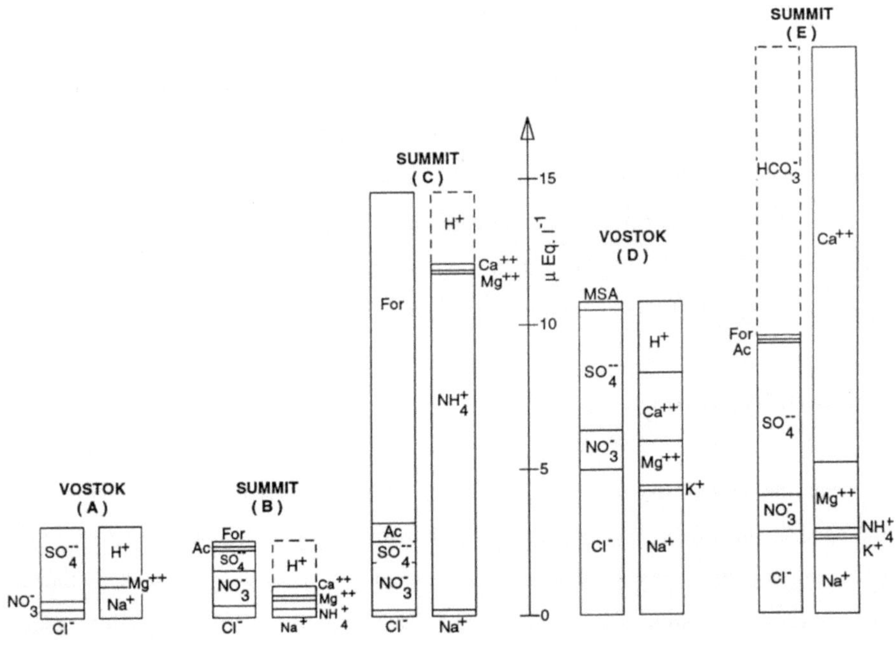

Fig. 1 Mean ionic composition of ice corresponding to present climate in Antarctica (A), Greenland (B), and to Glacial Age in Antarctica (D) and in Greenland (E). Dashed lines are used when acidity measurements are not yet available. Antarctic data are from Legrand et al. [1988] and Legrand et al. [1992b]. (C) is based on data from Legrand et al. [1992b] and corresponds to a sporadic event related to biomass burning plume reaching Greenland. Ac, For, and MSA abbreviate acetate, formate, and methanesulfonate respectively. Adapted from Legrand [1993].

Studies of the ionic budget of polar ice (see the overview reported in Figure 1) have permitted to examine which fraction of ions corresponds to the presence of an acid. The check of the ionic balance has served to demonstrate that nitrate is mainly present in antarctic precipitation as HNO_3, at least for present day warm climatic conditions [Legrand and Delmas, 1986]. It was proposed by Legrand et al. [1988] that in glacial Antarctic ice nitrate was present as a salt (neutral form). In Greenland ice deposited under present day conditions (Figure 1B), a main fraction of nitrate is likely in the form of HNO_3 but we cannot rule out the possibility that a small fraction of nitrate is also associated with ammonium. The non sea salt chloride (excess chloride) contribution, calculated by subtracting the sea salt contribution (using the reference bulk sea water Cl to Na ratio, 1.8 in weight and the sodium levels) to the chloride level, estimates the amount of HCl present in ice [Legrand and Delmas, 1988]. The very weak contribution of fluoride and light carboxylate to the ionic budget (see Figure 1), renders impossible a direct identification of chemical forms (salt or acid) of fluoride and carboxylates. Recently, De Angelis and Legrand [1994] suggested that the part of fluoride which is present in ice as HF can be estimated as the non soil contribution which is calculated by subtracting the soil contribution (using calcium levels and an estimation of the dust F/Ca ratio, 4 to 5 10^{-3} in weight) to the total fluoride level. With the exception of narrow Greenland snow layers exhibiting a large ammonium formate input (Figure 1C) related to high latitude biomass burning input [Legrand et al., 1992b], formate and acetate are assumed to be present as HCOOH and CH_3COOH [Legrand and De Angelis, 1995].

2.2. The ice records of the last two centuries

a) Greenland records. Greenland and Antarctic chemical profiles covering the last two centuries can provide useful information on the impact of human activities on the chemical composition of the atmosphere in the northern and southern hemispheres, respectively. Over such time periods, we can assume that no shift in meteorological parameters is large enough to significantly change the incorporation mechanisms at a given site. However, because of the acidification of the atmosphere following growing impact of human activities in particular in the northern hemisphere, incorporation modes may have been modified in particular for weak acids. In addition, information covering this time period is archived in upper snow and firn layers in which various processes of exchange can still take place with the atmosphere.

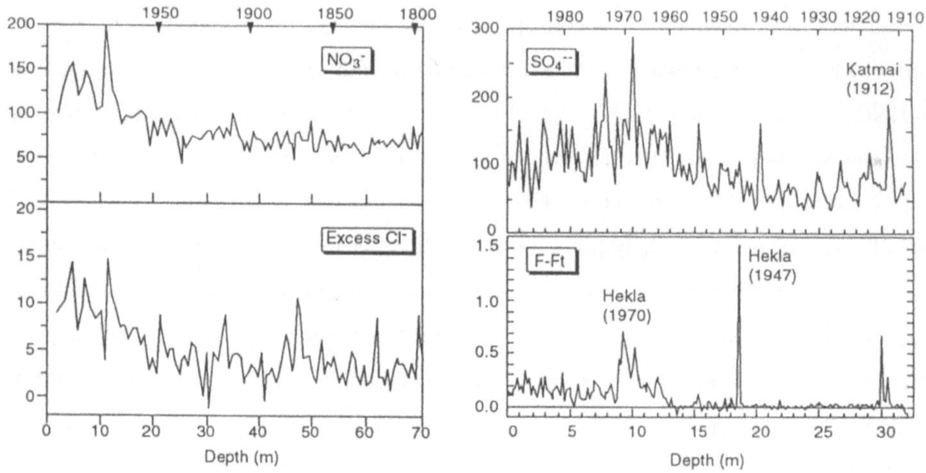

Fig. 2 Recent evolution of NO_3^-, excess-chloride since 1800 AD (adapted from Mayewski et al. [1993]), sulphate, and non soil fluoride (denoted F-F_t, see text) since the beginning of the century [De Angelis and Legrand, 1994] in Summit (Central Greenland) snow layers. Concentrations are reported in $ng.g^{-1}$.

At present the most comprehensive study of recent trends in acidic gases species has been achieved at Summit (central Greenland). In Figure 2, in addition to nitrate, we report the depth profiles of excess chloride, and non soil fluoride in snow layers deposited at Summit over the last hundreds of years. Enhanced levels are detected for these three species in recent Greenland snow layers. However, because of possible exchange between the atmosphere and these firn layers, valid interpretation of such records needs to be discussed on a case by case basis. To discuss whether these records reflect atmospheric signals or an artefact linked with post deposition effect in the firn, we must examine trends recorded in other Greenland sites. Although Summit has an accumulation rate nearly 2.5 times lower than that of Dye 3, which leads to a different depth-age relationship, similar temporal changes of nitrate are observed at Dye 3 and Summit (Figure 3). This indirect argument suggests that the recent NO_3^- trend in Greenland ice is atmospheric in origin, and not due to post depositional loss. This trend then reflects the increase of long range transport of NO_x pollution from middle northern latitudes [Neftel et al., 1985; Mayewski et al., 1990]. An excess-chloride trend is also revealed

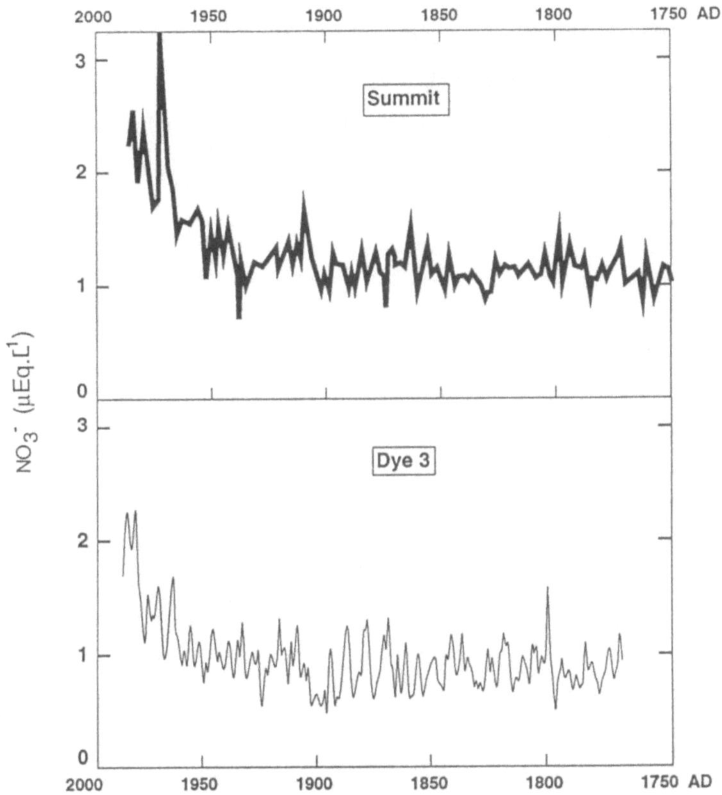

Fig. 3 Comparison of mean nitrate concentrations from Summit [Mayewski et al., 1993](upper trace) and Dye 3 [Mayewski et al., 1986](lower trace) ice over the last two centuries.

in firn from Summit (Figure 2) which may be related to growing direct man-made HCl emissions and/or to the increase of the acidity of the atmosphere leading to enhanced sea salt fractionation [Mayewski et al., 1993] following the reaction:

$$2\ NaCl\ (\text{sea salt}) + H_2SO_4 \ \text{->}\ \ Na_2SO_4 + 2\ HCl$$

However, for this halogenated species, available Greenland ice data remain more limited than in the case of nitrate and the interpretation of the Summit recent trend of HCl in terms of atmospheric signal needs further work. The recent increase of fluoride concentrations (Figure

2) may be due to growing coal-burning emissions [De Angelis and Legrand, 1994]. This being the only study of fluoride trend near the surface, it also remains difficult to discuss the validity of this record in terms of atmospheric changes.

In contrast to the preceding ions, the recent trends of light carboxylates exhibit a more complex picture with decreased formate and increased acetate levels in snow layers deposited

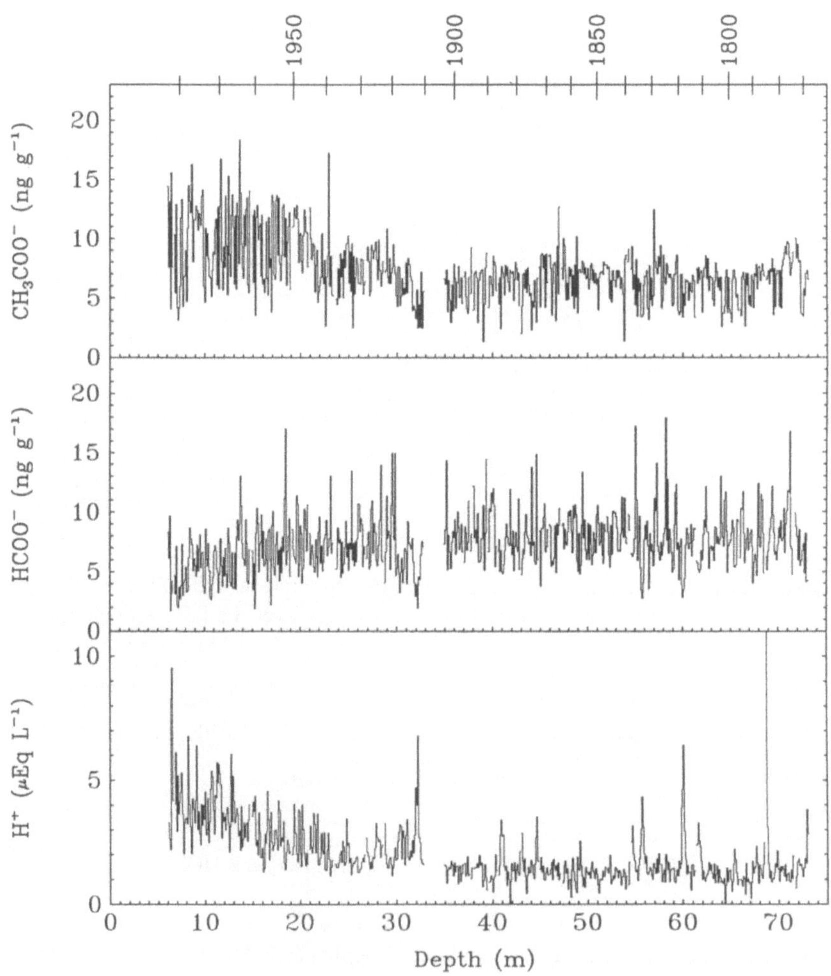

Fig. 4 Recent trend of CH_3COO^-, $HCOO^-$, and H^+ in Summit snow layers deposited since 1770. From Legrand and De Angelis [in press].

since 1940 (Figure 4). An additional complexity is introduced for these weak acids in the form of a possible role of the acidity of the atmosphere either on the incorporation mode or on the efficiency of post depositional effects. The acidity of the atmosphere over Greenland has been enhanced particularly after 1940, as depicted in Figure 4, as a result of enhanced levels of sulfate and nitrate (Figure 2) resulting from growing human emissions of SO_2 and nitrogen oxides. The role of pH on carboxylic acids content of Summit snow layers can be seen in Figure 5, where we report ratios (denoted $HCOO^-/HCOO^-_a$ and $CH_3COO^-/CH_3COO^-_a$) of formate and acetate concentrations in snow layers from the pre-industrial era containing enhanced acidity following volcanic activity to the respective background levels (denoted $HCOO^-_a$ and $CH_3COO^-_a$) as a function of the pH. Such a strong dependence of the formate level of Greenland snow deposits, which is in agreement with a pseudo-Henry's law equilibrium [Legrand and De Angelis, 1995], may reflect the role of pH during the incorporation of this weak acidic gaseous species into supercooled cloud droplets (see further

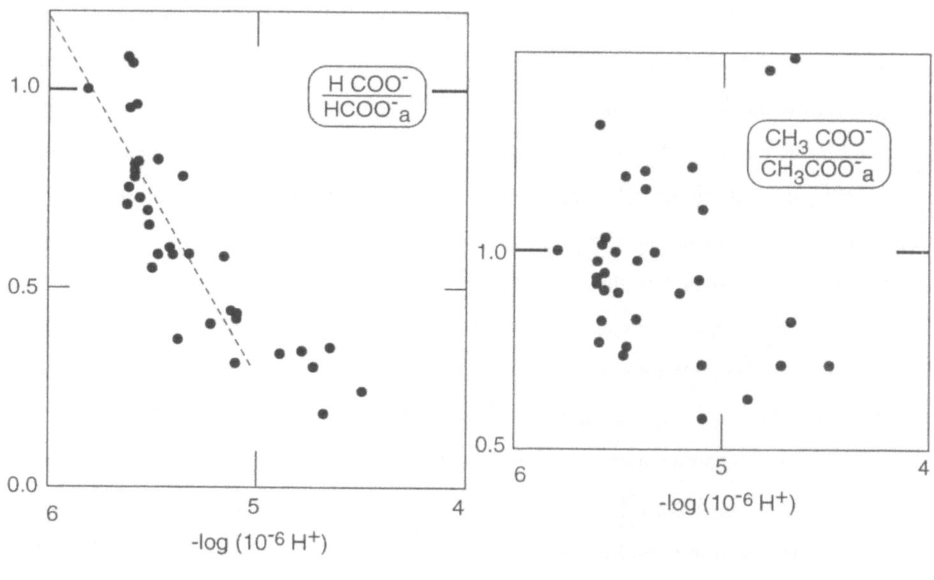

Fig. 5 Influence of the pH (i.e., - log (10^{-6} H^+), with H^+ expressed in micromoles per liter) on $HCOO^-$ and CH_3COO^- levels of Summit samples corresponding to the last 10,000 years. Ratios of disturbed carboxylic acid concentrations ($HCOO^-$ and CH_3COO^-) to the mean level ($HCOO^-_a$ and $CH_3COO^-_a$) are plotted versus pH. Adapted from Legrand and De Angelis [in press].

discussion in section 3) in the summer Summit atmosphere or into snow during post deposition phenomena. This relationship between the formate level and the acidity of Summit snow layers suggests that the expected increase of HCOOH in the atmosphere due to human related emissions of hydrocarbons (CH_4 and non-methane hydrocarbons) is counteracted by a simultaneous acidity increase related to growing fossil fuel combustion (HNO_3 and H_2SO_4) which leads to a less efficient uptake of HCOOH into the precipitation [Legrand and De Angelis, 1995] or a more intense post depositional effect in more acidic layers. Furthermore, as discussed by Legrand and De Angelis [1995], the very weak pH dependence of the acetate ice levels seen in Figure 5 is also in agreement with a pseudo-Henry's law equilibrium driven by a lower dissociation constant of the acetic acid with respect to the formic one.

b) Antarctic records. Over the two last centuries, the NO_3^- ice content at the South Pole remains fairly constant but an increase of the spring-summer maximum is seen in very recent layers (Figure 6A). It was proposed by Mayewski and Legrand [1990] that such a profile may be related to a recent enhanced denitrification of the lower stratosphere by sedimentation of polar stratospheric clouds (PSCs). As summarized by Wolff [1995], others workers [Legrand and Kirchner, 1988; Mulvaney and Wolff, 1993] also detected some spring peaks of nitrate in recent snow layers. The conclusion that these spring nitrate peaks provide valuable information on the polar stratospheric cloud activity in the past must be treated with some cautions because nitrate, initially trapped on and/or in snow flakes, may be subsequently redistributed in the firn during firnification processes leading to a smoothing of the seasonal signal with depth as seen for other volatile species like hydrogen peroxide [Sigg and Neftel, 1988]. Compared to the mean value of 100 $ng.g^{-1}$ which remains quite constant since 1955 at South Pole, cores from Dominion range, Vostok and Dome C (Figure 6) exhibit lower values before 1975 and a large increase in the shallow layers. It is unlikely that the difference by a factor of 5 between nitrate levels at Vostok and South Pole before 1975 reflects different atmospheric levels in these two stations of the high Antarctic plateau. It was proposed by several workers [Mayewski and Legrand, 1990; De Angelis and Legrand, 1995; Wolff, 1995] that low nitrate concentrations in deep layers of sites like Dome C, Vostok, and Dominian range reflect a reemission of nitrate from snow after deposition and that this phenomenon is more efficient in sites characterized by low accumulation rates of snow (34, 23, 30 $kg.m^{-2}.yr^{-1}$, respectively).

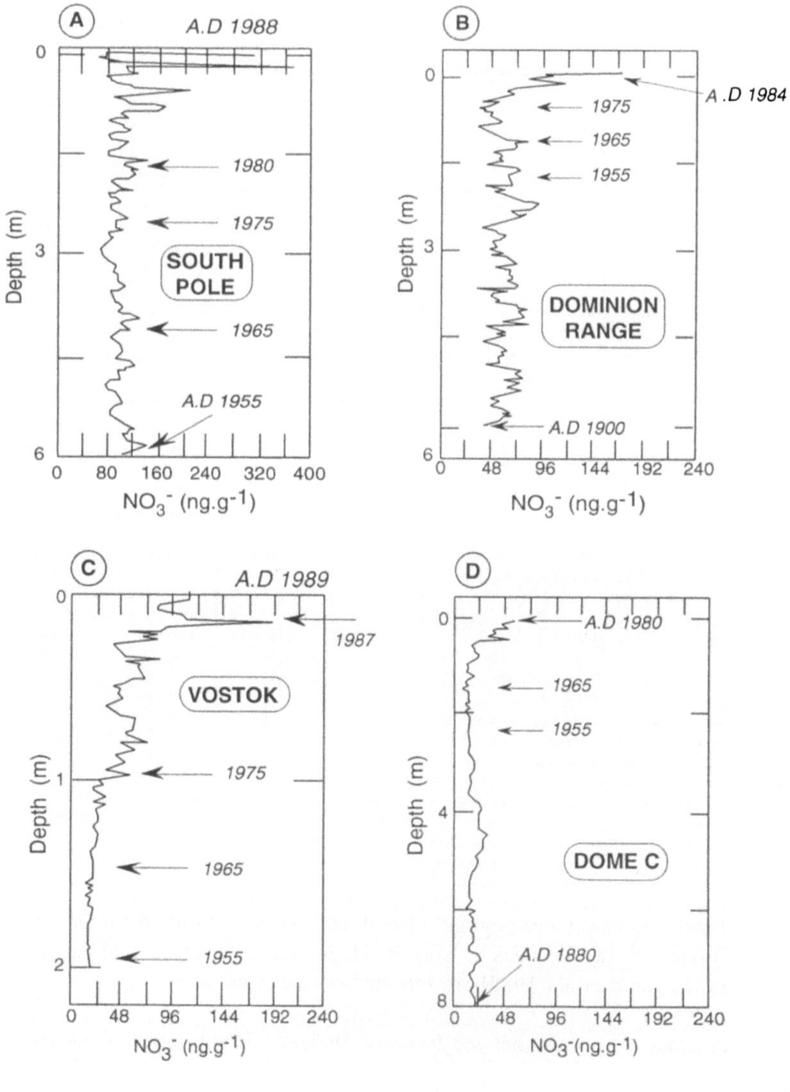

Fig. 6 Nitrate time series of the last decades from South Pole (A), Dominian Range (B), Vostok (C) and Dome C (D). Adapted from Mayewski and Legrand [1990].

A similar trend near the surface has been identified at Vostok for excess chloride as depicted in Figure 7. Similarly to nitrate, the excess chloride level at the South Pole is close to 10-20 ng.g^{-1} near the surface [Legrand and Delmas, 1984] remained rather unchanged in deeper layers [Kirchner and Delmas, 1988].

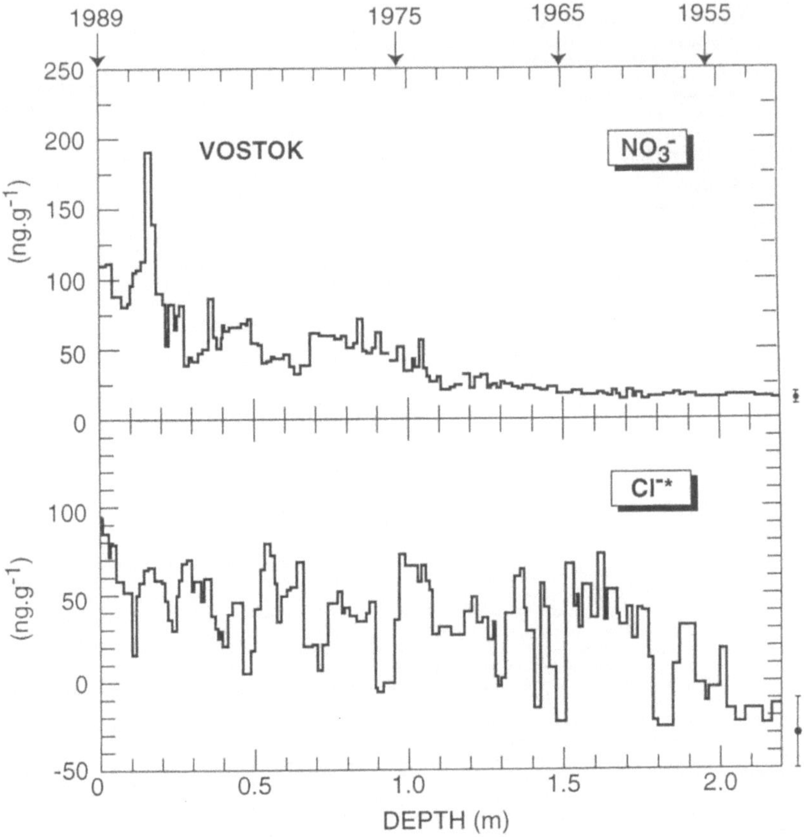

Fig. 7 Nitrate from Mayewski and Legrand [1990] and excess chloride from De Angelis and Legrand [1995] over the last decades at Vostok. Mean values of nitrate and excess chloride in deeper layers [Legrand et al., 1988] are reported on the right side.

In conclusion, it appears that the trend of HCl and HNO_3 detected in several East Antarctic shallow snow layers does not reflect changes in atmospheric concentrations of these species and represents "post depositional artefacts" which seem to be very efficient in these sites with low snow accumulation rates. In contrast, no trend of the mean value is detected for these two species in South Pole snow layers. Being in agreement with an expected absence of anthropogenic atmospheric trend in these remote regions of the Southern hemisphere, these south polar records suggest that snow accumulation rate in the order of 80 $kg.m^{-2}.yr^{-1}$ (eg 8 cm of equivalent water) is large enough to strongly limit post depositional artefacts. Under these specific conditions, we can speculate that changes of mean concentrations in ice corresponds to atmospheric changes. The atmospheric meaning of changing seasonal variations seen in south polar snow layers remains more questionable.

2.3. Long term ice records

Both deep ice cores extracted from Greenland and Antarctica covering the last great climate changes which occurred some 11,000 years ago have been scrutinized in order to investigate possible changes in the levels of gaseous acidic species in relation with the effect of the climate on their respective biogenic cycles. The richest glaciochemical studies covering the entire last climatic cycle are those of Vostok (East Antarctica, mean accumulation rate of 23 $kg.m^{-2}.yr^{-1}$, mean temperature of - 56 °C) and two deep ice cores extracted at Summit (230 $kg.m^{-2}.yr^{-1}$, mean temperature of - 32 °C) (United States GISP2 and European GRIP). At present, given our present poor knowledge of the air snow relationship, interpretation of the East Antarctic ice core remains highly speculative because of the very likely fast post deposition processes occurring near the surface (see section 2.2b). In Figure 8, we report depth profiles of nitrate and Cl^-/Na^+ along the Vostok core. The Cl^- to Na^+ ratio observed in Vostok ice remained close to the sea water ratio (1.8) during very cold climatic stages. The corresponding absence of sea salt fractionation at these time periods was interpreted by Legrand et al. [1988] in terms of limitation in the efficiency of sea salt alteration because of higher turbulence of the atmosphere and the resulting faster transport to inland Antarctica. As seen in Figure 8, the ratio is very low (0.5) during the Holocene period (0 to 13 kyrs BP) while during the preceding warm stage (116 to 140 kyrs BP) the ratio close to 3 is significantly higher than 1.8 suggesting the presence of HCl in this ice. From discussions in section 2.2b, we suspect that HCl present in the atmosphere with present day warm climatic conditions is not retained in the deep Vostok snow layers. One possible cause of its presence in ice deposited during the preceding warm stage may be related to the slight increase of the snow accumulation rate at this time [Jouzel et al., 1987].

The nitrate profile exhibits enhanced levels during the ice age (Figure 8). Based on a good correlation between calcium and nitrate concentrations along the Vostok core (Figure 9), Legrand et al. [1988] suggested that the nitrate increase corresponds to an additional input of neutral nitrate associated with the dust input which characterized the ice age. Our observation

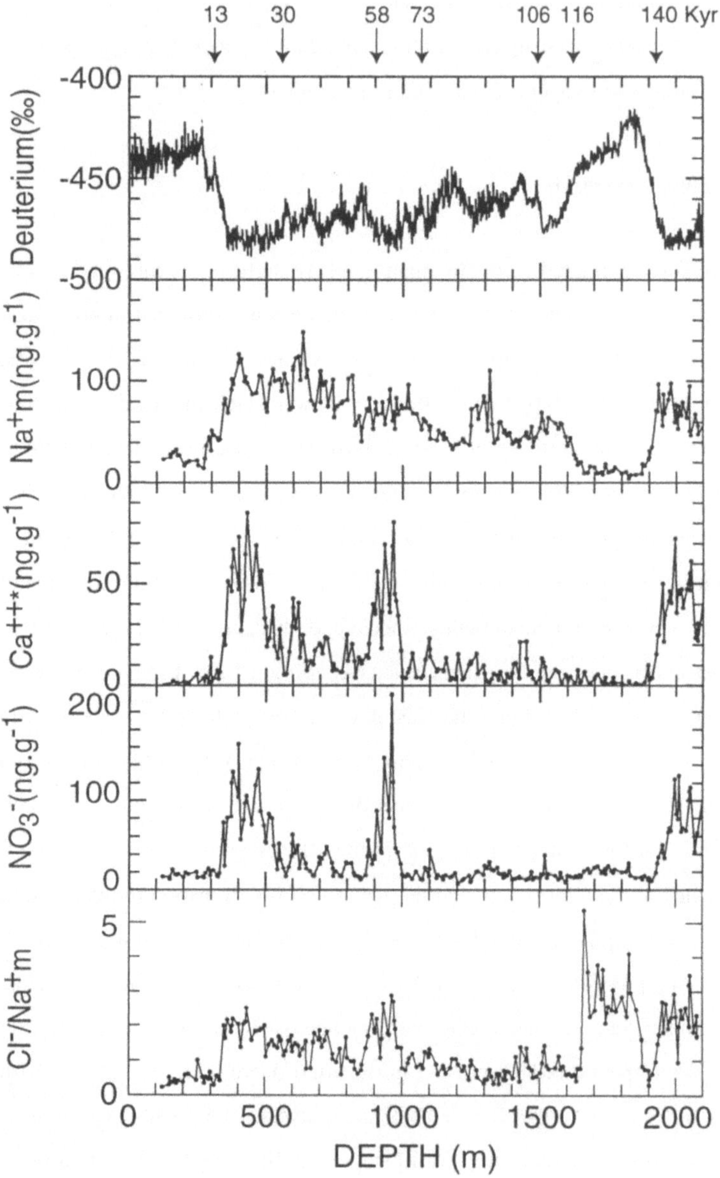

Fig. 8 Vostok ice core profiles (0 to 160,000 years): Deuterium isotope from Jouzel et al. [1987], marine sodium (Na^+_m), non sea salt calcium (Ca^{++*}), nitrate and Cl^-/Na^+_m from Legrand et al. [1988], Legrand et al. [1991].

of a strong post depositional effect in shallow snow layers at Vostok suggests an alternative explanation. Indeed, the enhanced dust amount in snow layers at this time may have limited exchange of HNO_3 between the gas phase and the solid precipitation in firn after deposition. It is therefore unclear if the observed nitrate increase during cold stages reflects higher atmospheric level of nitrate related to the scavenging of HNO_3 by dust particles or less efficient post depositional effect in relation with the enhanced presence of dust particles in/on the snow grains. The Greenland record which exhibits a rather weak nitrate changes to past climate conditions [Mayewski et al., 1994] is here not very useful to highlight the validity of nitrate record covering the past climate.

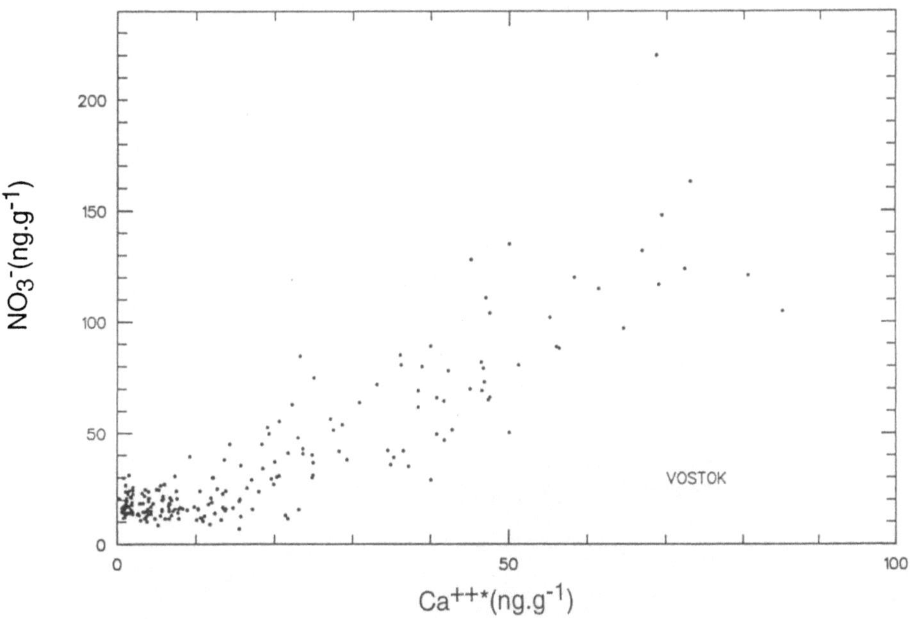

Fig. 9 Nitrate as a function of non sea salt calcium levels along the Vostok core [adapted from Legrand et al., [1988]).

The case of carboxylic acids in the Summit core is very interesting because most changes are clearly out of phase with climate variations, thus allowing the separation of effects due to changes in incorporation parameters from those due to source changes. Even considering the changes in meteorological conditions (temperature, wind speed at the snow surface) and in the acidity of the atmosphere (see Figure 10) which are expected to modify the air snow relationship of formate, the Summit formate record can be in a large extent interpreted in terms of atmospheric chemistry. For instance, the low formate levels (~ 2 ng.g^{-1}) during the warm stage C with respect to the Holocene values (~ 10 ng.g^{-1}) can be attributed to decreased atmospheric concentrations. Indeed, no drastic changes in

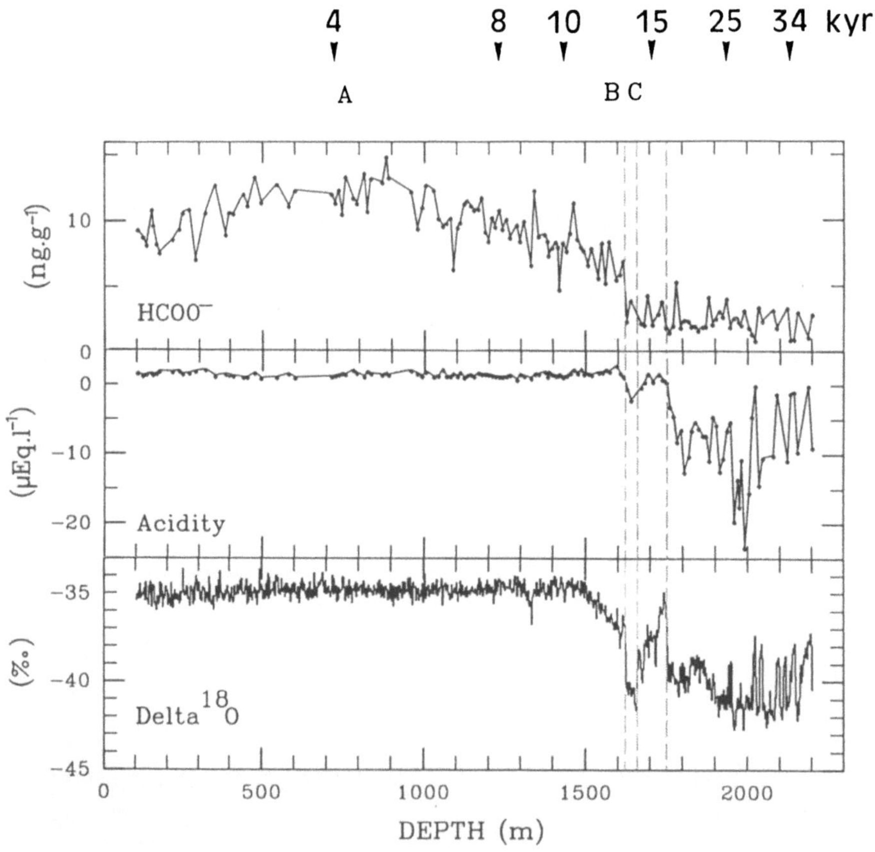

Fig. 10 Formate, acidity profiles down to 2200 m depth on the GRIP core, when considering only depth levels free of ammonium formate inputs (see section 2.1) along with the $\delta^{18}O$ values from Johnsen et al. [1992]. Vertical dashed lines refer to the successive climatic stages (denoted A to C): A is the Holocene (0 to 11,500 years BP), B the Younger Dryas (11,500 to 12,700 years BP) and C is Bolling and Allerod interstades (12,700 to 14,450 years BP). Adapted from Legrand and De Angelis [1995].

meteorological conditions nor in the acidity of the atmosphere had taken place between these two stages (Figure 10) and it is legitimate to assume the air snow relationship was unchanged. Furthermore, the increase between 11 and 5 kyrs BP and decrease between 5 kyrs BP and present time of formate ice levels (Figure 10) cannot be attributed to changes of the air snow relationship. As argued by Legrand and De Angelis [1995] such a record suggests changes in atmospheric levels in response to past climate, whose changes induced variations in the intensity of vegetation emissions from North America. Such emissions have been weakened during the ice age because of the large expansion of the Laurentide ice sheet over North America. During the last great climatic transition (some 13,000 years BP), the growth of vegetation over North America was delayed by about 5000 years, due to the slow retreat of continental ice.

We conclude that deep ice core records of acidic species from East Antarctica are mainly driven by the efficiency of post depositional effects which have likely changed over past climates. It is therefore unlikely that valid atmospheric information can be extracted from such records at present, considering our poor quantitative knowledge of the parameters involved in these post deposition effects. Greenland deep ice cores more likely contain informations on past atmospheric levels of these species. In spite of additional complexities in the form of influence of the acidity of the atmosphere, carboxylic acids records reveal changes which are out of phase with climate changes and are therefore more easy to translate in terms of relative changes in the chemical composition of the past atmosphere.

3. AIR SNOW RELATIONSHIPS

Field measurements of gas phase HNO_3 and HCl in the Summit boundary layer were first performed in the summer 1991 at ATM site using annular denuders coated with NaF [Silvente, 1993; Silvente and Legrand, 1995]. In summer 1993, mist chamber sampling techniques were used for HNO_3, HCOOH, and CH_3COOH by Dibb et al. [1994] and for HNO_3, HCl, HCOOH, and CH_3COOH by Léopold [1995]. As seen in Table 1, the three studies indicate HNO_3 mixing ratios in the range of a few tenths of pptv. Slightly higher HCl concentrations from 33 to 56 pptv are observed by Silvente [1993] and Léopold [1995]. Carboxylic acid mixing ratios are largely higher reaching a few hundreds of pptv. Although works from Léopold and Dibb having been both achieved in summer 1993, a large discrepancy appears for carboxylic acid results (Table 1). It still remains difficult to explain such large differences between the two studies and current work running in summer 1995 will certainly elucidate this problem. Formate, acetate, nitrate and chloride have also been measured in fresh snows deposited at Summit in summer (Table 1). In snow, the relative abundance of these four acidic species is very different. First, while rather similar atmospheric concentrations are observed for the two strong acids, nitric acid levels exceed those of excess chloride by almost one order of magnitude (Table 1). Second, weak acids are one order of magnitude more abundant than nitric acid in the atmosphere while the opposite is found in snow.

Table 1 Summary comparison of acidic gas mixing ratios (in pptv) at Summit in summer. Last line depicts average contents of fresh Summit snow in summer (in $ng.g^{-1}$).

HCOOH	CH_3COOH	HNO_3	HCl	References
235 ± 68	310 ± 123	30 ± 20	56 ± 30	this work, Léopold [1995]
1200 ± 685	783 ± 416	22 ± 15	-	Dibb et al. [1994]
-	-	15 ± 21	33 ± 50	Silvente [1993]
12 ± 8	13 ± 9	261 ± 233	30 ± 28	Léopold [1995] (snow)

Some aspects of the mechanisms of incorporation of these four acids have been discussed in recent papers (Legrand and De Angelis [1995] for carboxylic acids, Dominé et al. [1994], [1995] for HCl, Dominé and Thibert [1995] for HCl and HNO_3). Briefly, during the atmospheric phase of the snow, acids can be incorporated into growing snow crystals by dissolution into supercooled droplets followed by riming, or by co-condensation with H_2O, ie the sticking of the acid on the growing snow surface. Predictions of the snow composition according to both scenarios can be performed. Based on a unique set of visual examination of snow crystals collected at Summit in summer from Borys et al. [1992], Legrand and De

Angelis [1995] estimated that the mass fraction of the ice particles due to riming was about 4 10^{-3}. To examine wether the riming process is of importance in incorporating water soluble acidic gases, we have applied this value to the summer Summit atmosphere. Using atmospheric mixing ratios of 235, 310, 30, and 56 pptv for formic, acetic, nitric and hydrochloric acids respectively (Table 1) in the summer Summit boundary layer, we calculate that at - 15°C and for a present day pH of Greenland cloud droplets ranging from 3.5 to 4.3 [Léopold, 1995], fresh snowflake levels would be of 7.5, 10.7, 5.3, and 5.7 $ng.g^{-1}$ of formate, acetate, nitrate and chloride respectively (Figure 11). To estimate the expected concentration of acidic species in snowflakes due to a co-condensation process, we use the kinetic theory of gases which predicts that concentrations of the acidic species (X) in ice may be expressed as a function of sticking coefficient (α), molecular mass (M), and partial pressure (P) of the X and H_2O species following:

$$[X] = P_X \alpha_X (M_{H2O})^{0.5} / P_{H2O} \alpha_{H2O} (M_X)^{0.5}$$

Applying this equation to the summer Summit atmosphere (- 15°C, P_{H2O} = 165.7 Pa), with sticking coefficients of 0.5 for water [Haynes et al., 1992], 0.2 for HCl and HNO_3 [Dominé and Thibert, 1995], 0.1 and 0.15 for HCOOH and CH_3COOH respectively [Jayne et al., 1991], we calculate fresh snow levels of 30, 68, 9, and 13 $ng.g^{-1}$ for formate, acetate, nitrate and chloride respectively (Figure 11).

From these calculations, both riming and co-condensation processes seem able to explain the fresh snow content of HCl, HCOOH, and CH_3COOH. In contrast, for HNO_3, neither of these scenario is consistent with the high nitric acid content of fresh snow. Dominé and Thibert [1996] have measured the diffusion coefficient of HCl and HNO_3 in ice at - 15°C, which is close to the mean summer Summit temperature, and found values in the range of 10^{-12} and 10^{10} cm^2/s, respectively. Applying equilibrium solubilities measured by Dominé and Thibert [1995] to the Greenland HCl (56 ± 30 pptv, ie 3.8 ± 2 10^{-6} Pa) and HNO_3 (30 ± 21 pptv, ie 2 ± 1.4 10^{-6} Pa) atmospheric levels, we report in Figure 11 the expected amount of nitrate and chloride in snow if an equilibrium with the atmosphere has been thus reached. The diffusion coefficient for HNO_3 is large enough to allow equilibration of the falling snow flake with the atmosphere by solid phase diffusion [Dominé and Thibert, 1995]. In contrast, HCl diffusion in ice is too slow to allow the equilibrium value to be reached, and the HCl concentration is determined by riming or by the

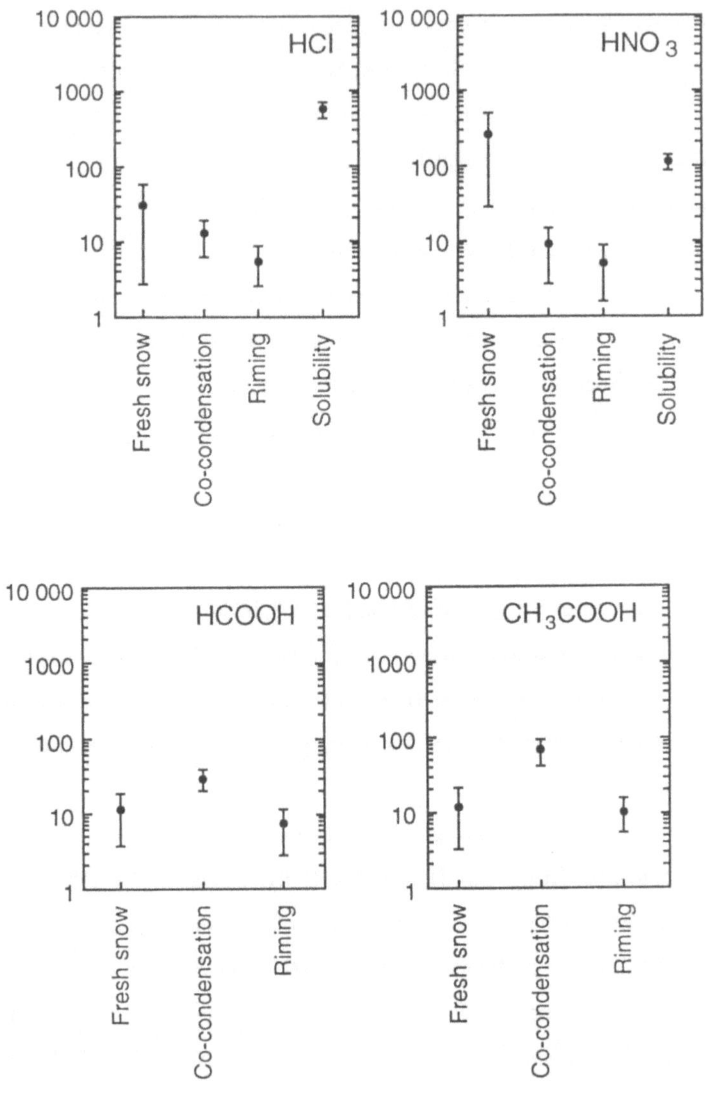

Fig 11. Comparison of formate, acetate, chloride and nitrate content of fresh snow at Summit in summer with expected concentrations calculated considering the riming, co-condensation processes at - 15°C using atmospheric concentrations from Léopold [1995]. For HCl and HNO_3, expected concentrations assuming a thermodynamic equilibrium are also reported.

kinetics of condensation rather than by thermodynamics. No laboratory measurements of the diffusion and solubility of carboxylic acids in ice are available. It is then not possible to determine the relationship between the measured values and equilibrium. An intuitive approach does not seem reasonable, because diffusion mechanisms of acids in ice are poorly known. For example, one would expect HNO_3 to diffuse very slowly in ice because of its size, but its diffusion is much faster than that of HCl. Thus, further experiments are necessary before we can begin to understand quantitatively carboxylic acids incorporation in snow.

4. CONCLUSIONS AND PERSPECTIVES

It is still difficult to provide straightforward interpretation of ice records of acidic gas species in terms of composition of our past atmosphere. With the exception of the ones from the South Pole, records extracted from the high Antarctic plateau do not reflect changes in the composition of the past atmosphere because of strong modifications of the atmospheric signals after deposition of snow at the ground. We believe that the Summit-Greenland ice cores have likely stored valid atmospheric informations. In order to assess the validity of these Greenland ice records, further glaciochemical studies achieved along ice extracted in other Greenland sites will be very valuable. It is obvious that field campains including overwinter with simultaneous measurements of atmospheric and snow composition are needed. Finally, a complete understanding of the air snow relationship will require more laboratory measurements.

ACKNOWLEDGEMENTS

The manuscript benefited from the helpful comments from the two reviewers. This work was supported by CNRS (Programmes "Phase Atmosphérique des Cycles Biogéochimiques" et "Programme National de Chimie Atmosphérique").

REFERENCES

Borys, RD, Del Vecchio, D, Jaffrezo, JL, Dibb, JE, Mitchell, DL (1992) Field observation, measurements and preliminary results from a study of wet deposition processes influencing snow and ice chemistry at Summit, Greenland. In "Precipitation Scavenging and atmosphere-surface exchanges", Schwartz and Slinn eds, 1705-1718

Conklin, MH, and Bales, RC (1993) SO_2 uptake on ice spheres: Liquid nature of the ice-air interface. J Geophys Res 98:16,851-16,855

Cragin, JM, Herron, MM, Langway, CC, and Klouda, G (1977) Interhemispheric comparison of changes in the composition of atmospheric precipitation during the late cenozoic era. Polar Oceans, proceedings of the polar oceans Conference, Montreal, 1974, Arctic Institute of North America, 617-631

De Angelis, M, and Legrand, M (1994) Origins and variations of fluoride in Greenland precipitation. J Geophys Res 99:1157-1172

De Angelis, M, and Legrand, M (1995) Preliminary investigations of post depositional effects on HCl, HNO_3 and organic acids in polar firn layers. In NATO ASI Ser "Ice core studies of global biogeochemical cycles", R Delmas ed, Springer-Verlag, 361-381

Dibb, JE, Talbot, RW, and Bergin, MH (1994) Soluble acidic species in air and snow at Summit, Greenland. Geophys Res Lett 21:1627-1630.

Dominé, F, Thibert E, Van Landeghem, F, Silvente, E and Wagnon, P (1994) Diffusion and solubility of HCl in ice: preliminary results. Geophys Res Lett 21:601-604

Dominé, F, Thibert, E, Silvente, E, Legrand, M, and Jaffrezo, JL (1995) Determining past atmospheric HCl mixing ratios from ice core analyses. J of Atmos Chem, in press.

Dominé, F, and Thibert, E (1995) Relationship between atmospheric composition and snow composition for HCl and HNO_3. In "Biogeochemistry of seasonally snow-covered catchments", IAHS Publ., 228: 3-10

Dominé, F, Thibert, E, and Chaix, L (1996) Interactions of gas phase HCl and HNO_3 with ice. In: NATO ASI Ser., "Chemical exchange between the atmosphere and polar snow", E Wolff and RC Bales eds, Springer-Verlag, this volume

Haynes, DR, Tro, NJ, and George, SM (1992) Condensation and evaporation of H_2O on ice surfaces. J of Phys Chem 96:8502-8509

Hammer, CU (1977) Past volcanism revealed by Greenland ice sheet impurities. Nature 270:482-486

Jayne, JT, Duan, SX, Davidovits, P, Worsnop, DR, Zahniser, MS, and Kolb, CE (1991) Uptake of gas phase alcohol and organic acid molecules by water surfaces. J of Phys Chem 95:6329-6336

Johnsen, SJ, Clausen, HB, Daansgard, W, Fuhrer, K, Gundestrup, N, Hammer, CU, Iversen, P, Jouzel, J, Stauffer, B, and Steffensen, JP (1992) Irregular glacial interstadials recorded in a new Greenland ice core. Nature 359:311-313

Jouzel, J, Lorius, C, Genthon, C, Barkov, NI, Kotlyakov, VM, and Petrov, VN (1987) Vostok ice core: a continuous isotope temperature record over the last climatic cycle (160,000 years). Nature 329:403-409

Kirchner, S, and Delmas, RJ (1988) A 1000 yr galciochemical study at the South Pole. Ann Glaciol 10:80-84

Legrand, M, and Delmas, RJ (1984) The ionic balance of Antarctic snow: A 10 yr detailed record. Atmos Environ 18:1867-1874

Legrand, M, and Delmas, RJ (1986) Relative contributions of tropospheric and stratospheric sources to nitrate in Antarctic snow. Tellus 38:236-249

Legrand, M, and Delmas, RJ (1988) Formation of HCl in the Antarctic atmosphere. J Geophys Res 93:7153-7168

Legrand, M, Lorius, C, Barkov, NI, and Petrov, VN (1988) Vostok (Antarctica) ice core : atmospheric chemistry changes over the last climatic cycle (160,000 years). Atmos Environ 22:317-331

Legrand, M, and Kirchner, S (1988) Polar atmospheric circulation and chemistry of recent (1957-1983) south polar precipitation. Geophys Res Lett 15:879-882

Legrand, M, Feniet-Saigne, C, Saltzman, ES, and Germain, C (1991) Ice-core record of oceanic emissions of dimethylsulphide during the last climate cycle. Nature 350:144-146

Legrand, M, Feniet-Saigne, C, Saltzman, ES, and Germain, C (1992a) Spatial and temporal variations of methanesulfonic acid and non sea salt sulphate in Antarctic ice. J Atmos Chem 14:245-260

Legrand, M, De Angelis, M, Staffelbach, T, Neftel, A, and Stauffer, B (1992b) Large perturbations of ammonium and organic acids content in the Summit-Greenland ice core: fingerprint from forest fires ? Geophys Res Lett 19:473-475

Legrand, M (1993) Ice core analysis in Arctic and Antarctic regions. In NATO ASI Ser., "The tropospheric chemistry of ozone in the polar regions", H Niki and KH Becker eds, 205-217

Legrand, M, and De Angelis, M (1995) Origins and variations of light carboxylic acids in polar precipitation. J Geophys Res 100:1445-1462

Legrand, M (1995a) Atmospheric chemistry changes versus past climate inferred from polar ice cores. In "Aerosol forcing of climate", Dahlem Konferenzen, RJ Charlson and J Heitzenberg eds, John Wiley and Sons Ltd, 123-151

Legrand, M (1995b) Sulphur-derived species in polar ice: A review. In NATO ASI Ser. "Ice core studies of global biogeochemical cycles", R Delmas ed, Springer-Verlag, 91-119

Legrand, M, and De Angelis, M (in press) Light carboxylic acids in Greenland ice: A record of past forest fires and vegetation emissions from the boreal zone. J Geophys Res

Léopold, A (1995) Contribution à l'étude de la fonction de transfert air-neige pour les gaz acides (Acides organiques, HNO_3, HCl). Thèse de doctorat de l'Université Joseph Fourier Grenoble I, pp 170

Mayewski, P, Lyons, WB, Spencer, MJ, Twickler, M, Dansgaard, C, Koci, B, Davidson, CI, and Honrath, RE (1986) Sulphate and nitrate concentrations from the south Greenland ice core. Science 232:975-977

Mayewski, PW, Lyons, B, Spencer, MJ, Twickler, M, Buck, CF, and Whitlow, S (1990) An ice core record of atmospheric response to anthropogenic sulphate and nitrate. Nature 346:554-556

Mayewski, PA, and Legrand, M (1990) Recent increase in nitrate concentration of antarctic snow. Nature 346:258-260

Mayewski, P, Meeker, LD, Morrison, MC, Twickler, MS, Whitlow, SI, Ferland, KK, Meese, DA, Legrand, M, and Steffensen, JP (1993) Greenland ice core "signal" characteristics: An expanded view of climate change. J Geophys Res 98:12,839-12,847

Mayewski, P, and 13 others (1994) Changes in atmospheric circulation and ocean ice cover over the North Atlantic during the last 41,000 years. Science 263:1747-1751

Mulvaney, R, Pasteur, EC, Peel, DA, Saltzman, ES, and Whung, PY (1992) The ratio of MSA to non sea salt sulphate in Antarctic Peninsula ice cores. Tellus 44:295-303

Mulvaney, R, and Wolff, EW (1993) Evidence for winter/spring denitrification of the stratosphere in the nitrate record of Antarctic ice cores. J Geophys Res 98:5213-5220

Murozumi, M, Chow, TJ, and Patterson, C (1969) Chemical concentrations of pollutant lead aerosols, terrestrial dusts and seasalts, Greenland and Antarctic snow strata. Geochim Cosmochim Acta 33:1247-1294

Neftel, A, Beer, J, Oeschger, H, Zurcher, H, and Finkel, F (1985) Sulphate and nitrate concentrations in snow from South Greenland 1895-1978. Nature 314:611-613

Neftel, A (1996) The record of gases and reactive species in ice cores, and problems of interpretation. In: NATO ASI Ser., "Chemical exchange between the atmosphere and polar snow", E Wolff and RC Bales eds, Springer-Verlag, this volume

Petit, JR, Briat, M, and Royer, A (1981) Ice age aerosol content from East Antarctic ice core samples and past wind strength. Nature 293:391-394

Sigg, A, and Neftel, A (1988) Seasonal variations in hydrogen peroxide in polar ice cores. Ann Glaciol 10:157-162

Silvente, E (1993) Contribution à l'étude de la fonction de transfert air-neige en régions polaires. Thèse de doctorat de l'Université Joseph Fourier Grenoble I, pp 157

Silvente, E, and Legrand, M (1995) A preliminary study of the Air-snow relationship for nitric in Greenland. In NATO ASI Ser. "Ice core studies of global biogeochemical cycles", R Delmas ed, Springer-Verlag, 225-245

Staffelbach, T, Neftel, A, Stauffer, B, and Jacob, D (1991) A record of the atmospheric methane sink from formaldehyde in polar ice cores. Nature 349:603-605

Waddington, ED, Cunningham, J, and Harder, S (1996) The effects of snow ventilation on chemical concentrations. In: NATO ASI Ser., "Chemical exchange between the atmosphere and polar snow", E Wolff and RC Bales eds, Springer-Verlag, this volume

Wolff, EW (1995) Nitrate in polar ice. In NATO ASI Ser. "Ice core studies of global biogeochemical cycles", R. Delmas ed, Springer-Verlag, 195-224

Wolff, EW (1996a) The record of aerosol deposited species in ice cores, and problems of interpretation. In: NATO ASI Ser., "Chemical exchange between the atmosphere and polar snow", E Wolff and RC Bales eds, Springer-Verlag, this volume

Wolff, EW (1996) Location, movement, and reactions of impurities in solid ice. In: NATO ASI Ser., "Chemical exchange between the atmosphere and polar snow", E Wolff and RC Bales eds, Springer-Verlag, this volume

THE RECORD OF GASES AND REACTIVE SPECIES IN ICE CORES, AND PROBLEMS OF INTERPRETATION

Albrecht Neftel
Physics Institute, University of Bern
CH-3012 Bern
Switzerland

Abstract

Since the start of the ice core drilling some 30 years ago, a whole zoo of species and parameter have been measured and reported in the literature. Those records per se are of minor relevance. Of great interest is the link to the corresponding atmospheric concentrations, which are related to basic parameter of the Earth System.

Ice, conserved well below the freezing point, knows two clearly distinguishable ways of carrying information of the past:

a) Composition of the gas enclosed in the bubbles

b) Trace species in the ice matrix

Once a good ice record is established it is still a long way to go until a relevant record for an "Earth system parameter" is gained. This path can be divided into different steps

$$c_{Atmosphere}(t) = \frac{c_{Atmosphere}}{c_{Atmospherelocal}} \cdot \frac{c_{Atmospherelocal}}{c_{Snow}} \cdot \frac{c_{Snow}}{c_{Firn}} \cdot \frac{c_{Firn}}{c_{Ice}} \cdot c_{ice}(t)$$

The reconstruction of an atmospheric record from the concentration versus depth profile gained from ice cores is similar to an odyssey through a labyrinth with many pitfalls ready to slur over enthusiastic students and researchers.

The paper focusses on the discussion of three „Earth system parameters“: CO_2, H_2O_2 and HCHO. It tries to show up some of the major difficulties on our odyssey.

NATO ASI Series, Vol. I 43
Chemical Exchange Between the Atmosphere and Polar Snow
Edited by Eric W. Wolff and Roger C. Bales

1) Introduction:

Over the last 30 years it has become well known that ice caps preserve information about past atmospheric composition. Changes in ice core and corresponding atmospheric compositions are always an indication of major changes in global environmental conditions. The temporal behaviour of parameters recorded in ice cores (PIC's) offer the possibility to get a better understanding of the cause and effect relationships of systematic changes to our environment. This information also serves as important scientific background for the consensus documents from the World Meteorological Organization and United Nations Environment Programm "Climate Change" (IPCC-Report,1990), which finally got a broader political discussion with the conference in Rio de Janeiro in 1992.

Ice caps and glaciers accumulate past precipitation. Cores are intentionally taken from places where a uniform deposition of snow occurs, which allows a good dating of the core. Counting years with seasonally varying parameters and converting depth scales to age scales is an old passion of the ice core community and sometimes a controversial area. Establishment of a depth to age scale is of course a prerequest for the interpretation of any ice core records. For the following discussion, let me assume that such a relation exists and is known.

The establishment of a property (mostly concentration) versus depth profile is already a great piece of work. Since the start of the ice core drilling some 30 years ago, a whole zoo of species and parameter (certainly over 50) have been measured and reported in the literature. Those records per se are of minor relevance. Of great interest is the link to the corresponding atmospheric concentrations, which are related to basic parameter of the Earth System.

Ice, that remains well below the freezing point, offers two clearly distinguishable ways of preserving information of the past:

a) Composition of the gas which is enclosed in bubbles at the firn - ice transition, thus in the depth interval where the density increases from 800 kg/m^3 to 840 kg/m^3 (Schwander and Stauffer, 1984)

b) Trace species in the ice matrix and/or at grain boundaries.

For the further discussion, let me assume that the reported results correspond to the real gas composition in the bubbles or the real concentration in the ice respectively. This is not a trivial assumption: many discrepancies in interpreting ice cores can be traced down to analytical artefacts.

Once a good ice record is established it is still a long way to go until a relevant record for an "Earth system parameter" is gained. This path can be divided into different steps, as we have proposed it already in an earlier contribution for a NATO-ARW (Neftel and Fuhrer,1993):

$$c_{Atmosphere}(t) = \frac{c_{Atmosphere}}{c_{Atmospherelocal}} \cdot \frac{c_{Atmospherelocal}}{c_{Snow}} \cdot \frac{c_{Snow}}{c_{Firn}} \cdot \frac{c_{Firn}}{c_{Ice}} \cdot c_{ice}(t)$$

Each of these steps will varie with time beginning on a diurnal scale.

This scheme will serve as a guideline for the further discussion. Let us have first a look at CO_2 and other non reactive trace gases such as methane (CH_4) and nitrous oxide (N_2O). These three gases are together with the CFC's the main greenhouse gases and are responsible for the predicted global warming.

2.) How reliable are the records of the Greenhouse gases taken from polar ice cores?

1990 I wrote in an other contribution to a NATO ARW: (Neftel, 1991) " *The statement that CO_2 is a "chemically boring" substance in snow and ice can be safely made". The same is true for CH_4 and N_2O. .. The main difficulty in interpreting "greenhouse data" from polar ice cores is the exact determination of the age of enclosed gas".*

In the mean time detailed CO_2 analysis of the newly drilled deep core in Greenland revealed that the story is not so simple, so that I have to revise my earlier statement.

The enclosure of stable gases deep in the firn at the firn ice transition is a slow process of several tens of years depending on the meteorological characteristics of the drill site (Schwander and Stauffer, 1984, Schwander this volume). Direct measurement of the firn - gas composition did not reveal a systematic chemical change of the gas composition under present day conditions. Minor changes of the concentration and isotopic composition are due to the effect of gravity and have been documented (Schwander, this issue). CO_2 measurements on both the GRIP and GISP2 core documented elevated CO_2 concentrations compared to results from Antarctica from samples having similar gas ages (Anklin et al, 1995; Barnola et al., 1995). There is not only a systematic shift in the level, but also elevated CO_2 concentrations in narrow layers. Such elevated concentrations were measured earlier on a core from Crete (Central Greenland), but have been attributed to a transport damage of the cores

(Anklin, 1991, Staffelbach et al., 1991a). The discrepancies between the Antarctic and Greenland ice cores can be summarised as follows:

a) CO_2 concentrations of Greenland deep cores in the so called Dansgaard-Oeschger events (the very rapid climatic fluctuations during the last glaciation) exactly parallels the $\delta^{18}O$ record as well as the Ca^{2+} record. Because the enclosure process of the gas in the bubbles acts as a low pass filter of several tens of years, the variation in the CO_2 signal cannot be of atmospheric origin (Staffelbach et al., 1988). Similar variations have not be seen in the Antarctic cores (Neftel et al., 1988; Anklin, 1994c)

b) Starting with a good match for pre industrial time (200 years B.P.) CO_2 levels between cores from both hemispheres, Greenland ice cores show an increasing CO_2 concentration with increasing gas age, up to a difference of 20 ppm (figure 1).

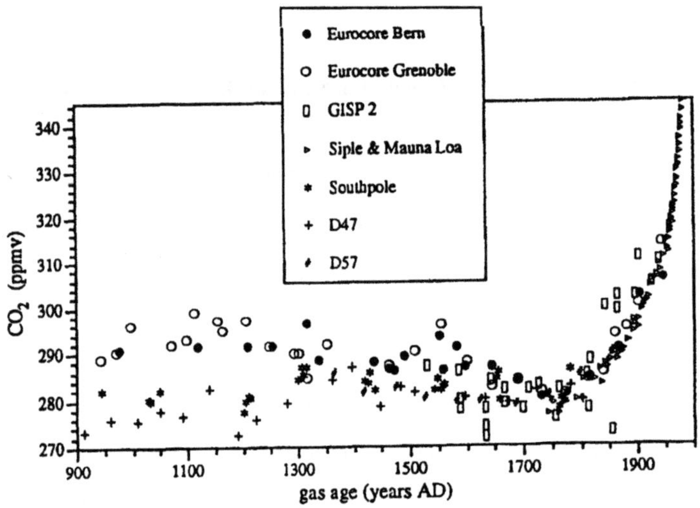

Figure 1: CO_2 data covering the past 1000 years from Greenland and Antarctic drill sites (from Anklin 1994c).

It is obvious, that these differences cannot be of atmospheric origin, because CO_2 is globally well mixed and the inter hemispheric difference due to different source and sink distribution can only be a few ppm (Pearman and Hyson, 1986). Such discrepancies have not be seen for the other two greenhouse gases CH_4 and N_2O (Blunier et al , 1993).

Table 1 summarizes the CO_2 concentration differences evaluated from Greenland and Antarctic ice cores.

Greenland			Antarctica				Difference Greenland-Antarctica
Gas age (Years AD)	CO_2 ppmV	σ ppmV	Gas age (Years AD)	CO_2 ppmV	σ ppmV		ppmV
1583	291	5.5	1605	282	3.2		9
1490	289	6.8	1489	283	2.2		6
1437	289	5.0	1416	279	2.7		10

Table 1a: Comparison of CO_2 for the last millenium derived from Antarctic ice cores (Siple and South Pole) and Summit Greenland.

Climatic stage	Age (years B.P.)		Summit	Dye 3	Byrd
	beginning	ending			
preboreal	11'500	10'000	295 ±16	331 ± 17	275 ± 7
younger dryas	12'700	11'500	251 ± 31	295 ± 24	258 ± 8
boelling-alleroed	14'450	12'700	267 ± 17	305 ± 4	238 ± 9
glacial maxima	23'000	17'000	205 ± 19	205 ± 13	197 ± 5
Dansgaard-Oeschger Events					
cold phases			220 ± 24	204 ± 11	<220
warm phases			261 ± 27	251 ± 12	<220

Table 1b: Comparison of CO2 data for the last glaciation between deep cores from Antarctica and Greenland.

Reaction of carbonates in the ice with acids are an additional source of CO_2 and could enrich CO_2 in the bubbles (Delmas, 1993). The same process could also lead to a depletion in alkaline ice, because dissolved CO_2 is a weak acid. Such an effect could only be of importance in alkaline ice which is present in Greenland ice from cold stages of the last glaciation. It would not explain the observed differences in the Holocene.

In his thesis Martin Anklin (1994c) suggests that carbonate dissolution is an additional source of CO_2 in the bubbles (Anklin et al, 1994b, Barnola et al., 1994). He suggests that the reaction takes place in the disordered surface layer at grain boundaries. Additional CO_2 is only produced if carbonate and acidic compounds coexists in the ice and survive down to the firn-ice transition. Such an explanation is reasonable, because it is more likely that dust particles containing carbonate survive the transport to Greenland, whereas they have most likely interacted while airborn with acidic compounds on the longer way to Antarctica.

Nevertheless, there are some potential inconsistencies in this explanation.

a) Does the conversion process go to completion or does it stops after a few hundred to thousand years?

It should be expected to invoke only impurities that are located at grain boundaries. Anklin et al. (1994b) made the statement that the availability of carbonate is the limiting factor for the CO_2 production. If strong acids are homogeneously distributed in the ice matrix, as I speculated from the electric conductivity measurements (ECM) (Neftel, 1991) the availability of strong acids could be the limiting factor. The supply of these acids in the reaction zones could be limited by diffusion of these acids. This puts us to the question of the distribution of impurities in the ice, which is addressed by Eric Wolff in this volume. Recently the Bern group build up a very sophisticated instrument to determine the low carbonate levels in polar ice (Fuhrer A., master thesis, 1994). Andreas Fuhrer could show, that there is still carbonate in Holocene ice from Summit up to levels of $0.4 \mu moll^{-1}$ which constitute a potential contribution of 60 ppm CO_2.

b) CO_2 production at grain boundaries requires three chemical steps:

$CaCO_3 \rightarrow HCO_3^- \rightarrow H_2CO_3 \rightarrow CO_2$

These scheme is taken from aqueous phase chemistry and it is supposed that a kind of aqueous phase chemistry occurs in the disordered region at grain boundaries. Mulvaney et al., (1989) showed that liquid veins enriched with H_2SO_4 exists at triple junction. I would expect, that in such chemically reactive zones H_2O_2 is destroyed more efficiently compared to H_2O_2 molecules located within an ice crystal. To my knowledge a link between CO_2 enhancement and H_2O_2 depletion has not been observed.

Because of the much higher carbonate level in Greenland ice the CO_2 records from Greenland ice cores are less reliable for the reconstruction of the past atmospheric CO_2 record compared to records from Antarctic cores.

Other important greenhouse gases are CH_4 and N_2O. From an analytical point of view CH_4 and N_2O behave as "friendly" gases. There are no potential precursors in the ice and their solubility in water is small, so that even liquid extraction systems do work.

Next on the list of chemical trace gases to be measured are CO and H_2. Because of their low concentrations, experimental difficulties in extracting the gas from the ice in sufficient amounts without affecting the concentration have thus to my knwledge prevented a successful measurements.

Looking at our transfer function scheme in establishing the transfer function for stable gases, the major difficulties arises from the last step, transformation within the ice. All other steps are rather simple, especially the relation between the local atmosphere and the global concentration. The enumerated stable gases have such a long lifetime in the atmosphere, that local measurements in a remote location far away from sources or sinks are significant measurements for hemispheric values.

3. Reversibly deposited species.

HCHO and H_2O_2 are the two trace gases which occur in the atmosphere in the low ppb to ppt range and that are highly soluble in water and that do not form major ionic species. Even if these species were perfectly conserved in the bubbles it is highly unlikely that they could be experimentally determined in the extracted air.

H_2O_2 and HCHO have been both determined in the ice matrix. The measurement of these species enlarged quite a bit the potential of information and also the speculation about changes in the atmospheric chemistry in the past. Let me first summarise the main features of them in polar snow, firn and ice.

3.1 H_2O_2

3.1.1. Pit studies

Pit studies show detailed features related to seasonality and initial post depositional processes. It is common to all pit studies, that I am aware off, that they have been taken during the summer season. H_2O_2 shows a clear seasonal variation in all locations that have mean temperature well below the freezing point. An example is shown in Figure 2.

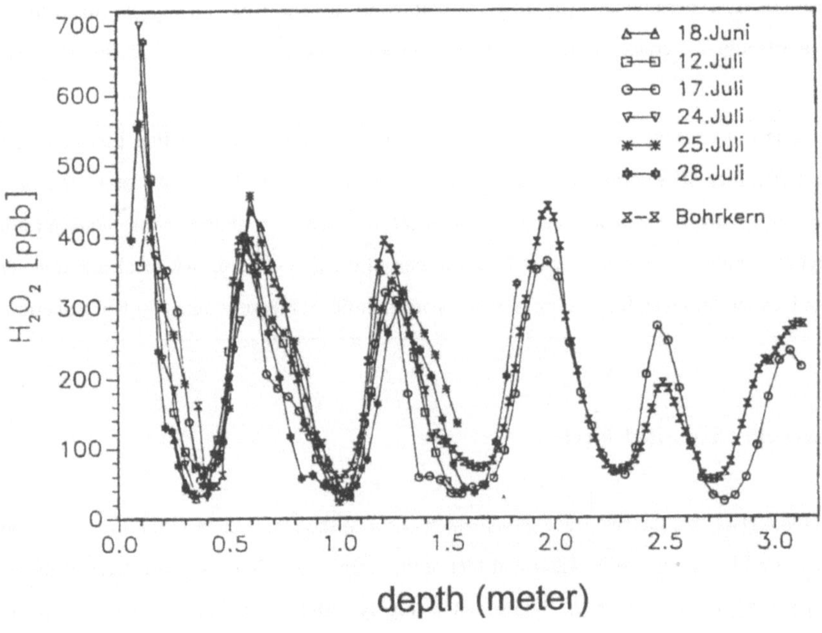

Figure 2: H_2O_2 pit studies from Central Greenland (from Sigg, 1990).

This seasonal variation corresponds to the expected seasonality in the atmospheric concentration. A phase shift between the H_2O_2 and $\delta^{18}O$ summer peak is seen in pits from Dye 3 and Summit and shallow cores from Central Greenland. Figure 3 shows the phase relation of $\delta^{18}O$ and H_2O_2 concentrations in the Greenland cores.

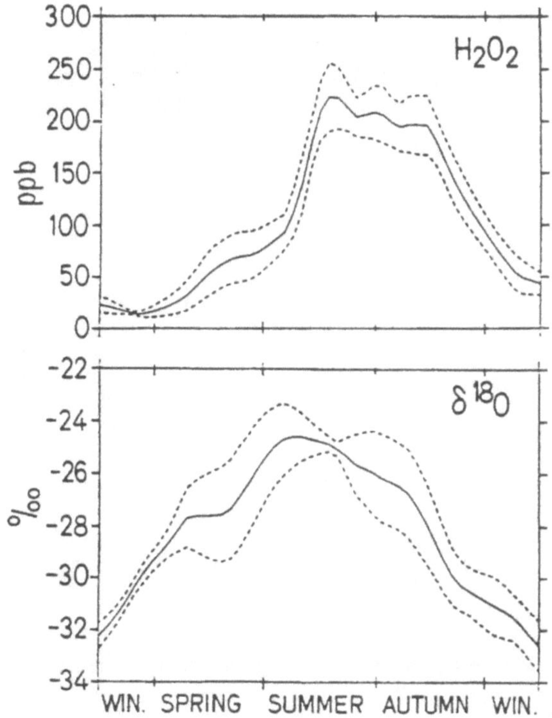

Figure 3: Phase shift $\delta^{18}O$ - H_2O_2 in a pit from Dye 3 (from Beer et al., 1991)

The H_2O_2 lags one or two month behind the $\delta^{18}O$ peak (Beer et al, 1991, Sigg 1990). In the Siple core, Antarctica, no clear shift between the H_2O_2 and $\delta^{18}O$ summer peak is seen (Sigg, 1990), whereas in the high accumulation core from Law Dome the H_2O_2 seems to preceed the $\delta^{18}O$ signal by approx. half a month (T. van Ommen, pers. comm.).

Almost all pit studies show a clear decrease from the actual surface value to the preceding summer peak, while no systematic trend is seen for the first few winter minima. Pit profiles from the Greenland plateau above 2400 m.a.s.l. show no systematic losses after the first year. A good overview is given in the paper by Anklin et al., 1994 where pits have been taken along a traverse from Summit to the West coast of Greenland (Figure 4).

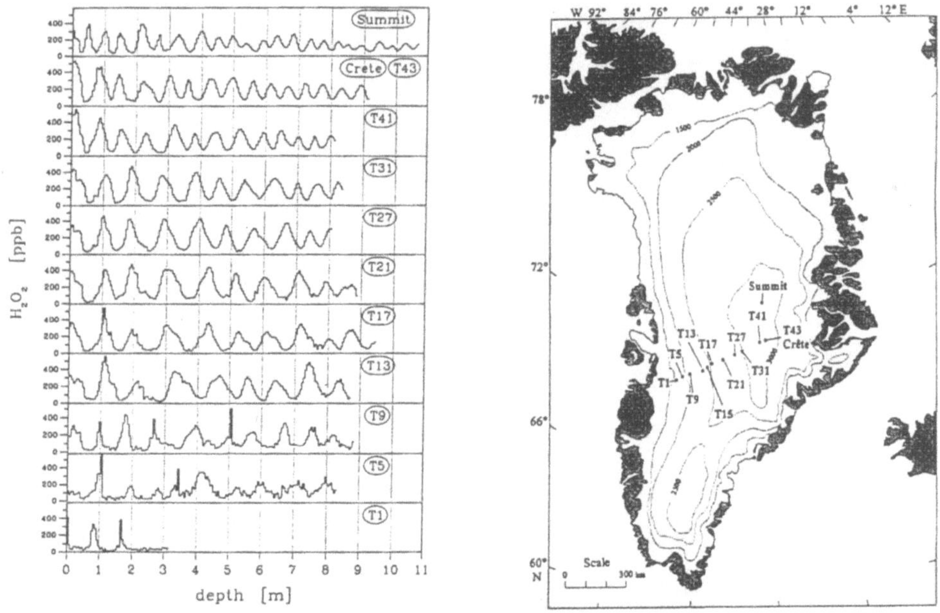

Figure 4: H_2O_2 profiles of 10 metres shallow cores along a western flow line starting from Central Greenland (Anklin, 1994c)

In Antarctica the situation is more complex. Station located in a narrow coastal band show similar pattern as for Greenland, whereas all station situated in the interior which is shielded by the circumpolar vortex show a strong and systematic loss of H_2O_2 (Neftel et al., 1995 K. Fuhrer, this volume).

3.1.2 Shallow cores

Continuous high resolution measurements of H_2O_2 revealed a more pronounced smoothing of the initial seasonal signal than occurs for $\delta^{18}O$. From the Siple and Summit records Sigg (1990) deduced a mean displacement length of 12 to 13 cm over the whole firnification. This displacement length is 50% larger than the one for $\delta^{18}O$.

The Dye 3 and Summit H_2O_2 records gave evidence for a systematic increase of the long-term mean value by 50% over the last 200 years (Sigg and Neftel, 1991). The largest fraction of this increase occurred over the last 30 years. It has to be noted, that the trend is relatively small compared to the year to year fluctuation.

Sigg (1990) showed that the mean annual concentration of the Summit and Dye 3 core are weakly correlated, whereas the cross correlation between Summit and Siple yielded no correlation. This has to be expected in view of the atmospheric lifetime of a few days of H_2O_2. Consequently atmospheric H_2O_2 concentrations in the Southern and Northern hemisphere are not connected.
The Siple record from Antarctica does not show an increase over the last 50 years. In the mean time a newer record should be available from Law Dome, a high accumulation site in Antarctica (Morgan, pers. communication) which will allow a re-evaluation of temporal trends in the H_2O_2 concentration.

3.1.3 Deep cores

H_2O_2 has been analysed continuously on only part of the new GRIP deep core. On previous deep cores (Camp Century, Dye 3 and Byrd) only a few samples were analyzed. The general picture established in 1986 for Greenland is confirmed by the Summit record. During the first 3000 years H_2O_2 decreases at an average of 2% per 100 years. From there on the concentration is stabilized until 9'000 years B.P., where a more rapid decay starts (Neftel et al, 1986, Sigg and Neftel, 1991, Fuhrer et al, 1993). The decrease of H_2O_2 occurs in small layers where sharp drops start to develop with a decay rate on the order of 0.1% a^{-1} (Figure 5). These holes must have developed in the ice, because they would have been smoothed out, had they have been there initially. They must be distinguished from some low H_2O_2 values associated with volcanic eruptions, which are already clearly seen in the firn concentration (Sigg and Neftel, 1991 Laj et al., 1990). During the Younger Dryas the H_2O_2 died out completely and recovered slightly in the Boelling-Alleroed. Deeper down the signal reappeared to values up to 2 ppb associated with NH_4^+ spikes (K. Fuhrer pers. comm.). H_2O_2 and Ca^{2+} concentration are basically anti correlated (Fuhrer et al, 1993). Where the Ca^{2+} level reaches a level above 1 µmolar the signal has died out.

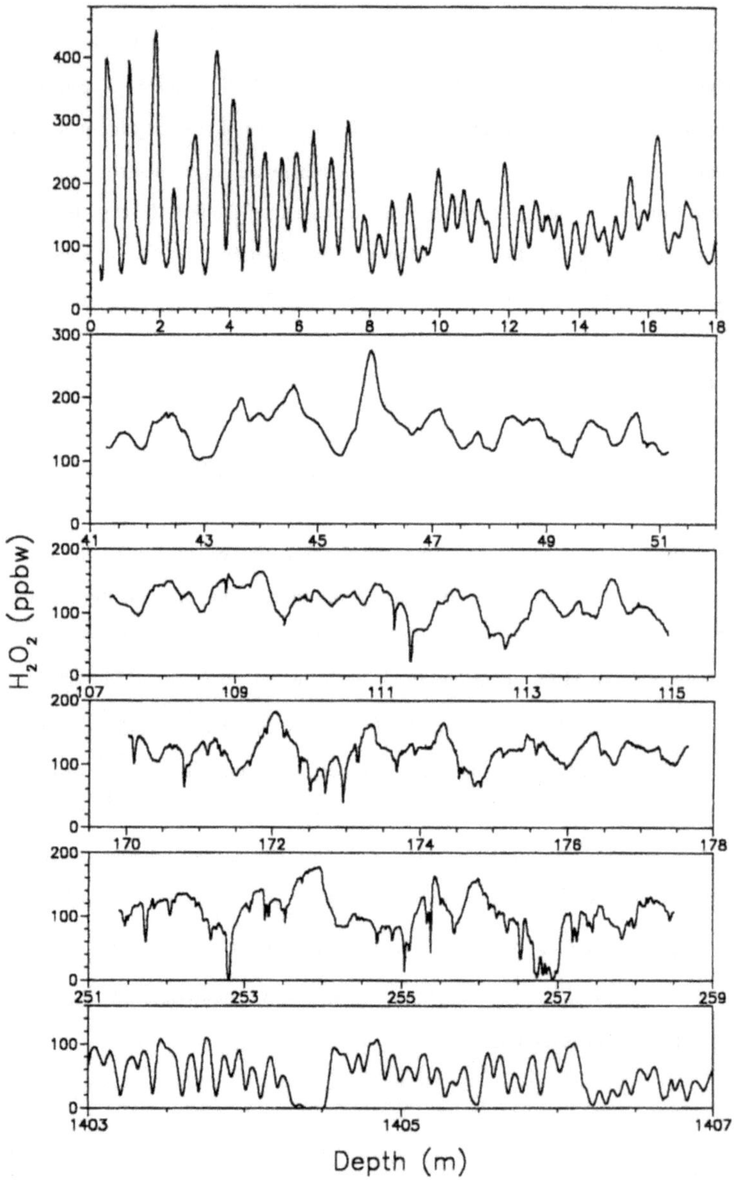

Figure 5: Developments of holes in the Summit H_2O_2 record

For Antarctica a continuous long-term record does not exist. A few measurements have been done along the Byrd core (Neftel et al., 1986; Neftel and Fuhrer, 1993).

The Byrd samples have a rather surprising elevated concentrations in the time period of the transition from glacial time to the Holocene between 15'000 and 8'000 years B.P. Detailed analysis even suggested that seasonal variations could be preserved for this time interval or that regularely appearance of wind crust switched off the smoothing process in the firn. A seasonal structure would only safely survive in case an accumulation rate double that of present day conditions was present. The drop into glacial time is less pronounced than in Greenland, which is consistent with the lower dust load for Antarctica compared to Greenland.

Interesting results are emerging from the „Australian" cores from Law Dome. Perfect seasonal cycles are unaffected at least 600 years back in time. This is remarkable especially in light of the high mean annual temperature of -22°C (Tas van Ommen, pers. comm.).

3.2 HCHO

3.2.1 Pit studies

Greenland pit studies showed a surprising behaviour. Lowest concentrations are found 10 cm below the surface which is followed by an increase to about 50% of the concentrations found in fresh snow fall (Staffelbach et al., 1991). These characteristics have been found in all pit studies from the Greenland plateau. Further down in the snow pack, the HCHO concentration is constant and does not show any seasonal variation.

3.2 Shallow core

On shallow cores from Dye 3 and Summit Staffelbach (1991) found a 50% increase in the HCHO concentrations over the last 100 years which parallels the CH_4 increase. Continuous measurements performed immediately after core recovery in Summer 1993 along a shallow core from Summit confirmed the increase (Fuhrer K. pers. comm.). Freshly measured ice cores show no seasonal variation of the concentration, changes in the ±20% range do occur over a time scale of tens of years. To my knowledge no HCHO measurements have been done on an Antarctic core covering the last few hundred years. Again the „Australian" core from Law Dome seems the most promising core material for such an analysis.

3.3 Deep ice cores

Staffelbach made a survey along the Dye 3 and Byrd core. He found a simple picture with constant values during the Holocene and lower value in glacial time (Staffelbach, 1990). The continuous measurement along the GRIP core, performed directly in the field showed a radically different picture (Fuhrer et al., 1993). HCHO concentration in cold stages of glacial period were up to ten times higher, than the Holocene values. HCHO parallels the Ca^{2+} content in the coarse features, the transitions from "dusty and cold" to "clean and warm" conditions. The levels in warm stages are comparable to the earlier Holocene values, but lower compared to the pre-industrial level. It is likely, that Staffelbach picked up ice from warm stages in the glaciation, or misinterpreted high values as contamination. (It has to be noted, that Staffelbach made the first HCHO measurements in polar ice).

4. Discussions of the H_2O_2 and HCHO records under the header "transfer function"

In the following I make an attempt to put up a series of hypothesis that are important to explain how concentration changes in the ice are related to changes in the atmosphere. I follow thereby the scheme of the transfer function with the 6 steps.

4.1 Step 1: Global atmosphere to local atmosphere relation.

The stable gases CO_2, CH_4 and N_2O are hemispherically well mixed and measurements at a remote location such as the polar ice sheets will be representative globally or at least hemispherically.
The atmospheric lifetimes of H_2O_2 and HCHO are determined primarily by the photolytical decay. In polar regions lifetimes are a few hours for HCHO to a few days for H_2O_2 in summer time and several days to weeks for both in winter time. The atmospheric summer HCHO and H_2O_2 concentrations over Central Greenland are locally to regionally determined respectively. Due to the absence of major emission sources close by they should reflect tropospheric background concentrations over Greenland. In winter time, long-range transport will determine the atmospheric concentrations over Greenland. The link between global mean values and atmospheric values derived from snow and ice concentration must be given by at least two dimensional atmospheric photochemical models in order that the lattitudinal dependence is taken into account.
From the recent field experiments devoted to the transfer function of H_2O_2 and HCHO carried out in Summer 93 and 94 (Bales et al, 1995, Bales et al. JGR in press, Fuhrer K. et al., this volume) the following hypothesis can be derived:

1.) Atmospheric surface concentrations of HCHO are enhanced by outgassing from the snow surface layer. These enhanced HCHO level in turn increase the atmospheric H_2O_2 concentrations via the photlytical decay of HCHO which produces HO_2 radicals. The recombination of HO2 radicals yields enough H_2O_2 to explain the differences between the measured diurnal maxima values and the modeled concentrations.

2.) Diel peak atmospheric H_2O_2 concentration as measured during the summer field campaigns are not influenced by a direct degassing of H_2O_2 from the snow pack.

4.2 Step 2: Atmosphere - Snow transfer

In case the snowflakes grow over the vapour phase the transfer function for H_2O_2 molecules is given by co-condensation (P.Jacob, 1986). Co-condensation means that in snowflakes the same ratio between H_2O_2 and H_2O molecules is established as in the gas phase. Concentration measurements on hoar frost, which grow overnight when a strong inversion has developed supported co-condensation as the appropriate transfer mechanism (Sigg, 1990). The co-condensation model implies, that H_2O_2 molecules are uniformly distributed into the ice. The snow resp. ice is then supersaturated with respect to the laboratory determined partitioning shown in Figure 6 (Conklin et al., 1993) which seems applicable for the conditions in central Greenland firn. Thus to reach equilibrium the deposited snow must release a large fraction of the H_2O_2 originally present. The snow is therefore a potential atmospheric H_2O_2 source. Attempts to determine H_2O_2 fluxes so far did not reveal a clear evidence for a systematic net exhalation of H_2O_2 (Bales et al, 1995, Bales et al. JGR in press).

For HCHO the transfer function can only be estimated. If we take the predicted model concentration as a reference (Neftel et al., 1995) and compare with concentration measued in fresh snow a lower concentration ratio than for H_2O_2 co-condensation emerges for Summit conditions p_{atm} = 700mbar:

	Air concentration (ppbV)	Fresh snow concen. (ppb)	K_D - ratio (Mol/liter)/atm
HCHO	0.1	10	4300
H_2O_2	2	600	12800

Table 2: K_D - ratio for HCHO and H_2O_2 during the initial snow formation in the atmosphere

If this difference of the K_D values between HCHO and H_2O_2 is real, this means a lower sticking probability of HCHO compared to H_2O_2 molecules on the growing snow.

4.3 Step 3: Transformation in the early lifetime of the snow

Pit studies reveal most of the problems we encountered when discussing the transfer puzzle. A consistent explanation of the different features does not yet exist. Therefore I try to review at least major inconsistencies and to give suggestions as to how they could be solved.

Observation 1): All H_2O_2 records from Greenland pit studies show only a clear decrease from the actual summer level to the previous one, whereas the winter layers do not show a significant trend

Explanation: The decrease in concentration in the actual summer layer is consistent with the idea, that snow adjusts its concentration from the supersaturated co-condensation value to a temperature dependent equilibrium concentration given by the K_D value from Conklin et al, 1993 (shown in figure 6).

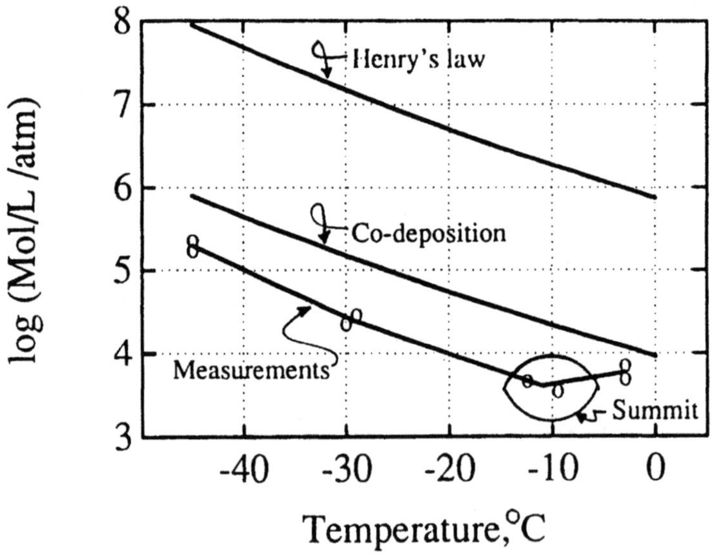

Figure 6: Comparison of various equilibrium transfer functions with experimental data. (Conklin. et al., 1993).

At constant temperature the initial snow concentration should decrease to 10% of its initial value in order to reach the equilibrium value. The actual change in the snow concentration will be less because the deposited snow layer will adjust to a lower temperature and thus a higher equilibrium concentration. The transfer coefficient strongly increases with decreasing temperature.

Pitfalls: It is unclear during which time period this loss occur. A detailed survey of surface snow concentration over two field seasons did not reveal a loss (Bales et al, 1995a and 1995b). Therefore the loss has to occur after the field season, between autumn and the next spring. In this time period the surface snow is getting colder and the distribution of H_2O_2 is shifted toward the solid phase. H_2O_2 molecules leaving the supersaturated summer layers will diffuse into snow layers above and below, which have much lower concentration and have therefore a high affinity for H_2O_2. We would thus expect to see an increase in the winter layers that is not observed.

Observation 2): Pits taken across the Antarctic vortex show an increasing loss with latitude of H_2O_2. (Fuhrer,pers. comm.). Initial surface snow concentrations are comparable across the whole Antarctica, consistent with atmospheric model calculation and assuming co-condensation.

Explanation: We speculated that losses were due to extremely low atmospheric winter values combined with an efficient outgassing and a high ventilation rate (Neftel and Fuhrer, 1993; Neftel et al.,1995). Within and outside the Vortex a periodic recharge of the surface with H_2O_2 could take place when air masses carrying H_2O_2 arrive from northern latitude.

Pitfall: If we accept the laboratory determined K_D value as an equilibrium constant, the atmospheric concentration is not allowed to exceed the values given in Table 3.
It seems questionable that the shielding effect of the Antarctic vortex is so strong, that such low atmospheric concentration are maintained in polar winter. During the polar winter the atmospheric lifetime is very much prolonged due to the absence of light.

Temperature	K_D Moll^{-1}atm^{-1}	c_{snow} (µmolar)	c_{atm} (natm)
-30	15'800	100	0.21
-40	100'000	100	0.033
-50	316'00	100	0.0104

Table 3: Upper limits of atmospheric concentrations needed to maintain a snow concentration of 100 µmolar at different temperatures.

Observation 3): HCHO snow concentrations show a minimum 1 cm to 20 cm below the surface. This has been repeatedly confirmed for Greenland. HCHO snow gas measurements performed during the field seasons 1993 & 1994 in Summit as well as on the high alpine research station at the Jungfraujoch by Katrin Fuhrer and Manuel Hutterli showed enhanced concentrations in the surface snow layer compared to atmospheric concentration.

Explanation: Previously Staffelbach et al. (1991) explained the low snow concentrations below surface by a photolytical decay. As already pointed out in this paper a photolysis could not occur if HCHO was in the glycol form. An oxidation within the snowpack leading to additional HCHO in the firn gas is highly unprobable. It would also be surprising the high firn gas concentrations parallel the minimum in the firn concentration.

A plausible explanation is as follows: HCHO is predominantly adsorbed on the surfaces of snowflakes rather than incorporated in the bulk ice. After deposition the specific surface of the snow is rapidly reduced and drives HCHO molecules into the gas phase. The snow therefore acts as an HCHO source and might explain why measured atmospheric concentrations exceed the calculated ones by more than 300%. It seems probable that HCHO is brought to the Greenland ice sheet within cloud systems.

HCHO molecules diffusing down into the snow pack are increasingly shielded from light and are gradually less degraded by photolytical decay. A kind of equilibrium between firn and firn gas will be established. The equilibrium concentration between the firn and the firn gas concentration has been determined to 50 Molatm^{-1}l^{-1}(C_{ice}/c_{air}) compared to a value of 16000 Molatm^{-1}l^{-1}(C_{ice}/c_{air}) for H_2O_2 (Conklin et al., 1993). By analogy with the discussion of the transfer function for H_2O_2 by Conklin et al (1993) it appears that the sticking coefficient for HCHO molecules on ice surfaces is considerably lower as the one for H_2O_2. As a consequence any variation in the firn gas will be smeared out over a larger depth interval. The ice will reach a concentration that is proportional to a long-term average of the atmospheric concentration.

The overall transfer ratio will be sensitive to the adsorption strength of HCHO on snow surfaces. It seems, that a higher impurity load reduced the loss process after deposition, which could explain the higher HCHO concentration measured in the cold stages in glacial times.

4.4 Step 4: Snow to Firn transfer.

During the whole firnification process, all chemical species which have a finite vapour pressure, will be smoothed out to a certain degree due to reevaporation and recondensation in the open pore space. Driving forces for this redistribution are: a) temperature gradients; b) pressure gradients; c) concentration gradients. The degree of redistribution decreases with increasing depth, because all gradients decrease with depth. Johnsen (1977) developed a simple algorithm to determine the mean displacement length for a water molecule integrated over the whole firnification process. He determined from the smoothing of $\delta^{18}O$ variations the mean displacement length for water molecules to be between 8 and 10 cm. As consequence a seasonal structure of $\delta^{18}O$ only survives in case the yearly accumulation exceeds 20 to 25 cm.w.equ.

Observation 4): Sigg applied the same analysis to the continuous H_2O_2 record from Summit and Siple and deduced a displacement length of 12 to 13 cm, thus 30 to 50% higher than for water molecules.

The consequences of this is that the minimum accumulation rate for a conservation of seasonal cycles of H_2O_2 is 30 cm. water. equivalent. Another consequence is that H_2O_2 molecules might diffuse out of the snowpack from layers down to roughly 1 meter. The longer displacement length of H_2O_2 molecules compared to H_2O molecules can be explained with a lower sticking probability on the surface. From laboratory studies Conklin et al., 1993 deduced the sticking coefficient of H_2O_2 to be 20% of the one of water molecules.

Pitfall: Consequently H_2O_2 released from the firn into the firn gas will stay longer in this phase compared to H_2O molecules and a relative enrichment in the open pore space of H_2O_2 should be established. Field observation gave no evidence for this (Bales et al., 1995).

Continuous HCHO measurements performed immediately after core recovery show an extremely constant signal with no indication of a seasonal variation (K. Fuhrer, pers com.). Changes occur with time constants of several years. This supports the hypothesis by Staffelbach et al. (1991b), that the

HCHO in the ice is proportional to a multiyear average of the atmospheric concentration over the drilling site.

4.5 Step 5: Firn to ice transition

The transition from firn to ice, thus the depth interval where the close off of the bubbles occurs is crucial for the enclosure of the gas. There is no apparent change in HCHO and H_2O_2 associated with this step. The fact that the connection to the atmosphere is canceled does not make any difference, because these molecules are too sticky to diffuse to the atmosphere.

4.6 Step 6: Chemical transformation within the ice sheet - $c_{ice}(t)$.

The discussion of CO_2 already made clear that polar ice is not as chemically inert as we perhaps thought a few years ago. The high resolution measurements of H_2O_2 clearly revealed that a slow decay occurs in narrow layers, with reaction rates on the order up to 0.01% a^{-1}, much too slow to be observed in a laboratory study. Catalytic decay of H_2O_2 on metal surfaces are known and described in the literature. It is likely, that a similar type of reaction occurs in the ice. Obviously movements of the reactants is very slow in the ice, which explains the very slow decay rates. It does not seems possible to describe the loss reaction with a simple first or second order reaction rate, since it is most probably transport limited. There is also only a qualitative relation between the Ca^{2+} concentration and the development of H_2O_2 holes.

Observation 5): The H_2O_2 decay process seems to be slowed down or stopped in the depth interval between 300 m.b.s. and 1600 m.b.s.

Explanation: Let us assume, that some of the H_2O_2 molecules are homogeneously distributed in the bulk ice, whereas the reminder are concentrated in grain boundaries. Those impurities that can act as catalysts for the H_2O_2 destruction are also concentrated at grain boundaries or at other "hot spots". Only those H_2O_2 molecules that diffuse to the catalysts will be destroyed. In case the diffusion across the ice matrix is slower then the destruction rate only the part of the H_2O_2 molecules sitting on grains sizes will be destroyed. At greater depths, the annual layers are thinner and the paths for H_2O_2 molecules to reach the next catalysts are smaller and smaller. The destruction rate is again higher. This can be seen in the Summit record between 1400 m and 1700 m.b.s. (Fuhrer et al, 1993).

Observation 6): HCHO values in cold stages of the glaciation shows a pattern typical of an aerosol related compound.

The persistence of the signal is determined by the depth resolution of the analytical system. It is therefore likely that already the initial HCHO signal was very noisy, a typical feature of an aerosol derived species.

Pitfall: HCHO is believed to be in the gas phase and incorporated in the ice in its gaseous form with a rather low sticking coefficient. This is in contradiction to the aerosol type of signal, as it is seen in the cold stages of the glaciation.

A possible way out is that a high impurity content would prevent the loss of HCHO from the surface snow after deposition.

5. Conclusions:

Even thought we have made substantial progress in our knowledge about the transfer process of reversibly deposited species and gaseous compounds since initialisation of the ICSI working group five years ago, we still lack a consistent picture of the extraction of atmospheric records from ice core records. For CO_2 the new findings on the GRIP and GISP2 core complicated the picture more than they simplified it. It is urgent that we understand the reasons for the observed inconsistencies within the different records to prevent discrediting of the basic information gained from ice cores regarding climatic changes.

For the "reversible" species H_2O_2 and HCHO a model is needed in order to put the story outlined in my contribution into a quantitative picture. Possible approaches and the state of the art will be reported by Roger Bales (this volume). But such models cannot be better than the story behind. If the wrong processes are modelled, the model output will not simulate what happened in the ice sheets.

In my contribution I tried to line up, what we know from field and laboratory data and in which direction the evaluation of the transfer process has to go. There is no claim for completeness nor a claim that the explanations given are reflecting the truth. They are intended to catalyse a better understanding of the transfer process, so that environmental scientists can take advantage of the ice core records.

In our present time human beings are intensifying all cycles. The modern Earth System Science allows to calculate all sort of projections in the future, which are not very pleasant. It is our hope, that an improved understanding of the earth system and a more accurate description might be in favour of a more reasonable behaviour of human beings.

Acknowledgements: I am grateful to Roger Bales, Tas van Ommen and an unknown reviewer for the many helpful suggestions. I thank Katrin Fuhrer, Manuel Hutterli, Thomas Staffelbach and Bernhard Stauffer for encouraging discussions. This work was partly supported by the EU-project "Transfer of aerosols and gases to the Summit of the Greenland ice sheet". Part of the work was carried out under a NATO Cooperative Research Grant awarded to Roger Bales.

References:

Anklin M. (1991) CO_2 - Analyse an einem Eisbohrkern aus Zentralgrönland und Bestimmung von Niederschlagsraten entlang einer Fliesslinie mittels H_2O_2 - Analyse
Lizentiatsarbeit, Physikalisches Institut Universität Bern

Anklin M., Stauffer B., Geis K. and Wagenbach D.(1994a) Pattern of annual snow accumulation along a west Greenland flow line - no significant change during the last decade observed.
(Tellus, in press)

Anklin M., Barnola J.M., Schwander J., Stauffer B. and Raynaud D. (1994b)
Processes affecting the CO_2 concentration measured in Greenland ice
(Tellus in press)

Anklin M. (1994c) Kohlenstoffdioxid Bestimmung in Luftproben aus einem neuen Tiefbohrkern von Summit (Grönland)
PhD thesis University of Bern

Bales R.C., McConnell J., Losleben M., Conklin M.H., Fuhrer K., Neftel A., Dibb J., Kahl J. and Stearns C.(in press JGR)
The diel variations of H_2O_2 in Greenland: A discussion of the cause and effect relationship.

Bales R.C., McConnell J., Losleben M., Fuhrer K. and Neftel. A.(1995)
H_2O_2 in Snow, Air and Open Pore Space in Firn at Summit, Greenland. Geophys Res Lett 22(10): 1261-1264

Barnola J.M., Anklin M., Porcheron J., Raynaud D., Schwander J. and Stauffer. B.(1995)
CO_2 evolution during the last millenium as recorded by Antarcic and Greenland ice. Tellus 47B: 264-272

Beer J., Finkel R.C., Bonani G., Gäggeler H., Görlach U., Jacob P., Klockow D., Langway Jr C.C.., Neftel A., Oeschger H., Schotterer U., Schwander J., Siegenthaler U., Suter M., Wagenbach D. and Wölfli W.(1991) Seasonal variations in the concentration of ^{10}Be, Cl^-, NO_3^-, SO_4^{2-}, H_2O_2, ^{210}Pb, ^3H, Mineral Dust and $\delta^{18}O$ in Greenland Snow Atmos Environ 25A: 899-904

Blunier T., Chappelaz J., Schwander J., Barnola J.M., Desperts F., Stauffer B. and Raynaud D.(1993) Atmospheric methane from a Greenland ice core from the last 1000 years Geophys Res Lett 20: 2219-2222

Conklin M. H., Sigg A., Neftel. A. and Bales R.C.(1993) Atmosphere-snow transfer function for H_2O_2: microphysical considerations J Geophys Res 98: 18367 - 18376

Delmas, R. (1993) A natural artefact in Greenland ice-core CO_2 measurements. Tellus 45B: 391-396

Fuhrer. K., Neftel. A, Anklin. M., Maggi. V. (1993) Continuous in situ measurements of chemical species along the new GRIP ice core from Summit, Central Greenland. Atmos Environt 27A: 1873-1880

Intergovernmental Panel on Climate Change (1990) Climate Change The IPCC Scientific Assessment Cambridge University Press ISBN 0 521 40720 6

Jacob P. (1987) Wasserstoffperoxid in der atmosphärischen Gas- und Flüssigphasenchemie: Laborstudien und Felduntersuchungen. Dissertation am Fachbereich Chemie der Universität Dortmund

Johnsen S.J. (1977);Stable Isotope Homogenization of Polar Firn and Ice. International Association of Hydrological Sciences Publication 118, Isotopes and Impurities in Snow and Ice, Grenoble, 210

Laj P., Drummey S. M., Spencer M.J., Palais J. M., and Sigurdsson H.(1990) H_2O_2 depletation in a Greenland ice core: implication for oxidation of volcanic SO_2. Nature 346: 45-48

Mulvaney R., Wolff E.W. and Oates K.(1989) Sulphuric acid at grain boundaries in Antarctic ice. Nature 331: 247-249

Neftel A.; Jacob P. and Klockow D. (1984) Measurements of hydrogen peroxide in polar ice samples Nature 311: 43-45

Neftel A.; Jacob P. and Klockow D. (1986) Long-term trend of Hydrogen Peroxide in polar ice cores Tellus 38B: 262-270

Neftel A. (1991) Use of snow and firn analysis to reconstruct past atmospheric compositionin: Seasonal Snowpacks - Processes of compositional changes. Editors: T.D. Davies, M. tranter and H.G. Jones. Springer Verlag Series G: Ecological Sciences Vol. 8

Neftel A., Oeschger H., Staffelbach T. and Stauffer B.(1988)CO_2 record in the Byrd ice core 50'000 - 5'000 years BP Nature, 331: 609-611

Neftel. A. and Fuhrer, K. (1993) A record of atmospheric oxidant from polar ice cores over the past 100'000 years - Dream or real possibility? NATO/ASI Series Vol 17 "The tropospheric chemistry of ozone in the polar regions Edited by H. Niki and K.H. Becker Springer Verlag

Neftel A., Bales R.C. and Jacob D.J.(1995) H_2O_2 and HCHO in polar snow and their relation to atmospheric chemistry, in Biogeochemical Cycles in Ice Cores, NATO ASI Series Vol .. Springer-Verlag

Pearman G.I. and Hyson P.(1986) Global transport and inter-reservoir exchange of carbon dioxid with particular reference to stable isotopic distribution. J of Atmos Chem 4: 81-124

Schwander J. and B. Stauffer (1984) Age difference between polar ice and the air trapped in its bubbles. Nature 311: 45-47

Sigg A. and Neftel A.(1988) Seasonal variations of hydrogen peroxide in polar ice cores. Annals Glaciol 10: 157-162

Sigg A. (1990) Wasserstoffperoxid-Messungen an Eisbohrkernen aus Grönland und der Antarktis und ihre atmosphärenchemische Bedeutung, Dissertation an der Universität Bern,

Sigg A. and Neftel A.(1991) Evidence for a 50% increase in H_2O_2 over the past 200 years from a Greenland ice core. Nature 351: 557-559

Sigg A., Staffelbach T., and Neftel A.(1992) Gas phase measurements of hydrogen peroxide in Greenland and their meaning for the interpretation of H_2O_2 records in ice cores J of Atmos Chem 14: 223-232

Staffelbach T., Stauffer B. and Oeschger H.(1988) A detailed analysis of the rapid changes in ice-core paramters during the last ice age Annals of Glaciology 10:
Staffelbach T., Stauffer, B., Sigg, A. and H. Oeschger (1991a) CO_2 measurements from polar ice cores more data from different sites. Tellus 43: 91-96

Staffelbach T., Neftel A., Stauffer B. and Jacob D.(1991) Formaldehyde in polar ice cores; a possibilty to characterize the atmospheric sink of methane in the past. Nature 349: 603-605

Thompson A.M., Owens M.A. ande Stewart R.W. (1989) Sensitivity of tropospheric hydrogen peroxide to global chemical and climate change. Geophys Res Lett 16: 53-56

Thompson A.M. (1992) The oxidizing ccapacity of the earth's atmosphere: Probable past and future changes Science 256: 1157-1165

ATMOSPHERIC TRANSPORT PATHWAYS FOR THE ARCTIC

Trond Iversen
Department of Geophysics
University of Oslo
P.O. Box 1022, Blindern
N-0315 Oslo
Norway

Introduction

Dispersion of a gravitationally neutral tracer in the lower 50 km of the atmosphere is generally governed by the macroscopic motion of the air itself. Molecular diffusion can be disregarded except for laminar layers of air with thickness on the order of a few centimetres close to rigid surfaces. The macroscopic perception of fluid flows are based on the concept that fluid parcels can be regarded as thermodynamically homogeneous. The total mass of each parcel is assumed to be constant; i.e. no molecular diffusion of total mass is permitted, and a unique velocity of the air-parcels can be defined as the derivative of parcel positions with respect to time.

In this paper we are discussing transport of constituents into the Arctic troposphere. These can be gases or particles which are typically transported over distances longer than 1000 km. For the understanding of Arctic air quality it is justified to assume that the species are gravitationally neutral, and the transport is therefore completely determined by the macroscopic velocity of the air parcel. Dispersion due to irregular molecular motions are negligible except for in very thin laminar layers in connection with interfacial processes such as deposition to the ground surface or to cloud and precipitation particles.

The velocity of all air-parcels encompass motions on scales ranging from millimetres to ten thousand kilometres in space and fractions of a second to months in time. Briefly, air motions can be separated into two basic types. Waves are characterized by their amplitude, wavelength and period, and closed circulations are described by amplitude, diameter and period. These two types of motion are commonly referred to as eddies. The scale of an eddy is defined in space as a quarter of a wavelength or radius of a circulation cell (denoted L horizontally and H vertically), and in time as a quarter of a period or the e-folding time for motions with increasing or decaying amplitudes (denoted T). The scale of a parameter is in some connections a typical average value and in others a typical amount of variation over L, H or T. The scales of atmospheric motions are crudely separated into three main classes: the **microscale** with $L \leq 1$ km, $H \leq 1$ km, and $T \leq 10^3$s; the **mesoscale** with $10\text{km} \leq L \leq 10^3$ km, $H \leq 10$ km, and 10^3 s $\leq T \leq 10^4$ s; and the **macroscale** with $L > 10^3$ km, $H = 10$ km, and 1 day $\leq T \leq$ 1 month. These broad scale-regimes are frequently connected with the patterns observed in the spectrum of kinetic energy; i.e. the kinetic energy of the eddies as a

NATO ASI Series, Vol. I 43
Chemical Exchange Between the Atmosphere and Polar Snow
Edited by Eric W. Wolff and Roger C. Bales

function of their size or periods. The mesoscale is then seen to cover a interval of scales for which there is a local minimum of kinetic energy, whilst there are local maxima on the micro- and macro-scales (e.g. Vinichenko, 1970; Atkinson, 1981).

On the microscale, kinetic energy is created by hydrodynamic instability caused by wind-shear and ground-surface roughness (mechanical turbulence) as well as convective overturning due to ground-surface heating (thermal turbulence). Microscale eddies are connected with unpredictable and very irregular flows, and are therefore termed turbulence. This turbulent flow is dominating in the lower 1000m of the atmosphere called the atmospheric boundary layer (ABL) due to the boundary conditions on the ground surface, but it is also important in the free troposphere in connection with deep moist convection and clear-air turbulence close to jet streams.

Macroscale kinetic energy is predominantly produced at mid-latitudes by baroclinic instability of the jet-stream systems. The term baroclinic stems from the large temperature-gradient on isobaric (constant pressure) surfaces which is driving the instability and conversion from potential to kinetic energy (see e.g. Charney, 1947; Eady, 1949; Eliassen, 1952; Charney and Stern, 1962, Simmons and Hoskins, 1978; Eliassen, 1984; Hoskins *et al.*, 1985). The macroscale eddies thus created are migrating cyclones and anticyclones embedded in the jet-stream systems. The jet-streams themselves are brought about by the combined action of the rotation of the planet and the atmospheric heating at low latitudes and the cooling at high latitudes. This heating/cooling pattern maintains the entropy of the atmosphere whilst heat is carried polewards by meridional (i.e. North-South) transport. Parts of this meridional transport are brought about by the eddies created by baroclinic instability as mentioned above, and parts by quasi-stationary planetary waves caused by the distribution of mountain ranges and ground surface types with heterogeneous thermodynamic properties, viz. oceans and continents. By quasi-stationary is meant periods longer than about 10 days. In comparison, the periods of migrating cyclones and anticyclones are typically between 2.5 and 6 days. Since the atmospheric motions are governed by non-linear laws of nature, these two eddy-regimes are not independent; the quasi-stationary eddies largely determine the paths of the migrating eddies, but the migrating eddies also influence the position and amplitudes of the quasi-stationary eddies.

An important difference between microcale and macroscale atmospheric motions is that the former is fully three-dimensional while the latter is almost horizontal and thus two-dimensional. The macroscale eddies are furthermore not considered as turbulence, since they can be predicted with deterministic skill up to 7-10 days when a reasonable dense measurement network (100 - 1000 km between radiosonde sites) is used to calculate initial conditions. This is also the case for most eddies on the mesoscale. The macroscale eddies is therefore considered to be known and resolved, while microscale eddies must be statistically treated. If only monthly or seasonally averaged meteorological data are explicitly known, however, also major parts of the macroscale eddies must be treated statistically as if being turbulence. In this respect, dispersion of atmospheric constituents by macroscale eddies can be termed turbulent diffusion.

The statistical treatment of turbulence is based on the concept that state-variables for the atmosphere (pressure, temperature, wind etc.) are stochastic variables; i.e. their values are not known exactly but must be given as probability distributions. A stochastic variable is said to be stationary if its probability distribution is constant in time, and homogeneous if it is spatially constant. Stationary and homogeneous turbulence are corresponding terms used for eddy flows in which probability distribution of any state variable is independent of time and space respectively. (Note the difference between stationary turbulence and a stationary eddy or wave which does not move with respect to the ground.) For most practical purposes a knowledge of the full probability distribution is not necessary, but only estimates of its first (mean values) and second (variances and co-variances) order moments.

Atmospheric Turbulent Diffusion

Theories for dispersion of plumes and puffs in homogeneous and stationary turbulence prescribe the dilution to be dependant on the time, t_L, elapsed since release into the atmosphere (the Lagrangian time). According to the similarity theory of Batchelor from the 1950s, eddies of similar size as a puff are the most efficient contributors to distorting the puff-shape and thus to the dilution of the material in it and the growth of the puff (see e.g. Pasquill and Smith, 1983). As long as the puff is in a size-range comparable to turbulent eddies with considerable kinetic energy, the size grows linearly with t_L, or even proportionally to $t_L^{3/2}$ if the kinetic energy is increasing with eddy-size in the size-range of the puff. After a sufficiently long time when all eddies are considerably smaller than the puff, the puff size is proportional to $t_L^{1/2}$ which is the same as obtained by molecular diffusion (with a much smaller coefficient of proportionality though).

Atmospheric turbulence is neither homogeneous nor stationary, except within limited intervals in space and time. For example there is usually a sharp jump in kinetic energy for microscale eddies between the ABL and the free troposphere above. The height of this discontinuity is normally taken to be the height of the ABL, or the mixing height. Turbulent diffusion of puffs released in the ABL can be assumed to approximately follow Batchelor's theory as long as its size is smaller than the mixing height, but the further diffusion will be much slower. Furthermore, the mixing height varies strongly both in space and time. In terrestrial regions without snow-cover during a summer day with clear skies, it may vary from less than 100m in the morning to 2000m in the afternoon. Heterogeneous thermodynamic properties of the ground surface may produce similar differences in the mixing height over distances of 10 km. Examples of the latter are coasts and borders of sea-ice. These heterogeneous and non-stationary characteristics of the turbulence seriously complicates the diffusion compared to the predictions of Batchelor's theory. A growing puff released in the ABL with decreasing mixing height, may suddenly be situated partly in the free troposphere above and partly inside the ABL. Vertical decoupling in the free troposphere may then provoke horizontal separation of the puff segments due to considerable vertical wind-shear (i.e. vertical changes in horizontal wind speed and direction).

Atmospheric dispersion on the macroscale and large mesoscale is normally not connected with diffusion unless available meteorological information does not resolve the macroscale eddies in time and space. If the eddies are adequately resolved, this part of the dispersion is called advection while the unresolved microscale turbulence is diffusive. The dispersion is then reduced to a combined

advection-diffusion process. In cases when only long-term or zonally (east-west) averaged meteorological data are available, also the dispersion by macroscale eddies must be accounted for in statistical terms. Due to so-called Stokes drift, parts of the effects of the macroscale eddies on the dispersion are of advective nature and the remaining is diffusive (Andrews and McIntyre, 1978; Plumb and Mahlman, 1987). Stokes drift is a process in multidimensional periodic flows for which there are systematic phase-shifts between the different velocity components. The effect is that air-mass trajectories yields a mean transport of mass, even if averages of the velocity components are vanishing. The macroscale baroclinic eddies transfer available potential energy associated with horizontal temperature gradients to kinetic energy whilst they simultaneously transport heat from middle to high latitudes. Their adiabatic dynamics provoke warm air to rise as it flows polewards and cold air to sink on its way equatorwards. Ground-based emission sources of pollutants at middle and low latitudes therefore contribute to elevated plumes entering the Arctic (see e.g. Iversen, 1993). In connection with the warm air rising, adiabatic expansion causes cooling and the water vapour in the air frequently becomes saturated. Hence, water soluble pollutants in plumes following these rising motions may be efficiently scavenged by precipitation.

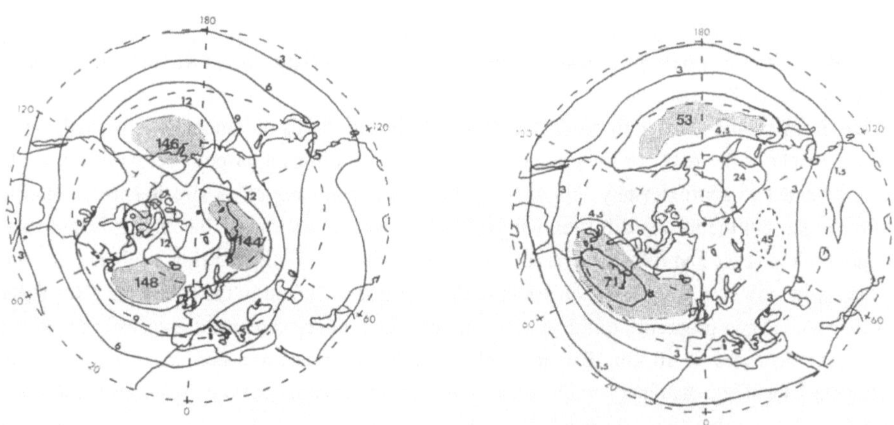

a) Low-pass filtered data emphasizing periods longer than 10 days. Equidistance 10 m.

b) Band-pass filtered data emphasizing periods between 2.5 and 6 days. Equidistance 5 m.

Fig. 1: Variance of 500 hPa geopotential height for 18 winter seasons after removal of the climatological mean annual cycle. Unit 10 m. (After Wallace and Blackmon, 1983) Shaded areas in a) are prone to onset and decay of quasi-stationary, macroscale eddies; shaded areas in b) experience frequent passage of migrating cyclones and anticyclones (baroclinic waveguides).

Blackmon (1976) investigated the occurrence of macroscale eddies of different wavelength and periods in the Northern Hemisphere, based on analyzed heights of the 500 hPa pressure surface from US-NMC. The 500 hPa-maps are excellent to apply for identification of macroscale eddies.But this does not imply that the major part of the transport of a pollutant released on the ground actually takes place in 500 hPa. The main mass of such pollutants is confined to below 2-3 km, also in the Arctic, and the air-flows in these lower levels are more irregular than the 500 hPa flow. The variance of the 500 hPa heights over 18 winter seasons after applying time-filters constructed to emphasize periods between 2.5 and 6 days (band-pass filtered) and periods longer than 10 days (low-pass filtered) were presented by Wallace and Blackmon (1983) (Figure 1). Areas dominated by the relatively short periodic propagating weather systems (Fig. 1b) are found to have maxima between 30°N and 60°N over the Pacific and Atlantic Oceans. These areas are interpreted as the main tracks of migrating cyclones and anticyclones, and named baroclinic waveguides after Wallace *et al.* (1988). Macroscale dispersion of gases and particles horizontally as well as vertically will be considerably more pronounced inside these areas than other places. Their lack of extension into the Arctic and the enhanced precipitation frequency for these areas hamper the transport into the Arctic. Areas dominated by quasi-stationary eddies (Fig 1.a) are found over the North Atlantic and North Pacific Oceans as well as over Northern Eurasia. The amplitude of the quasi-stationary variance is 2-3 times larger than for the variance connected with the baroclinic waveguides, and the areas extend considerably deeper into the Arctic as well. If the macroscale eddies are interpreted as turbulence, it is evident from Fig.1 that this turbulence is not homogeneous. Similar statistics for other seasons would reveal that the turbulence is non-stationary as well, since the amplitudes and position of the areas of large variance changes with season. Batchelor's theory of puff diffusion can therefore not be applied, and an important consequence for transport to the Arctic is that certain longitude sectors and certain seasons are more efficient for transport of mid-latitudinal air into the Arctic than others.

Pathways to the Arctic

The quasi-stationary eddies (Fig. 1a) can be understood as the result of interactions between long baroclinic waves, and the very long standing waves on the jet-streams caused by the major mountain ranges and thermal contrasts between continents and oceans (Charney and Eliassen, 1949). In the northern hemisphere, these contrasts in ground-surface properties are distributed in different longitudinal sectors, as opposed to in the southern where they are much more rotationally symmetric at middle and high latitudes. The so-called standing waves are not truly stationary because the strength and position of the westerly jets are varying with season in concordance with the seasonally changing temperature difference between the tropics and polar areas. Depending on the latitudinal variation of the westerly wind component averaged zonally, stationary wave-energy emanating from one source (e.g. Rocky Mountains) may occasionally be trapped to follow certain latitude bands and interfere constructively with wave energy from another source (e.g. The Himalayas) (Hoskins and Karoly, 1981; Held, 1983). Resonance can thus be observed, and certain regions may encounter persistent weather conditions for several weeks. These patterns appear to block the regular eastward propagation of baroclinic eddies and frontal cyclones and are traditionally termed regional blocking.

The weather forecasters' traditional view of a blocking is a more or less pronounced dipole feature in the geopotential height of the 500 hPa surface (Fig. 2), with a warm anticyclone (ridge) around 60°N and a cold cyclone (trough) at about 40°N and a split in the jets upstream (Rex, 1950; Lejenäs and Økland, 1983). In practice, the picture will often be dominated by the blocking anticyclone with a so-called Ω-circulation, and in some cases by the cold trough (Kanestrøm *et al.*, 1985). It is natural to associate the variance on periods longer than 10 days (Fig. 1a) with onset and decay of blocking. According to Kanestrøm *et al.* (1985) blocking over the North Atlantic and North Pacific regions can be dominated by warm ridges as well as cold troughs, whilst blocking over central Eurasia is almost exclusively occurring as a warm anticyclone. Shukla and Mo (1983) diagnosed the same three blocking regions as Wallace and Blackmon (1983) for all seasons by selecting episodes of positive anomalies for 500 hPa geopotential height with at least 7 days of persistence. The classical study of Rex (1950) and the later one by Lejenäs and Økland (1983) both included the dipole in their selection criteria, and therefore excluded the Eurasian blocking.

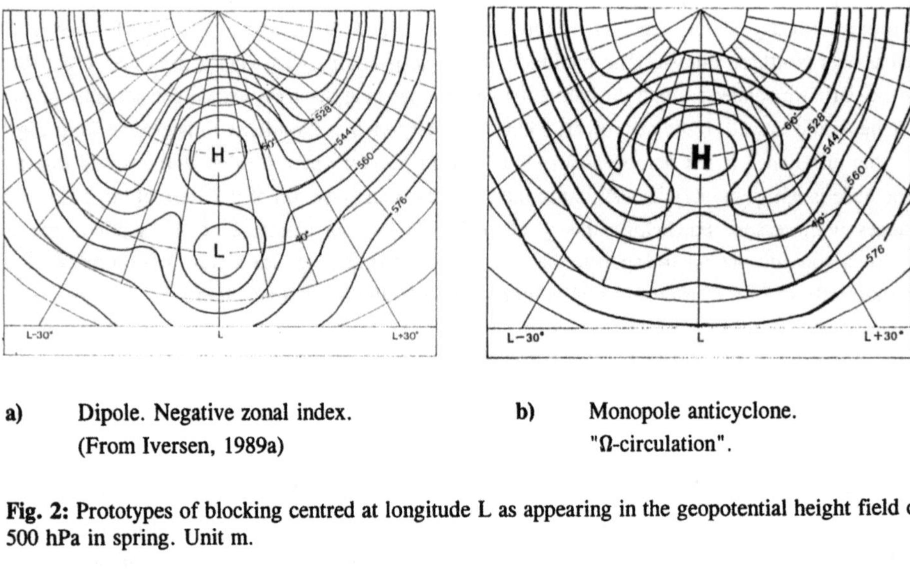

a) Dipole. Negative zonal index. (From Iversen, 1989a)

b) Monopole anticyclone. "Ω-circulation".

Fig. 2: Prototypes of blocking centred at longitude L as appearing in the geopotential height field of 500 hPa in spring. Unit m.

In connection with blocking, meridional exchange of air will be enhanced, and on the western rim of the anticyclone, conditions are optimal for transport of air from mid-latitudes into the Arctic. Under stagnant anticyclonic condition over the emission areas large amounts of pollutants may accumulate over several days, and then being advected swiftly polewards as the anticyclone slowly is moving eastwards (Raatz, 1983). Air originating from the ABL on the warm side, e.g. inside the blocking anticyclone, will have to rise considerably as it moves northwards through the baroclinic current. Air originating on the cold side, however, will tend to stay inside the ABL. This is now a widely agreed conceptual model for the transport of sulphate deep into the Arctic, including elevated haze layers, which are regularly observed in winter and spring (Carlson, 1981; Reiter, 1981; Raatz, 1983; Iversen and Joranger, 1985; Iversen, 1989a).

Fig. 3 shows relative frequencies of high meridional index between 60°N and 70°N brought about by the planetary wavenumbers 1 to 4 (Iversen, 1989a; Iversen, 1993). The planetary wavenumber gives the number of wavelengths along a latitude circle, and the meridional index is defined as high for air-flow with a component from south to north larger than 10 degrees of latitudes over 36 hours. These ultra-long wave-components are expected to determine a major part of the variance connected with quasi-stationary eddies with periods of weeks and longer (Blackmon, 1976; Hoskins and Karoly, 1983). The heterogeneous distribution with strong longitudinal variations shown in Fig. 1 is confirmed by the meridional index shown in Fig. 3. Furthermore, strong seasonal variations are also evident from Fig. 3, with maximum frequency of high meridional index in winter. Hence we will expect a seasonal variation in transport from middle to high latitudes, and that some areas will influence Arctic air quality considerably more than others.

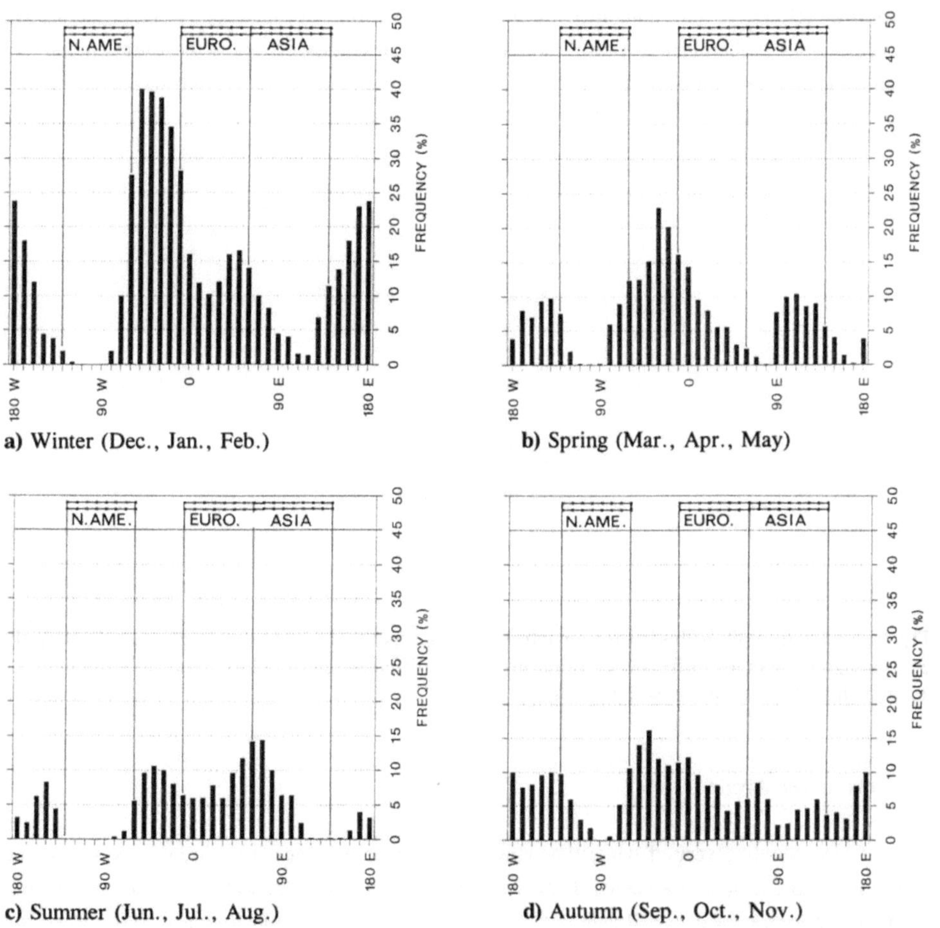

a) Winter (Dec., Jan., Feb.)

b) Spring (Mar., Apr., May)

c) Summer (Jun., Jul., Aug.)

d) Autumn (Sep., Oct., Nov.)

Fig. 3: Relative frequency of high meridional index (%) for every 10 degree longitude over a period of 2 years and 3 months. (Based on Iversen, 1989a.)

The figures can be used to identify main transport pathways of air into the Arctic; air is entering over the Atlantic Ocean, over Eurasia and over the Pacific Ocean. The relative importance of the entrance areas varies over the seasons, but the Atlantic pathway seems to be the most frequent. Europe and Eurasia seem to influence Arctic air quality much more than North-America. This statement is based on only a little more than two years though, and should be confirmed by statistics for many more years. The 18-years statistics of Blackmon (1976), however, supports the pathways sketched in Fig. 4. It is also consistent with model-calculations of sulphur. Barrie *et al.* (1989) estimated fluxes of sulphur through the Arctic circle, and found that Eurasia in 1979/80 contributed with about 95% and North America with 5%. Iversen (1989b) calculated budgets for areas north of 72.5°N. in March and July 1983, and arrived at similar numbers for the relative contributions in March 1983 (95% Eurasian 5% N.-American), but a larger N.-American fraction in July (14%). Not all entrance regions are responsible for increased concentration levels in the Arctic. The position of the pathway relative to mid-latitudinal source regions. Air-flow into the Arctic through the oceanic entrance regions predominantly cause lowered sulphur concentrations, while the Eurasian entrance region is the predominant pathway for elevated sulphur concentrations in the Arctic (Iversen, 1989a).

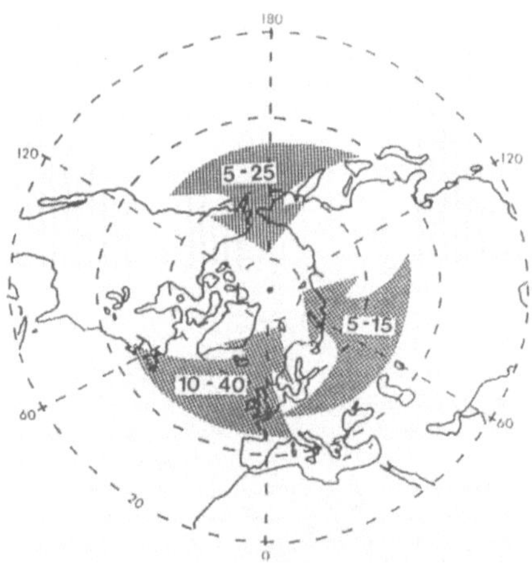

Fig. 4: Sketch of main pathways for air-intrusions from mid-latitudes to the Arctic. Numbers on the arrows signifies ranges in frequency of occurrence in different seasons: low frequencies in summer-autumn, high frequencies in winter (unit %).

Aerosols in The Arctic

After the discovery of wintertime episodic Arctic haze (Mitchell, 1957) it took a long time before it became generally accepted that distant anthropogenic sources were the main contributors. Both meteorological evidence (Reiter, 1981; Carlson, 1981; Iversen and Joranger, 1984; Raatz and Shaw, 1984; Barrie *et al.*, 1989; Iversen, 1989a and b), the analyzed composition of trace elements in aerosols (Heidam, 1981, 1984 and 1985; Lowenthal and Rahn, 1985; Maenhout *et al.*,1989, Barrie and Barrie, 1990), as well as isotopic analyses (Sturges and Barrie, 1989; Nriagu *et al.*, 1991) strongly emphasize long-range transport (LRT) from anthropogenic emission sources. Much evidence indicates that emissions in Eurasia, in particular of sulphur dioxide, are the major contributors to the total mass of the Arctic aerosol on an annual basis. High concentrations of particulate matter in the Arctic occur episodically and with pronounced seasonal cycles in monthly averages and variances of the concentrations (e.g. Barrie and Hoff, 1985; Iversen and Joranger, 1985; Shaw, 1985; Davidson *et al.*, 1989; Bodhaine, 1989; Iversen, 1989a). The life-time of aerosol particles in the Arctic is comparable to that of mid-latitudes in the warm and light season (a little less than a week), but considerably longer in the darkness of the Arctic winter. This is mainly due to very inefficient dry and wet deposition of the particles as well as their possible gaseous precursors, and the life-time in fact approaches the life-time of air-masses in the Arctic, i.e. a few weeks.

According to the emission inventory for sulphur published by Spiro *et al.* (1992), more than 80% of the northern hemispheric sulphur emissions are of anthropogenic origin. The two main contributors of natural origin are di-methyl-sulphide (DMS) from the oceans and SO_2-release from volcanoes. In the northern hemisphere in 1980, the two sources were estimated by the authors to be of the same importance. The DMS-emissions depend on the biological activity of certain algae-species and of the wind-speed providing conditions for a kinetically driven flux from the sea (e.g. Tarrasòn *et al.*, 1995). The contribution from DMS to long-range transported particulate sulphate depends on the rate of oxidation and the efficiency of turbulent transport away from the sea surface. Emissions from volcanoes probably give a more efficient production of particulate sulphate throughout the troposphere. Occasional powerful eruptions may even produce considerable amounts of sulphate aerosol in the stratosphere. Examples of the latter are Mt. St. Helens in 1980, El Chichon in 1982 and Mt. Pinatubo in 1991. Stratospheric sulphate aerosol has a lifetime of a couple of years. One sink is exchange of air with the underlying troposphere through mixing in tropopause-folding events (see e.g. Shapiro *et al.*, 1984). These are frequently occurring at middle and high latitudes causing, for example, a net flux of ozone into the troposphere of about 7×10^{10} molec. $cm^{-2}s^{-1}$ (Levy *et al.*, 1985). Therefore periods of increased particulate sulphate in the lower stratosphere probably have a significant impact on middle and upper tropospheric sulphate aerosol in the Arctic. Unfortunately, quantitative estimates of this has not been found published in the literature, but a maximum occurrence in the cold part of the year would be expected from the frequency of strong folding events. Recently, Trepte *et al* (1994) published measured stratospheric aerosol optical depths for the globe during volcanic calm periods and two years after the Mt. Pinatubo eruption. The optical depth is about an order of magnitude larger in the latter case, and the largest values are measured at high latitudes. The unpredictable nature of volcanic eruptions render it a difficult task to estimate a typical distribution of sulphur of volcanic origin. Volcanic emissions of a less eruptive nature occurs more regularly in certain areas. Of particular importance for the Arctic are Iceland, The Aleutian Islands and the western parts of the North Pacific Oceans. Calculations of global sulphur budgets (with a very crude resolution of the Arctic) made by Langner *et al.* (1992), indicate a pre-industrial level of 10 pptv in the Arctic during January and 25 pptv during July. The opposite seasonal variation than for anthropogenic Arctic aerosols are mostly due to the spring-summer maximum of DMS emissions.

Black Carbon (BC) is the light-absorbing part of an aerosol, mainly produced by incomplete combustion or pyrolysis of hydrocarbons. The chemical form of the light-absorbing part of the soot particles is microcrystalline graphitic carbon (Hansen and Rosen, 1985), but it is believed to be internally mixed with both organic particles and particulate sulphate. In spite of the hydrophobic nature of elemental carbon, BC in an aged aerosol may be scavenged by precipitation due to this mixing (Ogren, 1982). BC has been observed on several occasions in the high Arctic, on the ground (Heintzenberg, 1982; Rosen *et al.*, 1984; Clarke, 1989; Hopper *et al.*, 1994), and concentrations comparable to typical urban values ($\approx 1.7\ \mu g\, m^{-3}$) have been measured in thin layers by aircraft in the free troposphere (Hansen and Rosen, 1984; Rosen and Hansen, 1984; Hansen and Rosen, 1985). Hopper *et al.* (1994) reported similar episodicity and seasonal cycle for BC as for particulate sulphate. Based on observations taken in Alert, Canada, the monthly median values for BC can quite crudely be estimated from concentrations of SO_4^{-2} by multiplication with the factor 0.06. Hansen and Rosen (1985) found an abrupt increase of this ratio with height, however, and a ratio of 0.15 is possibly more representative for the aerosol column. BC may play an important role in climatic change, since it is an efficient absorber of solar radiation, and over an area with a large natural surface albedo, such as in the Arctic, the absorption may overrule the increased reflectivity due to increased aerosol content in general. Deposition of soot onto the Arctic snowpack has been estimated to have a similar impact on the Arctic radiation budget as the airborne haze has (Clarke and Noone, 1985). The emissions of BC have a large anthropogenic component, the main contributions being fossil fuel combustion and biomass burning. The average atmospheric residence-time at low and mid-latitudes is about a week which is twice as long as for sulphur due to the hydrophobic initial phase. The main transport routes to the Arctic probably are similar to anthropogenic sulphur, although the biomass burning is a more wide-spread source. There are presently only one published global emission-inventory of BC (Penner *et al.*, 1993).

In between episodes and during the "clean" summer and autumn half of the year, the composition of particles in Arctic air seems to point to natural origins, even though there are episodes of anthropogenic origin also during summer (Pacyna and Ottar, 1985; Nriagu *et al.*, 1991). Observations of natural particulate matter indicate crustal or sea-salt origins, with a significant but less pronounced contribution from photochemical oxidation of reduced sulphur, presumably DMS or mercaptans (Maenhout *et al.*, 1989; Pacyna and Ottar, 1989; Barrie and Barrie, 1990). These components are probably also present during "pollution" episodes, but in those cases they are efficiently masked by the much higher concentrations of anthropogenic origin. Components typical for aerosols of the sea-salt origin produced by bursting bubbles during windy conditions over sea, are Na, Mg, K, Cl, and Sr; for crustal aerosols typical elements are Si, Al, Fe, Sc, Ti, La, Ce, Sm, and Th; and for anthropogenic aerosols the dominating elements are NH_4^+, Pb, V, Ni, SO_4^{-2}, Cu, Zn, As and Se. Some elements occasionally originates from several sources in approximately similar relative amounts. For example, Mn can be indicative of both soil and anthropogenic emissions, and K can have important contributions from all three sources. At a measurement site at the southern tip of Greenland (Heidam, 1981), Ca is an almost equally important tracer for crustal, marine and anthropogenic aerosols, whilst in Thule at north-west Greenland it gave a clear signature of marine aerosols. Thus there are considerable uncertainties in these methods for source-apportionment, even though a few elements quite uniquely are connected with only one factor throughout.

Pacyna and Ottar (1989) gave examples for which loess-deposits in arid areas in Asia probably contributed significantly to observed aerosols in the Norwegian Arctic, in spite of the very long transport pathway. One would perhaps expect Saharan minerals to be transported more frequently, but Ottar and Pacyna (1989) were unable to select the Sahara as a separate source region due to lack of data on elemental composition of these minerals. Saharan dust is, however, known to be transported far away over the North Atlantic Ocean and Europe (Prospero and Carlson, 1972; Haeberli, 1977; File, 1986), and Mosher *et al.* (1993) discusses an episode of high concentration of particles of crustal origin at Dye 3, Greenland, which most probably was caused by transport from the Sahara. According to Jaenicke (1993), the earth's crust and cryosphere is now believed to be of at least the same importance as the oceans as contributors to the global content of aerosol particles. Crustal particles have been observed to be distributed over a much wider size-spectrum than earlier assumed (D'Almeida and Schütz, 1983), thus having a considerably larger potential for being long-range transported than believed before. In order to provide conditions favourable for LRT of mineral dust, local high wind-speeds (dust- and sand- storms) over the deposit regions are needed. Conditions provoking swift transport into the free troposphere and subsequent transport polewards in large-scale wind systems would be ideal for transport into the Arctic. D'Almeida (1986) made a thorough source-strength inventory of Saharan mineral dust based on measured correlations between aerosol turbidity and aerosol mass concentrations. His results indicated the Saharan annual source of particles with radii smaller than $5\mu m$ to be 8-9 10^7 t/a, and that 12% of the mass was model calculated to move northwards to Europe and 28% westward to the equatorial North Atlantic Ocean. Even though only relatively small amounts of Saharan aerosol particles reach the Arctic, major parts of air-concentrations and depositions of crustal origins over European parts of the Arctic probably are of Saharan origin. At least this is indicated by the model-calculations made by Joussaume (1990). Fig. 5 shows the model-calculated wet deposited dust together with mixing ratios in air at 500 and 200 hPa for February. The model is a general circulation model and the result is thus a pure model product with systematic and random errors. Such errors are discussed in the paper by Joussaume (1990). For February an important systematic error by the GCM used is an over-representation of blocking over western Europe and north-east Atlantic Ocean. This may lead to the effect that the western European and north Atlantic pathway into the Arctic is exaggerated. It is seen that the dust may take a route from Sahara over Europe and the North Atlantic Ocean to Greenland or from the Middle East and Asian deserts through the mid-Eurasian pathway into the Arctic. This dust is found at very high levels in the Arctic in accordance with the meridional temperature distribution of the troposphere, which provokes ground-level air in hot areas to rise and expand adiabatically as it moves to colder areas.

Transport to the Greenland glacier

Greenland is an obstacle which influences air-currents in a way which may hamper the airborne transport of certain air constituents to the glacier. With increasing static stability and decreasing upstream wind-speed, an increasing portion of air flows around in stead of over the obstacle. Implications of this can be seen in calculations of sulphur deposition during winter 1988 (Fig. 6), (Tarrasòn, 1992; Tarrasòn and Iversen, 1992). Sulphur fluxes to Greenland is probably not representative for most other Arctic areas, at least not those prone to significant pollution levels. There are also large differences between the situation in the northern parts (e.g. Summit) and the

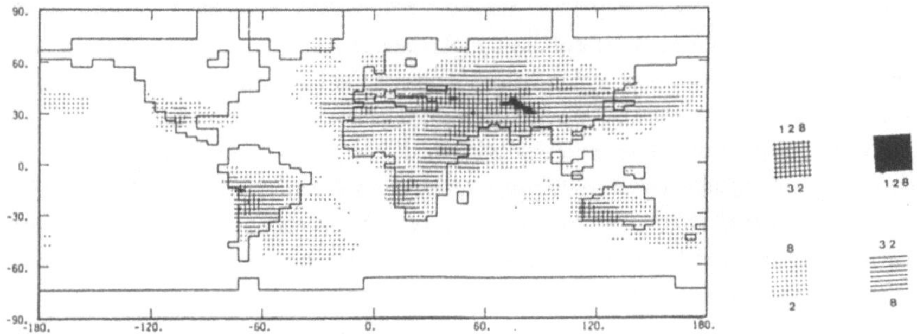

a) Wet deposition.

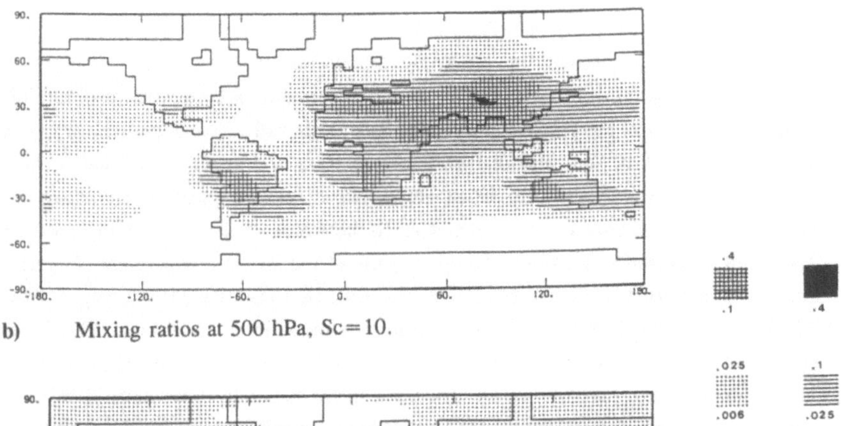

b) Mixing ratios at 500 hPa, Sc=10.

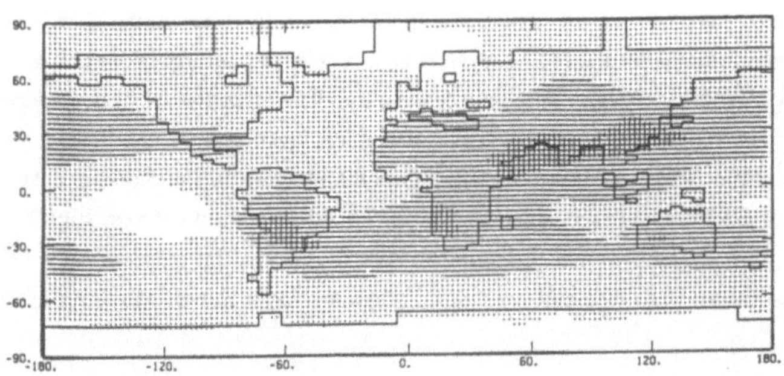

c) Mixing ratios at 200 hPa, Sc=209.

Fig. 5: Model-calculations of the dispersion of desert dust by a general circulation model for the month of February (From Joussaume, 1990). Unit for deposition: kg(AU) m^{-2} d^{-1}, and mixing ratio: kg(AU)/kg(air). (AU=arbitrary units: stems from the presetting of saturation mixing ratio at the surface in the source regions to the constant 1. The resulting dust concentrations thus are determined except for an arbitrary constant, and only patterns of variation in space and time can be inferred from the results. In b and c numbers are scaled with factor Sc to obtain readable maps.)

southern parts (e.g. Dye 3). Similar patterns can be seen for calculations of air concentrations of sulphate (not shown) and of BC (Penner *et al.*, 1993). The high altitude and the thermal properties of the massive glacier provides a blocking effect on the position of the atmospheric jet-stream and the storm tracks (Fig. 1). This is also supported by the trajectory analysis for back-trajectories arriving at the South-Greenland site Dye 3 published by Davidson *et al.* (1993), and shown in Fig. 7. The domain of influence for Dye 3 seems to be areas to the south and west (autumn), west (winter), south, west and north (spring). Unfortunately the trajectory statistics of Davidson *et al.* (1993) do not separate between trajectories originating in the ABL and those which are not. A considerable part of the trajectories originated from a longer distance than 2000 km, except during summer when shorter transport distances are frequent. The five-year trajectory statistics for Barrow in Northern Alaska by Miller (1981) also showed a considerable transport from beyond 2000 km, but the direction showed a bimodal distribution with one population of trajectories from the south and one population from the north. Even though an interannual variability in this bimodality was seen, the pattern was very persistent even throughout the different seasons. The pathway from the North to Barrow is a clear indication of possible pollution transport from Siberian emission sources over the North Pole, and is almost lacking in the regular data for Dye 3. The pattern in spring indicates, that there probably is occasional transport from the Eurasian sources to Dye 3 over the central polar basin, but very few 5-day trajectories actually originate in Eurasia. Since the life-time of Arctic winter aerosol is much longer than 5 days, one may argue that longer trajectories should have been used in the statistics. A full-year calculation of sulphur deposition (Tarrasòn, 1992) shows that Southern parts of Greenland

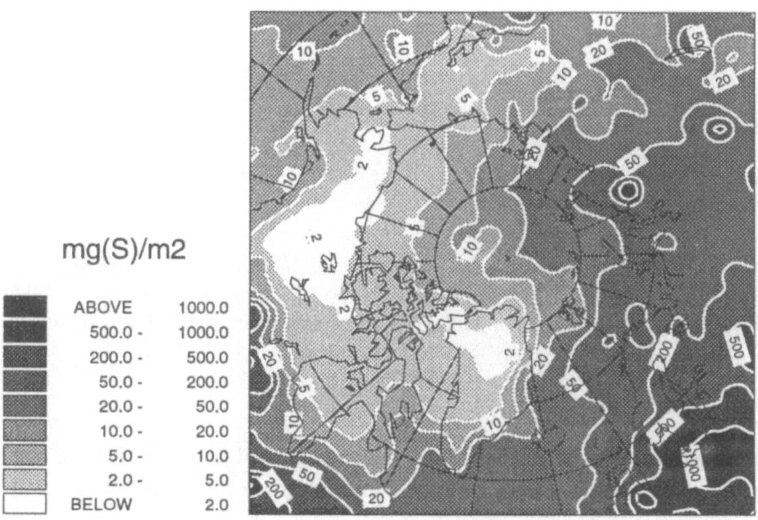

Fig. 6: Model-calculated deposition of sulphur for winter (January, February and December) 1988. (Calculations made by L. Tarrasón, The Norwegian Meteorological Institute.)

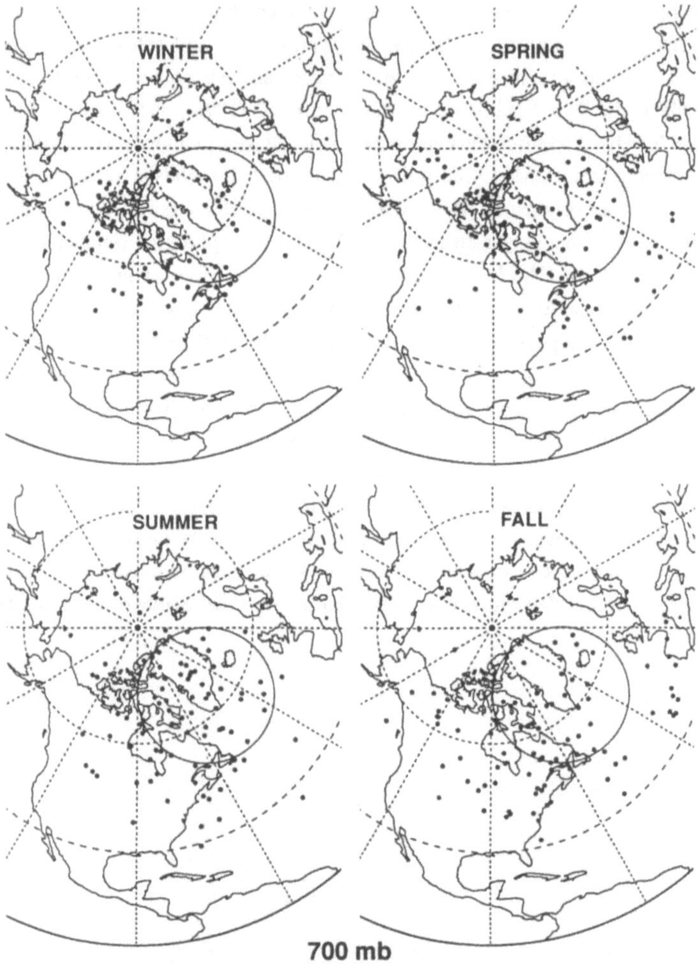

Fig. 7: Location of air-masses 5 days before reaching Dye 3 at 700 hPa (surface pressure in Dye 3 is normally varying between 700 hPa and 750 hPa) each day at 00 UT during the year Aug. 1. 1988 through July 31. 1989. The circles have radii of 2000 km and are centred in Dye 3. From Davidson *et al.* (1993).

("Dye 3") receive about equal mounts of sulphur from North America and Europe with about 5 % from Asia (including Siberia). Northern parts receive a little less than one half of the sulphur from Europe, one third from Siberia and about one fifth from North America. In central parts ("Summit") calculations show that about half of the deposition comes from Europe, about one third from North America, and one sixth from Siberia. In north Alaska, Siberian emissions are the only sources of importance.

Based on the indications of different pathways to the Greenland glacier, one will expect systematically different composition of the aerosol over inner parts of Greenland than encountered at other regions of the Arctic. This is particularly evident during winter. This may not be very serious for the interpretation of ice-core samples, since the relatively short atmospheric residence time of aerosols (except during Arctic winter) precludes a concept of a "typical" aerosol anyway. The calculations of Joussaume (1990) may indicate a rather opposite behaviour for crustal particles than for anthropogenic, since desert particles are transported at very high altitudes to the Arctic, and that there are local maxima of precipitation over southern parts of Greenland (see e.g. Fig. 7 of Davidson *et al.*,1985). A comparison between trace-element analyses made for Dye 3 (Mosher *et al.*, 1993) and other Arctic sites (e.g. Lowenthal and Rahn, 1985; Maenhout *et al.*, 1989) also indicates a stronger crustal factor over the Greenland glacier. However, this statement should be regarded as speculative at present; a unifying factor analysis of Arctic-wide elemental measurements seems to be missing in the literature.

Estimates of the transport during the last glacial maximum (LGM).

There are quite a few global models for the atmospheric circulation (AGCM = atmospheric general circulation models) being operated around the world, and some have been used for paleoclimatic research. A NATO ARW in 1987 concentrating on global scale atmospheric transport during past climatic periods resulted in comprehensive proceedings (Leinen and Sarnthein, 1989). Besides knowing the different astronomical parameters influencing solar radiation received by the earth, simulations of meteorological conditions during the last ice-age requires knowledge of extension of land-ice, sea-ice and sea-surface temperature, and the contents of greenhouse gases and aerosol particles. Furthermore, if one wishes to use the simulated wind circulation to estimate transport and chemistry of airborne constituents, an additional complicating factor is the emission sources. We are not able to discuss all these complexities in the present paper; examples of how such problems can be dealt with can be found in e. g. Manabe and Broccoli (1985), Kutzbach and Guetter (1986), Rind (1987), Lautenschlager and Herterich (1990), Mitchell (1993), and Valdes and Hall (1994). The figure 1 of Kutzbach and Guetter (1986) gives a nice overview of the changes in climatic forcing since the last glacial maximum (LGM, 18000 years before present). The solar radiation received in the northern hemisphere during LGM was similar to present, and with the same seasonal variation. The globally averaged annual sea-surface temperature was about 4 degrees lower than today, and the CO_2-concentrations were 200 ppmv as opposed to the modern, pre-industrial value of about 280 ppmv (1990-value: 353 ppmv; IPCC, 1990).

Judging from ice-core analyses from Greenland and the Antarctica, aerosol concentrations during LGM were much larger than present values. Joussaume (1993) tried to simulate the desert dust generation of crustal aerosols 18000 years BP using an AGCM with a prognostic equation for dust mass included. Typical February and August conditions were aimed at for the experiments. It is immediately understood how difficult it is to obtain accurate estimates of global emissions of particles which are small enough to be transported thousands of kilometres away. Potential source areas were defined by the condition that the soil water content averaged over the winter and summer seasons is less than 1.3% of the maximum soil water capacity. The actual emission release from these areas were parameterized by a bulk aerodynamic formula. Results from the calculations of deposition of dust onto the Greenland snowpack during LGM gave a factor of 4 increase compared to present depositions. On contrast to the present, maximum deposition was largest in August with a factor of 10 increase compared to present. However, this increase is significantly smaller by a factor of 10 to 20 than determined by ice-core observations. This underestimation of soil deposition may be due to erroneous atmospheric circulations in the present-day and LGM climates. The present day simulation probably overestimate the February transport to Greenland due to a larger North-Atlantic blocking activity than observed. Poor knowledge of dry areas as dust sources in Europe and Northern Asia is claimed to be the most probable reason for the underestimated LGM deposition to Greenland (Joussaume, 1983).

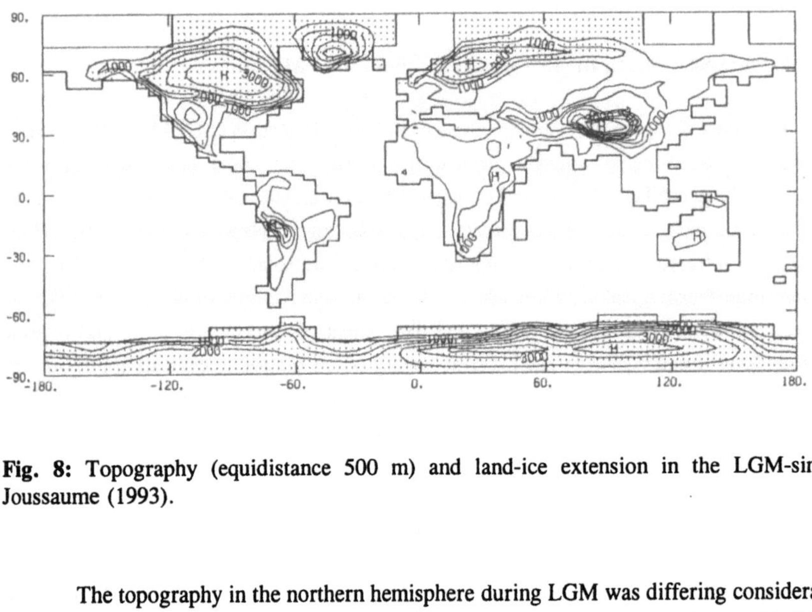

Fig. 8: Topography (equidistance 500 m) and land-ice extension in the LGM-simulations by Joussaume (1993).

The topography in the northern hemisphere during LGM was differing considerably from the present over North-America (the Laurentide ice cover) and over northern Europe (the Fennoscandian ice cover), see Fig. 8. These considerably altered large-scale topographies together with the different sea-surface temperature, sea-ice cover and land-ocean thermal contrast, provide conditions for a very different pattern of standing waves on the jet-streams in the northern hemisphere compared to present (Fig. 9). In the northern hemisphere, the main feature is a much more zonal and stronger jet-stream, extensive anticyclonic circulation over the Laurentide ice cover with a separate polar (or Arctic) jet to the north of it in February, and a removal of the present-day ridge with a south-westerly current over the North Atlantic Ocean. In the Pacific the differences are less pronounced, since the sea is still open and warm relative to the air in that area. Surface pressure maps (not shown here) indicate a decreased cyclonic activity over the Gulf stream in February due to the decrease in cold north-westerly winds brought about by the extensive ridge over the Laurentide ice cover. The more south-westerly warm winds relative to the sea-surface during LGM weakens the evaporation and supply of latent heat

a)
February, present day.

b)
February, ice-age (LGM).

c)
August, present day.

d)
August, ice-age (LGM).

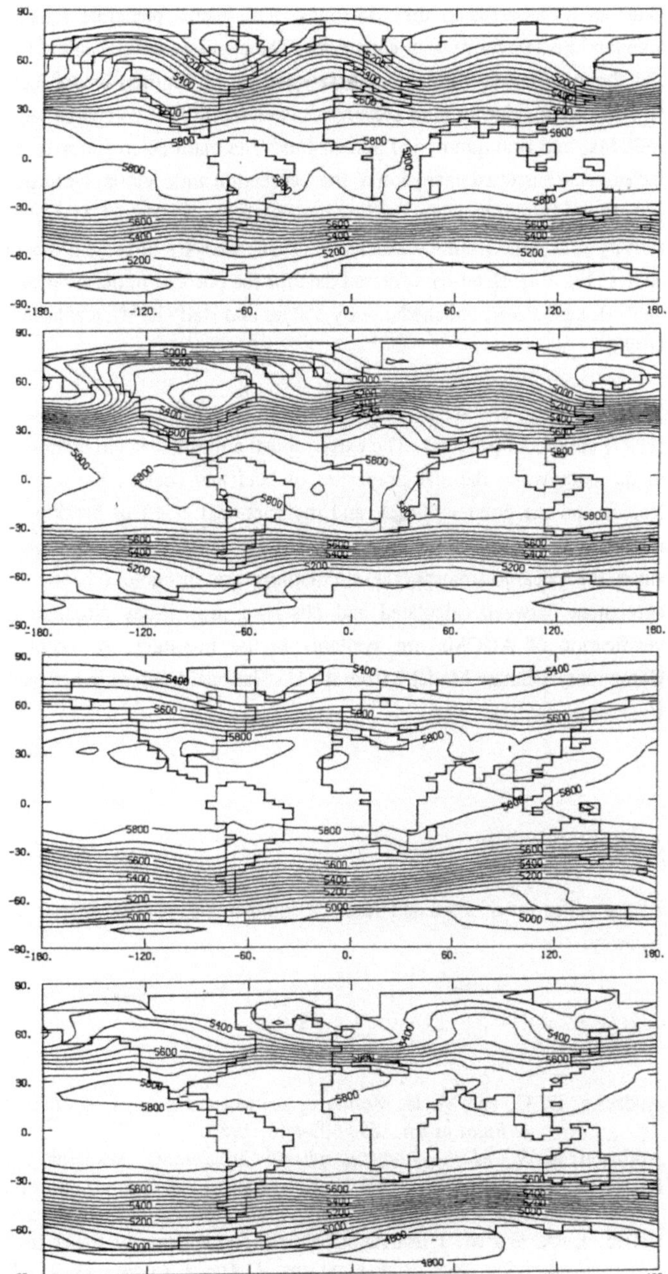

Fig. 9: Calculations of 500 hPa geopotential height (meters) for present day February (a), LGM February (b), present day August (c), and LGM August (d). From Joussaume (1993).

from the ocean. Furthermore, the storm tracks are extending eastwards along the sea-ice border rather than north-eastwards as they do today, and storms penetrate farther into the southern European mainland than under present conditions. The cyclonic activity in the Pacific is increased in the LGM, and the Aleutian low is increased in intensity as opposed to the Icelandic low which seems to be replaced by a ridge in both seasons. Joussaume (1993) compares her results with those from other AGCMs, and find quite good agreements. The main discrepancy by the present model is said to be the extreme eastward extension of the Laurentide anticyclone. Unfortunately, it has not been possible to find published statistics for low-pass variance for the LGM which can be assigned to blocking activity as is done from the Blackmon (1976) analyses of present-day climatology of quasi-stationary eddies (Fig. 1a). Band-pass filtered data for the purpose of defining storm-tracks (or baroclinic wave-guides), have been published e.g. by Valdes and Hall (1994), and agree with the results by Joussaume (1993)

Finally, one should be aware of the fact that also the present-day calculations of Joussaume (1993) are model products. The extratropical atmospheric circulation is among other things sensitive to the contrasting thermal properties of land and oceans, the temperature difference between the tropics and the polar latitudes, and the frictional effect of breaking stationary gravity waves over mountainous regions. These are all internal processes which may provoke complex feed-backs, and since the oceanic-atmospheric circulations are unstable (chaotic), one cannot expect complete agreement between calculated and observed present-day circulation patterns. Several papers on verification of AGCMs are available in the literature. A recent and interesting paper on the Community Climate Model 2 from the US National Center for Atmospheric Research is published by Hack *et al.* (1994).

Acknowledgements

I am grateful for detailed and valuable comments by Dr. L. Barrie, Canada, and Prof. K. Pedersen, Norway.

References

Andrews, D. G. and M. E. McIntyre, An exact theory of nonlinear waves on a Lagrangian mean flow. *J. Fluid Mech.*, **89**, 609-646, 1983.

Atkinson, B. W., *Meso-scale Atmospheric Circulations*. Academic Press, 495 pp., 1981.

Barrie, L. A., M. P. Olson and K. K. Oikawa, The flux of anthropogenic sulphur into the Arctic from mid-latitudes in 1979/80. *Atmos. Environ.*, **23**, 2505-2512, 1989.

Barrie, L. A. and M. J. Barrie, Chemical components of lower tropospheric aerosols in the high Arctic: Six years of observations. *J. Atmos. Chem.*, **11**, 211-226, 1990.

Barrie, L. A. and R. M. Hoff, Five years of air chemistry observations in the Canadian Arctic. *Atmos. Environ.*, **19**, 1995-2010, 1985.

Blackmon, M. L., A climatological spectral study of the 500 mb geopotential height of the northern hemisphere. *J. Atmos. Sci.*, **25**, 1607-1623, 1976.

Bodhaine, B. A., Barrow surface aerosol: 1976-1986. *Atmos. Environ.*, **23**, 2357-2370, 1989.

Carlson, T. N., Speculations on the movement of polluted air to the Arctic. *Atmos. Environ.*, **15**, 1473-1477, 1981.

Charney, J. G., The dynamics of long waves in a baroclinic westerly current. *J. Meteorol.*, **4**, 135-162, 1947.

Charney, J. G. and A. Eliassen, A numerical method for predicting the perturbations on the middle latitude westerlies. *Tellus*, **1**, 38-54, 1949.

Charney, J. G. and M. E. Stern, On the stability of baroclinic jets in a rotating atmosphere. *J. Atmos. Sci.*, **19**, 159-172, 1962.

Clarke, A. D., Aerosol light absorption by soot in remote environments. *Aer. Sci. Tech.*, **10**, 161-171, 1989.

Clarke, A. D. and K. J. Noone, Soot in the Arctic snowpack: a cause for perturbations in radiative transfer. *Atmos. Environ.*, **19**, 2045-2054, 1985.

D'Almeida, G. A., A model for Saharan dust transport. *J. Clim. Appl. Meteorol.*, **25**, 903-916, 1986.

D'Almeida, G. A. and L. Schütz, Number, mass and volume distribution of mineral aerosol and soils of the Sahara. *J. Clim. Appl. Meteorol.*, **22**, 233-243, 1983.

Davidson, C. I., S. Santhanam, R. C. Fortman and M. P. Olson, Atmospheric transport and deposition of trace elements onto the Greenland ice sheet. *Atmos. Environ.*, **19**, 2065-2082, 1985.

Davidson, C. I., J.-L. Jaffrezo, M. J. Small, P. W. Summers, M. P. Olson and R. D. Borys, Trajectory analysis of source regions influencing the south Greenland ice sheet during the Dye 3 gas and aerosol sampling program. *Atmos. Environ.*, **27A**, 2739-2749, 1993.

Eady, E. T., Long waves and cyclone waves. *Tellus*, **1**, 33-52, 1949.

Eliassen, A., Simplified dynamic models of the atmosphere designed for the purpose of numerical weather prediction. *Tellus*, **4**, 145-156, 1952.

Eliassen, A., The Charney-Stern theorem on barotropic-baroclinic instability. *PAGEOPH.*, **121**, 563-572, 1984.

File, R. F., Dust deposits in England on 9 November 1984. *Weather*, **41**, 191-195, 1986.

Hack, J. J., B. A. Boville, J. T. Kiehl, P. J. Rasch and D. L. Williamson, Climate statistics from the National Center for Atmospheric Research community climate model CCM2. *J. Geoph. Res.*, **99**, 20785-20813, 1994.

Haeberli, W., Sahara dust in the Alps - a short review. *Zeitschrift für Gletscherkunde und Glazialgeologie*, **13**, 206-208, 1977.

Hansen, A. D. A. and H. Rosen, Vertical distribution of particulate carbon, sulfur and bromine in the Arctic haze and comparison with ground-level measurements at Barrow, Alaska. *Geoph. Res. Lett.*, **11**, 381-384, 1984.

Hansen, A. D. A. and H. Rosen, Horizontal inhomogeneities in the particulate carbon component of Arctic haze. *Atmos. Environ.*, **19**, 2175-2180, 1985.

Heidam, N. Z., On the origin of the Arctic aerosol: a statistical approach. *Atmos. Environ.*, **15**, 1421-1427, 1981.

Heidam, N. Z., The components of the Arctic aerosol. *Atmos. Environ.*, **18**, 329-343, 1984.

Heidam, N. Z., Crustal enrichments in the Arctic aerosol. *Atmos. Environ.*, 2083-2098, 1985.

Heintzenberg, J. Size-segregated measurements of particle elemental carbon and light absorption at remote Arctic locations. *Atmos. Environ.*, **16**, 2461-2469, 1982.

Held, I. M., Stationary and quasi-stationary eddies in the extratropical troposphere: theory. In *Large-scale Dynamical Processes in the Atmosphere*, B. J. Hoskins and R. P. Pearce, eds., 127-168, 1983.

Hopper, J. F., D. E. J. Worthy, L. A. Barrie and N. B. A. Trivett, Atmospheric observations of aerosol black carbon, carbon dioxide, and methane in the high Arctic. *Atmos. Environ.*, **28**, 3047-3054, 1994.

Hoskins, B. J. and D. J. Karoly, The steady linear response of a spherical atmosphere to thermal and orographic forcing. *J. Atmos. Sci.*, **38**, 1179-1196, 1981.

Hoskins, B. J., M. E. McIntyre and A. W. Robertson, On the use of isentropic potential vorticity maps. *Q. J. Roy. Meteorol. Soc.*, **111**, 877-946, 1985.

IPCC, *Climate Change. The IPCC Scientific Assessment.* WMO/UNEP, University Press, Cambridge, U.K. 365 pp. 1990

Iversen, T., On the atmospheric transport of pollution to the Arctic. *Geoph. Res. Lett.*, **11**, 457-460, 1984.

Iversen, T., Some statistical properties of ground level air pollution at Norwegian Arctic stations and their relation to large scale atmospheric flow systems. *Atmos. Environ.*, **23**, 2451-2462, 1989a.

Iversen, T., Numerical modelling of the long range atmospheric transport of sulphur dioxide and particulate sulphate to the Arctic. *Atmos. Environ.*, **23**, 2571-2595, 1989b.

Iversen, T., Meteorology and transport of air masses in Arctic regions. In *The Tropospheric Chemistry of Ozone in the Polar Regions.*, H. Niki and K. H. Becker, eds., *NATO ASI Series*, **I7**, Springer-Verlag, 57-75, 1993.

Iversen, T. and E. Joranger, Arctic air pollution and large scale atmospheric flows. *Atmos. Environ.*, **19**, 2099-2108, 1985.

Jaenicke, R., Troposheric aerosols. In *Aerosol-Cloud-Climate Interactions.*, P. V. Hobbs, ed., *International Geophysics Series*, **54**, Academic Press, 1-31, 1993.

Joussaume, S., Three-dimensional simulations of the atmospheric cycle of desert dust particles using a general circulation model. *J. Geoph. Res.*, **95**, 1909-1941, 1990.

Joussaume, S., Paleoclimatic tracers: An investigation using an atmospheric general circulation model under ice age conditions 1. Desert dust. *J. Geoph. Res.*, **98**, 2767-2805, 1993.

Kanestrøm, I., K. Pedersen and H. Skåtun, Major stationary ridges and troughs at 500 mb. *Geophysica Norvegica*, **33**, 1-40, 1985.

Kutzbach, J. E. and P. J. Guetter, The influence of changing orbital parameters and surface boundary conditions on climate simulations for the past 18000 years. *J. Atmos. Sci.*, **43**, 1726-1759, 1986.

Langner, J., H. Rodhe, P. J. Crutzen and P. Zimmermann, Anthropogenic influence on the distribution of tropospheric sulphate aerosol. *Nature*, **359**, 712-716, 1992.

Lautenschlager, M and K. Herterich, Atmospheric response to ice age conditions - Climatology near the Earth's surface. *J. Geoph. Res.*, **95**, 22547-22557, 1990.

Leinen M. and M. Sarnthein, *Paleoclimatology and Paleometeorology: Modern and Past Patterns of Global Atmospheric Transport. NATO ASI Series*, **C282**, Kluwer Academic Publishers, 909 pp., 1988.

Lejenäs, H. and H. Økland, Characteristics of northern hemispheric blocking as determined from long time-series of observational data. *Tellus*, **35A**, 350-362, 1983.

Levy, H. B., J. D. Mahlman, W. J. Moxim and S. Liu, Tropospheric ozone: The role of transport. *J. Geoph. Res.*, **90**, 3753-3771, 1985.

Lowenthal, D. H. and K. A. Rahn, Regional sources of pollution aerosol at Barrow, Alaska during winter 1979-80 as deduced from elemental tracers. *Atmos. Environ.*, **16**, 2011-2024, 1985.

Maenhaut, W., P. Cornille, J. M. Pacyna and V. Vitols, Trace element composition and origin of the atmospheric aerosol in the Norwegian Arctic. *Atmos. Environ.*, **23**, 2551-2570, 1989.

Manabe, S. and A. J. Broccoli, The influence of continental ice sheets on the climate of an ice-age. *J. Geoph. Res.*, **90**, 2167-2190, 1985.

Miller, J. M., A five-year climatology of five-day back trajectories from Barrow, Alaska. *Atmos. Environ.*, **15**, 1401-1405, 1981

Mitchell, J. M. Jr., Visual range in the polar regions with particular reference to the Alaskan Arctic. *J. Atmos. Terr. Phys., Spec. Suppl., Pt. 1*, 195-211, 1957.

Mitchell, J. F. B., Modelling of Palaeoclimates: Examples from the recent past. CRTN 37, Hadley Centre, Met. Office, Bracknell, U.K., 21 pp. 1993.

Mosher, B. W., P. Winkler and J.-L. Jaffrezo, Seasonal aerosol chemistry at Dye 3, Greenland. *Atmos. Environ.*, **27A**, 2761-2772, 1993.

Nriagu, J. O., R. D. Coker and L. A. Barrie, Origin of sulphur in Canadian Arctic haze from isotope measurements. *Nature*, **349**, 142-145, 1991.

Ogren, J. A., Deposition of particulate elemental carbon from the atmosphere. In *Particulate Carbon: Atmospheric Life Cycle*, G. T. Wolff and R. L. Klimisch, eds., Plenum Press, 1982.

Pacyna, J. M. and B. Ottar, Transport and chemical composition of the summer aerosol in the Norwegian Arctic. *Atmos. Environ.*, **19**, 2109-2120, 1985.

Pacyna, J. M. and B. Ottar, Origin of natural constituents in the Arctic aerosol. *Atmos. Environ.*, **23**, 809-815, 1989.

Pasquill, F. and F. B. Smith, *Atmospheric Diffusion*, 3rd edition, John Wiley & Sons, 437 pp., 1983.

Penner, J. E., H. Eddleman and T. Novakov, Towards the development of a global inventory for black carbon emissions. *Atmos. Environ.*, **27A**, 1277-1295, 1993.

Plumb, R. A. and J. D. Mahlman, The zonally averaged transport characteristics of the GFDL circulation/transport model. *J. Atmos. Sci.*, **44**, 298-327, 1987.

Prospero, J. M. and T. N. Carlson, Dust concentration in the atmosphere of the equatorial North Atlantic Ocean. *J. Geoph. Res.*, **77**, 5255-5265, 1972.

Raatz, W. E., Arctic haze: Meteorological aspects of long-range transport. Ph.D.-thesis, Univ. of Alaska, Fairbanks, 1983.

Raatz, W. E. and G. E. Shaw, Long-range tropospheric transport of pollution aerosols into Alaskan Arctic. *J. Clim. Appl. Meteorol.*, **23**, 1052-1064, 1984.

Reiter, E. R., Planetary wave behaviour and Arctic air pollution. *Atmos. Environ.*, **15**, 1465-1471, 1981.

Rex, D. F., Blocking action in the middle troposphere and its effects upon regional climate - II. The climatology of blocking action. *Tellus*, **2**, 275-301, 1950.

Rind, D., Components of the ice age circulation. *J. Geophys. Res.*, **92**, 4241-4281, 1987.

Rosen, H. and A. D. A. Hansen, Role of combustion-generated carbon particles in the absorption of solar radiation in the Arctic haze. *Geoph. Res. Lett.*, **11**, 461-464, 1984.

Rosen, H., A. D. A. Hansen and T. Novakov, Role of graphitic carbon particles in radiative transfer in the Arctic haze. *Sci. Total Environ.*, **36**, 103-110, 1984.

Shukla, J. and K. C. Mo, 1983, Seasonal and geographical variation of blocking. *Mon. Wea. Rev.*, **111**, 388-402, 1983

Shapiro, M. A., R. C. Schnell, F. P. Parungo, S. J. Oltmans and B. A. Bodhaine, El Chichon volcanic debris in an Arctic tropopause fold. *Geoph. Res. Lett.*, **11**, 421-424, 1984.

Simmons, A. J, and B. J. Hoskins, The life cycles of some nonlinear baroclinic waves. *J. Atmos. Sci.*, **35**, 414-432, 1978.

Sturges, W. T. and L. A. Barrie, Stable lead isotope ratios in Arctic aerosols: evidence for the origin of Arctic air pollution. *Atmos. Environ.*, **23**, 2513-2520, 1989.

Spiro, P. A., D. J. Jacob and J. A. Logan, Global inventory of sulfur emissions with $1° \times 1°$ resolution. *J. Geoph. Res.*, **97**, 6023-6036, 1992.

Tarrasòn, L., Contributions to sulphur background deposition over Europe: Results for 1988. EMEP/MSC-W Note 5/92 The Norwegian Meteorological Institute, Oslo, Norway. 24 pp., 1992.

Tarrasòn, L. and T. Iversen, The influence of North American anthropogenic sulphur emissions over western Europe. *Tellus*, **44B**, 114-132, 1992.

Tarrasòn, L., S. Turner and I. Fløisand, An estimation of seasonal DMS fluxes over the North Atlantic Ocean and their contribution to European pollution levels. *J. Geoph. Res.*, in press, 1995.

Trepte, C. R., L. W. Thomason and G. S. Kent, Banded structures in stratospheric aerosol distributions. *Geoph. Res. Lett.*, 2397-2400, 1994.

Vinichenko, N. K., The kinetic energy spectrum in the free atmosphere - 1 second to 5 years. *Tellus*, **22**, 158-166, 1970.

Valdes, P. V. and N. M. J. Hall, Mid-latitude depressions during the last ice-age. In *Long-Term Climatic Variations*. J.-C. Duplessy and M.-T. Spyridakis, eds. *NATO ASI Series*, **I22**, Springer Verlag, 1994.

Wallace, J. M. and M. L. Blackmon, Observation of low-frequency atmospheric variability. In *Large-Scale Dynamical Processes in the Atmosphere.* B. J. Hoskins and R. P. Pearce, eds., 55-94, 1983.

Wallace, J. M., G.-H. Lim and M. L. Blackmon, Relationship between cyclone tracks and baroclinic waveguides. *J. Atmos. Sci.*, **45**, 439-462. 1988.

OCCURRENCE AND TRENDS OF POLLUTION IN THE ARCTIC TROPOSPHERE

Leonard A. Barrie
Atmospheric Environment Service
4905 Dufferin Street
North York
Ontario M3H 5T4
Canada

Introduction

The chemical composition of the troposphere (0 to ~8 km above msl) in the Arctic is distinctly different than it is in the Antarctic. As pointed out in previous reviews (Barrie, 1986; Barrie et al, 1992; Barrie, 1993; Barrie, 1995), the Arctic is surrounded by populated continents from which pollution is released to the atmosphere and is transported readily to the north. In contrast, the Antarctic region is entirely surrounded by the southern Pacific ocean and is remote from human activities. Thus, it comes as no surprise that the tropospheric concentration of many anthropogenic aerosols and gases is much higher in the Arctic than in the Antarctic. What may be less obvious is that atmospheric trace constitutents of natural origin are found to have a different chemical climatology in the Arctic than in the Antarctic. Substances derived from sea spray, wind blown dust, marine biogenic activity and volcanoes generally have different seasonal variations and concentrations in the Arctic compared to the Antarctic.

This is attributable to the uniquely different topography and geographical configuration of oceans and continents relative to the locations of measurement locations in the two regions (Figure 1). Topographically, the Arctic is very flat compared to the Antarctic (major exceptions- Greenland reaching 2 to 3 km above msl and the Rocky Mountains of eastern North America reaching ~ 6km). The northern troposphere is underlain by the Arctic ocean while the southern troposphere is bounded below by the massive Antarctic ice sheet that covers most of the continent from about 65 °S to the pole and reaches elevations of 3 to 4 km above msl.

Anthropogenic trace constitutents enter the Arctic atmosphere via transport after being released near the Earth's surface. They tend to be most concentrated in the first 2 km of the atmosphere declining to much lower "background values" in the upper

NATO ASI Series, Vol. I 43
Chemical Exchange Between the Atmosphere and Polar Snow
Edited by Eric W. Wolff and Roger C. Bales

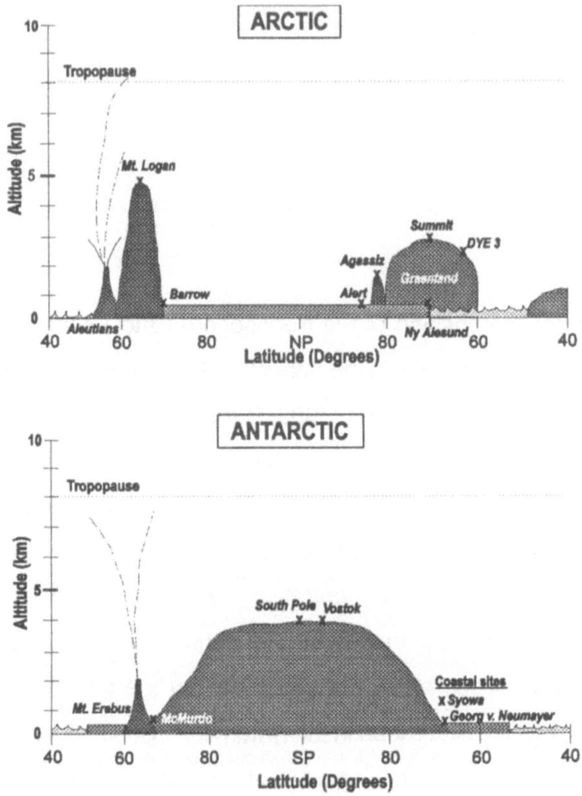

Figure 1 **A schematic contrasting the topographical and geographical differences of the two polar regions that play a major role in the composition of the atmosphere. Also shown are selected atmospheric and glacial ice chemistry measurement sites.**

troposphere. Natural constituents can enter the Arctic troposphere from sources within the Arctic such as volcanoes in the Aleutians or Iceland, dust erosion from exposed land surfaces around Greenland and in the Canadian Archipelago, sea spray production from whitecaps in open Arctic Ocean areas or from biogenic marine sources in the open or ice-covered Arctic ocean. They can also be transported into the region from similar sources located further south such as the Eurasian deserts in the case of dust (Welch et al, 1991;Franzen et al, 1995) or the North Atlantic or Pacific oceans in the case of marine constituents. Some natural constituents may enter the Arctic troposphere from the stratosphere such as cosmogenically produced Be^7 and Be^{10} (Dibb et al, 1994) or nitrates in the form of nitric acid in the gas phase or in ice crystals.

Table 1 Groups of gaseous and particulate trace constituents in the Arctic troposphere that are of regional, hemispheric or global interest.

GROUP	ENVIRONMENTAL SIGNIFICANCE IN ARCTIC
Sulphur Compounds	Climate Change, Visibility, Acidification of Ecosystems, Biogenic Marine Tracer
Black Carbon	Climate Change, Fossil Fuel Combustion Tracer of sources
Metals	Toxicity, Tracers of sources
Nitrogen oxide	Atmospheric Oxidant Chemistry, Ecosystem Acidification and Fertilization
Reactive Hydrocarbons	Atmospheric Oxidant Chemistry, Tracers of Natural and Anthropogenic Sources, Indicator of Chemical Reactions
Ozone	Climate Change, Atmospheric Chemical Stability
Halogens(Br, I)	Ozone depletion, Biogenic Marine Tracers
PAHs	Tracers of sources, Toxicity
Industrial Organics(PCBs)	Toxicity, Tracers of sources
Agricultural Pesticides	Toxicity, Tracers of sources
Greenhouse Gases	Climate Change

The abundance and main source areas of atmospheric trace constituents in the polar troposphere are very dependent on altitude. The wide range in elevation of air chemistry observatories and glacial ice coring sites in the polar regions from sea level to 4 km (Fig. 1) is very important in interpreting glacial ice core chemistry records in terms of atmospheric composition and chemical processes. Chemical groups of interest in the polar regions that are discussed in this paper are shown in Table 1. In this paper, the occurrence, origin and trend of each of these classes of compounds in the Arctic troposphere will be reviewed after a brief discussion of atmospheric aerosols. The emphasis is on comprehensive measurements at the Canadian baseline observatory at Alert which are representative of the lower troposphere in the North American Arctic (Barrie and Hoff, 1985). Few other sites have such a comprehensive set of long term observations.

Aerosols

Aerosols play a key role in the atmospheric cycles of many substances. Since the early 1970's, they have been studied in the Arctic, first out of concern for Arctic haze and its effects on visibility, climate and acidification and more recently for their involvement in arctic tropospheric chemistry (e.g. McConnell et al, 1992). Arctic haze is essentially the remnant of acid rain pollution from mid-latitudes (Barrie, 1986). Aerosols are not only produced as primary products by a variety of natural and anthropogenic processes at the Earth's surface but also by chemical and physical reactions. Chemical changes at polar sunrise including ozone depletion (Barrie et al, 1988; Niki and Becker, 1993, Kawamura et al, 1995) and the oxidation of sulphur dioxide to form particulate sulphates (Barrie and Hoff, 1984; Barrie et al, 1994a) are notable atmospheric chemistry phenomena involving aerosols. Once deposited in glaciers, aerosols yield an indication of the composition of the Earth's atmosphere, its general circulation and insight into a variety of environmental processes such as biomass burning, marine biological activity and volcanism.

The general characteristics of aerosols and mechanisms of deposition to glaciers

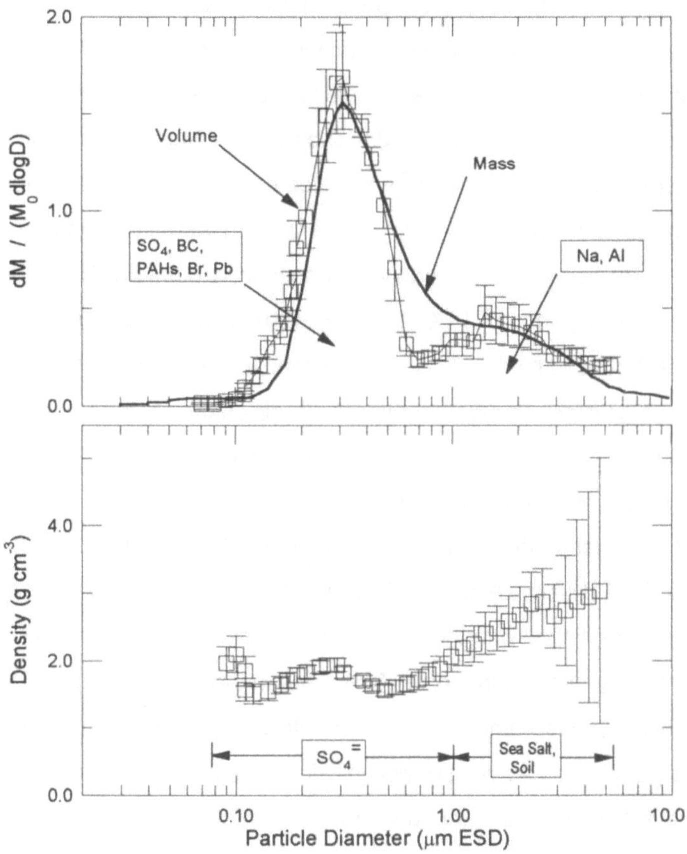

Figure 2 **The average mass and volume size distribution (top) and the distribution of density (bottom) of aerosol at Alert Canada for the period January to April 1992 (Staebler et al, 1994).**

have been reviewed by this author (Barrie,1985; Barrie, 1991). Spanning particle sizes from a few molecules in diameter to large assemblages of compounds reaching diameters of 100 μm, they are transported long distances before being removed by clouds, precipitation and dry deposition to the surface. Figure 2a shows the average mass and volume size distribution of aerosols (at about 14% relative humidity) measured at Alert, Canada during the polluted Arctic haze season. The main mass constituent of Arctic aerosols in the polluted winter is sulphate (Hillamo et al, 1993, Staebler et al, 1994) and it is distributed in the fine particle range peaking at 0.3 to 0.6 μm diameter depending on relative humidity and the definition of diameter. Variation in chemical composition with particle size is indicated by the size distribution of particle density (Fig. 2b) derived by dividing the aerosol mass distribution by the volume distribution. Particles < 1 μm equivalent scattering diameter (ESD) have a mean density at relative humidity of 14% of 1.52 ± 0.07 g cm^{-3} while those > 1 μm ESD are denser at 2.58 ± g cm^{-3}. This is consistent with a composition of fine particles being sulphuric acid or ammonium bisulphate as observed from chemical analysis (Barrie and Barrie, 1990) and the larger particles being a mixture of wind blown dust (2.6 to 2.75 g cm^{-3}) and sea salt particles (2.42 g cm^{-3}). The mass size distribution of sulphur particles measured at Dye III, on Greenland during March in air of relative humidity 55 to 90% (Hillamo et al, 1993) shows a peak at about 0.6 μm aerodynamic diameter. This is consistent with the 0.3 μm ESD observed at Alert if the difference in the ESD and aerodynamic diameter (~ 40%) and in the relative humidity (14% versus 55-90%) of the two measurements are taken into account since hygroscopic sulphate acid particles grow in size as relative humidity increases. Major aerosols constituents are mixed with traces of other compounds that are of interest. For instance, the fine particles dominated by sulphates contain black carbon, Pb and many other anthropogenic metals as well as PAHs.

Table 2. Components of winter aerosols (December to April) commonly observed in the Arctic (Barrie and Barrie, 1990; Maenhaut et al, 1993; Mosher et al, 1993; Barrie, 1995)

COMPONENT	ELEMENTS
1. Wind Blown Soil	Al, Mn, Mg, Sr
2. Sea Salt	Na^+, Cl^-, K^+, Sr, NO_3^-, Mg
3. Anthropogenic:primary	Pb NH^+_4, Ni, $SO^=_4$, V, H^+, Zn, Mn, Bl.-C
4. Anthropogenic:photochemical	Br, I, $SO_4^=$, H^+

For most aerosol constituent's, a large fraction of their total mass is attributable to source types shown in Table 2. The first two components are associated with sources that produce particles whose main mass is concentrated in diameters larger than 1 μm while the last two occur in fine particles smaller than 1 μm. Carbonaceous compounds are also present in Arctic aerosols as black carbon and as non-black organic compounds. At Alert in the polluted winter, the observed fraction of total aerosol mass that is carbonaceous (black carbon plus organics) is 2.8 to 11% (Fig.3) while that of the water soluble organic fraction ranges from 1.0 to 5.2%. Black carbon makes up part of the difference between carbon and water soluble organic carbon in the aerosols.

Sulphur and Black Carbon

In the Arctic troposphere, sulphur compounds play a prominent role in the reduction of visibility, perturbation of the radiation budget (Blanchet and Girard, 1995) and as markers of marine biogenic activity in the Arctic and northern oceans (Li and Barrie, 1993). The primary anthropogenic species are the gas SO_2 and its oxidation product particulate $SO_4^=$. Sulphate can also originate from sea salt and from the oxidation of dimethyl sulphide gas (DMS) released from the marine biogenic activity in the oceans (Restelli and Angeletti, 1993). Another product of DMS oxidation is methane sulphonic acid (MSA) which, like $SO_4^=$, is relatively involatile and therefore forms particulate matter in the atmosphere. Observations at Alert and on Greenland (Li and Barrie, 1993; Li et al, 1993a,b) and in Iceland (Prospero et al, 1995) show that the ratio of MSA/ $SO_4^=$ from oxidation of marine DMS in northern regions is approximately 0.3 (range 0.2 to 0.9). This is much higher than the ratio observed over temperate and tropical oceans (0.04 to 0.07). Thus, not only is MSA a tracer of biogenic marine activity but its ratio to $SO_4^=$ is an indicator of the latitude of the oceanic region in which it was formed.

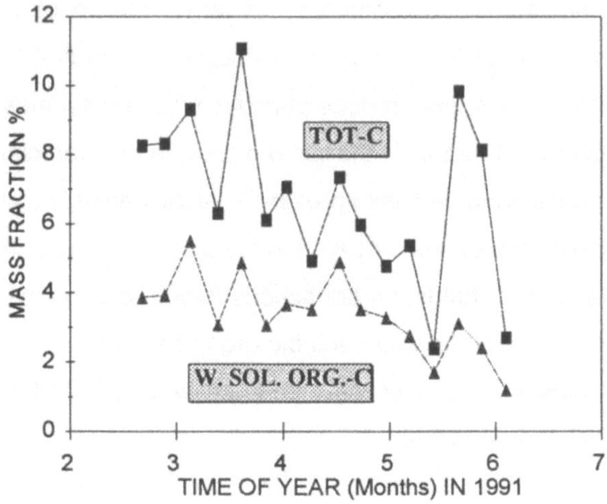

Figure 3 **The fraction of total aerosol mass at Alert, Canada that is comprised of carbon and the fraction that is comprised of carbon in water soluble organic carbon compounds (Kawamura et al, 1995a). Based on observations from February to June 1991.**

Particulate sulphate can be apportioned into anthropogenic, marine biogenic and sea salt components using the S^{34}/S^{32} ratio in $SO_4^{=}$ and simultaneously measured marine Na^+ concentrations as a tracer of sea salt sulphate. This was done on a seasonal basis at Alert, Canada by Li and Barrie (1993) who concluded that the sea salt fraction was <8 % throughout the year and that from June to August, 25-30% of $SO_4^{=}$ was marine biogenic and the rest anthropogenic. In contrast in the more polluted remainder of the year, the biogenic fraction was <14% of the total. Thus, even in the relatively unpolluted summer months, anthropogenic influences on the Arctic aerosol are at least as important as the marine ones.

Particulate sulphur and black carbon play a unique role in the energy budget of the Arctic atmosphere. The mean seasonal variation in concentration at Alert in 1989 and 1990 is shown in Figure 4. Sulphur and black carbon aerosols in the Arctic

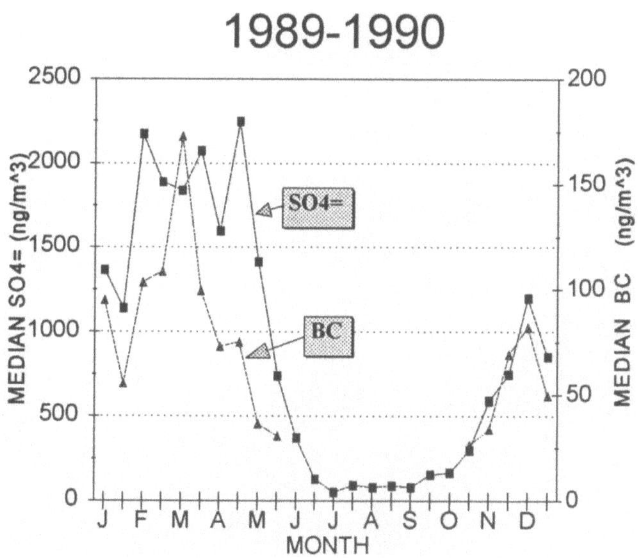

Figure 4 **The average seasonal variation of aerosol sulphate and black carbon at Alert for the period 1989 to 1990.**

are good cloud condensation nuclei. Their number concentration in the atmosphere affects the number of cloud droplets in optically thin marine stratus clouds and hence the whiteness of the cloud. A whiter cloud reflects more solar radiation back to space. Thus, sulphur aerosols play potentially a role in the Earth's radiation budget and thus possibly in climate (Charlson et al, 1992). Also, they play a similar role in cloud-free air as they decrease incoming solar energy by scattering sunlight back to space.

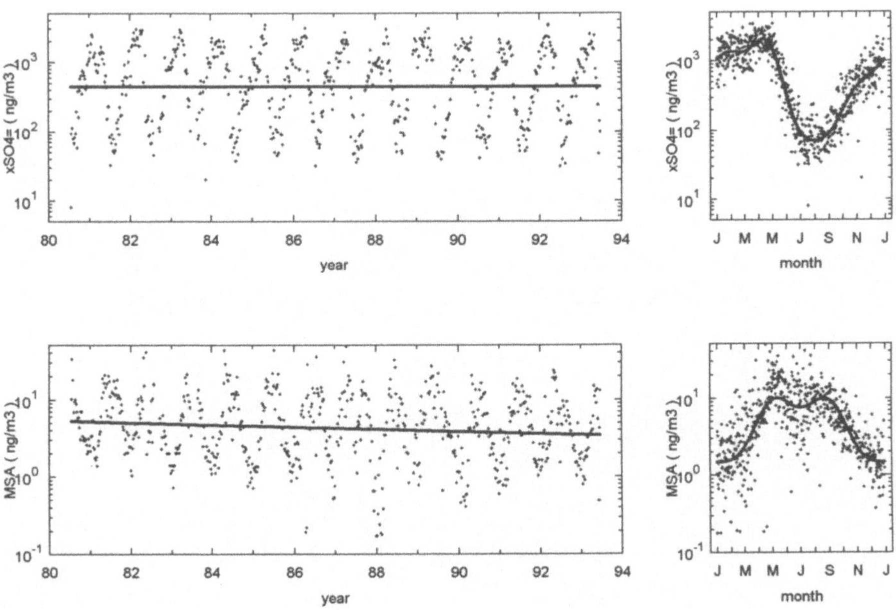

Figure 5 The long term trend and seasonal variation of $SO_4^=$ and MSA at Alert (Li et al, 1993; Barrie and Sirois, 1995).

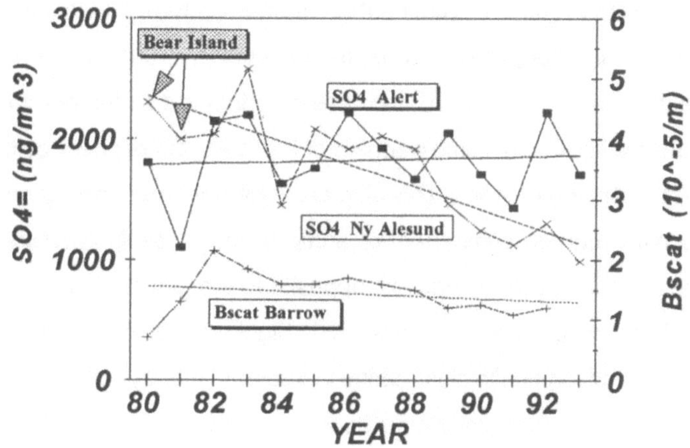

Figure 6 **A comparison of trends in arithmetic means of the March-April period for total $SO^{=}_4$ at Alert and Ny Ålesund with those of Bscat at Barrow Alaska (Bodhaine and Dutton, 1993)**

Globally, sulphur aerosols of anthropogenic and biogenic origin tend to mediate the global greenhouse warming effect. However, this global-average-cooling effect due to reduction of incoming solar radiation does not apply in the Arctic. Blanchet and Girard (1995) have proposed that the observed decrease in winter surface air temperature in the Arctic region over the past two decades (Kahl et al, 1993a,1993b; Bradley et al, 1992) which is contrary to global climate model predictions of strong warming in the polar regions due to greenhouse gas increases is actually due to a dehydration of the lower atmosphere caused by enhanced precipitation resulting from anthropogenic arctic haze aerosols. Since water vapour is a good greenhouse gas, its removal results in greater infrared radiative cooling than normal. In turn, this results in lower surface temperatures and deeper surface-based inversions. Furthermore, the same authors calculate that the observed 2 ° C increase in spring-time air temperature in the Arctic supports the predicted springtime warming of the lower atmosphere by the black carbon component of Arctic aerosols. This situation where the net absorption of solar radiation by the black carbon component of arctic haze aerosols exceeds the net scattering of solar radiation back to space is opposite to the global effect. The reason for this is that the underlying surface in the Arctic is highly reflective for solar radiation allowing black carbon absorption to significantly increase planetary absorption of solar energy while elsewhere on the globe over darker surfaces, the net effect of black carbon is to not significantly alter the planetary absorption. The long term trend and mean seasonal variation of $SO_4^=$ and MSA at Alert Canada between 1980 and 1993 are shown in Figure 5. There has been no significant trend in $SO_4^=$ for any season. In contrast to $SO_4^=$, MSA peaks in the period April to October reflecting its marine biogenic origins. It also shows a decline of 3%/yr.

Trends in anthropogenic haze particles in the Arctic have been noted but observed trends differ from region to region Bodhaine and Dutton (1993) report that between 1982 and 1992 at Barrow Alaska, surface aerosol scattering (Bscat) and aerosol optical depth have decreased significantly in the months of March and April but not at other times of year. Bscat is proportional to aerosol mass and hence should be driven mainly by $SO_4^=$, the predominant mass constituent. Aerosol $SO_4^=$ observations at Alert and Spitsbergen (a composite of Bear Island 1980-1981; Ny Ålesund 1982-1990; Zeppelin Mt. 1991-1993) are compared to those at Barrow for Bscat and to anthropogenic SO_x emission trends in Eurasia in Figures 6 ans 7 for the whole measurement period not just the period 1982 to 1992 selected by Bodhaine and Dutton (1993). No major change in $SO_4^=$ at Alert and in Bscat at Barrow is evident while at Spitsbergen, it has decreased by 50% from 1980 to 1989. At Spitsbergen, the decrease in SOx (sum of $SO_4^=$ and SO_2) parallels the trend in reported SOx emissions in Eurasia while that of $SO_4^=$ is slightly less than the reported

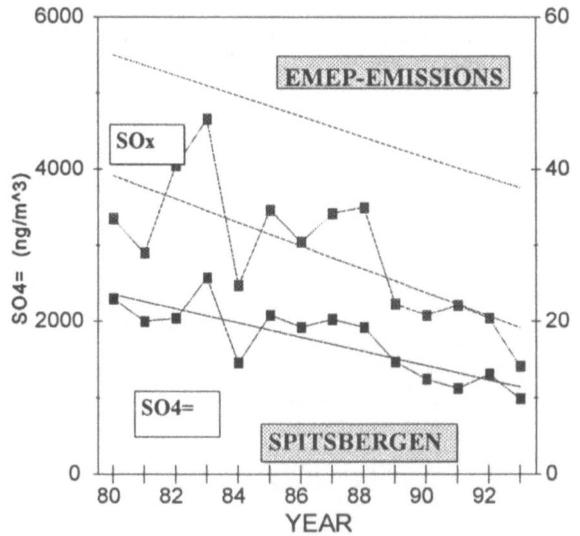

Figure 7 **A comparison of the trend in March/April mean atmospheric sulphur oxide concentrations at Spitsbergen(from NILU) with those in annual emissions of anthropogenic SO_2 from Europe and Russia(Mylona, 1993)**

SO_x emissions. The difference in trends may be due to the greater relative influence of the Russian smelting region of Norilsk on the North American Arctic than on the western European Arctic around Spitsbergen. Emissions of the former have changed less than the latter. To resolve this, chemical transport modelling studies are needed.

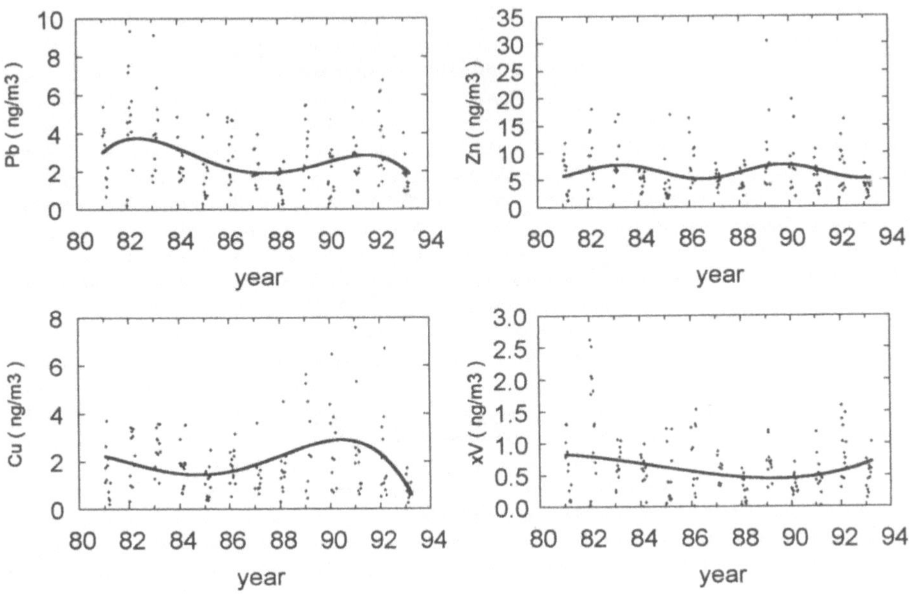

Figure 8 **Long term trends in weekly mean concentrations from January to March of anthropogenic metals in aerosols measured at Alert (Barrie and Sirois, 1995).**

Anthropogenic Heavy Metals

Heavy metals such as Pb, Cu, Zn and non-soil V (x-V) have been routinely measured in the Arctic. Downward trends since the late 1960s have been reported for the concentrations in glacial ice at Summit Greenland (Fig. 1) for Pb and Zn but not for Cu (Boutron et al, 1991). In glacial ice, Pb and Zn are definitely of anthropogenic origin while Cu is in part from soil. At Alert, the concentrations of these metals in air show a distinct seasonal variation peaking from January to March and minimizes in summer (Barrie and Sirois, 1995). Their long term trend at this high Arctic station during winter (Fig. 8) shows a marked decrease in Pb and in x-V between 1980 and 1993. In contrast, variations of Zn and Cu are less clear. Copper concentration decreased markedly in 1993 after peaking in 1990 and 1991. Whether this reflects the economic state of the smelting industry in Russia or whether it is due to meteorological variations remains an open question. Unlike the situation in Greenland snow, soil enrichment factors of Cu (relative to Al in crustal rock) in winter (January-March) aerosols are high (75 to 230) indicating that in the atmospheric aerosol in winter, this element is anthropogenic in origin.

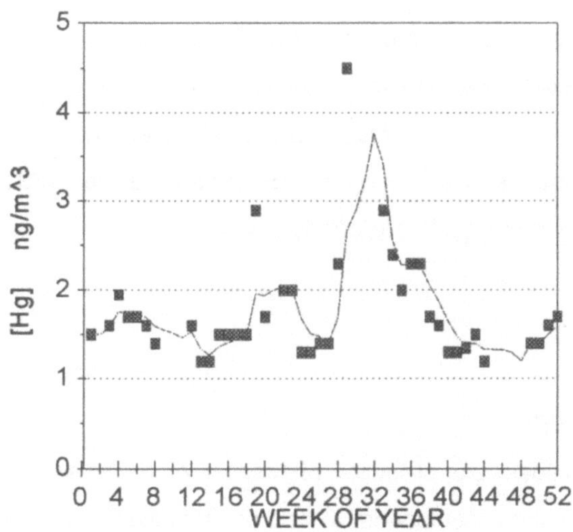

Figure 9 **The mean seasonal variation of total gas phase Hg in air at Alert based on measurements from August 1992 to July 1993 (adapted from Schroeder et al, 1994). The fitted line is a 5-point running mean of weekly mean observations.**

Atmospheric mercury which is mainly in the form of gas phase elemental mercury is increasing in the global environment at about 1% yr^{-1} (Lindqvist, 1994; EPRI, 1994). It is also increasing in sediments laid down in high Arctic lakes during the post-industrial era(Lockhart, 1993). The main anthropogenic sources of Hg are industrial processes such as chlor-alkali plants, waste incineration, coal combustion and metal production as well as natural gas combustion. Natural emissions from the Earth's crust can occur via the biosphere and oceans. Best estimates of the relative importance of anthropogenic sources to the global total input to the atmosphere range from 50 to 75% (EPRI, 1994). There are few measurements of atmospheric Hg in the Arctic. The first observations made routinely for a year at Alert are shown in Figure 9. Concentrations are approximately constant throughout the year at 1.5 ± 0.5 ng m^{-3} except for a slight peak in May to early June and a significant peak in August and early September at 2 to 4.5 ng m^{-3}. This compares with average concentrations observed in the North Atlantic in 1990 of 2.5 ng m^{-3} (Slemr and Langer, 1992). The processes that control this seasonal cycle are poorly understood.

NITROGEN OXIDES

Nitrogen oxides are products of fossil fuel combustion. They generally accompany acidic sulphur compounds in acid rain pollution and in Arctic haze pollution (Bottenheim et al, 1993). The chemistry of these compounds in the atmosphere is discussed in more detail by Bottenheim and Barrie (this volume). Suffice it to say that in contrast to lower latitudes, the predominant form of nitrogen oxides in the arctic troposphere is the organic gas peroxyacetylnitrate (PAN) forming 50 to 65% of the total. Alkylnitrate gases form 7 to 20% and the rest is HNO_3 and particulate NO^-_3 (Bottenheim et al, 1993; 1994; Muthuramu et al, 1994, Barrie et al, 1994a). During Polar Sunrise Experiment 1992 at Alert total inorganic nitrate was on average 74% particulate and the rest gaseous HNO_3 (Barrie et al, 1994a). Concentrations were about 50 pptv compared to 200 to 350 pptv for PAN. The seasonal variation of PAN and particulate NO^-_3 representing most of the airborne inorganic NO_3 are compared in Figure 10.

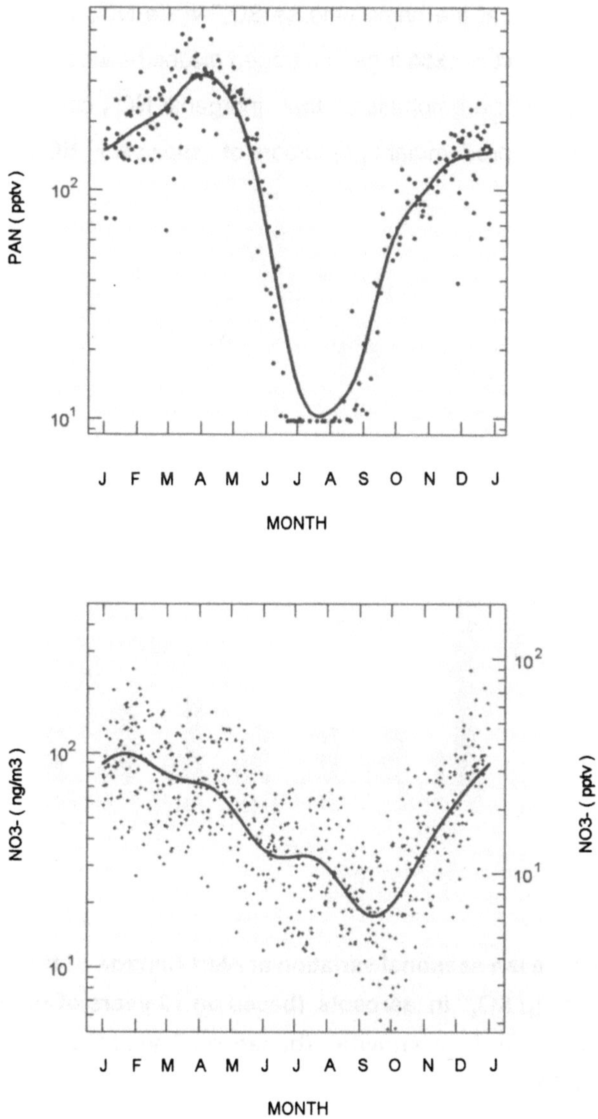

Figure 10 The mean seasonal variation of: (top) peroxyacetylnitrate (PAN) at Alert in the high Arctic based on measurements between 1987 and 1993 (Bottenheim et al, 1994) and (bottom) aerosol nitrate based on measurements from 1980 to 1993 (Barrie and Sirois, 1995).

PAN peaks in spring at the same time as $SO_4^=$ while NO_3 peaks in December to February. With respect to exchange of nitrogen oxides between the atmosphere and glacial snow it should be emphasized that inorganic NO^-_3 compounds (e.g HNO_3 , HONO) are likely the predominant precursors of snowpack NO^-_3.

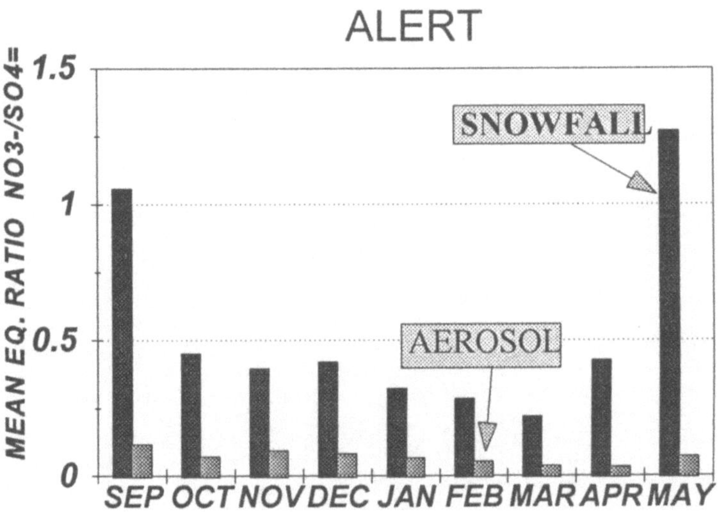

Figure 11. The mean seasonal variation at Alert Canada of the equivalents ratio of $NO_3^-/SO_4^=$ in aerosols (based on 13 years of observations, 1980-1993) and in snowfall (based on 3 years of observations, 1990-1993).

Figure 11 shows that the equivalents ratio of NO_3^- / $SO^=_4$ is much lower in ground level Arctic aerosols at Alert (0.05 to 0.2) than it is in snowfall (0.2 to 1.06).

The ratio in Alert snowfall is in the range of that observed in the Alaskan snowpack (Jaffe and Zukowski, 1993) as well as that seen in glacial snow and ice in Greenland (Mayewski et al, 1993). This marked enrichment of nitrates in snowfall is likely the result of more efficient scavenging of NO_3^- than of $SO_4^=$ compounds. The former tend to be on larger particles than the latter (Barrie et al, 1994b). Larger particles are more efficiently scavenged by falling ice crystals (Barrie, 1991). In addition, the uptake of HNO_3 which comprises, on average, 25% of total inorganic nitrate (Barrie et al, 1994a) in lower tropospheric air at Alert in winter is likely greater than that of SO_2. There is a seasonal variation in $NO_3^- / SO_4^=$ ratio in snowfall. It is highest at the beginning and end of the cold snowfall season in September and May, respectively. This possibly reflects a relative increase in the proportion of total nitrogen oxides that are in the form of particulate and gaseous inorganic nitrates since, at warmer temperatures, PAN thermally decomposes to NOx and hence to these products. PAN and alkyl nitrates are not very soluble or reactive with snow surfaces and, therefore, are not expected to contribute directly to snowfall nitrate.

REACTIVE HYDROCARBONS

Volatile organic hydrocarbons (VOCs) that are reactive in the atmosphere play a role in oxidant chemistry and hence in O_3 formation (Bottenheim and Barrie, this volume). They include alkanes, alkenes and alkynes. Routine measurements of numerous compounds have been made in the high Arctic at Alert and at Spitsbergen (Solberg et al, 1995). The average seasonal variation based on weekly grab samples from Alert in 1989 and 1990 (J. Rudolph, KFA Juelich, personal communication) of ethane (C_2H_6) and of acetylene (C_2H_2) is shown in Figure 12. In contrast to the Arctic, ethane and acetylene levels in Antarctica are much lower averaging annually 350 and 18 pptv, respectively (Rudolph et al, 1992). There is a strong seasonal variation in both constituents which is likely driven by seasonal variation in destruction by reaction with OH radical. Detailed observations of VOCs at Alert during Polar Sunrise Experiment 1992 (Jobson et al, 1994) showed that the relative composition of VOCs in the high Arctic is somewhat different than that observed at Spitsbergen and Barrow. Relative to iso-pentane, C_2-C_5 alkanes were more abundant at Barrow than at Alert. Furthermore, C_6 hydrocarbons at Spitsbergen are somewhat higher in concentration

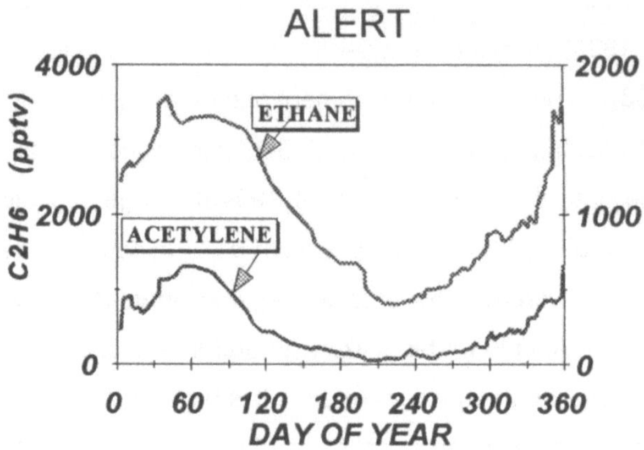

Figure 12 The average seasonal variation of lower tropospheric ethane and acetylene at Alert for the period 1989 to 1990 (J. Rudolph, KFA Juelich, personal communication).

than at Alert or Barrow indicating the proximity to mid-latitudinal pollution sources in Eurasia. The high Arctic air mass during winter and spring is relatively homogeneous for non-methane hydrocarbons with long atmospheric lifetimes, namely C_2-C_5 alkanes, acetylene and benzene.

In addition to reaction with OH radicals, many VOCs including ethane show evidence of reaction with Cl atoms at polar sunrise during surface ozone depletion events (Jobson et al, 1994; Solberg et al, 1995). Acetylene not only reactn with Cl atoms but also with Br atoms. A more detailed discussion of VOC chemistry in the Arctic is given elsewhere (Bottenheim and Barrie, this volume).

OZONE

Tropospheric ozone originates from the stratosphere and from photochemical reactions of VOCs and nitrogen oxides . In the Arctic troposphere, it has been decreasing in concentration since the late 1970's (Tarasick et al, 1994). Based on Be stable isotope observations and vertical ozone profiles Dibb et al (1994) have estimated that less than 10 to 15% of surface level ozone originates from stratospheric sources. However, this issue needs more research. The seasonal variation of mean ground level O_3 mixing ratio at Alert (Fig. 13) shows a peak in winter at 35 ppbv and a sharp decline in April to a minimum of 20 ppbv in May. Similar seasonal variations are observed at Barrow Alaska (Oltmans, 1994). The greatest variability in concentration occurs in April and May during periods of O_3 depletion in the Arctic boundary layer (Barrie et al, 1986) that are connected with elevated concentrations of gaseous and particulate Br and I (Barrie et al, 1994a, b).

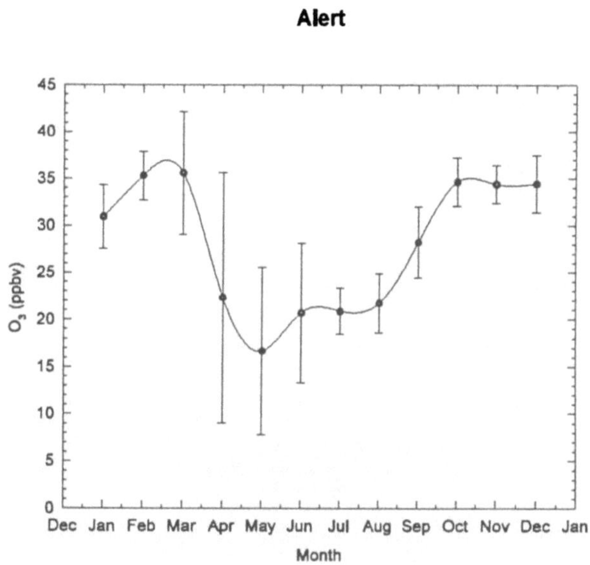

Figure 13 The average seasonal variation of ground level ozone at Alert for the period 1992 to 1993. Bars indicate the standard deviation of daily mean ozone mixing ratios (Anlauf, AES, personal communication)

HALOGENS (Br and I)

The mean seasonal variation of particulate Br and I at Alert for 13 years is compared to that at DYE 3 on southern Greenland for a year. Concentrations at Alert are much higher than at DYE 3 which is at a high elevation (Fig. 1). They show a peak in both Br and I in March to May and again in I in August to October. A similar seasonal variation is observed at Dye 3. It has been suggested that the spring peak is connected with photochemical reactions at polar sunrise that result in ozone depletion and end-product halogen formation and that the fall peak in particulate I is due to a late summer early autumn bloom in northern ocean production of iodine gases such as CH_3I and CH_2ICl (Sturges and Barrie 1988; Barrie and Barrie, 1990).

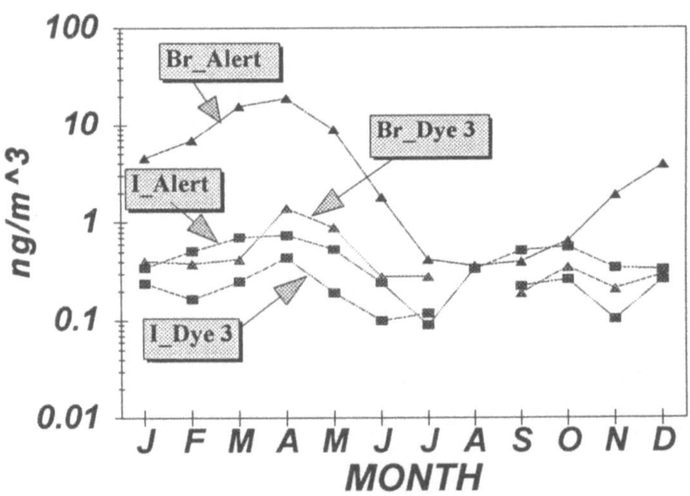

Figure 13 **A comparison of the mean seasonal variation of particulate Br and I at Alert for the period 1980 to 1993 (Barrie and Sirois, 1995) with that for DYE 3 Greenland from September 1988 to July 1989 (Mosher et al, 1993).**

POLYCYCLIC AROMATIC HYDROCARBONS(PAHs)

Many PAHs originate from the combustion of fossil fuel and from wood burning. They are released to the atmosphere in hot polluted mixtures of gases and particles. The mass of each PAH compound quickly partitions between the gas phase and particulate phase depending on volatility, temperature and aerosol surface area (Patton et al, 1991). This process leads to most of their particulate mass being on particles less than 1 μm diameter (Fig.2). Routine long-term measurements of the gas and particulate concentration of these compounds in the Arctic atmosphere have only recently been undertaken (Fellin et al, 1995) at Alert, Tagish-Yukon and Cape Dorset, Canada as well as at Dunay Island, Russia. Results for two years of measurements

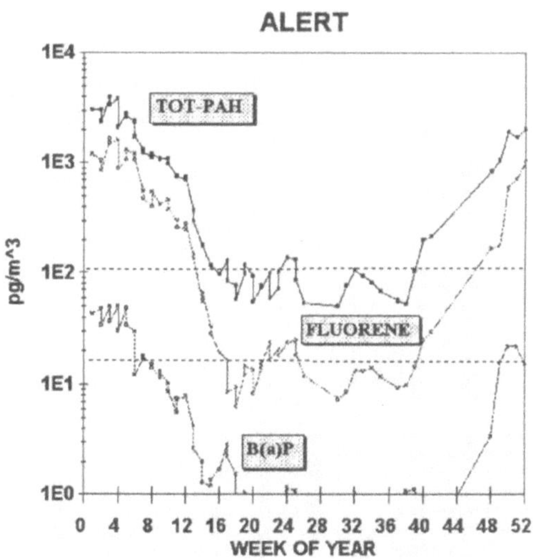

Figure 15 **The mean seasonal variation of total PAH (see Fig 16 for the compounds measured), a major constituent (fluorene) and a minor but highly reactive and toxic particulate PAH (B(a)P). Based on data at Alert Canada for 1992 and 1993 (Fellin et al, 1995).**

at Alert are shown in Figures 15 to 17. There is a strong seasonal variation in all PAHs (Fig.15) peaking during the winter months of December to March roughly in step with Arctic haze sulphates (Fig. 5). A comparison of the average concentration of selected PAHs that are mainly on particles observed at Alert during the peak concentration months of January and February with the second highest concentration observed at Dye3 by Jaffrezo et al (1993) shows that they are at least five times higher at Alert than at Dye3 and that the relative composition of the PAH mixtures are very different (Fig. 16). This underlines the point made earlier that due to differences

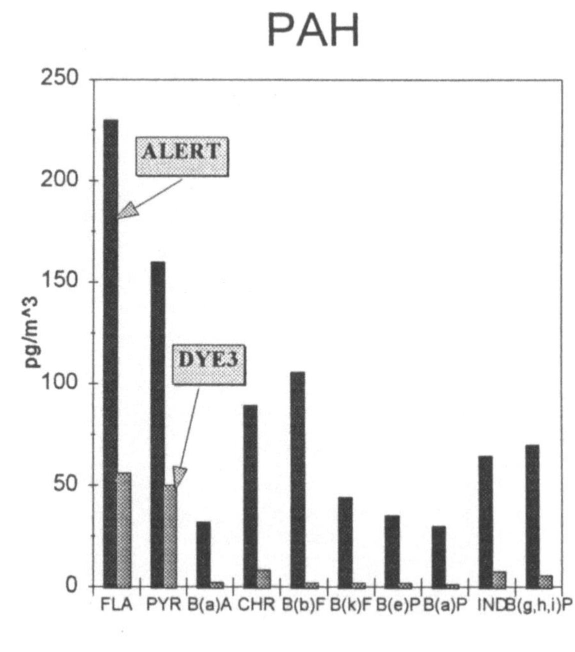

Figure16 **A comparison of mean tropospheric concentrations of selected PAHs that are mainly in particulate matter for Alert in January and February 1992 and 1993 with an observation made at Dye 3, Greenland (sample #6 of Jaffrezo et al, 1993) in spring 1991.**

of topography and geography (Fig. 1), the airshed for elevated Greenland sites is very different than that of Alert. At Alert, both gaseous and particulate PAHs were measured. The contribution of those compounds observed to the total mean PAH concentration for the period January to February of 1992 and 1993 (Fig 17) shows that fluorene, phenanthrene, fluoranthene and pyrene make up 60% of the total mass. However, as with trace metals in particulate matter the value of minor constituents is in the information they carry as tracers of sources and as indicators of chemical reactions. For instance, retene(RET) is relatively more abundant in wood smoke and biomass burning pollution than in fossil fuel combustion products. Some PAHs are more reactive than others. For instance B(a)P degrades in the environment more rapidly than its sister compound B(e)P.

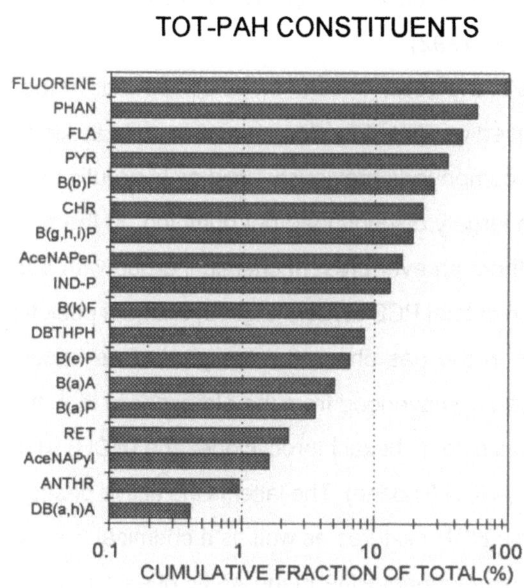

Figure 17 The mean relative composition of observed PAHs at Alert Canada for the most polluted winter period of January to February.

PERSISTENT ORGANIC POLLUTANTS (POPs)

The polar regions are susceptible to pollution from synthetic organic compounds that are persistent in the environment (Barrie et al, 1992; MacKay and Wania, 1995). These compounds may originate from industrial activities (e.g. polychlorinated biphenyls, PCBs) or from agricultural practices (herbicides and pesticides). The Arctic tends to receive a disproportionate fraction of the total mass of these compounds released to the environment over the past three decades. There are several reasons for this: (I) the increased solubility of gases in the colder Arctic water, (ii) the tendency of POPs to partition more onto particles at low temperatures and thereby be deposited more readily than through gas exchange with snow and (iii) for some POPs compounds(e.g. HCH) slower degradation in polar regions than elsewhere. They tend to bioaccumulate up the food chain in fats of animals. Thus, northern people who consume mainly wildfood receive a disproportionate amount of these substances in their diets (Kinlock et al, 1992).

Two examples of the occurrence of these compounds in the Arctic atmosphere will be given here based on recent routine measurements at Alert. First, PCBs are an industrial class of compounds previously used as hydraulic and transformer fluids. Their use has been largely discontinued but dumping in the environment and their persistence makes them an ever present chemical group with about 209 congeners. The seasonal variation of total PCB at Alert (Fig. 18) shows a peak from April to August. Most of the mass is in the gas phase but the particulate fraction is still important because it is more readily scavenged from the atmosphere than the gaseous fraction.

The second example is hexachlorocyclohexane (HCH) which has two isomers of note, α-HCH and γ-HCH (lindane). The latter is the active pesticide while the former is a byproduct in some HCH mixtures as well as a chemical transformation product of γ-HCH. It is the most abundant POP in the Arctic ocean (Bidleman et al, 1995). A mass budget of HCH for the Arctic ocean surface layer (Barrie et al, 1992) shows that the atmosphere is a major source and that the mean residence time of HCH is about 20

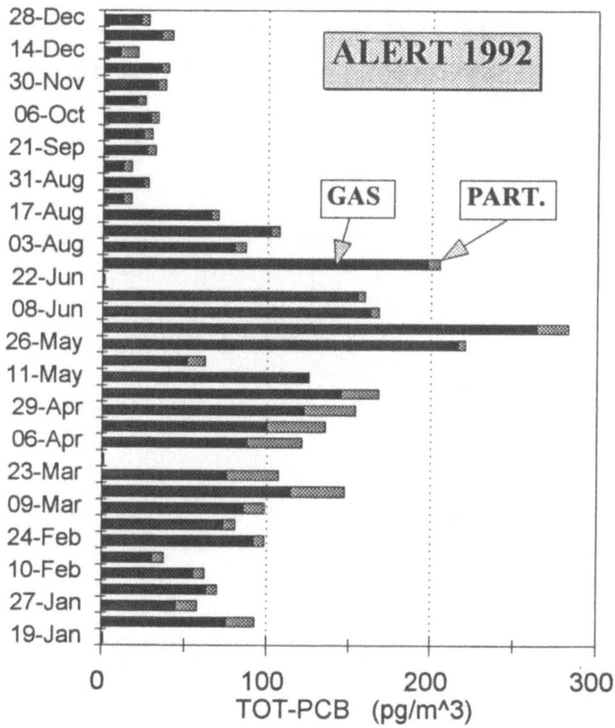

Figure 18 The seasonal variation of total PCBs at Alert in 1992 (Stern et al, 1995)

years as it is limited by the residence time of water rather than chemical degradation rates which are much slower. It is an agricultural herbicide widely used in the world today (China, India, Russia, Turkey, Spain France and Mexico are major users) although in the 1980s a shift in use of the wasteful technical HCH mixture that is mostly the inactive α-HCH and only a small fraction γ-HCH (8%) to pure γ-HCH has reduced emissions to the environment. Between 1980 and 1993, it has been observed that the atmospheric concentration of this constituent in the Arctic decreased by a factor of 10. Accompanying this decrease has been a shift in air-to-ocean exchange from into-the-ocean in 1980 to out-of-the-ocean in 1993(Bidleman et al, 1995).

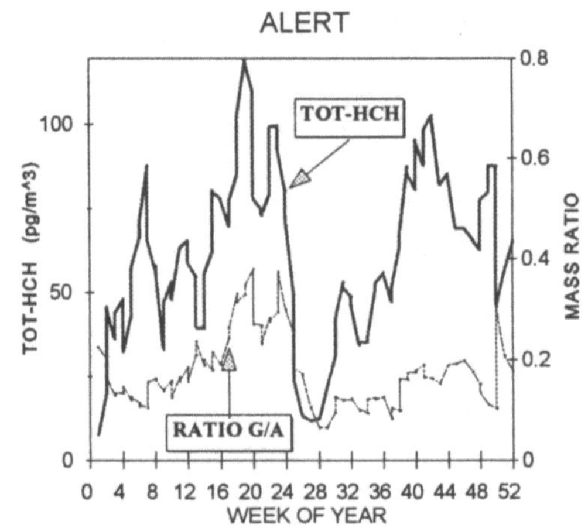

Figure 19 **The seasonal variation of tot-HCH and its γ-HCH/ α-HCH ratio (G/A) at Alert. Based on data at Alert Canada for 1992 and 1993 (Fellin et al, 1995). Data are smoothed with a three point running mean.**

The seasonal variation of HCH isomers in the high arctic atmosphere has only recently been measured(Fig. 19). There is a peak in the spring months of tot-HCH that is marked by a peak in the γ-HCH/ α-HCH ratio. Since current applications are mostly pure γ-HCH and the background is enriched in α-HCH, this indicates that the spring peak is due to relatively fresh emissions of HCH probably associated with loss to the atmosphere from soil during spring tilling when evaporation is high and vegetation is not covering the surface.

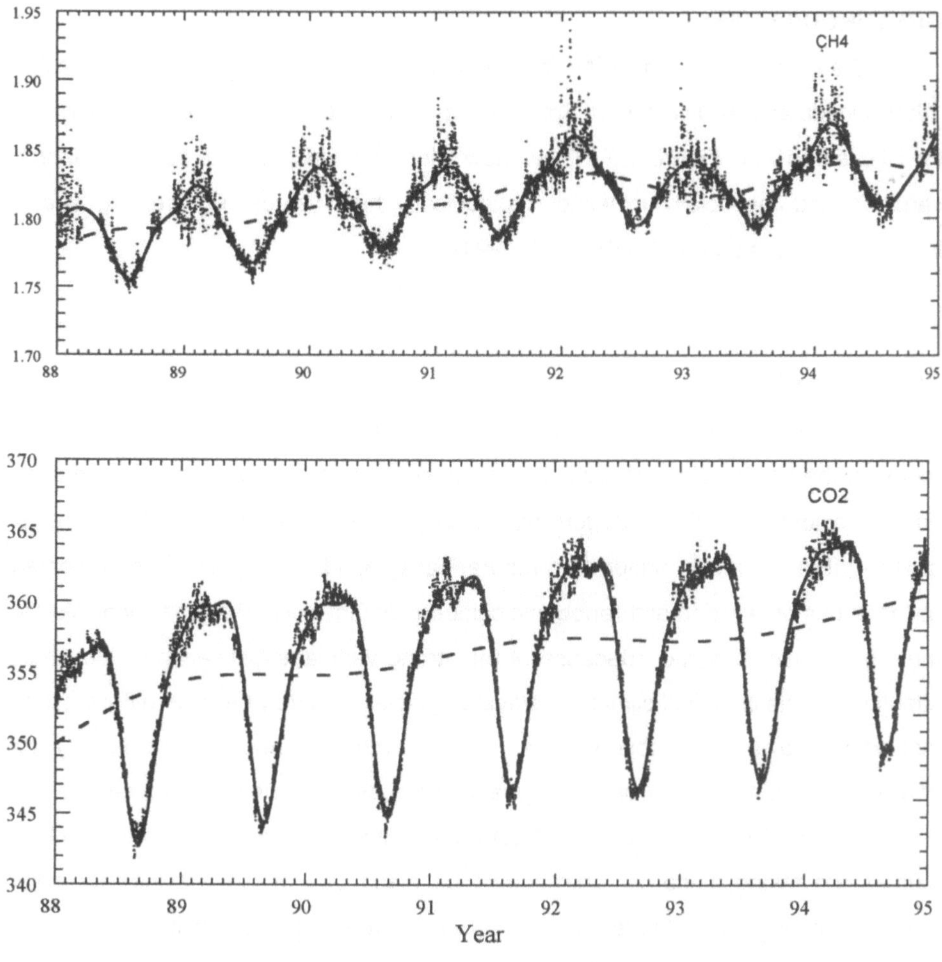

Figure 20 Observed concentrations (ppmv) of carbon greenhouse gases at Alert Canada (Worthy et al, 1994). (Top-methane and bottom-carbon dioxide). Dashed lines represent inter-annual variations.

GREENHOUSE GASES

Greenhouse gases in the Arctic are being routinely monitored at Barrow Alaska, Alert Canada and at Ny Ålesund, Spitsbergen. The time series at Alert for CO_2 and for CH_4 is shown in Figure 20. They show a long term increase , a strong seasonal variation and, especially in winter, episodically high values representing strong transport from the south (Worthy et al, 1994).

CONCLUSION

The composition of the arctic troposphere is showing clearly that this remote region of the Earth's atmosphere is being influenced by human activity. Greenhouse gases, acidic aerosols, nitrogen oxides and persistent organic pollutants are all present at levels much higher than natural background. This paper has presented observations of selected anthropogenic compounds that were taken mainly at Alert in the high Canadian Arctic. Because of the strong vertical stratification of the Arctic atmosphere, the chemical climate at elevated glacial sites may be quite different than that indicated by sea level observations. Unfortunately, there exists to date very few year-round measurements of atmospheric composition at elevated glacial sites on Ellesmere Island, Greenland or in the Rocky Mountains where major ice core chemistry studies are taking place. This should be made one of the priorities for polar atmospheric research on the relationship between atmospheric composition and glacial ice chemistry. It is likely that the lower tropospheric observations reported here for sea level observations are typical of large areas of the high Arctic atmosphere up to 2 km above mean sea level. Thus, low altitude glacial sites on Ellesmere Island (1.6 km) are affected by the atmosphere described herein. The other sites on Greenland and the Rocky Mountains likely experience middle to upper tropospheric air that is much less polluted and from a different air shed.

ACKNOWLEDGMENTS

Thanks are due to K. Anlauf, D. Worthy, N. Trivett and J. Rudolf for providing ozone and carbon gas observations at Alert and to K. Hayden for assistance in drafting some of the figures. Also thanks to the Norwegian Institute For Air Research(NILU) for providing long term sulphur observations at Spitsbergen.

REFERENCES

Barrie, L.A. and R.M. Hoff, 1984, The oxidation rate and residence time of sulphur dioxide in the Arctic atmosphere, *Atmos. Envir., 18*, 1995-2010.

Barrie, L.A. and R.M. Hoff, 1985, Five years of air chemistry observations in the Canadian Arctic, *Atmos. Envir. 18*, 1459-1469.

Barrie, L.A., 1985, Atmospheric particles: their physical/chemical characteristics and deposition processes relevant to the chemical composition of glaciers, *Ann. Glaciog. 7*, 100-108.

Barrie, L.A., 1986, Arctic air pollution: an overview of current knowledge, Atmos. Environ., 20, 643-663.

Barrie, L.A., J.W. Bottenheim, R.C. Schnell, P.J. Crutzen and R.A. Rasmussen, 1988, Ozone destruction and photochemical reactions at polar in the lower Arctic atmosphere, *Nature, 334*, 138-141.

Barrie, L.A. and M.L. Barrie, 1990, Chemical components of lower tropospheric aerosols in the high Arctic: Six years of observations, J. Atmos. Chem., 11, 211-226

Barrie, L.A., J.W. Bottenheim, R.C. Schnell, P.J. Crutzen and R.A. Rasmussen, 1988, Ozone destruction and photochemical reactions at polar in the lower Arctic atmosphere, *Nature, 334*, 138-141.

Barrie, L.A. and J.W. Bottenheim, 1991, Sulphur and nitrogen pollution in the Arctic atmosphere, Chapt.6, 155-184, in Ed. W. Sturges, *Pollution of the Arctic Atmosphere*, Elsevier Press, pp334.

Barrie, L.A., 1991, Snow formation and processes in the atmosphere that influence its chemical composition, NATO ASI Series, Vol. G28, Seasonal Snowpacks, Ed. T. Davies. et al., 1-20.

Barrie, L.A., D. Gregor, B. Hargrave, R. Lake, D. Muir, R. Shearer, B.Tracey and T. Bidleman, 1992, Arctic contaminants: sources, occurrence and pathways, *Sci. of the Total Envir., 122*, 1-74.

Barrie, L.A., 1993, Features of polar regions relevant to tropospheric ozone chemistry, in the *Tropospheric Chemistry of Ozone in the Polar Regions,* eds. H. Niki and K.H. Becker, Springer Verlag, 3-24.

Barrie, L.A., S.M. Li, D.L. Toom, S. Landsberger and W. Sturges, 1994a, Lower tropospheric measurements of halogens, nitrates and sulphur oxides by denuder and filter systems during Polar Sunrise Experiment 1992, *J. Geophys. Res. D.*, *99*,25453-25469.

Barrie, L.A., R. Staebler, D. Toom, B. Georgi, G. den Hartog, S. Landsberger, D.Wu, 1994b, Arctic aerosol size-segregated chemical observations in relation to ozone depletion during Polar Sunrise Experiment 1992, *J. Geophys. Res. D.*, *99*, 25439-25453.

Barrie, L.A., 1995, Arctic aerosols: composition sources and transport, *NATO Advanced Studies Workshop on Ice Core Studies of Global Biogeochemical Cycles, Ed. R.Delmas*, NATO ASI Series 1: Global Environmental Change, 30, 1-23, Springer Verlag, Berlin, New York.

Barrie. L.A. and A. Sirois, 1995, Arctic aerosols trends at Alert Canada:1980 to 1993, Atmos. Envir., submitted.

Bidleman, T.F., L.M. Jantunen, R.L. Falconer, L.A. Barrie and P. Fellin, 1995, Decline of hexachlorocyclohexane in the arctic atmosphere and reversal of air-sea gas exchange, *Geophys. Res. Lett., 22*, 219-222.

Blanchet, J. P. and Girard, E. , 1995, Water vapour feedback in the formation of continental Arctic air: its implications for climate, *Sci. Tot. Environ., 160/161*, 793-802.

Bodhaine, B.A. and E.G. Dutton, 1993, Long-term decrease in Arctic haze at Barrow Alaska. *Geophys. Res. Lett., 20,* 947-950

Bottenheim, J.W., L.A. Barrie, E. Atlas, L.E. Heidt, H. Niki, R.A. Rasmussen and P.B. Shepson, 1990, Depletion of lower tropospheric ozone during Arctic Spring: The polar sunrise experiment 1988, *J. Geophys. Res., 9*, 101

Bottenheim, J.W. L.A. Barrie and E. Atlas, 1993, The partitioning of nitrogen oxides in the lower Arctic troposphere during spring 1988, *J. Atmos. Chem.,17*, 15-27.

Bottenheim, J.W., L.A. Barrie and A. Sirios, 1994, Analysis of seven years of continuous observations of peroxyacetyl nitrate (PAN) at the high Arctic station Alert, paper #225, Joint meeting on global atmospheric chemistry, CACGP/IGAC, Fuji-Yoshida, 5-9 Sept 1994.

Boutron, C.F., U. Görlach, J.P. Candelone, M.A. Bolshov and R.J. Delmas, 1991, Decrease in anthropogenic lead, cadmium and zinc in Greenland snows since the late 1960's, *Nature, 353*, 153-157.

Bradley, R.S., F.T. Keimig and H.F. Diaz,1992 Climatology of surface-based inversions in the North American Arctic, *J. Geophys. Res., 97,* 15699-15712.

Charlson, R.J., S.E. Schwartz, J.M. Hales, R.D. Cess, J.A. Coakley, J.E. Hansen and D.J. Hofman, 1992, Climate forcing by anthropogenic aerosols, *Science, 255,* 423-430

Dibb, J.E., L.D. Meeker, R.C. Finkel, J.R. Southon, M.W. Caffee and L.A. Barrie, 1994, Estimation of stratospheric input to the Arctic troposphere: ^{7}Be and ^{10} Be in aerosols at Alert, Canada, *J. Geophys. Res, 99D,* 12855-12864

EPRI, 1994, Mercury Atmospheric Processes:A Synthesis Report , EPRI/TR-104214, Proc. Workshop in Tampa Florida, Sept. 1994.

Franzén, L.G., M. Hjlemroos, P. Kållberg, E. Brorström-Lunden, S. Juntto and A.-L. Savolainen, 1994, The 'yellow snow' episode of northern fenoscandia, March 1991- a case study of long-distance transport of soil, pollen and stable organic compounds, *Atmos. Envir. , 28*, 3587-3605.

Fellin, F., L.A. Barrie, D. Dougherty, D. Toom, D. Muir, B. Grift, L. Lockhart and B. Billeck, 1995, Air monitoring in the Arctic: selected results for OC and PAH compounds for 1992, *Envir. Toxic and Chem. J.,* in press.

Hopper, J.F., D.E.J. Worthy, L.A. Barrie and N.B.A. Trivett, 1994, Atmospheric observations of aerosl black carbon, carbon dioxide, and methane in the high Arctic, *Atmos. Envir. 28*, 3047-3054.

Hillamo, R.E., V.-M. Kerminen, W. Maenhaut, J.-L. Jaffrezo, S. Balachandra and C.I. Davidson, 1993, Size distributions of atmospheric trace elements at Dye3

Geenland-I. Distribution characteristics and dry deposition velocities., *Atmos. Envir., 27A*, 2787-2803.

Jaffe, D.A. and M.D. Zukowski, 1993, Nitrate deposition to the Alaskan snowpack, *Atmos. Envir., 27*, 2935-2934.

Jaffrezo, J.L., P. Masclet, M.P. Clain, H. Wortham, S. Beyne and H. Cachier, 1993, Transfer function of polycyclic aeromatic hydrocarbons from the atmosphere to the polar ice-I. Determination of atmospheric concentrations at Dye 3, Greenland, *Atmos. Envir., 27A*, 2781-2786.

Jobson, B.T., H. Niki, Y. Yokouchi, J. Bottenheim, F. Hopper, R. Leaitch, 1994, Measurements of C_2-C_6 Hydrocarbons During the Polar Sunrise 92 Experiment, *J. Geophys. Res. 99D*, 25355-25368.

Kahl, J.D., M.C. Serreze, R.S. Stone, S. Shitani, M. Kisley, and R. C. Schnell, 1993a, Tropospheric temperature trends in the Arctic, *J. Geophys. Res., 98D*, 12825-12838.

Kahl, J.D., D.J. Charlevoix, N.A. Zaitseva, R.C. Schnell and M.C. Serreze, 1993b, Absence of evidence for greenhouse warming over the Arctic Ocean, *Nature, 236*, 335-337.

Kawamura, K., H. Kasukabe and L. A. Barrie, 1995a, Concentration changes of total organic carbon and nitrogen, water soluble organic carbon, and polar organic compounds in the arctic aerosols at polar sunrise, *Atmos. Envir., 29*, submitted.

Kawamura, K., H. Kasukabe and L. A. Barrie, 1995b, Production of dicarboxylic acids in the arctic atmosphere at polar sunrise, *Geophys. Res. Lett., 22*, 1253-1256.

Kinlock D., H. Kuhnlein, and D.C.G. Muir, 1992, Inuit foods and diet: a preliminary assessment of benefits and risks, *Sci. Tot. Envir. 122*, 247-278.

Li, S.M. and L.A. Barrie, 1993, Biogenic sulphur aerosols in the Arctic troposphere: I Contributions to sulphate, *J. Geophys. Res., 98D*, 20613-20622.

Li, S.M., L.A. Barrie and S. Sirois, 1993, Biogenic sulphur aerosols in the Arctic troposphere, *J. Geophys. Res. 98D*, 20623-20631.

Li, S.M., R.W. Talbot, L.A. Barrie, R.C. Harriss, C.I. Davidson and J.-L. Jaffrezo, 1993, Seasonal and geographic variations of methane sulphonic acid in the arctic troposphere, *Atmos. Envir. 27A,* 3011-3024

Lindqvist, O. 1994, Atmospheric cycling of mercury: an overview, 181-185 *in Mercury Pollution:Integration and Synthesis*, Eds. Watras and Huckabee, Lewis Publishers.

Lockhart, W.L., 1993, Depositional trends: lake and marine sediments, pp46-56, *Synopsis of Research Conducted Under 1992/1993 Northern Contaminants Program*, Eds. Murray and Shearer, DIAND ISBN 0-662-21037-9, Ottawa, Canada.

Mackay, D. and F. Wania, 1995, Transport of contaminants to the Arctic: partitioning, processes and models, *Sci. Tot. Environ., 160/161*, 25-38.

Maenhaut, W., G. Ducastel and J.M. Pacyna, 1993, Atmospheric aerosol at Ny Ålesund, during winter 1989: multielement composition, receptor modelling and comparision with data from previous winters, *in Proc. of 5th Int. Symp. on Arctic Air Chem*, 8-10 Sept. Copenhagen, NERI Tech. Rep. 70, Danish Min. of Envir. P.O Box 358, Frederiksborgvej 399, DK-4000 Roskilde, Denmark.

Mayewski, P.A., L.D. Meeker, M.C. Morrison, M.S. Twickler, S. Whitlow, K.K. Ferland, M.R. Legrand, and J.P. Steffenson,1993, Greenland ice core "signal" characteristics offer expanded view of climate change, *J. Geophys. Res., 98,* 12839-12847.

Mosher, B.W., P. Winkler and J.L. Jaffrezo, 1993, Seasonal aerosol chemistry at Dye3, Greenland, *Atmos. Envir., 27A*, 2761-2773.

Muthuramu, M., P.B. Shepson, J.W. Bottenheim, B.T. Jobson, H. Niki, K.G. Anlauf, 1994, Relationships between organic nitrates and surface ozone destruction during Polar Sunrise Experiment 1992, *J. Geophys. Res., 99*, 25369-25378.

Mylona, S., 1993, Trends of sulphur dioxide emissions, air concentrations and depositions of sulphur in Europe since 1880, EMEP/MSC-W Rep. #2/93, The Norweg. Met. Instit., P.O. Box 43-Blindern, N-0313 Oslo 3, Norway.

Niki, H. and K.-H. Becker, 1993, *The Tropospheric Chemistry Of Ozone In The Polar Regions,* NATO ASI Series I:Global Environmental Change, Vol. 7, Springer Verlag, 425 pp.

Otlmans, S.J. 1993, Climatology of Arctic and Antarctic tropospheric ozone, in *The Tropospheric Chemistry Of Ozone In The Polar Regions,* eds Niki and Becker,

NATO ASI Series I:Global Environmental Change, Vol. 7, Springer Verlag, 25-40.

Patton, G.W., M.D. Walla, T.F. Bidleman and L.A. Barrie, 1991, Polycyclic aromatic and organochlorine compounds in the atmosphere of northern Ellesmere Island, Canada, *J. Geophys. Res., 96D,* 10867-10877.

Prospero, J.M., D.L. Savoie, R. Arimoto, H.Olafson, and H. Hjartarson, Sources of aerosol nitrate and non-sea salt sulphate in the iceland region, *Sci. Tot. Environ., 160/161*, 181-192.

Restelli, G. and G. Angelletti, 1993, *Dimethylsulphide: Oceans, Atmosphere and Climate,* Kluwer Acad. Pub., Dordrecht, 399 pp.

Rudolph, J., A. Khedim, T. Clarkson and D. Wagenbach, 1992, Long term measurements of light alkanes and acetylene in the Antarctic troposphere, *Tellus, 44B*, 252-261.

Schroeder, W. H. R. Ebinghaus, M. Shoeib, K. Timoschenko and L. A. Barrie, 1995, Atmospheric mercury measurements in the northern hemisphere from 56° to 82.5° N latitude, *Water Air Soil Poll,* 80: 1227-1236.

Slemr, F. and E. Langer, 1992, Increase in global atmospheric concentrations of mercury inferred from measurements over the Atlantic Ocean, *Nature, 355*, 434-437.

Solberg, S., O. Hov, N. Schmidbauer, A. Semb and F. Stordal, 1995, Tropospheric ozone depletion in the Norwegian Arctic in spring, J. Atmos. Chem. submitted May 1995 (NILU rep U-93093).

Staebler, R.M., G. den Hartog, B. Georgi and T. Düsterdiek, 1994, Aerosol size distributions during the Polar Sunrise Experiment 1992, *J. Geophys. Res, 99D,* 25429-25439.

Stern, G., L. Barrie, P. Fellin and D. Muir, 1995, Polychlorinated biphyenls in air in the high Arctic:1992-1993, *Envir. Sci. Techn.*, submitted.

Sturges, W.T. and L.A. Barrie, 1988, Chlorine, bromine and iodine in Arctic aerosols, *Atmos. Envir., 22,* 1179-1194.

Tarasick, D. W., D.J. Wardle, J.B. Kerr, J.J. Bellefleur and J. Davies, 1995, Tropospheric ozone trends over Canada:1980-1993, *Geophys. Res. Lett., 22,* 409-412.

Welch, H.E., D.C.G. Muir, B.N. Billeck, W.L. Lockhart, G.J. Brunskill, H.J.Kling, M.P. Olson and R.M. Lemoine, 1991, Brown snow. a long range transport event in the Canadian Arctic, *Envir. Sci. and Tech., 25,* 280-286.

Worthy, , D.E.J., N.B.A. Trivett, J.F. Hopper, J.W. Bottenheim and I. Levin, Analysis of long range transport events at Alert, Northwest Territiories, during Polar Sunrise Experiment 1992, *J. Geophys. Res., 99D*, 25329-25344.

ONE YEAR'S CONTINUOUS AEROSOL SAMPLING AT SUMMIT IN CENTRAL GREENLAND

Peter Wåhlin
National Environmental Research Institute
Division of Emissions and Air Pollution
Frederiksborgvej 399
DK-4000 Roskilde
Denmark

Abstract—Aerosol samples were collected continuously in two size fractions from June 1993 to June 1994 at a site 6 km north of the GRIP camp at Summit in central Greenland. The battery-driven sampling equipment was designed for automatic operation and low energy consumption. The samples were analyzed for elements using Particle Induced X-ray Emission (PIXE). The elements Al, Si, S, Cl, K, Ca, Ti, Mn, Fe, Zn, Br, Sr and Pb were measured well above the detection limits in many samples. The contents of most of these elements can be attributed to three sources, a crustal, a sea salt and an anthropogenic source. Additional contributions from other sources seem to be of importance for S and especially for Br. The crustal source had a peak incidence in the spring, while the anthropogenic source was present all year, strongest in winter and spring. The sea salt signal was very weak and difficult to resolve from the data, but was strongest in the winter. The results from Summit are compared with published data from Dye 3 on the southern part of the Greenland Ice Sheet, and with measurements from Station Nord, a sea level site in northeastern Greenland.

INTRODUCTION

Snow deposition studies are very important for the interpretation of icecore data. Measurements of chemical species in air compared with measurements of the same species deposited simultaneously in snow are of obvious importance for studies of how icecore data can be used as a record of atmospheric conditions in the past. A great effort is made in these years to retrieve such data in the Summit region in central Greenland, where the GISP 2 and GRIP deep core drillings have been carried out. Determination of sources and studies of deposition mechanisms depend on long continuous time series to cover all seasonal variation. While doing measurements on winter snow is hardly more difficult than doing measurements on summer snow, this is not the case for the corresponding atmospheric measurements because of the harsh climatic conditions and the resulting short summer field seasons. The data presented here are results from work done in association with GRIP in an attempt to do all-year aerosol sampling near the GRIP camp using an unattended, automatic sampler. The sampling began in the summer 1989 but has until now not been very successful and, although an all-year air concentration record can now be presented for a series of elements in two size fractions, some reservations must be made for possible errors that are due to imperfect sampling.

EXPERIMENTAL

The samples are collected with a battery-driven two-stage discrete streaker in miniature designed to provide small samples suitable for Particle Induced X-ray Emission (PIXE)

NATO ASI Series, Vol. I 43
Chemical Exchange Between the Atmosphere and Polar Snow
Edited by Eric W. Wolff and Roger C. Bales

analysis (Kemp and Wåhlin, 1992). The sampler, shown schematically in Fig. 1, is encased in a cylinder 35 cm in length and with a diameter of 15 cm. Particles separated at about 2 μm diameter are collected sequentially in small spots on a sampling wheel with two coaxial discs. Coarse particles are deposited by means of a rectangular impaction nozzle on a Apiezon coated mylar foil, while fine particles are collected on a 0.4 μm Nuclepore filter. The size of the sample spots matches the cross-section of the proton beam in the PIXE apparatus and is small enough to allow 64 samples to be collected on the same foil/filter. A comprehensive set of blank samples are obtained by skipping every eight sample position. The air flow is ~15 cm^3/s corresponding to ~ 6 m^3/week (STP) at Summit conditions. The power consumption is less than 4 W which makes it possible to use batteries (600 kg zinc/air cells) as the power source for a whole year. The main power consumer is a small pump. The sampling wheel is controlled by a small computer and is automatically turned a 5° angle once a week (Monday at 0 GMT) which allows the collection of weekly samples in more than one year. The computer also logs time, wheel position, air flow and temperature.

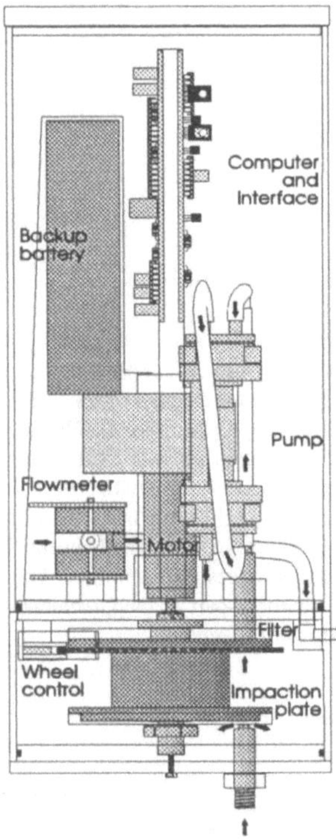

Fig. 1. Schematic drawing of the streaker.

The sampling site is situated 6 km north of the GRIP camp at Summit (72°35' N, 37°38' W, 3220 m a.s.l.) The deployment of the streaker on the icecap is shown in Fig. 2. As the ambient temperature may range below -60 °C the instrument is placed inside a large aluminum box insulated with polystyrene foam. With this geometry the energy dissipated internally in the compact instrument can maintain a temperature of about 30 °C above that of the surroundings. The box is placed in a 2 m pit and covered with snow, which ensures that ambient temperature is constantly close to the annual average of -30 °C at Summit. A 2 W heating element is automatically turned on if the temperature inside the streaker falls below 0 °C. The air intake is constructed to avoid blocking by blowing snow. It is equipped with an ordinary Frisbee disc to minimize turbulence around the sampling point (Hall and Upton, 1988).

The first years of the sampling, which began in 1989, can retrospectively be considered as prototype testing. Different incidents of error were encountered such as computer malfunctions and too high pump voltage. The most serious problem, however, has been condensation of rime and blocking with ice in the air inlet, although this was not realized in the

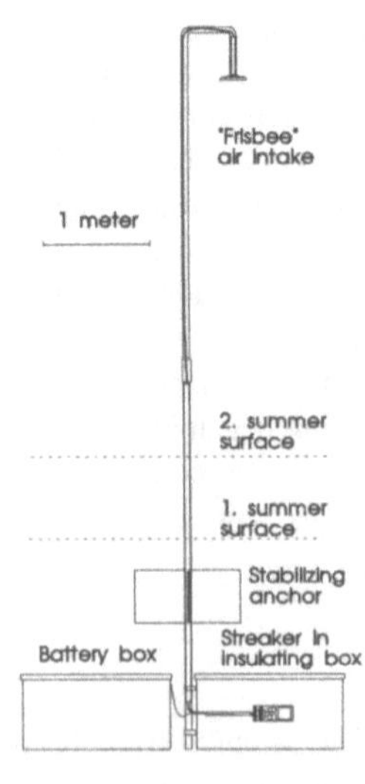

Fig. 2. The experimental set-up at the sampling site.

Table 1. Detection limits in ng/m^3 for a series of elements.

	Fraction	
	Fine	Coarse
Al	4	3
Si	3	2
S	1.5	1.5
Cl	0.8	2
K	0.4	0.4
Ca	0.2	2
Ti	0.05	0.06
V	0.03	0.04
Cr	0.05	0.04
Mn	0.02	0.02
Fe	0.1	0.07
Ni	0.015	0.02
Cu	0.015	0.02
Zn	0.01	0.3
Ga	0.007	0.008
As	0.015	0.015
Se	0.01	0.007
Br	0.07	0.015
Rb	0.03	0.015
Sr	0.02	0.015
Zr	0.04	0.02
Pb	0.05	0.04

beginning. The condensation takes place in the subsurface part of the inlet, and it seems to be a summer phenomenon caused by the relatively high vapor content, when the temperature of air is higher than that of the equipment below in the snow. Steps have been taken to prevent blocking of the inlet by progressively widening the subsurface part of the inlet, and the sampling season from June 1993 to June 1994 was the first successful in the sense that the inlet was never blocked and the streaker had operated satisfactorily. Nevertheless, as discussed later, there are still indications for some occurrence of condensation processes leading to losses of particles.

The air flow, recorded as float positions in the variable area flowmeter, was calculated and rescaled from Summit conditions (average air pressure 0.68 atm) to STP (1 atm, 0 °C) using the laboratory calibration values, the Bernoulli equation and the law of ideal gases. The flowmeter is read with an uncertainty of about 10% by means of an array with 11 sets of LED/phototransistors.

The samples were analyzed by means of PIXE using a Van de Graaff accelerator in Copenhagen. With the filter/foil used the PIXE provides detection limits down to 10^{-10} - 10^{-11} g for medium atomic weights. Concentrations down to 0.01 ng/m^3 could be measured for several elements (Table 1). Detection limits for some elements were rather high due to impurities in the foil/filter materials.

OBSERVATIONS

The continuous series of 51 coarse fraction samples and 51 fine fraction samples ranged over almost one year in the period from 21-6-93 to 8-6-94 with one week's sampling of each (except the last one in each series) beginning on Mondays at 0 GMT. The elements Al, Si, S, Cl, K, Ca, Ti, Mn, Fe, Zn, Br, Sr and Pb were measured well above the detection limits in many samples. The air volume ranged from 4.5 to 6 m^3 (STP).

The annual variation of Al, Si, K, Ca, Ti, Mn, Fe, Zn, Sr and Pb

The concentration data for all elements except S, Cl and Br could be expressed in reduced form using a constrained multivariate receptor model.

The model

$$x_{ij} \cong \sum_k a_{ik} f_{kj} \quad (\mathbf{X} \cong \mathbf{AF})$$

Table 2. Summit mean values June 1993 - June 1994. Unit: ng/m^3 (STP). The source contributions are determined only for the fine fraction data.

	Fine fraction	Soil	Sea	Combustion	Coarse fraction
Al	1.4±0.3	100%			1.3±0.3
Si	6.2±0.6	65%		35%	2.8±0.3
S	46±4	-	-	-	7.7±0.7
Cl	2.8±0.3	-	-	-	1.2±0.2
K	1.26±0.12	35%	5%	60%	0.47±0.05
Ca	1.21±0.11	70%	5%	25%	0.7±0.2
Ti	0.090±0.009	85%		15%	0.064±0.008
Mn	0.034±0.004	60%		40%	0.022±0.003
Fe	1.24±0.11	70%		30%	0.90±0.08
Zn	0.092±0.008	~0%		100%	0.03±0.02
Br	0.061±0.008	-	-	-	0.0019±0.0013
Sr	0.013±0.002	70%	15%	15%	0.0084±0.0015
Pb	0.075±0.008	~0%		100%	0.005±0.003

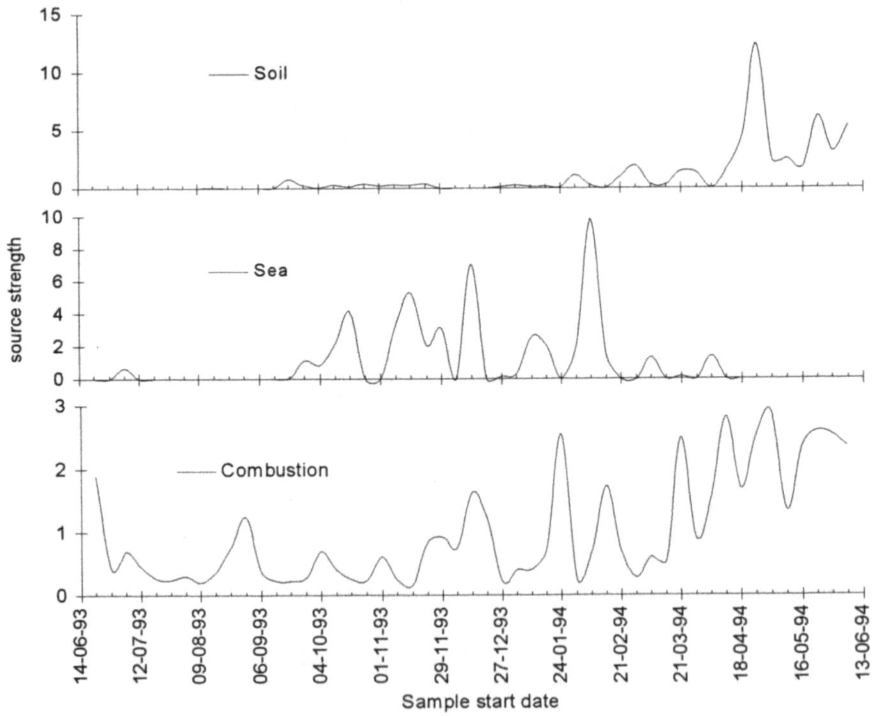

Fig. 3. Normalized strengths (mean value = 1) of the soil, sea and combustion sources.

expresses the measured concentrations x_{ij} as linear combinations of constant source profiles a_{ik}. The coefficients f_{kj} represent the source strengths found in the individual samples. Index i refers to the chemical elements, index j to the samples and index k to the sources. An iterative method is used in which the elements of the **F** matrix and selected elements of an initial **A** matrix are fitted by turns, and the model is confined to permit only non-negative elements of **A** and **F** (Wåhlin, 1992). Those elements of **A** which are not allowed to change are marked with zeros in a form matrix with the same number of rows and columns as **A**. The model has a hybrid character. It can be characterized as a chemical mass balance model with a built-in ability to fine tune source profiles. On the other hand, if all elements of **A** are allowed to adjust freely (permitting a confusion of more or less mixed up sources), the method has resemblance to factor analysis. The uncertainties of the measurements, σ_{ij}, are handled in an appropriate manner in the fitting process, in which the chi-square,

$$\chi^2 = \sum_j \sum_i \frac{(x_{ij} - \sum_k a_{ik} f_{kj})^2}{\sigma_{ij}^2},$$

is minimized within the limits imposed by the constraints. This also allows data below the detection limit to be put on a par with other data.

Three sources were recognized: A prominent crustal (soil) source, a prominent anthropogenic (combustion) source, and a third, rather indistinct, marine salt (sea) source. Only the fine fraction data was used in the model, because the coarse fraction data was considerably closer to detection limits. The reference values of Sverdrup et al. (1942) were used in the sea profile, while the composition of the soil source and the combustion source were allowed to change freely except for Zn and Pb, which in the soil profile were fixed to Fe in the very small ratios found in igneous rock (Kaye and Laby, 1959). A fairly good total fit ($\chi^2 = 611$, $\nu = 387$) was achieved, in which only the analytical uncertainties (without air volume uncertainties) were used as weights. The mean values are shown in Table 2 along with the fractions assigned by the model to each source. The annual variation of the source strengths are shown

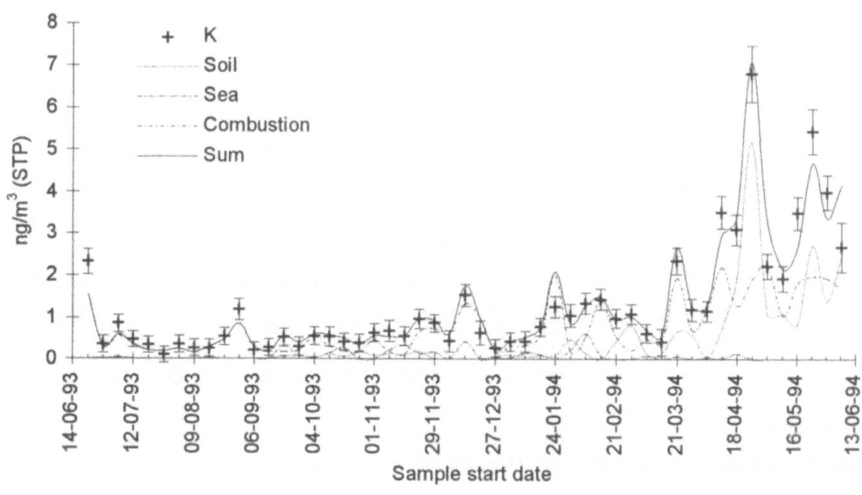

Fig. 4. Concentrations of K in the fine fraction samples fitted with contributions from three sources.

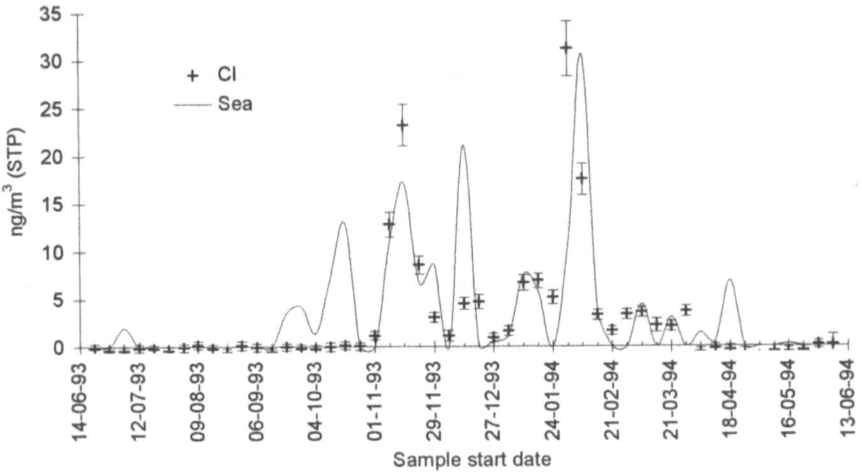

Fig. 5. Fine fraction concentrations of Cl compared with the estimate of the contribution from sea salt.

in Fig. 3. The source contributions to the individual fine fraction samples can be constructed by multiplying the normalized source strengths (in Fig. 3) by the mean values and the source fractions (in Table 2). The determination of the sea source strengths is uncertain because it relies on small contributions from only three elements K, Ca and Sr. (Only 5%, 5% and 15%, respectively, of the annual mean concentrations are assigned to the sea source). In Fig. 4 the fit is shown for K to which all three sources contribute.

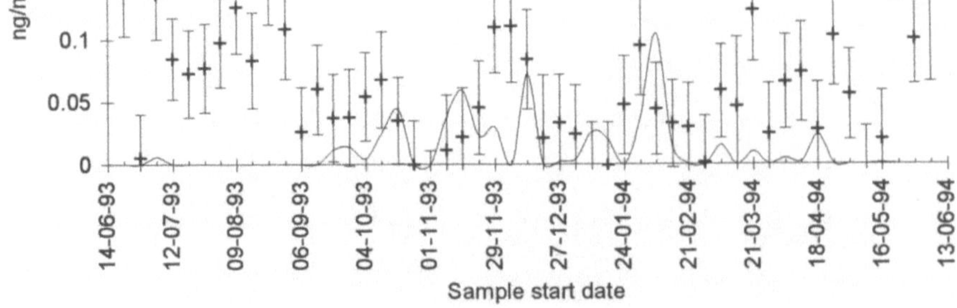

Fig. 6. Fine fraction concentrations of Br compared with the estimate of the contribution from sea salt.

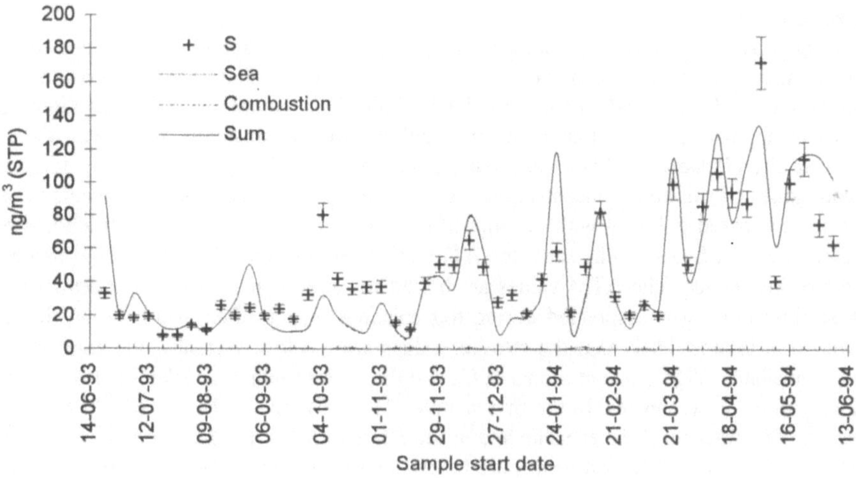

Fig. 7. Fine fraction concentrations of S compared with model. The sea salt contribution is negligible.

The sea salt components Cl and Br

In Fig. 5 the fine fraction concentrations of Cl are compared with the expected contribution from sea salt using the fitted source strength values mentioned in the preceding section. Halogen depression associated with volatilization by the effect of H_2SO_4 in filters from sea level sampling sites in the Arctic has been reported by several authors (cf. Shaw, 1991), but it is not observed with certainty in these samples. The agreement between the experimental values and the corresponding model values is not good in all samples, but, since the mean values agree well, this is probably due to the high uncertainty in the sea source strength values. This indicates that the sulfate aerosol at Summit is well aged and neutralized.

The fine fraction concentrations of Br are in the same manner compared with expected sea salt contributions in Fig. 6. Although both the Br data and the details in the estimate are uncertain, it is possible to recognize a rise in Br concentration in the spring and in the summer which is not related to sea salt. High spring Br has previously been observed in the Arctic (cf. Br measurements at Station Nord presented later in this paper) and comes possibly from other marine sources (Barrie and Barrie, 1990).

Sulfur

The fine fraction concentrations of S are shown in Fig. 7. The measurements are compared with the sum of the (negligible) sea salt contribution and a contribution from the combustion source that is normalized to 100% of the measured fine fraction mean value 46 ng/m^3 (STP). The agreement is reasonably good, except in the fall when the model predictions are systematically too low for a period of about two months. An explanation for the lack of agreement could be the delayed oxidation of sulfurdioxide and special scavenging mechanisms in the atmosphere or impact from a source region with different elemental profile composition, but it might also be influence from other sulfur sources, e.g. marine dimethylsulfide production or volcanic activity. Measurements of methanesulfonic acid (MSA) in the summer aerosol and snowpit samples at Summit (Jaffrezo et al., 1994) show that the marine MSA contribution to the sulfur budget is not negligible, but, on the other hand, MSA seems to peak in the summer rather than in the fall.

Particle losses

Comparison with other aerosol measurements at Summit indicates that particles are lost in the streaker inlet. Some preliminary results are available of sulfate and Ca measurements in the summer 1993 and the summer 1994 with daily high volume sampling on filters at the ATM camp, which were situated on two slightly different positions 30-45 km SW of the streaker site (Jaffrezo, 1994). It is possible, on the basis of these data, to calculate weekly mean values of sulfate-S concentrations, which overlap the streaker measurements. The values are compared in Fig. 8 with the sum of the fine fraction and the coarse fraction streaker measurement of S. As can be seen, the ATM values are between 25% and 200% higher than the streaker values. The ATM values are biased towards higher daytime concentrations because the sampling is suspended during fog episodes, which often occur during the nights. Even so, it is most likely that the streaker values are too low. Actually, the comparison between available ATM concentrations of Ca and the corresponding streaker values shows even more serious discrepancies. In the first sample (21-06-94) the ATM value is 26 times higher than the fine fraction streaker value and in the 5 last samples the ratios are between 3 and 9. (The streaker coarse fraction values are below detection limit because of high Ca content in the impaction foil.) This strongly supports the hypothesis that particles are lost in the inlet by filtering in rime, which will preferably sort out the larger particles. Cascade impactor sampling at Dye 3 (Hillamo et al., 1993) has shown that particles of anthropogenic origin and of crustal origin exhibit almost unimodal size distributions, but with very different geometric mean aerodynamic diameters (respectively 0.6 μm and 2 μm). A speculative explanation of the particular large deviations in the first samples could be that condensation occurs in the summer and, since only a little amount of rime was observed when the sampler was retrieved in early June 1994, that the rime evaporates again during the winter.

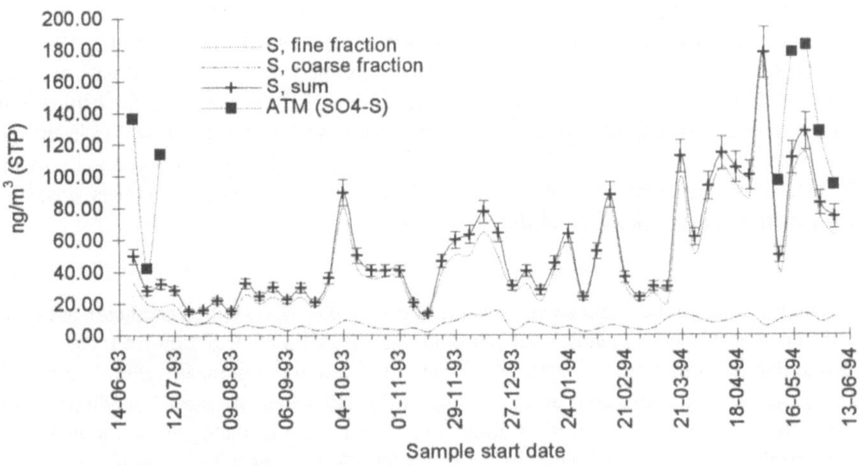

Fig. 8. Total concentrations of S compared with high volume sampler measurements of sulfate.

COMPARISON WITH OTHER ALL-YEAR DATA FROM GREENLAND

Dye 3 on the southern dome of the Greenland Ice Sheet.

The Dye 3 Gas and Aerosol sampling Program (DGASP) was conducted during August 1988 – July 1989 at the Dye 3 radar base on the Greenland Ice Sheet. The location relative to Summit is approx. 850 km SSW at an altitude of 2480 m. Airborne concentrations of SO_4^{2-}, gases, radionuclides and several trace elements were measured during a whole year. Both SO_4^{2-} and the crustal elements had strong springtime peaks, while the sea salt elements had maxima in the winter (Davidson et al., 1993a). This is in agreement with the Summit measurements. The Summit measurements of elemental sulfur are compared with sulfate-S values from Dye 3 in Table 3. The numbers are not directly comparable, due to different sampling times and averaging methods (arithmetic versus geometric means), but estimates of arithmetic means at Dye 3, assuming a log-normal number distribution (last column in Table 3), seem to show that the levels of sulfur are relatively higher at Dye 3 in spring and fall.

Table 3. Summit arithmetic mean concentration of S compared to Dye 3 geometric mean concentrations of sulfate-S with geometric standard deviations in parentheses (Davidson, 1993a). The numbers in parentheses in the last column are estimated arithmetic means assuming a log-normal number distribution. Unit: ng/m^3 (STP).

	Summit arith. means	Dye 3 geom. means	Dye 3 arith. means
spring	88	83(3.4)	(176)
summer	35	18(1.7)	(21)
fall	37	32(3.6)	(72)
winter	52	11(5.8)	(52)

Table 4. Three year's arithmetic mean values at Station Nord in northeastern Greenland. Unit: ng/m^3.

	Soil	Sea	Combustion	Other sources
Al	19		3	
Si	57		7	
S		13	345*	
Cl		270†		
K	6	5	4	
Ca	14	6	4	
Ti	1.48		0.17	
V	0.04		0.18	
Cr	0.05		0.10	
Mn	0.23		0.21	
Fe	15		4	
Ni	0.02		0.09	0.03‡
Cu	0.02		0.09	0.17‡
Zn	0.02		1.09	
As			0.14	
Se			0.019	
Br		0.93		1.11§
Sr	0.04	0.19	0.06	
Zr	0.05			
Pb			0.79	

*Sum of S in sulfurdioxide (189 ng/m^3) and sulfate (156 ng/m^3).

†Due to Cl depletion the measured value was only 42 ng/m^3

‡Metal source

§Non sea salt spring Br

Comparison between annual arithmetic mean values of trace element concentrations measured at both places shows that the Dye 3 values (Mosher et al.) are about 5-10 times higher than the Summit streaker values. This does not necessarily reflect any real difference between the levels at Summit and at Dye 3. It is more probably a consequence of the loss of particles which was observed for Ca and consequently is expected for other elements in the coarse particle mode.

Sea level sampling in Greenland.

The first aerosol measurements in Greenland of trace elements were performed in the years 1979-83 at five different locations round the Greenland coast line. A distinct anthropogenic component was found at all locations with a pronounced winter maximum and a summer minimum (Arctic haze). The highest levels of

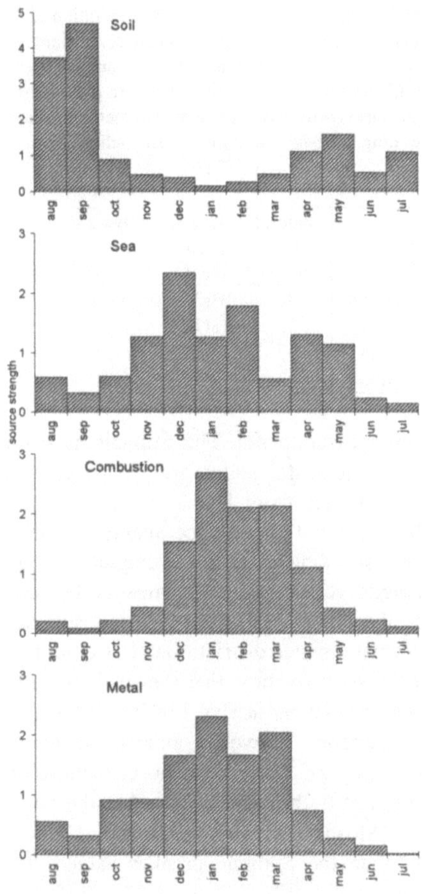

Fig. 9. Seasonal variation of normalized source strengths expressed as monthly means. The values summarize results from three year's continuous aerosol sampling at Station Nord in northeastern Greenland (Aug. 90 - July 93).

the anthropogenic metals Cr, Cu, Zn and Pb were found at the northernmost site Station Nord (Heidam, 1984).

A continuous monitoring at Station Nord of gases and particles was initiated in the summer 1990. Weekly samples of particulate matter were collected on membrane filters together with gases on impregnated filters in filterpacks. Unpublished results from the first 3 years (Aug. 90 - July 93) are summarized in Fig. 9 and Table 4. The source types found at Summit are also recognized at Station Nord (soil, sea, combustion, non sea salt bromine), but in addition to these is also found a metal component with extra contributions of Cu and Ni. The combustion signal and the metal signal have nearly the same characteristic seasonal variation with a winter maximum in January/February, only the metal signal appears a little earlier in the fall/winter which spots the polluters to be situated in the northern part of the main source region. The winter appearance of anthropogenic pollution (Arctic Haze) is a well known phenomenon at sea level monitoring sites around the Arctic Sea, so it is noticeable that the situation in the winter is very different at Summit and at Dye 3, where the anthropogenic signal is much weaker and first peaks in April/May. Two factors, especially, are believed to be responsible for the difference. First, the Greenland Ice Sheet is at an elevation well above the Arctic Haze which, due to subsidence and surface radiational cooling in the winter, presumably is confined to the lower part of the Arctic air (cf. Barrie and Hoff, 1984). Second, the southern part of Greenland is much more influenced by midlatitude Atlantic pathways than sites within the Arctic Basin, where transport over the Pole from Eurasia is most important (Davidson et al., 1993b).

The soil source has a peak in April/May both at Station Nord and at Summit, and the sea salt source has maximum in the winter at both places. The very high soil signal in August/September at Station Nord is probably of regional/local origin and connected with the lack of snow cover in the summer.

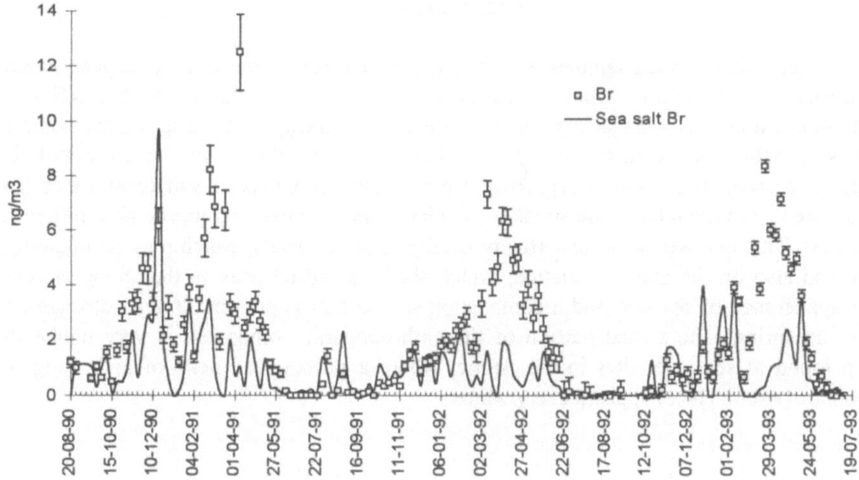

Fig. 10. Measurements of particulate Br at Station Nord compared with model calculation of sea salt Br. High values not related to sea salt are measured in the spring.

The appearance in the spring of filterable non sea salt bromine (Fig. 10) has been observed at many different places in the Arctic and has been connected with the destruction of tropospheric ozone at Polar Sunrise (Bottenheim et al., 1990).

Because particulate sulfur did fit badly, it was not included in the mathematical analysis which led to the results summarized in Fig. 9 and Table 4, but it is demonstrated in Fig. 11 that the sum of particulate sulfur and sulfurdioxide fits fairly well with the combustion source. The concentrations found at Station Nord of sulfurdioxide and sulfate, as well as the high fractions of sulfurdioxide in the Arctic night, are similar to the levels found at other places around the Arctic Sea (cf. Barrie and Hoff, 1984).

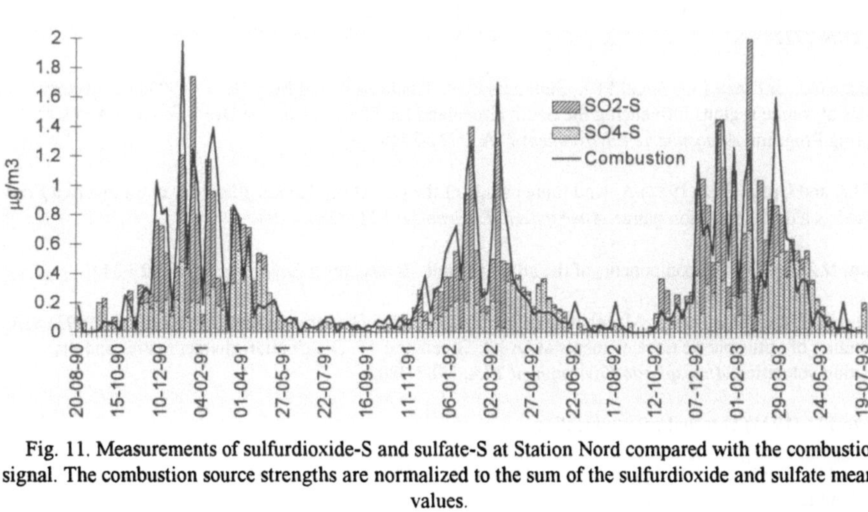

Fig. 11. Measurements of sulfurdioxide-S and sulfate-S at Station Nord compared with the combustion signal. The combustion source strengths are normalized to the sum of the sulfurdioxide and sulfate mean values.

CONCLUSION

The elemental content in the samples from Summit, which cover almost a whole year, can be apportioned to three source types, a crustal source, a sea salt source and a combustion source. In addition to this, a rise is observed in the Br concentration in the spring and summer which is probably connected to photochemical reactions and the ozone depletion at Polar Sunrise. The comparison with independent summer measurements of sulfate indicates that particles are lost in the inlet of the streaker sampler in the summer, and maybe also in the rest of the year. This imposes some uncertainty on the absolute levels, which may be underestimated, and also on the annual variation, which shows a crustal peak in the spring, a weak winter appearance of sea salt, and a combustion source that is present all year, strongest in winter and spring. The annual pattern of the anthropogenic component is very unlike the pattern found at sea level sites in the Arctic, where a pronounced peak of anthropogenic compounds (Arctic Haze) appears every winter.

REFERENCES

Barrie L.A. and Hoff R.M. (1984) The oxidation rate and residence time of sulphur dioxide in the Arctic atmosphere. *Atmospheric Environment* **18**, 2711-2722.

Barrie L.A. and Barrie M.J. (1990) Chemical components of lower tropospheric aerosols in the high Arctic: Six years of observation, *J. Atmos. Chem.* **11**, 211-226.

Bottenheim J.W., Barrie L.A., Atlas E., Heidt L.E., Niki H., Rasmussen R.A. and Shepson P.B. (1990) Depletion of lower tropospheric ozone during Arctic spring: The Polar Sunrise Experiment 1988, *J. Geophys. Res.*, **95**, 18555-18568.

Davidson C.I., Jaffrezo J.L., Mosher B.W., Dibb J.E., Borys R.D., Bodhaine B.A., Rasmussen R.A. , Boutron C.F., Gorlach U., Cachier H., Ducret J., Colin J.L., Heidam N.Z., Kemp K. and Hillamo R. (1993a) Chemical constituents in the air and fresh snow at Dye 3, Greenland—I. Seasonal variations. *Atmospheric Environment* **27A**, 2709-2722.

Davidson C.I., Jaffrezo J.L., Small M.J., Summers P.W., Olson M.P. and Borys R.D. (1993b) Trajectory analysis of source regions influencing the South Greenland Ice Sheet during the Dye 3 Gas and Aerosol Sampling Program. *Atmospheric Environment* **27A**, 2739-2749.

Hall D.J. and Upton S.L. (1988) A wind tunnel study of the particle collection efficiency of an inverted Frisbee used as a dust deposition gauge. *Atmospheric Environment* **22**, 1383-1394.

Heidam N.Z. (1984) The components of the arctic aerosol. *Atmospheric Environment* **18**, 329-343.

Hillamo R.E., Kerminen V.M., Maenhaut W., Jaffrezo J.L., Balachandran S. and Davidson C.I. (1993) Size distribution of atmospheric trace elements at Dye 3, Greenland—I. Distribution characteristics and dry deposition velocities. *Atmospheric Environment* **27A**, 2787-2802.

Jaffrezo J.L. (1994) Personal communication.

Jaffrezo J.L., Davidson C.I., Legrand M. and Dibb J.E. (1994) Sulfate and MSA in the air and snow on the Greenland Ice Sheet. *J. Geophys. Res.*, **99**, 1241-1253.

Kaye G.W. and Laby T.H. (1959) Chemical and physical constants. Longman Green, London.

Kemp K. and Wåhlin P. (1992) Aerosol sampler for unattended all-year sampling at remote sites. Proceedings of the Fifth International Symposium on Arctic Air Chemistry, September 8-10, Copenhagen.

Mosher B.W., Winkler P. and Jaffrezo J.L. (1993) Seasonal aerosol chemistry at Dye 3, Greenland. *Atmospheric Environment* **27A**, 2761-2772.

Shaw G.E. (1991) Aerosol chemical components in Alaska air masses, 2. Sea salt and marine product. *J. Geophys. Res.* **96**, 22369-22372.

Sverdrup H.U., Johnson, M.W. and Flemming R.H. (1942) The oceans. Prentice-Hall, New York.

Wåhlin P. (1992) A multivariate receptor model with a physical approach. Proceedings of the Fifth International Symposium on Arctic Air Chemistry, September 8-10, Copenhagen.

CENTRAL ANTARCTICA: ATMOSPHERIC CHEMICAL COMPOSITION AND ATMOSPHERIC TRANSPORT

Barry A. Bodhaine
National Oceanic and Atmospheric Administration
Climate Monitoring and Diagnostics Laboratory
325 Broadway
Boulder, Colorado 80303, U.S.A.

Introduction

Central Antarctica, by virtue of its remoteness and the lack of local anthropogenic pollution sources, can probably be considered the cleanest location on earth. It is a true pristine background environment in the most fundamental sense. Indeed, it may be the *last* true pristine background environment. Yet, airborne materials of anthropogenic origin are detected there, albeit at very small concentrations, thanks to long-range transport and mixing processes in the atmosphere. The Antarctic troposphere plays an important role in global climate because it links the global atmosphere and oceans with the Antarctic ice sheet and stratosphere. Tropospheric transport processes occur on fairly short time scales, of the order of days to weeks; whereas stratospheric processes (e.g., interhemispheric exchange processes) may occur on the order of years.

Antarctica has many unique features that qualify it as a valuable laboratory for studying and understanding our global environment. It is surrounded by a massive, biologically active ocean, and is covered by a glacial ice sheet that contains an historic record of atmospheric composition and climate. It is subject to a day-night period that can be described as a year-long diurnal cycle, making it ideal for some studies of photochemical processes.

The Antarctic continent has an area of about 14 million km^2, surprisingly about twice the size of Australia. It is the highest continent with an average altitude of about

NATO ASI Series, Vol. I 43
Chemical Exchange Between the Atmosphere and Polar Snow
Edited by Eric W. Wolff and Roger C. Bales

1.9 km, is completely surrounded by oceans, and contains about 90% of the world's ice (about 25 million km^3). The high polar plateau of East Antarctica reaches an altitude of about 3.5 km; however, the highest mountain peaks, located in the Transantarctic Range reach altitudes of nearly 4.6 km. The South Pole, located on the greater polar plateau, is at an altitude of about 2.8 km. A nice map of Antarctica, shown here as Figure 1, was provided by *Weller et al.* [1987]. An introduction to a series of review articles was provided by *Meriwether* [1988], and an especially interesting first person account of scientific activities in Antarctica was written by *Bouton* [1981]. Another interesting article by *Satchell* [1983] describes the experiences of women wintering over with all-male crews at the South Pole.

The Amundsen-Scott Station was formally established at the geographic South

Figure 1. Map of Antarctica, showing major geographic features and some of the scientific stations (from *Weller et al.*, 1987).

Pole in 1956 for the International Geophysical Year (IGY). A diagram of the original station (now referred to as Old Pole) was provided by *Miller* [1974]. Operated by the U. S. National Science Foundation (NSF), a large number of research programs have been conducted there for many years. In 1972 the Geophysical Monitoring for Climatic Change Program (GMCC, now the Climate Monitoring and Diagnostics Laboratory (CMDL)) of the National Oceanic and Atmospheric Administration (NOAA) established a monitoring program at Amundsen-Scott Station to measure gases, aerosols, solar radiation, and meteorological variables [*Pack,* 1973]. The mission of this program was to measure the concentrations and characteristics of these atmospheric quantities, to detect any trends in the concentrations of these species caused by anthropogenic or natural sources, and to assess possible climatic effects that may result from these changes. It is especially important to understand natural sources and sinks in order to allow detection of anthropogenic effects. Three other background stations – located at Barrow, Alaska; Mauna Loa, Hawaii; and American Samoa – were established as part of the GMCC program to create a four-station network.

Atmospheric chemistry measurements were made from the Dobson hut during the austral winter of 1972. In 1973 the Dobson hut was moved to avoid local contamination. Then in 1974 measurements were made from the Aurora Tower (built above an existing structure to extend above the snow surface) using a new aerosol-gas sampling apparatus [*Komhyr,* 1983]. The new South Pole station (known for its distinctive geodesic dome structure) was occupied during 1975, but a temporary Clean Air Facility (CAF) was constructed below the surface of the snow and was used during 1975 and 1976. In late 1976 a new CAF building was constructed on stilts above the snow and was occupied during the 1977 austral winter. This configuration is shown in Figure 2. The aerosol-gas sampling apparatus was moved from the old Aurora Tower to the temporary CAF, and then to the new CAF. Atmospheric chemistry measurements have continued to the present at the new CAF. A new CAF is in the planning stages and should be completed by about 1996-1997.

The CAF is located upwind from the main station and 330°-110° has been designated as the clean air sector. Winds are from the clean air sector about 98% of the time, the most persistent directions being N-NNE. The annual average wind speed is about 5 m s^{-1} and a typical maximum wind speed is about 15 m s^{-1} in winter. The average pressure is about 680 hPa with a maximum of about 700 hPa and a minimum of about 650 hPa. The annual average air temperature is about −50 C with a maximum of about −20 C in December-January and a minimum of about −78 C in July-August. Although there is only a trace of annual precipitation at the South Pole, blowing and drifting snow causes accumulation around the buildings.

Several excellent review articles have been written on Antarctic atmospheric chemistry research. Only a few of the most recent ones will be mentioned here; of course, each of these contains numerous references to previous Antarctic research. The author apologizes for those not mentioned here. *Washburn* [1980] discussed the direction of future research in Antarctica. A special issue of the *Journal of Geophysical Research* [Vol. 87: 4269-4318, 1982] was devoted to papers presented at the Symposium on Progress in Antarctic Meteorology, held at Canberra, Australia, 4-6

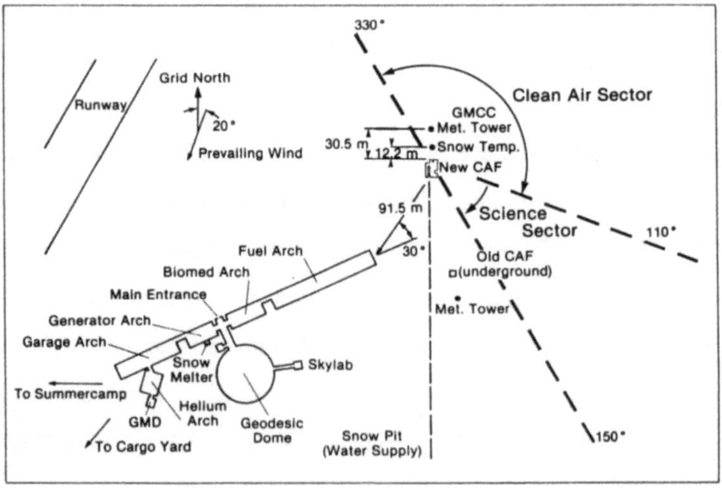

Figure 2. Amundsen-Scott Station at the South Pole [from *Bodhaine and Harris,* 1982].

December 1979. A review of polar meteorology by *Robinson* [1983] included a discussion of Antarctic atmospheric chemistry research, and *Robinson et al.* [1988] reviewed air quality monitoring at the South Pole. Several papers concerning Antarctic atmospheric chemistry appeared in a special issue of the *Journal of Atmospheric Chemistry* [Vol. 3: 3-201, 1985] as a result of the Conference on the Scientific Application of Baseline Observations of Atmospheric Composition, 7-9 November 1984, in Aspendale, Australia. Some of the problems encountered when measuring atmospheric quantities at background locations were discussed by *Heintzenberg* [1985]. *Weller et al.* [1987] discussed the Antarctic environment in terms of a huge natural laboratory. An issue of *Reviews of Geophysics* [Vol. 26: 41-207, 1988] was devoted to atmospheric sciences in Antarctica. In June 1991 a Symposium on the Tropospheric Chemistry of the Antarctic Region was held in Boulder, Colorado, U.S.A. [*Bodhaine et al.*, 1992], and a collection of papers from this meeting was published in two special issues of *Tellus* [Vol. 44B: 249-442, 1992; Vol. 45B: 89-207, 1993].

Aerosol Measurements

Condensation nuclei

The first aerosol measurements on the Antarctic continent appear to have been made in 1965 at the Russian Mirnyy station on the Antarctic coast by *Voskresensky* [1968]. However, he states in his paper that he made a limited number of measurements earlier in 1962. His instrument was of the Sholtz design, using an expansion cloud chamber operating at a supersaturation of 200-300%, constructed at the Main Geophysical Observatory in Leningrad (St. Petersburg). The first reference to aerosol measurements at the South Pole is by *Roberts* [1969] who reports that Aitken nuclei measurements were established during the austral winter of 1968. He didn't document what instrument was used but it was probably a Gardner condensation nucleus (CN) counter [*Rich*, 1955]. Note that the name Gardner Counter derives from the manufacturer, Gardner Associates, Schenectady, New York. *Warburton* [1973] used a Gardner CN counter during the 1971-1972 austral summer at the South Pole and reported a concentration of 150 ± 20 cm^{-3}, a value very much in line with present day values, but at the lower limits of sensitivity of the Gardner counter. Routine CN measurements were established by the GMCC program at the South Pole in December 1971 using a Gardner counter. An excellent review of Antarctic aerosol research was presented by *Shaw* [1988].

A Pollak CN counter [*Metnieks and Pollak,* 1959] was installed at the South Pole in January 1974 by Austin W. Hogan of the State University of New York at Albany. This same instrument is still in regular operation and has produced an outstanding long-term data set consisting of twice daily observations for the entire time period. At the time of this writing, this CN record is more than 20 years long. These data were first presented by *Hogan* [1975], and more recently by *Hogan and Bodhaine* [1995]. Transport processes were discussed by *Hogan and Barnard* [1978], *Hogan* [1979], and *Hogan et al.* [1982]. *Samson et al.* [1990] presented the long-term CN concentration record and performed a harmonic analysis to emphasize the annual cycle. They found a decrease in springtime aerosol concentrations over the 1977-1985 period. A detailed review of CN and meteorological measurements at the South Pole was provided by *Hogan et al.* [1993].

Continuous CN measurements were established at the South Pole in 1974 using a General Electric (G.E.) automatic CN counter [*Skala,* 1963]. This instrument and its operation at GMCC stations were described by *Bodhaine* [1983] and *Bodhaine and Murphy* [1980]. The on-site Pollak CN counter was accepted as the standard, and the automatic CN counter was adjusted to agree with it. In this way the continuous CN record, available as hourly means, was tied to the Pollak counter's twice daily observations. In 1987 a TSI 3020 alcohol-based CN counter was operated for one year at the South Pole and it was found to give somewhat higher values than the G.E. counter, possibly because of the mass flow controller built into the TSI instrument. In 1989 a TSI 3760 CN counter was permanently installed at the South Pole and it too gave consistently higher values than the G.E. counter. However, the TSI 3760 CN counter uses a critical orifice to control the sample flow, giving a somewhat lower mass flowrate at the South Pole than it has at sea level. Because of these differences between various CN instruments, the Pollak CN record has been maintained through the present, and the other CN data sets may be compared with the Pollak counter data set if desired. The long-term CN and aerosol scattering (σ_{sp}) data for the South Pole, in addition to the references given above, were presented by *Murphy and Bodhaine* [1980], *Bodhaine and Shanahan* [1990], and *Bodhaine* [1992]. The long-term record of CN and σ_{sp} data is shown in Figure 3.

Aerosol Chemistry

Apparently the first aerosol chemistry measurements at the South Pole station were made during the 1970-1971 austral summer, as reported by *Duce et al.* [1971],

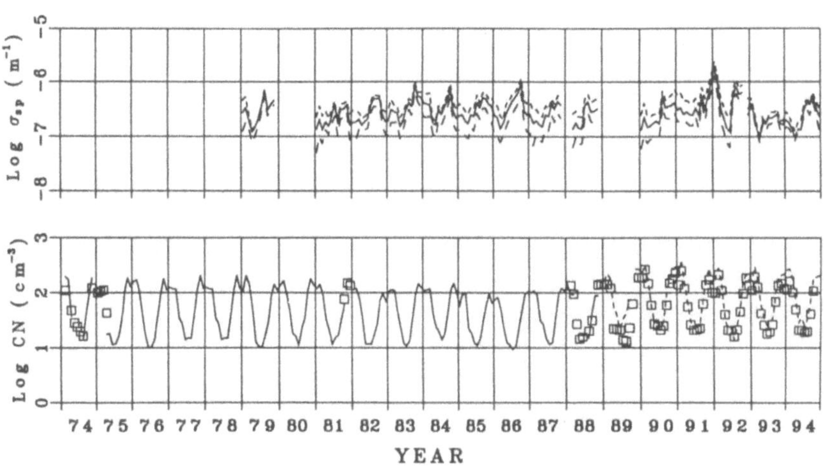

Figure 3. South Pole monthly geometric mean CN concentration (bottom); and σ_{sp} (top) for 450 (dotted), 550 (solid), and 700 nm (dashed). For CN concentration the solid line shows G.E., the dashed line shows TSI, and the squares show Pollak CN data.

Duce [1972], *Duce et al.* [1973], *Zoller et al.* [1974], and *Duce et al.* [1975]. They took filter samples for subsequent analysis by neutron activation and atomic absorption techniques. Aerosol chemistry measurements were resumed at the South Pole during the 16 December 1974 to 4 February 1975 austral summer by *Maenhaut and Zoller* [1977] and *Maenhaut et al.* [1979]. *Cunningham et al.* [1979] identified sulfate in South Pole aerosol samples taken during November 1975 to January 1976 and November 1977 to January 1978 using the Raman microprobe technique. Samples taken during austral winters 1975 and 1976 were discussed by *Cunningham and Zoller* [1981]. Aerosol samples collected between February 1979 and November 1983 were discussed by *Tuncel et al.* [1989]. This total piece of work, beginning with *Duce et al.* [1971] and ending with *Tuncel et al.* [1989], will stand as one of the outstanding classical scientific data sets. An example of the seasonal cycles of marine and crustal aerosol at the South Pole is shown here in Figure 4. Of primary interest are the concentrations of elements in summer and winter aerosol in Tables 1 and 2 of *Tuncel*

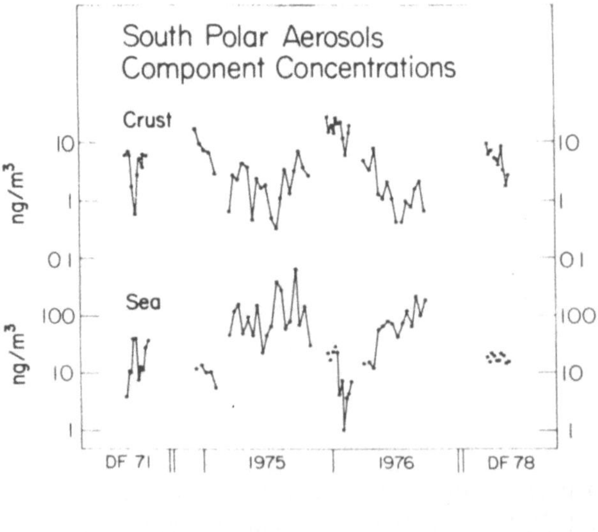

Figure 4. Seasonal variation of crustal and marine aerosol components at the South Pole in 1971 and 1975-1978 [from *Cunningham and Zoller,* 1981].

et al. [1989]. A summary of their crustal, marine, and sulfate data is shown here in Table 1. A comprehensive review of the results of that work was given by *Shaw* [1988] and therefore will not be repeated here. It is important to note that this work was followed through from beginning to end by one scientist: Dr. William H. Zoller, who is now serving as Professor of Chemistry at the University of Washington in Seattle.

Table 1. Average concentrations (ng m^{-3}) and standard deviations of crustal, marine, and sulfate components in the South Pole atmosphere during the summer and winter of 1979-1983 [*Tuncel et al.*, 1989].

	Crustal	Marine	Sulfate
Summer	6.65 ± 4.97	27 ± 16	126 ± 11
Winter	3.02 ± 2.32	95 ± 87	21 ± 13

The first black carbon (BC) measurements at the South Pole were made during December 1985 to February 1986 by *Warren and Clarke* [1990]. They found an atmospheric concentration of about 1.4 ng m^{-3}. BC data for an entire year (December 1986 to November 1987) were reported by *Hansen et al.* [1988], and are shown here in Figure 5. These data showed an annual cycle with a maximum concentration of about 2 ng m^{-3} in December 1986 and a minimum of about 0.1 ng m^{-3} in May 1987. These data were also presented by *Bodhaine* [1995] along with two more years of data and a more detailed analysis, including aerosol absorption coefficient and single scatter albedo.

An interesting study by *Warren and Clarke* [1990] examined the amount of black carbon (BC) in the atmosphere and deposited in the snow around the camp at South Pole. They found a well-defined plume of BC deposited in the snow downwind of the station due to the burning of diesel fuel. However, the soot concentration in the snow (about 0.2 ng g^{-1}) was too small to affect snow albedo significantly, suggesting that surface radiation measurements made near Antarctic stations are representative of the surrounding area.

Aerosol scattering

In January 1979 a four-wavelength nephelometer similar in design to that of *Ahlquist and Charlson* [1969] was installed by the author at the South Pole. Except for one year (1980), this instrument has operated continuously to the present, and has produced an excellent data record for 1979-present. A nephelometer measures the scattering extinction coefficient of suspended particles (σ_{sp}), an important parameter in climate modeling. The dependence of multiwavelength σ_{sp} measurements on particle size distribution was discussed by *Thielke et al.* [1972]. The first year of data was presented by *Bodhaine and Bortniak* [1981] and *Parungo et al.* [1981], and is shown here in Figure 6. They found a clear annual cycle in σ_{sp} that is out of phase with the cycle in CN concentration. It is important to note that CN concentration is simply the

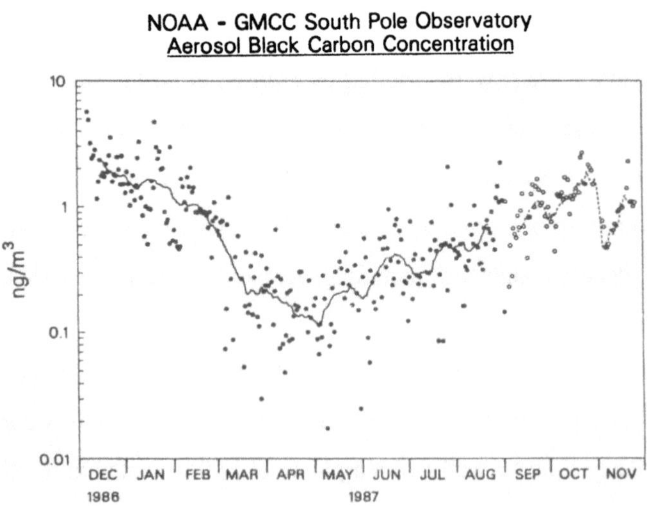

Figure 5. Daily means of BC aerosol at the South Pole during 1986-1987 [from *Hansen et al.*, 1988].

total number of particles, regardless of size or chemical composition, whereas σ_{sp} depends on particle size (and index of refraction) and responds strongly to larger particles. A late-winter maximum in σ_{sp} is caused by sea salt events transported over long distances from the surrounding oceans. The largest σ_{sp} value ever recorded at the South Pole (1 August 1979) coincided with a large NaCl event [*Parungo et al.*, 1981]. An aerosol chemistry experiment during the 1982 austral winter, by *Bodhaine et al.* [1986, 1987] using the PIXE method of analysis, showed conclusively that large cyclonic storm systems located off the coast of Antarctica cause rapid transport of sea-salt laden air from lower latitudes to the vicinity of the Pole, appearing as a warming in the midtroposphere. The annual cycle in CN concentration corresponded closely with the annual cycle in S, and the annual cycle in σ_{sp} corresponded closely with the annual cycle in NaCl (see Figure 7). The possibility of transport of sea salt to the South Pole under such conditions during the austral winter was suggested by

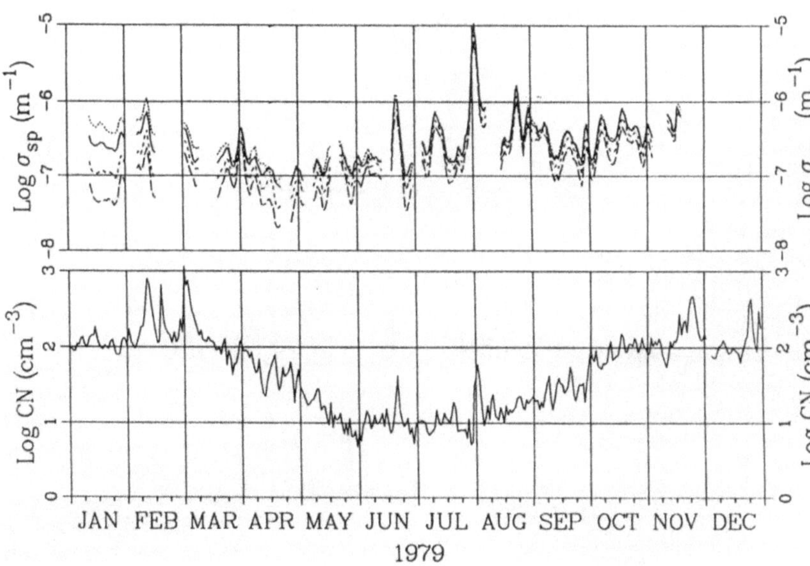

Figure 6. South Pole daily geometric mean CN concentration (bottom), and σ_{sp} (top) for 450 (dotted), 550 (solid), 700 (dashed), and 850 nm (long-dashed).

Cunningham and Zoller [1981] and *Bigg* [1980]. A warm transport event marked by a σ_{sp} event is shown in Figure 8.

An aerosol chemistry experiment conducted during 1987 [*Bodhaine and Harris*, 1992] found results similar to the 1982 experiment. However, the 1987 experiment included additional BC measurements using an aethalometer, and size distribution measurements using a diffusion battery with four size cuts based on Nuclepore filters. The BC measurements from this experiment had been published by *Hansen et al.* [1988]. In addition to daily means of the aerosol data, *Bodhaine and Harris* [1992] presented time-height cross sections of temperature and twice daily back trajectories for the entire year.

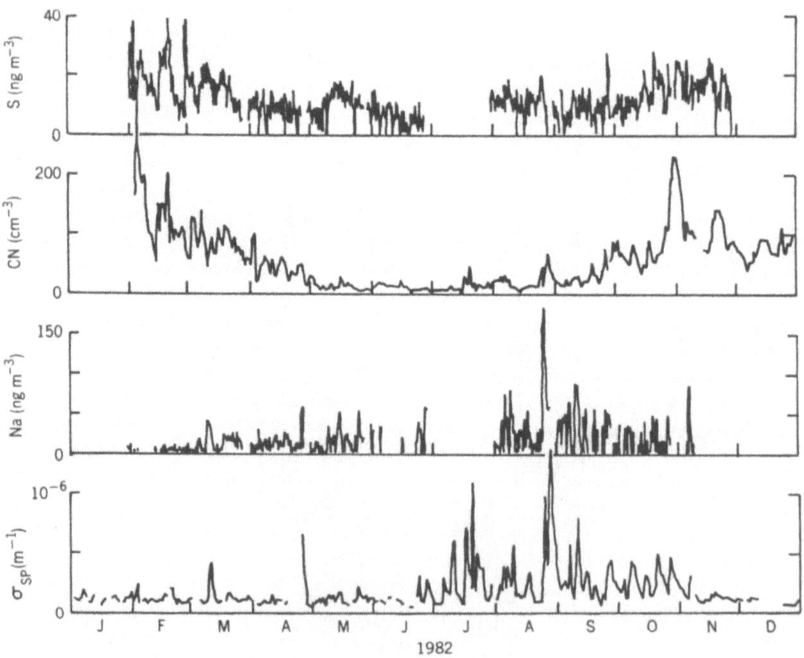

Figure 7. South Pole daily geometric means of CN concentration and σ_{sp}, and 8-hr means of S and Na concentration during 1982 [from *Bodhaine et al.*, 1986].

Single particle analyses of South Pole aerosols were performed by *Parungo et al.* [1979] and *Bigg* [1980]. However, both of these studies involved samples taken during the austral summer. *Parungo et al.* [1981] performed single particle analyses of samples taken during the austral winter. The origin and transport of Antarctic aerosols was discussed by *Ito and Iwai* [1981], *Ito et al.* [1984], *Bigg et al.* [1984], and *Ito* [1985].

Aerosol size distribution

Apparently the first study of individual particles sampled at the South Pole was performed by *Parungo et al.* [1979]. They examined particles deposited on Nuclepore filters during November 1977 using a transmission electron microscope (TEM) and a scanning electron microscope (SEM) with energy dispersive x-ray analysis (EDXA). They analyzed 100 particles and found a single mode size distribution which peaked at about 0.08 μm diameter with a total concentration of about 100 cm^{-3}. Half of the

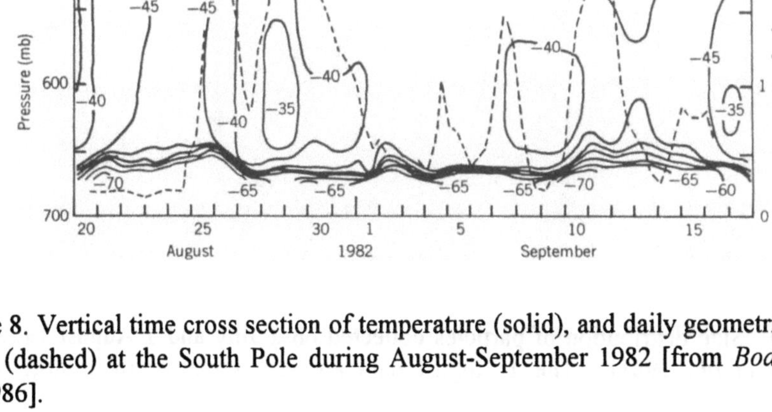

Figure 8. Vertical time cross section of temperature (solid), and daily geometric means of σ_{sp} (dashed) at the South Pole during August-September 1982 [from *Bodhaine et al.*, 1986].

particles with d < 0.2 μm contained S. A subsequent study by *Parungo et al.* [1981] examined Nuclepore filter samples taken once per month during February 1979 to January 1980. They analyzed 100 particles on each of the 12 filters. Fortunately, the 1 August 1979 sample coincided with a large sea salt event (see Figure 9). They found the expected annual cycle in particle concentration. However, the percentage of particles containing S remained fairly constant throughout the year with the exception of the August sample which showed a smaller percentage containing S and a larger percentage containing Na and Cl. The percentage of particles containing Na and Cl showed an annual cycle with a maximum in winter. The February sample contained mostly S, while the August sample contained mostly Na and Cl.

At the time of installation of the Pollak CN counter in January 1974, a diffusion battery apparatus of the *Rich* [1966] parallel plate design was also installed. When routine Pollak CN observations were performed, additional routine measurements were

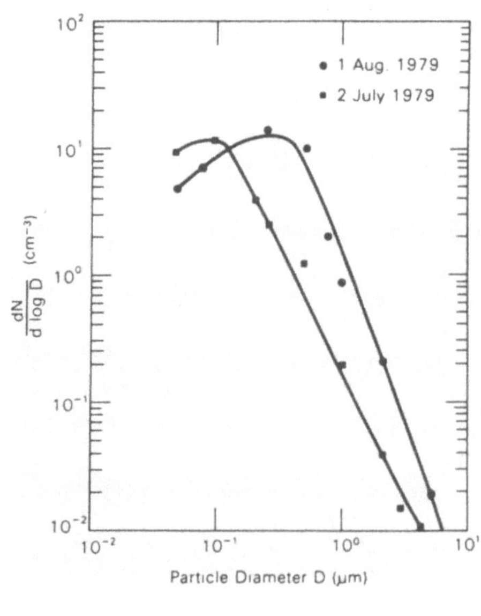

Figure 9. Size distribution of particles collected on 2 July and 1 August 1979 [from *Parungo et al.*, 1981]. The 1 August data are representative of a large sea-salt episode.

made using the diffusion battery. *Hogan* [1975], using this apparatus, reported that during January 1974 CN concentrations were about 40-80 cm^{-3} with a geometric mean diameter of about 0.02-0.03 μm. *Hogan and Barnard* [1979] used the diffusion battery technique to infer the aerosol size distribution in the 0.01-1.0 μm diameter range during December 1978. They found about 100 cm^{-3} with a number concentration peak at about 0.06-0.08 μm diameter and volume concentration peak at about 1 μm. *Hogan et al.* [1984a] studied South Pole aerosol sampled in January 1993 using diffusion battery and SEM with EDXA techniques. They also inferred aerosol mass using σ_{sp} data. However, they did not show size distributions. A warm event that occurred at the South Pole on 13 September 1983 was studied by *Murphey et al.* [1991]. They found an increase in CN concentration and an increase in aerosol size during this event. *Riley* [1987] presented a summary of diffusion battery data for the years 1974-1983, that clearly showed the warm large-particle events.

The 1987 aerosol chemistry experiment conducted by *Bodhaine and Harris* [1992] included a diffusion battery experiment. This experiment used six air-flow paths, selected by motor-driven ball valves, consisting of four paths through 8-, 5-, 2-, and 1-μm pore-diameter Nuclepore filters, one path through an absolute filter, and one straight-through path. Each path was sampled for 10 minutes by a continuously operating TSI 3020 CN counter, giving a complete cycle every hour. Interpretation of these diffusion battery data in terms of aerosol size distribution was described by *Bodhaine et al.* [1989]. An entire year (1987) of South Pole hourly data was obtained in this way. Using the penetration fraction through the 2-μm pore Nuclepore filter, they estimated the aerosol median diameter as about 0.02 μm during October (monthly minimum), 0.04 μm during January (monthly maximum) and as large as 0.06 μm during an individual sea salt event. These diffusion battery data show clearly that the penetration fraction increases dramatically whenever NaCl and σ_{sp} events occur. A more detailed study by *Bodhaine et al.* [1988] for July 1987 hourly data showed the correspondence between diffusion battery penetration fraction and σ_{sp} data (see Figure 10).

Transport

Antarctic meteorology has been studied extensively and was reviewed by *Radok* [1979] and *Robinson* [1983]. *Schwerdtfeger* [1984] provided the authoritative textbook on the subject, so that it is not necessary to repeat that information here. *Shaw* [1988] also provided a brief summary of Antarctic transport processes. Rather, it is the purpose of this review to consider individual transport studies related to

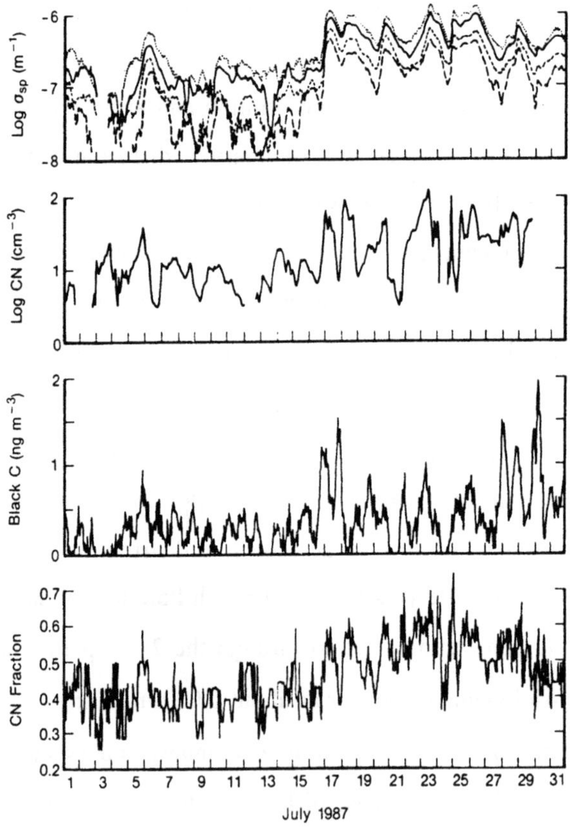

Figure 10. South Pole hourly mean aerosol data for July 1987. From the top, σ_{sp} at 450 (dotted), 550 (solid), 700 (dashed), and 850 nm (long-dashed); CN concentration; BC concentration; and CN penetration fraction through a 2-μm pore-diameter Nuclepore filter [from *Bodhaine et al.,* 1988]. A fraction of 0.5 corresponds to an aerosol median diameter of approximately 0.03 μm.

specific measurement programs. For example, *Mroz et al.* [1989] conducted an atmospheric tracer study using heavy methane to determine transport paths over the Antarctic continent. They released heavy methane at various locations around the continent during January, June, and October, and performed measurements at various coastal locations and at the South Pole. They found rapid transport to the plateau during spring and summer. However, transport was inhibited during winter by the winter circumpolar vortex. Other studies, such as back trajectory cluster analyses by *Harris* [1992], found the most vigorous long-range transport paths to the South Pole during July-October.

Stearns and Wendler [1988] reviewed results derived from a network of Antarctic automatic weather stations. That network consisted of 25 units sited at various locations on the continent to support research on katabatic winds, barrier winds, surface heat fluxes, and climate. Surface winds over the Antarctic continent were reviewed by *Parish* [1988], who used detailed topographic maps and surface temperatures to estimate the katabatic, or drainage, wind flow off the ice sheet. *Bromwich* [1988] reviewed precipitation over Antarctica.

One of the first descriptions of the influence of meteorology on particle measurements at South Pole was provided by *Hogan and Barnard* [1978]. They found that variations in particle concentration and size accompany meteorological events. For example, during winter months when temperature decreases and inversion strength increases, particle concentrations decreased to <10 cm^{-3}. Winter warming events are accompanied by an increase in particle concentrations in general and large particles in particular. *Hogan* [1975] also discussed the origin of particles measured at South Pole but at that time his observations were limited to the austral summer of January 1974. *Hogan* [1979] performed aircraft observations in the vicinity of South Pole and discussed the origins of particles, but his observations were again limited to the austral summer. *Hogan et al.* [1979] discussed meteorological transport and the effect on particle concentration and size for measurements during November 1978. More detailed studies of transport to the Pole were carried out for aerosol events in both summer and winter by *Hogan et al.* [1982, 1984b]. An interesting study by *Raisbeck and Yiou* [1981] showed the viability of using $^{10}Be/^{7}Be$ ratios in surface-based aerosol samples at the South Pole to infer stratospheric input. January (summer) samples showed significant stratospheric input whereas June (winter) samples did not.

Meteorological transport was referred to by *Bodhaine and Bortniak* [1981] and *Parungo et al.* [1981] when discussing the presence of Na and Cl in South Pole aerosol samples. The aerosol chemistry experiment of *Bodhaine et al.* [1986], along with CN and σ_{sp} measurements, allowed sufficient time resolution (8-hr) to identify individual events and relate them to calculated back trajectories. For example the large sea salt event of 1 August 1979 was found to originate in the W-SW sector, whereas a large 11 September 1982 event originated in the south (Ross Sea) sector. The study by *Harris* [1992] showed a climatology of 5-day trajectories calculated backwards from the South Pole for 1985-1989 using clustering techniques. Cyclonic westerly flow patterns were the strongest and most frequent, and easterly flow patterns were the least frequent and usually anticyclonic. Isentropic trajectories from the west usually ascended from the marine boundary layer along the coast to about 5 km above the pole (see Figure 11). Monthly cluster means showed strongest transport in October, and weakest transport in January. Some year-to-year variability was observed. For example 1987 showed northerly transport to be more vigorous than westerly transport. It seems clear, however, that strong transport can occur from the north, west, or south, whereas strong transport from the east is rare. *Harris* [1992] also showed that, during July 1985-1989, vigorous westerly flow corresponded to 500-hPa temperatures about 3°C warmer than easterly flow, consistent with enhanced warm air advection from the west.

Global circulation models (GCMs) have been used to estimate the distribution of tracers over the Antarctic continent. *Law et al.* [1992] used CO_2 as a tracer and deduced transport and distribution over Antarctica. They found realistic distributions and used the model to predict transport times with reference to tracer growth in Antarctica. They found a speedup when the tracer entered the baroclinically active regions of southern midlatitudes. *Genthon* [1992] used a GCM to study the features of desert dust and sea-salt distribution over the Antarctic continent. His most important conclusion was that surface based measurements do not provide the necessary information for understanding transport processes aloft. The vertical structure is important because, in the case of the seasonal cycle for example, what is seen at the surface may be the opposite of that seen in the column. Sea salt and desert dust sources complement each other for modeling studies because they have different seasonal production and vertical distribution. The GCM of *Yamazaki and Chiba* [1993] was used to simulate 2-yr transport for sources in Japan, Europe, North America, and the Amazon. Once the Japanese tracer reached the equatorial regions, it took about 3 months more to reach Antarctica and showed maximum values in June. In the Antarctic middle and upper troposphere, the minimum density appeared over the top of the east Antarctic ice sheet, whereas at the surface the distribution was uniform.

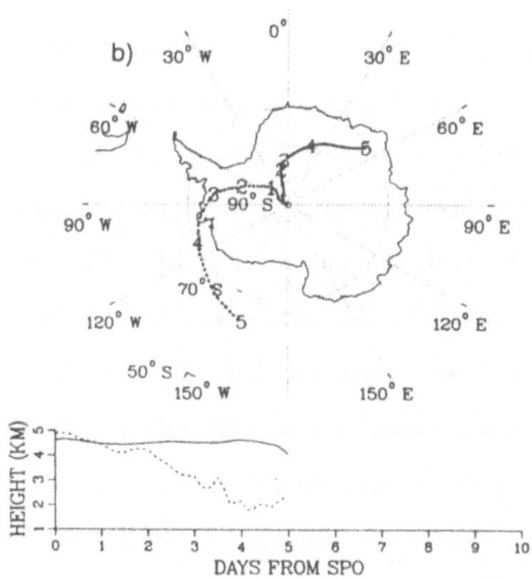

Figure 11. Isentropic back trajectories on the 280-K surface arriving at the South Pole on 21 August 1989 at 00 Z (solid) and 12 Z (dashed). Numerals along the trajectories show 1-day intervals. The plot below the trajectories shows trajectory heights [from *Harris*, 1992].

Over the interior of the continent, the density decreased with altitude, indicating downward transport of stratospheric air.

CMDL Monitoring Program

Carbon dioxide (CO_2) measurements were first made at the South Pole by *Keeling et al.* [1976]. Those measurements continue to the present and were recently presented by *Keeling et al.* [1989]. Continuous analyzers have been operated at the South Pole as part of the CMDL program [*Gillette et el.*, 1987]. Flask sample data were presented by *Conway et al.* [1988, 1994]. The *Conway et al.* [1994] publication will surely stand as one of the classic data sets in the field, presenting long-term data from 46 global sites, including 14 shipboard data sets. Continuous and flask CO_2 data for the South Pole have been presented in the series of CMDL Annual Reports [e.g., *Peterson and Rosson,* 1993]. Because the global CO_2 growth is well known, and the possible role of CO_2 in global environmental change is a popular topic in its own right, further discussion of the CO_2 issue should be considered beyond the scope of this paper. However, the South Pole CO_2 data provide an anchor point for global models that describe the CO_2 distribution and its effect on global climate.

Methane (CH_4) has been measured at the South Pole since 1983 as part of a global flask sampling network [*Dlugokencky et al.*, 1994] consisting of 53 sites (including 17 shipboard stations). The first South Pole CH_4 measurements were made by *Steele et al.* [1987, 1989]. Because CH_4 has been increasing in the global atmosphere, and because it is considered an important greenhouse gas, it has a direct effect on climate forcing in the troposphere and stratosphere. Although the CH_4 budget has been assessed in numerous studies, the actual causes of the global increase are unknown. No significant sources exist in the high latitudes of the southern hemisphere so that CH_4 is well mixed and there is little variability among sites. The average increase in CH_4 concentration was about 11 ppb yr^{-1} over the 10-yr period of record, starting with a concentration of about 1550 ppb in 1983.

Chlorofluorocarbons (CFCs) 11 and 12 have been measured at the South Pole since 1978 [*Peterson and Rosson*, 1993; *Elkins et al.*, 1993]. Because of the potential for CFCs to destroy ozone in the stratosphere, their widespread use in refrigeration, aerosol sprays, and foams, and the Montreal Protocol, which called for the elimination of CFCs, trends in CFC concentrations have been closely monitored. *Elkins et al.* [1993] predict that if growth rates continue to slow in line with industrial emissions, global values will peak before the turn of the century, and then decline.

Surface ozone was measured at the South Pole in the early 1960s, as summarized by *Oltmans and Komhyr* [1976]. *Oltmans and Levy* [1994] present surface ozone data for the period 1975-1992 that show a significant long-term decrease, particularly in the summer values. They attribute this decrease to two factors: greater penetration of UV radiation to the surface leading to surface ozone loss, and enhanced transport of ozone-depleted air from the coast to the Pole. These complex circumstances were discussed in more detail by *Schnell et al.* [1991]. Although it is not the purpose of this paper to discuss stratospheric ozone and the ozone hole, *Solomon* [1988] and *Hofmann* [1988] presented excellent review articles on this phenomenon, and related it to tropospheric effects.

Measurement of trace gas concentrations in the Antarctic atmosphere using an infrared spectrometer was accomplished by *Murcray et al.* [1988]. Ground-based spectra were obtained at the South Pole during November-December 1986. The system was capable of measuring the column abundances of a number of trace gases including HNO_3, NO, NO_2, HCl, H_2O, CH_4, and N_2O.

The first solar radiation measurements at the South Pole were made by *Viebrock and Flowers* [1968], as part of the IGY. They used the 1960-1965 data to examine the effects of the eruption of Mt. Agung on the radiation budget at the South Pole. After a 10-yr gap, the CMDL program initiated new solar radiation measurement program starting in 1976. This program was described in detail by *Dutton et al.* [1991].

Conclusions

Despite the size of Antarctica and the interest in atmospheric chemistry, few measurements are made there except at a few coastal locations and at the South Pole itself. By virtue of its remoteness the continent has not been studied as extensively as its counterparts elsewhere. In June 1991, following the Symposium on the Tropospheric Chemistry of the Antarctic Region [*Bodhaine et al.,* 1992], a one-day workshop was held to assess the current state of Antarctic research and to address the necessary future directions of this research. Approximately 65 researchers from 15 countries took part in this workshop [*Barrie et al.,* 1991]. An entire section of this report was devoted to air-snow exchange and glacial chemistry research. It was strongly suggested that the incorporation processes of ^{18}O, ^{2}H, soluble and insoluble particles, reactive gases, organic species, and the gas precursors of NO_3, SO_4, and MSA be studied. It was recommended that air-snow exchange be studied at Vostok, Dome-C, Plateau, and South Pole, and that the snow chemistry should be studied at Byrd and Law Dome.

Antarctica will surely continue to be one of the last true frontiers for atmospheric chemistry research and an ideal location for air-surface exchange studies. Antarctic ice core studies are essential and will be crucial for the understanding of global climate.

Acknowledgments

I wish to thank the many observers who wintered over at the Amundsen-Scott Station and operated the large array of instrumentation that is responsible for the many excellent data sets that exist today. I also wish to give a special tribute to William H. Zoller who was responsible for the excellent aerosol chemistry data set that will stand as a classic background data set.

References

Ahlquist NC, Charlson RJ (1969) Measurement of the wavelength dependence of atmospheric extinction due to scatter. *Atmos Environ 3:* 551-564

Barrie LA, Bodhaine BA, Schnell RC, Shaw GE (1991) *Antarctic Tropospheric Chemistry Research, A Report of a Workshop Held in Conjunction with The International Symposium on the Tropospheric Chemistry of the Antarctic Region,* 3-6 June 1991, Boulder, Colorado.

Bigg EK (1980) Comparison of aerosol at four baseline atmospheric monitoring stations. *J Appl Meteorol 19:* 521-533

Bigg EK, Gras JL, Evans C (1984) Origin of Aitken particles in remote regions of the southern hemisphere. *J Atmos Chem 1:* 203-214

Bodhaine BA (1983) Aerosol measurements at four background sites. *J Geophys Res 88:* 10753-10768

Bodhaine BA (1992) The U.S. aerosol monitoring program in Antarctica. In *Conference Proceedings Vol. 35, 4th Workshop Italian Research on Antarctic Atmosphere,* Colacino M, Giovanelli G, Stefanutti L (eds) Italian Physical Society, Bologna: 15-25

Bodhaine BA (1995) Aerosol absorption measurements at Barrow, Mauna Loa, and South Pole. *J Geophys Res,* submitted

Bodhaine BA, Bortniak JC (1981) Four wavelength nephelometer measurements at South Pole. *Geophys Res Lett 8:* 539-542

Bodhaine BA, Harris JM (1982) *Geophysical Monitoring for Climatic Change No. 10, Summary Report 1981,* Air Resources Laboratory, Boulder, Colorado: 158 pp

Bodhaine BA, Harris JM (1992) Aerosol measurements at the South Pole during 1987. *NOAA Data Rep ERL CMDL-9,* Climate Monitoring and Diagnostics Laboratory, Boulder, Colorado: 120 pp

Bodhaine BA, Murphy ME (1980) Calibration of an automatic condensation nuclei counter at the South Pole. *J Aerosol Sci 11:* 305-312

Bodhaine BA, Shanahan MK (1990) Condensation nucleus and aerosol scattering extinction measurements at the South Pole Observatory. *NOAA Data Rep ERL CMDL-1,* Climate Monitoring and Diagnostics Laboratory, Boulder, Colorado: 148 pp

Bodhaine BA, DeLuisi JJ, Harris JM, Houmere P, Bauman S (1986) Aerosol measurements at the South Pole. *Tellus 38B:* 223-235

Bodhaine BA, DeLuisi JJ, Harris JM, Houmere P, Bauman S (1987) PIXE analysis of South Pole aerosol. *Nucl Instr Methods B22:* 241-247

Bodhaine BA, Dutton EG, DeLuisi JJ, Harris JM, Shaw GE, Hansen ADA, Novakov T (1988) South Pole aerosol measurements during 1987. In *Atmospheric Aerosols and Nucleation,* Wagner PE, Vali G (eds) Springer-Verlag, Heidelberg.

Bodhaine BA, Dutton EG, DeLuisi JJ, Herbert GA, Shaw GE, Hansen ADA (1989) Surface aerosol measurements at Barrow during AGASP-II. *J Atmos Chem 9:* 213-224

Bodhaine BA, Barrie LA, Schnell RC, Shaw GE, McKie JK (1992) Symposium on the tropospheric chemistry of the Antarctic region. *Tellus 44B:* 250-251

Bouton K (1981) A reporter at large, South of 60 degrees south. *New Yorker,* 23 March 1981: 42-122

Bromwich DH (1988) Snowfall in high southern latitudes. *Rev Geophys 26:* 149-168

Conway TJ, Tans P, Waterman LS, Thoning KW, Masarie KA, Gammon RH (1988) Atmospheric carbon dioxide measurements in the remote global troposphere, 1981-1984. *Tellus 40B:* 81-115

Conway TJ, Tans PP, Waterman LS, Thoning KW, Kitzis DR, Masarie KA, Zhang N (1994) Evidence for interannual variability of the carbon cycle from the National Oceanic and Atmospheric Administration/Climate Monitoring and Diagnostics Laboratory Global Air Sampling Network. *J Geophys Res 99:* 22831-22855

Cunningham WC, Zoller WH (1981) The chemical composition of remote area aerosols. *J Aerosol Sci 12:* 367-384

Cunningham WC, Etz ES, Zoller WH (1979) Raman microprobe characterization of South Pole aerosol. *Microbeam Analysis,* Newbury D (ed) San Francisco Press, San Francisco, California, U.S.A.: 148-154

Dlugokencky EJ, Steele LP, Lang PM, Masarie KA (1994) The growth rate and distribution of atmospheric methane. *J Geophys Res 99:* 17021-17043

Duce RA (1972) Trace substances in the Antarctic atmosphere. *Proceedings of the Colloquium on Conservation Problems of Antarctic and Circumpolar Waters,* Parker B (ed) Allen Press, Lawrence, Kansas, U.S.A.: 27-42

Duce RA, Zoller WH, Jones AG (1971) Atmospheric particle and gas sampling at McMurdo and South Pole stations. *Ant J United States 6:* 133-134

Duce RA, Zoller WH, Moyers JL (1973) Particulate and gaseous halogens in the Antarctic atmosphere. *J Geophys Res 78:* 7802-7811

Duce RA, Hoffman GL, Zoller WH (1975) Atmospheric trace metals at remote northern and southern hemisphere sites: Pollution or natural? *Science 187:* 59-61

Dutton EG, Stone RS, Nelson DW, Mendonca BG (1991) Recent interannual variations in solar radiation, cloudiness, and surface temperature at the South Pole. *J Climate 4:* 848-858

Elkins JW, Thompson TM, Swanson TH, Butler JH, Hall BD, Cummings SO, Fisher DA, Raffo AG (1993) Decrease in the growth rates of atmospheric chlorofluorocarbons 11 and 12. *Nature 364:* 780-783

Genthon C (1992) Simulations of desert dust and sea-salt aerosols in Antarctica with a general circulation model of the atmosphere. *Tellus 44B:* 371-389

Gillette DA, Komhyr WH, Waterman LS, Steele LP, Gammon RH (1987) The NOAA/GMCC continuous CO_2 record at the South Pole, 1975-1982. *J Geophys Res 92:* 4231-4240

Hansen ADA, Bodhaine BA, Dutton EG, Schnell RC (1988) Aerosol black carbon measurements at the South Pole: Initial results, 1986-1987. *Geophys Res Lett 15:* 1193-1196

Harris JM (1992) An analysis of 5-day midtropospheric flow patterns for the South Pole: 1985-1989. *Tellus 44B:* 409-421

Heintzenberg J (1985) What can we learn from aerosol measurements at baseline stations? *J. Atmos Chem 3:* 153-169

Hofmann DJ (1988) Balloon-borne measurements of middle atmosphere aerosols and trace gases in Antarctica. *Rev Geophys 26:* 113-130

Hogan AW (1975) Antarctic aerosols. *J Appl Meteorol 14:* 550-559

Hogan AW (1979) Meteorological transport of particulate matter to the south polar plateau. *J Appl Meteorol 18:* 741-749

Hogan AW, Barnard S (1978) Seasonal and frontal variation in Antarctic aerosol concentrations. *J Appl Meteorol 17:* 1458-1465

Hogan AW, Barnard S (1979) Physical properties of the aerosol at the South Pole. *Geophys Res Lett 6:* 845-848

Hogan AW, Bodhaine BA (1995) The twenty year aerosol record at South Pole. Paper presented at the 75th AMS Annual Meeting, Fourth Conference on Polar Meteorology, Dallas, Texas, USA, 15-20 January 1995

Hogan AW, Barnard S, Bortniak J (1979) Physical properties of the aerosol at the South Pole. *Geophys Res Lett 6:* 845-848

Hogan A, Barnard S, Samson J, Winters W (1982) The transport of heat, water vapor and particulate material to the south polar plateau. *J Geophys Res 87:* 4287-4292

Hogan A, Samson J, Kebschull K, Townsend R, Barnard S, Murphey B (1984a) On the interaction of aerosol with meteorology. *J Rech Atmos 18:* 41-67

Hogan A, Kebschull K, Townsend R, Murphey B, Samson J, Barnard S (1984b) Particle concentrations at the South Pole, on meteorological and climatological time scales; is the difference important? *Geophys Res Lett 1:* 850-853

Hogan AW, Riley D, Murphey BB, Barnard SC, Samson JA (1993) Variation in aerosol concentration associated with a polar climatic iteration. In *Antarctic Meteorology and Climatology: Studies based on automatic weather stations, Antarctic Research Series 61:* 175-199

Ito T (1985) Study of background aerosols in the Antarctic troposphere. *J Atmos Chem* 3: 69-91

Ito T, Iwai K (1981) On the sudden increase in the concentration of Aitken particles in the Antarctic atmosphere. *J Meteorol Soc Japan 59:* 262-271

Ito T, Ono A, Iwai K (1984) On the origin and nature of Antarctic aerosols. In *Proceedings of the Fourth Symposium on Polar Meteorology and Glaciology,* National Institute of Polar Research, Tokyo, Japan: 289-296

Keeling CD, Adams JA, Ekdahl CA, Guenther PR (1976) Atmospheric carbon dioxide variations at the South Pole. *Tellus 28:* 552-564

Keeling CD, Bacastow RB, Carter AF, Piper SC, Whorf TP, Heimann M, Mook WG, Roeloffzen H (1989) A three-dimensional model of atmospheric CO_2 transport based on observed winds, 1, Analysis of observational data. In *Aspects of Climate Variability in the Pacific and the Western Americas, Geophys Monogr 55,* Peterson DH (ed) American Geopysical Union, Washington, DC

Komhyr WD (1983) An aerosol and gas sampling apparatus for remote observatory use. *J Geosphys Res 88:* 3913-3918

Law R, Simmonds I, Budd WF (1992) Application of an atmospheric tracer model to high southern latitudes. *Tellus 44B:* 358-370

Maenhaut W, Zoller WH (1977) Determination of the chemical composition of the South Pole aerosol by instrumental neutron activation analysis. *J Radioanal Chem 37:* 637-650

Maenhaut W, Zoller WH, Duce RA, Hoffman GL (1979) Concentration and size distribution of particulate trace elements in the South Pole atmosphere. *J Geophys Res 84:* 2421-2431

Meriwether JW Jr (1988) Atmospheric sciences in Antarctica. *Rev Geophys 26:* 41-43

Metnieks AL, Pollak LW (1959) Instruction for use of photo-electric condenstaion nucleus counters. *Geophys. Bull. No. 16,* School of Cosmic Physics, Dublin Institute for Advanced Study, Dublin, Ireland

Miller JM (1974) In *Geophysical Monitoring for Climatic Change No. 1, Summary Report 1972* Environmental Research Laboratories, Boulder, Colorado: 10

Mroz EJ, Alei M, Cappis JH, Guthals PR, Mason AS, Rokop DJ (1989) Antarctic atmospheric tracer experiments. *J Geophys Res 94:* 8577-8583

Murcray FJ, Murcray FH, Murcray DG (1988) Infrared measurements of column abundances of several trace gases in the Antarctic atmosphere. *Rev Geophys 26:* 81-88

Murphey BB, Hare T, Hogan AW, Lieser K, Toman J, Woodgates T (1991) Vernal atmospheric mixing in the Antarctic. *J Appl Meteorol 30:* 494-507

Murphy ME, Bodhaine BA (1980) The South Pole automatic condensation nuclei counter: Instrument details and five years of observations. *NOAA Tech Memo ERL ARL-82,* Air Resources Laboratories, Silver Spring, Maryland, 88 pp

Oltmans SJ, Komhyr WD (1976) Surface ozone in Antarctica. *J Geophys Res 81:* 5359-5364

Oltmans SJ, Levy H II (1994) Surface ozone measurements from a global network. *Atmos Environ 28:* 9-24

Pack DH (1973) Geophysical monitoring for climatic change. *Ant J United States 8:* 253-254

Parish TR (1988) Surface winds over the Antarctic continent: A review. *Rev Geophys 26:* 169-180

Parungo F, Ackerman E, Caldwell W, Weickmann HK (1979) Individual particle analysis of Antarctic aerosols. *Tellus 31:* 521-529

Parungo F, Bodhaine B, Bortniak J (1981) Seasonal variation in Antarctic aerosol. *J Aerosol Sci 12:* 491-504

Peterson JT, Rosson RM (1993) In *Climate Monitoring and Diagnostics Laboratory No. 21, Summary Report 1992,* NOAA Environmental Research Laboratories, Boulder, Colorado

Radok U (1979) Polar meteorology and climatology 1975-78. *Rev Geophys Sp Phys 17:* 1772-1781

Raisbeck GM, Yiou F (1981) Cosmogenic $^{10}Be/^{7}Be$ as a probe of atmospheric transport processes. *Geophys Res Lett 8:* 1015-1018

Rich TA (1955) A photo-electric nucleus counter with size discrimination. *Geofis Pura Appl 31:* 60-65

Rich TA (1966) Apparatus and method for measuring the size of aerosols. *J Rech Atmos 2:* 79-86.

Riley DC (1987) *A study of the relationship between the variation of aerosol and meteorological parameters at South Pole on synoptic, seasonal and climatological time scales.* Masters thesis submitted to the State University of New York, Albany, New York, 141 pp

Roberts CL Jr (1969) ESSA's Antarctic meteorological program. *Ant J United States 4:* 224-225

Robinson E (1983) Polar meteorology and climatology 1979-1982. *Rev Geophys Sp Phys 21:* 1048-1064

Robinson E, Bodhaine BA, Komhyr WD, Oltmans SJ, Steele LP, Tans P, Thompson TM (1988) Long-term air quality montioring at the South Pole by the NOAA program Geophysical Monitoring for Climatic Change. *Rev Geophys 26:* 63-80

Samson JA, Barnard SC, Obremski JS, Riley DC, Black JJ, Hogan AW (1990) On the systematic variation in surface aerosol concentration at the South Pole. *Atmos Res 25:* 385-396

Satchell M (1983) Women who conquer the South Pole. *Parade Magazine,* 5 June 1983: 16-17

Schnell RC, Liu SC, Oltmans SJ, Stone RS, Hofmann DJ, Dutton EG, Deshler T, Sturges WT, Harder JW, Sewell SD, Trainer M, Harris JM (1991) Decrease of summer tropospheric ozone concentrations in Antarctica. *Nature 351:* 726-729

Schwerdtfeger W (1984) *Weather and Climate of the Antarctic,* Elsevier, New York, 261 pp

Shaw GE (1988) Antarctic aerosols: A review. *Rev Geophys 26:* 89-112

Skala GF (1963) A new instrument for the continuous measurement of condensation nuclei. *Anal Chem 35:* 702-706

Solomon S (1988) The mystery of the Antarctic Ozone "Hole." *Rev Geophys 26:* 131-148

Stearns CR, Wendler G (1988) Research results from Antarctic automatic weather stations. *Rev Geophys 26:* 45-61

Steele LP, Fraser PJ, Rasmussen RA, Khalil MAK, Conway TJ, Crawford AJ, Gammon RH, Masarie KA, Thoning KW (1987) The global distribution of methane in the atmosphere. *J Atmos Chem 5:* 125-171

Steele LP, Lang PM, Martin RC (1989) Atmospheric methane in Antarctica. *Ant J United States 24:* 239-241

Thieke JF, Charlson RJ, Winter JW, Ahlquist NC, Whitby KT, Husar RB, Liu BYH (1972) Multiwavelength nephelometer measurements in Los Angeles smog aerosols II. Correlation with size distributions, volume concentrations. *J Colloid Inter Sci 39:* 252-259

Tuncel G, Aras NK, Zoller WH (1989) Temporal variations and sources of elements in the South Pole atmosphere 1. Nonenriched and moderately enriched elements. *J Geophys Res 94:* 13025-13038

Voskresensky AI (1968) Condensation nuclei in the Mirny region. *Tr Sov Antarkt Eksped 38:* 194-198

Warburton JA (1973) Surface measurments of Aitken nuclei at McMurdo, Siple, Byrd, and South Pole stations. *Ant J United States 8:* 236

Warren SG, Clarke AD (1990) Soot in the atmosphere and snow surface of Antarctica. *J Geophys Res 95:* 1811-1816

Washburn AL (1980) Focus on polar research. *Science 209:* 643-652

Weller G, Bentley CR, Elliot DH, Lanzerotti LJ, Webber PJ (1987) Laboratory Antarctica: Research contributions to global problems. *Science 238:* 1361-1368

Yamazaki K, Chiba M (1993) A 3-D global simulation of the advective transport of passive tracers from various northern hemisphere sources. *Tellus 45B:* 160-178

Zoller WH, Gladney ES, Duce RA (1974) Atmospheric concentrations and sources of trace metals at the South Pole. *Science 183:* 198-200

COASTAL ANTARCTICA: ATMOSPHERIC CHEMICAL COMPOSITION AND ATMOSPHERIC TRANSPORT

Dietmar Wagenbach
Institut für Umweltphysik
University of Heidelberg
INF 366
69120 Heidelberg
Germany

1 Introduction

Snow fields of the coastal Antarctic region are situated intermediately between the vast environmental regimes of the high antarctic plateau and that of the South Polar Ocean. Since they are directly exposed to moist maritime air masses, a relatively high snow accumulation rate, associated with the deposition of mainly marine derived aerosol species, is expected here. These characteristics make this region unique to recover high resolution ice core records that reveal the environmental and climatic history of the South Polar Ocean. Indeed, several extensive ice core studies have been or are currently performed in coastal antarctic regions (see Figure 1). To "calibrate", in particular, their glacio-chemical information in terms of the corresponding atmospheric changes, representative long term records of the coastal antarctic aerosol chemistry (including the relevant precursor gases) are needed.

Although most of the antarctic overwintering stations are arranged along the rim of the continent (see Figure 2) no year round observation of the near surface aerosol chemistry (except for single particle analyses) is available before 1983. Before that time, year round monitoring of fission products (i.e. ^{90}Sr, ^{137}Cs) along with the naturally produced radioisotopes ^{222}Rn, ^{210}Pb and ^{7}Be started in the sixties at the Antarctic Peninsula (Feely et al., 1988) and at the French station Dumont D'Urville (Polian et al., 1986). From these data, the first information was gained on the transport of stratospheric and continental species to the antarctic coast. Later on, continuous observations of the number concentration of condensation nuclei (CN) were performed at Syowa and Mawson (Bigg et al., 1984) showing, similar to South Pole data, regular, strong summer maxima which suggest photochemical production of new particles during that season.

NATO ASI Series, Vol. I 43
Chemical Exchange Between the Atmosphere and Polar Snow
Edited by Eric W. Wolff and Roger C. Bales

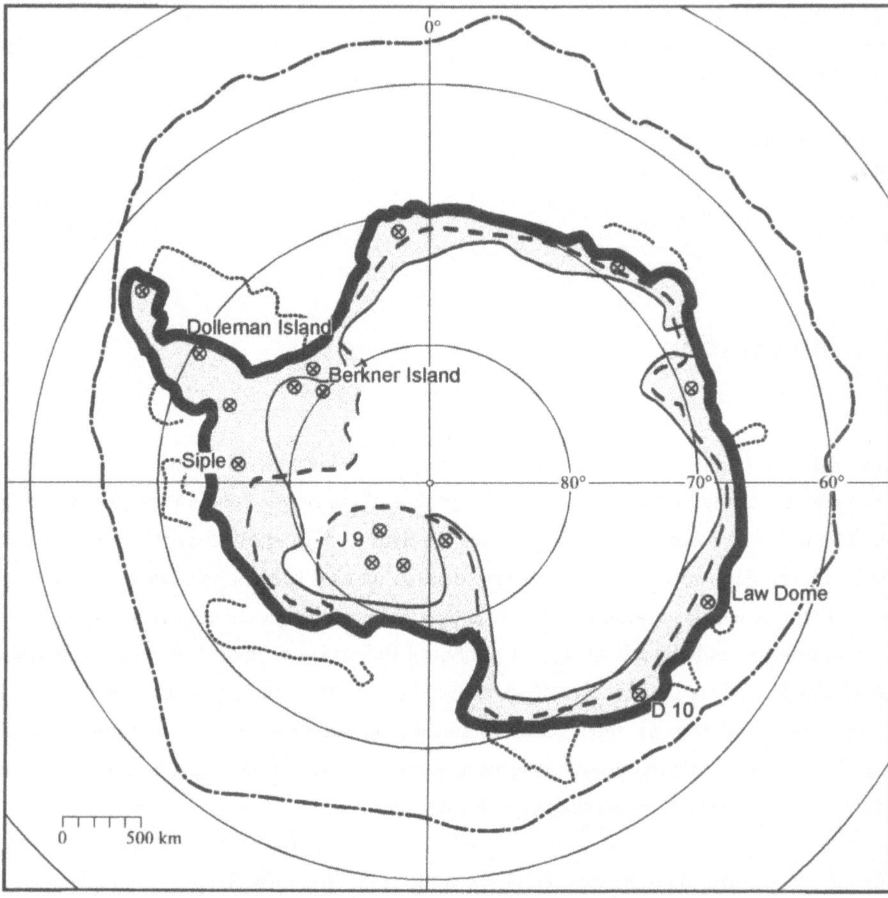

Figure 1: Sketch of the southern boundary areas of coastal Antarctica are marked by the 20 cm surface balance (fine line) and the 1500 m elevation (dashed line) contours. Important ice core drill sites in coastal regions (see e.g. review of Mulvaney and Wolff, 1994) are marked by crosses. Lines over ocean areas show the 50 % sea ice coverage during winter and summer, respectively (adapted from Giovinetto et al., 1990).

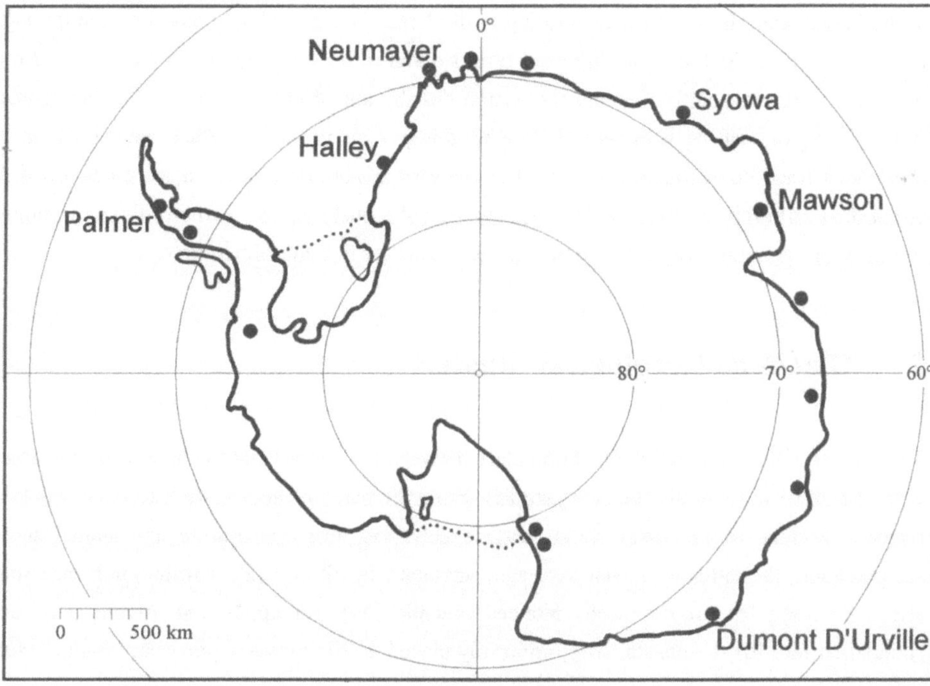

Figure 2: Map of all overwintering stations in coastal Antarctica. Those providing year-round surface aerosol records, are indicated by their names.

Continuous, year round observational programs on the major chemical components and on radioactive isotopes of the coastal antarctic aerosol were placed in operation by Wagenbach et al. (1988b) as late as 1983 at Neumayer Station. Here, aerosol sampling was performed in a contamination protected clean air station established by the University of Heidelberg. A similar year round program (with respect to high volume aerosol sampling and analyses) was started in 1987 at Mawson by Prospero et al. (1991) and by Savoie et al. (1992). It was extended in 1990 to Palmer Station and King George Island (not considered here) by Savoie et al. (1993). Gillett et al. (1993) supplemented the Mawson aerosol sampling program by investigating particular sulphur components in fine and large particle fractions. Within a collaborative effort between French, British and German research groups, continuous observations of the ionic aerosol composition were also performed since 1991 at Halley and Dumont D'Urville (E.

Wolff and M. Legrand, personal communication) including nitric acid and ammonia sampling at the French and black carbon monitoring at the British station. Mainly based on the findings of these new investigations and on recent results obtained by Wagenbach and co-workers from Neumayer, the seasonal cycles and the spatial distribution of chemical aerosol in the coastal Antarctic region will be summarised. Emphasis is put on the near surface concentration of short lived trace constituents for which the analyses in Antarctic firn and ice cores are well established. Therefore, long lived trace gases (like methane or carbon dioxide), upper tropospheric soundings or surface ozone observations are not considered here.

2 Glacio-meteorological situation

There is certainly not a well defined boundary between the coastal and the inner antarctic areas with respect to snow or air chemistry studies. From the main distinguishing features of the two regions, distance to the open water, surface elevation, firn temperature and annual snow accumulation, the latter is chosen here as a surrogate for the overall maritime influence (but also accounting for orographically induced precipitation). An upper limit of the snow accumulation rate for the coastal strip, arbitrarily placed at 20 cm water per year, would, if following the compilation by Giovinetto et al. (1990), constrain the other parameters to:

1) an elevation lower than ~1800 m a.s.l.,
2) a firn temperature higher than -27 ^{0}C, and
3) a mean annual distance to the open water with 50% sea ice coverage of less than ~1600 km.

Some justification for the chosen accumulation rate criterion shown in Figure 1 arises from various isotopic and chemical investigations of surface snow samples in the above defined coastal area. Compared to the interior they all show considerably higher sea salt concentrations (Herron and Langway, 1979; Clausen and Langway, 1990; Minikin et al., 1994) (suggesting a more direct impact of the marine aerosol body) and relatively low deuterium excess values indicating a significant contribution of high latitudes moisture sources to the precipitation in these escarpment and ice shelf regions (Petit et al., 1991; Graf et al., 1994; Peel and Mulvaney, 1992). Including the large ice shelves (although accumulating in their central parts less than 20 cm per year), the coastal antarctic regions, as defined above, would contribute roughly one third to the total area of the continent. As shown in Figure 1, this coastal part of Antarctica has been the target region for many important ice core studies like for example at Law Dome (Etheridge et al., 1992), at Siple (Mosley-Thompson et al., 1991), on the Antarctic Peninsula (Peel and Mulvaney, 1992) and currently within the Berkner Island and the West Antarctic ice core drilling programs. Notably, the glacio-meteorological conditions of these drill sites with respect to the firn temperature and the snow accumulation rate are quite similar to those commonly found in Central Greenland.

As shown in Figure 2, most stations in coastal Antarctica providing atmospheric records are situated close to sea level, and, for logistical reasons, relatively near to the shore or to the ice edge of an ice shelve. Among others, the following specific meteorological features at these positions are relevant for the interpretation of short lived atmospheric species:

- the influence of katabatic winds draining the source free interior from air masses which are loaded with aged aerosol (Shaw, 1988) contrary to the frequent advections of moist, maritime polar air associated with the passage of frontal systems,
- the persistence of low level stratus clouds and relatively high precipitation rates that is expected to diminish the atmospheric residence time of aerosol particles and water reactive trace gases,
- the relatively weak or frequently disappearing surface inversion that enables a regular replenishment of the boundary layer air also during polar night (Phillpot and Zillman, 1970),
- the immediate influence of the annual sea ice cycle changing the distance to vast areas of open sea water from some tens to up to 1000 km (Zwally, 1984).

3 Sources and transport pathways of atmospheric trace constituents to coastal antarctic sites

Short lived atmospheric trace constituents mainly investigated in antarctic ice core studies are various sea salt components (Na^+, Cl^-, Mg^{2+} ...) , nitrate, biogenic and vulcanogenic sulphur species, ammonium, mineral dust, heavy metals and radioisotopes (Oeschger and Langway, 1989). The principal source and pathways of these trace constituents into the near surface atmosphere of coastal Antarctica are illustrated in Figure 3. Important atmospheric compartments contributing in various ways to the atmospheric aerosol load in coastal regions are:

- the free troposphere over the northward continents fed by deep convection with continental aerosols (and precursor gases) like mineral dust, ammonium sulphate, nitrate and (anthropogenic) trace elements,
- the stratosphere constitutes a (potential) source for nitrate, biogenic background sulphate and for volcanic sulphate to the antarctic troposphere,
- the free marine troposphere over the subantarctic circle and the adjacent mid latitudinal ocean areas which is strongly influenced by the evasion of dimethylsulfide (DMS) from the

ocean surface during summer. Since the main particulate end products of the photochemical DMS oxidation chain are sulphuric and methanesulphonic acids (MSA) a large fraction of the total aerosol load in this layer (besides the ubiquitous sea particles) is made up by these biogenic sulphur components (Saltzman et al., 1983),

- the local boundary layer offshore and in the vicinity of the stations. This reservoir may be loaded with marine derived atmospheric species during the open water season but, potentially, also be contaminated by anthropogenic emissions, by ammonia from animal life and by mineral dust mobilised from bare soil areas.

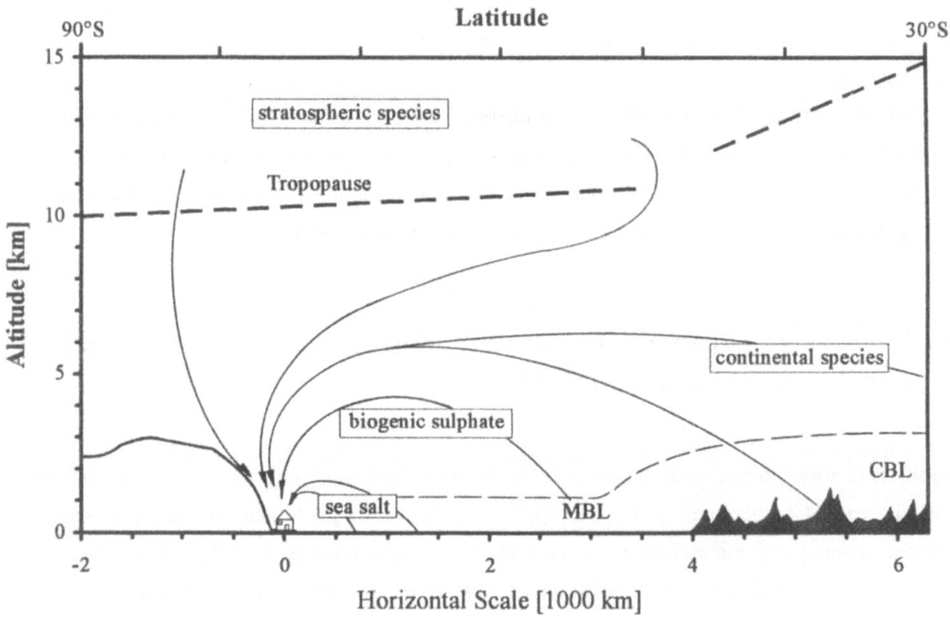

Figure 3: Principle long range transport pathways of short lived atmospheric constituents to coastal antarctic sites.

The transport of trace constituents from the above atmospheric reservoirs to coastal antarctic stations obviously covers considerably different spatial scales and, hence, may operate with variable efficiencies during different seasons.

3.1 Meridional long range transport

Based on world-wide atmospheric ^{85}Krypton and recently also SF_6 observations (Levin and Hesshaimer, 1995 and references therein) it has been showed conclusively that the characteristic time for interhemispheric exchange is around one year. It is, thus, by more than one order of magnitude larger than the typical tropospheric aerosol residence time. This makes the northern hemisphere utterly unimportant as a source for the antarctic aerosol body if large injections into the stratosphere, e.g. volcanic eruptions, are ignored.

The meridional aerosol transport from southern mid latitudes appears to operate most effectively in the upper tropospheric level where the life time of particles and water reactive precursor gases is expected to be significantly longer than in the moist marine boundary layer. The poleward air mass transport at this higher tropospheric level is considerably reduced, however, by the circumpolar vortex, established between autumn and late winter (Schwerdtfeger, 1984; Mroz et al., 1989). Observational findings provided by long term records of ^{210}Pb and ^{222}Rn at Dumont D'Urville corroborate this picture (Lambert et al. 1990). Both radioisotopes ^{210}Pb and its mainly terrestrial precursor gas ^{222}Rn, show a weak but concurrent annual cycle with levels generally enhanced by about 50% over the annual means from November till February. A similar pattern for ^{210}Pb is observed at Mawson (Larsen and Sanderson, 1990) and Neumayer (at the latter station, unexpectedly, no regular ^{222}Rn summer maxima occur, Winkler et al., 1992).

A much more pronounced seasonal cycle of continental aerosol shows up at Neumayer in the mineral dust records with a mean summer level exceeding the winter background by more than a factor of three (see section 4.1). It is open to discussion, however, how much the stronger surface inversion over the stations and the enhanced cyclonic storm activities in the subantarctic region during the winter season may contribute to the observed annual cycles of long range transported continental species (Genthon, 1992). Comparing the typical coastal antarctic ^{210}Pb and mineral dust levels with those observed by Feely et al. (1988) for ^{210}Pb and Adams et al. (1980) for background mineral dust at the Bolivian high elevation site Chacaltaya (serving here as a reference for the upper continental troposphere) reveals an overall depletion by a factor of 15 for ^{210}Pb and more than a factor of 60 for mineral dust particles. This large contrast in the depletion factors clearly illustrates the different behaviour of submicron particles supported by a precursor gas (here, ^{222}Rn with a lifetime of 5.5 days) and large sized primary aerosol particles during long range transport to Antarctica over source free areas.

3.2 Downward mixing from the upper troposphere and the stratosphere

During the polar winter seasons when the circumpolar vortex is well established and temperature gradients across the tropopause are weak, one may expect an enhanced subsidence of stratospheric air (Schwerdtfeger, 1984). What is observed, however, at coastal and interior antarctic sites is that stratospheric aerosol markers such as fission products and the cosmic ray produced ^{7}Be arrive at the surface during the austral summer month (Feely et al., 1988; Wagenbach et al., 1988b). This may be again partly the result of the weaker surface inversion during that time of the year. But also at southern mid latitudes, elevated ^{7}Be concentrations are found during summer by Feely et al. (1988). ^{7}Be, whose equilibrium concentration steadily increases with height, may, on the other hand, be only a poor tracer for stratospheric aerosol, perhaps indicating downward mixing from the upper troposphere only. Indeed mean ^{7}Be levels at Mawson (Larsen and Sanderson, 1990) and Dumont D'Urville (Lambert et al., 1990), where the influence of katabatic downslope winds is strong, are higher by about a factor of 1.5 - 2 compared to Neumayer Station (situated on an ice shelf which is less influenced by air masses descending from the high plateau). To extract the net stratospheric aerosol input arriving at Neumayer, the ^{7}Be record was normalised, therefore, to ^{210}Pb (which subsides also from above) and, in a second step, ^{7}Be was combined with ^{10}Be measurements. ^{10}Be offers a much higher stratosphere/troposphere contrast than ^{7}Be due to its negligible radioactive decay rate. The results shown in Figure 4 suggest that the input of stratospheric aerosol at Neumayer (as indicated by relatively high ^{7}Be/^{210}Pb and ^{10}Be/^{7}Be ratios) is strongest from late summer through autumn. During that time the stratospheric ^{10}Be fraction appears to be nearly twice as high than during the other seasons (Wagenbach et al., 1988a).

All these findings are evidently in contradiction to the observations of stratospheric tritium (Teegarden, 1967) in monthly precipitation samples at Halley Bay (IAEA), in surface snow samples at South Pole (Jouzel et al., 1979) and in water vapour at Neumayer. At all those sites a winter/spring maximum in tritium was found. At Halley Bay tritium regularly peaks in spring showing only a secondary maximum in summer (Figure 4) whereas a distinct summer and a spring peak are observed in the Neumayer water vapour record.

It is not clear *a priori*, if the stratospheric aerosol subsides directly into the Antarctic troposphere (possibly by seasonal adjustment of the tropopause height (Sanak et al., 1985) or by mountain wave induced injections (Robinson et al., 1983)) or, instead, enters the troposphere via the tropopause gaps at mid latitudes. The winter/spring tritium, however, may originate exclusively from the Antarctic stratosphere, if sedimenting polar stratospheric clouds (PSC's), reach the winter Antarctic troposphere. A possible link between the occurrence of PSC's and tritium observed in the antarctic winter precipitation was already mentioned by

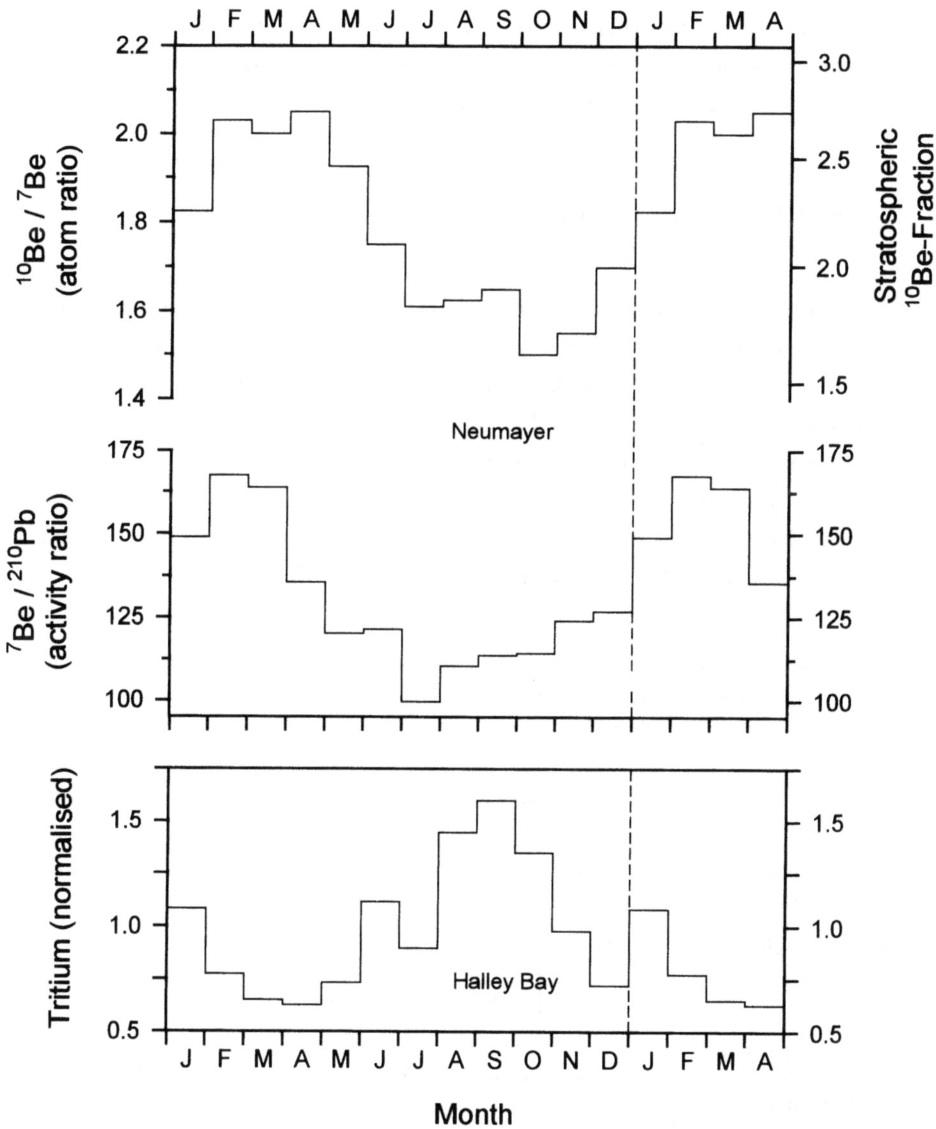

Figure 4: Seasonal pattern of arrival of stratospheric aerosol tracers at Neumayer Station, indicated by enhanced $^{10}Be/^{7}Be$ (1983 - 1985) or $^{7}Be/^{210}Pb$ (1983 - 1993) ratios. Based on a stratospheric and tropospheric aerosol residence time of 15 and one months, respectively, the apparent stratospheric ^{10}Be-input flux, normalised to the tropospheric ^{10}Be production rate, is displayed on a non-linear scale on the right hand ordinate. Further shown, are the tritium values in monthly precipitation (normalised to the annual means) as reported by IAEA at Halley during 1983 - 1993.

Jouzel et al. (1979) and Mroz et al. (1989). It is worth mentioning that, in contrast to the sub-micron stratospheric aerosol particles, no substantial air mass exchange across the tropopause would be required to detect PSC-tritium remnants at the surface of Antarctica.

4 Long term observations of chemical records

At Neumayer Station for which the longest records (1983-1993) are available, the relative aerosol composition is found to differ greatly between mid summer and mid winter. For example in July, the dry aerosol mass concentration (1.1 $\mu g/m^3$) consists of more than 90% sea salt with the remainder mainly composed of non sea salt (nss) sulphate and nitrate in a one to one ratio. The situation is dramatically changed in January when approximately half of the total mass (1.3 $\mu g/m^3$) is associated with sea salt and the remainder predominately made up by nss-sulphate and MSA. Except for the relative contribution of sea salt particles, this picture does not vary much throughout the coastal stations (see Table 1 for Mawson and Neumayer).

Table 1: Multy year observations of the ionic composition in bulk aerosol samples at Mawson (68°S / 63°E) (Savoie et al., 1993) and Neumayer (70°S / 8°W) (Wagenbach et al., 1988b and recent results)

	Sodium [ng/m^3]	nss-sulphate [ng/m^3]	MSA [ng/m^3]	nitrate [ng/m^3]	ammonium [ng/m^3]
Mawson					
Feb. 1987 - Feb. 1991	100	90	30	28	18.5
Dec. - Jan. (mean)	-	264	59	-	58
Jun. - Jul. (mean)	-	11	2	-	3
Neumayer					
Mar. 1983 - Mar. 1994	280	147	36 *	38	6.8
Dec. - Jan. (mean)	-	345	114 *	-	9.8
Jun. - Jul. (mean)	-	32	1.9 *	-	2.5
*: corrected for sea salt sulphate depletion (see text) -: no number given for Mawson					

4.1 Continental Aerosol

At Neumayer Station, which provides the only year round records of mineral dust and heavy metals at coastal sites (Wagenbach et al., 1988b), emission of the northward continents is seen mainly during the austral summer months, although in very limited quantities (mean mineral dust level in summer: 16 ng/m^3). As shown in Figure 5 the mineral dust (calculated from Manganese and Aluminium analyses) exhibits an annual cycle quite similar to that reported by Tuncel et al. (1989) from South Pole (despite a somewhat higher dust level at Neumayer). The only other atmospheric mineral dust concentration from coastal Antarctica which can be compared with the Neumayer data are short term measurements during summer by Dick and Peel (1985) and Pszenny et al. (1989), giving mean values of 9.5 and 21 ng/m^3, respectively.

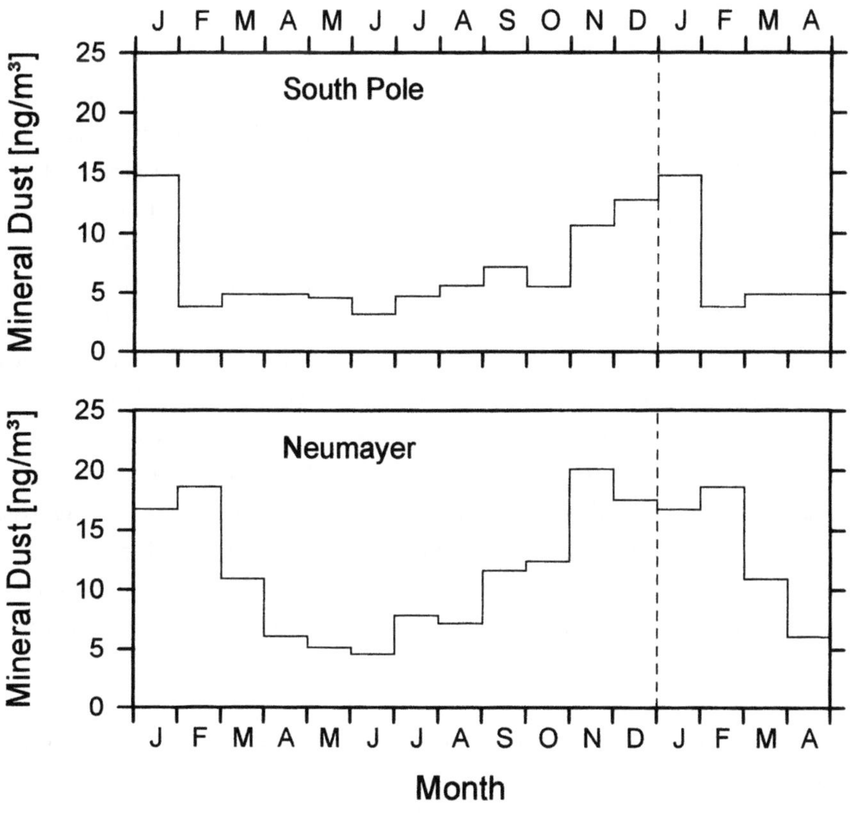

Figure 5: Mean annual cycle of atmospheric mineral dust concentrations at Neumayer (1983 - 1988) and at South Pole (1970 - 1979). South Pole values adapted from monthly mean Scandium concentrations, reported by Tuncel et al. (1989).

The only non-radioactive, particulate species for which a long term anthropogenic impact on the Antarctic atmosphere could be conclusively confirmed is lead (snow pit study at Coats Land performed by Wolff and Suttie, 1994). Atmospheric lead concentrations measured during two years at Neumayer gave a mean level of 11 pg/m^3, which is in excellent agreement with the above mentioned snow analyses, if the scavenging ration for ^{210}Pb at Neumayer is taken in account (Wagenbach et al., 1988b). Also Dick and Peel (1985) reported a mean lead concentration obtained during a short summer campaign at the Antarctic Peninsula lower than 9 pg/m^3. On the other side, erroneously high values in the range of 100-300 pg/m^3 were observed by Völkening et al. (1988) during ship-board measurements in the Weddell Sea region. These data and, although difficult to assess, as well the South Pole mean lead level of 76 pg/m^3 reported by Maenhaut and Zoller (1979) may, due to unsolved local contamination problems, not give a reliable picture of atmospheric lead in Antarctica.

4.2 Sea Salt

Near surface sea salt concentrations (deduced from Na) clearly show the largest spatial and temporal variability among all chemical aerosol components measured so far at coastal sites (see Table 1). While a mean sea salt level of 0.062 µg/m^3 is reported by Tuncel et al. (1989) for South Pole, 0.27 µg/m^3 are found by Savoie et al. (1993) at Mawson which may be typical for sites close to the ice sheet margin where downslope winds prevail. Going from the Neumayer ice shelf station (being less influenced by the katabatic wind system) to the north of the Antarctic Peninsula the sea salt levels rise from 0.7 to 3.2 µg/m^3 at the Palmer Station (Savoie et al., 1993). It is likely, however, that all coastal sea-salt data are significantly underestimated due to particle losses at the air intake, and, as in the case of Neumayer, due to selective sampling with respect to wind speed and wind direction. Interestingly, the seasonal change of the atmospheric sea salt load at Mawson, as well as at Neumayer is relatively weak and does not mirror the seasonal sea ice cycle. Minimum concentrations were seen at Neumayer consistently during mid summer and weak maxima in autumn and spring (see Figure 6). Savoie et al., (1993) observed a mean sea salt cycle at Mawson which resembles that of South Pole in showing its maximum in spring, however, no obvious seasonal changes were reported from the Palmer Station.

On the synoptic time scale, short term sea salt peaks (sometimes 100-fold above background) were observed throughout the year at Mawson, Neumayer and Dumont D'Urville (Legrand, unpublished). This poses the question about the source regions and pathways of sea salt particles during the winter-spring season when large areas north to the stations are covered with sea ice. Likely candidates during that time are a rapid long range advection of sea salt particles from large, vigorously stirred, open water areas, followed by entrainment into the downslope winds (Parish, 1988), as well as the transport from local polynias (patches of ice free water) temporarily opened by heavy storms and seaward winds. The relative contributions of both sources to the winter atmospheric sea salt load in coastal areas are currently unknown, but a clear link of the Neumayer sea salt record to the local storm activity is observed.

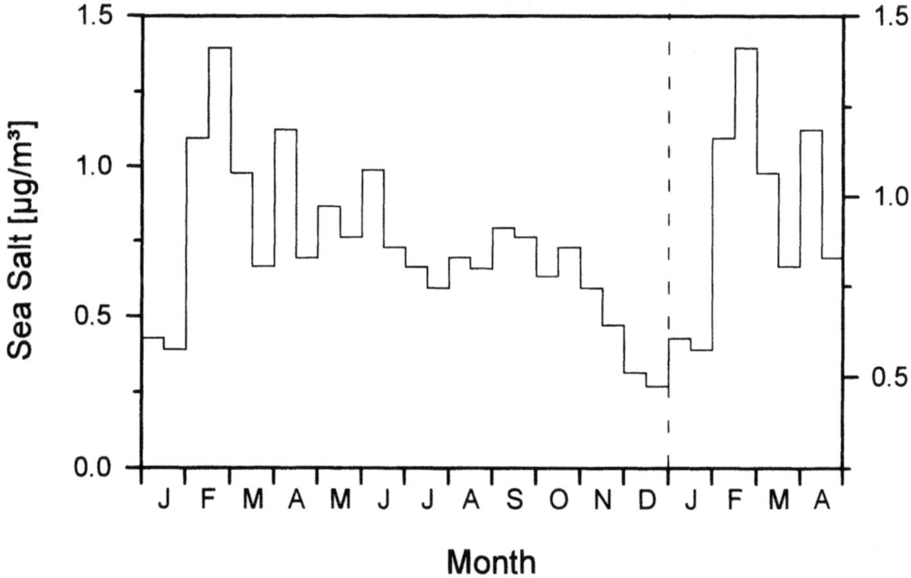

Figure 6: Mean annual cycle of atmospheric sea salt concentration at Neumayer (1983 - 1993).

4.3 Sulphur species (nss-sulphate, MSA)

The seasonal cycle in the number concentration and in the properties of the size distribution of condensation nuclei (CN) as well as the chemical morphology of individual particles have been extensively studied throughout Antarctica by several investigators (Shaw, 1988 and references therein). Considering the findings of this work, most authors agree in suggesting that the antarctic aerosol body is dominated during summer by photochemically produced particles most of which contain sulphur or sulphuric acid, probably of biogenic origin (DMS oxidation). Recently, Gras (1993) was able to confirm this qualitative picture at Mawson by comparing the CN-volume-concentration (estimated from the particle size distribution) with the atmospheric MSA and nss-sulphate records. In this key paper he convincingly demonstrated that almost all of the well-known seasonal CN changes are quantitatively explained by the respective annual cycle of nss-sulphate and MSA. Indeed, all atmospheric records of these species available from coastal sites show a remarkably regular, high amplitude seasonal cycle. There are significantly increasing MSA summer levels starting from Mawson, via Neumayer to Palmer Station (Savoie et al., 1993). The MSA and nss-sulphate seasonal records (shown in Figure 7 for the Neumayer Station) are perfectly in phase which holds true as well for the comparison with the annual DMS cycle reported by Gillett et al. (1993) from the sub-antarctic Macquairie Island (no year round DMS records at Antarctic sites are currently available).

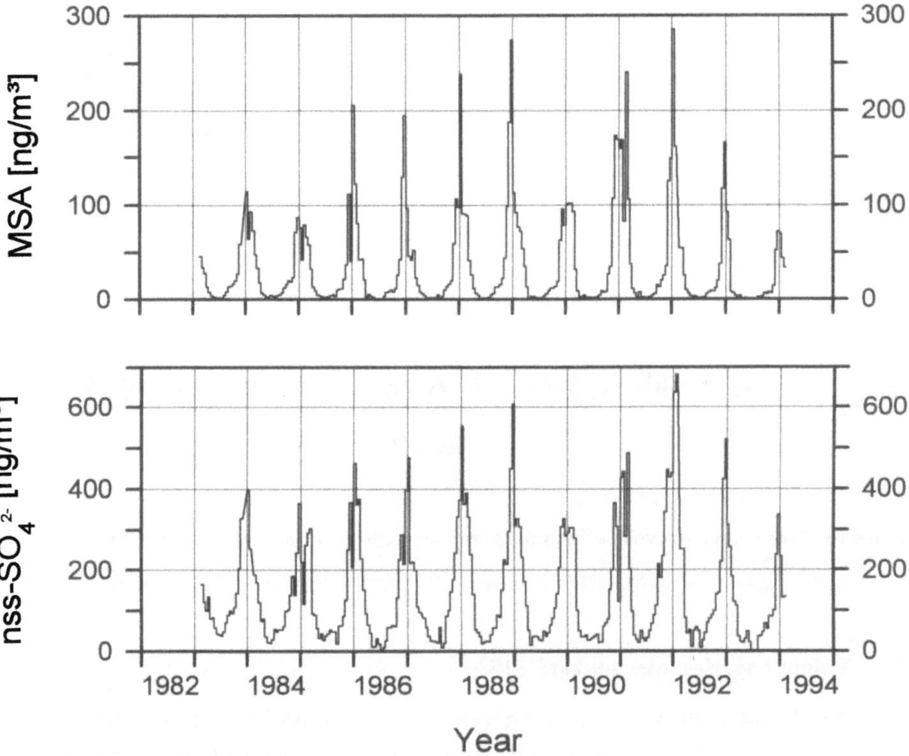

Figure 7: MSA and nss-sulphate records observed at Neumayer between 1983 - 1993. Individual data points represent the typically two week sampling period.

Since the only significant MSA source is the photochemical oxidation of biogenic DMS (Ayers et al., 1991), almost all antarctic nss-sulphate may also be attributed to this process. Further evidence that DMS emissions from the marine biota are responsible for the majority of coastal antarctic nss-sulphate arises from prelimarily $\delta^{34}S$ analyses of Neumayer sulphate and MSA samples (MSA-analyses by F. Pichlmayer, ÖFZ-Seibersdorf) showing in summer only a slight depletion of about 5‰ with respect to sea water sulphate. Virtually the same isotopic sulphur property was observed by Calhoun et al. (1991) in nss-sulphate from ship-board sampling during February-March in the South-Pacific.

The temporal pattern of the seasonal MSA cycle exhibits a spectacular regularity from year to year (in particular during the polar night period). At Neumayer for example 10 out of 11 annual cycles show absolute minima and maxima of the bi-weekly mean concentrations centered around the 18th of July (± 5 days) and around the 20th of January (± 8 days), respectively. The only relevant cycles operating with a similar year to year precision in this region are the solar insolation at the surface and (lagging approximately three months) the large scale sea ice coverage (Zwally, 1984). This suggests that the seasonality of the biogenic sulphur particles observed in coastal Antarctica is predominantly driven by the well-fixed interplay between these two extremal cycles. The outstanding feature of atmospheric MSA in coastal Antarctica is the remarkably large seasonal contrast. The mean MSA level during January-February exceeds, for example at Neumayer, the means in June-July by a factor of 75. As shown in Figure 8 similar annual MSA cycles are observed at 41 S in marine baseline air of Cape Grim (CSIRO) albeit with a lower amplitude arising from the less prominent summer peaks there.

The situation in coastal Antarctica is less clear with respect to nss-sulphate, however, and consequently with its ratio to MSA. This ratio is widely debated in view of its climatological and air chemical implications and is going to be increasingly used in antarctic ice core studies (Mulvaney et al., 1992; Legrand et al., 1992). Due to high sea salt loads at coastal sites, the corrections of measured total sulphate concentrations for the contribution of sea salt sulphate are substantial and frequently lead to negative nss-sulphate values if applied to aerosol or precipitation data of the winter half year. This effect is too large that it could be explained by the (systematical) analytical uncertainties involved. Although commonly not considered, it is observed in all year round aerosol records as well as in various ice cores at coastal sites (as e.g. by Aristarain et al., 1982 and Mulvaney and Wolff, 1993). This apparent depletion of the sulphate to sodium (or chloride) ratio with respect to the corresponding sea water ratio was discussed by Gjessing (1989). He proposed that a sulphate/sodium fractionation may occur during freezing of large sea water droplets or during rime formation. No fractionation in sea water spray could be simulated, however, in the laboratory by Keck (1994).

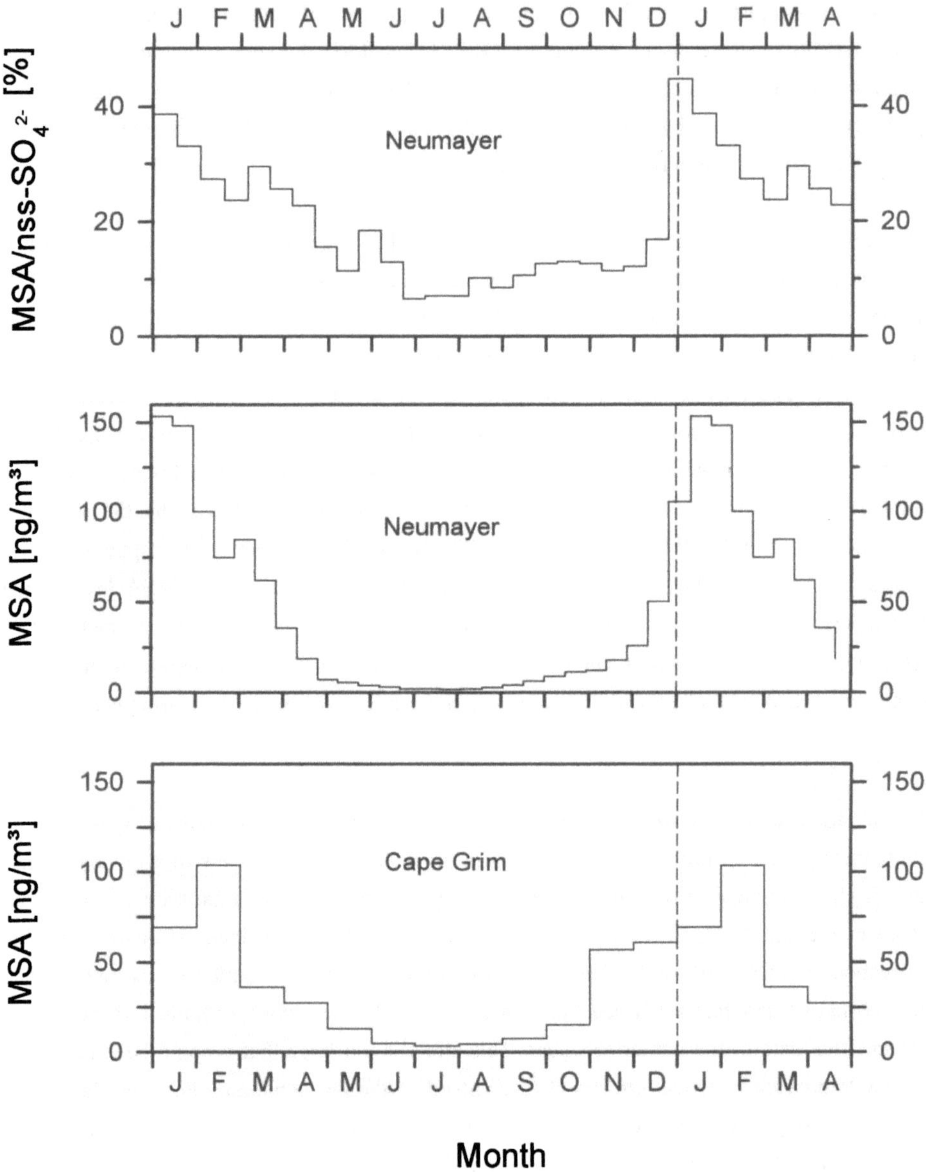

Figure 8: Mean annual cycle of MSA and the MSA to nss-sulphate ratio at Neumayer (1983 - 1993), above and mean annual MSA cycle observed in bulk aerosol samples at Cape Grim. The Cape Grim monthly means are based on data reports from CISRO Baseline Series (1989-1991).

The Neumayer nss-sulphate data were empirically corrected, therefore, for the apparent sulphate fraction by requiring that the correlation coefficient between the winter nss-sulphate and the respective sodium values approaches a minimum (commonly zero) (Minikin et al., 1994). This procedure suggests, at least during winter, a dramatic depletion of the sulphate to sodium ratio in air-borne sea salt particles by a factor of 4.3, approximately. If this method is applied to fresh snow at Neumayer and to high resolution firn cores from various sites on the Ekstrøm and Ronne/Filchner ice shelves similar results are obtained by Minikin et al. (1994), Wagenbach et al. (1994) and Trefzer (1992). Therefore, „uncorrected" nss-sulphate concentrations from coastal Antarctica appear to be significantly underestimated depending on season, and the marine influence at the site investigated. Among others, this would introduce important complications in the use of MSA to nss-sulphate ratios, especially with respect to its seasonal and spatial variability but also in view of its particle size dependence. For example, ship-board studies of atmospheric sulphur species performed by Pszenny et al. (1989) and by Berresheim et al. (1987) during March/April 1986 in the northern Weddell Sea showed anomalously low nss-sulphate but typical MSA concentrations for this season. Possibly as a consequence, Berresheim et al. (1987) could not balance the estimated local DMS evasion flux by the total deposition flux of nss-sulphur species.

Nevertheless, the MSA to nss-sulphate ratios observed during the mid summer season at Mawson and Neumayer are not seriously affected by the ambiguity in the sea salt sulphate corrections. Indeed, they are virtually the same during that time although no correction was made on the Mawson data with respect to a sea salt sulphate depletion (Neumayer: arithmetric mean: 33%, Mawson: 36% obtained by multiregression analyses and assumed to be seasonally invariant (Savoie et al., 1993)). These summer ratios are, however, by about one third higher than at remote marine sites of the mid southern latitudes, as for instance observed in the sub-micron aerosol fraction by Ayers et al. (1991) at Cape Grim.

The possible reasons for a systematical latitudinal change of the ratio (such as a temperature dependence of the MSA-sulphur dioxide branching ratio or a different removal efficiency for MSA and nss-sulphate particles) have been discussed in detail by Saltzmann (1994) and Bates et al. (1992).

During mid winter months, when the MSA level at coastal antarctic sites is as low as 2 ng/m^3, the DMS associated nss-sulphate fraction becomes an unmeasurably low quantity (also if no sea salt sulphate fractionation would occur). At that time the MSA to nss-sulphate ratio at Neumayer is consistently low and in the range of 7-13 % which is obviously due to an unknown surplus of nss-sulphate not related to marine DMS. Fortuitously, quite similar low ratios are observed at southern mid latitudes in winter by Ayers et al. (1991) and at south tropical sites throughout the year by Saltzman et al. (1984). Given the uncertainties involved, a typical range of nss-sulphate background concentration (possibly not associated with marine DMS) of 15-30 ng/m^3 can be deduced from the current record at Neumayer. Savoie et al. (1993) assumed a continental origin of the non DMS sulphate fraction and arrived at 21 ng/m^3 from multi regression analyses with MSA and nitrate (the latter was hypothesised to be continental as well). The above pictures are in reasonable agreement with the mean winter level of total sulphur (10 ng/m^3) and sodium (31 ng/m^3) at South Pole reported by Tuncel et al. (1989). However, there is currently no straightforward explanation for the relatively low MSA to nss-sulphate ratios around 10-15% which persist consistently into early summer and frequently until mid of December at Neumayer (see Figure 8) and Dumont D'Urville (M. Legrand unpublished data). At this time, the MSA and nss-sulphate concentrations are already increased over their winter background level by, at least, one order of magnitude making a significant reduction of their ratio by non DMS sulphate very unlikely.

4.4 Nitrogen components (nitrate, ammonium)

There is currently no unequivocal source apportionment available for background nitrate (aerosol nitrate and nitric acid) observed in the remote marine troposphere of the southern hemisphere (Levy et al., 1980; Prospero and Savoie, 1989). This is also true for antarctic nitrate which controlling source (mid-latitudes and tropical troposphere versus stratosphere) has been the subject of controversial debate for many years (Levy et al., 1980; Laird et al., 1988; Legrand and Delmas, 1986; Laird et al., 1988; Legrand and Delmas, 1988; Wolff, 1995). Also nitrite photolysis in highly productive surface sea water was considered by Zafiriou and True (1979) as a potential source of background nitrate in the marine troposphere. Mean nitrate concentrations in shallow ice cores from the Ronn/Filchner Ice Shelves are found by Minikin et al. (1994) to steadily increase with distance to the ice edge by roughly a factor two over 600 km, however. This makes a significant contribution of marine derived nitrate very unlikely.

More recently, several authors brought up the intriguing suggestion that the occurrence of elevated antarctic near surface nitrate may be connected (nowadays) to the spring denitrification of the antarctic stratosphere (Mayewski and Legrand, 1990; Legrand and Kirchner, 1990; Mulvaney and Wolff, 1993; Laird et al., 1988; Pfeilsticker and Wagenbach, 1990). This would imply, however, that the fraction of polar stratospheric cloud particles, which is associated with nitric acid trihydrate would grow large enough to reach the antarctic troposphere by sedimentation (Solomon, 1988; Mc Elroy, 1989).

At coastal sites (presently providing the only atmospheric nitrate records in Antarctica) the annual nitrate cycle stands out in showing its annual maximum during October-November. This is seen at Mawson (Savoie et al., 1992), Neumayer (see Figure 9) and Dumont D'Urville (M. Legrand unpublished data) without any large intersite differences. As stated by Savoie et al. (1992) and Wagenbach et al. (1988b), due to the relatively high atmospheric sea salt loads, these data are presumed to represent total atmospheric nitrate. Equal inspections of the seasonal change of nitrate in high resolution ice core records of coastal regions, commonly reveal an annual nitrate maximum in summer in close coincidence with nss-sulphate (Minikin et al. 1994; Mulvaney and Wolff, 1993; Wagenbach et al., 1994; Zeller et al., 1986 and Herron and Langway, 1979). Only in rare cases a secondary maximum is observed in winter snow layers (Moser et al., 1990). Part of this mismatch with the atmospheric observations may be attributed to stratigraphical noise but even in year round sampled fresh snow, no straight forward coherence with the seasonality of atmospheric nitrate could be observed by Trefzer (1992) at Neumayer (see Figure 9) and Wolff (unpublished data) at Halley.

Savoie et al. (1993) and Savoie et al. (1992) hypothesised that their nitrate signal above background is mainly linked to continental aerosol due to a similar temporal pattern of the mean seasonal nitrate and ^{210}Pb cycles at Mawson. This suggestion is not confirmed, however, by the records of mineral dust (see Figure 6) and ^{210}Pb at Neumayer and by the ^{210}Pb seasonallity at Dumont D'Urville (Lambert et al., 1990) all showing a significant phase lag to the nitrate peak region.

Some evidence for a stratospheric origin of the early summer nitrate peak arises from the annual tritium cycle (shown in Figure 4) which is the only one also providing a clear annual maximum in spring. Further hints for a non tropospheric source for antarctic nitrate emerged from the $^{15}N/^{14}N$ isotopic ratio which was found to be extremely depleted in Neumayer nitrate samples during the nitrate enhancement period every year (Pichlmayer and Wagenbach, 1995). As shown in Figure 9, during the high nitrate period, the $\delta^{15}N$ values are clearly outside the range ever observed in tropospheric nitrate or ammonium but they remain consistently within the typical tropospheric range during background conditions. This isotopic irregularity of the antarctic aerosol nitrate, however, does not automatically suggest an extra tropospheric source since Freyer et al. (submitted) report an inverse isotopic effect from central antarctic ice core nitrate which they found to be extremely enriched in ^{15}N. Such a strong ^{15}N enrichment is most likely caused by fractionating processes during the remobilisation of nitric acid from the firn which in turn would lead to a significant flux of ^{15}N depleted nitrate to the antarctic near surface troposphere. Indeed, observational evidence for a significant net loss of nitrate from the summer snow surface near Neumayer are reported by Neubauer and Heumann (1988). This is in agreement with extraordinary high nitrate concentrations in year round sampled fresh snow found by Wolff (personal communications) at Halley and by Trefzer (1992) at Neumayer (see Figure 9).

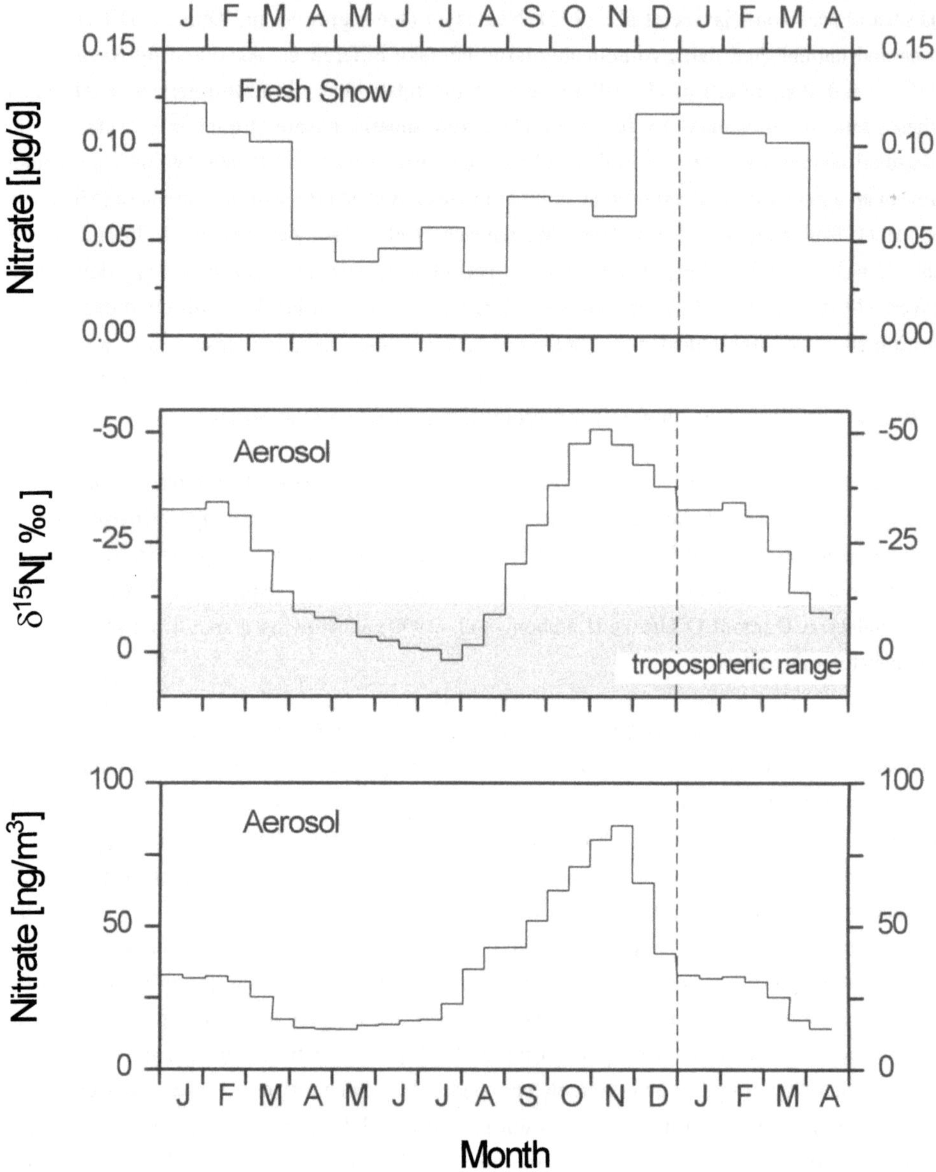

Figure 9: Mean annual cycles of nitrate (1983 - 1993) and δ^{15}N-Nitrate (1986 - 1992) in bulk aerosol samples at Neumayer. Also shown are the monthly mean nitrate concentrations, observed at Neumayer in fresh snow during 1983 and 1990.

Atmospheric ammonia is commonly derived from biogenic sources of the continents. However, in very remote marine regions where the influence of continental air masses is negligible, also the ocean is believed to be a net ammonia source to the marine troposphere (Quinn et al., 1987). At coastal antarctic sites, year round ammonium records are available from Mawson and Palmer (Savoie et al., 1993), Neumayer (see Figure 10) and Dumont D'Urville (Legrand unpublished data). Note that, especially during the summer months when the mineral acid load of the exposed aerosol filter is relatively high a significant fraction of gaseous ammonia may have contributed to the observed ammonium concentrations.

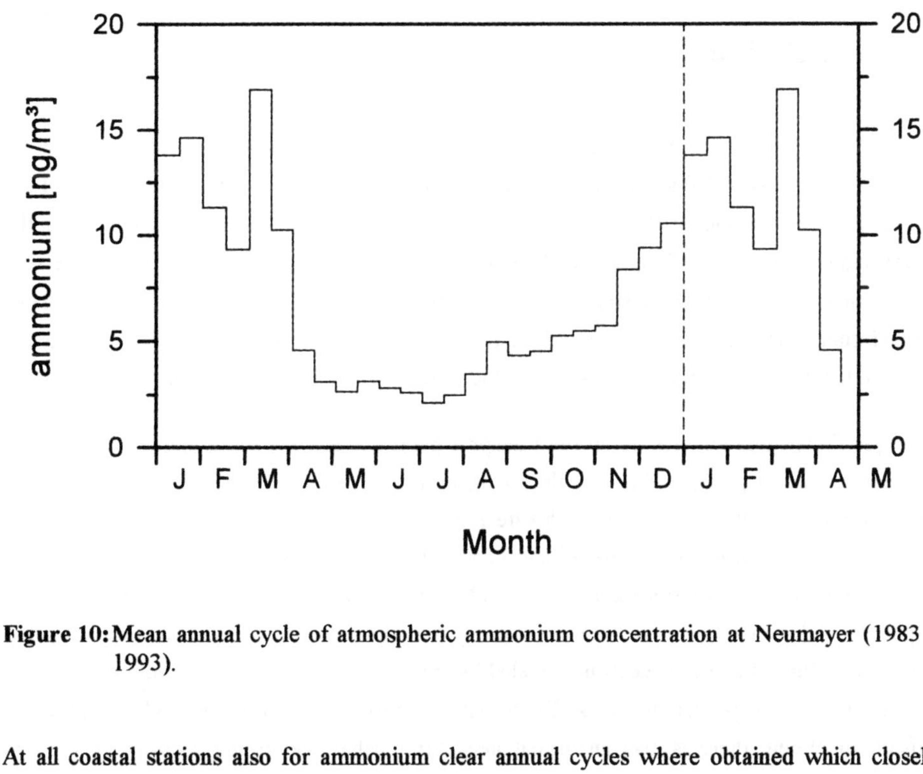

Figure 10: Mean annual cycle of atmospheric ammonium concentration at Neumayer (1983 - 1993).

At all coastal stations also for ammonium clear annual cycles where obtained which closely follow the temporal pattern of the biogenic sulphur species (Figure 10). Savoie et al. (1993) argue, therefore, that a marine biogenic source is mainly contributing to the ammonium concentration observed at Mawson and Palmer. There is, however, a large intersite variability of the summer means (ranging from 12 ng/m^3 at Neumayer via ~58 ng/m^3 at Mawson to about 150 ng/m^3 at Dumont D'Urville) which is not easily explained in view of the otherwise moderate differences of biogenic sulphur concentrations. As discussed by Silvente and Legrand (1993) to some depth, avoiding post sampling artefacts is extremely critical in obtaining reliable atmospheric ammonium measurements in polar regions. However, no significant ammonia contamination of artificially loaded blank filters have been observed by Savoie et al. (1993). On the other hand, Legrand (personal communication) attributed the extremely high ammonium values of about 150 ng/m^3 he found during November-December, to local contaminations by the large penguin population around Dumont D'Urville. Although the representative atmospheric ammonium level remains still somewhat unclear, the annual ammonium cycles are consistently observed to follow roughly the temporal MSA pattern. This makes the marine biota of the polar ocean (or in rotting sea ice) likely to be the dominating source for ammonium found in antarctic aerosol and snow samples.

5 Conclusions

Observational programs launched in coastal Antarctica during the last decade have considerably improved our knowledge on the chemical composition and on the seasonal variability of the near surface aerosol body in that region. Important progress was achieved in demonstrating the overwhelming influence of biogenic DMS emissions from the South Polar Oceans on the seasonally changing aerosol chemistry. Also the precise timing of the annual cycles of the major aerosol components appears now to be relatively well established which contributes significantly to the interpretation of glacio-chemical ice core records from these areas.

New questions came up such as on the transport pathways and the representative ionic composition of sea salt particles during polar night, on the driving forces responsible for the specific seasonality of the MSA to nss-sulphate ratio as well as on the source and the mean atmospheric level of ammonium aerosol and background nss-sulphate. Also in the coastal antarctic troposphere total nitrate continues to offer the most enigmatic feature, not only with respect to its dominating source but now also regarding its preservation in snow. Perhaps some of these questions may be successfully tackled by extending the current year round observation also to the relevant precursor gases (DMS, sulphur dioxide and nitrogen oxides). Although difficult to obtain, there is also an urgent need for reliable year round records on the gas phase/aerosol partition of nitrate, chloride and reduced nitrogen species as well as on the size distribution of the major chemical aerosol components.

Coastal antarctic ice core studies will, in the end, greatly profit from such an improved knowledge of the recent atmospheric situation in narrowing their actual interpretation deficits. Conversely, only such sophisticated ice core studies enable us to recover useful air chemical information from many other interesting sites in the vast regions of coastal Antarctica but, most important, also from the recent and earliest pre-observational eras.

6 Acknowledgements

I wish to thank the Alfred Wegener Institute for Polar and Marine Research for logistical and financial support with respect to the air chemical observation at Neumayer station. Special thanks are going to the Neumayer overwintering crews for invaluable commitment, to Dr. F. Pichlmayer for performing the $\delta^{15}N$ and $\delta^{34}S$ analyses as well as to P. Saey for the figures and drawings and the manuscript editing.

7 References

Adams F, Van Graen M, Van Espen P and Andreuzzi D (1980) The elemental composition of atmospheric aerosol particles at Chacaltaya, Bolivia, *Atmos Env 14:* 879-893

Aristarain AS, Delmas RJ and Briat M (1982) Snow chemistry on James Ross Island (Antarctic Peninsula), *J Geophys Res 87:* 11004-11012

Ayers GP, Ivey JP and Gillett RW (1991) Coherence between seasonal cycles of dimethyl sulphide, methanesulphonate and sulphate in marine air, *Nature 239:* 404-406

Bates TS, Calhoun JA and Quinn PK (1992) Variations in the methansulfonate to sulfate molar ratio in submicrometer marin aerosol particles over the South Pacific Ocean, *J Geophys Res:* 9859-9865

Berresheim H (1987) Biogenic sulfur emissions from Antarctic waters, *J Geophys Res 92:* 13245-13262

Bigg EK, Gras JL and Evans C (1984) Origin of Aitken particles in remote regions of the southern hemisphere, *J Atmos Chem 1:* 203-214

Calhoun JA, Bates TS and Charlson RJ (1991) Sulfur isotope measurements of submicrometer sulfate aerosol particles over the Pacific Ocean, *Geophys Res Lett 18:* 1877-1880

Clausen HB and Langway CC Jr (1989) The ionic deposits in polar ice cores, *Dahlem Konferenzen: The Environmental Record in Glaciers and Ice Sheets*, edited by Oeschger H and Langway CC Jr, pp 225-247, John Wiley, New York

CSIRO (1989-1991) Baseline Series, *Ann. Rep. Baseline Atmospheric Program (Australia)* Burea of Meteorology, Melbourne 3000, Australia

Dick AL and Peel DA (1985) Trace elements in Antarctic air and snowfall. *Ann. Glaciol. 7:* 12-19

Etheridge DM, Pearman GI and Fraser PJ (1992) Changes in tropospheric methane between 1841 and 1978 from a high accumulation-rate Antarctic ice core, *Tellus 42B:* 282-294

Feely HW, Larsen RJ and Sanderson CG (1988) Annual report of the surface air sampling program, *EML 497*

Freyer HD, Kobel K, Delmas RJ, Kley D and Legrand MR (1995) First results of $^{15}N/^{14}N$ ratios in nitrate from alpine and polar ice cores (submitted)

Genthon C (1992) Simulations of desert dust and sea-salt aerosols in Antactica with a general circulation model of the atmosphere, *Tellus 44B:* 371-389

Gillett RW, Ayers GP, Ivey JP and Gras JL (1993) Measurement of dimethylsulfide, sulfurdioxide, methanesulfonic acid and non sea salt sulfate at the Cape Grim baseline station in *Restelli G and Angeletti G (eds) DMS: ocean atmosphere and climate*: 117-128

Giovinetto MB and Waters NM (1990) Dependence of Antarctic surface mass balance on temperature, elevation, and distance to open ocean, *J Geophys Res 95:* 3517-3531

Gjessing Y (1989) Excess and deficit of sulfate in polar snow, *Atmos Env 23:* 155-160

Graf W, Moser H, Reinwarth O, Kipfstuhl J, Oerter H, Minikin A and Wagenbach D (1993) Snow accumulation rates and isotopic contents (^{2}H, ^{3}H, ^{18}O) of near surface firn of the Filchner-Ronne Ice Shelf, Antarctica. *Annals of Glaciology*, 1994: 121-128

Gras JL (1993) Condensation nucleus size distribution at Mawson, Antarctica: microphysics and chemistry, *Atmos Env 27A:* 1427-1434

Herron MM and Langway CC Jr (1979) Dating of Ross Ice Shelf cores by chemical analysis, *J Glaciology 24:* 345-356

International Atome Energy Agency (IAEA), Environmental isotope data No 8 (1986) and 9 (1990) Technical Reports Series 264 and 311 and personal communication by K. Rozanski

Jouzel J, Merlivat L, Mazaudier D, Pourchet M and Lorius C (1982) Natural tritium deposition over Antarctica and estimation of the mean global production rate, *Geophys Res Lett Vol 9:* 1191-1194

Jouzel J, Merlivat L, Pourchet M and Lorius C (1979) A continuous record of artificial tritium fallout at the South Pole, *Earth Planet Sci Lett 45:* 188-200

Keck L (1994) Experimentelle Simulation einer möglichen Sulfatabreicherung in polaren Seesalzaerosolen, MS-Thesis, Institut für Umweltphysik, Univerversity Heidelberg

Laird CM, Zeller EJ, Armstrong TP and Parker BC (1982) Solar activity and nitrate deposition in South Pole snow, *Geophys Res Lett 9:* 1195-1198

Laird CM, Zeller EJ and Dreschhoff GAM (1988) Comment on "Relative contributions of tropospheric and stratospheric sources to nitrate in Antarctic snow" by M. R. Legrand and R. J. Delmas, *Tellus 40B:* 233-236

Lambert G, Ardouin B and Sanak J (1990) Atmospheric transport of trace elements toward Antarctica, *Tellus 42B:* 76-83

Larsen RJ and Sanderson CG (1990) Annual report of the surface air sampling program, *EML 524:* 1-34

Legrand MR, Fenient-Saigne C, Saltzman ES and Germain C (1992) Spatial and temporal variations of methanesulfonic acid and non sea salt sulfate in Antarctic ice, *J Atmos Chem 14:* 245-260

Legrand MR and Delmas RJ (1986) Relative contributions of tropospheric and stratospheric sources to nitrate in Antarctic snow, *Tellus 38B:* 236-249

Legrand MR and Delmas RJ (1988) Reply to C. M. Laird, E.J. Zeller and G. A. M. Dreschhoff, *Tellus 40B:* 237-240

Legrand MR and Kirchner S (1990) Origins and variations of nitrate in South Polar precipitation, *J Geophys Res 95:* 3493-3507

Levin I and Hesshaimer V (1995) Refining of atmospheric transport model entries by the globally observed passiv tracer distribution of 85Krypton sulphur hexafluoride (SF6). To be submitted to J. Geophys. Res.

Levy II H, Mahlman JD and Moxim WJ (1980) A stratospheric source of reactive nitrogen in the unpolluted troposphere, *Geophys Res Lett 7:* 441-444

Maenhaut W and Zoller WH (1979) Determination of the chemical composition of the South Pole aerosol by instrumental neutron activation analysis, *J Radioanal Chem 37:* 637-650

Mayewski PA and Legrand MR (1990) Recent increase in nitrate concentration of Antarctic snow, *Nature 346:* 258-260

McElroy MB (1989 Studies of Polar Ice: Insights for Atmospheric Chemistry, *Dahlem Konferenzen: The Environmental Record in Glaciers and Ice Sheets, John Wiley, New York*: 363-377

Minikin A, Wagenbach D, Graf W and Kipfstuhl J (1994) Spatial and seasonal of the snow chemistry at the central Filchner -Ronne Ice Shelf, Antarctica, *Ann Glaciology*

Moser K, Wagenbach D and Münnich KO (1990) Nitrate in coastal Antarctic shallow firn cores: comparison of seasonal pattern and total fluxes, *Annal Glaciol 14:* 349

Mosley-Thompson E, Dai J, Thompson LG, Grootes PM, Arbogast JK and Paskievitch JF (1991) Glaciological studies at Siple Station (Antarctica): potential ice-core paleoclimatic record, *J Glaciol 37:* 11-22

Mroz EJ, Alei M, Cappis JH, Guthals PR, Mason AS and Rokop DJ (1989) Antarctic atmospheric tracer experiments, *J Geopys Res 94:* 8577-8583

Mulvaney R, Pasteur EC, Peel DA, Saltzman ES and Whung P-Y (1992) The ratio of MSA to non-sea-salt sulphate in Antarctic Peninsula ice cores, *Tellus 44B:* 295-303

Mulvaney R and Wolff EW (1993) Evidence for winter/spring denitrification of the stratosphere in the nitrate record of Antarctic firn cores, *J Geophys Res 98:* 5213-5220

Mulvaney R and Wolff EW (1994) Spatial variability of the major chemistry of the Antarctic ice sheet, *Annals of Glaciology*, 1994: 440-447

Neubauer J and Heumann KG (1988) Nitrate trace determination in snow and firn core samples of ice shelves at the Weddell Sea, Antarctica, *Atmos Env 22:* 537-545

Oeschger H and Langway Jr CC (eds) (1989) Dahlem Konferenzen: The Environmental Record in Glaciers and Ice Sheets, John Wiley, New York

Parish TR (1988) Surface winds over the Antarctic continent: A review, *Rev Geophys 26:* 169-180

Peel DA and Mulvaney R (1992) Time-trends in the pattern of ocean-atmosphere exchange in an ice core from the Weddell Sea sector of Antarctica, *Tellus 44B:* 430-442

Petit JR, White JWC, Young NW, Jouzel J and Korotkevich YS (1991) Deuterium excess in recent Antarctic snow, *J Geophys Res 96:* 5113-5122

Pfeilsticker K and Wagenbach D (1990) The seasonal cycle of nitrate in the Antartic troposphere, paper presented at Symposium on the Chemistry of the Global Atmosphere CACGP

Phillpot HR and Zillman JW (1970) The surface temperature over the Antarctic continent, *J Geophys Res 75:* 4161-4169

Pichlmayer F. and Wagenbach D. (1995) An EA / MS method for source characterization of airborn nitrates, utilizing the stable isotope ratio of nitrogen, presented at 46th Pittsburg Conference, New Orleans, LA, March 5 - 10, 1995, *Report OEFS-A-3369*, pp 14, Austrian Research Center, Seibersdorf, Austria

Phillpot HR and Zillman JW (1970) The surface temperature over the Antarctic continent, *J Geophys Res 75:* 4161-4169

Polian G, Lambert G, Ardouin B and Jegou A (1986) Long-range transport of continental radon in Subantarctic and Antarctic areas, *Tellus 38B:* 178-189

Prospero JM, Savoie DL, Saltzman ES and Larsen R (1991) Impact of oceanic sources of biogenic sulphur on sulphate aerosol concentrations at Mawson, Antarctica, *Nature 350:* 221-223

Prospero JM and Savoie DL (1989) Effect of continental sources on nitrate concentrations over the Pacific Ocean, *Nature 339:* 687-689

Pszenny AA, Castelle AJ, Galluway JN and Duce RA (1989) A study of the sulfur cycle in the Antartic marine boundary layer, *J Geophys Res 94:* 9818-9380

Quinn PK, Charlson RJ and Zoller WH (1987) Ammonia, the dominant base in the remote marine troposphere: a review, *Tellus 39B:* 413-425

Rehfeld S and Heimann (1994) Three-dimensional atmospheric transport simulation of the radioactive tracers ^{210}Pb, ^{7}Be, ^{10}Be and ^{90}Sr, Max-Planck-Institut für Meteorologie, Hamburg (FRG) Report No144

Robinson E, Clark D, Cronn DR, Bamesberger WL and Hogan AW (1983) Stratospheric-tropospheric ozone exchange in Antarctica caused by mountain waves, *J Geophys Res 88:* 10,708-10,720

Saltzman ES, Savoie DC, Zika RG and Prospero JM (1983) Methansulfonic acid in the marine atmosphere, *J Geophys Res 88:* 10897-10902

Saltzman ES, Savoie DC, Prospero JM and Zika RG (1984) Methansulfonic acid and non-seasalt sulfate in pacific air: regional and seasonal variations, *J Atmos Chem 4:* 227-240

Saltzman ES (1994) Ocean/atmosphere cycling of dimethylsulfide: sensitivity to climate change, *NATO ARW* on GBCs in polar ice

Sanak J, Lambert G and Ardouin B (1985) Measurement of stratosphere-to-troposphere exchange in Antarctica by using short-lived cosmonuclides, *Tellus 37B:* 109-115

Savoie DL, Prospero JM, Larsen RJ and Saltzman ES (1992) Nitrogen and sulfur species in aerosols at Mawson, Antarctica, and their relationship to natural radionuclides, *J Atm Chem 14:* 181-204

Schwerdtfeger W (1984) *Weather and Climate of the Antarctic,* Elsevier, New York, 261 pp

Shaw GE (1988) Antarctic aerosol: A review. *Rev Geophys 26:* 89-112

Silvente E and Legrand M (1993) Ammonium to sulphate ratio in aerosol and snow of Greenland and Antarctic regions, *Geophys Res Lett:* 687-690

Solomon S (1988) The mystery of the Antarctic Ozone "Hole", *Rev Geophys 26:* 131-148

Teegarden BJ (1967) Cosmic-ray production of deuterium and tritium in the Earth's atmosphere, *J Geophys Res 72:* 4863-4867

Trefzer U (1992) Vergleich der ionischen Zusammensetzung von Aerosol- und Neuschneeproben aus dem Bereich der Georg-von-Neumayer Station, Antarktis. MS-Thesis, Institut für Umweltphysik, Univerversity Heidelberg

Tuncel G, Aras NK, Zoller WH (1989) Temporal variations and sources of elements in the South Pole atmosphere 1. Nonenriched and moderately enriched elements, *J Geophys Res 94:* 13025-13038

Völkening J, Baumann H and Heumann KG (1988) Atmospheric distribution of particulate lead over the Atlantic Ocean from Europe to Antarctica, *Atmos Environ 22:* 1169-1174

Wagenbach D, Münnich KO, Beer J and Wölfli W (1988a) Time pattern of natural radionuclides in Antarctic aerosol and snow (Abstract), *Chemical Geology 70:* 105

Wagenbach D, Görlach U, Moser K and Münnich KO (1988b) Coastal Antarctic aerosol: the seasonal pattern of its chemical composition and radionuclide content, *Tellus 40B:* 426-436

Wagenbach D, Graf W, Minikin A, Trefzer U, Kipfstuhl J, Oerter H and Blindow N (1994) Reconnaissance of chemical and isotopic firn properties on top of Berkner Island, Antarctica, *Annals of Glaciology 20: 307-312*

Winkler P, Brylka S and Wagenbach D (1992) Regular fluctuations of surface ozone at Georg-von-Neumayer station, Antarctica, *Tellus 44B*: 33-40

Wolff EW (1995) Nitrate in polar ice, *NATO ASI Series, Vol I 30, Ice Core Studies of Global Biogeochemical Cycles,RJ Delmas (ed.)*: 91-119

Wolff EW and Suttie ED (1994) Antarctic snow record of southern hemisphere lead pollution, *Geophys Res Lett 21:* 781-784

Zafiriou OC and True MB (1979) Nitrite Photolysis as a source of free radicals in productive surface waters, *Geophys Res Lett 6:* 81-84

Zeller EJ, Dreschhoff GAM and Laird CM (1986) Nitrate flux on the Ross Ice Shelf, Antarctica and its relation to solar cosmic rays, *Geophys Res Lett 13:* 1264-1267

Zwally HJ (1984) Observing polar-ice variability, *Ann of Glaciology 5:* 191-198

CHEMICAL REACTIONS IN THE POLAR TROPOSPHERE RELEVANT TO C, S, AND N COMPOUNDS

Jan W. Bottenheim and Leonard A. Barrie
Atmospheric Environment Service
4905 Dufferin Street
Downsview, Ontario, Canada
M3H 5T4

INTRODUCTION

The chemistry of the atmosphere is driven by solar radiation which can give rise to photodissociation. The amount of radiation reaching the troposphere is attenuated by scattering and absorption processes in the upper layers of the atmosphere. In polar regions, the path to the troposphere is longer than at other parts of the globe, and hence less radiation is available for photochemical processes. In fact, during the winter months the lower parts of the polar troposphere are entirely excluded from receiving direct solar radiation. In addition, temperature regimes are different. In the winter they drop to levels of -25 to -50^{0}C in the Arctic and lower in the Antarctic. Consequently the absolute humidity is very low. Finally, due to the low level of incoming solar radiation and loss of energy due to infrared cooling at the surface the lower troposphere is stratified and vertical mixing is hindered. These natural differences between the polar troposphere and other regions of the globe cause changes in chemical reactions that take place.

The absence of solar radiation means that in the winter photodissociation does not supply the spark for chemical reactions. At the same time the lower temperatures will have an effect on the rate of collisionally-driven reactions. Those requiring a certain amount of activation energy will slow down or even become negligible. Combined with the decrease in vertical mixing these considerations will lead to a different atmospheric composition from other regions: the abundance of *photochemically reactive compounds* will be high and the level of *photochemical products* will be low. Concentrations of compounds that normally would react with photochemically produced species will rise as they are transported further from Southern source regions to the polar troposphere.

When sunlight returns after the long polar night the atmosphere contains elevated concentrations of reactant and lower levels of product species. Thus from a purely photochemical reaction perspective, spring will be the season of greatest chemical change.

NATO ASI Series, Vol. I 43
Chemical Exchange Between the Atmosphere and Polar Snow
Edited by Eric W. Wolff and Roger C. Bales

However, since temperatures, absolute humidity and rate of vertical mixing are still generally low, the mechanisms of atmospheric chemical transformations are often different than at lower latitudes.

Summer and fall are quite different in nature than winter and spring. Solar radiation is present 24 hours per day in the summer. Temperatures are higher and since the polar regions are surrounded by oceans the presence of stratus clouds is widespread. The lower troposphere is much more photochemically processed and therefore contains largely product compounds. To the extent that these can be removed by precipitation processes their concentrations are very low. Transport of air from midlatitude regions, however, still occurs in the mid- to upper troposphere.

In summary, it can be anticipated that in the polar troposphere, unique chemical reactions will take place in the winter and spring. In the following sections, a short overview of what drives chemical reactions in the unperturbed troposphere is followed by a discussion of principal reaction chains involving carbon, sulphur, and nitrogen compounds that are currently considered important in the polar troposphere. As we indicate whether these reactions have been confirmed by observation, it will become clear that this is an area of research that is still in its infancy. The potential impact of what happens in the polar troposphere in relation to the global climate is uncertain. This region is sufficiently different from other parts of the globe, however, that a determination of its chemistry constitutes a crucial test of the quality and scope of understanding of atmospheric chemistry in general. Such knowledge is the first step in the process of unraveling the link between atmospheric composition and glacial snow and ice composition.

KEY DRIVER OF CHEMICAL REACTIONS IN THE TROPOSPHERE

Chemical reactions in the troposphere center around a set of highly reactive free radicals, OH and HO_2. New OH radicals, which are often described as the 'vacuum cleaners' of the atmosphere, are mostly due to the photodissociation of ozone

$$O_3 + h\nu \rightarrow O(^1D) + O_2 \quad (\lambda < 320 \text{ nm}) \qquad (1)$$

followed by reaction of the excited oxygen atom $O(^1D)$ with water molecules

$$O(^1D) + H_2O \rightarrow 2\,OH \qquad (2)$$

It is immediately clear that there are some limitations to this process in polar regions. Firstly in the winter without solar radiation this route of OH formation is excluded. Furthermore the fact that this process only occurs with light of wavelength less than ca. 320 nm makes it quite sensitive to the solar zenith angle. Demerjian *et al.* (1980) calculated a rate constant for reaction (1) of 0.8, 0.26, and 0.06 ($*10^{-5}$ s^{-1}) for z = 60^0, 70^0 and 78^0 respectively. This process is further limited in the spring by the low absolute humidity due to the prevailing low temperatures, thus reducing the efficiency of reaction (2).

Other photochemically induced primary processes exist that yield OH radicals. The major reactions are photolysis of nitrous acid:

$$HONO + h\nu \rightarrow OH + NO \quad (\lambda < 380\ nm) \quad (3)$$

and formaldehyde:

$$HCHO + h\nu \rightarrow H + HCO \quad (\lambda < 340\ nm) \quad (4)$$
$$H + O_2\ (+M) \rightarrow HO_2 \quad (5)$$
$$HCO + O_2\ (+M) \rightarrow HO_2 + CO \quad (6)$$

Normally these processes are more identified with polluted air masses than tropospheric background air. However, the arctic troposphere in winter and spring is really a polluted background air mass and we will show later that recent observations in the Arctic suggest that these processes are in fact quite important radical formation routes in the spring. This is not necessarily true for the Antarctic.

In the absence of solar radiation another mechanism of formation exists for these radicals, although its efficiency is much lower. This mechanism involves the formation of the $NO_3^\bullet$ radical species from reaction of ozone with NO_2.

$$NO_2 + O_3 \rightarrow NO_3^\bullet \quad (7)$$

followed by reaction with a compound that has an abstractable hydrogen atom, e.g.

$$NO_3^\bullet + HC \rightarrow HNO_3 + R \quad (8)$$
$$R + O_2\ (+M) \rightarrow RO_2 \quad (9)$$

Subsequent reactions of the RO_2 radical then lead eventually to the OH/HO_2 couple.

Due to their high reactivity, the concentration of these species is rather low (global averages are estimated to be ca. 10^5 to 10^6 molecules/cm^{-3} for OH (Spivakovsky *et al.*, 1990) and about an order of magnitude more for HO_2). However, their efficiency is enhanced by the fact that their chemical reactions involve small chains through which they are regenerated:

OH	+	CO	$\rightarrow CO_2 + HO_2$	(10)
H	+	O_2 (+M)	$\rightarrow HO_2$	(11)
HO_2	+	NO	$\rightarrow NO_2 + OH$	(12)
HO_2	+	O_3	$\rightarrow 2\,O_2 + OH$	(13)

It is interesting to note that reaction (12) creates ozone via the follow up reactions

NO_2	+	$h\nu$	$\rightarrow NO + O(^3P)(\lambda < 400$ nm)	(14)
$O(^3P)$	+	O_2 (+M)	$\rightarrow O_3$	(15)

while reaction (13) will destroy ozone (e.g. Liu and Trainer, 1988; Ridley, 1991). Reaction (12) dominates as long as the NO concentration is larger than ca. $2.5 * 10^7$ molec.cm^{-3} (in mixing ratio units ca. 10 pptv), a condition that is often not attained in the polar troposphere (e.g. Honrath and Jaffe, 1992).

Eventually the chain reactions involving OH and HO_2 will terminate, leading to the formation of nitric acid and hydrogen peroxide:

OH	+	NO_2 (+M)	$\rightarrow HNO_3$	(16)
HO_2	+	HO_2 (+M)	$\rightarrow H_2O_2 + O_2$	(17)
HO_2	+	OH	$\rightarrow H_2O + O_2$	(18)

Halogen atoms are well known to be important reactive compounds in stratospheric chemistry. Their participation in tropospheric chemistry is not well understood (Graedel and Keene, 1995). Although widely speculated to be participants in the chemistry of marine atmospheres, observational evidence is limited (e.g. Pszenny *et al.*, 1993). However, we will show later in this paper that in the arctic troposphere in the spring they do play a role. How the halogen atoms are initially produced is not known, although many potential sources have been proposed. Direct photodissociation of halocarbon compounds is now believed to be unimportant (Moortgat *et al.*, 1993). Speculation is currently focused on inorganic photochemistry involving aerosols as summarized in Table 1. Once produced these atoms may set up a chain reaction that leads to ozone destruction:

$$X + O_3 \rightarrow XO + O_2 \quad (19)$$

$$XO + XO \rightarrow \rightarrow 2X + O_2 \quad (20)$$

Table 1. Possible primary formation routes for halogen atoms in the polar troposphere.

	Proposed chemical reaction	Reference
1	gas phase reaction of N_2O_5 with seasalt particles N_2O_5 + NaX -> NO_2X + $NaNO_3$ NO_2X + hν -> NO_2 + X	Finlayson-Pitts *et al*, 1990 Vogt and Finlayson-Pitts, 1994 Behnke *et al.*, 1995
2	photoinduced conversion involving dissolved organic materials or transition metals $4 Br^- + 4 H^+ + O_2 + h\nu \rightarrow Br_2 + H_2O$	McConnell *et al.*, 1992
3	during SO_2 oxidation on seasalt particles high pH: HSO_5 + Br^- -> SO_4^{2-} + HOBr HOBr + Br^- + H^+ -> Br_2 + H_2O low pH: Br_2^- + HO_2 -> Br_2 + HO_2^-	Mozurkewich, 1995
4	reduction of nitrate ions $NO_3^- + 2 Br^- + 3 H^+ \rightarrow Br_2 + HONO + H_2O$	Li, 1994

Halogen atoms will efficiently react with compounds that have an abstractable hydrogen atom similar to reactions of OH. However, this will lead to chain termination rather than propagation in contrast to the OH/HO_2 chain. Thus the overall chain of reactions involving the halogen atoms is difficult to sustain in the gas phase only; it is believed to involve reactions in the liquid or solid phase involving aerosols or ice crystals. A hypothetical diagram of the halogen chemistry in the polar troposphere is shown in Figure 1.

In this summary we focus primarily on the gas phase reactions. Reactions not occurring in the gas phase can have an important impact. In the winter and spring due to the low humidity, chemistry in liquid cloud water is not expected to be important. There is, however, usually an abundance of ice clouds, ice crystal showers, and aerosols on which chemical reactions can take place. The snow-covered surface at the lower boundary of the troposphere is another potentially important medium for such reactions. Ice crystal surface area in the arctic troposphere is highly variable and cannot be neglected relative to surface snow area as a potential host for chemical reactions (Barrie, 1993).

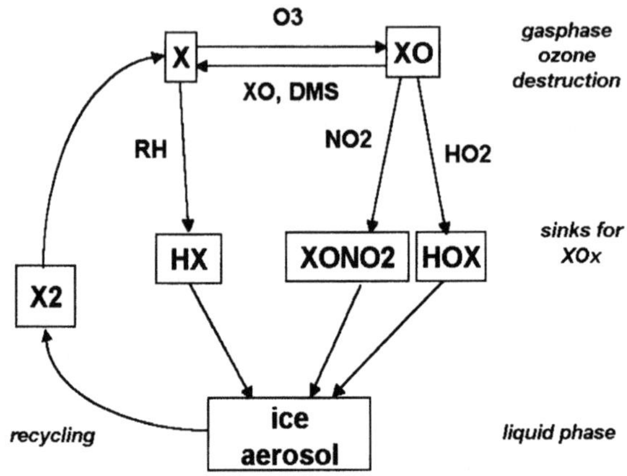

Figure 1 Diagram of halogen chemistry in polar regions (X=Cl, Br or I)

Several factors besides chemical reactivity influence the net rate of such chemical transformations (e.g. Table 2). Consequently, the efficiency of the actual chemical reaction is not necessarily the rate limiting step of a chemical transformation but rather the physico-chemical state of the condensed phase (Molina, 1993; Hanson *et al.*, 1994; Ravishankara *et al.*, 1995). Reactants have to be able to stick to the surface of a heterogeneous medium and preferably adsorb on or dissolve in the condensed phase before chemistry can occur. This limits the importance of such reactions for most organic compounds because of their low solubility in water. This includes the organic nitrates such as PAN (peroxyacetyl nitrate) and the class of alkylnitrates ($RONO_2$). Exceptions are organic acids and carbonyls such as formaldehyde (HCHO) and acetone (CH_3COCH_3). The inorganic acids, hydrogen peroxide (H_2O_2) and the peroxyhydroxyl radical (HO_2) are highly soluble. Dissolution of acids will tend to lower the pH and this will have an impact on the extent to which compounds of moderate to low solubility dissolve and undergo chemical transformation such as SO_2 (oxidation of S(IV) to S(VI)), O_3 (destruction) and NO_2 (NO_2^- formation).

Table 2. Potential rate limiting processes for heterogeneous chemical transformations in the atmosphere (after Lelieveld and Crutzen, 1991)

1	transfer of gases from the gas phase to the droplet/ice crystal surface
2	exchange across the gas-liquid/solid surface
3	dispersion of dissolved gases through the droplet/particle
4	attainment of aqueous-phase equilibria (e.g. ionization)
5	chemical reaction

Chemical reactions that occur in the liquid phase are more complex than in the gas phase although they bear some resemblance. Much of the oxidation chemistry probably again involves the (photochemically produced) HO_2 radical, either by absorption from the gas phase or by formation in the liquid phase. Thus, Chameides and Davis (1983) have proposed that O_3 destruction as well as formic acid (HCOOH) formation in the aqueous phase may occur via the following scheme:

$$HO_2 \rightarrow H^+ + O_2^- \quad (21)$$
$$O_2^- + O_3 \; (+ H^+) \rightarrow OH + 2\,O_2 \quad (22)$$
$$OH + CH_2(OH)_2 \rightarrow CH(OH)_2 + H_2O \quad (23)$$
$$CH(OH)_2 + O_2 \rightarrow HCOOH + HO_2 \quad (24)$$

This would probably be followed by

$$HCOOH \leftrightarrow HCOO^- + H^+ \quad (25)$$
$$OH + HCOO^- \; (+ O_2) \rightarrow CO_2 + H_2O + HO_2 \quad (26)$$

(note that HCHO will hydrate upon dissolving in water to form $CH_2(OH)_2$). In the next section we will indicate that there are also gas phase processes that may lead to production of organic acids such as HCOOH and CH_3COOH (acetic acid).

REACTIONS OF CARBON COMPOUNDS

We mentioned earlier that the efficiency of the OH radical is at least partly driven by its recycling in chain reactions such as in the reactions (10)-(13). Hydrocarbons play an important role in these processes. Not only are they oxidized by OH, but during that process invariably a peroxy radical, generically indicated with the formula RO_2, is produced. RO_2 is the alkyl equivalent of the HO_2 radical and its reactions are comparable to those of HO_2. Thus they will usually oxidize an NO molecule into NO_2 (thus leading to ozone production, reactions 14-15) or react with other peroxy radicals to produce peroxides. In the case that the alkyl group of the RO_2 radical contains 3 or more C atoms, reaction with NO will also produce alkylnitrates ($RONO_2$). Except in this latter case, several other hydrocarbon compounds are formed in these reactions and the ultimate product is CO_2. A good example of hydrocarbon degradation is the path along which methane (CH_4) is transformed as shown in Figure 2. Figures 3 and 4 show the general pattern for hydrocarbons with more than one carbon atom, and routes to some key intermediates, in particular carbonyl compounds.

CH_4	+ OH	$\xrightarrow{O_2}$	CH_3O_2	+ H_2O		
	+ Cl	$\xrightarrow{O_2}$		+ HCl		
CH_3O_2	+ NO	$\xrightarrow{O_2}$	CH_2O	+ HO_2	+ NO_2
	+ HO_2	$\longrightarrow$	CH_3O_2H	+ O_2	
	+ CH_3O_2	$\xrightarrow{O_2}$	2 CH_2O	+ 2 HO_2	+ O_2
		$\longrightarrow$	CH_3OH	+ CH_2O	+ O_2
CH_2O	+ OH	$\xrightarrow{O_2}$	HO_2	+ CO	+ H_2O
	+ Cl	$\xrightarrow{O_2}$			+ HCl
	+ Br	$\xrightarrow{O_2}$			+ HBr
	+ NO_3	$\xrightarrow{O_2}$			+ HNO_3
	+ $h\nu$	$\xrightarrow{O_2}$	2 HO_2	+ 2 CO	
CO	+ OH	$\xrightarrow{O_2}$	HO_2	+ CO_2	

Note:

CH_2O	+ HO_2	$\rightarrow\rightarrow\rightarrow$	$HCOOH$
HCO	+ O_2	$\rightarrow\rightarrow\rightarrow$	not $HCOO_2$

Figure 2. Diagram of the transformation of methane to CO_2.

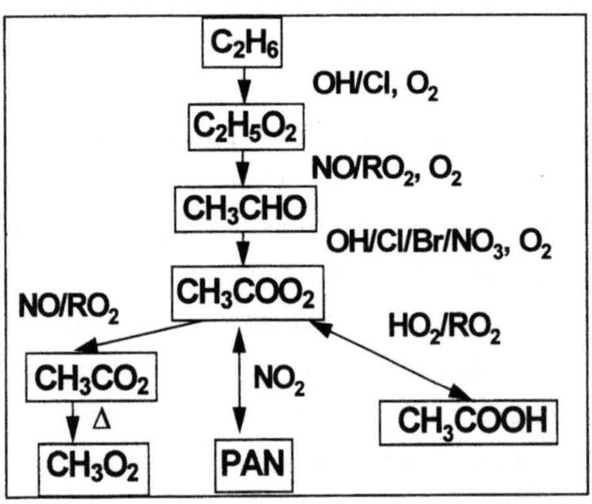

Figure 3. Partial diagram of ethane (C_2H_6) transformation

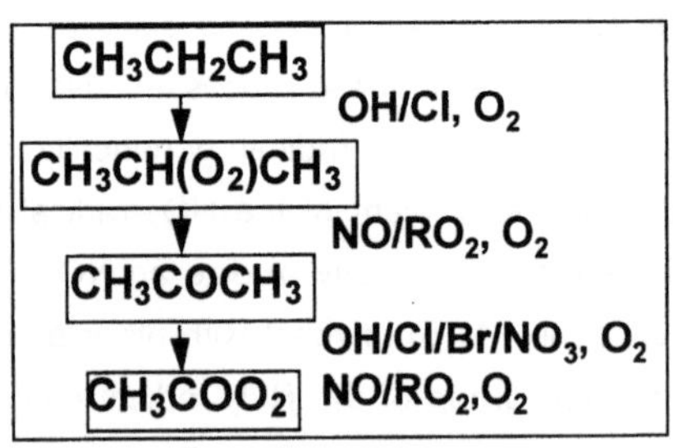

Figure 4. Partial diagram of propane (C_3H_8) transformation

The examples of Figures 2-4 are representative for the chemical transformation of alkanes which have the generic chemical formula C_nH_{2n+2}. Alkenes (hydrocarbons that have a double bond in their chemical structure, C_nH_{2n}) will have additional reaction pathways. Reactive intermediates such as OH and $NO_3^{\bullet}$ may add to the double bond rather than abstract a H atom. Furthermore alkenes can react with ozone. Overall the transformation routes of alkenes are more complex than those of the alkanes but the basic idea is the same: formation of an RO_2 radical followed by reaction either with NO leading to ozone and carbonyl species production or another RO_2 radical to form a peroxide.

Figures 2 and 3 indicate that HCHO and CH_3CHO are precursors to gas phase formation of the associated acids HCOOH and CH_3COOH. In the case of HCOOH this requires addition of HO_2 to HCHO as the initial step. Since this is a slow reaction this process is unimportant in the atmosphere. Instead, it is currently assumed that the addition of O_3 to ethene (C_2H_4) is the predominant gas phase route to HCOOH formation (Calvert *et al.*, 1978; Horie *et al.*, 1994). On the other hand, formation of CH_3COOH from CH_3CHO is a viable process (Madronich *et al.*, 1990). The key difference is that here the initial step is H atom abstraction from CH_3CHO by OH followed immediately by O_2 addition to yield the peroxyacetyl radical CH_3COO_2 (Figure 3). Reaction of the peroxyacetyl radical with other peroxy radicals such as HO_2 and CH_3O_2 is fast (Moortgat *et* al., 1989) and known to yield CH_3COOH (Niki *et al.*, 1985). As indicated in Figure 2 a similar mechanism is not possible in the case of HCOOH since the equivalent $HCOO_2$ radical is too unstable to be of importance.

The reaction rate coefficients for the hydrocarbon reactions with OH are variable depending on the reactant as well as the hydrocarbon itself. This variation will have an impact on the relative concentration of a suite of hydrocarbons. After emission into the atmosphere the concentration ratio of two hydrocarbons will remain constant if no chemical transformation takes place, and selective removal by wet and dry deposition or vertical diffusion can be ignored (Jobson *et al.*, 1994). This is especially true for polar regions, well removed from emission regions (see McKeen and Liu (1993) for a discussion on why this is less valid closer to source regions). Therefore a *change* in the concentration ratio may be interpreted in terms of chemical reactions of the hydrocarbons. To illustrate how this takes place, consider two hydrocarbons HC_1 and HC_2, both reacting with a reactive intermediate, say OH. In chemical kinetics the rate of change of each hydrocarbon is expressed as

$$\frac{d[HC_i]}{dt} = k_i \cdot [HC_i] \cdot [OH] \tag{27}$$

where $[HC_i]$ denotes the concentration of the hydrocarbon, [OH] the concentration of OH and k_i the rate coefficient of the reaction between HC_i and OH. Equation (27) is of course incomplete since it ignores atmospheric dilution of HC_i. However, the change in the concentration ratio between HC_1 and HC_2, R_{12}, will be independent of this effect; its rate of change can be expressed as function of [OH] and the difference in the rate coefficients with OH ($\Delta k_{12} = k_1 - k_2$)

$$\frac{d[R_{12}]}{dt} = \Delta k_{12} \cdot [R_{12}] \cdot [OH] \tag{28}$$

High precision measurements of hydrocarbons in the Arctic have allowed the use of this property to show that OH driven chemistry is indeed absent in the winter months. Figure 5 shows a comparison of the temporal variation of CH_4 and C_2H_6 from January to April at the

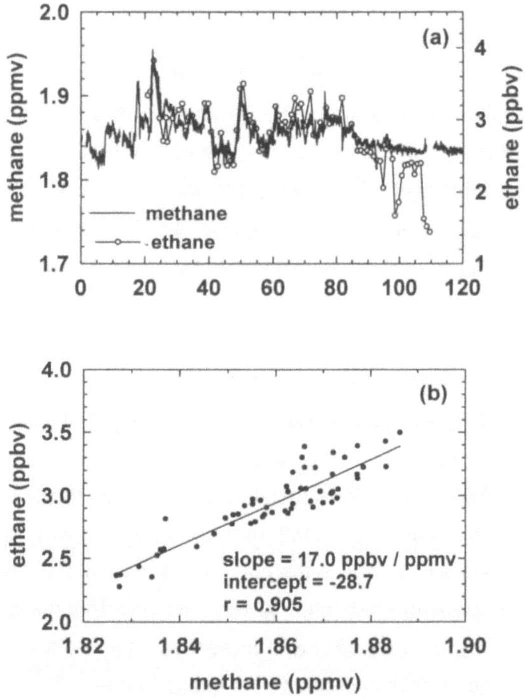

Figure 5 Relation between methane and ethane observed at Alert, N.W.T. (a) temporal trend from January to April 1992, (b) linear regression of data from January 21 to March 24, 1992 (from Jobson *et al.*, 1994)

high arctic station Alert. Rates of reaction of these hydrocarbons with OH are different by at least an order of magnitude and yet before sunrise in the high Arctic, these two hydrocarbons show an excellent linear correlation. Note that the temporal change in each compound is due to variations in origin and travel time from midlatitude to arctic regions which therefore would have led to a nonlinear relationship, if OH chemistry had occurred. Figure 6 compares CH_4 with HCHO measured at the same location. HCHO not only reacts much more efficiently with OH than CH_4, but most hydrocarbon reactions involving OH (including CH_4) will ultimately produce HCHO. The strong correlation between CH_4 and HCHO therefore implies that reactions with OH have not occurred in the observed air mass in its recent past. Below we will address the question whether reactions with non-photochemically produced $NO_3^{\bullet}$ radicals can be discerned.

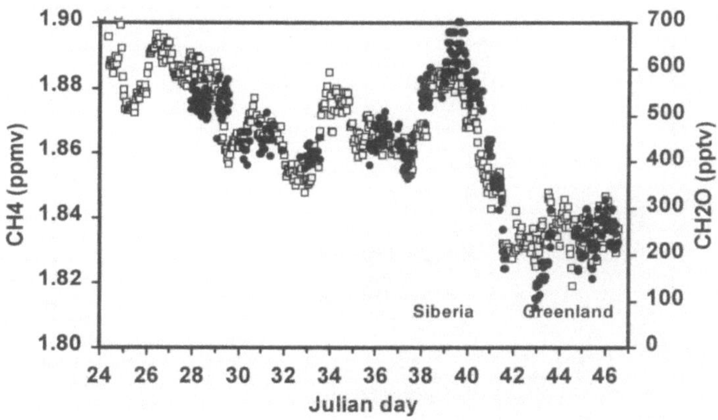

Figure 6. Comparison of HCHO (dots) and CH_4 hourly means (squares) during January/February 1992 at Alert, N.W.T. (from DeServes, 1994)

Recently it was reported that after sunrise in the Arctic widespread O_3 depletion is observed in the lower troposphere (Barrie *et al.*, 1988, Oltmans *et al.*, 1989, Bottenheim *et al.*, 1990). The chemical reactions involved in this process are not well understood, but it is widely speculated that it does involve attack by halogen atoms, in particular Br but possibly also Cl (e.g. McConnell and Henderson, 1993) and I (Solomon *et al.*, 1994). If this were the case then it should be possible to discern signals of halogen chemistry in the concentration ratios of hydrocarbons, making use of equation (28), replacing OH and the associated reaction rate coefficients by those for Cl or Br (reactions with I are most likely too slow to be of importance, although little is know about their kinetics in this respect). A particularly striking example of this is shown in Figure 7. Here the ratio [isobutane]/[n-butane] is compared with the ratio [isobutane]/[propane]. The reaction rate coefficients, k_i, for reactions of these hydrocarbons with OH or Cl are such that if reaction with OH occurred in the atmosphere then the former ratio should remain constant since $\Delta k \approx 0$, but the latter ratio should be variable depending on the age of the observed air mass. On the other hand if reaction were due predominantly to reaction with Cl then the opposite would be true. A similar analysis searching for Br driven chemistry has not been possible with the available data. This is due to the generally lower efficiency of reactions with Br atoms, and the evidence for Br chemistry from hydrocarbon data analysis is presently less compelling (Jobson *et al.*, 1994).

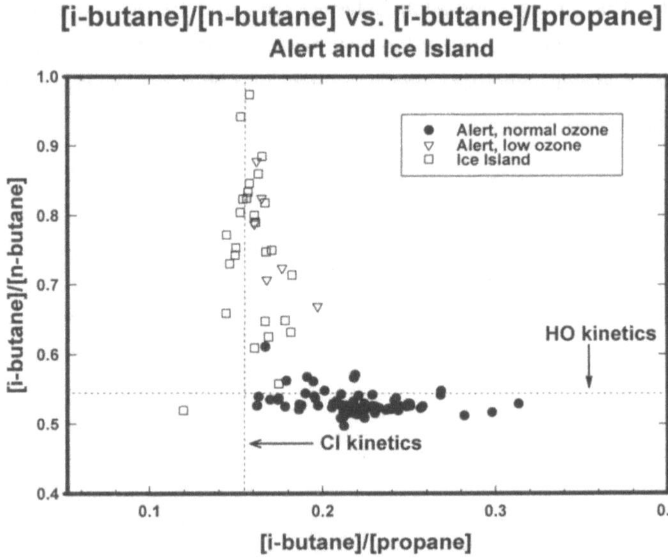

Figure 7 Plot of [i-butane]/[n-butane] versus [i-butane]/[propane] at Alert, N.W.T. Solid lines indicate approximately how data would align under OH oxidation (horizontally) and Cl atom oxidation (vertically). (from Jobson *et al.*, 1994)

The strong evidence for chemical reactions involving Cl atoms in the arctic troposphere in the spring has many implications for the overall understanding of polar tropospheric chemistry that have hardly been explored. We note that the evidence is not limited to the Canadian studies; similar results have been obtained in the Norwegian Arctic (F. Stordahl (Norwegian Institute for Air Research), personal communication, 1994). Furthermore, it has been speculated for some time that such reactions play an important role in the marine boundary layer outside polar regions but clear evidence has not been reported (see e.g. Graedel and Keene, 1995).

One example of the poorly understood implications of the Cl chemistry involves the carbonyl compounds. During episodes of O_3 depletion, which correspond with the times that Cl driven chemistry has been identified, a remarkable *increase* was observed for both HCHO (DeServes, 1994) and CH_3COCH_3 (Yokouchi *et al*, 1994) (these are the only two carbonyl compounds that have been measured simultaneously with O_3 depletion events). Although reactions of many hydrocarbons with Cl atoms would produce HCHO, it also *reacts* very efficiently with Cl atoms. Chemical reaction mechanisms that have been proposed to describe the polar tropospheric chemistry in the spring therefore predict a *decrease* in HCHO during O_3 depletion episodes (Barrie *et al.*, 1988; McConnell *et al.*, 1992). The amount of CH_3COCH_3 production is similarly difficult to understand with current chemical reaction mechanisms.

Finally it should be emphasized that the sparse measurements of H_2O_2 in the Arctic troposphere are in agreement with the general picture described above. In the winter virtually no H_2O_2 is observed in accordance with the absence of OH chemistry. In the spring during O_3 depletion episodes its concentration appears to decrease (DeServes, 1994).

REACTIONS OF SULPHUR COMPOUNDS

Chemical reactions of sulphur compounds in the troposphere revolve around the oxidation of reduced sulphur compounds originating from biological processes (e.g. dimethyl sulphide, DMS) and sulphur dioxide (SO_2), a major gaseous product of biogenic sulphur decomposition as well as of anthropogenic activities such as fossil fuel burning and smelting. The products are aerosol inorganic sulphates (SO_4^{2-}) as well as organic aerosol methane sulfonic acid (MSA). SO_2 also originates also from volcanic eruptions but on a global basis this source is small in comparison to that from anthropogenic and marine biogenic origin. However, in pre-industrial times at elevated glacial sites in the polar regions this can be a dominant natural source, although highly episodic in nature. Since SO_4^{2-} particles are effective scatterers of solar radiation and form cloud condensation nuclei, they play an important role in the radiative balance of the atmosphere (Barrie, this issue). In the Arctic the sulphur chemistry is dominated by SO_2 and SO_4^{2-} from anthropogenic sources in Europe and Asia (Russia) (Barrie and Bottenheim, 1991).

A diagram summarizing the main transformations is presented in Figure 8. The oxidation of the biogenic sulphur compound DMS has been recently reviewed by Turnipseed and Ravishankara (1993). It appears that it is largely initiated by reaction with OH, either through addition, or abstraction of an H atom. Within the polar regions in winter and spring this is not expected to be of importance. However, on the periphery of polar regions in the spring and summer it is important, affecting the nature of the source which is then transported long distances to the pole. In the winter chemical attack by $NO_3^\bullet$ radicals would be possible; the products of this reaction which is postulated to involve mainly addition, are unknown. Reaction with halogen atoms is known to be fast but normally considered not important. However the identification of important Cl chemistry in the arctic spring (see previous section) might require a reevaluation. As with $NO_3^\bullet$, addition of halogens leads to unknown products.

H atom abstraction is also possible and this will initiate the same chain of reactions following H abstraction by OH. Reaction of DMS with IO (iodine-monoxide) radicals is now known to be unimportant (Barnes *et al.*, 1991) but the same authors determined that BrO (bromine-oxide) efficiently reacts with DMS to produce ultimately dimethylsulfoxide (DMSO). BrO is produced when Br atoms react with ozone, and hence if ozone depletion in the arctic troposphere is resulting from Br atom reactions then the BrO + DMS reaction could be an important degradation route for DMS at that time (Toumi, 1994).

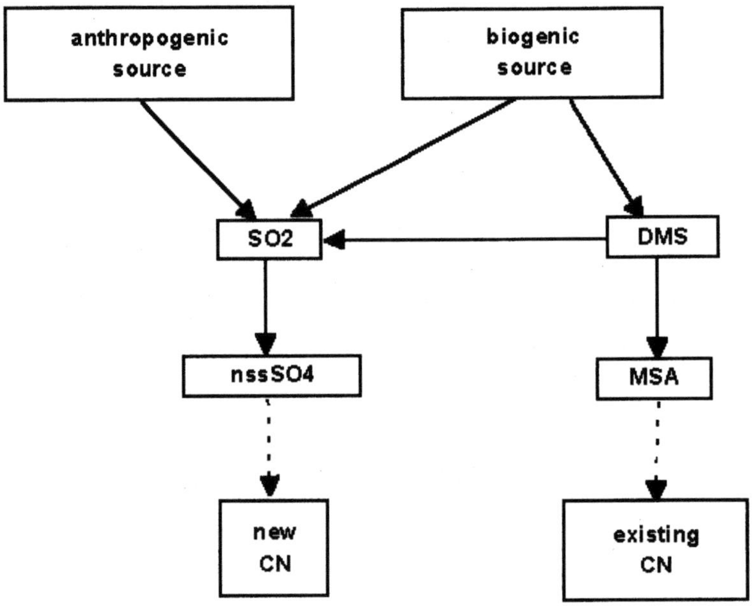

Figure 8. Summary diagram of sulphur chemistry

The chain of reactions that follows the abstraction of an H atom from DMS is not yet well known. It is a multi-step process which currently is postulated to involve largely oxidation by ozone or NO_2. The process leads either to the production of MSA, or SO_2 (or SO_3) and hence SO_4^{2-}. The distinction is important since MSA tends to condense on *existing* particles while SO_4^{2-} may lead to the production of *new* particles (Pandis *et al.*, 1994). It appears that certain steps in the route to SO_2 (or SO_3) formation involve substantial activation energy (e.g. Hynes *et al.*, 1986; Turnipseed and Ravishankara, 1993). Therefore the ratio of [MSA]/[SO_4^{2-}] resulting from DMS oxidation is expected to decrease with temperature and be larger in polar regions than at midlatitude and tropical regions. This is consistent with the observations of Li and Barrie (1993) in the summer at the arctic station Alert. They found an average ratio of 0.3 which is much higher than that found in the tropics. In the winter however, when DMS sources are lowest, they found a much lower ratio in the Arctic which suggests that different chemical transformation processes may have played a role.

The oxidation of SO_2 to SO_4^{2-} cannot be explained solely by reaction with OH in the gas phase. Other mechanisms such as oxidation by H_2O_2 and ozone in clouds, or heterogeneous oxidation by heavy metals are possible. In all cases the oxidation rate increases with solar radiation. This explains the observation that in the Arctic the degree to which SO_2 is oxidized to SO_4^{2-} increases dramatically between January and April (Barrie and Hoff, 1984), see for example Figure 9. As was stated above this oxidation will probably largely occur in the air mass before it reaches the high Arctic.

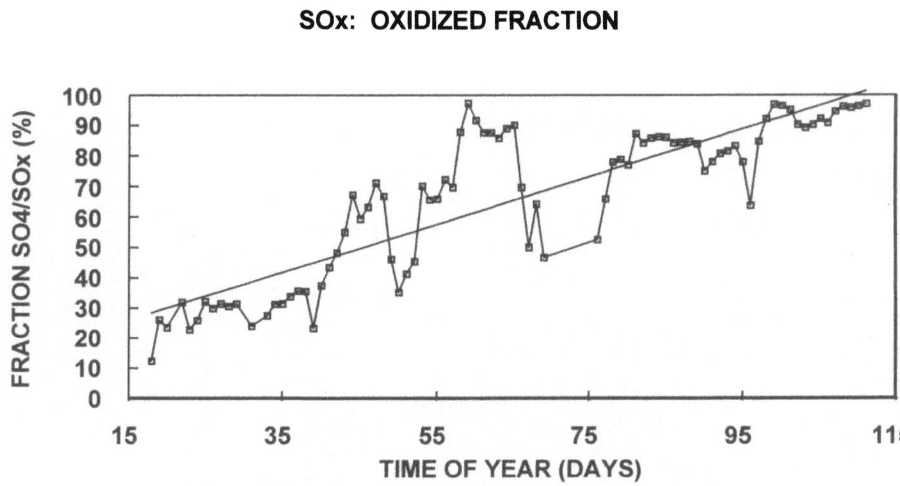

Figure 9. Ratio SO_4/SO_x, observed in 1992 at Alert, N.W.T. (from Barrie *et al.*, 1994)

REACTIONS INVOLVING NITROGEN OXIDE COMPOUNDS

The chemistry of nitrogen oxides is intimately related to the abundance of ozone in the troposphere. As discussed above, the conversion of NO to NO_2 as in reaction (12) leads to ozone production via reactions (14) and (15). This is in contrast with the fast reaction

$$O_3 + NO \rightarrow NO_2 + O_2 \qquad (29)$$

which together with reactions (14) and (15) indicates that under solar radiation the ozone concentration is determined by the so-called photostationary state equation

$$[O_3] = \frac{k_{14} \cdot [NO_2]}{k_{29} \cdot [NO]} \tag{30}$$

There are several other nitrogen oxides in the troposphere besides the pair NO and NO_2 (the sum of NO + NO_2 is usually denoted by the term NO_x; the total of all nitrogen oxides excluding N_2O on the other hand is denoted by NO_y while in recent years the term NO_z is gaining acceptance as representing the difference NO_y - NO_x). Figure 10 shows a diagram of the main components of NO_y and the chemical reactions that are involved in the

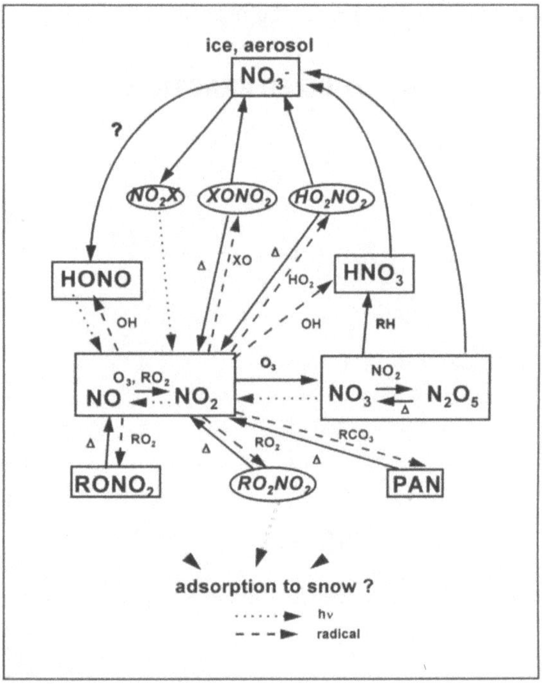

Figure 10. Diagram of nitrogen oxide chemistry in polar regions.

interconversions between them. We have indicated which reactions are only occurring when sunlight is present (either direct photochemical reactions indicated with dotted lines, or those involving photochemically produced radicals shown with dashed lines). Some reactions require a large amount of thermal energy and are slow at subzero temperatures; the symbol Δ is used for denoting this. Furthermore, compounds in square boxes have been observed in the Arctic. These distinctions help to explain the main features of chemical transformations involving the nitrogen oxides in the polar troposphere.

Due to the absence of sunlight, organic compounds such as PAN and $RONO_2$ will not be produced in the winter and their presence will depend on transport from midlatitude regions. The central chain in Figure 12

$$\{NO <-> NO_2\} -> \{NO_3^{\bullet} -> N_2O_5\} \quad (31)$$

will be shifted to the right and reactions involving $NO_3^{\bullet}$ might be expected to occur. As a result chemical reactions will lead predominantly to the formation of *inorganic* oxidation products: HNO_3 and NO_3^- particles are expected to maximize, and this is in fact observed (Figure 11).

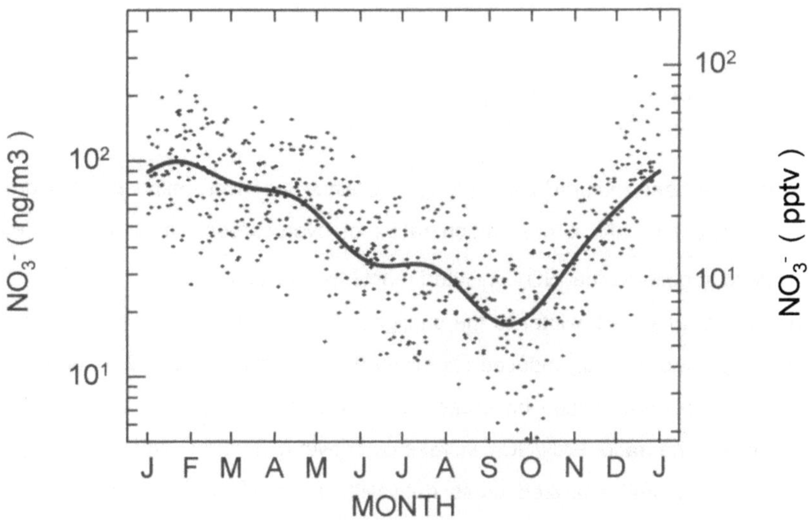

Figure 11. Annual cycle of total nitrate (from weekly averaged high volume samples) at Alert N.W.T.

A further consequence of these considerations is that in the winter the ratio [PAN]/[nitrate] is lower than in the spring (Figure 12).

The question arises whether concentration ratios of hydrocarbons indicate that $NO_3^{\bullet}$ reactions did occur, similar to what was outlined earlier with respect to halogen atom reactions. The efficiency of the reactions with $NO_3^{\bullet}$ is substantially lower, and polar regions would be an ideal region to probe for such processes. Analysis of hydrocarbon data collected over the winter at Alert is inconclusive in this respect (Bottenheim, unpublished results). Penkett *et al.* (1993) have reported hydrocarbon measurements collected over the Atlantic ocean in air masses that appeared to have originated form the Arctic and hence were

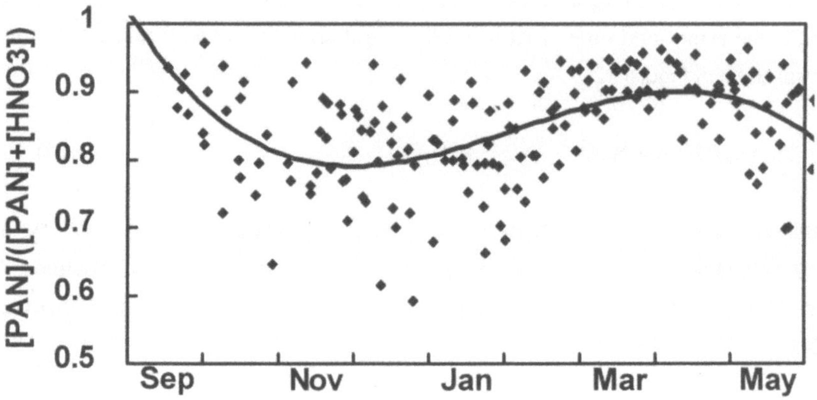

Figure 12. Ratio of [PAN]/{[PAN] + [HNO_3]} from 7 years data at Alert, N.W.T. (from Bottenheim *et al.*, 1994)

assumed to have been matured over a long time since they were exposed to hydrocarbon sources. Comparison with concentration ratios of hydrocarbons in air near to emission sources indeed suggested that $NO_3^\bullet$ initiated reactions had occurred.

The transition from the dark winter to the sunlight irradiated spring will signal the increasing importance of photochemically induced chemistry as outlined in Figure 10. The result is that organic nitrogen oxide formation becomes more important. In fact, as with the SO_4^{2-}/SO_x ratio (Figure 9), the ratio PAN/NO_y increases between January and April (Figure 13). Bottenheim *et al.* (1994) analyzed these data with a simplified chemical mechanism and showed that they are consistent with OH driven chemistry. The analysis also

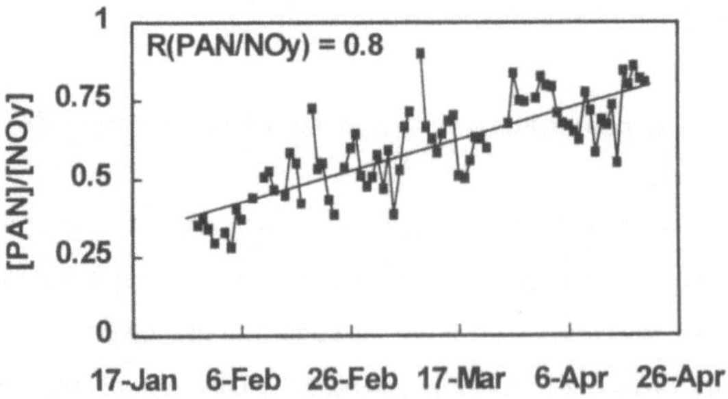

Figure 13. Ratio of [PAN]/[NO_y] observed in 1992 at Alert, N.W.T. (from Bottenheim*et al.*, 1994)

Table 3. Average OH concentrations, estimated from PAN and NO_y observations at arctic station Alert, N.W.T. (from Bottenheim *et al.*, 1994)

February 1992	2.9 e+04	molec.cm^{-3}
April 1992	1.1 e+05	molec.cm^{-3}

yielded an estimate of the average OH radical concentration (see Table 3) during the oxidation process. These concentrations are higher than expected in the Arctic if photolysis of O_3 was the main source of OH radicals (reactions 1 and 2). This might suggest that the formation of PAN occurred well before the air reached the Arctic, but we will indicate below that other sources within the Arctic may also play a role.

One recent addition to the suite of nitrogen oxides measured in the Arctic is gaseous (presumably nitrous acid, HONO) and particle nitrite (NO_2^-) (Li, 1993, 1994). The observed concentrations, shown in Figure 14, are higher than expected from the simple gas phase

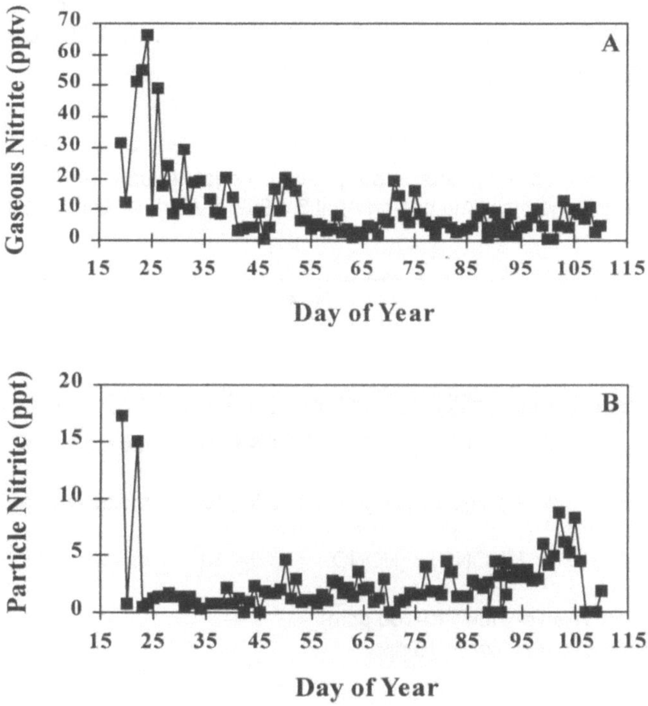

Figure 14. Time series of gas phase (diagram A) and particle phase (diagram B) nitrite, as observed at Alert, N.W.T. in 1992. From Li, 1994.

formation route:

$$OH + NO \rightarrow HONO \qquad (31)$$

Li (1994) proposed several potential reactions on acidic aerosol particles in the arctic troposphere. He deduced that the observed gaseous/particle nitrite ratio could only be understood if in the aerosol nitrite was not in the form of an NO_2^- ion, but rather as a hydrated NO^+ ion (H_2ONO^+). This would seem possible in view of the apparent high acidity of the aerosol particles. We note that due to the analytical techniques employed, H_2ONO^+ would still be measured as an NO_2^- ion.

The presence of HONO, as well as HCHO discussed earlier, implies that formation mechanisms for OH other than photolysis of O_3 exist in the Arctic troposphere, namely reactions (3) and (4)-(6). Li (1994) used the observed concentrations at Alert to estimate the relative importance of these OH formation routes. The results are reproduced in Table 4. As can be seen HONO photolysis is the dominant formation mechanism. Interestingly the predicted OH concentration is in very good agreement with the number obtained by Bottenheim *et al.* (1994) from the analysis of the PAN data. The implications of these results have not been explored in comprehensive model calculations to date.

Table 4. Estimated noontime steady state concentration of OH radicals from photochemical sources on April 5, 1992 at Alert, N.W.T. (after Li, 1994)

$$[OH]_{ss} = \frac{\sum R_{OH}}{k[CH_4] + k[CO] + k[O_3] + k[NO_2]}$$

$[O_3] \cong 40$ ppb, $[CH_2O] \cong 100$ ppt, $[HONO] \cong 10$ ppt, $[CH_4] \cong 1.8$ ppm, $[NO_2] \cong 30$ ppt, $[CO] \cong 170$ ppb

$R(OH, O_3)$	$\cong$	$\alpha * j(O_3) * [O_3]$	$\cong$	$1.4 * 10^3$	$molec.cm^{-3}.s^{-1}$
$R(OH, CH_2O)$	$\cong$	$\beta * j(CH_2O) * [CH_2]$	$\cong$	$2.8 * 10^4$	$molec.cm^{-3}.s^{-1}$
$R(OH, HONO)$	$\cong$	$\gamma * j(HONO) * [HONO]$	$\cong$	$3.0 * 10^5$	$molec.cm^{-3}.s^{-1}$

α is low because $[H_2O]$ is very low (10-100 ppm)
β is low because [NO] is very low (< 10 ppt)

$k[CH4] + k[CO] + k[O_3] + k[NO_2] \cong 1\ s^{-1}$

$$[OH]_{ss} = 3 * 10^5\ molec.cm^{-3}$$

CONCLUDING REMARKS

We have attempted in this paper to describe the central aspects of what drives tropospheric chemical reactions, and then indicated how the unique features of the polar troposphere might be expected to result in specific trends, different from other regions of the globe. The focus of attention therefore was on the winter and spring. These seasons are especially interesting since in the former no direct solar radiation is present, while in the latter chemical transformations will occur at reduced tropospheric temperatures in an atmosphere that is primed for photochemistry after the extended absence of sunlight.

Winter observations support the general thesis that in the polar troposphere variations in the chemical composition are controlled by chemical reactions at mid-latitude sources followed by transport modulated differences in the concentrations. Photochemically produced trace compounds such as carbonyl compounds and peroxides are at a minimum. Evidence for the occurrence of $NO_3^{\bullet}$ reactions, expected to occur largely in the absence of sunlight, is as yet circumstantial.

The spring observations are revealing in several aspects. Many of the observations described above were made during studies undertaken to determine the mechanism of ozone depletion in the lower arctic troposphere (Bottenheim, 1993; Barrie *et al.*, 1994). In view of the central role of ozone in the composition of the troposphere this phenomenon needs to be understood before it can be assumed that our description of the chemistry of the atmosphere is satisfactorily complete. Several measurements are in agreement with predictions from chemical models of the atmosphere, e.g. the seasonal trends in NO_y compounds, the increase in oxidation products of SO_2, and the decrease in hydrocarbons reactive with photochemically generated reactants. New features have been identified that were unexpected. For instance, halogen atom driven chemistry, and the apparent efficient formation of HONO which should create much higher OH radical concentrations than anticipated. Furthermore, measurements of carbonyl compounds are of concern since their trends during ozone depletion events seem contrary to what is expected. There remains also little doubt that chemical transformations on the surface of ice and aerosol particles are an important aspect of the polar troposphere just as they are in the stratosphere.

The study of chemical transformations in the polar troposphere is largely in its infancy. The observations that were discussed here are mostly from very recent studies and many require confirmation from other locales. However, it is clear that much is to be learned from these studies that will go beyond the confinement of polar regions. Chemical transformations of a photochemical and heterogeneous nature that are similar to those taking place in the atmosphere, very likely will also occur in the surface snowpack after substances are deposited by wet and dry deposition processes and as the snowpack is ventilated (Waddington, this issue) and resuspended in blowing snow (Pomeroy, this issue).

REFERENCES

Barrie L.A. and Hoff R.M. (1984), The oxidation rate and residence time of SO_2 in the arctic atmosphere. *Atmos. Environ.*, 18: 2711-2722.

Barrie L.A., Bottenheim J.W., Schnell R.C., Crutzen P.J. and Rasmussen R.A. (1988), Ozone destruction and photochemical reactions at polar sunrise in the lower arctic atmosphere. *Nature,* 334: 138-141.

Barrie L.A. and Bottenheim J.W. (1991), Sulphur and nitrogen pollution in the arctic atmosphere. In: *Pollution of the arctic atmosphere*, pp. 155-184, W. Sturges, Ed., Elsevier Press, London, 335 pp.

Barrie L.A. (1993), Features of polar regions relevant to tropospheric ozone chemistry. In: *The tropospheric chemistry of ozone in polar regions*, pp. 3-24, H. Niki and K.H. Becker, Ed., NATO ASI series I: Global Environmental Change, Vol. 7, Springer verlag, Berlin.

Barrie L.A., Bottenheim J.W. and Hart W.R. (1994), Polar sunrise experiment 1992 (PSE 1992); Preface. *J. Geophys. Res.*, 99D: 25313-25314.

Barrie L.A. (1996), Occurrence and trends of pollution in the Arctic troposphere. *This volume.*

Barnes I., Bastian B., Becker K.H. and Overath R.D. (1991), Kinetic studies of the reactions of IO, BrO and ClO with dimethylsulfide. *Int. J. Chem. Kinet.*, 23: 579-591.

Behnke et al. (1995), presented at the European Geophysical Society general meeting, Hamburg.

Bottenheim J.W., Barrie L.A., Atlas E., Heidt L.J., Niki H., Rasmussen R.A. and Shepson P.B. (1990), Depletion of lower tropospheric ozone during arctic spring: the polar sunrise experiment 1988. *J. Geophys.Res.,* 95D: 18555-18568.

Bottenheim J.W. (1993), Polar sunrise studies. In: *The tropospheric chemistry of ozone in polar regions*, pp. 41-57, H. Niki and K.H. Becker, Ed., NATO ASI series I: Global Environmental Change, Vol. 7, Springer verlag, Berlin.

Bottenheim J.W., Barrie L.A. and Sirois A. (1994), Analysis of seven years of continuous observations of peroxyacetyl nitrate (PAN) at the high arctic station Alert. Paper #225, *Joint meeting on global atmospheric chemistry, CACGP/IGAC*, Fuji-Yoshida, 5-9 Sept.

Calvert J.G., Su F., Bottenheim J.W. and Strausz O.P. (1978), Mechanism of the homogeneous oxidation of sulfur dioxide in the troposphere. *Atmos. Environ.*, 12:197-226.

Chameides W.L. and Davis D.D. (1983), Aqueous-phase source of formic acid in clouds. *Nature*, 304: 427-429.

Demerjian K.L., Schere K.L., and Peterson J.T. (1980),Theoretical estimates of actinic (spherically integrated) flux and photolytic rate constants of atmospheric species in the lower troposphere. Adv. Environ. Sci. Technol., 10: 369-460.

DeServes C. (1994), Gas phase formaldehyde and peroxide measurements in the arctic atmosphere. *J. Geophys. Res.*, 99D: 25391-25398.

Finlayson-Pitts B.J., Livingstone F.E. and Berko H.N.(1990), Ozone destruction and bromine photochemistry in the arctic spring. *Nature*, 343: 622-625.

Graedel T.E. and Keene W.C. (1995), Tropospheric budget of reactive chlorine. *Global Biogeochem. Cycles*, 9: 47-77.

Hanson D.R., Ravishankara A.R., and Solomon S. (1994), Heterogeneous reactions in sulfuric acid aerosols: a framework for model calculations. *J. Geophys. Res.*, 99D, 3615-3629.

Honrath R.E. and Jaffe D.A. (1992), The seasonal cycle of nitrogen oxides in the arctic troposphere at Barrow, Alaska. *J. Geophys. Res.*, 97D: 20615-20630.

Horie O., Neeb P., Limbach S. and Moortgat G.K. (1994), Formation of formic acid and organic peroxides in the ozonolysis of ethene with added water vapour. *Geophys. Res. Lett.*, 21:1523-1526.

Hynes A.J., Wine P.H. and Semmes D.H. (1986), Kinetics and mechanism of OH reactions with organic sulfides. *J. Phys. Chem.*, 90: 4148-4156.

Jobson B.T., Niki H., Yokouchi Y., Bottenheim J., Hopper F. and Leaitch R. (1994), Measurements of C_2-C_6 hydrocarbons during the polar sunrise 1992 experiment: Evidence for Cl atom and Br atom chemistry. *J. Geophys. Res.*, 99D: 25355-25368.

Lelieveld J. and Crutzen P.J. (1991), The role of clouds in tropospheric photochemistry. *J. Atmos. Chem.*, 12: 229-267.

Li S.-M. (1993), Particulate and snow nitrite in the spring arctic troposphere. *Atmos. Environ.*, 27A: 2959-2968.

Li S.-M. (1994), Equilibrium of particle nitrite with gas phase HONO: Tropospheric measurements in the Arctic during polar sunrise. *J. Geophys. Res.*, 99D: 25469-25478.

Li S.-M. and Barrie L.A. (1993), Biogenic sulfur aerosol in the arctic troposphere: 1. Contributions to total sulfate. *J.Geophys. Res.*, 98D: 20631-20622.

Liu S.C. and Trainer M. (1988), Response of the tropospheric ozone and odd hydrogen radicals to column ozone change. *J. Atmos. Chem.*, 6: 221.

Madronich S., Chatfield R.B., Calvert J.G., Moortgat G.K., Veyret B. and Lesclaux R. (1990), A photochemical origin of acetic acid in the troposphere. *Geophys. Res. Lett.*, 17:2361-2364.

McConnell J.C., Henderson G.S., Barrie L., Bottenheim J., Niki H., Langford C.H. and Templeton E.M.J. (1992), Photochemical bromine production implicated in arctic boundary layer ozone depletion. *Nature*, 355: 150-152.

McConnell J.C. and Henderson G.S. (1993), Ozone depletion during polar sunrise. In: *The tropospheric chemistry of ozone in polar regions*, pp. 89-104, H. Niki and K.H. Becker, Ed., NATO ASI series I: Global Environmental Change, Vol. 7, Springer verlag, Berlin.

McKeen S.A. and Liu S.G. (1993), Hydrocarbon ratios and photochemical history of air masses. *Geophys. Res. Lett.*, 20: 2363-2366.

Molina M. (1993), Chemical interactions of tropospheric halogens on snow/ice. In: *The tropospheric chemistry of ozone in polar regions*, pp. 273-279, H. Niki and K.H. Becker, Ed., NATO ASI series I: Global Environmental Change, Vol. 7, Springer verlag, Berlin.

Moortgat G.K., Veyret B. and Lesclaux R. (1989), Kinetics of the reaction of HO_2 with $CH_3C(O)O_2$ in the temperature range 253-368 K. *Chem. Phys. Lett.*, 160: 443-447.

Moortgat G.K., Meller R. and Schneider W. (1993), Temperature dependence (256-296 K) of the absorption cross sections of bromoform in the wavelength range 285-360 nm. *Ibid.*, pp. 359-370.

Mozurkewich M. (1995), Mechanisms for the release of halogens from sea salt particles by free radical reactions. *J. Geophys. Res.*, 100D: 14199-14208.

Niki H., Maker P.D., Savage C.M. and Breitenbach L.P., FTIR study of the kinetics and mechanism for Cl-atom initiated reactions of acetaldehyde. *J. Phys. Chem.*, 89: 588-591.

Oltmans S.J., Schnell R.C., Sheridan P.J., Peterson R.E., Li S.M., Winchester J.W., Tans P.P., Sturges W.T., Kahl J.D. and Barrie L.A. (1989), Seasonal surface ozone and filterable bromine relationship in the high Arctic. *Atmos. Environ.*, 23: 2431-2441.

Pandis S.N., Russell L.M. and Seinfeld J.H. (1994), The relationship between DMS flux and CCN concentration in remote marine regions. *J. Geophys. Res.*, 99D: 16945-16957.

Penkett S.A., Blake N.J., Lightman P., Marsh A.R.W., Anwyl P. and Butcher G. (1993), The seasonal variation of nonmethane hydrocarbons in the free troposphere over the North Atlantic Ocean: Possible evidence for extensive reaction of hydrocarbons with the nitrate radical. *J. Geophys. Res.*, 98D: 2865-2886.

Pomeroy J. (1996), Wind-blown snow: sublimation, transport and changes to polar. *This volume.*

Pszenny A.A.P., Keene W.C., Jacob D.J., Fan S., Maben J.R., Zetwo M.P., Springer-Young M. and Galloway J.N. (1993), Evidence of inorganic chlorine gases other than hydrogen chloride in the marine surface air. *Geophys. Res. Lett.*, 20: 699-702.

Ravishankara A.R., Hanson D.R. and Lovejoy E.R. (1996), Processes at ice surfaces: physical uptake and reactions. *This volume.*

Ridley B.A. (1991), Recent measurements of oxidized nitrogen compounds in the troposphere. *Atmospheric Environment*, 25A: 1905-1926.

Solomon S., Garcia R.R. and Ravishankara A.R., On the role of iodine in ozone depletion. *J. Geophys. Res.*, 99D: 20491-20500.

Spivakovsky C.M., Yevich R., Logan J.A., Wofsy S.C., McElroy M.B. and Prather M.J. (1990), Tropospheric OH in a three-dimensional chemical tracer model: An assessment based on observations of CH_3CCl_3. *J. Geophys. Res.*, 95D: 18441-18471.

Toumi R. (1994), BrO as a sink for dimethylsulphide in the marine atmosphere. *Geophys. Res. Lett.*, 21: 117-120.

Turnipseed A.A. and Ravishankara A.R. (1993), The atmospheric oxidation of dimethylsulfide: Elementary steps in a complex mechanism. In: *Dimethylsulphide: Oceans, atmosphere and climate*, pp. 185-195, G. Restelli and G. Angeletti, Ed. Kluwer Academic Publishers, Dordrecht.

Vogt R. and Finlayson-Pitts B.J. (1994), Tropospheric HONO and reactions of oxides of nitrogen with NaCl. *Geophys. Res. Lett.*, 21: 2291-2294.

Waddington E.D., Cunningham J. and Harder S. (1996), The effects of snow ventilation on chemical concentrations. *This volume.*

Yokouchi Y., Akimoto H., Barrie L.A., Bottenheim J.W., Anlauf K. and Jobson B.T. (1994), Serial gas chromatographic/mass spectrometric measurements of some volatile organic compounds in the arctic atmosphere during the 1992 polar sunrise experiment. *J. Geophys. Res.*, 99D: 25379-25390.

MODELING FRAMEWORK FOR ATMOSPHERIC TRACE GAS MEASUREMENTS AT THE AIR-SNOW INTERFACE

Anne M. Thompson
NASA/Goddard Space Flight Center
Laboratory for Atmospheres
Greenbelt, Maryland 20771
USA

1. Introduction

There are two principal aims of this chapter. (1) To present a framework for the use of field observations, including transfer coefficients among snow-ice-atmosphere, in defining current atmospheric chemical processes. (2) To present a framework for interpreting ice core concentrations of key species in the past with a focus on the deduction of pre-industrial global hydroxyl (OH) radical concentration. The highly reactive hydroxyl radical determines the lifetime of major C, N, S and halogen species. These frameworks define sets of measurements that must be made to constrain atmospheric models for interpreting oxidant (O_3, OH, and H_2O_2) and other trace gas levels in present-day and past atmospheres.

In the first instance we see the use of models in assimilating measurements and guiding experimental design. In the case of atmospheric trace gases in a given environment, a standard set of species, reactions and rates of exchange processes (scavenging and deposition) is assumed to determine concentrations of gases that are typically measured. The major biogeochemical cycles of interest in the polar atmosphere are reviewed by Bottenheim and Barrie [1996] in this volume, along with examples of key species measurements in the Arctic. Bottenheim and Barrie show that a number of species (e.g. O_3, CO, hydrocarbons, NO_x [=NO+NO_2], alkyl nitrates, PAN [=peroxyacetylnitrate], halogenated hydrocarbons, DMS [=dimethyl sulfide]) measured at the surface or in the mixed layer occur at the concentrations expected from climatology and standard sets of chemical reactions. Exceptions occur during polar sunrise when ozone is destroyed rapidly within a matter of days, apparently due to interactions with halogens. Bromine and chlorine-containing constituents appear to play a role, though it is not yet known which particular species are responsible [Jobson et al., 1994]. This leads to the use of models to test new mechanisms and to predict other species

NATO ASI Series, Vol. I 43
Chemical Exchange Between the Atmosphere and Polar Snow
Edited by Eric W. Wolff and Roger C. Bales

concentrations that might be present but are not currently being measured.

It is a challenge to explain exceptions to simple gas-phase photochemical equilibrium that occur for what are referred to as "reversibly deposited" species, e.g. H_2O_2 and HCHO [Neftel, 1996; Bales and Choi, 1996]. These species interact for a limited time with snow, ice, and within pore space in firn, so that we cannot make the standard assumption in a photochemical model, that once removed by contact with snow or ice, the constituent has disappeared from the atmosphere. As in the case of polar sunrise ozone depletion, this is another problem for which photochemical models of the polar atmosphere are needed to test mechanistic assumptions against observations.

For modeling local processes one-dimensional, box or point models are adequate. In this chapter, an example of simulating polar gas phase chemistry is given by applying the one-dimensional Goddard Space Flight Center (GSFC) tropospheric photochemical model to observations from the 1992 Polar Sunrise Experiment (PSE-92) (December 1994 Issue of Journal of Geophysical Research - Atmospheres). A framework is given to complement the framework for exchange processes at snow and ice, given by Neftel [1996] and Bales and Choi [1996]. The first step in extrapolating from surface measurements at high latitudes to the global atmosphere is to relate surface or near-surface measurements to the mixed layer and local free troposphere, which may be probed by aircraft. PSE-92 was an airborne [Leaitch et al, 1994] as well as ground-based experiment and the use of high-frequency aircraft data in a photochemical model gave additional atmospheric information. This is further illustrated with a non-polar, high-latitude data set from ABLE-3B (Arctic Boundary Layer Experiment) [Harriss et al., 1994]. Of particular interest are distributions of the oxidants: H_2O_2, OH, and O_3. The local concentration of OH can be determined from a comprehensive set of O_3/NO_x/PAN/hydrocarbon/uv and CO measurements. Likewise, the net rate of ozone production (or destruction) can be calculated, which is another parameter that characterizes the local oxidizing environment.

The second use of atmospheric photochemical models for polar data is to deduce the composition of past atmospheres from ice-core measurements of trace gases. This requires a more complicated framework than for explaining a given set of trace gas measurements because of disparate lifetimes of gases measured, ranging from a few hours to tens of years. For gases with short lifetimes, polar concentrations may not be representative of most regions, or of a global burden. In the case of OH, it is the latter parameter that is of interest in assessing global change. Thus, we ask: what is the relationship between global composition and a gas measured in polar ice? I will illustrate with methane, which is one of the "simpler" gases in terms of transfer function (and, therefore, interpretation of the ice core measurement), atmospheric chemistry, and nearly uniform global distribution. However, even this "simple" gas is subject to many feedbacks, both in source - related to its biogenic origins - and in sink. In climates different from today's, the interaction of temperature (radiative forcing) and chemical composition is a driving issue. In other words, a cooler climate affected sources of CH_4 (nearly all biogenic during pre-industrial times) as well as the other species that determined the OH radical (the CH_4 sink). In the absence of definitive source data, modelers make varying assumptions of CH_4 sources and sinks. However, depending on the assumptions, a wide range of CH_4 fluxes and values for "global OH" have been derived [Thompson, 1992; 1995; Thompson et al, 1993a].

A final set of modeling issues is critical for more reactive constituents and is central to the interests of this NATO Workshop. How are localized, even regional, abundances determined in field experiments, or extracted from ice cores at a small number of locations, scaled up to global burden or distribution? For example, the OH radical, which is the only way of removing CH_4 from the atmosphere, is central to the biogeochemical cycles for hydrocarbons, N, S and halogenated constituents [Bottenheim and Barrie, 1996]. However, it has a lifetime of seconds and, in turn, it changes according to the concentrations of methane, CO, non-methane hydrocarbons (NMHC) and NO_x. Because the global abundance of OH is so critical to global change, a modeling and experimental framework is given for approaching the problem of deducing OH from ice-core measurements.

2. Modeling Atmospheric Photochemistry in Polar Environment

Figure 1 illustrates a conceptual framework for relating atmospheric measurements over snow or ice to surface processes as well as the partitioning of constituents within snow, firn and ice. Although our experimental focus is primarily on polar regions, the diagram is constructed to indicate that air-snow-ice relationships are needed at all latitude ranges. This is because species concentrations and key interactions among them have a large regional variation.

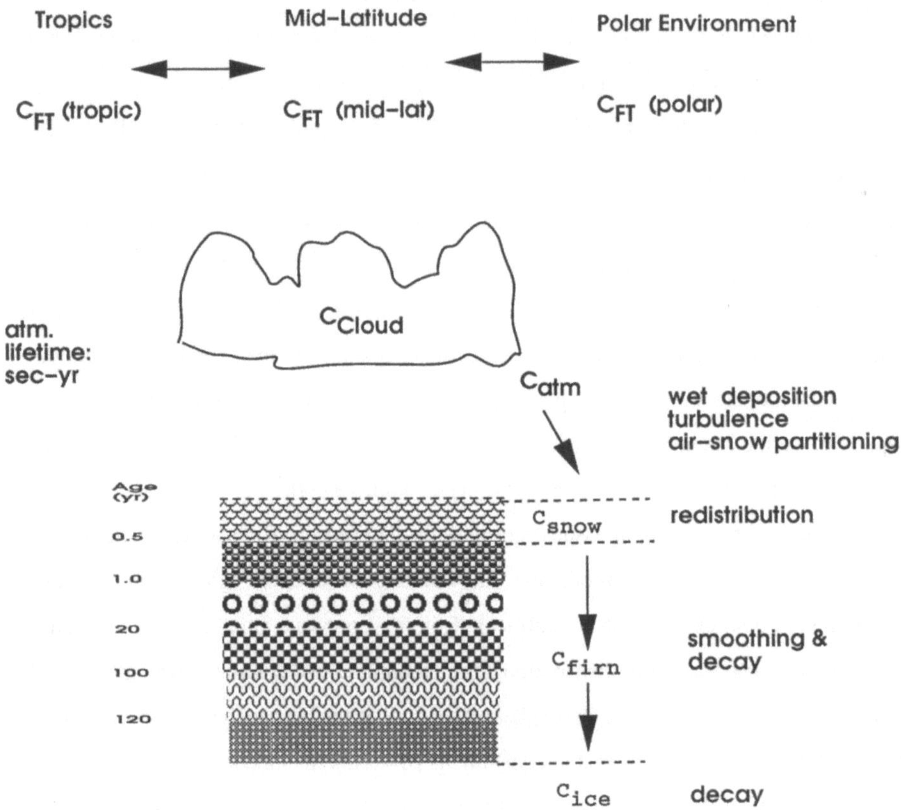

Fig. 1. Conceptual Framework for atmospheric modeling, with connection to processes at atmosphere-snow interface [cf Neftel et al, 1995; Bales and Choi, 1996].

To interpret atmospheric trace species concentrations, "c_{atm}" in Figure 1, we use a model that solves a series of continuity equations for the major gas-phase constituents. In the following section, a model of gas-phase photochemical reactions is applied to measurements made in PSE-92, conducted at Alert, Northwest Territories, in March-April 1992.

2.1 Photochemical Model Calculations for PSE-92

The general form of the continuity equation for a given species, c_i, is given by:

Flux Div.	+	Prod.	-	Loss(i)	=	$\partial [c_i]/ \partial t$
(Transport, Eddy diffusion)		(Photochem. Production)		(Photochem. Loss)		

The Production term consists of all processes that form species with mixing ratio, c_i, and Loss(i) represents all processes that destroy species i. In the steady-state version of the model, the right-hand side = 0. Eddy diffusion is typical for the formulation of vertical transport in a one-dimensional model. Strictly speaking, a one-dimensional model is valid only for one location and time and is best used for an air mass that is horizontally homogeneous. This is often a reasonable assumption for measurements made at a given polar sampling site.

The one-dimensional GSFC photochemical model [Thompson and Cicerone, 1982, 1986; Thompson et al, 1993b] is used to compute diurnally averaged profiles for conditions of the PSE-92 after a periodic 24-hour behavior is achieved. The period chosen for simulation is late March, when daylight is present but ozone depletion has not progressed significantly. Table 1 shows mean computed and measured concentrations for key species measured ~ 3m above the surface at Alert, Northwest Territories, Canada (NWT, 82°28'N, 62°30'W).

The model vertical domain spans 0-15 km with 24 altitude grid points. Vertical transport is prescribed by eddy diffusion; grid spacing is 1 km except in the lowest km near the surface, which is at higher resolution to allow prescription of a distinct surface layer below 100 m (smallest layer, 6.25 m). In the current formulation of the model, there are 37

Table 1. Mixing ratios for near-equinox (pre-ozone depletion) period
Polar Sunrise Experiment - 1992

Species	Model Mix. Ratio (24-hr mean)	Observed Mix. Ratio	Reference
O_3	38 ppbv	20-40 ppbv	Anlauf et al., 1994
NO_x	25 pptv	10-40 pptv	Leaitch et al., 1994 (airborne)
CH_4	1.8 ppmv	1.8 ppmv	Jobson et al., 1994
C_2H_6	1.6 ppbv	1.5-2.0 ppbv	Jobson et al., 1994
HCHO	34 pptv	30-600 pptv	de Sevres, 1994
PAN	400 pptv	150-600 pptv	Muthuramu et al., 1994
NO_y	500 pptv	600 pptv	Muthuramu et al., 1994
C_3H_6	31 pptv	30 pptv	Jobson et al., 1994
C_3H_8	330 pptv	300-1000 pptv	Jobson et al., 1994
C_2H_4	11 pptv	10-20 pptv	Jobson et al., 1994
toluene	.32 ppbv	.2-.3 ppbv, C_6H_6	Jobson et al., 1994
CH_3OOH	120 pptv	10-40 pptv	de Serves, 1994
initrate	0.1 pptv	10-15 pptv	Muthuramu et al., 1994
pnitrate	0.9 pptv	3 pptv	Muthuramu et al., 1994

continuity equations (Table 2) with an additional ten species solved assuming photochemical equilibrium. The reaction scheme is based on the JPL-1992 recommended rate coefficients for the 128 reactions shown in Table 3, except for some NMHC reactions which have been updated to the JPL-1994 recommendation. NMHC selected for model simulation include major lower molecular-weight NMHC - C_2H_6, C_3H_8, C_2H_4, and C_3H_6 - the oxidation of which contributes to aldehydes, organic peroxides, PAN and CO. The formation and destruction of the ethyl, 1-propyl and 2-propyl nitrates are also included; mechanisms and rate coefficients are given by Roberts [1990]. Temperature and relative humidity are prescribed using the mean profiles for 75° latitude in the US Std Atmosphere [1966]. Profiles of CH_4 and H_2 are held constant; the CH_4 mixing ratio is fixed at 1.8 ppmv.

Surface fluxes are expressed as the boundary conditions in the transport term. These consist of influxes for those species injected from the surface or the stratosphere as well as

Table 2. Species in GSFC 1D Photochemical Model (1995 Version, update from Thompson et al, 1993b)

Solved for in Continuity Equations:
O3, $O(^3P)$, NO, NO2, NO3, HNO_3, HNO_4, N_2O_5, PAN
H, OH, HO_2, H_2O_2, toluene, HCP, MPN ($CH_3O_2NO_2$),
CH_3O_2, CH_3OOH, H_2CO, C_2H_6, $C_2H_5O_2$, CH_3CHO, CH_3CO_3, C_2H_5OOH, CO,
C_3H_6, $C_3H_6OHO_2$, CH_2O_2, $C_2H_4O_2$, C_3H_8, $NC_3H_7O_2$, $IC_3H_7O_2$,
$C_2H_5ONO_2$, 1-propyl nitrate ("PNITRATE"),
2-propyl nitrate ("INITRATE"), C_2H_4, $C_2H_4OHO_2$

Fixed Profiles assumed for: Molecular density (M), N_2, O_2, H_2O, H_2, CH_4

Transients computed from photochemical equilibrium:

HCO, $O(^1D)$, C_2H_5O, CH_3O, CH_3, C_3H_6OHO, CH_2O_2, $C_2H_4O_2E$ (E = excited Criegee intermediate), CH_2O_2E, NC_3H_7O, IC_3H_7O

deposition for those species (e.g. HNO_3, PAN, organic nitrates) that are assumed to be removed irreversibly. Influxes represent actual injections or are surrogates for transport of moderate lifetime constituents from mid-latitudes (e.g. for CO, hydrocarbons, NO_x). For irreversibly deposited species, the transfer ratio c_{atm}/c_{snow}, is reflected in the first-order removal coefficients, i.e. deposition velocity. A similar parameter would also be included for irreversible loss to ice particles or fog removal. In the GSFC model, surface deposition is assumed for O_3, the peroxides, aldehydes, and all N-containing species except NO.

Photodissociation rates are computed for clear sky conditions at 80°N with an above-troposphere total O_3 column of 320 DU, which is typical for mid-high northern latitudes at equinox. A surface albedo of 0.8 is assumed, in accordance with the highly reflecting snow-ice site. Solar flux is appropriate for equinox. All mixing ratios given in Table 1 are based on 24-hour averages in which "diurnally averaged" rate coefficients and photodissociation rates are used. Table 1 shows that the constituents with poorest agreement between computed mixing ratios and those observed are HCHO, isopropyl nitrate, and CH_3OOH. Disagreement in the case of HCHO is taken as a sign that this species is formed in snow or ice. Models are

Table 3: Reaction Scheme in GSFC 1D Tropospheric Model (1995 version)

N#	REACTION
	Photodissociations
1	O3 + hv = O2 + O
2	O3 + hv = O2 + O(1D)
3	NO2 + hv = NO + O
4	HNO3 + hv = OH + NO2
5	H2O2 + hv = OH + OH
6	NO3 + hv = NO + O2
7	NO3 + hv = NO2 + O
8	H2CO + hv = H + HCO
9	CH3OOH + hv = OH + CH3O
10	HNO4 + hv = HO2 + NO2
11	CH3CHO + hv = CH3 + HCO
12	N2O5 + hv = NO2 + NO3
13	H2CO + hv = H2 + CO
14	C2H5OOH + hv = C2H5O + OH
15	PAN + hv = CH3CO3 + NO2
16	C2H5ONO2 + hv = C2H5O + NO2
17	PNITRATE + hv = NC3H7O + NO2
18	INITRATE + hv = IC3H7O + NO2
	Free-radical Reactions
19	O + O2 + MMM = O3 + M
20	O + O3 = O2 + O2
21	O + O + MMM = O2 + M
22	O(1D) + N2 = O + N2
23	O(1D) + O2 = O + O2
24	NO + O3 = NO2 + O2
25	NO2 + O = NO + O2
26	NO2 + O3 = NO3 + O2
27	NO + NO3 = NO2 + NO2
28	NO + O + MMM = NO2 + M
29	NO2 + NO3 + MMM = N2O5 + M
30	N2O5 = NO2 + NO3
31	O(1D) + H2O = OH + OH
32	O(1D) + CH4 = OH + CH3
33	O(1D) + CH4 = H2 + H2CO
34	O(1D) + H2 = OH + H
35	H + O3 = OH + O2
36	H + O2 + MMM = HO2 + M
37	OH + O3 = HO2 + O2
38	HO2 + O3 = OH + O2 + O2
39	OH + O = H + O2
40	HO2 + O = OH + O2
41	H2O2 + O = OH + HO2
42	OH + CH4 = H2O + CH3
43	HO2 + NO = OH + NO2
44	OH + CO = CO2 + H
45	OH + H2 = H2O + H
46	OH + NO2 + MMM = HNO3 + M
47	HNO3 + OH = NO3 + H2O
48	OH + H2O2 = H2O + HO2
49	OH + HO2 = H2O + O2
50	OH + OH = H2O + O
51	OH + H2CO = H2O + HCO
52	HO2 + HO2 = H2O2 + O2
53	OH + OH + MMM = H2O2 + M
54	H + HO2 = H2 + O2
55	H + HO2 = H2O + O
56	H + HO2 = OH + OH
57	H2CO + O = OH + HCO
58	CH3 + O2 + MMM = CH3O2 + M
59	CH3O2 + NO = CH3O + NO2
60	CH3O2 + HO2 = CH3OOH + O2
61	CH3OOH + OH = CH3O2 + H2O
62	CH3O + O2 = H2CO + HO2
63	H2 + O = OH + H
64	HO2 + NO2 + MMM = HNO4 + M
65	HNO4 = HO2 + NO2
66	HNO4 + OH = H2O + O2 + NO2
67	HCO + O2 = CO + HO2
68	C2H6 + OH = C2H5O2 + H2O
69	C2H5O2 + NO = C2H5O + NO2
70	C2H5O + O2 = CH3CHO + HO2
71	C2H5O2 + HO2 = C2H5OOH + O2
72	CH3CHO + OH = CH3CO3 + H2O
73	CH3CO3 + NO = CH3 + CO2 + NO2
74	CH3CO3 + NO2 + MMM = PAN + M
75	PAN = CH3CO3 + NO2
76	CH3OOH + OH = CH2OOH + H2O
77	C2H5OOH + OH = C2H5O2 + H2O
78	C2H5OOH + OH = C2H4OOH + H2O
79	N2O5 + H2O = HNO3 + HNO3
80	CH3O2 + CH3O2 = CH3OH + H2CO + O2
81	CH3O2 + CH3O2 = CH3O + H2CO + O2
82	CH3O2 + CH3O2 = CH3OOCH3 + O2
83	CH3CHO + NO3 = CH3CO3 + HNO3
84	H2CO + NO3 = HCO3 + HNO3
85	C3H6 + OH + MMM=C3H6OHO2 + H2O + MMM
86	C3H6OHO2 + NO = C3H6OHO + NO2
87	C3H6OHO + MMM = H2CO + CH3CHO
88	C3H6OHO2 + HO2 = C3H7O3H + O2
89	C3H6OHO2+C3H6OHO2=C3H6OHO+C3H6OHO+O2
90	C3H6OHO2 + CH3O2 = CH3O + C3H6OH + O2
91	C3H6 + O3 = CH2O2E + CH3CHO
92	CH2O2E + MMM = CH2O2
93	CH2O2E + MMM = H2 + CO2
94	CH2O2E + MMM = H2O + CO
95	CH2O2E + MMM = HCOOH
96	CH2O2 + H2O = HCOOH + H2O
97	C3H6 + O3 = C2H4O2E + H2CO
98	C2H4O2E + MMM = C2H4O2
99	C2H4O2E + MMM = CH4 + CO2
100	C2H4O2E + MMM = CH3OH + CO
101	C2H4O2E + MMM = CH3COOH
102	C2H4O2 + H2O = CH3COOH
103	C3H8 + OH = NC3H7O2 + H2O
104	NC3H7O2 + NO = NC3H7O + NO2
105	NC3H7O + O2 = PALD + HO2
106	NC3H7O2 + HO2 = NC3H7OOH + O2
107	NC3H7O2 + CH3O2 = NC3H7O + H2CO + O2
108	NC3H7O2 + NC3H7O2 = PALD + PALD + O2
109	C3H8 + OH = IC3H7O2 + H2O
110	IC3H7O2 + NO = IC3H7O + NO2
111	IC3H7O2 + NO = INITRATE + O2
112	IC3H7O + O2 = CH3COCH3
113	IC3H7O2 + HO2 = IC3H7OOH + O2
114	IC3H7O2 + CH3O2 = IC3H7O + H2CO + O2
115	C2H5O2 + NO = C2H5ONO2
116	NC3H7O2 + NO = PNITRATE
117	OH + C2H5ONO2 = PROD
118	OH + PNITRATE = PROD
119	OH + INITRATE = PROD
120	C2H4 + OH + MMM = C2H4OHO2 + H2O + M
121	C2H4OHO2 + NO = C2H4OHO + NO2
122	C2H4OHO2 + HO2 = C2H4O3H + O2
123	C2H4OHO2 + CH3O2 = CH3O + C2H4OH + O2
124	TOL + OH = HCP + HCP
125	HCP + NO = CH3CHO + HO2 + NO2
126	HCP + HO2 = C3H7O2H
127	CH3O2 + NO2 + MMM = MPN + MMM
128	MPN = CH3O2 + NO2

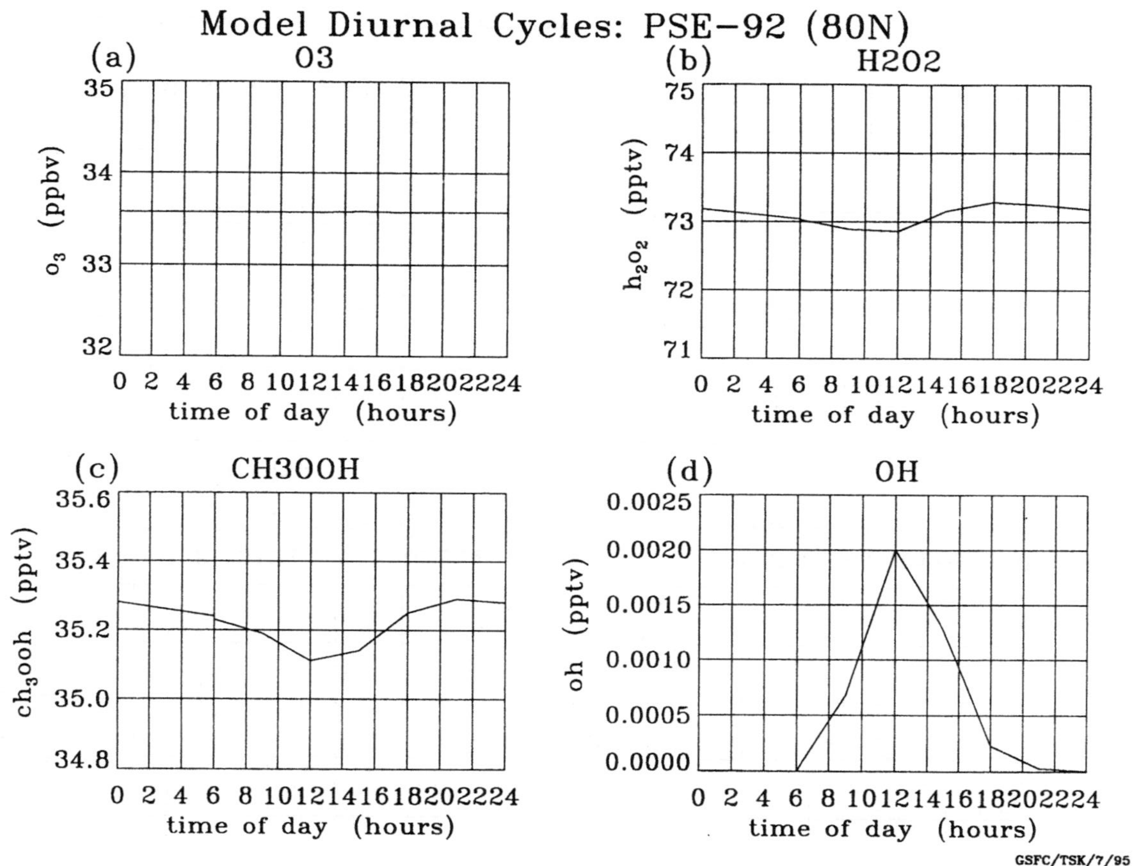

Fig. 2. Diurnal pattern of oxidants computed with GSFC 1D photochemical model for 80°N conditions (equinox) to simulate mean chemical concentrations of pre-ozone loss Polar Sunrise Experiment-92 (a) O_3; (b) H_2O_2; (c) CH3OOH; (d) OH.

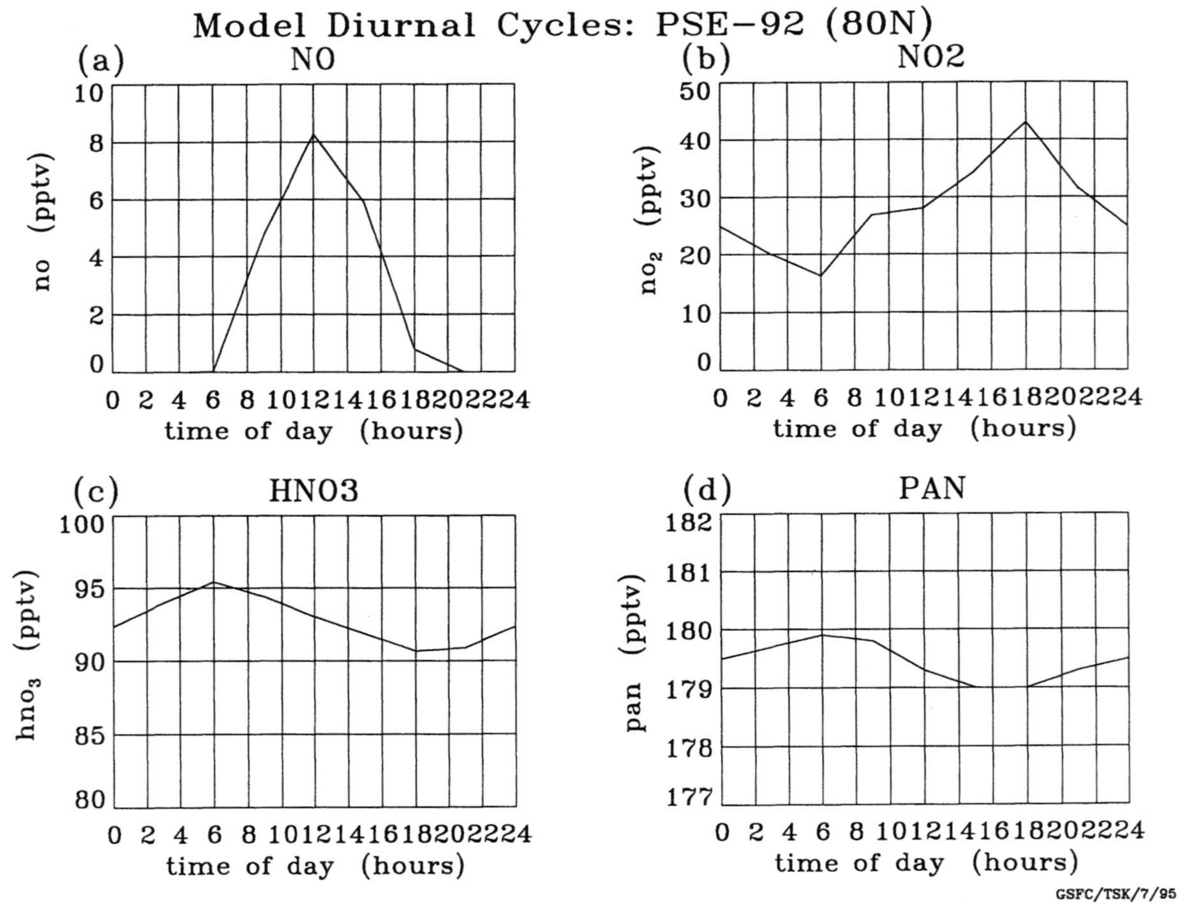

Fig. 3. Same as Fig. 2 except for odd nitrogen constituents.

often too low in HCHO in polar regions [Neftel et al, 1995] and non-polar regimes as well [Jacob et al, 1995].

Diurnal cycles for the constituents simulated are in two categories. Sunlight initiated free radicals appear only between sunrise and sunset with a peak at noon, e.g. OH and NO in Figures 2d and 3a, respectively. Note that OH reflects the very high solar zenith angle at Alert. Computed mixing ratios of OH and H_2O_2 at Summit, Greenland (72°N), at the same equinoctal conditions and boundary conditions illustrate a sharp latitude gradient. They are, respectively, factors of 7.5 and 4 higher than at Alert. Computed H_2O_2 mixing ratios for Greenland, 275 pptv, agree with observations [Bales and Choi, 1996; McConnell and Bales, personal communication, 1995]. Longer-lived species with little diurnal variation include O_3, the peroxides and PAN (Figures 2, 3). Intermediate between this are NO_2 and HNO_3 (Figures 3b, c) which vary with transformation rates to other reactive nitrogen species: NO_3 and N_2O_5. Although our NO_y agrees with observations, computed HNO_3 is too high [J. Bottenheim, personal communication, 1995]; a too low simulation of organic nitrogen species may account for some of the difference.

2.2 Modeling Framework for Atmospheric Measurements at Polar Sites

Relationships among transfer coefficients for the various processes occurring on the snow and ice (Figure 1) mean that as many processes as possible need to be measured simultaneously to determine partitioning and mechanisms. Bales and Choi [1996] treat experimental approaches to determine the coefficients for transfer among atmosphere, firn, pore-space and ice.

In designing experiments, we are concerned with the question: Assuming that we know partitioning between snow, c_{snow}, and the atmosphere in contact with it, c_{atm}, how do we determine the relationship of the surface atmospheric layer to the free troposphere, c_{FT}? The answer is by coupling an aircraft experiment to a ground-based experiment in a framework that relates c_{FT} and c_{atm}, with simultaneous measurements. Disparity between the two suggests that these regimes are separated from one another, just as a complex relationship among c_{snow}, c_{atm}, and c_{firn}, for example, indicates that chemical processes are occurring within the snow or ice. Likely explanations are a transport barrier between surface layer and free troposphere, as well as heterogeneous processes, e.g. removal in snow, ice crystals, precipitation. Thus, in addition to concentrations in the gas phase, concentrations of a given species, c, in clouds, fog or snow, are required to complete the framework.

Interpretation of aircraft data, i.e. a set of measurements of c_{FT}, can be made with a photochemical model as described above. With aircraft samples from a fairly homogeneous air mass, e.g. taking a set of vertical profiles, a 1D or box model can give meaningful results. Over a farther free tropospheric range, an aircraft is likely to intercept multiple air masses. To characterize distinct layers, high-frequency instrumentation is required and the age and origin of air masses can be determined by trajectory analysis, ratios of key compounds, tracer dilutions [McKeen and Liu, 1993]. A photochemical point model (an instantaneous steady-state model) can be used to assimilate high-frequency data from multiple air masses provided sufficient numbers of species are measured [Davis et al, 1995; Jacob et al, 1995]. Ideally, the aircraft has the same complement of instrumentation as the ground-based experiment.

An example of high-frequency aircraft data collected at high latitude is given in Figure 4. These observations were made during summertime (21 July 1990) on the Arctic Boundary Layer Experiment (ABLE 3B) using the NASA Electra in Manitoba, Canada. The air masses sampled mostly represented background nonpolluted conditions although biomass burning plumes, typical of the region, were also encountered [Blake et al, 1994; Wofsy et al, 1994]. Note hydrocarbon spikes in the ethene example in Figure 4c. Using these data in a merged set, calculations were made with the GSFC "point model." The point model uses the same reaction set as in Table 3.

Enough species were measured to characterize ozone photochemical production and loss and to constrain the calculation of OH (Figure 5a) quite tightly [Fan et al. 1994, Mauzerall et al., 1995]. Without the prescribed input of O_3/NO/CO/hydrocarbons/uv/H_2O/HNO_3/PAN, uncertainties in computed OH could exceed 50% [Thompson and Stewart, 1991]. Note the large (> factor of 10) variability in instantaneous OH for air masses sampled on this flight. The net rate of O_3 formation is computed assuming instantaneous photochemical equilibrium at each measurement point (Figure 5b). Note that most of the environment is net O_3-destroying. Only in a few air masses with elevated NO is there net O_3 formation. The fact that CO and hydrocarbon spikes (note correlation in Figure 6) denoting biomass burning-affected plumes only occasionally coincide with elevated NO suggests that the plumes are relatively well-aged.

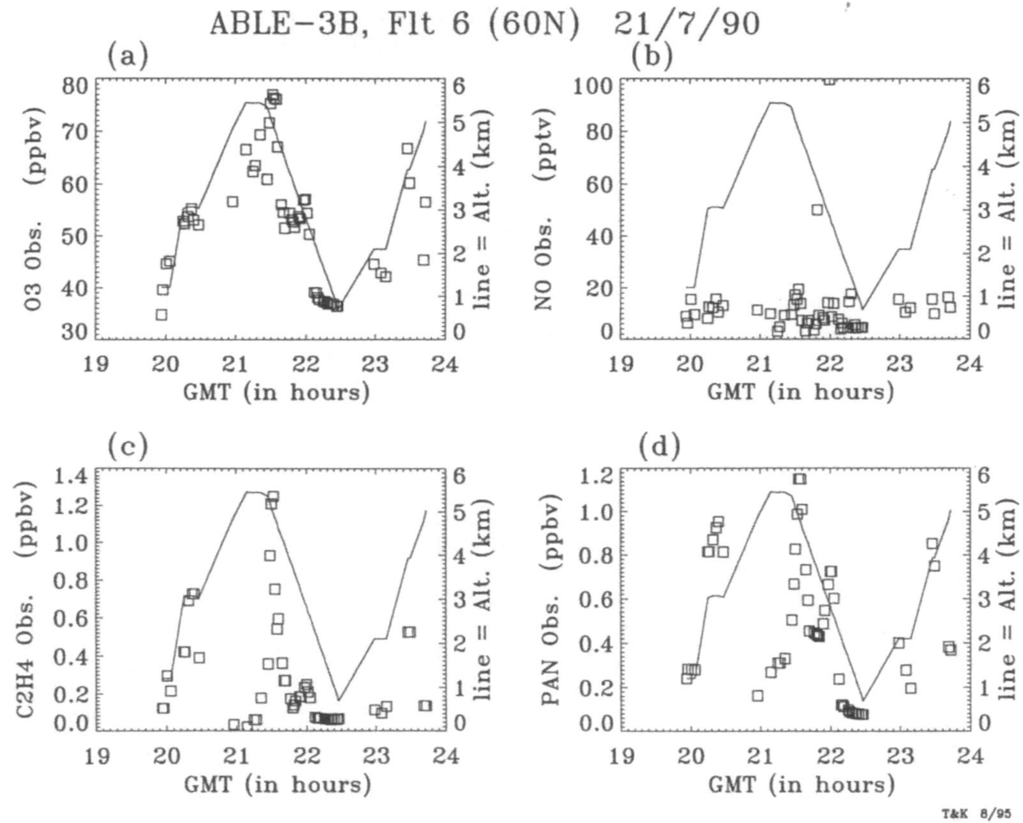

Fig. 4. High-frequency data collected from NASA Electra during ABLE 3B Flight 6, 21 July 1990, out of Churchill, Manitoba. Solid line shows altitude. (a) O_3; (b) NO; (c) C_2H_4; (d) PAN.

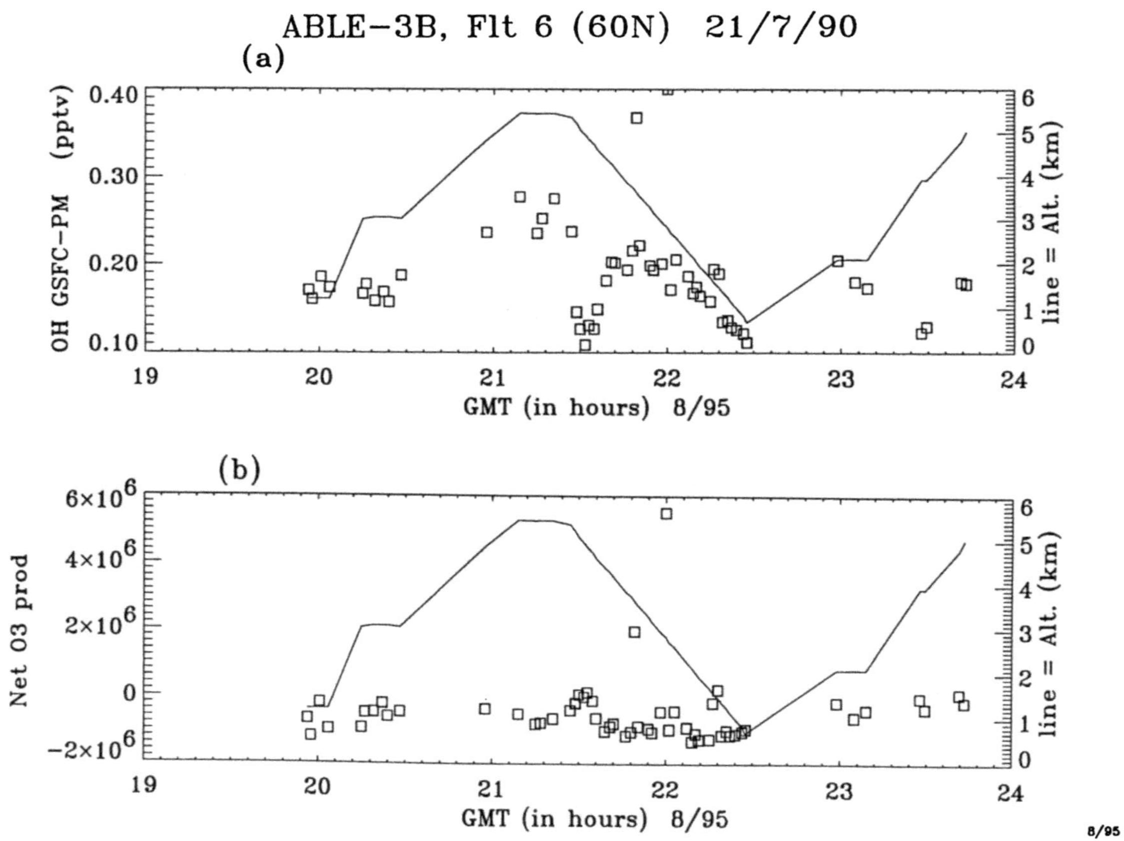

Fig. 5. Instantaneous (a) OH and (b) net O_3 formation rate computed by GSFC point model using input from ABLE 3B with O_3/CO/NO/NMHC/PAN/uv/HNO_3 constraints.

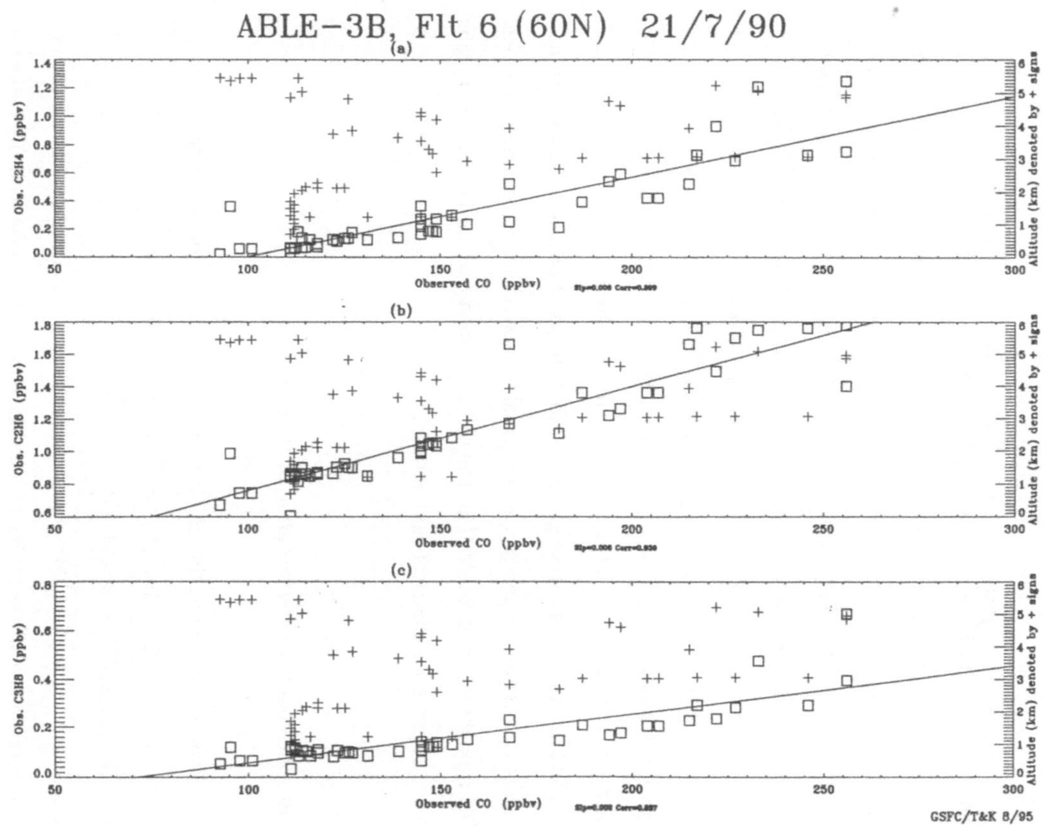

Fig. 6. Correlations between NMHC and CO from ABLE 3B Flight 6. High correlation indicates common source, with low values representing clean background air and higher values suggestive of moderately aged pollution from wildfires.

3. Modeling Framework for Interpreting Ice Core Measurements

There are several challenges in using models to interpret ice core data when one considers the processes illustrated in Figure 1. First is that the mechanisms, processes and transfer coefficients one measures in a contemporary experiment might be different in a regime with different temperature, moisture, for example. The only species for which this appears not be a problem is a relatively inert constituent like methane for which the concentration extracted from the ice core is a good proxy for the global atmospheric abundance. Methane is not soluble and its present-day (PD) lifetime is 8-10 years, yet interpretation of how and why ambient levels are changing is not simple (Section 3.1). The second challenge is that even if the transfer coefficients of today are valid, the problem of scaling from localized measurements to global distributions for trace gases measured in ice cores is formidable. Section 3.2 presents a conceptual framework for determining the global burden of OH from the ice-core record. This can serve as a model for the derivation of global information on the past trace chemical environment.

3.1 Modeling Methane for Interpretation of Ice Core Data

The problem in interpreting the ice-core methane record in terms of perturbed chemical processes is apparent in the CH_4 continuity equation:

$$\text{Flux Div.} - k^*[CH_4][OH] = \partial [CH_4]/ \partial t$$

where the flux divergence term includes transport as well as fluxes from a variety of sources (Table 4). The predominant physical sink is soil uptake (included mathematically in the flux divergence) and reaction with OH is the primary photochemical removal process.

First, even with a good present-day record of CH_4 at global monitoring sites, it has been necessary to conduct model sensitivity studies to decompose the contemporary CH_4 budget [Fung et al, 1991]. Second, perturbed climatic conditions affect the rate at which

Table 4. Methane sources and sinks	
Natural sources -	
Wetlands	
Termites	Clathrates
Wildfires	Ocean
Sources affected by human activity -	
Animals (ruminants)	
Gas leaks	
Rice paddies	
Biomass & wood fuel burning	
Coal mining	
Landfills	
Fossil fuel combustion	
Physical sink: soil uptake	Chemical sink: OH reaction

natural processes produce CH_4. The temperature sensitivity of biogenic sources is obvious, but other factors are indirect. For example, a perturbed climate will have different lightning frequency [Price and Rind, 1994], with presumably, altered occurrence of natural wildfires (Table 4). Moisture levels affect wetlands CH_4 emissions. Estimates of PD, pre-industrial Holocene and Last Glacial Maximum (LGM, 18K years ago) methane sources, with parameterized formulation of wetlands emissions under altered climatic conditions, were made by Chappellaz et al. [1993].

The complexity in evaluating whether methane is perturbed by modified sources or a different OH level is that OH levels have probably not remained constant. Multiple, nonlinear interactions of the hydroxyl radical with methane itself and with other atmospheric trace species that have changed over time, imply that OH in past atmospheres could have been higher or lower than at present (Table 5). Model studies indicate a range of possible changes in global OH since the LGM [Thompson et al, 1993a]. Most suggest that global OH is 20-30% less now than in the pre-industrial Holocene and that during the LGM, OH was probably somewhat higher than in pre-industrial times [Pinto and Khalil, 1991]. This "loss of OH" is the result of varying assumptions on present-day and pre-industrial distributions for NO_x, CO and hydrocarbon sources. In general, only methane data have been used as a constraint. This is not very rigorous because methane fluxes parameterized to match pre-industrial and LGM conditions cover almost a factor of 2 [Thompson et al, 1993a].

Hence, there is interest in finding ice-core data for non-methane constituents, including those species that are markers for atmospheric oxidizing capacity.

3.2 An Approach to Modeling Historic Oxidant Levels

Several models have used H_2O_2 and HCHO ice-core data and found that it is not possible to simultaneously reproduce all the observations [Thompson et al., 1993a] without making assumptions about highly perturbed ultraviolet conditions during the LGM [Staffelbach et al, 1993]. Thus, a new framework for interpretation of ice cores is warranted.

The challenge to modeling oxidation changes is that OH is itself affected by CH_4 changes as well as by altered emissions of hydrocarbons, CO and NO_x. Present-day models frequently use published emission inventories for global distributions of CH_4, CO, hydrocarbons and NO_x, but there are no inventories for past atmospheres. We suggest a framework whereby measurements of the snow-air relationship and of species deposited in ice and snow can be used to deduce a regional OH level. A regional approach is required because of highly diverse NO_x/CO/CH_4/O_3/OH environments. The global OH burden can be derived from adding regional OH amounts. The regions studied include the tropics where OH may be at highest column amount due to a higher tropopause (i.e. more mass in the troposphere) and more uv and moisture from which to convert O_3 to OH. Under present-day conditions, biomass burning, industrialization in the tropics, and possibly upper tropospheric lightning (giving rise to NO_x and through photochemical reaction to O_3) tend to promote OH formation. Regional models, however, require more reliable CH_4, CO, NO_x and NMHC budgets than we have at present. Sensitivity studies should be done for all of these sources. Soluble gases, which end up in snow and ice, can be useful constraints for resolving present-day chemical budgets. With contemporary anthropogenic sources identifed within a region, OH can be computed from a model and natural contributions to the budgets will be better isolated. Examples for two constituents follow:

<u>NO_x</u>: Combustion dominates NO_x sources in mid- and high latitude industrialized regions [Logan, 1983; Dignon, 1992]. In the tropics three sources potentially more important than industrial activity are biomass burning, lightning and soil/biogenic sources. Tropical snow-atmosphere relationships and snow nitrate seasonality from, for example, Peruvian ice could help distinguish regional NO_x sources. Thompson et al. [1995] report that nitric acid in snow

Table 5. Interaction of CH_4/CO and NMHC/NO with O_3 and OH photochemistry

Methane Oxidation:

$OH + CH_4 (+ O_2) \longrightarrow CH_3O_2 + H_2O$
$CH_3O_2 + NO \longrightarrow NO_2 + CH_3O$
(Form formaldehyde: $CH_3O (+ O_2) \longrightarrow HCHO$)
$HCHO + h\nu \longrightarrow H_2 + CO$ <-- formation of CO
$HCHO + h\nu (+ O_2) \longrightarrow HO_2 + CHO$

OH/HO2 Balance Modified by CO Rxn
$OH + CO (+ O_2) \longrightarrow CO_2 + HO_2$

* With fixed NO_x, increases in CH_4, CO and NMHC ==> lower OH

Conversion of NO to NO_2:

$CH_3O_2 + NO \longrightarrow NO_2 + CH_3O$
$RO_2 + NO \longrightarrow NO_2 + RO$ (or any RO_2 from NMHC)
$HO_2 + NO \longrightarrow NO_2 + OH$

Formation of O_3:
$NO_2 + h\nu \longrightarrow O + NO$
$O + O_2 (+ M) \longrightarrow O_3 + M$ CO/NO_x/HC favor O_3

* With fixed NO_x, increases in CH_4, CO and NMHC ==> more O_3

* Increases in NO_x (up to 1 ppbv) ==> more O_3, more or less OH, depending on CH_4/CO/NMHC changes

Formation of OH:
$O_3 + h\nu \longrightarrow O(^1D) + O_2$
$O(^1D) + H_2O \longrightarrow OH + OH$

When NO_x levels > 1 ppbv, adding more NO_x may suppress OH through formation of nitric acid: HNO_3

Formation of H_2O_2:
$HO_2 + HO_2 \longrightarrow H_2O_2 + O_2$

is greatest during the season of greatest precipitation (November-January). Two explanations appear plausible: enrichment by biomass burning combustion (or biogenic) products, eg insoluble but oxidizable organic-nitrogen compounds; and HNO_3 formed from oxidation of lightning NO. Nitric acid in the upper troposphere over South America and the South

Atlantic Basin is high during the dry season [Talbot et al., 1995], when biomass burning is at its peak. Deep convection followed by long-range transport is the transport mechanism for getting nitrogen compounds to the middle and upper troposphere. However, deep convection is accompanied by lightning. Surface and lightning NO_x should be distinguishable by correlating nitric acid in snow with tracers, e.g. soot or organic acids. Indeed, soluble organics are expected to differ depending on whether the hydrocarbon origins are biogenic or combustion. With present-day NO_x sources better resolved, historic NO_x budgets should be better constrained. This is just one example of how to use atmosphere-snow relationships and non-polar snow/ice measurements to infer chemical budget information for both present and past atmospheres.

CO: Regional CO budgets should be easier to construct than NO_x because of the longer lifetime of this constituent. However, the photochemical CO source (from oxidation of NMHC, both anthropogenic and biogenic) is uncertain, partly due to the complex nature of oxidation pathways. Direct measurement of CO in ice cores would be a big step forward for constraining OH. Not only is it the major sink term for OH, but it is expected to have relatively straightforward atmosphere-snow partitioning. Note that carbon monoxide, like NO_x and CH_4, has large tropical sources - from biomass burning, plant/soil emissions, oxidation of vegetative hydrocarbon emissions. CO budgets can be refined in regional models of the present day atmosphere when sufficient observations of concentrations and fluxes, with correlative tracer measurements, are available. Field measurements need to include temperature dependence of biogenic sources. For example, in isoprene emissions from vegetation, model prescriptions are usually parameterized according to temperature.

With sources and regional sinks (OH) better defined for present-day conditions, climate model parameters can be used to prescribe historic sources for CH_4, NO, NMHC and CO as in the study of Chappellaz et al [1993]. Besides perturbed sources for gases with sources that respond to different temperature forcing, transport and other factors like precipitation, moisture and cloud cover, affected by climate, will feed back on OH.

4. Summary

The framework described above is summarized as shown in Figure 7. To constrain regional (and global) OH take simultaneous measurements of multiple species in the atmosphere and in snow and ice cores -- CH_4, peroxides, acids. Use correlations and process models to infer regional sources and sinks. Scale up from regional to global to determine atmospheric oxidizing capacity present and past. It is important to take advantage of as many soluble species measurements as possible in resolving sources.

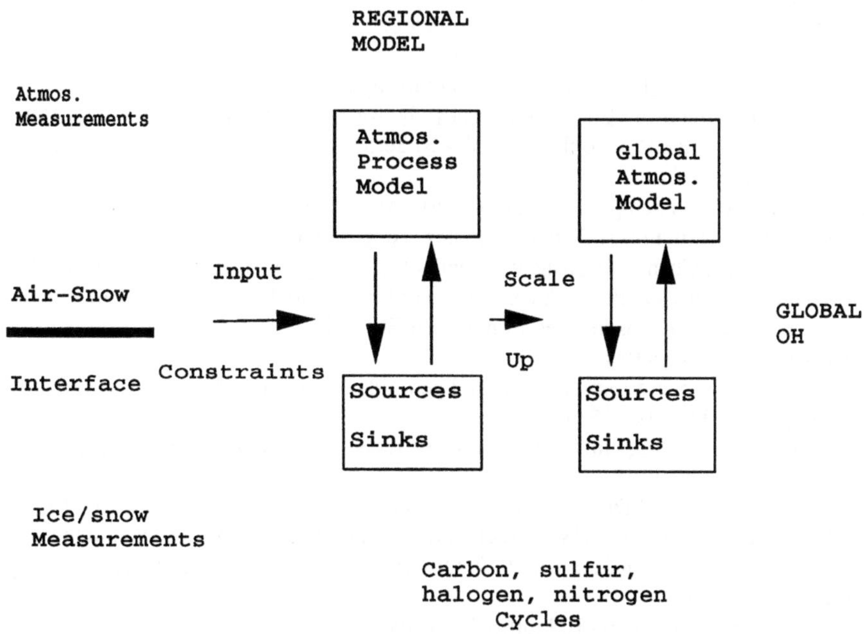

Fig. 7. Schematic for relating measurements from snow and ice to atmospheric measurements with models as integrating device.

Acknowledgments. This work was supported by the NASA EOS and Tropospheric Chemistry Programs. Thanks to T. L. Kucsera (Applied Research Corporation at GSFC) for modeling and graphics assistance and to D. L. Mauzerall and D. J. Jacob of Harvard for the

ABLE 3B data set. R. C. Bales and J. McConnell (Univ. of Arizona), D. P. McNamara (Applied Research Corporation at GSFC) and J. Bottenheim (AES, Downsview) made helpful comments on the manuscript.

References

Anlauf KG, Mickle RE, Trivett NBA (1994) Measurement of polar ozone during Polar Sunrise Experiment 1992, J Geophys Res, **99:**25345-25353.

Bales RC and Choi J (1996) Conceptual framework for interpretation of exchange processes, in Chemical exchange between the atmosphere and polar snow, ed. E. W. Wolff and R. C. Bales, NATO ASI Series, Springer-Verlag, Heidelberg, (This Volume).

Blake DR, Smith TW, Chen T-Y, Whipple WJ, and Rowland FS (1994) Effects of biomass burning on summertime nonmethane hydrocarbon concentrations in the Canadian wetlands, J Geophys Res, **99:** 1699-1720.

Bottenheim JW, Barrie LA (1996) Chemical reactions in the polar troposphere relevant to C, S, and N compounds, in Chemical exchange between the atmosphere and polar snow, ed. E. W. Wolff and R. C. Bales, NATO ASI Series, Springer-Verlag, Heidelberg, (This Volume).

Chappellaz JA, Fung IY, Thompson AM (1993) Atmospheric methane increase since the Last Glacial Maximum. 1. Source estimates, Tellus, **45B:**228-241.

Davis DD, Crawford J, Chen G, Bradshaw J, Sandholm S, Gregory G, Anderson B, Blake D, Rowland F, Sachse G (1995) An assessment of the photochemical O_3 tendency in the western North Pacific as inferred from GTE/PEM-WEST-A observations during the Fall 1991, J Geophys Res, in press.

de Serves C (1994) Gas phase formaldehyde and peroxide measurements in the Arctic, J Geophys Res, **99**:25391-25398.

Dignon J (1992) NO_x and SO_x emissions from fossil fuels: A global distribution, Atmos Environ **26A:**1157-1163.

Fan SM, Jacob DJ, Mauzerall DL, Bradshaw JD, Sandholm ST, Blake DR, Singh HB, Talbot RW, Gregory GL, Sachse GW, (1994) Photochemistry of reactive nitrogen in the subarctic troposphere in Summer 1990: Observations and modeling, J Geophys Res, **99:** 16867-16906.

Fung I, John J, Lerner J, Matthews E, Prather M, Steele LP, Fraser PJ (1991) Three-dimensional model synthesis of the global methane cycle, J Geophys Res, 96:13033-13065.

Harriss RC, Wofsy SC, Hoell JM Jr, Bendura RJ, Drewry JW, McNeal RJ, Pierce D, Rabine V, Snell RL, The Arctic Boundary Layer Expediton (ABLE 3B): July-August 1990, J Geophys Res, 99:1635-1644.

Jacob DJ, Heikes BG, Fan S-M, Logan JA, Mauzerall DL, Bradshaw JD, Singh HB, Gregory GL, Talbot RW, Blake DR, Sachse GW (1995) The origin of ozone and NO_x in the tropical troposphere: A photochemical analysis of aircraft observations over the South Atlantic Basin, J Geophys Res submitted.

Jobson BT, Niki H, Yokouchi Y, Bottenheim JW, Hopper JF, Leaitch WR (1994) Measurements of C_2-C_6 hydrocarbons during the Polar Sunrise 1992 Experiment:

Evidence for Cl atom and Br atom chemistry, J Geophys Res, **99:**25355-25368.
Leaitch WR, Barrie LA, Bottenheim JW, Li S-M, Shepson PB, Muthuramu K, Yokouchi Y (1994) Airborne observations related to ozone depletion at polar sunrise, J Geophys Res, **99**:25499-517
Logan JA (1983) Nitrogen oxides in the troposphere: Global and regional budgets, J Geophys Res, **88**:10785-10471.
Mauzerall DL, Jacob DJ, Fan S-M, Bradshaw JD, Gregory GL, Sachse GW, Blake DR (1995) Origin of tropospheric ozone at remote high northern latitudes in summer, J Geophys Res, in press.
McKeen SA, Liu SC (1993) Hydrocarbon ratios and photochemical history of air masses Geophys. Res. Lett. **20**:2363-2366.
Muthurama K, Shepson PB, Bottenheim JW, Jobson BT, Niki H, Anlauf KW (1994) Relationships between organic nitrates and surface ozone destruction during the Polar Sunrise Experiment 1992, J Geophys Res, **99:** 25369-25378.
Neftel A. (1996) The record of gases and reactive species in ice cores, and problems of interpretation, in Chemical exchange between the atmosphere and polar snow, ed. E. W. Wolff and R. C. Bales, NATO ASI Series, Springer-Verlag, Heidelberg, (This Volume).
Neftel A, Bales RC, Jacob DJ (1995) H_2O_2 and HCHO in polar snow and their relation to atmospheric chemistry. in Ice Core Studies of Global Biogeochemical Cycles, ed. R. J. Delmas, NATO ASI Series, vol I 30, Springer-Verlag, Heidelberg, pp 249-264.
Pinto JP, Khalil MAK (1991) The stability of tropospheric OH during ice ages, interglacial epochs and modern times, Tellus, 43B:136-151.
Price C, Rind D (1994) The impact of a $2XCO_2$ climate on lightning-caused fires, J. Climate, **7:**1484-1494.
Roberts JM (1990) The atmospheric chemistry of organic nitrates, Atmos Environ **24A**:243-287.
Staffelbach T, Neftel A, Stauffer B and Jacob D (1991) A record of the atmospheric methane sink from formaldehyde in polar ice cores, Nature, 349, 603-605.
Talbot RW et al (1995) Chemical characteristics of continental outflow over the tropical south Atlantic Ocean from Brazil and southern Africa, J Geophys Res submitted.
Thompson, AM (1992) The oxidizing capacity of the Earth's atmosphere: Probable past and future changes, Science, **256**, 1157-1168.
Thompson AM (1995) Photochemical modeling of chemical cycles: Issues related to the interpretation of ice core data, in Ice Core Studies of Global Biogeochemical Cycles, ed. R. J. Delmas, NATO ASI Series, vol I 30, Springer-Verlag, Heidelberg, pp 265-297.
Thompson AM and Cicerone RJ (1982) Clouds and wet removal as causes of variation in the the composition of the unpolluted troposphere, J Geophys Res, **87:**8811-8826.
Thompson AM and Cicerone RJ (1986) Possible perturbations to tropospheric CO, CH_4, and OH, J Geophys Res, **91:**10853-10864.
Thompson AM, Stewart RW (1991) The effect of kinetics imprecisions on constituents calculated in a tropospheric photochemical model. J Geophys Res, **96**:13089-13109.
Thompson AM, Chappellaz JA, Fung IY, Kucsera TL (1993a) Atmospheric methane increase since the Last Glacial Maximum. 2. Interactions with oxidants, Tellus, **45B:**242-257.
Thompson AM et al. (1993b) SAGA-3 ozone observations and a photochemical model analysis of the marine boundary layer during SAGA-3, J Geophys Res, **98**:16955-16968.

Thompson LG, Mosley-Thompson E, Davis ME, Lin P-N, Henderson KA, Cole-Dai J, Bolzan JF, Liu K-b (1995) Late glacial stage and holocene tropical ice core records from Huascaran, Peru, Science **269**: 46-50.

Wofsy SC, Fan S-M, Blake DR, Bradshaw JD, Sandholm ST, Singh HB, Sachse GW, Harriss RC (1994) Factors influencing atmospheric composition over subarctic North America during summer, J Geophys Res **99**: 1887-1898.

OVERVIEW OF FIELD DATA ON THE DEPOSITION OF AEROSOL-ASSOCIATED SPECIES TO THE SURFACE SNOW OF POLAR GLACIERS, PARTICULARLY RECENT WORK IN GREENLAND

Jack E. Dibb
Glacier Research Group
Institute for the Study of Earth, Oceans and Space
University of New Hampshire
Durham, New Hampshire 03824, USA
(also affiliated with the Department of Earth Sciences)

ABSTRACT

This contribution presents a review of recent field experiments investigating the relationship between the composition of snow falling onto polar glaciers and the composition of aerosols in the overlying atmosphere. The limited data exisiting prior to the late 1980s indicated that aerosol removal processes should cause fractionation between the composition of aerosols and snow. Uncertainties regarding the relative importance of ice-nucleation scavenging, in-cloud riming of snow flakes and dry deposition over polar ice sheets precluded assessment of the impact the likely fractionation would have on efforts to reconstruct temporal variations in aerosol chemistry from ice core chemistry records. Two large international experiments on the Greenland ice sheet after 1988 focused on these, and other, issues central to understanding air-snow exchange processes. These experiments confirmed that ice-nucleation scavenging is the major process incorporating aerosol-associated species into polar snow. This process enriches snow in large aerosols relative to the aerosol population aloft. Dry deposition was found to be of minor importance at the present time, but also enriches the snow in large aerosols and the species associated with them. Since the large aerosols over Greenland are predominantly derived from sea-salt and dust, while $SO_4^{=}$, NH_4^{+} and several pollutant trace metals are concentrated in submicron aerosols, snow chemistry presents a biased view of aerosol composition. However, it would appear to be possible to incorporate such a bias into efforts to reconstruct aerosol chemistry records from ice cores, were it not for several complicating factors. Spatial variability in snow chemistry will likely impose an inherent limit on the temporal resolution that will be possible in such

NATO ASI Series, Vol. I 43
Chemical Exchange Between the Atmosphere and Polar Snow
Edited by Eric W. Wolff and Roger C. Bales

reconstructions. Variations in the seasonal pattern of snow accumulation over time have the potential to greatly complicate interpretation of ice core chemistry records. Unfortunately, it is not clear if seasonal changes in snow accumulation could be recognized and quantified from ice core records, if they have occurred. Deposition of aerosols and associated species by fog droplets was found to be of greater importance than expected. This mechanism does not fractionate the atmospheric aerosol population to the same extent as snowfall and dry deposition. Variations in the relative contributions of fog versus snow over time (on seasonal to millenial scales) would thus alter the relationship between the composition of aerosols and the snowpack. Here again, it is uncertain whether such changes could be recognized in an ice core and accounted for in the reconstruction of past aerosol composition and loading.

INTRODUCTION

Polar glaciers are, without doubt, the premier repositories of information on the past composition of the atmosphere. Examples of the records obtained from ice cores collected in Greenland and Antarctica are presented and discussed by Wolff (this volume) and Neftel (this volume). However, this meeting (and volume) serves as a reminder that the chemical composition of glacial snow and iee is a proxy for the chemical composition of the overlying atmosphere (see also Bales et al., 1992; Jaffrezo et al., in press). The relationships between the composition of air and snow may not be that well understood, even for the present day. Reconstruction of atmospheric composition at earlier times from ice core chemical records requires not only that we improve our understanding of the transfer processes operating now, but that we become able to predict how these processes would have changed under different climatic conditions in the past.

This paper focuses on "irreversibly" (*sensu* Bales et al., 1992) deposited species, i.e., non-volatile components of the atmosphere that are not felt to partition between the condensed and vapor phases after deposition to snow. Insoluble particles like crustal dust or volcanic ash represent the "irreversible" end members contrasting highly "reversible" compounds like H_2O_2 (Fuhrer, this volume). Insoluble particles have no appreciable interactions with gaseous species in the atmosphere (other than providing surfaces for heterogeneous reactions and perhaps becoming coated with products of such reactions). The composition of sea-salt aerosols can be modified by interactions with gaseous acids during their atmospheric lifetimes, but they have been considered essentially chemically inert once deposited on cold, dry snow. Similarly, trace metals can be emitted to the atmosphere as vapor (by volcanoes, smelting, and high temperature combustion), but soon condense into particulate forms in which they remain until and after deposition. The situation is less clear for the present day major components of remote free tropospheric aerosols, namely H_2SO_4, NH_4HSO_4 and $(NH_4)_2SO_4$. Gaseous precursors (NH_3 and SO_2) are the dominant atmospheric sources of these species, with gas to aerosol conversion and ongoing exchange between the phases important issues in tropospheric chemistry. However, NH_4^+ and $SO_4^=$ records in snow and ice have usually been discussed in terms of aerosol removal processes, so this paper will include them as mainly "irreversible" species. It should be borne in mind that the concept of reversible vs. irreversible deposition is a convenient way to think about some of the important issues in air-snow exchange, but that transitional behavior between the idealized end members is likely. For example, Delmas (pers comm) suggests that Cl^- can be liberated from sea-salt aerosols after deposition to snow.

The fundamental issue regarding air-snow exchange of irreversibly deposited species to glacier surfaces can be formulated as a simple question: Does the composition of accumulating snow record an unbiased picture of the composition of the aerosols in the air masses in which the snow flakes formed? If it does, then explanations for downcore (temporal) changes in the concentrations and fluxes of these species can be sought in varying source-receptor relationships (source types and strengths interacting with transport mechanisms) over time. Even if the composition of snow presents a biased view of aerosol composition (due to fractionation between aerosol and snow by the removal processes) interpretation of downcore variations may still focus on source-receptor arguments, if it can be shown that the bias between air and snow composition is stationary over the relevant time scales. Most, if not all, attempts to interpret ice-core records of aerosols and aerosol-associated species have implicitly assumed either no bias, or a stationary bias, between the composition of aerosols and snow (see Wolff this volume).

This paper will address two main tasks. First, published work demonstrating that aerosol removal processes operating over polar glaciers must fractionate the aerosol population on the basis of size and composition will be briefly reviewed. These studies also suggest that the bias between aerosol and snow composition could vary significantly in response to relatively small changes in atmospheric conditions (mainly temperature and humidity). In the second task, results from recent investigations of these issues on the Greenland ice sheet will be summarized. Particular attention will be paid to assessing the possible bias imparted to snow chemical records of aerosol-associated species by air-snow exchange processes, examining whether any bias might vary over time, and considering how changes in the species specific efficiency of aerosol removal processes might be identified in ice core records and separated from changes in aerosol loading in the overlying atmosphere.

BACKGROUND

Aerosol Fractionation by Snow

Rahn and McCaffrey (1979) pointed out an apparent contradiction between the chemical composition of aerosols collected at Barrow, Alaska in the late 1970s and analyses of contemporary Alaskan snow and preanthropogenic firn and ice from Greenland reported by Weiss et al. (1978). Enrichment factors calculated from the aerosol data indicated a large impact of pollution reaching the Arctic basin, which had been presumed to be pristine due to its remoteness. (The importance of pollution-derived species and elements as a component of the winter Arctic basin aerosol has been confirmed by many later investigations (e.g., Rahn and McCaffrey, 1980; Barrie, 1986; Ottar, 1989; Shaw and Khalil, 1989; Barrie and Barrie, 1990).) Surprisingly, the composition of Alaskan snow falling out of this polluted air was quite similar to that falling on Greenland before 1900, showing only modest enrichments in V and Mn. Concentrations of all other elements determined in these samples could be explained by the natural sources of sea-salt aerosols and crustal dust. Rahn and McCaffrey (1979) suggested that nucleation of snow crystals on clay minerals "could fractionate snow chemically relative to aerosol", preferentially carrying down soil-derived over pollution-derived aerosols.

At about the same time, Boutron and Lorius (1979) compared trace metal concentrations determined in snow collected near South Pole with concentrations in aerosol samples collected at the South Pole station (Maenhaut et al., 1979) (with the samples not covering same time periods). In this case no discrepancies were found between the relative concentrations of the various species in the aerosol and snow. Dick and Peel (1985) collected simultaneous samples of aerosols and snow at several sites on the Antarctic Peninsula and found that the ratio (concentration in air (C_a)/concentration in snow (C_s)) had similar, roughly constant, values for Al (felt to be crustally derived) and Na, K and Ca (felt to reflect sea-salt inputs). Similar values were found for the C_a/C_s ratios of the trace metals Cd, Cu, Pb and Zn (though the aerosol and snow samples for these elements were not from exactly overlapping time periods). These findings were taken to indicate that aerosol removal processes did not cause significant fractionation between the sea-salt, crustal dust, and trace metal components over much of Antarctica, contrary to the situation in the Arctic. However, Dick (1988) later reported that simultaneous aerosol and snow sampling, using improved protocols, at another site slightly further north on the Antarctic Peninsula suggested that scavenging by snow was most effective for sea-salt aerosols, followed by crustal dust, with trace metals least efficiently incorporated into snow.

Davidson et al. (1985) conducted simultaneous aerosol and snow sampling at Dye 3 (on the southern part of the Greenland ice sheet) and calculated scavenging ratios (defined as $C_s\rho_a/C_a$, where ρ_a is the density of the air) for several snow storms. A suite of crustal elements plus Cd and Cu had comparable, high, scavenging ratios (> 1200) while the ratios for $SO_4^=$ and Pb were always < 200. Nucleation of snow crystals on crustal dust was suggested to be the major scavenging process, a mechanism that would account for the large scavenging ratios for the crustal species (Davidson et al., 1985). The low scavenging ratios calculated for Pb and $SO_4^=$ were hypothesized to reflect their association with submicron aerosols which do not serve as good ice nuclei (IN). The large scavenging ratios for Cd and Cu were surprising, as size distribution data from other sites indicated that Cd and Cu are usually found on these small particles as well (Davidson et al., 1985). Only a limited amount of qualitative size distribution data for the various elements in the aerosol at Dye 3 was obtained by SEM during the experiments.

A series of experiments in winter storms on mountain peaks at lower latitudes have also provided insight into fractionation of aerosols during removal by snow (Borys et al., 1983, 1988; Mitchell and Lamb, 1989). Simultaneous collection of cloudwater, snow flakes and interstitial aerosol demonstrated that snow flakes scavenged only a small fraction of the available aerosols. Specifically, ice nucleation followed by diffusive growth into snow flakes mainly incorporated only the initial IN, which were most often a crustal dust particle since the crystal structure of these particles is conducive to ice nucleation. On the other hand, condensation of water vapor into cloudwater and growth of these droplets by coalescence resulted in a more concentrated condensed phase with a composition quite similar to the initial aerosol. Under certain conditions the growing snow flakes can incorporate significant amounts of super-cooled cloudwater in a process termed riming, which differs from diffusive growth of snowflakes in that the transfer of water from cloudwater to snow does not include the distillation effect of evaporating the cloudwater drops. As a result, rimed snow flakes incorporate more of the chemical impurities present in the much more abundant cloud condensation nuclei (CCN) than do unrimed flakes. The CCN are mainly soluble salts of NH_4^+ and $SO_4^=$, often enriched in anthropogenic trace metals (Borys et al., 1983, 1988; Mitchell and Lamb, 1989). As the degree of snow flake riming increased in these cold alpine storms, the differences in the composition of snow relative to cloudwater and aerosols decreased. The riming process has been shown to increase the scavenging efficiency of $SO_4^=$ by snow by an order of magnitude or more in regions outside the mountains as well (e.g., Scott, 1981; Barrie, 1985).

Borys (1983) argued that the conditions required for significant in-cloud riming of snow flakes, specifically high enough liquid water concentrations to allow growth of numerous cloudwater droplets to sizes > 10 μm before ice nucleation begins, should be quite rare in the Arctic under present conditions. This hypothesis expands on the earlier suggestion of Rahn and McCaffrey (1979), and implies that the composition of Greenland snow should generally be biased toward those aerosols that are good IN (dust and sea-salt) and against those aerosols that are readily activated as CCN (ammonium sulfates and associated trace species). Riming would seem even less likely in the colder and drier air over Antarctica (or Greenland during a colder climatic stage), suggesting that the aerosol-snow fractionation reported by Dick (1988) and Davidson et al., (1985) is more the rule than the earlier results from Antarctica which found no evidence for such fractionation.

If riming is now rare over the polar glaciers, and even rarer during colder climates, a significant bias between aerosol and snow composition should be expected in these regions. Specifically, polar snow must be enriched in IN and associated species relative to CCN-associated species when compared to the aerosol population in the overlying atmosphere. On the other hand, this bias would not seem likely to change greatly over time. Temporal changes in the concentration of any aerosol-associated species in snow or ice could then be considered a direct reflection of changing concentration of the aerosols overhead (Hammer, 1985), though the scaling factors would be different for sea-salt vs. crustal vs. CCN-related species.

However, Davidson et al. (1987, 1989) suggested that riming of snow crystals might be a significant scavenging mechanism over south Greenland in summer months. This would imply that the type and extent of fractionation between aerosol and snow in south Greenland currently varies widely over the course of the year. (The previously noted similarity of snowpit chemistry all over Greenland would suggest that the same might be true for a much wider region (Mayewski et al., 1990; Q. Yang, unpublished data).) If large seasonal changes in scavenging do occur now, relatively small changes in temperature over longer time scales could also alter aerosol-snow fractionation through changes in cloud liquid water content. The resultant changes in the composition of snow might then be hard to partition between variations in aerosol loading in the atmosphere (the objective) and variations in the removal processes. It should be noted that the hypothesis of significant riming of snow falling on south Greenland during summer was developed by comparing detailed snow chemistry records from pits near Dye 3 to long-term aerosol concentration records from several sites in the Arctic basin. Davidson et al. (1987, 1989) cautioned that the aerosol records from lower altitude sites might not be representative of the air masses reaching the crest of the Greenland ice sheet, but argued that the extremely limited aerosol data available from Dye 3 seemed to support the use of the available records from similar latitudes for such comparisons.

Aerosol Fractionation by Dry Deposition

Davidson (1989) reviewed the current understanding of the micro-physical processes actually removing aerosols from the atmosphere to polar snow and showed that dry deposition should be more important for the same types of aerosols that are preferentially removed by nucleation scavenging (sea-salt and dust). Dry deposition velocities estimated for crustally-derived aerosol-associated elements from limited size distribution data at Dye 3 compared well with estimates obtained by measuring increasing concentrations of the same elements in snow exposed at the surface during dry periods (Davidson et al., 1985). Applying these estimated dry deposition velocities to a number of sites in Greenland indicated that up to 35% of the annual flux of crustal elements may be due to dry deposition at low accumulation rate sites in north Greenland. The estimated potential importance of dry deposition decreased to as little as 10% of the annual flux for high accumulation sites like Dye 3. Dry deposition would appear to account for a negligible fraction of the annual flux of those species and elements felt to be primarily associated with submicron aerosols even at the lowest accumulation rate sites considered (Davidson et al., 1985). During periods of reduced accumulation of snow, both seasonally and over longer periods of time, the relative importance of dry deposition would increase for all aerosols, but proportionately more for the larger particles (again, sea-salt and dust), such that dry deposition could become the dominant process delivering supermicron aerosols to the surface of the ice sheet.

Source Apportionment from Snow and Ice Chemical Records

Most attempts to interpret snow and ice records of aerosol-associated species have not considered the potential impact of any biases introduced by aerosol removal processes (see Delmas, this volume). If the likely biases outlined above are valid, they would not impact the large number of examples where the finest discrimination attempted was defining the relative role of marine, crustal and anthropogenic sources. Likewise, source apportionments inferred entirely from changes in concentration, or relative concentrations, of species that are only associated with those aerosols that are preferred IN would not be greatly impacted by changes in the relative importance of ice nucleation, riming and dry deposition. Interpretations based on analysis of insoluble particles (mineralogy, trace element ratios, isotopic ratios) would probably not be affected at all (e.g., Gaudichet et al., 1992; Grousset et al., 1992). However, changes in the concentrations of species associated with CCN ($SO_4^=$, NH_4^+, trace metals) along an ice core could be partially the result of variations in the efficiency of processes removing these small particles from the atmosphere (particularly extent of riming) in addition to any changes in atmospheric loading. By extension, any conclusions that are based on changes in the relative concentrations of IN-associated vs. CCN-associated species may need to be carefully assessed.

RECENT EXPERIMENTS IN GREENLAND

Beginning in summer 1988, a series of experiments have been conducted at Dye 3 and Summit, Greenland. These studies share the common objective of developing process-level understanding of air-snow exchange to facilitate more rigorous interpretation of the records recovered from ice cores. The Dye 3 Gas, Aerosol and Snow Sampling Program (DGASP) was the first, and so far only, campaign to make measurements through the winter season on the Greenland ice sheet (August, 1988 - July, 1989). Many DGASP results have been compiled in a special issue of *Atmospheric Environment* (vol. 27A, # 17/18, December, 1993). Similar experiments, involving an ever growing number of investigators with specific expertise, have been conducted each summer since 1989 near the GISP2 and GRIP deep drilling sites. The work at Summit is still in progress, but many significant results have already appeared in the literature. This section will highlight the findings from the investigations at both Dye 3 and Summit that are particularly relevant to "irreversibly" deposited species, but some of these results are germane to reconstruction of atmospheric chemistry signals from snow and ice records more generally.

Aerosol-Snow Fractionation

A surprising result of DGASP was the finding that the aerosol loading in air masses reaching the Greenland crest at Dye 3 was nearly unrelated to the situation prevailing in the adjacent and much better studied Arctic basin. The Arctic haze phenomenon, where exceptionally high concentrations of most aerosol-associated species blanket the basin for several winter-spring months each year, was conspicuously absent at Dye 3 during the DGASP year (Jaffrezo and Davidson, 1993; Davidson et al., 1993a,b; Dibb and Jaffrezo, 1993; Mosher et al., 1993). In fact, concentrations of most aerosol-associated species, except those in sea-salt, were at their lowest in the winter when Arctic haze peaks at lower altitudes. Snowpit sampling at the end of DGASP revealed similar composition vs. depth profiles to numerous previous investigations, suggesting that this was not an unusual year.

It appears that the impact of Arctic haze on the composition of aerosol and snow at Dye 3 is restricted to brief episodes near the end of the Arctic haze season, each lasting only a few days. Most of these episodes occurred in mid-April, suggesting that they may make a major contribution to the "spring" peaks in dust, crustal species and $SO_4^=$ reported for a number of snowpits and ice cores from south Greenland (e.g., Finkel et al., 1986; Mayewski et al., 1987; Davidson et al., 1987, 1989; Beer et al., 1991). Year round aerosol and snow sampling at other sites on the Greenland ice sheet will be needed to confirm a similar episodic impact from Arctic haze, but similarities in seasonal patterns of snow chemistry throughout much of interior Greenland (Mayewski et al., 1990; Q. Yang and P. Mayewski, unpublished data) suggest that this finding probably applies to most high elevation locations on the ice sheet.

Lack of a pronounced "Arctic haze" seasonality in aerosol concentrations over Greenland calls the need to invoke large summer-time increases in aerosol scavenging efficiency (through riming) (Davidson et al., 1987, 1989) into question. Examination of snowflake replicas collected throughout the DGASP year and during several summer seasons at Summit revealed very low numbers of very lightly rimed flakes (Borys et al., 1992, 1993). It is thus unlikely that seasonal variations in the extent of in-cloud riming of snow flakes significantly alters aerosol-snow fractionation over these time scales at present. As noted earlier, in-cloud riming of snowflakes over Greenland should be even less frequent during colder climate episodes, since colder air masses would be drier and the liquid water content of clouds would not be high enough.

However, cloudwater, with its characteristic enrichment in CCN-associated species, may influence snow composition through deposition directly to the surface. Bergin et al. (1993, in press A, in press B) have demonstrated that frequent night-time radiation fog events at Summit carry down a significant amount of soluble ionic species, despite constituting a very small or negligible net water flux. A large fraction of all aerosols > 0.5 μm diameter are activated as CCN during these events and many of them grow large enough to be deposited before the fog dissipates during the following morning. These super-cooled water droplets are greatly enriched in all soluble ionic species, relative to snow falling at Summit. However, the fluxes of $SO_4^=$ and MSA (primarily associated with sub-micron aerosols, see below) due to fog have been estimated to account for nearly 1/3 of their total depositional flux during the summer season (with snow accounting for most of the rest and very small contributions from dry deposition) while fog contributes < 1/5 of the seasonal flux of ionic species derived from sea-salt or crustal sources (Bergin et al., in press B). It seems likely that fog also contributes disproportionately to the deposition of other submicron-aerosol-associated species, including anthropogenic trace metals, but chemical analyses to confirm this hypothesis have not been made to date. Changes in the amount of fog deposits retained in the snowpack over time (seasonal or longer time scales) could result in variations in overall aerosol-snow fractionation that would hinder reconstruction of records of atmospheric composition It is not clear how, or even whether, variations in the frequency and intensity of fog events could be identified and accounted for in snowpit or ice core records.

Vertical mixing between near-surface air and that aloft has been found to exert important, if not dominant, control on the concentration of aerosols just above the snow surface on the Greenland ice sheet (where all aerosol sampling has been conducted) (Dibb, 1990, 1993; Dibb et al. 1992; Dibb and Jaffrezo, 1993; Mosher et al., 1993). Since snow often forms in clouds well above the surface (Borys et al., 1992, 1993), direct comparison between aerosol and snow samples can be misleading if the connection (or lack of one) between air at cloud height and that at the surface is not considered on a case by case basis. The differing results obtained in the various comparisons between Antarctic aerosol and snow may be partially explained by such boundary-layer exchange effects.

The data base on the size distributions of several aerosol-associated ionic species and elements in the air over the Greenland ice sheet has been greatly expanded through sampling at Dye 3 and Summit. These results largely confirm expectations based on similar investigations in more accessible locations, with $SO_4^{=}$, MSA, NH_4^+, S, Pb, Zn, Br, and Ni predominantly found on submicron aerosol and Na^+, K^+, Ca^{2+}, Mg^{2+}, Si, K, Ca, Mn, Fe, Ti, Al, Na, Cl and Mg residing on supermicron particles (Hillamo et al., 1993; Jaffrezo et al., in press; Bergin et al., in press B). Sulfate, and sometimes MSA, can have a second mode in the 1-2 μm range which may be large enough to occasionally dominate the mass distribution. Similarly, a secondary submicron mode is sometimes apparent for Si, K, Ca, Mn, Fe and Ti. More work is needed to fully understand the processing during long-range transport to Greenland that creates short term variability in size distributions of aerosol-associated species. However, this expanding data base confirms many of the assumptions Davidson et al. (1985) made from limited data at Dye 3 and extrapolation from size-distributions measured at other locations. These data also confirm that the distinct differences expected between the composition of IN and CCN, that should result in aerosol-snow fractionation, do exist over Greenland. It should be noted that deposition velocities calculated from the size distributions measured by Hillamo et al. (1993) are 3- to 5-fold greater for crustal species than those found mainly in the accumulation mode.

Sources of Aerosols and Associated Species Reaching Greenland

A variety of tracers and modeling approaches have been used to determine the natural and anthropogenic sources of aerosol-associated species entering the Arctic basin, particularly during the Arctic haze season (e.g., Rahn and McCaffrey, 1980; Heidam, 1981; Barrie et al., 1989; Li and Winchester, 1989). Unfortunately, this wealth of information can not be applied directly to the situation over Greenland, since the high altitude air masses over the ice sheet and the low altitude air masses in the Arctic basin appear to be decoupled. The question of where the aerosols preserved in the glacier come from is an important one, but it has received relatively little attention in the recent experiments at Summit and Dye 3.

Davidson et al. (1993c) presented a summary of 5-day back trajectories arriving at Dye 3 once per day through the DGASP year. This analysis suggested a seasonal progression of most likely source regions that was in qualitative agreement with observed seasonal changes in aerosol and snow composition. However, trajectory models have well known shortcomings (e.g., Kahl, 1993) that are exacerbated by sparse sounding data in remote regions like those over the oceans and both poles. Verification of trajectory model output can be accomplished through examination of tracer transport (e.g., Harris and Kahl, 1990), but such tests were not conducted for the DGASP trajectories. Climatologies of trajectories arriving at Dye 3 and Summit for the past 10 years are now being calculated (with a different model) and will be verified to the extent possible with the available chemical data from aerosol and snow sampling at both sites.

A variety of multi-variate statistical techniques have been used to apportion Arctic aerosols to a small number of dominant sources on the basis of variations in composition over time (e.g., Heidam, 1981; Rahn, 1981; Lowenthal and Rahn, 1985; Li and Winchester, 1989; Barrie and Barrie, 1990). Regional scale discrimination of sources is often possible, particularly if enough tracers of anthropogenic emissions are included in the aerosol composition data set (e.g., Lowenthal and Rahn, 1985). Samples of aerosol and snow have been collected throughout the experiments in Greenland for the express purpose of generating multi-element data sets tailored for existing source-receptor apportioning techniques. So far, only a few of these samples have been chemically analyzed and no multi-variate statistical tools have yet been used to address the issue of sources of natural and anthropogenic aerosols reaching Greenland.

The isotopic composition of Pb in Arctic haze aerosol has been used to assess the relative importance of industrial and automotive Pb emissions from North American, European, and Eurasian sources (Sturges and Barrie, 1989; Sturges et al., 1993) Changing concentrations and isotopic composition of Pb in Atlantic Ocean surface water have been related to reduced use of Pb additives in gasoline in the US (Shen and Boyle, 1987, 1988; Sherrell et al., 1992). Application of these apparently powerful tools to the question of the source of Pb in snow at Summit (by two different groups) has resulted in divergent conclusions regarding the relative importance of North American versus European and Eurasian sources of Pb reaching Summit (Boutron et al., 1991; Rosman et al., 1993; Boyle et al., 1994; Sherrell et al., in review).

The source(s) of the dust reaching Summit is also still an open question. Exploratory analyses to examine the potential of natural tracers (pollen, spores, mineralogy, rare earth element and Nd/Sr isotopic signature) are being conducted on snow and ice samples collected at Summit. Several of the groups making these measurements in snow samples are also interested in analyzing aerosol samples. Preliminary tests to determine whether archived filters from sampling at Dye 3 and Summit might be suitable for these purposes are also in progress.

Other Complications Impacting Reconstruction of Atmospheric Chemistry

Non-steady accumulation of snow. Snowfall and accumulation are episodic processes, such that depth in a snowpit or core can not be perfectly related to the steady passage of time. It is generally assumed that precipitation occurs frequently enough at most snow and ice sampling sites in Greenland to justify assuming a linear relation between depth and time for the depth intervals between independently established time lines (these time lines are usually annual signals in pit and shallow core studies, but can be 100's of years apart in the deeper sections of long cores). However, several lines of evidence suggest that winter accumulation is presently underrepresented in the snowpack at Dye 3 and Summit, and perhaps much of the interior of the Greenland ice sheet. This evidence is summarized below, followed by a brief discussion of the problems non-uniform accumulation of snow may create in the interpretation of snow and ice chemistry records.

Snow accumulation is not easily measured during the winter on the Greenland ice sheet, even at a manned station, so all of the information is indirect to varying degrees. Colder air carries less moisture, so reduced precipitation onto the ice sheet might be expected during winter, an expectation that has long been confirmed for the Arctic basin (e.g., Rahn and McCaffrey, 1980; Barrie, 1986). Long-term (but probably not quantitative) measurements by station personnel at Dye 3 do indicate less precipitation in winter than other seasons, consistent with observations and stake measurements during DGASP (Jaffrezo et al., 1994). Correlations between detailed stable isotope records from snowpits and satellite-borne passive microwave surface temperature estimates suggest that winter-spring is the time of least snow accumulation at Summit (Shuman et al., in review). These findings are also supported by several arguments based on the depth profiles of aerosol-associated species in the snowpack.

The short (53 day) half-life of ^{7}Be means that radioactive decay should reduce its activity below detectable levels within < 1 year. In several snowpits sampled at Summit, high ^{7}Be activities have been detected at depths where the snow is "too old" if age is estimated by assuming a linear relation to depth between well established annual markers (Dibb et al., 1992; Dibb, 1993). Concentrations of MSA in the aerosol at Dye 3 and Summit peak in spring, then again in late summer. These peaks are recorded in the upper few years in the snowpack, but the 4-5 months between spring and summer in a given year are consistently separated by greater depth in the snow than are the 7-8 months between a summer peak and the spring peak in the following year (Jaffrezo et al., 1994). Sea-salt aerosols were very abundant in the atmosphere over Dye 3 in December-January of the DGASP year and Na^+ and Cl^- also reached their highest concentrations in surface snow at this time (Davidson et al., 1993a; Mosher et al., 1993) However, the concentrations of sea-salt derived species in their "winter" peak, identified in numerous pits from the Dye 3 region (Mayewski et al., 1987; Davidson et al., 1987, 1989; Beer et al., 1991), are always subordinate to the concentrations of $SO_4^=$ despite the higher concentrations of sea-salt-derived species in the winter aerosol (Fig. 1). Any aerosol-snow fractionation should exaggerate the dominance of Na^+ and Cl^- over $SO_4^=$ in winter, but there is no clear signal preserved in the snow. All of these observations can be explained if the winter season is represented by considerably less than 1/4 the thickness of an annual layer of snow.

Under-representation of a certain season in the snowpack can impact interpretation of snow and ice records in several ways. Assuming that depth in the snow is a linear function of time may very well frustrate efforts to understand the causes of "seasonal" variations in snow chemistry in terms of seasonal variations in atmospheric chemistry. Variations in the concentration of sea-salt and dust in Greenland snow provide a particularly clear illustration of this possibility. Concentrations of Na^+ (representing sea-salt) and Ca^{2+} (reflecting dust inputs) have reproducible annual oscillations in snow all over Greenland (Mayewski et al., 1990; Q. Yang, unpublished data), with the sea-salt peak usually a few cm deeper (earlier). Often the two peaks partially overlap (Fig. 2, bottom panel). Comparison to the annual oscillations of $\delta^{18}O$ suggests that the sea-salt peak occurs in "late-winter" while most of the dust is delivered in "late-winter to early spring". To explain this from the viewpoint of aerosols in the atmosphere, one must invoke a large influx of sea-salt with a pulse of dust arriving at most a few weeks later, with both events occurring in a few weeks to perhaps 2 months somewhere near February or March (though some might argue that spring does not arrive in Greenland until April or May). We have already seen that the peak sea-salt concentrations in the aerosol over Greenland occur in the dead of winter, while peak dust concentrations during DGASP were not observed until mid-April (Davidson et al., 1993a; Mosher et al., 1993). Surface snow sampling the past three seasons at Summit has shown that Ca^{2+} concentrations remain near the "spring" peak levels well into May, if not June (Fig. 2, upper panels), yet the annual peaks of Ca^{2+} and Na^+ in the snowpack partially overlap nearly every year (Fig. 2, bottom panel). Thus the snow chemistry record in Greenland indicates near simultaneity of major annual atmospheric signals that are actually at least 3, and maybe as much as 6, months apart, if it is not recognized that the relationship between snow depth and age may not be continuous.

An equally problematic aspect of uneven snowfall and accumulation is also illustrated by the example just described. If proportionally more winter snow fell onto Greenland (and was preserved) in a given year (or for a number of years), the annual flux and mean annual

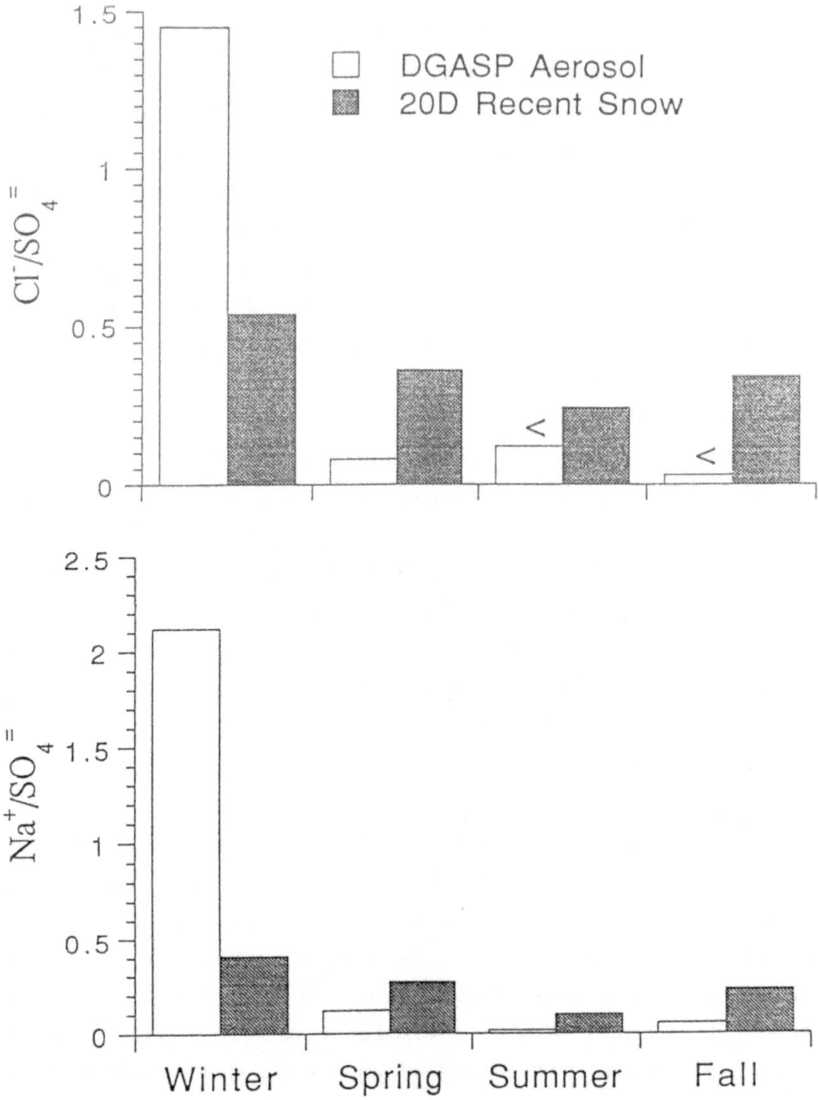

Figure 1. Seasonally resolved ratios of Cl^- and Na^+ over $SO_4^=$ in recent (1940-1984) 20D snow and in surface-level aerosol samples collected at Dye 3 from August, 1988, through July, 1989. Concentrations of Cl^- were often below detection limit on aerosol filters collected June - November during the DGASP year (Davidson et al., 1993a; Mosher et al., 1993). The abundance of sea-salt in aerosols December - February is not well reflected in "winter" snow. Note that the $Na^+/SO_4^=$ and $Cl^-/SO_4^=$ equivalence ratios in sea-salt are 8.3 and 9.7, respectively, so there is considerable "excess" $SO_4^=$ in aerosol and snow throughout the year. (20D is 40 km southwest of Dye 3, and was the site of a shallow coring operation in 1984 (Dibb et al., in review).

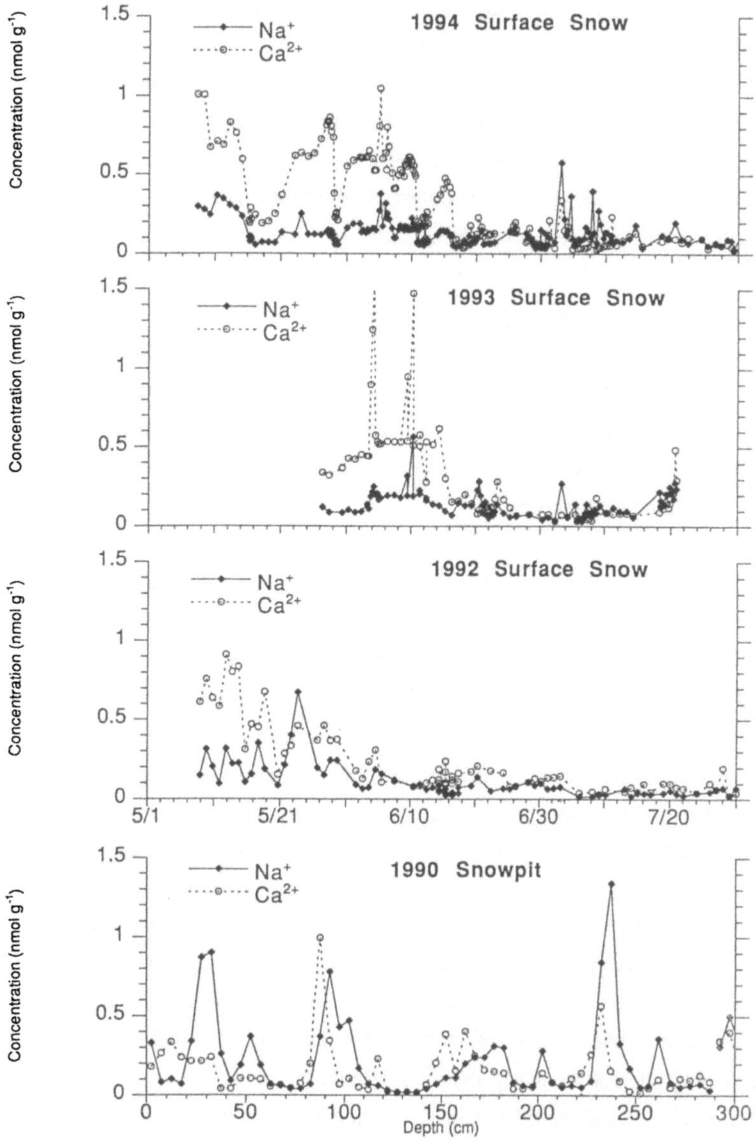

Figure 2. Comparison between the concentrations of Na^+ and Ca^{2+} in surface snow collected in 3 summer seasons and representative snowpit profiles of the same species (see Table 1 for description of the surface snow samples). All sampling was conducted at the remote sampling camp 30 km south west of the GISP2 drill camp. Note that surface snow concentrations of Ca^{2+} remain comparable to the elevated concentrations characterizing the "spring" peak in the snowpack until late May or early June each season. (J. Dibb, V. Hart, J.-L. Jaffrezo, M. Bergin and H. Kuhns, unpublished data)

concentration of sea-salt associated species would increase appreciably, with no change in the atmosphere concentration or seasonal abundance of sea-salt aerosols required. (Of course, extended periods of modified precipitation patterns would be accompanied by circulation changes which would probably also impact aerosol concentrations and their seasonal variations, so the problem is even more complicated.) Similar arguments would apply for any season characterized by exceptionally high or low concentrations of one or more aerosol-associated species (i.e., more late-summer snow would dilute the mean annual concentrations of all aerosol-associated species except NH_4^+, and perhaps $SO_4^=$, because atmospheric concentrations of sea-salt and dust are so low at this time of year (Bergin et al., in press B)). In theory, careful examination of the physical stratigraphy of snow and ice could provide indications of changes in the relative amounts of winter and summer snow over time, but such analyses would be very labor intensive and somewhat subjective. As precipitation fields calculated by general circulation models improve and are verified, results from these models might be used to incorporate changing seasonal patterns of snow accumulation into the interpretation of ice chemistry time series.

Spatial variability of snow composition and accumulation. Experiments to quantify the impacts of dry deposition, sublimation and exchange of reversible species between air and snow often involve repeated sampling of a given layer of snow over time-scales ranging from hours to several years (e.g., Davidson et al., 1985; Dibb, 1989; Dibb and Jaffrezo, 1993). Temporal changes in the composition of surface and near-surface snow are then interpreted as reflecting these post-deposition process. It is assumed that the initial composition of the layer is spatially uniform over the scales that will be covered by repeated sampling during a given experiment. Spatial uniformity of snow composition is also assumed when direct air-snow comparisons are made, as well as when reconstruction of past atmospheric composition is attempted from an ice core chemical record. Departures from strictly uniform composition of snow are most critical for the post-deposition experiments, since the changes that are expected can be quite small. However, snow sampling for various purposes at Summit and Dye 3 has revealed spatial variability over a range of scales, particularly for aerosol-associated species, that can be large enough to impact nearly any interpretation of the data.

Preservation of Chernobyl fallout in Greenland snow is extremely uneven (Dibb, 1989). The variability of the ^{137}Cs signal from this accident in a 150 x 150 km^2 region around Summit exceeds the range previously reported for the continental scale survey, probably just because more pits are included for the Summit region (Fig. 3). This is an extreme example, since the cloud of radioactive debris was patchy, deposition occurred in only one or two storms and the labeled snow was probably redistributed by drifting (Dibb, 1989). However, a fresh snow sampling traverse conducted within hours after the snow stopped revealed substantial variability in the deposition of two natural radionuclide tracers of

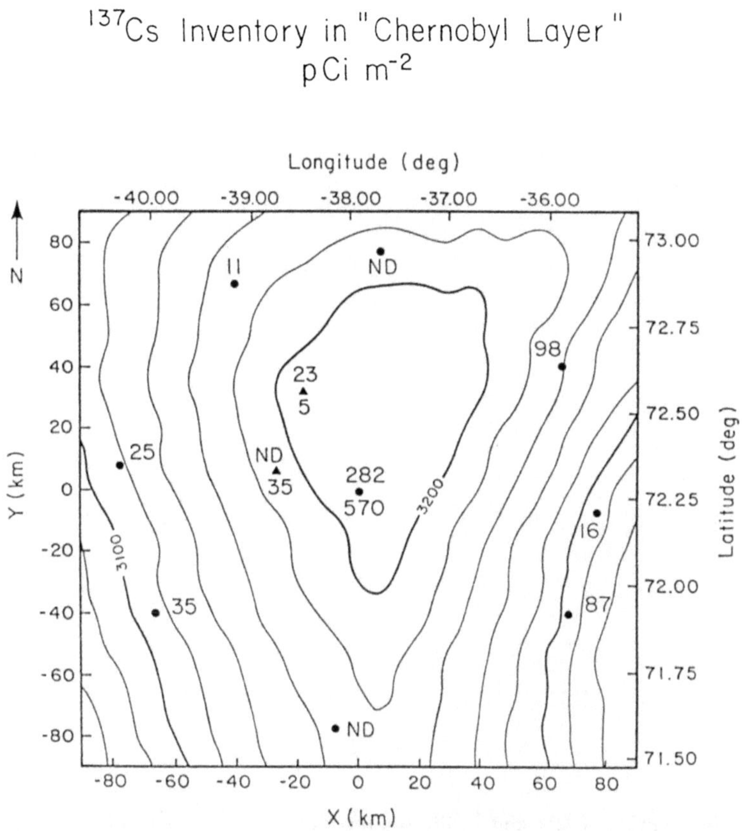

Figure 3. Uneven preservation of Chernobyl fallout in the Summit, Greenland region as reflected by extreme variability in the ^{137}Cs inventory in the Chernobyl layer in a series of snowpits sampled between 1987 and 1993. Note that two pits were sampled within 100-200 m of each other at each of three sites in the central portion of the region. The smallest difference for any of these pairs is > 2. No evidence of Chernobyl fallout was found in 3 of the pits (ND). (J. Dibb, unpublished data.)

submicron aerosols (Fig. 4). The atmospheric distribution of ^{7}Be and ^{210}Pb is presumably quite smooth over 60 km scales at a given elevation, so this example largely reflects variability imparted by air to snow transfer of submicron aerosol plus spatial variability in the depth of snow that accumulated from this storm at the different spots that were sampled.

The extreme spatial variability shown in Figures 3 and 4 are from singular (though probably not unusual) events, but some degree of temporal averaging is accomplished during the firnification process, and often by the sampling resolution chosen for a pit or core study. Such temporal smoothing may reduce the importance of initial spatial variability. However,

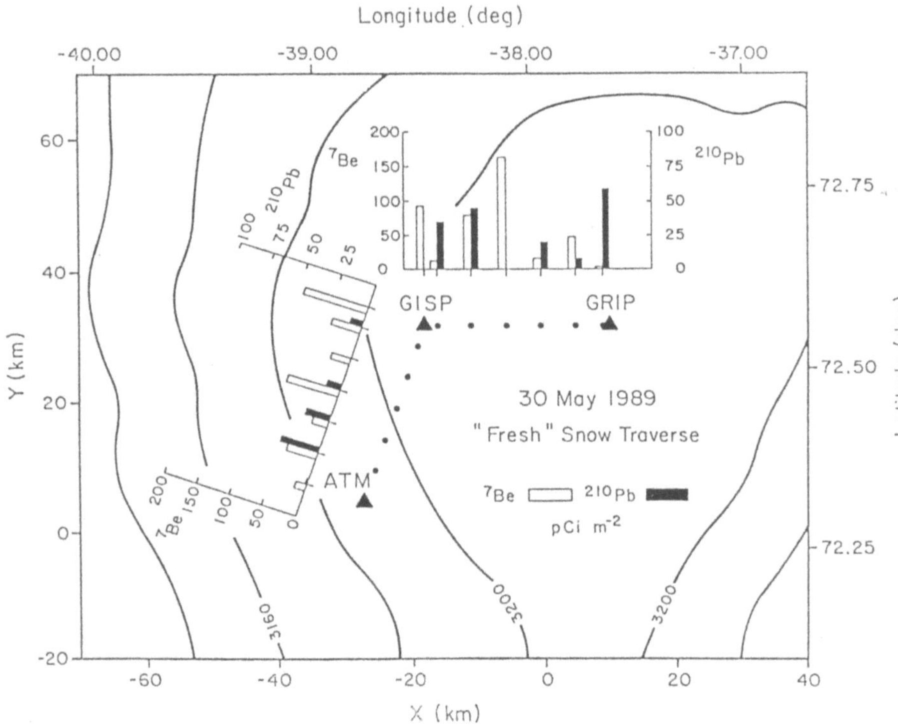

Figure 4. Inventories of ^{7}Be and ^{210}Pb in a layer of new snow that fell at Summit and was subjected to some drifting during the course of the storm (from Dibb, 1990).

signals felt to reflect seasonal, or longer, intervals can also vary widely between sites from a fairly small region (e.g., Zielinski et al., 1992). A recent attempt to quantify changes in the concentrations and inventories of soluble ionic species in summer snow one year after deposition provides another example. Snow that accumulated at Summit in the 1993 summer season was marked by non-particulating, inert strings between four poles defining a 10 x 10 m^2 plot. The mean concentration of each species was similar on the three occasions that the layer was sampled, but their inventories varied nearly 3-fold due to the different thickness of the summer layer preserved on 3 different sides of the plot (Fig. 5).

Daily, and more frequent, surface snow sampling at Summit has shown that the concentrations of aerosol-associated ions in adjacent replicates collected in succession (3-6

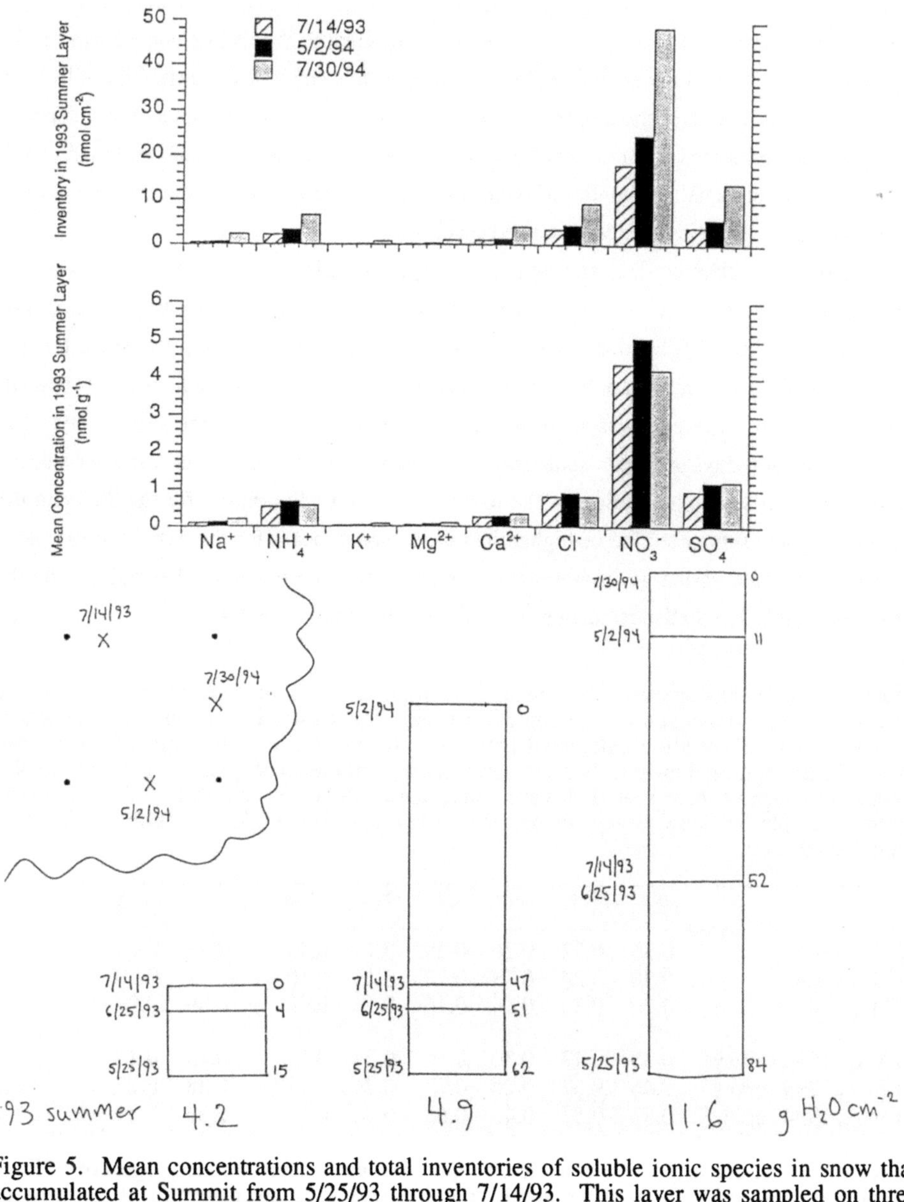

Figure 5. Mean concentrations and total inventories of soluble ionic species in snow that accumulated at Summit from 5/25/93 through 7/14/93. This layer was sampled on three different occasions at locations separated by about 10 m. The accumulation of snow varied by nearly a factor of three between the sites. Note that at the site with the greatest accumulation of snow over the 1993 summer season, the markers set out on 6/25 and 7/14 were found in the same horizon. (J. Dibb, unpublished data)

samples from an area < 0.2 m^2) from a single stratigraphic layer can vary by as much as a factor of two (Fig. 6) Some of the synchronous large increases and decreases in the coefficient of variation for all species can be related to specific events like wind storms or fog deposition, but detailed understanding of the causes of this small-scale variability is lacking. On a daily basis, and over the length of a summer season, the concentrations of ionic species with gaseous precursors are less variable than those associated with sea-salt and dust (Table 1). (The relatively small variability of NH_4^+ and $SO_4^=$ is one indication that they may not be so "irreversibly" deposited as IN-associated species.)

Efforts to characterize, and understand, spatial variability of snow composition at Summit are ongoing. In 1994 three surveys were conducted on a 1 km^2 sampling grid designed to facilitate application of newly developed and evolving spatial statistical tools to delineate spatial patterns of variability over scales ranging from the cm separation of replicate samples to 1 km. The grid was sampled after a big wind event, after the surface layer had been exposed to 6 days of sunny, settled weather with night-time fog events, and after a heavy snowfall in nearly calm conditions. The mean coefficients of variation for the 300 samples collected in each survey tend to be higher than the seasonal means for the past three summers, but the small-scale variability represents a significant fraction of the total variability on the km^2 scale (Table 1). Statistical analyses of these results are in progress.

Table 1. Coefficients of variation (standard deviation/mean) in the concentrations of soluble ionic species in summer snow at Summit, Greenland. The seasonal means are the averages of the coefficient of variation calculated from each sample set, not the ratio of the overall standard deviations and means. For comparison, the coefficients of variation for 300 samples collected in each of three spatial surveys conducted in 1994 are also presented. Separation between samples in these surveys ranged from a few cm to over a km (J. Dibb and V. Hart, unpublished data).

	Na^+	K^+	Mg^{2+}	Ca^{2+}	NH_4^+	$SO_4^=$	Cl^-	NO_3^-
1992 Season[1]	0.36	0.32	0.26	0.28	0.19	0.10	0.13	0.08
1993 Season[1]	0.30	0.30	0.18	0.22	0.13	0.10	0.11	0.09
1994 Season[1]	0.20	0.20	0.14	0.16	0.12	0.08	0.09	0.07
23 May, 1994 survey[2]	0.35	0.47	0.30	0.16	0.24	0.26	0.13	0.17
23 June, 1994 survey[3]	0.46	0.67	0.26	0.27	0.24	0.17	0.28	0.27
28 June, 1994 survey[4]	0.53	0.52	0.45	0.63	0.20	0.19	0.24	0.21

1 A sample set consisted of 3-6 adjacent replicates, each approximately 100 cm^2. Each sample set was collected from the single, dominant, surface layer present at the time of sampling. At least one, but up to seven, sample sets were collected per day. A total of 86, 87 and 196 sample sets were collected in the 1992, 1993 and 1994 seasons, respectively.
2 Sampled after wind storm.
3 Layer aged 6 days in fine weather.
4 Fresh snow fall.

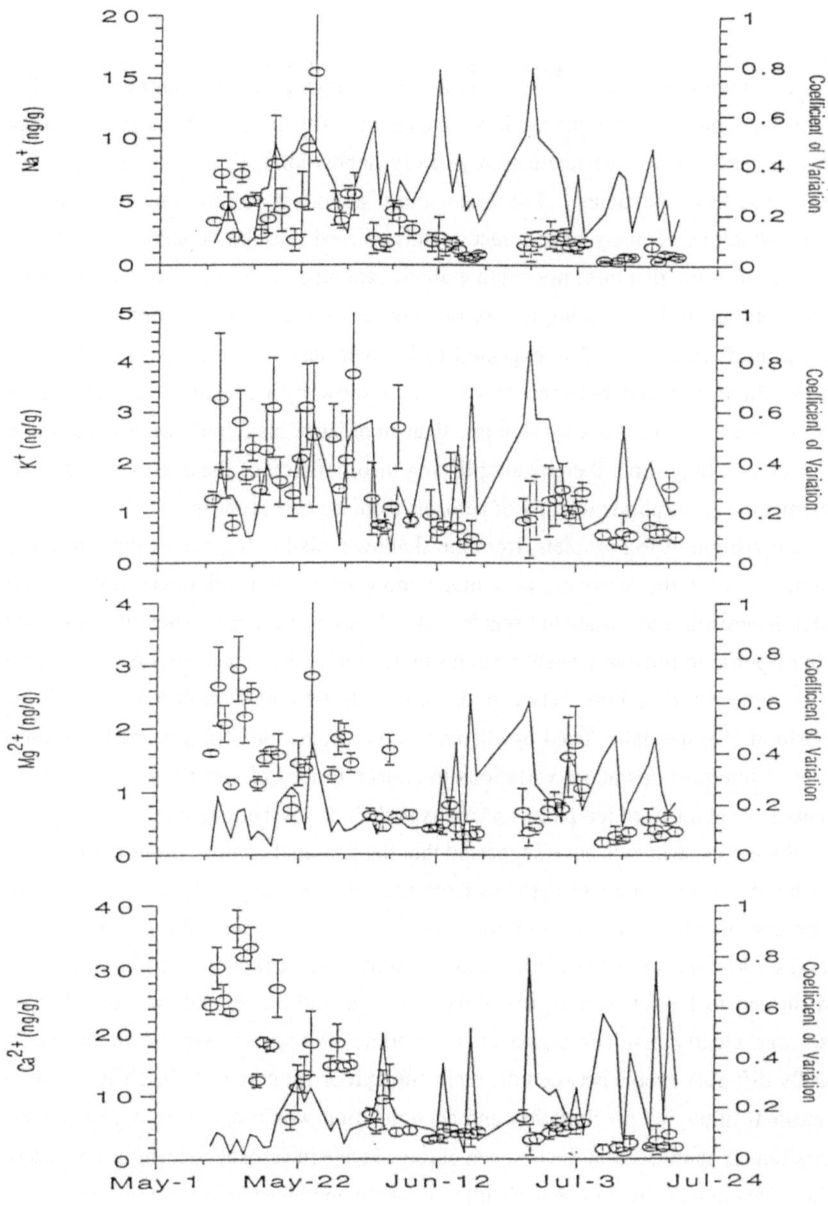

Figure 6. Concentrations and coefficients of variation, for 4 soluble ionic species felt to be aerosol-associated, in daily samples of surface snow at Summit in 1992. The ellipses and error bars represent the mean and standard deviation, respectively, of 5 replicates collected on each occasion. The line shows the temporal variation of the coefficient of variation (standard deviation/mean). (J. Dibb, unpublished data.)

CONCLUSIONS

This review has shown that the processes removing aerosols, and associated species, from the atmosphere over polar glaciers must result in fractionation between the composition of aerosols and snow. Ice nucleation and dry deposition will selectively remove larger, mainly crystalline, particles. The much more abundant submicron ammonium sulfate aerosols, which carry many of the trace metals emitted into the atmosphere, will be under-represented in snow that does not capture significant amounts of cloudwater through riming. Under current conditions, riming is very rare during snow events over high elevation sites on the Greenland ice sheet and is expected to be even less important over Antarctica. The fractionation introduced between aerosols and snow by nucleation scavenging and dry deposition is expected to vary by well less than an order of magnitude, and perhaps as little as a factor of 2-3, over time, thereby simplifying incorporation of these biases into attempts to reconstruct the atmospheric loading of aerosols from ice core records.

Deposition of fog droplets, from the shallow radiation fog events that presently occur frequently during the summer, is a much more efficient mechanism for removing the submicron aerosols and associated species. It is likely that the frequency of fog events could vary in response to relatively small changes in meteorological conditions over any given core site. A time varying bias between the concentration of submicron aerosols and the composition of snow, introduced by changes in the deposition of fog droplets, could make it difficult to interpret downcore variations in concentrations of species associated with these small aerosols in terms of temporal variations in their atmospheric loading.

Recent experiments have suggested that the potential complications for reconstructing atmospheric concentrations of aerosols from ice cores introduced by chemical fractionation may be completely overshadowed by complications introduced through purely physical processes. Variability in the composition of snow over a range of spatial scales has been demonstrated to be substantial, particularly for aerosol-associated species. Such spatial "noise" can clearly result in seasonal, or shorter, atmospheric signals being recorded as markedly different events between different snowpit or ice core sampling sites. Variation of the seasonal timing of precipitation and accumulation would result in large changes in the composition of snow even in the absence of any change in aerosol concentration patterns over the site. This possibility will be difficult to confirm and identify in ice core records, but may be the largest impediment to quantitative reconstruction of changing atmospheric composition from ice core chemistry records.

ACKNOWLEDGMENTS

I would like to thank all of the participants in the recent field experiments in Greenland. Discussions with Jean-Luc Jaffrezo, Cliff Davidson, Roger Bales, and Mike Bergin over the years of struggling with air-snow exchange issues are much appreciated. Many of the GISP2 and GRIP investigators assisted in various aspects of field work at Summit, while support from the steering committees of both drilling programs made it possible for us to undertake the investigations. The Polar Ice Coring Office provided logistic support for operations at Dye 3 and Summit. Financial support from NSF OPP is also acknowledged and much appreciated. This manuscript benefited from the comments of M. Bergin, C. Davidson, S. Whitlow and an anonymous referee.

REFERENCES CITED

Bales, R. C., J. E. Dibb and A. Neftel (1992) Processes of snow-atmosphere chemical exchange related to interpretation of GISP2 ice core; Report of the Workshop at Durham, NH, 6-7 March, 1992. Workshop organized, and report available from, the working group on Snow-Atmosphere Chemical Exchange of the International Commission on Snow and Ice.

Barrie, L. A., (1986) Arctic air pollution: An overview of current knowledge; *Atmos. Environ.*, 20, 643-663.

Barrie, L. A., (1985) Scavenging ratios, wet deposition and in-cloud oxidation: an application to the oxides of sulphur and nitrogen; *J. Geophys. Res.*, 90, 5789-5799.

Barrie, L. A., and M. J. Barrie (1990) Chemical components of lower tropospheric aerosols in the high Arctic: Six years of observations; *J. Atmos. Chem.*, 11, 211-226.

Barrie, L. A., M. P. Olson and K. K. Oikawa (1989) The flux of anthropogenic sulphur into the Arctic from mid-latitudes in 1979/80; *Atmos. Environ.*, 23, 2505-2512.

Beer, J., R. C. Finkel, G. Bonani, H. Gaggeler, U. Gorlach, P. Jacob., D. Klockow, C. C. Langway, Jr., A. Neftel, H. Oeschger, U. Schotterer, J. Schwander, U. Siegenthaler, M. Suter, D. Wagenbach and W. Wolfli (1991) Seasonal variations in the concentration of ^{10}Be, Cl^-, NO_3^-, SO_4^{2-}, H_2O_2, ^{210}Pb, ^3H, mineral dust and $\delta^{18}O$ in Greenland snow; *Atmos Environ.*, 25A, 899-904.

Bergin, M. H., J.-L. Jaffrezo, C. I. Davidson, R. Caldow and J. Dibb (1994) Fluxes of chemical species to the Greenland ice sheet at Summit by fog and dry deposition; *Geochim. Cosmochim. Acta*, 58, 3207-3215.

Bergin, M. H., C. I. Davidson, J.-L. Jaffrezo, J. E. Dibb, R. Hillamo and H. D. Kuhns (in press A) The contributions of wet, fog, and dry deposition to the summer SO_4^{2-} flux at Summit, Greenland; in, ed., R. Delmas, Ice Core Studies of Global Biogeochemical Cycles, NATO Advanced Science Institutes Series.

Bergin, M. H., J.-L. Jaffrezo, C. I. Davidson, J. E. Dibb, S. N. Pandis, R. Hillamo, W. Maenhaut, H. D. Kuhns and T. Makela (in press B) The contributions of snow, fog, and dry deposition to the summer flux of anions and cations at Summit, Greenland; *J. Geophys. Res.*

Borys, R. D., (1983) The interaction of aerosol particles, cloud water and ice crystal and their relationship to precipitation scavenging in the Arctic; in, ed., H. R. Pruppacher, R. G. Semonin and W. G. N. Slinn, Precipitation Scavenging, Dry Deposition and Resuspension, vol. 2, pp. 1281-1288, Elsevier, Amsterdam.

Borys, R. D., E. E. Hindman and P. J. Demott (1983) The significance of snow crystal and mountain surface riming to the removal of atmospheric trace-constituents from cold clouds; in, ed., H. R. Pruppacher, R. G. Semonin and W. G. N. Slinn, Precipitation Scavenging, Dry Deposition and Resuspension, vol. 1, pp. 181-190, Elsevier, Amsterdam.

Borys, R. D., E. E. Hindman and P. J. Demott (1988) The chemical fractionation of atmospheric aerosol as a result of snow crystal formation and growth; *J. Atmos. Chem.*, 7, 213-239.

Borys, R. D., D. Del Vecchio, J.-L. Jaffrezo, J. Dibb and D. L. Mitchell (1992) Field observations, measurements, and preliminary results from a study of wet deposition processes influencing snow and ice chemistry at Summit, Greenland; in, ed., S. E. Schwarz and W. G. N. Slinn, Precipitation Scavenging and Atmosphere-Surface Exchange Processes, vol. 3, pp. 1693-1702, Hemisphere Publishing Corporation.

Borys, R. D., D. Del Vecchio, J.-L. Jaffrezo, C. I. Davidson and D. L. Mitchell (1993) Assessment of ice particle growth processes at Dye-3, Greenland; *Atmos. Environ.*, 27A, 2815-2822.

Boutron, C. F., and C. Lorius (1979) Trace metals in Antarctic snows since 1914; *Nature*, 277, 551-554.

Boutron, C. F., U. Gorlach, J.-P. Candelone, M. A. Bolshov and R. J. Delmas (1991) Decrease in anthropogenic lead, cadmium and zinc in Greenland snows since the late 1960's; *Nature*, 353, 153-156.

Boyle, E. A., R. M. Sherrell and M. P. Bacon (1994) Lead variability in the western North Atlantic Ocean and central Greenland ice: Implications for the search for decadal trends in anthropogenic emissions; *Geochim. Cosmochim. Acta*, 58, 3227-3238.

Davidson, C. I., (1989) Mechanisms of wet and dry deposition of atmospheric contaminants to snow surfaces; in, ed., H. Oeschger and C. Langway, The Environmental Record in Glaciers and Ice Sheets, pp. 29-51, Wiley and Sons, New York, N.Y.

Davidson, C. I., S. Santhanam, R. C. Fortmann and M. P. Olson (1985) Atmospheric transport and deposition of trace elements onto the Greenland ice sheet; *Atmos. Environ.*, 19, 2065-2081.

Davidson, C. I., R. E. Honrath, J. B. Kadane, R . S. Tsay, P. A. Mayewski, W. B. Lyons and N. Z. Heidam (1987) The scavenging of atmospheric sulfate by Arctic snow; *Atmos. Environ.*, 21, 871-882.

Davidson, C. I., J. R. Harrington, M. J. Stephenson, M. J. Small, F. P. Boscoe and R. E. Gandley (1989) Seasonal variations in sulfate, nitrate and chloride in the Greenland ice sheet: Relation to atmospheric concentrations; *Atmos. Environ.*, 23, 2483-2494.

Davidson, C. I., J.-L. Jaffrezo, B. W. Mosher, J. E. Dibb, R. D. Borys, B. A. Bodhaine, R. A. Rasmussen, C. F. Boutron, U. Gorlach, H. Cachier, J. Ducret, J.-L. Colin, N. Z. Heidam, K. Kemp and R. Hillamo (1993a) Chemical constituents in the air and snow at Dye 3, Greenland-I. Seasonal variations; *Atmos. Environ.*, 27A, 2709-2722.

Davidson, C. I., J.-L. Jaffrezo, B. W. Mosher, J. E. Dibb, R. D. Borys, B. A. Bodhaine, R. A. Rasmussen, C. F. Boutron, F. M. Ducroz, H. Cachier, J. Ducret, J.-L. Colin, N. Z. Heidam, K. Kemp and R. Hillamo (1993b) Chemical constituents in the air and snow at Dye 3, Greenland-II. Analysis of episodes in April 1989; *Atmos. Environ.*, 27A, 2723-2737.

Davidson, C. I., J.-L. Jaffrezo, M. J. Small, P. W. Summers, M. P. Olson and R. D. Borys (1993c) Trajectory analysis of source regions influencing the south Greenland ice sheet during the Dye 3 gas and aerosol sampling program; *Atmos. Environ.*, 27A, 2739-2749.

Dibb, J. E., (1989) The Chernobyl reference horizon (?) in the Greenland ice sheet; *Geophys. Res. Lett.*, 16, 987-990.

Dibb, J. E., (1990) ^{7}Be and ^{210}Pb in the atmosphere and surface snow over the Greenland ice sheet in the summer of 1989; *J. Geophys. Res.*, 95, 22,407-22,415.

Dibb, J. E., (1993) Further insights into air-snow relationships at Summit, Greenland from beryllium-7 and lead-210; in, ed., N. Z. Heidam, Proceedings of the 5th International Symposium on Arctic Air Chemistry, Copenhagen, Denmark, September 8-10, 1992, National Environmental Research Institute, Technical Report no. 70.

Dibb, J. E., and J.-L. Jaffrezo (1993) Beryllium-7 and lead-210 in aerosol and snow in the Dye 3 gas, aerosol and snow sampling program; *Atmos. Environ.*, 27A, 2751-2760.

Dibb, J. E., J.-L. Jaffrezo, and M. Legrand (1992) Initial findings of recent investigations of air-snow relationships in the Summit region of Greenland; *J. Atmos. Chem.*; 21, 167-180.

Dibb, J. E., P. A. Mayewski, M. S. Twickler, S. I. Whitlow and G. Holdsworth (in review) Recent Temporal Trends in the Soluble Ionic Burdens of Snow and Ice in the Yukon Territory and South Greenland: Implications for Atmospheric Composition; *submitted to J.Geophys. Res.*

Dick, A. L., and D. A. Peel (1985) Trace elements in Antarctic air and snowfall; *Ann. Glaciol.*, 7, 12-19.

Dick, A. L., (1988) Trace elements in simultaneously sampled aerosol and snow from the Antarctic Peninsula; (abst.) *Ann. Glaciol.*, 10, 201.

Finkel, R. C., C. C. Langway, Jr., and H. B. Clausen (1986) Changes in precipitation chemistry at Dye 3, Greenland; *J. Geophys. Res.*, 91, 9849-9855.

Fuhrer, K., (1995) Overview of field data for reversibly deposited species; *this volume.*

Gaudichet, A., M. De Angelis, S. Joussaume, J. R. Petit, Y. S. Korotkevitch and V. N. Petrov (1992) Comments on the origin of dust in East Antarctica for present and ice age conditions; *J. Atmos. Chem.*, 14, 129-142.

Grousset, F. E., P. E. Biscaye, M. Revel, J. R. Petit, K. Pye, S. Joussaume and J. Jouzel (1992) Antarctic (Dome C) ice-core dust at 18 k.y. B.P.: Isotopic constraints on origins; *Earth Planet. Sci. Lett.*, 111, 175-182.

Hammer, C. U. (1985) The influence on atmospheric composition of volcanic eruptions as derived from ice-core analysis; *Ann. Glaciol.*, 7, 125-129.

Heidam, N. Z., (1981) On the origin of the arctic aerosol: A statistical approach; *Atmos. Environ.*, 15, 1421-1427.

Hillamo, R. E., V.-M. Kerminen, W. Maenhaut, J.-L. Jaffrezo, S. Balachandran and C. I. Davidson (1993) Size distributions of atmospheric trace elements at Dye 3, Greenland-I. Distribution characteristics and dry deposition velocities; *Atmos. Environ.*, 27A, 2787-2802.

Jaffrezo, J.-L., and C. I. Davidson (1993) The Dye 3 gas and aerosol sampling program (DGASP): An overview; *Atmos. Environ.*, 27A, 2703-2707.

Jaffrezo, J.-L., C. I. Davidson, M. Legrand and J. E. Dibb (1994) Sulfate and MSA in the air and snow on the Greenland ice sheet; *J. Geophys. Res.*, 99, 1241-1254.

Jaffrezo, J.-L., J. E. Dibb, R. C. Bales and A. Neftel (in press) Current status of atmospheric studies at Summit (Greenland) and implications for future research; in, ed., R. Delmas, Ice Core Studies of Global Biogeochemical Cycles, NATO Advanced Science Institutes Series.

Kahl, J. D., (1993) A cautionary note on the use of air trajectories in interpreting atmospheric chemistry measurements; *Atmos. Environ.*, 27A, 3037-3038.

Li, S.-M., and J. W. Winchester (1989) Resolution of ionic components of late winter Arctic aerosols; *Atmos. Environ.*, 23, 2387-2399.

Lowenthal, D. H., and K. A. Rahn (1985) Regional sources of pollution aerosol at Barrow, Alaska during winter 1979-80 as deduced from elemental tracers; *Atmos. Environ.*, 19, 2011-2024.

Maenhaut, W., W. H. Zoller, R. A. Duce and G. L. Hoffman (1979) Concentration and size distribution of particulate trace elements in the South Polar atmosphere; *J. Geophys. Res.*, 84, 2421-2431.

Mayewski, P. A., M. J. Spencer, W. B. Lyons and M. S. Twickler (1987) Seasonal and spatial trends in south Greenland snow chemistry; *Atmos. Environ.*, 21, 863-869.

Mayewski, P. A., M. J. Spencer, M. S. Twickler and S. Whitlow (1990) A glaciochemical survey of the Summit region, Greenland; *Ann. Glaciol.*, 14, 186-190.

Mitchell, D. L., and D. Lamb (1989) Influence of riming on the chemical composition of snow in winter orographic storms; *J. Geophys. Res.*, 94, 14,831-14,840.

Mosher, B. W., P. Winkler and J.-L. Jaffrezo (1993) Seasonal aerosol chemistry at Dye 3, Greenland; *Atmos. Environ.*, 27A, 2761-2772.

Neftel, A. (1996) The record of gases and reactive species in ice cores, and problems of interpretation; *this volume.*

Ottar, B., (1989) Arctic air pollution: A Norwegian perspective; *Atmos. Environ.*, 23, 2349-2356.

Rahn, K. A., (1981) The Mn/V ratio as a tracer of large-scale sources of pollution aerosol for the Arctic; *Atmos. Environ.*, 15, 1457-1464.

Rahn, K. A., and R. J. McCaffrey (1979) Compositional differences between Arctic aerosol and snow; *Nature*, 280, 479-480.

Rahn, K. A., and R. J. McCaffrey (1980) On the origin and transport of the winter Arctic aerosol; *Annals N.Y. Acad. Sci.*, 338, 486-503.

Rosman, K. J. R., W. Chisolm, C. F. Boutron, J.P. Candelone and U. Gorlach (1993) Isotopic evidence for the source of lead in Greenland snows since the late 1960s; *Nature*, 362, 333-335.

Scott, B. C., (1981) Sulfate washout ratios in winter storms; *J. App. Met.*, 20, 619-625.

Shaw, G. E., and M. A. K. Khalil (1989) Arctic haze; in, ed., O. Hutzinger, The Handbook of Environmental Chemistry, vol 4/Part B, pp. 69-111, Spinger-Verlag, Berlin Heidelberg.

Shen, G. T., and E. A. Boyle (1987) Lead in corals: Reconstruction of historical fluxes to the surface ocean; *Earth Planet. Scie. Lett.*, 82, 289-304.

Shen, G. T., and E. A. Boyle (1988) Thermocline ventilation of anthropogenic lead in the western North Atlantic; *J. Geophys. Res.*, 93, 15,715-15,732.

Sherrell, R. M., E. A. Boyle and B. Hamelin (1992) Isotopic equilibration between dissolved and suspended particulate lead in the Atlantic Ocean: Evidence from Pb-210 and stable Pb isotopes; *J. Geophys. Res.*, 97, 11,257-11,268.

Sherrell, R. M., E. A. Boyle and K. K. Falkner (in review) Temporal variability of Pb, Cd and Pb isotope deposition in Central Greenland snow; *submitted to Earth Planet. Sci. Lett.*

Shuman, C. A., R. B. Alley, S. Anandakrishan, J. W. C. White, P. M. Grootes and C. R. Stearns (in review) Temperature and accumulation at the Greenland summit: Comparison of high-resolution isotope profile and satellite passive microwave brightness temperature trends; *submitted to J. Geophys. Res.*

Sturges, W. T., and L. A. Barrie (1989) Stable lead isotope ratios in Arctic aerosols: Evidence for the origin of Arctic air pollution; *Atmos. Environ.*, 23, 2513-2519.

Sturges, W. T., J. F. Hopper, L. A. Barrie and R. C. Schnell (1993) Stable lead isotope ratios in Alaskan Arctic aerosols; *Atmos. Environ.*, 27A, 2865-2872.

Weiss et al. (1978) as cited by Rahn and McCaffrey, 1979.

Wolff, E. W., (1996) The record of aerosol deposited species in ice cores and problems of interpretation; *this volume.*

Zielinski, G. A., J. E. Dibb, R. J. Fiacco, P. A. Mayewski, S. Whitlow, M. Twickler, M. S. Germani and P. Grootes (1992) Variability in the Chemical Signal of the 1982 Eruption of El Chichon as Recorded in Greenland Snow Pits; (abst.) AGU Chapman Conference on Climate, Volcanism and Global Change, p. 20.

THE DEPOSITION OF PARTICLES AND GASES TO ICE SHEETS

Cliff I. Davidson, Michael H. Bergin, and Hampden D. Kuhns
Departments of Civil & Environmental Engineering, and Engineering & Public Policy
Carnegie Mellon University
Pittsburgh, Pennsylvania 15213, USA

Introduction

The glacial record has proven to be a most valuable source of information about climate change, major geologic events, and variations in global cycling of chemical species (Barnola et al., 1987; Dansgaard et al., 1989; Mayewski et al., 1990; Taylor et al., 1993). This information has been obtained by noting variations in chemical concentrations and in ice properties with depth in ice sheets worldwide (Delmas and Legrand, 1989). Virtually all of the chemical constituents found in the world's glaciers, aside from those near bedrock, originally came from the atmosphere. Yet published studies of ice core data have seldom incorporated information about the atmosphere or air-to-snow transfer into their interpretations.

If we are to make maximum use of the wealth of glacial record data available, we need to understand the processes by which these constituents reach their final position embedded in glacial ice. There are generally considered to be three major steps involved. First, particles and gases are emitted from sources and carried through the atmosphere to the air over the ice sheet. Second, the constituents are transported from the atmosphere to the surface of the ice sheet by a combination of deposition processes. This step includes processes occurring during precipitation as well as those occurring as the surface ages, such as atmosphere-surface exchange of particles and gases. Third, the constituents may experience modification as the surface becomes buried by successive layers of snow. This paper focuses on the second step with examples of data obtained on the Greenland Ice Sheet. The fundamental processes described here apply to ice sheets in the Arctic and Antarctic, and to some extent to mid-latitude glaciers.

The deposition of particles and gases to ice sheets can occur by dry deposition, by wet deposition during formation and subsequent precipitation of snow, and by fog deposition. Each of these three mechanisms is discussed here. The paper is essentially an update on descriptions of mechanisms presented in earlier reviews (Davidson, 1989; Davidson et al., 1991), with special attention given to recent advances in our understanding of deposition from the work of Barrie (1992), Bergin et al. (1994, 1995 a,b), Borys et al. (1992, 1993), Dibb et al. (1992), Dibb and Jaffrezo (1993), and others. The paper focuses on irreversibly depositing chemical species.

We begin by presenting descriptions of the deposition mechanisms, with separate sections for dry, wet, and fog deposition. Then we discuss experimental data illustrating these

NATO ASI Series, Vol. I 43
Chemical Exchange Between the Atmosphere and Polar Snow
Edited by Eric W. Wolff and Roger C. Bales

mechanisms for irreversible deposition. Finally, we summarize these findings and briefly discuss implications for ice core research.

Dry Deposition

Although wet deposition and fog deposition occur only during discrete time intervals when meteorological conditions dictate, dry deposition occurs continuously. Thus the term is used to describe the collection of mechanisms that continue to transport particles and gases to the earth's surface even in the absense of clouds or fog. The dry deposition flux of a chemical species in the form of particles or gases can be described mathematically as:

$$F_d = V_d(z)C_a(z) \tag{1}$$

where F_d is the dry deposition flux (g/cm^2 sec), V_d is the dry deposition velocity (cm/sec), and C_a is the airborne concentration of the chemical species (g/cm^3). Both V_d and C_a are functions of height, although the flux is assumed to be uniform near the ground (the "constant flux" layer).

The deposition process is assumed to occur in three major steps, namely aerodynamic transport, boundary layer transport, and interactions with the surface (Davidson and Wu, 1990). If particles large enough to be influenced by gravity are included, then sedimentation is considered an additional step in parallel with the other three. Each of these steps is associated with a resistance to transport, which is essentially the inverse of the dry deposition velocity:

$$V_d = \frac{1}{r_t} = \frac{1}{r_a + r_b + r_c} + \frac{1}{r_g} \tag{2}$$

where r_a, r_b, r_c, and r_g are the aerodynamic, boundary layer, surface, and gravitational resistances, respectively, and r_t is the total resistance (sec/cm). Figure 1 illustrates the various resistances using an analogy to an electrical circuit.

Aerodynamic transport occurs as chemical constituents are carried from the free atmosphere to the region of relatively quiescent air just above the surface, often called the viscous sublayer. For snow, the viscous sublayer is likely to extend a few millimeters or less over the surface. The principle mechanism of transport in this region is eddy diffusion, where particles and gases are carried by wind eddies from regions of relatively high concentration to regions of lower concentration. If the particles and gases are removed from the atmosphere when they reach the snow surface, the snow is considered to be a perfect sink and the concentration is modeled as zero at the surface. There is thus a perpetual concentration gradient between the free atmosphere and the snow which drives the eddy diffusion process. Note that if the surface is a source of a constituent, the concentration gradient may be in the opposite direction; this can result in a continual flow

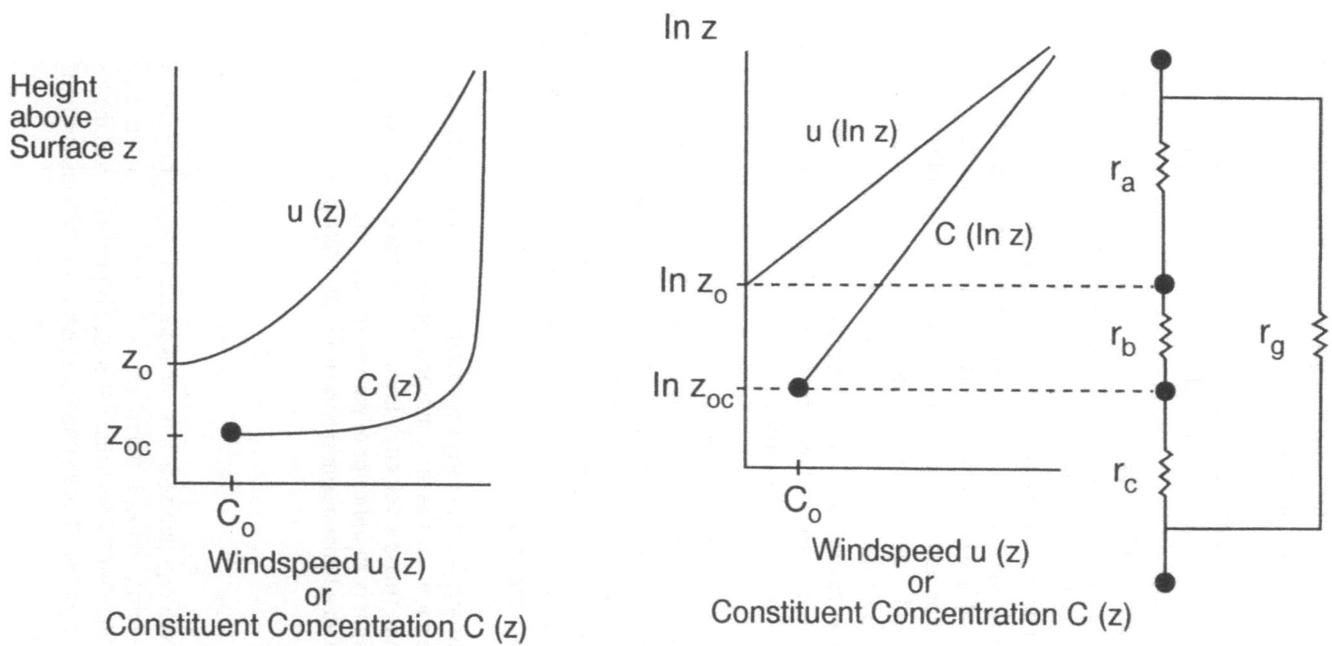

Figure 1. Resistances to transport, following an analogy to an electrical circuit. Plots of windspeed and chemical constituent concentration as functions of height are also shown. Reprinted from Davidson and Wu (1990).

of the chemical species from the surface into the atmosphere by eddy diffusion. Such flow may occur when gases desorb from the surface.

Boundary layer transport, the second step, involves the movement of particles and gases across the viscous sublayer to the surface. Several mechanisms are applicable as illustrated in Figure 2. Brownian and eddy diffusion refer to transport of small particles and gases along a concentration gradient due to random thermal motion and turbulent wind eddies. Interception occurs when a particle on an air streamline passes within one particle radius of the surface. Inertial impaction refers to capture of a particle on an air streamline that rapidly changes direction upon encountering roughness elements on the surface (e.g., ice needles). Turbulent inertial deposition is similar to impaction, except that the particle is influenced by a turbulent eddy which may be moving normal to the surface rather than an air streamline. Sedimentation applies to particles larger than a few tenths μm that are influenced by gravity. Finally, gas sorption may occur directly onto the snow crystals or onto a layer of liquid water on the crystals. Besides these major mechanisms, movement across the viscous sublayer can occur along a temperature gradient (thermophoresis) or a humidity gradient (diffusiophoresis), although the effects of such phoretic forces are usually small compared with other mechanisms.

The third step determines whether a constituent remains on the surface or returns to the mainstream airflow after penetrating the viscous sublayer. Particles can stick to the snow or rebound, depending on snow conditions and particle characteristics. For example, a soluble particle encountering a warm snow surface where there is appreciable liquid water may dissolve into the liquid layer. In contrast, a particle reaching a dry snow surface may rebound across the viscous sublayer. Soluble gases may sorb onto the snow, while insoluble, unreactive gases may not.

The relative rates of transport for these steps determine the overall rate at which particles and gases deposit on the snow. Just as the windspeed profile is an indicator of the transport of air momentum (mass x velocity) down to the surface, the profile of concentration versus height is an indicator of particle or gas transport. The expression for the windspeed profile in an adiabatic atmospheric is (Stull, 1988, p. 378):

$$u(z) = \frac{u_*}{k} \ln \frac{z}{z_o} \tag{3}$$

where u_* is the friction velocity (cm/sec), k is von Karman's constant equal to 0.4, z is the height above the surface (cm), and z_o is the roughness height, also known as the momentum sink (cm). Equation 3 shows that when u(z) is plotted versus the logarithm of height, the result is a straight line. The analogous equation for the concentration profile is:

$$C(z) - C_o = \frac{F_d}{ku_*} \ln \frac{z}{z_{oc}} \tag{4}$$

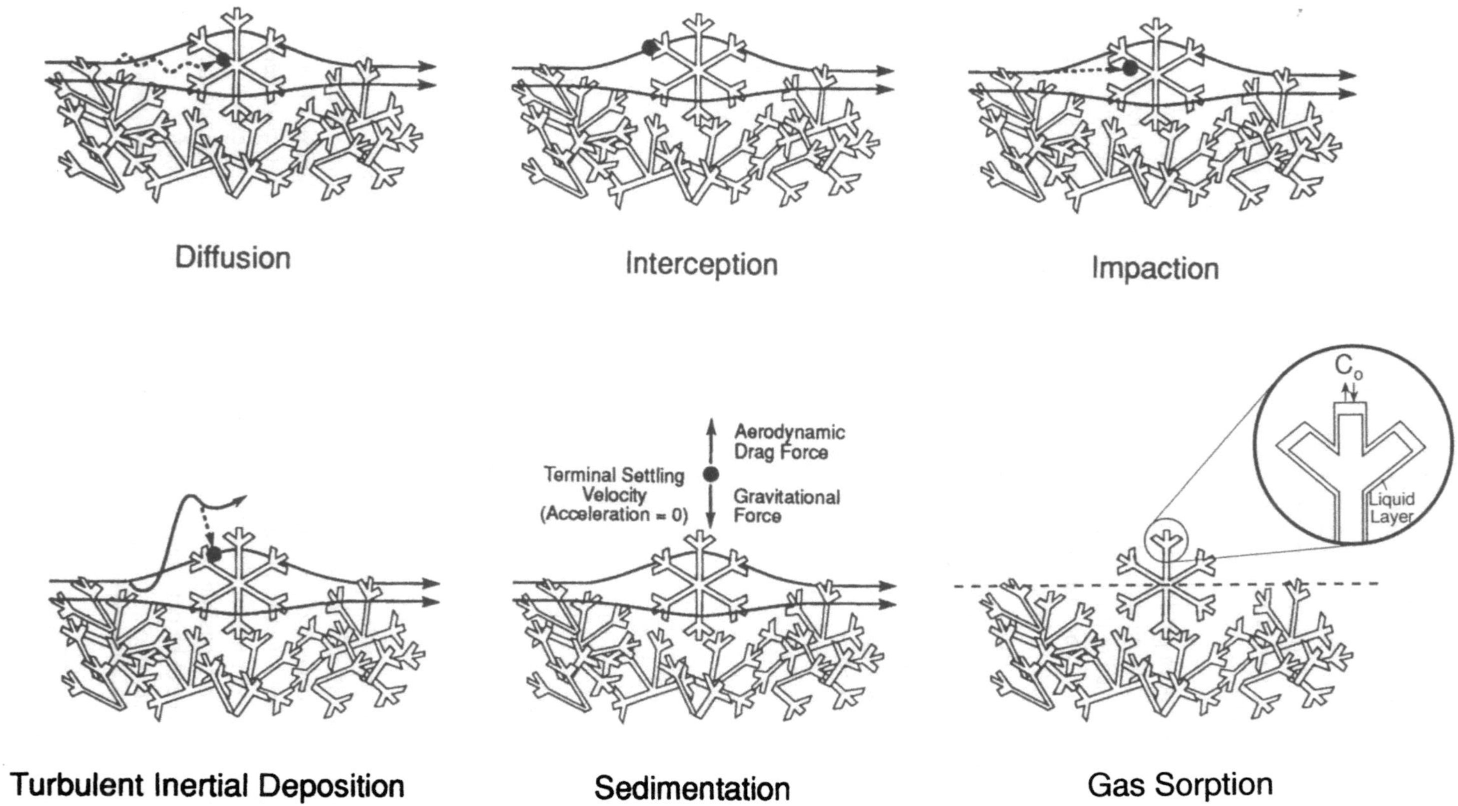

Figure 2. Mechanisms of particle deposition to snow. All of the mechanisms refer to transport across the viscous sublayer except for gas sorption, which involves interaction with the surface. This figure has been modified from Davidson et al. (1991).

where z_{oc} is the height above the surface at which the concentration reaches its minimum value. This value of concentration is zero for a perfect sink surface, which is the case for a chemical species that is depositing irreversibly. For a species whose deposition is reversible, there is exchange with the atmosphere resulting in an imperfect sink surface; this is described mathematically by assigning a nonzero value for C_o. Difficulties in attempting to quantify net deposition for reversible species are discussed later in this paper. Corrections to these equations for nonadiabatic conditions are discussed by Davidson and Wu (1990).

Equations 3 and 4 have been plotted in Figure 1. The figure implies that any of the transport steps may be rate-limiting in the overall dry deposition process. Particles or gases efficiently transported to the surface can have negligible deposition velocities unless they remain with the snow. For cases where the constituents do not stay on the surface, we define r_c as infinite so that there is essentially an open circuit in the resistance diagram of Figure 1. Conversely, particles and gases that would remain on the surface after reaching it may have small deposition velocities if transport down to the surface is sufficiently slow, indicating a large value of $r_a + r_b$. In the case of very large particles, high settling velocities result in a low value of r_g that short-circuits the other transport processes. These situations can also be seen by referring to the appropriate mathematical expressions for each resistance:

$$r_a = \frac{u(z)}{u_*^2} = \frac{1}{ku_*} \ln \frac{z}{z_o} \tag{5}$$

$$r_b = \frac{1}{ku_*} \ln \frac{z_o}{z_{oc}} \tag{6}$$

$$r_c = \frac{C_o}{F_d} \tag{7}$$

$$r_g = \frac{1}{V_s} = \frac{18\mu}{(\rho_p - \rho_a) g K d_p^2} \tag{8}$$

where μ is the dynamic viscosity of air, ρ_p is the particle density, ρ_a is the air density, g is the acceleration of gravity, K is the slip correction factor, and d_p is the particle diameter. The concept of a mass concentration sink z_{oc} is useful for visualizing the shapes of concentration profiles, but it is difficult in practice to calculate a value for this parameter. Rather, the boundary layer resistance r_b can be computed by considering expressions for each of the mechanisms responsible. For example, Ibrahim et al. (1983) considered the boundary layer resistance to be composed of three resistances in parallel, i.e., for Brownian diffusion, interception, and impaction, respectively:

$$\frac{1}{r_b} = \frac{1}{r_{diffusion}} + \frac{1}{r_{interception}} + \frac{1}{r_{impaction}} \tag{9}$$

$$r_{diffusion} = \frac{1100}{u(30cm)} Sc^{0.8} \tag{10}$$

where u(30 cm) is the windspeed at 30 cm, and Sc is the Schmidt number equal to ν/D, where D is the Brownian diffusivity.

$$r_{interception} = \frac{1}{\eta_{interception} u_* A_n N} \tag{11}$$

where $\eta_{interception}$ is the efficiency of collection by interception, equal to $2d_p/(d_n\sqrt{\rho_p})$. Here A_n is the vertical projected area of an ice needle, d_n is the diameter of an ice needle, and N is the number of needles per unit area of the ground.

$$r_{impaction} = \frac{0.4}{\eta_{impaction}} \frac{u(z)}{u_*^2} \tag{12}$$

where $\eta_{impaction}$ is the efficiency of collection by impaction, equal to 10^{-3}/St, where St is the Stokes number defined by:

$$St = \frac{(\rho_p - \rho_a) u_* K d_p^2}{9 \mu d_n} \tag{13}$$

These expressions can be used to determine r_b in Equation 2 to find the overall dry deposition velocity. Expressions for V_s and St in equations 8 and 13, respectively, have been modified from Friedlander (1977, pp. 35 and 96).

Figure 1 illustrates an important difference between the windspeed and concentration profiles. The aerodynamic resistance r_a is defined only for $z > z_o$, and the windspeed increases gradually with height over the region applicable to r_a. This is consistent with the fact that all of the resistance to transport of air momentum is associated with aerodynamic delivery. In contrast, most of the gradient in concentration occurs much closer to the surface; this indicates that the boundary layer resistance may be dominant for chemical constituents. In actuality, any of the three resistances may be rate-limiting for a given set of conditions. However, the situation illustrated in Figure 1 is commonly observed over many natural surfaces (Davidson and Wu, 1990), suggesting that the transport of chemical species is often limited by what happens in the viscous sublayer.

Equation 2 has been used with the various component resistances to predict V_d as a function of d_p for conditions over the Greenland Ice Sheet. Expressions for several of the mechanisms described above have been taken from the literature and combined with data for windspeed and snow surface conditions. It has been assumed that $r_c = 0$ so that all particles stick to the surface. The calculations have been performed for stable (Monin-Obukhov length = 100m), unstable (Monin-Obukhov length = -100m), and adiabatic atmospheric conditions. Results are shown in Figure 3. Note that V_d follows a characteristic curve that reflects high rates of transport for very small particles and for large particles, but not for the mid-size range. Small particles have high Brownian diffusivities while large particles have appreciable sedimentation and inertial influences. Deposition velocities for particles of approximately 0.1 μm diameter are low since none of the mechanisms are efficient for this size of particle.

The curves of Figure 3 do not include the possible effects of wind pumping which may deliver chemical constituents to porous snow below the surface. Cunningham and Waddington (1993) and Waddington et al. (1996) report that such effects may occur in areas where irregularities on the surface, such as sastrugi, create pressure gradients in the snow. The importance of snow and firn microstructure on permeability of the surface has been discussed by Albert et al. (1996). It is difficult to quantify these effects with the limited current understanding.

In addition, complications caused by blowing and drifting snow have been ignored in this analysis. Pomeroy et al. (1991) and Pomeroy and Jones (1996) have shown that the chemistry of snow can be modified by disturbance of surface layers. Chemical constituents that have been previously deposited may be resuspended; alternatively, dry deposition to blowing snow may be enhanced by movement of snow crystals, and sublimation of the snow may increase chemical species concentrations. The overall effects are unknown.

The issue of particle deposition is further complicated by differences between concentrations near the surface and those aloft. Recent temperature profile measurements over the Greenland Ice Sheet suggest that the air near the surface is on occasion isolated from the free troposphere where most of the long range transport of chemical species is occurring (Dibb et al., 1992). During these isolation events, the airborne concentrations of chemical constituents just above the ice sheet may be very small, and dry deposition is expected to be negligible. The frequency of occurrence of such events over extended time periods has not been investigated. These events might obscure the ice core record since free tropospheric air is advected to the surface somewhat sporadically.

Wet Deposition

Clouds in warm climates contain water droplets that are formed when hygroscopic particles serve as cloud condensation nuclei, and droplet growth occurs by water vapor diffusion. In cold regions, clouds may contain ice crystals formed when particles are activated as ice nuclei and grow by water vapor diffusion. Water droplets may coalesce to form raindrops, and ice crystals may aggregate to form snowflakes. These four components of clouds -- liquid droplets, ice crystals, raindrops, and snowflakes -- make up the components of clouds. The distinction between the first two terms and the last two is usually on the basis of size: a raindrop or snowflake is large enough to settle out of the cloud despite updrafts that may tend to keep smaller droplets and ice crystals suspended.

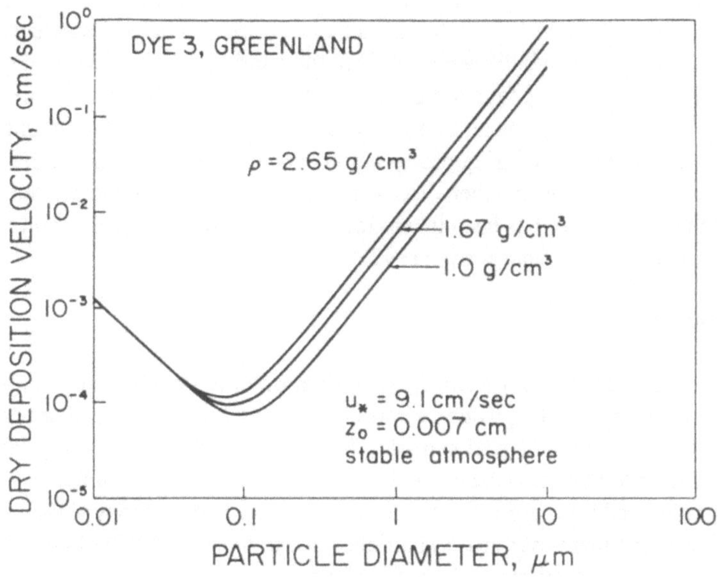

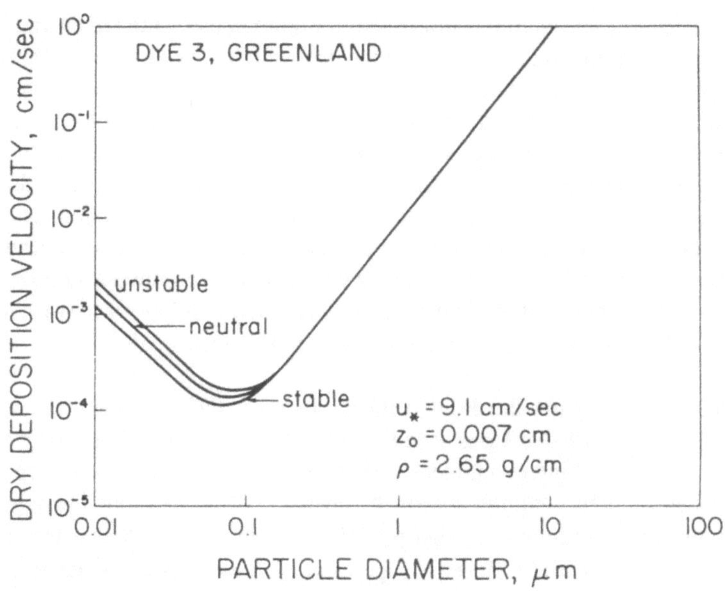

Figure 3. Dry deposition velocities as a function of particle diameter for a snow surface, based on the model of Ibrahim et al. (1993). The reference height is 10 cm and snow conditions are for Dye 3, Greenland. All curves assume no hygroscopic growth in the region just above the snow. Reprinted from Davidson (1989).

Removal of particles and gases by precipitation involves several mechanisms. Nucleation scavenging is the process by which particles are activated to serve as cloud condensation nuclei or ice nuclei and are eventually removed from the atmosphere. Particles and gases can also be captured by any of the four components of clouds as winds enhance mixing within the cloud. Alternatively, particles and gases can be scavenged by falling raindrops and snowflakes below the clouds during precipitation events. In this section, we describe each of these major mechanisms and then discuss formulations developed to predict scavenging. Much of the theory has been taken from Slinn (1984), Pruppacher and Klett (1980), Davidson et al. (1991), and Barrie (1992).

Nucleation Scavenging

Different categories of particles play a role in the formation of liquid droplets as opposed to the formation of ice crystals. Hygroscopic particles such as sulfates and sodium chloride are often the first to become condensation nuclei. In contrast, particles with a crystal lattice structure similar to that of ice, such as soil dust, make the best ice nuclei. Nevertheless, in cold conditions some fraction of hygroscopic particles may serve as ice nuclei and may even dominate over crustal particles, due to the abundance of the hygroscopic particles such as sulfate in the atmosphere.

The process of nucleation scavenging is illustrated in Figure 4, using the Greenland Ice Sheet as an example. Greenland is surrounded by ocean which provides the water vapor that results in snowfall on the ice sheet. In the first step, water vapor condenses on cloud condensation nuclei over the ocean. The nuclei can be natural particles, such as salt crystals from seaspray or sulfate emitted from biological activity in the ocean, or they can be anthropogenic particles that have been transported from mid-latitude sources. Some crustal particles may also serve as condensation nuclei. Note that condensation can occur well below $0^{\circ}C$, depending on the chemical composition of the nuclei. This is because the resulting droplets, e.g., those with salt nuclei, may be highly concentrated solutions with freezing points depressed below that of pure water. The cloud droplets reach diameters as large as 20-30 μm (Borys et al., 1992), although they can be much smaller.

After formation, winds may carry the droplets inland, rising as they follow the elevation of the ice sheet. The temperature decreases as the cloud rises, and eventually some of the droplets begin to freeze resulting in a mixed-phase cloud, shown as the second step in Figure 4. The frozen droplets continue to grow by diffusion of water vapor and achieve sizes of 100 μm or more. Furthermore, other particles present are activated as ice nuclei and grow by water vapor diffusion. Because supersaturation with respect to ice is always greater than that with respect to liquid water, some of the liquid droplets evaporate and contribute water vapor to the growing ice crystals. The particles which served as nuclei for these evaporating droplets are then released back into the atmosphere. They can again serve as nuclei, or they may be captured by existing cloud components as discussed below.

The ice crystals then aggregate to form snowflakes, shown as the third step in Figure 4. Some of the remaining liquid droplets collide with the ice crystals and snowflakes, freezing

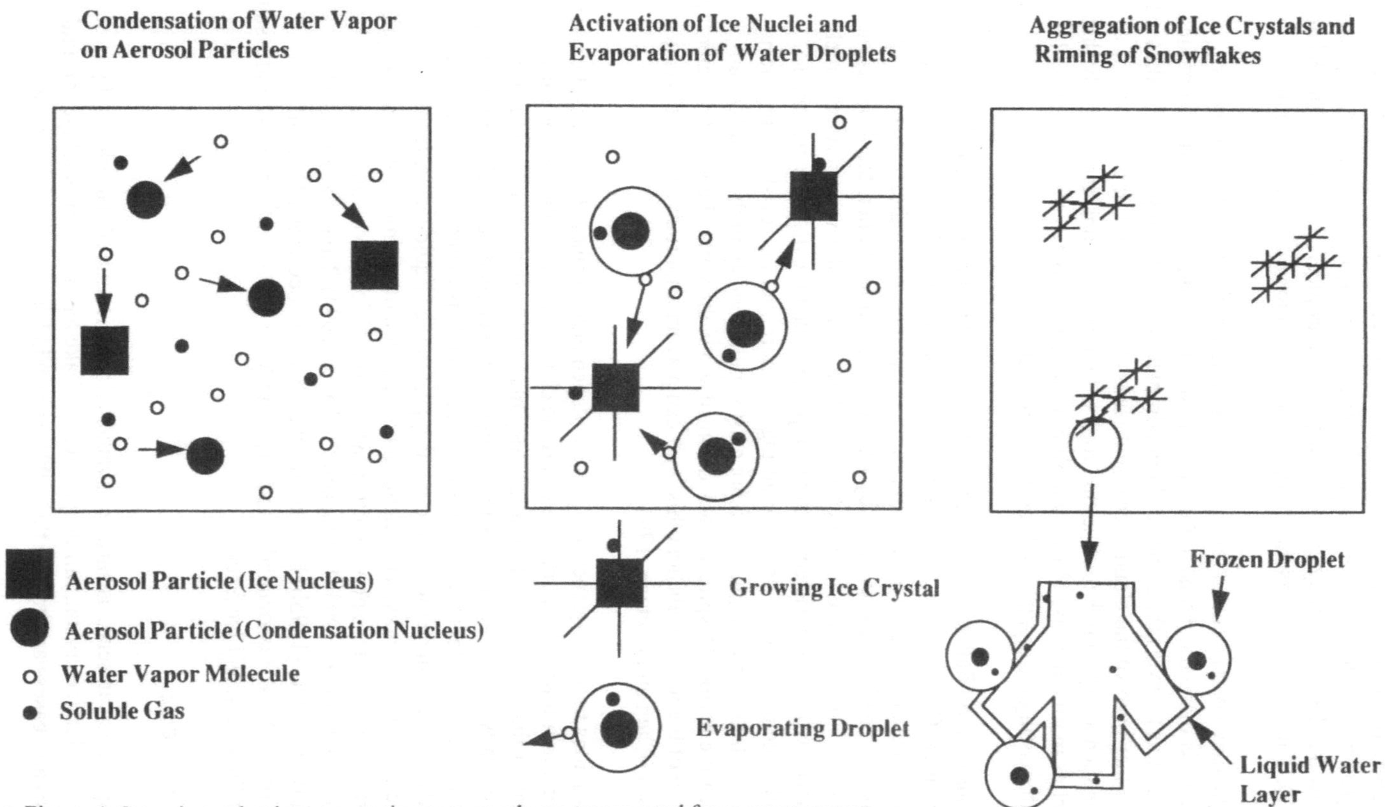

Figure 4. Steps in nucleation scavenging as aerosols are transported from ocean areas surrounding Greenland to the atmosphere over the ice sheet.

on contact. Strictly speaking, the term *ice crystals* refers only to components of clouds that have grown by water vapor diffusion; the term *rimed ice and snow crystals* refers to ice and snow that have grown by collisions with liquid droplets (Pruppacher and Klett, 1980, p. 27). The liquid droplets that have frozen on contact are known as *rime*.

A snowflake may contain only a few ice crystals or hundreds of crystals, rimed or unrimed. This is marked contrast to raindrops in warmer climates that result from the coalescence of as many as thousands of liquid droplets, or by melting of snowflakes that formed in upper parts of the cloud. Because it takes at least several minutes or tens of minutes for such a large number of liquid droplets to form raindrops, it is unlikely that liquid droplets in mixed-phase clouds will form rain; rather most of the droplets will collide with existing ice crystals and result in rimed snow.

Observations of snowflakes at Dye 3, Greenland suggest that riming sometimes occurs but is not ubiquitous. Observations at Summit show little evidence of riming, even in mid-summer (Borys et al., 1993). The lack of riming in summer may indicate that the liquid droplets are too small to be scavenged by ice crystals as rime. Rather, the droplets may evaporate or be carried off the ice sheet. Concentrations of water vapor are extremely low at this location, which is consistent with the hypothesis that the droplets are small. Alternatively, the temperature decrease during transport over the ice sheet may occur quickly so that liquid and ice phases do not exist simultaneously in the cloud for an appreciable time period. Despite the lack of rimed snow, there may be a liquid water layer at the surface of the ice crystals, and this layer can play an important role in scavenging soluble gases (Chen and Crutzen, 1994).

Several variations to this sequence are possible depending on characteristics of the nuclei and weather conditions. For example, even if the cloud droplets remain at the same elevation rather than rising after formation, they may still freeze upon continued growth because eventually the salt solution comprising the droplet becomes sufficiently dilute to allow freezing. If the initial temperatures are cold enough as expected in winter, snow may form directly from ice nucleation without the appearance of liquid water, resulting in unrimed ice crystals and snow. The chemical composition of unrimed snow can be quite different from that of rimed snow (Scott, 1981; Barrie, 1985; Davidson et al., 1989; Borys et al., 1993; Collett et al., 1993; Kalina and Puxbaum, 1994). Note that Kuhns et al. (1995) and Bergin et al. (1995b) have shown that there are occasionally significant concentrations of SO_4^{2-} (associated with NH_4^+) and MSA that could serve as condensation nuclei over the Greenland Ice Sheet.

In-Cloud Scavenging

Cloud components can serve as surfaces available for collection of particles and gases. Each cloud droplet, ice crystal, raindrop, and snowflake has a viscous sublayer around it, and particles and gases can traverse this sublayer as winds move through clouds.

The same transport steps as described for dry deposition are important here. Aerodynamic transport refers to delivery by wind to the top of the viscous sublayer. Boundary layer transport refers to movement across the sublayer by diffusion, interception, impaction, turbulent inertial deposition, sedimentation, and possibly other mechanisms such as phoretic forces. Finally the particle or gas interacts with the liquid water or ice, and is either retained or released back into the air. In parallel with these three steps, sedimentation can result in particle collisions with the cloud components.

Below-Cloud Scavenging

Falling raindrops and snowflakes can collect particles and gases during precipitation. The same mechanisms are important here, since once again there is a viscous sublayer around each raindrop and snowflake. Note that large particles may be collected by differential sedimentation, the difference between the settling velocities of the particles and those of the falling precipitation. When the settling velocities are essentially the same, there may be negligible collection.

Most particles in a cloud have taken up water due to the high relative humidities present. This can assist in-cloud scavenging by increasing particle size and by increasing retention of gases and other particles. In contrast, particles below clouds may be dry and may not have experienced growth by uptake of water, so below-cloud scavenging may be less efficient.

Combining the Mechanisms

All of the liquid water and ice in a cloud contain chemical species. However, cloud droplets and ice crystals generally remain airborne, while falling precipitation transfers chemical species from the atmosphere to the earth's surface. Therefore we confine our attention in this section to chemical species contained in snow. Deposition by fog, discussed below, addresses the special case where cloud droplets reach the surface.

We begin by writing an expression for the flux of chemical species with snowfall, or wet deposition flux F_W (g/cm^{2}sec), analogous to the dry deposition flux F_d defined by Equation 1:

$$F_w = \rho_w P C_s \qquad (14)$$

where ρ_W is the density of snow meltwater, set equal to 1 g/cm^3, P is the snow precipitation rate (cm of snow meltwater/sec), and C_S is the concentration of chemical species in the snow (g chemical constituent/g snow). Accounting for the various scavenging mechanisms that determine C_S is a difficult task. One approach is to define a scavenging rate ψ (sec^{-1}) such that the airborne concentration C_a is diminished as a first-order process by scavenging:

$$\frac{dC_a}{dt} = -\psi C_a \tag{15}$$

It is then possible to relate C_s and C_a according to:

$$C_s = \frac{1}{\rho_w P} \int_0^\infty \psi(z) C_a(z) dz \tag{16}$$

where the scavenging rate and airborne concentration are both allowed to be functions of height. Slinn (1984) suggests a simple function for ψ that accounts for the efficiencies of collection as air containing chemical species passes by snowflakes or as snowflakes fall to the ground:

$$\psi = \int_0^\infty \eta_s(L) A(L) V_s(L) N_s(L) dL \tag{17}$$

where η_s represents the total efficiency of collection by a snowflake of characteristic size L, equal to the collision efficiency multiplied by the retention efficiency, A is the effective cross sectional area of the snowflake (cm^2/snowflake), V_s is the settling velocity of the snowflake (cm/sec), and N_s is the number distribution function of snowflakes per unit volume of air (snowflakes/(cm x cm^3air)). With this formulation, the surface resistance is taken into account as part of the efficiency. If a single characteristic snowflake size L is assumed for the cloud, then the scavenging rate can be expressed as $\psi = \alpha\eta_s P/L$, where α is a constant of order unity.

Estimating ψ may be difficult, so many investigators have chosen to relate C_s to C_a empirically through a scavenging ratio W. Two forms of this parameter have been used in the literature:

$$W_{mass} = \frac{\rho_a C_s}{C_a} \tag{18}$$

$$W_{volume} = \frac{\rho_s C_s}{C_a} \tag{19}$$

W_{mass} has dimensions (g chemical/g snow) ÷ (g chemical/g air), and thus it is sometimes termed the "mass basis" scavenging ratio. In contrast, the "volume basis" scavenging ratio W_{volume} has dimensions (g chemical/cm^3 snow meltwater) ÷ (g chemical/cm^3 air). The two forms have values that differ by an amount ρ_s/ρ_a, or about a factor of 1000. Both are found in the literature, often without subscripts, so it is necessary to closely examine the defining relations to determine which is being used by an author.

Fog Deposition

The occurrence of fog reflects the presence of water in the form of liquid droplets at the earth's surface. Ice fog reflects the presence of water in the ice crystal phase. Either form of fog can occur on polar ice sheets, and it can have a significant effect on the transfer of particles and gases from the atmosphere to an ice sheet. In fact, it has been well-documented that concentrations of chemical species in fog can exceed those in snow (e.g., Borys et al., 1988; Davidson et al., 1989; Kalina and Puxbaum, 1994).

Figure 5 illustrates the process of scavenging of particles and gases by fog for the case of liquid water droplets, which frequently occurs in summer in the polar regions. The water is usually supercooled, existing at temperatures well below freezing. Water vapor condenses on particles to form fog droplets, and the droplets can capture gases and other particles which is essentially in-cloud scavenging as described earlier. The droplets grow to between 10 μm and 30 μm in diameter (Borys et al., 1992), and have settling velocities much greater than the dry deposition velocities of the original aerosol particles (Bergin et al., 1995a). A significant fraction of the particles scavenged by fog are deposited to the surface snow. Those droplets not reaching the surface can evaporate, releasing gases and particles back into the air. In some cases, gases react irreversibly in the fog droplets releasing particles having a different chemical composition than the pre-fog aerosol.

Number concentration data for aerosols with diameters in the range 0.5-3 μm before, during, and after a typical fog event are also shown in Figure 5. The data were obtained with a Laser Particle Counter (LPC - Thermo Systems Inc. model 3755). The data show that particles of this size range were effectively scavenged by fog. Slowly increasing concentrations after the fog event are attributed to release of fog nuclei due to evaporation as well as downward mixing of particle-laden air from aloft (Bergin et al., 1995b).

A simple model for fog scavenging has been developed. The flux of particulate chemical species by fog F_f (g/cm^2sec) can be expressed in a form analogous to Equations 1 and 14 for dry and wet deposition fluxes, respectively (Bergin et al., 1994):

$$F_f(t) = V_s(t) C_f(t) \tag{20}$$

where $V_s(t)$ is the settling velocity of the fog droplets and $C_f(t)$ is the concentration of aerosol chemical species in particles that have been incorporated into the fog droplets (g/cm^3 water). Equation 20 illustrates that the flux of chemical species with fog is not constant with time (as is the case for dry deposition). The model applies only to particles and assumes that nucleation is the only scavenging mechanism; there is no uptake of additional particles as the fog droplets settle. Data from the literature suggest that virtually all particles with diameters greater than about 1 μm and more than half the mass of particles with diameters smaller than this value serve as nuclei for fog droplets (Noone et al., 1992).

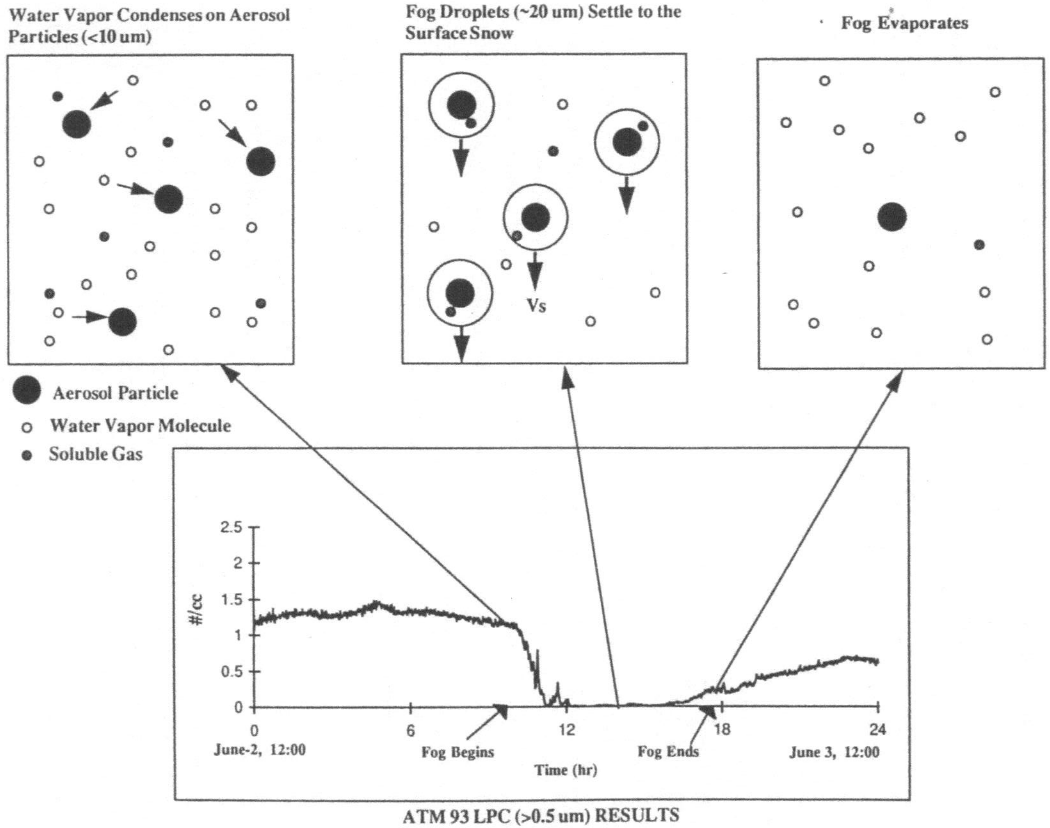

Figure 5. Steps in the removal of particles and gases from the atmosphere over the Greenland Ice Sheet by fog. Laser Particle Counter (LPC) data from a typical fog event are also shown, after Bergin et al. (1995b).

It is convenient to define the inventory of a chemical species for a fog event (mass of chemical species deposited to the surface/unit area) as follows:

$$I_f = \int_0^{t_{fog}} V_s(t) C_f(t) dt \qquad (21)$$

where $F_f(t) = \frac{dI_f}{dt}$. We can write an expression for $C_f(t)$ as:

$$C_f(t) = RC_{ai} \exp(-V_s t / h_f) \qquad (22)$$

where R is the dimensionless fraction of the initial chemical species concentration incorporated into the fog droplets ((g/cm^3 air that is contained in fog droplets) ÷ (g/cm^3 air just before fog formation)), C_{ai} is the initial airborne chemical species concentration just before fog formation, and h_f is the height of the fog (Bergin et al., 1994).

If we assume that V_s does not vary with time, we can combine Equations 21 and 22 to give the following expression for the fog inventory of an aerosol chemical species, I_f:

$$I_f = h_f RC_{ai}\left(1 - \exp\left(-\frac{V_s t_{fog}}{h_f}\right)\right) \qquad (23)$$

Developing a comparable fog scavenging model for a gas phase chemical species requires consideration of mass transport to the droplets, and partitioning of the gas between the air and the fog droplets. This partitioning depends on several factors including pH of the fog droplets, the Henry's law constant of a gas, and liquid water content of the fog. Such a model is the objective of current research but its discussion is outside the scope of this paper.

Note that it is also possible to express chemical species in snow resulting from dry deposition and from wet deposition in terms of inventories. It is convenient to report surface snow data in terms of inventories, since the inventory of a chemical species is independent of changes in snow density created by water vapor flux in and out of the snow pack.

Applying the Theory: Estimating the Deposition in Greenland

A number of experiments have been performed at Dye 3 in southern Greenland and at Summit in central Greenland over the past several years. These experiments have focused on the deposition of aerosols and gases as well as on post deposition processes that affect surface snow chemistry. Results from some of these experiments are presented in this section. First, dry deposition velocities for several chemical species are estimated from particle size distribution measurements at Dye 3 and Summit. Next, snow scavenging ratios based on chemical species concentrations in air and snow are presented for both locations. Following this, the fog model presented above is used to estimate the seasonal fog inventories of several aerosol chemical species at Summit, and the modeled inventories are compared with field data. Finally, the relative contributions of dry, wet, and fog deposition to the summer inventories of several chemical species are presented for Summit.

Dry Deposition

Table 1 shows dry deposition velocities for several aerosol chemical species. The table lists arithmetic mean values of V_d computed from a number of separate size distribution measurements. The standard deviation of the set of V_d values for each species is also listed. The deposition velocities are estimated using the theory of Ibrahim et al. (1983) for aerosols depositing to snow, assuming no hygroscopic growth in the humid layer above the surface. The values for Dye 3 have been taken from Hillamo et al. (1993) using impactor data obtained during the DGASP, the Dye 3 Gas and Aerosol Sampling Program (Jaffrezo and Davidson, 1993). The values for Summit have been taken from Bergin et al. (1995b) using impactor data obtained during the 1993 field season. Chemical analysis methods include ion chromatography (IC) and proton-induced x-ray emission (PIXE). Note that these values represent the results of limited sets of experiments; the true variability is probably much greater than indicated by the standard deviations in Table 1.

The table shows that crustal particles generally associated with the coarse mode (Ca, K, Mg, Na) have greater deposition velocities than the other chemical constituents. MSA, SO_4^{2-}, and S have much of their mass in the accumulation mode but with some fraction in the coarse mode. NO_3^- is associated primarily with the coarse mode in these samples. NH_4^+ exists only in the accumulation mode. For each chemical species, the largest airborne particles in a given distribution are responsible for most of the mass dry deposition flux of the species. It must be emphasized that these calculations are based on only a limited number of size distribution measurements. Furthermore, the estimates do not include the fraction existing in the gas phase, which may be significant for NO_3^-.

The higher values for Dye 3 reflect the greater fraction of coarse particles in these distributions relative to the distributions at Summit. This may be due in part to the different seasons of sampling: the Dye 3 data were obtained in March while the Summit data were obtained during late May - mid July. Airborne particles are likely to be larger in the earlier time period when there is more frequent transport from the Arctic Basin; smaller particles are associated with winds from the south which are more likely to have experienced precipitation scavenging, as evidenced with optical and scanning electron microscopy data (Davidson et al., 1985). The differences in distributions may also be partly due to differences in sampling equipment. The Dye 3 data were obtained using a 7-stage impactor from PIXE International Corporation, while the Summit data were obtained with a new 12-stage Hillamo-Maenhaut impactor (Maenhaut et al., 1995).

Table 1. Dry deposition velocities calculated from size distribution data obtained in Greenland. Dye 3 data are taken from Hillamo et al. (1993), while Summit data are taken from Bergin et al. (1995b). Calculations assume no hygroscopic growth.

Species	Dye 3 V_d, mean cm/s	Dye 3 Std. dev. cm/s	Summit V_d, mean cm/s	Summit Std. dev. cm/s
IC Analysis				
MSA	----	----	0.024	0.023
SO_4^{2-}	----	----	0.021	0.017
NO_3^-	----	----	0.073	0.053
NH_4^+	----	----	0.017	0.011
PIXE Analysis				
S	0.02	0.01	0.022	0.016
Ca	0.35	0.16	0.11	0.021
K	0.17	0.12	0.064	0.017
Mg	0.70	0.95	0.078	0.016
Na	0.16	0.12	0.067	0.015

Wet Deposition

Table 2 shows snow scavenging ratios computed on a mass basis (W_{mass}) for Greenland. These values have been determined empirically from equation 18 using measurements of C_s and C_a; predicting W from equations 15-17 has not been attempted due to lack of information on cloud parameters.

The values for Dye 3 are taken from Davidson et al. (1985), and represent the arithmetic mean and standard deviation of scavenging ratios from three snowstorms where clouds were at or very close to the surface. Values for Summit are for eight snowstorms with clouds at various heights. For both sets of data, scavenging ratios for species associated with the accumulation mode (SO_4^{2-}, Pb) are smaller than those for crustal species which are primarily in the coarse mode (Ca, K, Mg, Na). This is probably related to more efficient scavenging of larger particles (Miller and Wang, 1989; Mitra et al., 1990; Schumann, 1991). Uptake of particles as nuclei is likely to be the dominant wet deposition mechanism for these species. Chemical analysis methods include AA (atomic absorption spectrophotometry), IC, and PIXE.

Although the data in Table 2 are consistent with our understanding of scavenging as a function of particle size, other data in the literature show greater variability. For example, Davidson et al. (1993b) have used data reported by Dibb and Jaffrezo (1993) and Mosher et al. (1993) from the Dye 3 Gas and Aerosol Sampling Program (DGASP) to estimate scavenging ratios for several chemical species. Airborne concentrations for 1-2 day sampling periods were used with concentrations in snow that fell during or close to these periods to estimate W_{mass}. Results show values <100 for ^{7}Be and ^{210}Pb, with values for SO_4^{2-} and trace metals of <200-2000. There is considerable variability in the data, attributed to differences in airborne concentration between the surface and cloud level as well as to the fact that air sampling was not conducted simultaneously with snow events.

Snow scavenging ratios for other sites in the polar regions and the mid-latitudes have also been reported in the literature. For Antarctica, Lambert et al. (1983) report values of W_{mass} of roughly 500-700 for ^{210}Pb and roughly 900-1500 for ^{90}Sr. Engelmann (1970) reports values of 100-2700 for ^{131}I and gross beta activity for snow in Germany. Scott (1981) and Barrie (1985) each report values ranging from <200 to several thousand for SO_4^{2-} in North America, with higher values associated with rimed snow. Barrie et al. (1985) report an average value of 240 for acidity based on airborne concentrations from a routine monitoring program and concentrations measured in a snowpit in the Canadian Arctic. Overall, these studies suggest that a complexity of factors affects scavenging, and that predictions of W_{mass} based on fundamental meteorological parameters is difficult. It is therefore tenuous to extrapolate these empirical results to obtained over short time periods to other times of the year or other locations.

Table 2. Scavenging ratios calculated from concentrations of chemical constituents in air and snow over the Greenland Ice Sheet. Dye 3 values are taken from Davidson et al. (1985), while values for Summit are previously unpublished. Analyses were conducted by IC (SO_4^{2-}), PIXE (Ca), and AA (all other species).

	Dye 3		Summit	
Species	W_{mass}, mean	Std. dev.	W_{mass}, mean	Std. dev.
SO_4^{2-}	180	120	220	129
Pb	160	70	----	----
Ca	----	----	840	400
K	1200	620	----	----
Mg	1800	1100	----	----
Na	2000	1200	----	----

Fog Deposition

For each of the fog events occurring during the 1993 field season at Summit, the fog inventory of an aerosol chemical species can be estimated using Equation 23. Table 3 shows the estimated seasonal fog inventories for MSA, SO_4^{2-}, NH_4^+, Na^+, K^+, and Ca^{2+}. These values are obtained by summing the estimated fog inventories for each of the fog events occurring during the field season. For each event, C_{ai} values are taken from daily filter samples reported for the field season by Bergin et al. (1995b); it is assumed that each time-averaged value is representative of the concentration just before the onset of fog. For MSA and SO_4^{2-}, the R values are taken to be 0.7 +/- 0.2 based on SO_4^{2-} impactor results before and after fog events. For NH_4^+, which is associated with smaller particles, R is assumed to be 0.5 +/- 0.2. Na^+, K^+, and Ca^{2+} are expected to be more readily incorporated into fog droplets due to the larger particle sizes, and R is assumed to be 1.0 for these species. The duration of each fog event is 7 hr +/- 2 hr on the basis of observations from the 1993 field season. A settling velocity of 0.7 cm/s +/- 0.3 cm/s has been assumed, corresponding to a fog droplet diameter of 15 μm based on Borys et al. (1992). The fog height is assumed to be 150m +/- 50m based on tethersonde data for temperature, pressure and relative humidity profiles. This fog height is consistent with those predicted using a version of a fog model developed by Pandis and Seinfeld (1989) modified for the conditions at Summit (Bergin et al.,1995b).

Table 3 shows that the estimated fog inventories are in general agreement with measured values. For NH_4^+, the aerosol generally having the smallest mass mean diameter (MMD), the ratio of the estimated to measured seasonal inventory is 0.58 +/- 0.17. For Ca^{2+}, a coarse mode aerosol having the largest MMD of these six species, the ratio is 0.93 +/- 0.33. The results suggest that the R values for the aerosols having significant accumulation mode mass fractions are underestimates of actual R values. It is also possible that V_s and h_f have not been accurately estimated for fogs at Summit. Note that the geometric standard deviations in the table represent the variability in the estimated inventories for these specific fog events; the variability for a larger number of events is likely to be somewhat greater.

Total Deposition of Chemical Species at Summit

During the 1993 summer field season, experiments were conducted to estimate the relative contributions of dry, wet, and fog deposition to the total seasonal flux of several chemical species. Snow and fog samples were collected from nearly every event occurring from May 24 to July 13. Most of the samples were collected in trays positioned above the snow surface, although roughly 10% of the samples were collected directly from the snow surface at times when high winds prevented effective collection by the trays. The chemical species inventories were estimated for each fog and snow event by multiplying the fog or snow inventory (g/cm^2) by the chemical species concentration (ng/g), following the procedure of Bergin et al. (1995a,b).

Table 3. Comparison between the estimated and measured seasonal fog inventories for Summit, Greenland. All data were obtained using ion chromatography.

	Estimated Geo. Mean Seaonal Fog Inventory (GSD) ng/cm^2	Estimated Seasonal Fog Inventory/ Measured Seasonal Fog Inventory
MSA	0.5 (1.2)	0.64 +/- 0.16
SO_4^{2-}	51 (1.1)	0.67 +/- 0.11
NH_4^+	7.9 (1.1)	0.58 +/- 0.17
Na^+	2.5 (1.2)	0.85 +/- 0.39
K^+	1.1 (1.2)	0.84 +/- 0.34
Ca^{2+}	7.3 (1.1)	0.93 +/- 0.33

To complete the inventory, the dry deposition fluxes were estimated using dry deposition velocities from Table 1 along with aerosol concentrations estimated from daily filter samples (Bergin et al., 1995b).

Results are shown in Table 4. Wet deposition is the dominant process contributing to the summer chemical inventories for all species. For NO_3^-, a chemical species that has a significant fraction in the gas phase, wet deposition accounts for 93% of the summer inventory. This suggests efficient scavenging of vapor-phase HNO_3 by ice crystals (Silvente and Legrand, 1995). It is also possible that an additional NO_y species contributes to the NO_3^- in surface snow (Dibb et al., 1994). For the other chemical species, the mean contributions from wet, fog, and dry deposition are 64%, 21%, and 15%, respectively. For SO_4^{2-} and NH_4^+, which have significant mass fractions in the accumulation mode, the summer fog inventory is much greater than the dry deposition inventory. Although MSA also has a significant mass fraction in the accumulation mode, dry deposition and fog deposition contribute approximately the same amount to the inventory due to the relatively low concentrations of MSA during fog events (Bergin et al., 1995b). For the coarse mode aerosols (Ca^{2+}, K^+, and Na^+), the fog and dry deposition inventories are generally similar, but in this case it is due to their relatively high dry deposition velocities. It is clear that all three deposition processes may need to be considered when using ice core chemistry to provide information on past atmospheric concentrations.

The results in Table 4 apply only to the period late May to mid-July during 1993 at Summit. Marked differences exist in precipitation, frequency of fog, and atmospheric stability in different seasons and from year-to-year (Davidson et al., 1985; Dibb, 1996; Jaffrezo et al., 1995). Furthermore, estimates of the total deposition year-round must account for airborne concentrations at different times of the year, which are known to vary significantly over the ice sheet at Dye 3 (Davidson et al., 1993a; Dibb and Jaffrezo, 1993; Jaffrezo et al., 1993, 1994; Mosher et al., 1993) and at Summit (Wahlin, 1996).

There are numerous uncertainities in these estimates, such as the effect of hygroscopic growth in the viscous sublayer during dry deposition, the effect of wind pumping, complications caused by blowing snow, uncertainties in the scavenging ratios, and various assumptions in the model used to estimate deposition by fog. All of the key parameters must be examined for each site of interest; for example, dry deposition is likely to be far more important in Northern Greenland and in some regions of Antarctica where annual snowfall is low. Overall, while we have made some progress in recent years, our understanding of air-snow transfer of irreversibly deposited species is still weak in many areas.

There are numerous chemicals in the ice core that are deposited via vapor-snow transfer, and many of these species may experience atmosphere-surface transport in both directions before reaching a final concentration in the ice. However, lack of knowledge of processes affecting these reversibly depositing gases poses additional problems. Formaldehyde, hydrogen peroxide, sulfur dioxide, nitric acid, hydrochloric acid, acetic acid, and formic acid are all seen in both the snow and gas phase at Summit, Greenland (Dibb et al., 1994;

Table 4. Seasonal chemical inventory contributions due to fresh snow, fog, and dry deposition at Summit, Greenland, taken from Bergin et al. (1995b). All data were obtained by ion chromatography.

	MSA	NO_3^-	SO_4^{2-}	Na^+	NH_4^+	K^+	Ca^{2+}
Wet (Fresh Snow)							
Seasonal Inventory							
Geo. Mean (ng/cm^2)	1.1	1036	178	17	60	5.6	46
Geo. Std. dev.	1.7	1.1	1.1	1.4	1.1	1.2	1.1
%, Mean	45	93	61	71	76	69	63
%, Std. dev.	13	1	5	8	4	6	3
Fog							
Seasonal Inventory							
Geo. Mean (ng/cm^2)	0.8	70	76	3.0	14	1.4	8.3
Geo. Std. dev.	1.2	1.1	1.1	1.6	1.3	1.5	1.4
%, Mean	32	6	26	13	19	19	13
%, Std. dev.	9	1	3	6	4	6	4
Dry							
Seasonal Inventory							
Geo. Mean (ng/cm^2)	0.6	7.9	34	3.8	4.4	1.0	12
Geo. Std. dev.	1.3	1.2	1.2	1.1	1.1	1.2	1.1
%, Mean	23	1	13	16	5	12	18
%, Std. dev.	7	1	6	4	1	3	2

Bales, 1996; Fuhrer et al., 1996), and possibly experience such exchanges (Conklin and Bales, 1993; Dominé et al., 1995; Neftel, 1996). Formulations developed in this chapter for irreversibly depositing species are inadequate for such cases.

Summary

In this paper, we have considered several mechanisms by which chemical constituents are transported from the atmosphere to ice sheets where they can be eventually recovered in ice cores.

Dry deposition is considered a three-step process. In aerodynamic transport, the particles and gases are brought from the free atmosphere down to the viscous sublayer region immediately adjacent to the surface. They are then carried by boundary layer transport across the viscous sublayer. Finally, they experience chemical and physical interactions with the surface. Particles large enough to be influenced by sedimentation can short-circuit these steps and reach the surface directly under the force of gravity.

Wet deposition includes precipitation scavenging which can occur by one or more processes. Nucleation scavenging refers to the mechanism by which particles serve as cloud condensation nuclei or ice nuclei in the formation of clouds. In-cloud scavenging refers to collection of particles and gases by water droplets, ice crystals, raindrops, and snowflakes as winds sweep through clouds. Below-cloud scavenging occurs as raindrops and snowflakes collect particles and gases while falling to the earth's surface.

Fog deposition can be an important process when particles serve as condensation nuclei in the formation of fog, or as particles and gases are scavenged by fog droplets. Not all chemical constituents are removed permanently from the atmosphere by fog, however: some of the droplets evaporate before they reach the surface, releasing particles and gases back to the atmosphere.

Experiments at Summit, Greenland suggest that wet deposition is responsible for the bulk of the mass deposition of several chemical species onto the ice sheet. For MSA, SO_4^{2-}, Na^+, NH_4^+, K^+, and Ca^{2+} , wet deposition accounts for an average of 64% of the total deposition during the summer 1993 field season. For NO_3^-, which has an appreciable fraction in the gas phase, the corresponding value is 93%. It is apparent that dry and fog deposition are significant for many of these species, and must be considered along with wet deposition when estimating the total seasonal flux to the ice sheet.

Proper interpretation of ice cores requires an understanding of the deposition processes discussed here. The finding that wet deposition is a dominant process suggests that chemical constituents transported in precipitating clouds may be responsible for a considerable fraction of the chemicals in the cores. The fact that dry and fog deposition can be important at times suggests that trajectories transporting chemical species to ice sheets during periods of dry weather must also be accounted for, and that frequency of fog may be an important variable. These results also underscore the need for year-round research on constituents in the atmosphere, snow, and fog at ice coring sites.

Finally, the difficulty in quantifying deposition of reversible species indicates the need for research on atmosphere-surface exchange of such chemical species, as well as work on physical and chemical changes that occur as surface snow ages. Only through continued research on these processes can we develop a satisfactory interpretation of the wealth of ice core data becoming available.

Acknowledgments

We would like to thank Jack Dibb and Sallie Whitlow at the University of New Hampshire for advice concerning analysis of samples and interpretation of our data. We also thank Jean-Luc Jaffrezo of LGGE for advice on all aspects of the research presented in this paper. In addition, we thank Risto Hillamo and Timo Makela of the Finnish Meteorological Institute, as well as Willy Maenhaut at The University of Gent, for analysis and interpretation of impactor data. Discussions with Len Barrie, Paolo Laj, and Hans Pruppacher on several aspects of the material were very helpful. The dry deposition calculations for Figure 3 were performed by Richard Honrath. This work was supported by NSF Grants DPP-9123082, OPP-9321642, and OPP-9423410.

References

Albert, M.R., E.M. Arons, and R.E. Davis, Firn properties affecting gas exchange at Summit, Greenland: Recent results on near-surface ventilation possibilities, In: '*Processes of Chemical Exchange between the Atmosphere and Polar Snow,*' E.W.Wolff and R.C. Bales (ed), NATO ASI Series I., Springer Verlag, Berlin, in press, 1996 (this volume).

Bales, R.C., Conceptual framework for interpretation of exchange processes, In: '*Processes of Chemical Exchange between the Atmosphere and Polar Snow,*' E.W.Wolff and R.C. Bales (ed), NATO ASI Series I., Springer Verlag, Berlin, in press, 1996 (this volume).

Barnola, J.M., D. Raynaud, Y.S. Korotkevich, and C. Lorius, Vostok ice core provides 160,000-year record of atmospheric CO_2, *Nature, 329*, 408-414, 1987.

Barrie, L.A., Scavenging ratios, wet deposition, and in-cloud oxidation: an application to the oxides of of sulphur and nitrogen, *J. Geophys Res, 90,* 5789-5799, 1985.

Barrie, L.A., D. Fisher, and R.M. Koerner, Twentieth century trends in Arctic air pollution revealed by conductivity and acidity observations in snow and ice in the Canadian high Arctic, *Atmos. Environ., 19,* 2055-2063, 1985.

Barrie, L.A., Scavenging ratios: Black magic or a useful scientific tool?, In: '*Precipitation, Scavenging and Atmospheric Surface Exchange*', S.E. Schwartz and W.G.N. Slinn (ed), Hemisphere Publishing Corp., Vol. 3, pp. 403-419, 1992.

Bergin, M.H., J.L. Jaffrezo, C.I. Davidson, R. Caldow, and J.E. Dibb, Fluxes of chemical species to the Greenland Ice Sheet at Summit by fog and dry deposition, *Geochem. Cosmochim. Acta, 58,* 3207-3215, 1994.

Bergin, M.H., C.I. Davidson, J.L. Jaffrezo, J.E. Dibb, R. Hillamo, H.D. Kuhns, and T. Makela, The contributions of wet, fog, and dry deposition to the summer flux of SO_4^{2-} at Summit, Greenland, In: '*Ice Core Studies of Global Biogeochemical Cycles*', R. Delmas (ed), NATO ASI Series, Springer Verlag, Berlin, in press, 1995a.

Bergin, M.H., J.L. Jaffrezo, C.I. Davidson, J.E. Dibb, S.N. Pandis, R. Hillamo, W. Maenhaut, H.D. Kuhns, and T. Makela, The contributions of snow, fog, and dry deposition to the summer flux of anions and cations at Summit, Greenland, *J. Geophys. Res.,* in press, 1995b.

Borys, R.D., E.E. Hindman, and P.J. Demott, The chemical fractionation of atmospheric aerosol as a result of snow crystal formation and growth, *J. Atmos. Chem., 7,* 213-239, 1988.

Borys, R.D., D. Del Vecchio, J.L. Jaffrezo, J.E. Dibb, and D.L. Mitchell, Field observations, measurements and preliminary results from a study of wet deposition processes influencing snow and ice chemistry at Summit, Greenland, In: '*Precipitation, Scavenging and Atmospheric Surface Exchange*', S.E. Schwartz and W.G.N. Slinn (ed), Hemisphere Publishing Corp., Vol. 3, pp.1693-1702, 1992.

Borys, R.D., D. Del Vecchio, J.L. Jaffrezo, C.I. Davidson, and D.L. Mitchell, Assessment of ice particle processes at Dye-3, Greenland, *Atmos. Environ., 27*A, 2815-2822, 1993.

Chen, J.P. and P.J. Crutzen, Solute effects on the evaporation of ice particles, *J. Geophys. Res., 99,* 18,847-18,859, 1994.

Collett, J., B. Oberholzer, and J. Staehelin, Cloud chemistry at Mt. Rigi, Switzerland: Dependence on drop size and relationship to precipitation chemistry, *Atmos. Environ., 27*A, 33-42, 1993.

Conklin, M.H., and R. Bales, SO_2 uptake on ice spheres: liquid nature of the ice-air interface, *J. Geophys. Res., 98,* 16,851-16,855, 1993.

Cunningham, J. and E.D. Waddington, Air flow and dry deposition of non-sea salt sulfate in polar firn: paleoclimatic implications, *Atmos. Environ., 27*A, 2943-2956, 1993.

Dansgaard,W., J.W.C. White, and S.J. Johnsen, The abrupt termination of the Younger Dryas climate event, *Nature*, 339, 532-534, 1989.

Davidson C.I., Mechanisms of wet and dry deposition of atmospheric contaminants to snow surfaces, In *'The Environmental Record in Glaciers and Ice Sheets'*, H. Oeschger and C.C. Langway, Jr. (ed), Dahlem Workshop Report, Wiley and Sons, Chichester, pp. 29-51, 1989.

Davidson, C.I., S. Santhanam, R.C. Fortmann, and M.P. Olson, Atmospheric transport and deposition of trace elements to the Greenland Ice Sheet, *Atmos. Environ., 19*, 2065-2081, 1985.

Davidson, C.I., J.R. Harrington, M.J. Stephenson, M.J. Small, F.P. Boscoe, and R.P. Gandley, Seasonal variations in sulfate, nitrate, and chloride in the Greenland Ice Sheet: Relation to atmospheric concentrations, *Atmos. Environ., 23*, 2483-2493, 1989.

Davidson, C.I., and Y.L. Wu, Dry deposition of particles and vapors, In *'Acidic Precipitation,'* Vol. 3, S.E. Lindberg, A.L. Page, and S.A. Norton (ed), Springer Verlag, Berlin, pp. 103-216, 1990.

Davidson. C.I., Jaffrezo J.L., and P.A. Mayewski, Arctic air pollution as reflected in snowpits and ice cores, In *'Pollution of the Arctic Atmosphere'*, W.T. Sturges (ed), Elsevier Science Publishers, London, pp. 43-95, 1991.

Davidson, C.I., J.L. Jaffrezo, B.W. Mosher, J.E. Dibb, R.D. Borys, B.A. Bodhaine, C.F. Boutron, U. Gorlach, H. Cachier, J. Ducret, J.L. Colin, N.Z. Heidam, K. Kemp, and R. Hillamo, Chemical constituents in the air and snow at Dye 3, Greenland. I: Seasonal variations, *Atmos. Environ., 27*A, 2709-2722, 1993a.

Davidson, C.I., J.L. Jaffrezo, B.W. Mosher, J.E. Dibb, R.D. Borys, B.A. Bodhaine, C.F. Boutron, F.M. Ducroz, H. Cachier, J. Ducret, J.L. Colin, N.Z. Heidam, K. Kemp, and R. Hillamo, Chemical constituents in the air and snow at Dye 3, Greenland. II: Analysis of episodes in April 1989, *Atmos. Environ., 27*A, 2723-2737, 1993b.

Delmas, R.J., and M. Legrand, Long-term changes in the concentrations of major chemical compounds (soluble and insoluble) along deep ice cores, In *'The Environmental Record in Glaciers and Ice Sheets'*, H. Oeschger and C.C. Langway, Jr. (ed), Dahlem Workshop Report, Wiley and Sons, Chichester, pp. 319-341, 1989.

Dibb, J.E., Overview of field data on the deposition of aerosol-associated species to the surface snow of polar glaciers (emphasizing recent work in Greenland), In: *'Processes of Chemical Exchange between the Atmosphere and Polar Snow,'* E.W.Wolff and R.C. Bales (ed), NATO ASI Series I., Springer Verlag, Berlin, in press, 1996 (this volume).

Dibb, J.E., J.L. Jaffrezo, and M. Legrand, Initial findings of recent investigations of air-snow relationships in the Summit region of the Greenland Ice Sheet, *J. Atmos. Chem., 14,* 167-180, 1992.

Dibb, J.E. and J.L. Jaffrezo, Beryllium-7 and Lead-210 in aerosol and snow in the Dye 3 Gas and Aerosol Sampling Program, *Atmos. Environ., 27*A, 2751-2760, 1993.

Dibb J. E., Talbot R. W., and Bergin M. H. Soluble acidic species in air and snow at Summit, Greenland. *Geophys. Res. Lett., 21,* 1627-1630, 1994.

Dominé F., Thibert E., Silvente E., Legrand M., and Jaffrezo J.L., Determining past atmospheric HCl mixing ratios from ice core analysis, submitted for publication, 1995.

Engelmann, R.J., Scavenging predictions using ratios of concentrations in air and precipitation, In: *Precipitation Scavenging (1970),* R.J. Engelmann and W.G.N. Slinn (ed), CONF-700601, National Technical Information Service, Springfield, Virginia, pp. 475-486, 1970.

Friedlander, S.K., *Smoke, Dust, and Haze,* Wiley and Sons, New York, 1977.

Fuhrer, K., M. Hutterli, and J.R. McConnell, Overview of field data for reversibly deposited species, In: *'Processes of Chemical Exchange between the Atmosphere and Polar Snow,'* E.W. Wolff and R.C. Bales (eds.), NATO ASI Series I, Springer Verlag, Berlin etc., 1996 (this volume).

Hillamo, R.E., V.M. Kerminen, W. Maenhaut, J.L. Jaffrezo, S. Balachandran, and C.I. Davidson, Size distributions of atmospheric trace elements at Dye 3, Greenland I. Distribution characteristics and dry deposition velocities, *Atmos. Environ., 27*A, 2787-2802, 1993.

Ibrahim, M., L.A. Barrie, and F. Fanaki, An experimental investigation of the dry deposition of particles to snow, pine trees, and artificial collectors, *Atmos. Environ., 17*, 781-788. 1983.

Jaffrezo, J.L. and C.I. Davidson, The Dye 3 Gas and Aerosol Sampling Program (DGASP): An Overview, *Atmos. Environ., 27*A, 2703-2707, 1993.

Jaffrezo, J.L., R.E. Hillamo, C.I. Davidson, and W. Maenhaut, Size distributions of atmospheric trace elements at Dye 3, Greenland II. Sources and transport, *Atmos. Environ., 27*A, 2803-2814, 1993.

Jaffrezo, J.L., C.I. Davidson, M. Legrand, and J.E. Dibb, Sulfate and MSA in the air and snow on the Greenland Ice Sheet, *J. Geophys. Res., 99,* 1241-1253, 1994.

Jaffrezo, J.L., J.E. Dibb, R.C. Bales, and A. Neftel, Current status of atmospheric studies at Summit (Greenland) and implications for future research, In: *'Ice Core Studies of Global Biogeochemical Cycles'*, R. Delmas (ed), NATO ASI Series, Springer Verlag, Berlin, in press, 1995.

Kalina, M.F. and H. Puxbaum, A study of the influence of riming of ice crystals on snow chemistry during different seasons in precipitating continental clouds, *Atmos. Environ., 28,* 3311-3328, 1994.

Kuhns, H.D., C.I. Davidson, M.H. Bergin, J.L. Jaffrezo, J.E. Dibb, C. Stearns, and R. Caldow, Summertime NH_4^+, SO_4^{2-}, and MSA aerosol at Summit, Greenland, to be submitted for publication, 1995.

Lambert, G., B. Ardouin, and A. Mesbah-Bendezu, Atmosphere to snow transfers in Antarctica. In: *'Precipitation Scavenging, Dry Deposition, and Resuspension,'* H.R. Pruppacher, R.G. Semonin, and W.G.N. Slinn (ed), Elsevier, New York, pp. 1353-1360, 1983.

Maenhaut, W., R. Hillamo, T. Makela, J.L. Jaffrezo, M.H. Bergin, and C.I. Davidson, A new cascade impactor for aerosol sampling with subsequent PIXE analysis, submitted for publication, 1995.

Mayewski, P.A., W.B. Lyons, M.J. Spencer, M.S. Twickler, C.F. Buck, and S. Whitlow, An ice core record of atmospheric response to anthropogenic sulphate and nitrate, *Nature, 346,* 554-556, 1990.

Miller, N.L., and P.K. Wang, Theoretical determination of the efficiency of aerosol particle collection by falling columnar ice crystals, *J. Atmos. Sci., 46,* 1656-1663, 1989.

Mitra, S.K., U. Barth, and H.R. Pruppacher, A laboratory study of the efficiency with which aerosol particles are scavenged by snowflakes, *Atmos. Environ., 24*A, 1247-1254, 1990.

Mosher, B.W., P. Winkler, and J.L. Jaffrezo, Seasonal atmospheric chemistry at Dye 3, Greenland, *Atmos. Environ., 27*A, 2761-2772, 1993.

Neftel, A., The record of gases and reactive species in ice cores, and problems of interpretation, In: *'Processes of Chemical Exchange between the Atmosphere and Polar Snow,'* E.W.Wolff and R.C. Bales (ed), NATO ASI Series I., Springer Verlag, Berlin, in press, 1996 (this volume).

Noone, K.J., J.A. Ogren, A. Hallberg, J. Heintzenberg, J. Ström, H.-C. Hansson, B. Svenningsson, A. Wiedensohler, S. Fuzzi, M.C. Facchini, B.G. Ahrends, and A. Berner, Changes in aerosol size-and phase distributions due to physical and chemical processes in fog, *Tellus, 44*B, 489-504, 1992.

Pandis, S.N., and J.H. Seinfeld, Mathetical modeling of acid deposition due to radiation fog, *J. Geophys. Res., 94,* 12,911-12,923, 1989.

Pomeroy J. W., Davies T. D., and Tranter M. The impact of blowing snow on snow chemistry, *'Seasonal Snowpacks,'* NATO ASI Series, Vol. 28, Springer Verlag, Berlin, pp. 71-113, 1991.

Pomeroy, J.W. and H.G. Jones, Windblown snow: sublimation, transport, and changes to polar snow, In: *'Processes of Chemical Exchange between the Atmosphere and Polar Snow,'* E.W.Wolff and R.C. Bales (ed), NATO ASI Series I., Springer Verlag, Berlin, in press, 1996 (this volume).

Pruppacher, H.R. and J.D. Klett, *Microphysics of Clouds and Precipitation*, D. Reidel, Dordrecht, Holland, 1980.

Schumann, T., Aerosol and hydrometeor concentrations and their chemical composition during winter along a mountain slope-3. Size differentiated in-cloud scavenging efficiencies, *Atmos. Environ., 25*A, 809-824, 1991.

Scott, B.C., Sulfate washout ratios in winter storms, *J. App. Met., 20,* 619-625, 1981.

Silvente, E., and M. Legrand, A preliminary study of the air-snow relationship for nitric acid in Greenland, In: '*Ice Core Studies of Global Biogeochemical Cycles*', R. Delmas (ed), NATO-ASI Series, Springer Verlag, Berlin, in press, 1995.

Slinn, W.G.N., Precipitation scavenging, In '*Atmospheric Sciences and Power Production*', D. Randerson (ed), Technical Information Center, Oak Ridge, TN: U.S. Department of Energy, 1984.

Stull, R.B., *An Introduction to Boundary Layer Meteorology*, Kluwer Academic Publishers, Dordrecht, Holland, 1988.

Taylor, K.C., G.W. Lamorey, G.A. Doyle, R.B. Alley, P.M Grootes, P.A. Mayewski, J.W. White, and L.K. Barlow, The 'flickering switch' of late Pleistocene climate change, *Nature, 361*, 432-436, 1993.

Waddington, E.D., J. Cunningham, and S.L. Harder, The effects of snow ventilation on chemical concentrations, In: '*Processes of Chemical Exchange between the Atmosphere and Polar Snow,*' E.W.Wolff and R.C. Bales (ed), NATO ASI Series I., Springer Verlag, Berlin, in press, 1996 (this volume).

Wahlin, P., One year's continuous aerosol sampling at Summit in central Greenland, In: '*Processes of Chemical Exchange between the Atmosphere and Polar Snow,*' E.W.Wolff and R.C. Bales (ed), NATO ASI Series I., Springer Verlag, Berlin, in press, 1996 (this volume).

OVERVIEW OF RECENT FIELD EXPERIMENTS FOR THE STUDY OF THE AIR-SNOW TRANSFER OF H_2O_2 AND HCHO

Katrin Fuhrer and Manuel Hutterli
Physics Institute
University of Bern
3012 Bern
Switzerland

Joseph R. McConnell
Department of Hydrology and Water Resources
University of Arizona
Tucson
Arizona 85721
USA

1. INTRODUCTION

The analysis of hydrogen peroxide and formaldehyde in ice cores offers the potential to reconstruct the oxidation capacity of the atmosphere in the past [Neftel and Fuhrer, 1993], since these species are closely linked to the odd H and the odd O budget of the atmosphere [Thompson,1992]. These species are so-called reversibly deposited on the snow surface since they show a strong exchange with the gas phase after deposition and their ice concentrations are dominated to a large extent by postdepositional changes in the firn. Therefore, the transfer from the atmosphere into the snow is an important step in the overall transfer function of these species [Neftel, this issue]. To get a better understanding of this step several field experiments were carried out during the 1993 and 1994 summer seasons at Summit in Central Greenland, in the 1993/94 Austral summer in Antarctica along the Swedish traverse (SWEDARP), and at South Pole in November and December, 1994. We will give an overview of the data and point out possible contradictions and knowledge gaps.

2. OBSERVATIONS AT SUMMIT, GREENLAND

H_2O_2 gas phase measurements were performed in summers 1990, 93 and 94 at Summit [Sigg et al., 1992; Bales et al., 1995a]. The peak concentrations of 2 to 3 ppbv, which were mostly found in the afternoon, are

NATO ASI Series, Vol. I 43
Chemical Exchange Between the Atmosphere and Polar Snow
Edited by Eric W. Wolff and Roger C. Bales

significantly higher than the value of 0.7 ppbv calculated with a homogenous gas-phase model for the summer solstice at noontime [Neftel et al., 1995] when the H_2O_2 concentrations are at maximum. Even the mean concentration of 1.5 ppbv as found during a 2 month period in 1994 are more than double the modeled concentration. HCHO gas-phase measurements were also performed. Concentrations between 0.2 and 0.6 ppbv were found during the 1993 and 94 seasons (Figure 1), with the average of 0.3 ppbv being 3 times that expected from photochemical model calculations [Neftel et al., 1995].

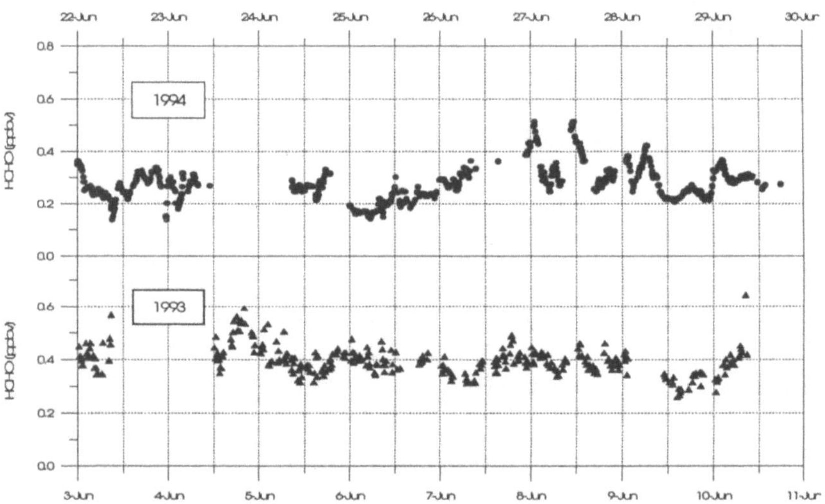

Fig. 1 Two seven-days period of HCHO gas phase concentrations measured at Summit in June 1993 and 1994.

Arguments against a potential H_2O_2 surface source

A surface source of H_2O_2 to explain the high atmospheric H_2O_2 mean concentration through much of the day is considered to be unlikely for several reasons. First, firn gas measurements between 0.3 and 1.6 m depth show concentrations that are always below atmospheric levels, suggesting the firn is not a source for atmospheric H_2O_2 [Bales et al., 1995a]. Second, gradient measurements (together with vertical profiles of eddy diffusivity calculated from balloon soundings) which could give a quantitative insight into fluxes caused by degassing are below the detection limit during all days when the modeling suggests that fluxes should reach steady state [Bales et al., 1995b]. Measured gradients on the order of 0.1 ppbv/m (Figure 2) (only few gradient data are available so far) are not in contradiction with days with low eddy diffusivity when modeling suggests that fluxes do not reach steady state. For these days, a model with a fixed eddy diffusivity profile in time and fixed spatial boundaries can not be used to estimate gradients. The measured gradients may, however, also result from the release of H_2O_2 in the morning and deposition to the snow surface in the evening as suggested by the changing direction of the measured gradients during the day (Figure 2). This diel exchange is the most likely explanation for the daily variation in the atmospheric H_2O_2 concentration, although only the peak concentrations and not the daily mean concentrations can be explained in this manner. Third, the amount needed to double the atmospheric H_2O_2 concentration, 2-20 $ngm^{-2}s^{-1}$ as suggested by Bales et al. [1995b], should lead to a significant depletion of surface snow (3-30 μM in a 5 cm surface sample after a 10 day period without snowfall). Surface snow monitoring during the summer season did not show losses between snowfall events [Bales et al., 1995b] and so makes a net degassing at these rates unlikely. However, sampling was not specifically designed to detect losses.

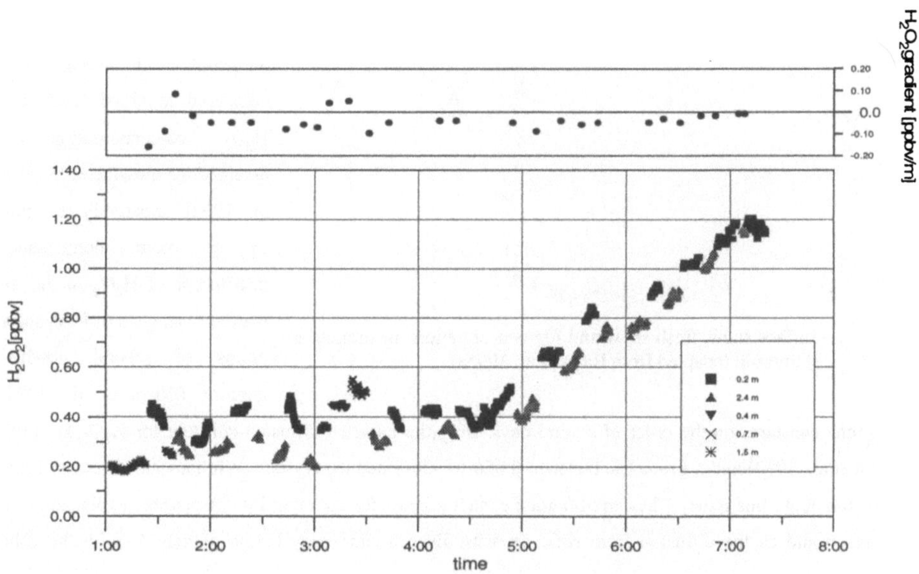

Fig. 2 H_2O_2 measurements performed at different heights between 0.2 and 2.4 m above the snow surface on the 19.6.94. Also the mean gradients in the evening and morning from the 9. and the 10. June 94 are indicated.

Degassing of HCHO and H_2O_2 after snowfall

Degassing from the snow surface has been observed after snowfall for both H_2O_2 and HCHO. Fresh snow H_2O_2 concentrations decrease within a few hours by 50% as shown in Figure 3 (adapted from Bales et al. [1995a]) by the lower concentrations of the surface snow compared to fresh snow. Fresh snow HCHO concentrations decrease within a few hours by 90% [Hutterli, 1996]. This leads to the often observed increase in atmospheric concentrations after snowfall, at least for H_2O_2. For HCHO, an atmospheric increase cannot yet be confirmed but available data suggest that an increase of more than 200% can be excluded [Hutterli, 1996].

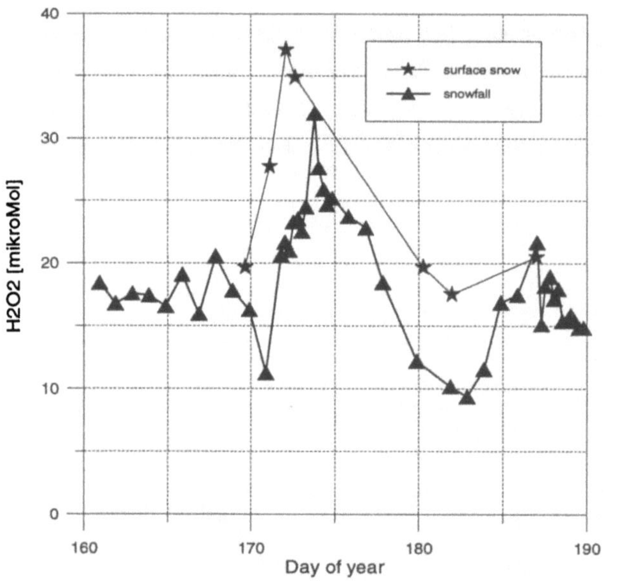

Fig. 3 Surface snow, fresh snow and fog concentrations in summer 93 at Summit (adapted from Bales et al. 1995a).

Degassing experiments by Staffelbach [1990] suggest that HCHO is preferentially adsorbed at the snow crystal surface rather than distributed homogenously throughout the ice matrix. Therefore, the fast decrease after snowfall can be explained by desorption in response to the higher temperatures at ground level compared to cloud level. For H_2O_2 co-condensation is assumed for snowflakes [Sigg et al., 1992], apparently resulting in a more homogenous distribution of H_2O_2 in the ice matrix. Diameters of fresh snow grains of several hundred microns [Borys et al., 1993] lead to time constants on the order of several days using the modeled diffusion constant for H_2O_2 at -11°C [Conklin et al., 1993] and a geometric factor of 0.056 for elongated ice crystals [Whillans and Grootes, 1985; Waddington et al., this issue]. Thus equilibration cannot account for the rapid loss. A possible explanation for the losses would be the diffusion from rime deposits. Even a rime contribution of only 1 % to the solid deposition at Summit, as estimated by Borys et al. [1993], would lead to a contribution of up to 50 % of the H_2O_2 in surface snow from rime assuming Henry's law equilibrium at a temperature of -10°C (by extrapolating the Henry's law constant according to Lind and Kok [1994]) and a 1 ppbv gas phase concentration (concentration of more than 1000 µM in rime droplets). A typical droplet diameter of 20 µm leads to a characteristic diffusion time of less than an hour (corresponding to the time needed to halve the initial concentration in a sphere [Crank, 1975]). The high concentration would allow for large fluxes, making it possible that a significant portion of the loss in fresh snow is caused by diffusion from rime deposits during the equilibration process. Observations of rime deposits in single snow falls with H_2O_2 losses in fresh snow is needed to test this idea. Recrystallisation may also contribute to a fast partial equilibration after snowfall, however, no data are available on the change in surface area during the first few hours after deposition. Even if significant degassing after snowfall can be confirmed for both H_2O_2 and HCHO, these events would not be able to maintain the high atmospheric concentrations over a longer time period considering the short atmospheric lifetime of these species over Greenland.

Simple point model calculations for the Summit atmosphere

Instead of an H_2O_2 surface source, an HCHO surface source could offer a consistent picture for both the H_2O_2 and HCHO gas phase concentrations at Summit. Calculations using the point model ACUCHEM [Braun et al., 1988] showed that observed HCHO concentrations could explain the high atmospheric H_2O_2 concentrations with typical atmospheric conditions for the high Arctic atmosphere [Fuhrer et al., 1996]. The photolytical destruction of HCHO leads to H and HCO radicals. Both of these will react with O_2 to produce HO_2 radicals. The reaction of two HO_2 radicals represents the production path of H_2O_2. Firn gas measurements [Hutterli, 1996] show elevated HCHO concentrations in the upper firn layers, indeed suggesting a potential surface source of HCHO. Firn gas concentrations reach up to 4 times the atmospheric value right below the surface and decrease to atmospheric concentrations at a depth of 3 m (Figure 4; from Hutterli, 1996). Atmospheric H_2O_2 profiles will still show gradients above the surface if an HCHO surface source is the reason for the high atmospheric H_2O_2 concentrations. Assuming a penetration depth of UV light in snow of 20 cm [Warren, 1982] and the firn gas HCHO concentration of 4 times the atmospheric concentration right below the surface, one would also expect a smoothed maximum of H_2O_2 in the upper firn air. This is not observed, however, suggesting that either H_2O_2 destruction in the firn air occurs or that HCHO initiated H_2O_2 production does not take place in the firn pore space (possibly due to the large surface area).

HCHO firn and firn gas measurements

The emission rate calculated with the ACUCHEM model for a 100 m mixed layer in the atmosphere is about 100 times that provided from the snow pack since the surface snow is usually 1-2 ppbw higher than the minimum right below the surface (Figure 4) [Hutterli, 1996]. Deeper snow layers cannot increase the atmospheric HCHO concentration, as indicated by the direction of the firn gas gradient. The ACUCHEM model may have overestimated the flux needed to increase the HCHO air concentration up to 0.3 ppbv (e.g. by using a too large mixed layer). However, as estimated from losses from fresh snow [Hutterli, 1996], a flux of this magnitude is expected after snowfall so, at that time, the atmosphere should be much more enriched with HCHO than has been observed. If destruction of HCHO after snowfall does not occur then another mechanism must be found to provide the observed high atmospheric H_2O_2 concentrations [Hutterli, 1996] (e.g. photochemical production).

The firn and firn gas profiles show opposite trends, with the firn concentration exhibiting a minimum close to the surface (Figure 4) [Staffelbach et al., 1991; Hutterli, 1996]. The downward directed flux of HCHO, as implied from the firn gas gradients, can be provided from the surface snow using the same argument as above (i.e. the surface snow is usually 1-2 ppbw higher than the minimum right below the surface). However, this flux would not be able to raise the firn concentration from less than 2 to approximately 5 ppbw. The 5 ppbw mean concentration of deeper snow layers therefore suggests either: i) that a photochemical source of HCHO exists at the surface or ii) that surface snow concentrations are higher during winter and/or they do not decrease to the minimum value found in pits during summer. A highly temperature dependent uptake capacity of snow for

HCHO, as was found for H_2O_2, could explain these observations. There could be a slower desorption or equilibration with fresh snow at lower temperatures. Advection of HCHO-rich air masses would also lead to high fresh snow concentration in winter, which is possible considering the lifetime of HCHO in winter of 180 days [DeServes, 1995]. Concentrations comparable to those measured at Summit in summer were found in air masses traveling from Greenland during the polar night [DeServes, 1995].

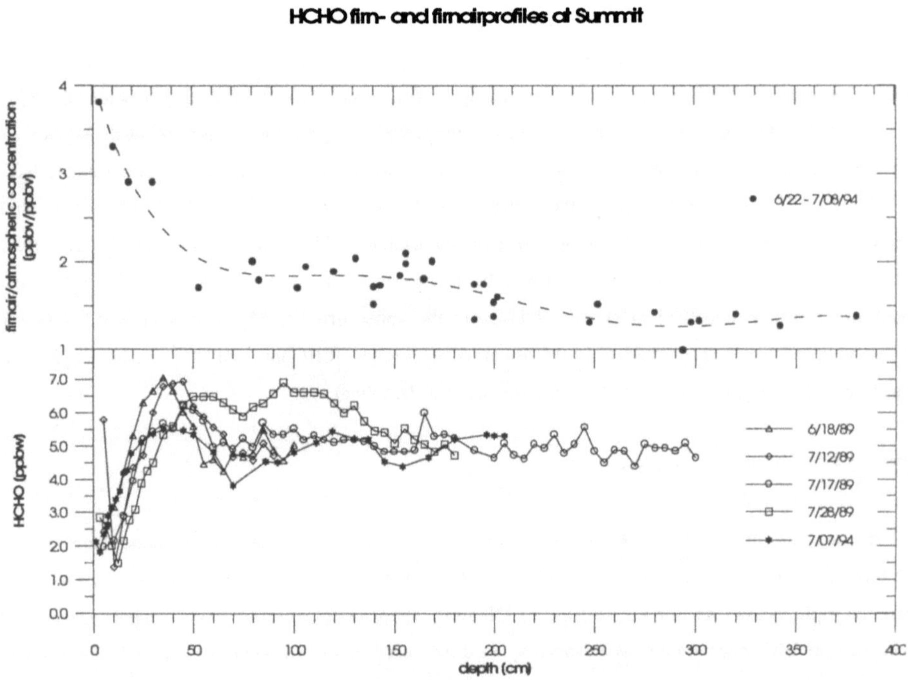

Fig. 4 On the bottom several pits analysed for HCHO between 1989 and 1994, during the summer season. On the top HCHO firn gas measurements performed in summer 1994 (from Hutterli, 1996)

H_2O_2 firn gas measurements

Conklin et al. [1993] calculated an equilibration time for H_2O_2 in snow layers in the order of weeks to months. Bales et al. [1995a] showed that the firn to firn air ratios cluster around the equilibrium value for the annual mean temperature of -30°C, with average summer values above this line and average winter values below. They concluded that the equilibration process is still going on at least down to a 2 m depth. However, some open questions remain. The fact that summer firn air concentrations are lower than atmospheric must be due to either a sink in the firn or to a slow exchange of firn gas with the atmosphere combined with a slow equilibration of the snow layers. In the latter case, the firn gas would therefore still reflect winter atmospheric concentrations. Summer firn layers should release H_2O_2 during the equilibration process. Winter firn layers will only be a sink for H_2O_2 if the surrounding firn air is higher than the equilibrium value for winter snow concentrations. This is not the case according to Figure 5 (adapted from Bales et al.,1995a) where most winter values lie above the equilibrium line for the respective temperature implying that these layers release H_2O_2. Waddington et al. [this issue] calculate a characteristic time constant for the snow-firn gas exchange to be on the order of a month in the upper meter of the snowpack. Equilibrium between snow and firn air would lead to firn air concentration between 500 and 2000 pptv according to the laboratory determined equilibrium line, using the temperatures measured at the time of the firn gas sampling. This is not observed (Figure 5), which could be explained by the firn air reflecting equilibrium with the snow layers for lower temperatures earlier in the year. However, diffusion and ventilation have characteristic time constants on the order of a day, using the exchange rate estimated by Albert et al. [this issue], and the diffusion coefficient from Schwander et al. [1988] for the upper meter of snow. Therefore the upper meter should be well mixed with regard to the atmosphere, with a significant sink in the firn being required to maintain the low firn air concentrations. We therefore suggest that low firn air concentrations imply that a destruction of H_2O_2 occurs in the firn air, that the equilibrium concentration for Greenland snow is different from that determined in the laboratory (and winter layers would provide a sink for H_2O_2), or the loss factor in the sampling tubes [Bales et al., 1995a] for the 1993 firn air measurements was underestimated. The latter seems to be unlikely because similar firn gas concentrations were found in 1994 [J. Winterle, personal communication]. The fact that the firn air shows a weak seasonal structure would instead argue against destruction. More detailed firn air measurements are needed.

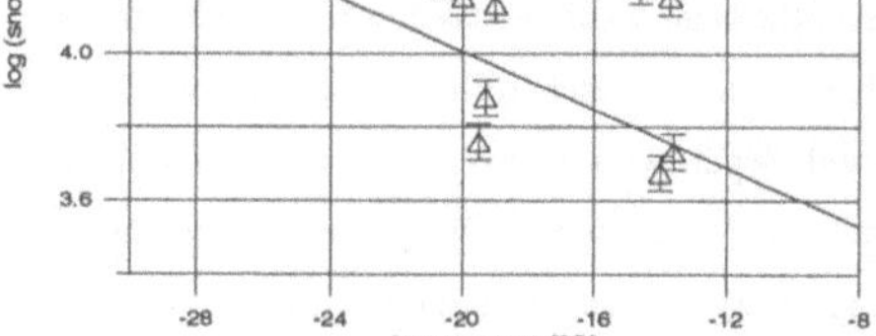

Fig. 5 Firn to firn air ratios measured in summer 1993 at Summit compared to the Co-condensation line and the laboratory determined Kd equilibrium line (adapted from Bales et al., 1995a).

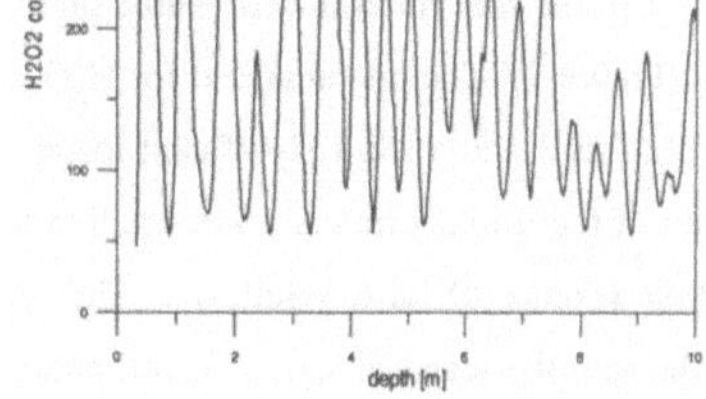

Fig. 6 H_2O_2 concentrations along Eurocore (1989), adapted from Sigg and Neftel [1991]

Bales et al. [1995a] pointed out that since winter concentrations in buried snow do not change with time over the last few years, the equilibration process must be very slow. This implies also that winter layers still reflect co-condensation at the temperature of snow formation, leading to winter atmospheric concentrations of 5 pptv at -40°C. This is not in contradiction with the only peroxide measurements during polar night at high Arctic latitudes that have been made [DeServes, 1995]. In addition, we point out that after one year the summer layers stay essentially unchanged for several years (Figure 6, adapted from Sigg and Neftel, 1991). A likely explanation for the first year summer loss is that with a strongly temperature dependent diffusion constant, a partial equilibration occurs for the first summer layer. Also, recrystallisation will be faster under higher temperature gradients near the surface, enhancing the equilibration process. Certainly, if the first year's summer layer decrease is related to temperature, it should be observed to occur during the summer season. Surface snow monitoring may not be sufficient to observe decreases in concentrations since the exposure time between snow fall events is too short. A monitoring of 10 cm deep pits would give better insight in to changes of the same layer over a longer period. It cannot be excluded that significant changes also occur for winter layers during the first half year after deposition, eventually related to enhanced ventilation rates during winter time. Winter layers would then no longer represent co-condensation conditions.

3. OBSERVATIONS IN DRONNING MAUD LAND, ANTARCTICA

Losses of H_2O_2 in Antarctic pits

Antarctic H_2O_2 gas phase measurements from Dronning Maud Land between 73 and 76°S show maximum concentrations of 0.5 ppb [Fuhrer et al., 1996], similar to the H_2O_2 gas phase concentrations measured at South Pole (Figure 7). These values are in agreement with measurements from Georg v. Neumayer Station [Jakob and Klockow, 1993]. Note the sharp drop in H_2O_2 concentration at South Pole on day 326 (Figure 7). While investigation of this event is incomplete, it is correlated in time with a change in air mass back-trajectory. HCHO gas phase concentrations measured on the traverse in the same area are between 50 and 80% of those at Summit (Figure 7). The pits analysed for H_2O_2 (Table 1) show a completely different picture from Greenland (Figure 8). Pits 3 and 4 show significant losses of H_2O_2 with depth, with the average concentrations dropping to a tenth of the initial summer concentration within less than 1 meter depth. This phenomena is observed at South Pole (Figure 8) [also Whitlow et al., 1992], and at Georg v. Neumayer station [Jakob and Klockow, 1993]. Because the seasonal signal is well preserved a few 100 km from the coast (Pit 1, Figure 8), the coastal loss may be due to a different mechanism, possibly related to the high seasalt content. Further studies are needed. The low accumulation rate in the Antarctic interior cannot be the dominant reason for the loss with increasing latitude since in North Central, Greenland, no losses were observed with similar accumulation rates and temperatures [Neftel,1984]. Also, the accumulation rates in Pits 2 and 3 seem to be similar as inferred from the H_2O_2 profile (Figure 8), and far lower losses are observed in Pit 2, in contrast to Pit 3. From the gas phase data it can be stated that the loss is not due to a difference in summer atmospheric concentrations. The difference in the near-surface summer snow concentrations between pits can be explained by co-condensation. A change in temperature from -10°C to -20°C from Pit 1 to Pit 2, which is reasonable based on year round temperature data from coastal and interior antarctic sites [Schwerdtfeger, 1984], leads to a change in concentration from 250 ppb to 500 ppb using the measured air concentrations of 0.4 ppbv. This is in the range of the observed concentrations.

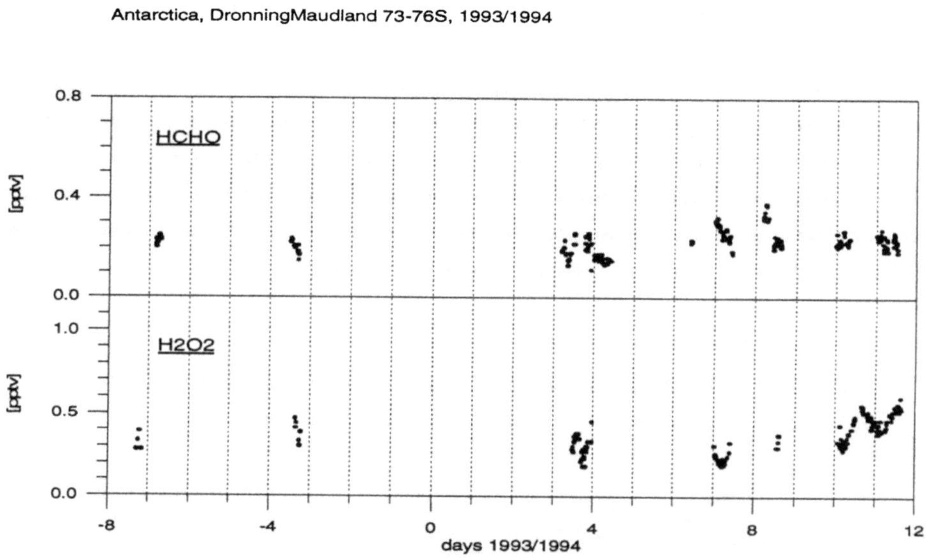

Fig. 7 HCHO and H_2O_2 gas phase measurements performed during the Austral summer 93/94 between 73 and 76°S in Dronning Maud Land, Antarctica (adapted from Fuhrer et al., 1996). On the bottom H_2O_2 gas phase concentrations for a 20 day period in 1994 measured at South Pole.

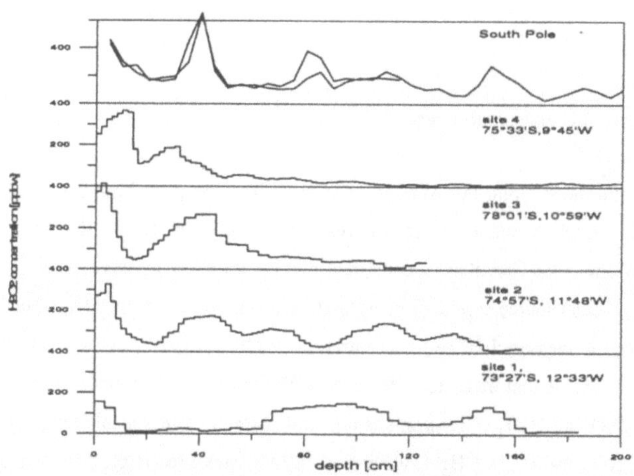

Fig. 8 H_2O_2 pits performed during the Austral summer 93/94 between 73 and 76°S in Dronning Maud Land, Antarctica. On top pits analysed at South Pole in November and December 1994.

The hypothesis made by Neftel [this issue] is that a loss with increasing latitude is due to the shielding effect of the polar vortex in the winter season, maintaining low atmospheric H_2O_2 concentrations in the interior Antarctic during winter time. However, even with lower atmospheric winter concentrations in the Antarctic interior, the equilibration process must be very different from Summit. In Summit, losses in pits were only observed for the first summer layer, and winter layers at a depth of 20 cm show the same concentrations as deeper buried winter layers. The same is true for the Antarctic pits 1 and 2. Pits 3 and 4 instead show significant losses below a depth of 40 cm, suggesting that equilibration with the atmosphere reaching deeper down in depth. Waddington et al. [this issue] calculated a dominating impact of the topographic ventilation on the equilibration process with the atmosphere down to a depth of several decimeters for typical conditions of low accumulation sites, in contrary to high accumulation sites. If we suggest ventilation as the predominant process which accounts for the observed losses, then this implies that the topography and/or windspeed are different for pits 2 from pit 3 since they hardly differ in accumulation rate.

Table 1: Locations and the 10 m bore hole temperature [E. Isaksson, unpublished data] (reflecting the mean annual temperature at the site) for the pit sites in Dronning Maudland, Antarctica.

Date	latitude	longitude	altitude	10 m temperature	site nr
24.12.93	73°27'47"S	12°33'70"W	800 m	-22.7°C	1
28.12.93	74°57'69"S	11°48'07"W	2150 m	-33.4°C	2
3.1.94	78°01'52"S	10°59'21"W	2151 m		3
8.1.94	75°33'16"S	9°45'92"W	2400 m	-37.4°C	4

Simple point model calculations for the Antarctic atmosphere

We also measured HCHO air concentrations along the Antarctic traverse. They ranged within 0.2-0.3 ppbv (Figure 7). We performed the same model calculation as for the Summit atmosphere also for the Antarctic atmosphere [Fuhrer et al., 1996]. Whereas for Greenland a consistent picture arises for measured HCHO and H_2O_2 concentration when a surface HCHO source is added, in Antarctica seventy-five percent of the Summit emission rate is needed to reproduce the measured HCHO concentration. Such an HCHO concentration then leads to an H_2O_2 concentration that is double that measured. However, with the ACUCHEM model there is no means to adjust to the measured HCHO and H_2O_2 concentrations and still maintain a reasonable range of the important species (O_3, H_2O, HNO_3, NO_x, PAN, CO, CH_4) [Fuhrer et al., 1996]. Measured H_2O_2 concentrations can only be reproduced with atmospheric HCHO concentrations below 0.1ppbv.

4. CONCLUSIONS

We find that this kind of qualitative discussion reveals several contradictions in the current understanding of the postdepositional processes affecting of H_2O_2 and HCHO. No consistent picture is yet found to explain the high H_2O_2 and HCHO gas phase concentration over Greenland since a surface source for H_2O_2 seems unlikely. For a final conclusion on possible diel uptake and degassing, daily water vapour fluxes will be needed to separate fluxes caused by degassing from fluxes caused by coevaporation and codeposition. The latter fluxes will follow the temperature profile, i.e. changing sign after the temperature maximum is reached whereas former fluxes will be at maximum when the highest temperature is reached where also highest vertical eddy diffusivities occur and where the lowest ice-air partitioning coefficients are found [Conklin et al., 1993]. Mass balance of H_2O_2 fluxes should be carried out to decide if positives fluxes dominate. In addition, the profiles of both H_2O and H_2O_2 should be compared to decide if they are strongly correlated.

From the H_2O_2 firn air measurements we conclude that there is a lack of agreement between the equilibrium line determined by Conklin et al. [1993] in the laboratory, and the field observations.

Point model calculations show that high H_2O_2 concentrations are compatible with the high HCHO concentrations found at Summit. High HCHO firn air concentration suggest a HCHO surface source, however, a consistent quantitative analysis considering fluxes into the firn and the atmosphere is missing. Further, the same model calculation does not lead to a compatible picture for the Antarctic atmosphere, suggesting that important parameters are not yet included in the conceptual model. Antarctic H_2O_2 pit studies show a completely different picture from Summit, suggesting that the large losses occurring during the winter season down to a depth of several decimeters in the Antarctic interior are most likely caused by the enhanced impact of ventilation on the equilibration process of low accumulation sites. In contrast, significant concentration changes are only observed in pits at Summit for the first summer. These losses are presumed to occur during the first few month after deposition. Low winter atmospheric concentrations in the Antarctic interior are not sufficient to explain the observed losses below 40 cm depth.

ACKNOWLEDGEMENTS

The data from Greenland were obtained within the frame of the european TAGGSI program. The data from Antarctica were gained during the SWEDARP expedition 1993/94. Studies by the University of Arizona were supported by National Science Foundation grants (DPP-9123025 and DPP-9224192) and a NASA Graduate Fellowship for Global Change (NGT30269). South Pole studies were conducted under a cooperative arrangement with the NOAA Climate Montitoring and Diagnostics Laboratory.

REFERENCES

Albert MR, Arons EM, Davis RE (this issue) Firn properties affecting gas exchange at Summit, Greenland: Ventilation possibilities.

Bales RC, McConnell JR, Losleben MV, Fuhrer K, Neftel A (1995a) H_2O_2 in snow, air and open pore space in firn at Summit, Greenland. Geophys. Res. Letters 22: 1261-1264

Bales RC, McConnell JR, Losleben MV, Conklin MH, Fuhrer K, Neftel A, Dibb JE, Kahl JDW, Stearns ChR (1995b) The diel variations of H_2O_2 in Greenland: A discussion of the cause and effect relationship. J. Geophys. Res. 100: 18,661-18,668

Borys RD, DelVecchio D, Jaffrezo JL, Davidson CI, Mitchell DL (1993) Assessment of ice particle growth processes at Dye-3, Greenland. J. Atm. Env. 17/18: 2815-2822

Braun W, Herron JG, Kahaner DK, (1988) Acuchem: A computer program for modelling complex chemical reaction systems. Int. J. of Chem. Kinet. 20: 51-61

Conklin MH, Sigg A, Neftel A, Bales R (1993) Atmosphere-snow transfer function of H_2O_2: Microphysical considerations. J. Geophys. Res. 98: 18,367-18,376

Crank J (1975) The mathematics of diffusion. Brunel University Uxbridge

DeServes C (1995) Gas phase formaldehyde and peroxide measurements in the Arctic atmosphere. J. Geophys. Res. 99: 25,391-25,398

Fuhrer K, Staffelbach Th, Neftel A, Hutterli M (in preparation) A comparison of Greenlandic and Antarctic HCHO and H_2O_2 gas phase data.

Hutterli M (in preparation) Approach of a quantitative understanding of the atmosphere-snow transfer of formaldehyde.

Jakob P, Klockow D (1993) Measurements of hydrogen peroxide in Antarctic ambient air, snow and firn cores. J.Anal.Chem. 346: 429-434

Lind A, Kok GL (1994) correction to "Henry's law constant for aequous solutions of hydrogen peroxide, methylperoxide, and peroxyacetic acid. J. Geophys. Res. 99: 21,119

Neftel A (1984) Measurements of hydrogen peroxide in polar ice samples. Nature 311:43-45

Neftel A, Bales RC, Jacob D (1995) H_2O_2 and HCHO in polar snow and their relation to atmospheric chemistry. In: Ice core studies of Global Biochemical Cycles, Delmas R (ed), NATO-ASI Series, Springer Verlag, Berlin

Neftel A (this issue) The record of gases and reactive species in ice cores and problems of interpretation.

Neftel A, Fuhrer K (1993) Reconstruction of the oxidation capacity of the atmosphere: Dream or possible reality ? NATO-ASI Series, Springer Verlag, Berlin

Schwander J, Stauffer B, Sigg A (1988) Air mixing in firn and the age of the air at pore close-off. Annals of Glaciology 10: 141-145

Schwerdtfeger W (1984) Weather and Climate of the Antarctic. Developments in atmospheric sciences 15, Elsevier

Sigg A, Neftel A (1991) Evidence of a 50% increase in H_2O_2 over the past 200 years from a Greenland ice core. Nature 351: 557-559

Sigg A, Staffelbach Th, Neftel A (1992) Gas phase measurements of hydrogen peroxide in Greenland and their meaning for the interpretation of H_2O_2 records in ice cores. J. Atm.Chem. 14: 223-232

Staffelbach Th, Neftel A, Stauffer B, Jacob D (1991) A record of the atmospheric methane sink from formaldehyde in polar ice cores. Nature 63: 10,603-10,605

Staffelbach Th (1990) Formaldehyd und Ammonium Messungen an Schnee-Firn-und Eisbohrkernen aus polaren und alpinen Regionen. PhD thesis, University of Bern

Thompson A (1992) The oxidizing capacity of the earth's atmosphere: Probable past and future changes. Science 256: 1157-1165

Waddington ED, Cunningham J, Harder SL (this issue) The effects of snow ventilation on chemical concentrations.

Warren StG (1982) Optical Properties of Snow. Rev.Geophys.and Space Phys.1: 67-89

Whillans IM, Grootes PM (1985) Isotopic diffusion in cold snow and firn. J. Geophys. Res. 90: 3910-1918

Whitlow S, Mayewski PA, Dibb JE(1992) A comparison of major chemical species: Seasonal concentration and accumulation at the South Pole and Summit, Greenland. Atm Environ. 23: 2045-2054

CONCEPTUAL FRAMEWORK FOR INTERPRETATION OF EXCHANGE PROCESSES

Roger C. Bales and Jungyill Choi
Department of Hydrology and Water Resources
The University of Arizona
Tucson, Arizona 85721
U. S. A.

1 Introduction

Polar firn and ice contain the most detailed, long-term, complete record of past climatic conditions on the Earth. Gas enclosed in bubbles and released from ice when it is brought to atmospheric pressure are samples of ancient atmospheres. Impurities in the ice matrix are related to atmospheric composition at or shortly after the time the snow that forms the ice was deposited. Climate reconstruction from polar ice cores offers the only possible indication of historical concentrations for many atmospheric chemical species that are critical to understanding how the Earth's atmosphere responds to changes in emissions of the various carbon, nitrogen and sulfur species. These historical records provide the data against which to evaluate chemical/climate models that are designed to simulate how the Earth's atmosphere will respond to changing patterns of anthropogenic emissions.

Ice core records of temperature and methane (CH_4) have provided the main data constraints on attempts at photochemical modeling for the pre-industrial Holocene and last glacial maximum [*Thompson*, 1992]. Somewhat arbitrary assumptions must be made for species for which there are no data. Use of additional records that would constrain the dominant atmospheric oxidant, hydroxyl radical (OH), and the feedbacks involving OH (e.g. with carbon monoxide (CO), ozone (O_3), CH_4, oxides of nitrogen (NO_x) and non-methane hydrocarbons) should yield a definite improvement in model accuracy and uncertainty [*Thompson*, 1995]. The major impediment to using these records is that we do not know the transfer function relating concentrations in the atmosphere to those in the ice.

First, it is useful to distinguish between chemical species that are reversibly versus irreversibly deposited to snow. Concentrations of hydrogen peroxide, formaldehyde, organic acids and HNO_3 in snow change in the days and months after the snow is deposited. That is, they are reversibly deposited to snow. Both decreases and increases in surface-snow concentration (and mass) have been observed.

Non-volatile atmospheric particles such as volcanic tephra, soil dust, soot carbon, H_2SO_4 aerosols and most anthropogenically produced trace species appear to be essentially immobile (in a chemical sense) once deposited in snow, and hence will tend to remain associated with

NATO ASI Series, Vol. I 43
Chemical Exchange Between the Atmosphere and Polar Snow
Edited by Eric W. Wolff and Roger C. Bales

a given layer of snow throughout firnification. This implies that wet and dry deposition are irreversible, and the "transfer function" needed for making interpretations of these species in ice cores is an understanding of how fluxes from the air to the snow depend on changing climatic conditions [*see Davidson et al.*, this volume].

In general, the chemical record of reversibly deposited species at a specific site depends on seasonal and annual patterns of accumulation and temperature and on the intensity of air exchange between the snowpack and the atmosphere [*Neftel et al.*, 1995]. During both rapid growth of ice by vapor deposition [*Valdez et al.*, 1989] and rapid freezing of liquid [*Iribarne and Pyshnov*, 1990], impurity concentrations in the ice should reflect concentrations in the cloud rather than those in equilibrium between ice and atmosphere. A volatile species in snow that is physically in contact with air that undergoes exchange with the atmosphere should eventually equilibrate with atmospheric levels. In high-accumulation areas there may be insufficient air exchange for equilibrium to be reached before subsequent snowfalls bury snow from the previous event. In low-accumulation areas there may be insufficient burial of the snow for it to retain concentrations characteristic of the season in which it fell. The key to understanding the relation between concentrations in snow and in those in the atmosphere then becomes a question of relating diurnal and seasonal changes in near-surface atmospheric processes to processes in the snow.

This paper addresses two central questions. First, what do we know about the processes governing the atmosphere-snow transfer for species that are "reversibly deposited" to snow? Reversibly deposited refers to species that have a significant presence in both the gas and condensed phases, and can undergo reversible exchange between snow or firn and the overlying atmosphere. Second, what more should we learn about the atmosphere-snow exchange process and subsequent changes in the snow/firn/ice in order to extract relevant climatic information about these species from ice-core records? This paper focuses on hydrogen peroxide (H_2O_2) as a model for the behavior of reversibly deposited species, though the conceptual model should also describe the behavior of formaldehyde (HCHO), nitric acid (HNO_3) and light organic acids (e.g. formic (HCOOH) and acetic (CH_3COOH)).

2 Observations

It has generally been observed that H_2O_2 concentrations in the surface layer of snow at Summit, Greenland exceed those in previous, buried summer layers [*Sigg*, 1990] (Figure 1). *Whitlow et al.* [1992] reported that at South Pole, H_2O_2 concentrations in buried summer layers dropped to one-half of the surface value in the top meter, and to about 25 percent of the surface value below about 2.5 m (Figure 2). Thus snow-pit samples from both high and low accumulation sites show an apparent loss in H_2O_2 after the first year. However, losses are greatest at the low-accumulation site in central Antarctica, i.e. South Pole.

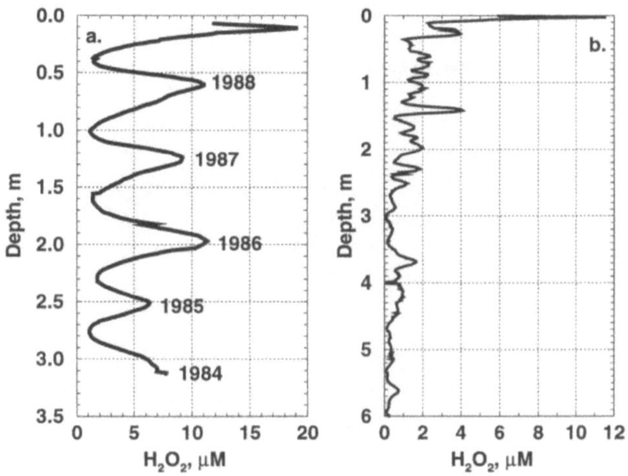

Figure 1: a) Typical depth profile of H_2O_2 in snow/firn from Summit, Greenland. Concentrations shown here are an average of 6 pits and one hand-augered core taken over a period of 40 days in June-July, 1989 [*Sigg*, 1990]. b) Depth profile of H_2O_2 in snow/firn from South Pole, as reported by *Whitlow et al.* [1992].

There is considerable variability in concentrations of H_2O_2 and other volatile species deposited by individual events. At Summit, surface layers deposited in individual snowfalls and altered by fog and diamond dust (clear sky precipitation) deposition range from under 10 to over 30 μM [*Bales et al.*, 1995*a*]. The average surface snow concentrations measured over 3-month periods in 1991 and 1992 were only slightly higher than the peaks measured in the same layer in a pit that was sampled in May 1993 (Figure 3). While these two years of data fail to show large H_2O_2 mass losses from the pack during the first year after snow is deposited, they do show some losses. Rather, the more striking feature on Figure 3 is the amount of post-depositional smoothing that occurs during the first year. Both mixing of snow from different events and exchange of H_2O_2 with the atmosphere are thought to contribute to the smoothing. A possible explanation for the higher first-year snow-pit concentrations (Figure 1) is that samples simply reflect high-concentration events before any smoothing (mixing) occurred.

Three points suggest that more surface snow data are needed to establish the range of first-year losses of H_2O_2 and other species from the snow/firn. First, the peak 1992 concentration in the 1993 snow pit (Figure 2) was higher than most previous buried years. Second, average surface snow values on Figure 2 show a definite increase in going from 1991 to 1993. While this trend is not unlike earlier patterns observed in pits and in the Eurocore, the question of possible analytical bias should be checked. Different laboratories did the summer 1991-92 versus 1993 analyses shown on Figure 2. Third, peak summer H_2O_2 values in consecutive years vary by

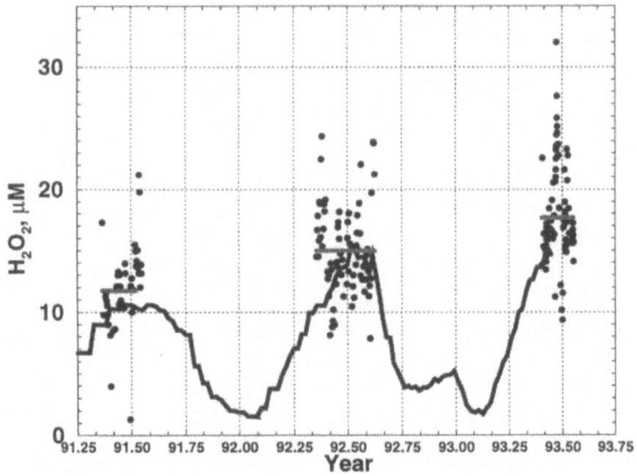

Figure 2: Comparison of H_2O_2 concentrations measured in the surface layer of snow at Summit, Greenland in 1991-1993, with concentrations measured in a snow pit sampled in 1993. Surface data (•) are daily averages; in most cases concentrations in five samples were reported. The three horizontal lines during summers show average of all surface samples for each year. Surface data for 1991 from *Fuller* [1995]; surface data for 1992 from *Drummey* [1993]; surface data for 1993 and snow pit data from *Bales et al.* [1995*a*]. Matching of the snow-pit concentration line (continuous solid line) to the time scale is only approximate. It was done by assuming that water accumulation was uniform throughout the year, and that the peak 1991 and 1992 concentrations in the snow pit were centered on mid year.

more than a factor of two, as can be seen in the 1989 snow pit data (Figure 1, *c.f.* 1985 versus 1986). One sees the same pattern of year-to-year differences in the Eurocore from Summit [*Sigg and Neftel*, 1991] (Figure 4). Comparing the H_2O_2 and $\delta^{18}O$ records, two qualitative features are apparent: i) the two exhibit similar seasonal patterns, and ii) years with higher peak H_2O_2 generally have a higher peak $\delta^{18}O$ value, suggesting a warmer summer temperature (Figure 4). More water vapor should be in the air at warm temperatures, resulting in more H_2O_2 production but a higher H_2O to H_2O_2 ratio in snow. Since between -10 and -30°C a 4° temperature decrease should result in a doubling of H_2O_2 in snow in equilibrium with a constant atmospheric concentration by either co-condensation or partitioning [*Conklin et al.*, 1993], year-to-year differences in peak H_2O_2 cannot be attributed to significant differences in summer temperature. This is quite different from our finding that within a single summer, H_2O_2 in surface snow is correlated with temperature and specific humidity [*Bales et al.*, 1995*b*].

Sigg and Neftel [1988] observed that H_2O_2 diffuses to a greater extent in snow/firn than do water molecules. $\delta^{18}O$ diffusion is less than that of H_2O_2 because gas-phase H_2O molecules striking an ice surface stick with a probability of 1.0, whereas the probability for an H_2O_2 molecule is on the order of 0.2-0.5. HCHO diffuses even further, as it has even less affinity for the ice surface. Figure 5 is a summary of mass accommodation coefficients, or sticking coefficients, of various molecules on water and supercooled water. These values give an indication of the

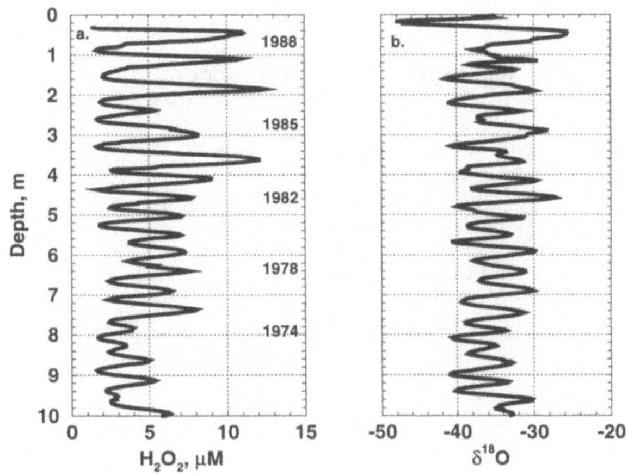

Figure 3: H_2O_2 concentrations and $\delta^{18}O$ in top 10 m of Eurocore [*Sigg et al.*, 1992]. $\delta^{18}O$ from S Johnsen, University of Copenhagen.

relative tendency of molecules to diffuse in snow/firn. Note that at -6°C the sticking coefficient for H_2O_2 is 0.3, versus 0.2 for HNO_3, 0.06 for HCHO and 0.02 for HCOOH. The implication is that smoothing of HNO_3 and HCOOH in the ice should be greater than for H_2O_2, and these species should be more-readily degassed from surface snow.

In principle, the sticking coefficient should also affect the relative concentration of an impurity found in fresh snow grown from the vapor phase, versus in the atmosphere. The growth of a particle from the co-condensation of two gaseous species should depend on the rate of surface

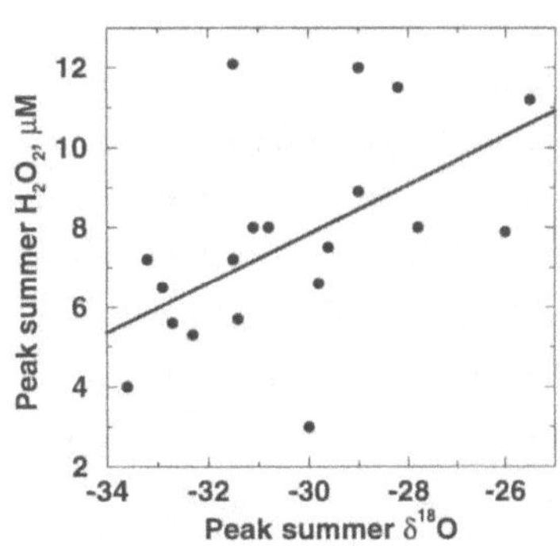

Figure 4: Peak summer values for $\delta^{18}O$ and H_2O_2, from Figure 3. Solid line is regression. Standard error of estimate is 2.17.

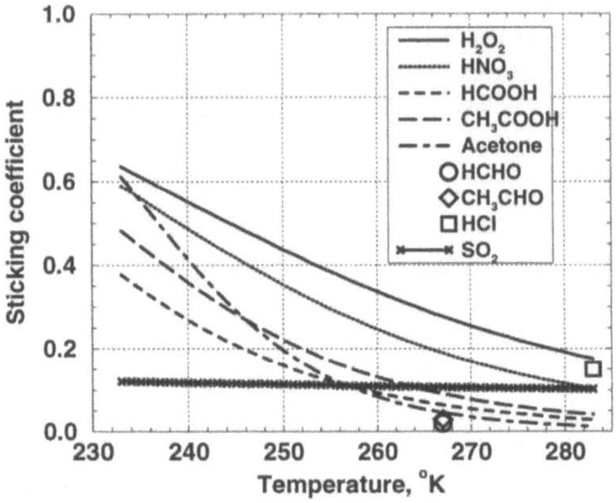

Figure 5: Sticking coefficients for various gases on water and supercooled water. Lines are fits to measurements over the temperature range of -10 to +20 °C. H_2O_2 and SO_2 from *Worsnop et al.* [1989]; HNO_3, HCOOH and CH_3COOH from *Jayne et al.* [1991]; HCHO and CH_3CHO from *Jayne et al.* [1992] acetone from *Duan et al.* [1993]; and HCl from *Van Doren et al.* [1990].

reaction of adsorbed reactants [*Harrison and MacKenzie*, 1990], i.e. on the sticking coefficient. So for a sticking coefficient of 0.5, co-condensation of H_2O and H_2O_2 should lead to an ice-to-air concentration ratio of 7.3×10^4 M atm^{-1} at -30° and 4.7×10^3 M atm^{-1} at -6° (for a sticking coefficient of 0.3) [*Conklin et al.*, 1993].

3 Conceptual Model

Relating concentrations of species in ice cores to those in the global atmosphere involves consideration of processes at the global to local microphysical scales. For the current discussion, we focus on the relations between concentrations of soluble gaseous species in the mixed layer of the lower troposphere and the upper few meters of the snow/firn. Other authors in this volume consider the larger-scale transport processes that bring species to this point.

For a given seasonal accumulation pattern, the depth of air-snow exchange will determine the extent to which previously deposited snow equilibrates with the atmosphere. Gas exchange with the atmosphere is controlled by the physical properties of the snow and its pore space [*see Davis et al.* this volume]. Both macroscale transport and exchange at the ice-air interface control transport of volatile species in the pack and uptake to or release from the snow/firn. Adsorption and absorption at the ice-air interface and partitioning into the interior of a snow grain enhance retention in near-surface layers and retard diffusion of gases to deeper layers.

3.1 Snowpack-Scale Conceptual Model

Macroscale transport is determined by processes of diffusion, wind pumping and longer-term atmospheric pressure changes (Figure 6). A general equation for exchange of chemical species between phases in a snowpack is:

$$\theta_a \frac{\partial C_a}{\partial t} + \theta_i \frac{\partial C_i}{\partial t} = \theta_a \frac{\partial}{\partial x_k} \left[D_k \frac{\partial C_a}{\partial x_k} \right] - \theta_a u_k \frac{\partial C_a}{\partial x_k} \tag{1}$$

Terms on the left are the changes in concentration in the air (C_a) and ice (C_i) phases, respectively. C_i includes material both on the surface and in the bulk ice grain, i.e. everything not in the air phase. On the right, we consider advection and dispersion; x_k is a coordinate direction (k = 1,2,3). Advection induced by pressure pumping is position dependent. Velocity induced by long-term atmospheric pressure changes should be largely one dimensional, whereas a three-dimensional description is needed to fully characterize the velocity field resulting from wind pumping. Dispersion includes both molecular diffusion, which is assumed to be homogeneous and isotropic within a given layer, and mechanical dispersion, which depends on velocity.

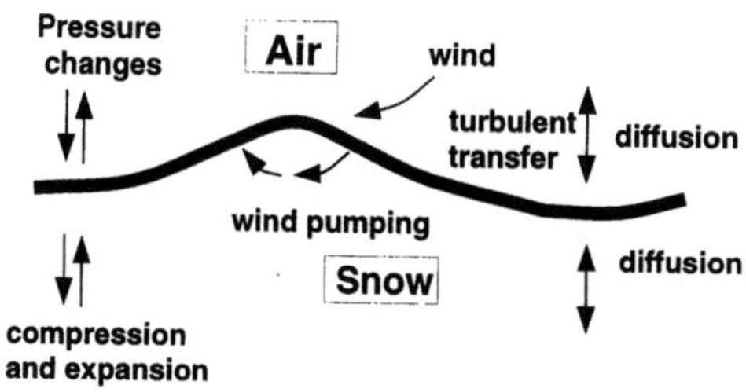

Figure 6: Processes contributing to post-depositional alteration of concentrations of volatile species in snow/firn.

Mathematically, the dispersion coefficient (D_k) in direction k is the sum of the molecular diffusion coefficient (D_0) and the dispersivity (α_d) times velocity (u_k). Diffusion must also consider an effective tortuosity of the medium (which for spheres reaches a maximum for a face-centered cubic packing). The volume fractions of ice and air (θ_a, θ_i) depend on depth and on the age of the snowpack. For a discussion of diffusion in firn, see *Schwander* [this volume].

While the process of gas mixing over depth is primarily caused by molecular diffusion, changes in atmospheric pressure force air exchange between the void volume of the snow and the atmosphere. In terms of equation 1, this barometric wind pumping influences u_k. Though barometric pressure changes may cause air to penetrate several meters into a porous medium [*Massmann and Farrier*, 1992], overall mass transport should be small. Pressure pumping from surface winds blowing over uneven topography is known to disrupt the thermal regime in snow packs [*Clarke and Waddington*, 1991; *Albert and McGilvary*, 1992], and can be responsible for considerable mass transport as well. At South Pole, topographic wind pumping is thought to account for much of the dry deposition of non-sea salt sulfate to snow [*Cunningham and Waddington*, 1993], with barometric pumping being insignificant. For a model of wind pumping, see *Waddington et al.* [this volume].

The second equation needed to solve for C_a and C_i comes from a general expression for slow air-ice transfer:

$$\theta_i \frac{\partial C_i}{\partial t} = \theta_a k_f C_a - \theta_i k_b C_i = \theta_i k_b \left[C_a K_D - C_i \right] \qquad (2)$$

where k_f and k_b are pseudo-first-order rate coefficients, or mass-transfer coefficients (s^{-1}) and K_D is the air-to-ice equilibrium partition coefficient ($cm_a^3\,cm_i^{-3}$). In terms of the conceptual model of Figure 7, k_f is an overall mass-transfer coefficient, and K_D is an overall partition coefficient (i.e. sum of ice surface and bulk concentrations), equal to $\frac{k_f \theta_a}{k_b \theta_i}$. The θ_a and θ_i parameters are given as constants in equations (1) and (2), but actually change as metamorphism and firnification proceed.

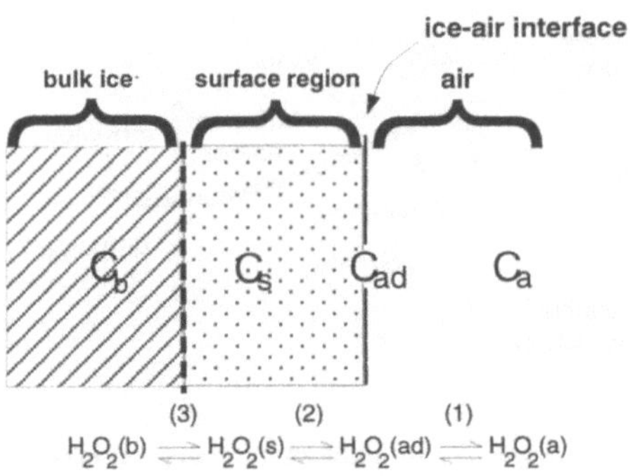

Figure 7: Conceptual model of partitioning at ice-air interface. Any of the three reversible reactions illustrated: i) adsorption, ii) partitioning to surface disordered region, and iii) partitioning to bulk ice, can be rate limiting. In the presence of liquid water, equilibrium between C_{atm} and C_{snow} would be expressed by Henry's law. C_{atm} can also reflect codeposition [*Conklin et al.*, 1993]. Reprinted with permission.

Temperature gradients in the snow/firn drive water-vapor movement [*Colbeck*, 1983], and should also enhance transport of volatile impurities. *Colbeck* [1983] formulates an expression for the rate of grain growth in snow as a function of both vapor diffusion, i.e. the rate of supply of water vapor from an adjacent layer with a higher vapor density, and surface kinetics, i.e. the rate of vapor deposition at a grain surface, which depends on the local excess vapor density.

He notes that because the difference in equilibrium vapor density between a source and sink is significantly greater than the local excess vapor density, diffusion rather than surface kinetics always controls growth rate. The difference in equilibrium water-vapor density depends on the temperature gradient in the snow/firn. Because ice conducts heat better than air, temperature gradients between grains exceed those across a grain, and grain growth proceeds by hand-to-hand movement at a rate that exceeds what would occur just due to one-dimensional vapor flow. This hand-to-hand vapor transport, which occurs in the upper portion of the snow and firn where significant temperature gradients exist, would also enhance release of impurities such as H_2O_2 from the ice matrix due to the continual movement of the ice-air interface. In polar snow, both annual and diel temperature cycles give rise to temperature gradients that influence vapor deposition. The vapor deposition rates, while quite small in winter due to cold temperatures, are sufficiently fast to cause complete recrystallization of snow grains in a few days during late summer to early fall [*Colbeck*, 1989]. It is during this time that depth hoar forms in polar snow [*Alley*, 1987]. See *Davis et al.* [this volume] for a discussion of metamorphism in firn the effect on physical properties, and the effect of hand-to-hand vapor transport on release of impurities from the ice matrix.

Whillans and Grootes [1985] showed that smoothing of seasonal $\delta^{18}O$ signals by diffusion of H_2O in firn should be limited by diffusion in the open pore space rather than by solid diffusion in ice grains of 7-10 mm diameter. Smoothing times are sensitive to both grain size and temperature, which determines the diffusion coefficient. For impurities in ice, the diffusion coefficient may be smaller than the self-diffusion coefficient for water.

3.2 Grain-Scale Conceptual Model

At the microscale, one can consider the exchange of a volatile species such as H_2O_2 between air and ice to be a three-step process involving physical adsorption of H_2O_2 at the interface, followed by either: i) incorporation of the H_2O_2 into the surface region where water molecules are also being deposited/exchanged (co-deposition), or ii) partitioning of H_2O_2 into an aqueous surface region in accord with Henry's law equilibrium (Figure 7). The third step is incorporation of H_2O_2 into the bulk ice. At short time scales the second step should control the rate and extent of uptake. At longer time scales, diffusion into and out of the bulk ice will be important.

More specifically, the change in concentration in the ice is that due to adsorption (C_{ad}), accumulation in the surface region (C_s), partitioning into bulk ice (C_b), and chemical reaction (C_{ch}):

$$\frac{\partial C_i}{\partial t} = \frac{\partial C_{ad}}{\partial t} + \frac{\partial C_s}{\partial t} + \frac{\partial C_b}{\partial t} + \frac{\partial C_{ch}}{\partial t} \tag{3}$$

Equation (3) then applies to the rate limiting process, or combination thereof. The adsorbed contribution is expressed as a reversible reaction with equilibrium given by a linear isotherm. The assumption for species such as H_2O_2, HCHO, HNO_3 and other acids under conditions at Summit is that surface adsorption is small relative to both the total surface area available for adsorption, and to total accumulation in the ice. While "adsorption" of both polar and non-polar species on ice has been reported [*Goss*, 1993], experimental measurements of uptake reflect both adsorption at the air-surface region interface and absorption in the surface region. Further theoretical and experimental work is needed to separate these processes.

The contribution due to accumulation in the surface region (also called a transition layer, quasi-liquid layer or surface premelting) can be expressed as a reversible reaction, with equilibrium given by Henry's law:

$$C_s = K_H C_a \tag{4}$$

At temperatures near 0°C liquid water is present; at lower temperatures the volume of surface region can be expressed as an equivalent volume of liquid water [*Conklin and Bales*, 1993]. Impurities incorporated into the bulk ice during growth of ice crystals in the cloud will translocate to the surface of snow grains, to grain boundaries or to inter-grain crystal boundaries during metamorphism. Dissolution of soluble but non-volatile salts will result in liquid water at grain boundaries or at the ice-air interface on snow grains having a high ionic strength, approaching that of brine solutions [*Brimblecombe and Conklin*, this volume].

Henry's coefficients for temperatures at or below zero are sorely lacking. Those few that are available are discussed by *Brimblecombe and Conklin* [this volume]. To get a more general picture of the relative behavior of the species of interest, we have extrapolated K_H values measured at or near 25°C to temperatures below 0°C (Figure 8a).

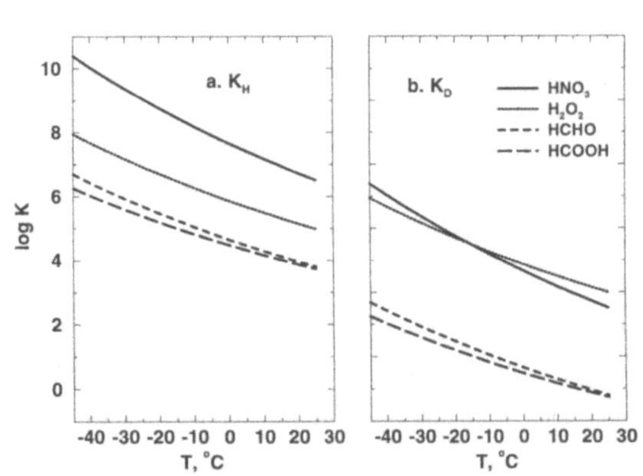

Figure 8: a) Henry's coefficients extrapolated to temperatures below 0°C, in units of $\mathrm{M\,atm^{-1}}$. K_H and $\Delta H°$ values from *Hoffmann and Calvert* [1985], for 25°C. b) calculated K_D values based on product of K_H and assumed K_{iw} (units of $\mathrm{M\,atm^{-1}}$).

Several investigators have probed the thickness and volume of liquid water at the ice-air interface. There is both thermodynamic evidence [*Fletcher*, 1968; *Gubler*, 1982] and experimental evidence [*Golecki and Jaccard*, 1978; *Clifford*, 1975] to support the existence of such a layer. For example, adsorption of CO_2 [*Ocampo and Klinger*, 1982] and SO_2 [*Conklin et al.*, 1993] onto ice have been observed to increase with temperature from -35°C to near 0°C. It is thought that this increase is due to trapping of solute molecules at the surface by mobile water molecules (i.e. a surface liquid-like layer) as the temperature approaches the melt point. This is consistent with the observed increase in surface roughening for crystals grown above about -10°C [*Colbeck*, 1985]. *Chen and Crutzen* [1994] recently reviewed the thermodynamic properties of this transition layer as part of a study of solute effects on ice-particle evaporation in clouds. They express the thickness of the transition layer as:

$$h = \max(h_{eq}, \frac{n_{w,min} M_w}{A \rho_w}) \tag{5}$$

where h_{eq} is the thickness in equilibrium with bulk ice and the vapor phase, $n_{w,min}$ is the minimum amount of water that must be maintained in a transition layer containing solutes, M_w is the molecular weight of water, A is the surface area of the ice particle, and ρ_w is the density of water in the transition layer. Following *Kuroda and Lacmann* [1992], *Chen and Crutzen* [1994] give h_{eq} as:

$$h_{eq} = -a + \left[\frac{n a^n \Delta\sigma_\infty V_m}{kT \ln \frac{e_w}{e_i}} \right]^{\frac{1}{n+1}} \tag{6}$$

where a is the radius of action of intermolecular forces (1 nm), n is a constant equal to 2, $\Delta\sigma_\infty$ is the interfacial energy introduced by the transition layer (0.025 J m^{-2}), k is the Boltzmann constant, T is temperature, V_m is the volume of a water molecule (~27Å^3), and e_i and e_w are the saturation vapor pressures over pure ice and water, respectively. Below about -1°C, h_{eq} is only a few monolayers thick. The minimum amount of water is a function of solute content:

$$n_{w,min} = n_s \frac{a_{w,min}}{1 - a_{w,min}} \tag{7}$$

where n_s is the equivalents of solute in the transition layer and $a_{w,min}$ is the minimum activity of water in the transition layer, equal to $\frac{e_w}{e_i}$. Since at equilibrium most solutes should largely be excluded from the ice matrix, n_s should just be the total solute content of the system. For solute concentrations in bulk snow on the order of 10 μM, h_{eq} gives the larger h.

Partitioning into bulk ice can also be given by an equilibrium expression:

$$C_b = K_{iw} C_s \tag{8}$$

where K_{iw} is an ice-water partition coefficient. This assumes a liquid-like surface region. Relatively few measurements have been made of distribution coefficients, K_{iw} for the ice/water system. The equilibrium distribution of alkali-metal fluorides and chlorides gives K_{iw} values on the order of $10^{-4} - 10^{-3}$ [*Gross*, 1968; 1977], but depend on counterion concentration. Values for SO_4^{2-} and NO_3^- are probably on the order of 10^{-4}. NH_4F has a K_{iw} of 0.13 [*Gross*, 1977], and antifreeze glycoprotein has a reported K_{iw} of 0.8 [*Gross et al.*, 1987]. *Sigg et al.* [1990] reported a K_{iw} of 0.013 for H_2O_2 at 0°C; *Conklin et al.* [1993] suggest a value on the order of 5×10^{-3} for -10°C. While a consistent picture of the values of K_{iw} for various species has yet to emerge, evidence suggests that most acid anions have values on the order of $10^{-4} - 10^{-3}$, with species such as H_2O_2 having values on the order of $10^{-2} - 10^{-1}$. Clearly, there is a need for further experimental work in this area.

Alternatively, such a liquid-like surface region need not be present. In either event equations (4) and (8) can be combined by eliminating C_s:

$$C_b = K_{iw} K_H C_a = K_D C_a \tag{9}$$

Conklin et al. [1993] report experimental K_D's for H_2O_2 on the order of 4×10^3 M atm^{-1} at -10°C and 2.5×10^4 M atm^{-1} at -30°C. Applying values of K_{iw} of 0.1 for HF, 0.01 for H_2O_2, and 0.0001 for HCOOH, CH_3COOH and HCHO gives the predicted pattern of K_D's shown on Figure 8b. Note in particular the similarity in the relative values on Figure 8b versus Figure 5.

The species concentration in the ice phase will, in general, not be homogeneous during periods of post-depositional change, at even the grain scale. So the average ice-phase concentration (C_i) given in equations (2) and (3) is the grain-averaged value for the ice phase. The concentration C_b in equations (3), (5) and (6) is also a grain-scale average for the non-surface part. A more-complete description of the impurity concentration at the grain scale has been given by describing slow uptake and release at the grain scale with a spherical diffusion model [*Conklin et al.*, 1993]. Assuming spherical geometry:

$$C_i = \frac{3}{a} \int_0^a r C_r \mathrm{d}r \tag{10}$$

where C_r is the local impurity concentration at any point in the ice sphere, assumed to be radially homogeneous; a is the average sphere radius (10^{-4} m). Impurity transport in an ice sphere is governed by the spherical diffusion equation:

$$\frac{\partial C_r}{\partial t} = \frac{D_i}{r^2} \frac{\partial}{\partial r} \left[r^2 \frac{\partial C_r}{\partial r} \right] \tag{11}$$

where D_i is the impurity diffusion coefficient in the ice. Boundary conditions for diffusion in the ice spheres are assumed to be:

$$C_r(x, a, t) = K_D C_a(x, t) \tag{12}$$

$$\frac{\partial C_r(x, 0, t)}{\partial r} = 0 \tag{13}$$

In general, both diffusion in the ice matrix and along grain boundaries should be considered. An effective value of D_i could reflect both processes.

For bulk-ice diffusion being rate limiting, D_i can be related to the mass-transfer coefficient k_f [*Conklin et al.*, 1993]:

$$k_b = \frac{D_i}{0.0441\, a^2} \tag{14}$$

where the constant 0.0441 is for spherical geometry. Grains in near-surface layers of polar snow are far from spherical (e.g. *Davis et al.*, this volume). Thus specification of an equivalent spherical radius would be required for the calculations. The sphere with an equal surface area to volume ratio may provide a better mean ice diffusion distance than would a simple linear measurement of grain size. Measured k_b values for H_2O_2 uptake on ice gave D_i's in the range of 3×10^{-16} to 4×10^{-14} $m^2 s^{-1}$ for -60° to -3°C, respectively. The estimated self-diffusion coefficient for H_2O in ice is on the order of 10^{-15} $m^2 s^{-1}$at $-$ 5°C and 2×10^{-10} $m^2 s^{-1}$ at -30°C [*Hobbs*, 1974].

Grain growth and loss due to vapor diffusion in response to temperature gradients would result in $\frac{\partial C_r}{\partial t}$ being larger than given by equation (8) due to the movement of the ice-air interface. Since vapor deposition and loss is not uniform around a snow grain, addition of an advection term to (8) would not be appropriate. Rather, enhanced mass transfer due to inter-grain vapor flux could be accounted for by a larger effective D_i in equation (8).

We assume that the loss by chemical reaction can be given by a pseudo-first-order rate expression, though this has not been verified. The production rate (r_p) also depends on the precursor species.

$$\frac{\partial C_{ch}}{\partial t} = k_1 C_s + r_p \tag{15}$$

where the rate coefficient for destruction (k_1) depends on such things as light penetration, temperature, and the presence of other species. Similar equations can be written for other phases in which reactions occur. See the contributions by *Neftel* [this volume], *Ravishankara et al.* [this volume], and *Hoffmann* [this volume] for a discussion of chemical reactions in snow, firn and ice.

4 Examples

The example of H_2O_2 at Summit, Greenland will be used for illustration. Observations at Summit have thus far been limited to the "summer" period, i.e. the months of April through September and show no apparent seasonal trend [*Bales et al.*, 1995*a*]. Significant losses in H_2O_2 concentrations in surface and near-surface snow have yet to be documented during this period [e.g. *Bales et al.*, 1995*b*]. While buried summer layers of snow/firn have lower H_2O_2 concentrations than does surface snow, it appears that averaging and loss processes occur over weeks to months [*Bales et al.*, 1995*a*]. However, our understanding of these processes is limited by the lack of observations on both the surface "initial" conditions during much of the year and the rate of change at the surface.

4.1 Equilibrium Deposition

Since H_2O_2 is photochemically produced, atmospheric concentrations follow the annual cycle of solar radiation. Calculations of atmospheric H_2O_2 using the Harvard photochemical model give lower summer concentrations than actually observed in air right above the snow [*Neftel et al.*, 1995], possibly due to degassing of H_2O_2 from surface snow. However the calculated seasonal trend from the model does closely follow solar radiation and can be assumed to indicate the trend one would observe if longer-term monitoring in the free troposphere were done. Figure 9a shows the annual cycle, repeated for 5 years, interpolated from calculations done for the summer solstice and three days between the summer and winter solstices [*A. Sigg*, personal communication]. No transport or heterogeneous losses were included in the calculation.

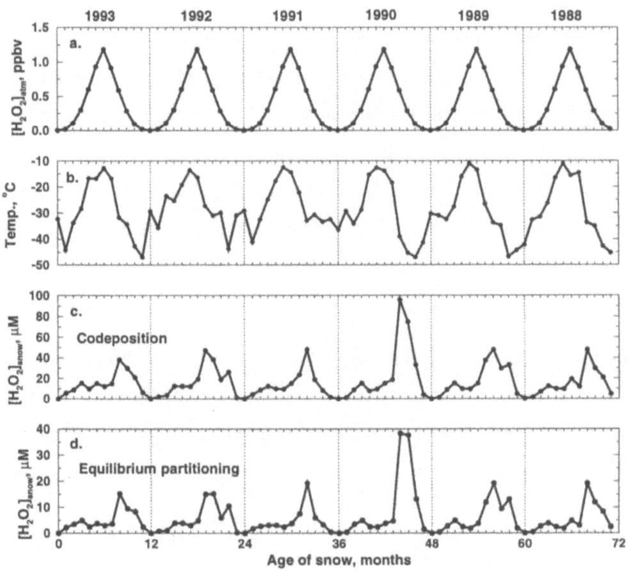

Figure 9: Model calculation for H_2O_2 signal expected in snow based on equilibrium partitioning. a) Seasonal variation in atmospheric H_2O_2, interpolated from photochemical model calculations. b) Monthly mean temperature at Summit, from *C. Stearns* [unpublished], c) Placement of atmospheric signal into the snow based on co-deposition of H_2O and H_2O_2 at the mean monthly temperature. d) Placement of atmospheric signal into the snow based on equilibrium partitioning of H_2O_2 into ice at the mean monthly temperature.

Using an equilibrium assumption (equation 9), the concentration of H_2O_2 in snow depends only on the gas-phase concentration and on the temperature [*Conklin et al.*, 1993]. Monthly mean temperatures at Summit for 1988-93 are shown on Figure 9b. Besides the equilibrium partitioning given by the linear equation (9), it has been suggested that fresh snow contains higher than equilibrium concentrations due to codeposition of H_2O and H_2O_2 [*Sigg*, 1990]. In this case, the molar H_2O to H_2O_2 ratio in the snow would be the same as that in saturated air. For this example, we assume the ratio of H_2O_2 to H_2O in ice is $\frac{1}{2}$ of the value in saturated air, to account for the lower sticking coefficient of H_2O_2 relative to H_2O. However, co-deposition of H_2O and H_2O_2 gives H_2O_2 concentrations in snow (Figure 9c) that are significantly higher than the 15-30 μM typically observed during the summer at Summit. Using the temperature-dependent "equilibrium" partition coefficient reported by *Conklin et al.* [1993] gives concentrations in the snow (Figure 9d) that are near those observed. Note that using this equilibrium model, the peak H_2O_2 concentration in snow is predicted to occur 1-2 months before the summer solstice. This occurs because the partition coefficient is quite temperature sensitive, and temperature apparently drops off faster than does solar radiation.

The results shown on Figure 9d are compared to snow-pit observations in Figure 10. It was assumed that water accumulation was uniform throughout the year, an assumption supported by the small seasonal variation seen in a correlation of changes in passive microwave brightness temperatures with the $\delta^{18}O$ record in snow pits [*Schuman et al.*, 1995]. Using the average monthly precipitation pattern from a model of precipitation variations for 1963-1988 [*Bromwich et al.*, 1993], which suggests a summer peak and winter minimum in precipitation, gave a result that is nearly indistinguishable. Both the depths and total H_2O_2 mass in the profile in Figure 10a were adjusted to match those of the two snow pits in Figure 10b. Qualitatively the patterns on the two figures show one similarity: the rather steep drop in concentration in the fall versus the slower rise in spring. They also exhibit some key differences: i) the summer peak is broader in the data vs. model, and ii) the model profile has a broad spring plateau that is missing in the data. It is also apparent that the monthly time step used in the model is too coarse to match the data, and diffusional smoothing of the model curves is needed.

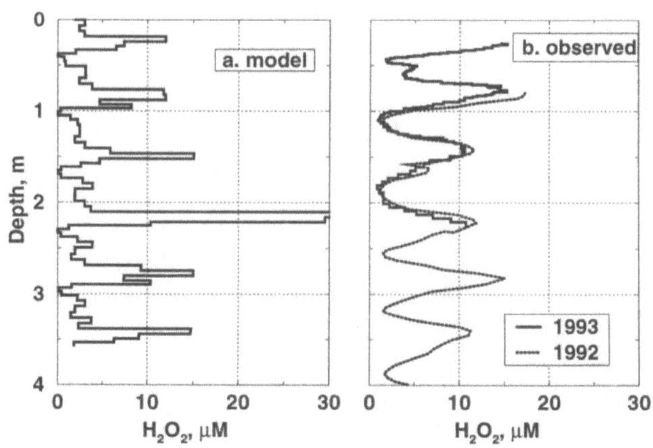

Figure 10: Comparison of H_2O_2 profiles: a) based on features of Figure 9d, using equal monthly water accumulation and b) data from 1992 [*Neftel et al.*, 1995] and 1993 [*Bales et al.*, 1995*a*] pits at Summit. Concentrations in Figure 9d were multiplied by 0.8 to give the same H_2O_2 mass as in part b) of this figure.

4.2 Post-Depositional Diffusion and Advection

Using the features of profile on Figure 10 to define an "initial condition", one can then calculate post-depositional changes due to diffusion and advection in the snow/firn, and examine the effect of limitations by slow grain-scale mass transfer. For illustration, we show changes after one year for equilibrium conditions at the grain scale. For every mole of H_2O_2 in the air, there is 3×10^5 mole in the ice. A value of $10^{4.4}\,\mathrm{M\,atm^{-1}}$ was used for K_D (-30°); other parameters were $D_a = 1.9 \times 10^{-5}\,\mathrm{m^2s^{-1}}$, $\theta_i = 0.35$ and $\theta_a = 0.65$. We assumed $a = 1$ mm and $k_b = 10^{-3}\,\mathrm{s^{-1}}$; in this case the ice and air are in equilibrium over the time scale considered, and the respective profiles (Figure 11) have the same pattern. For $k_b \leq 10^{-6}$, the profile in air becomes dampened out relative to that in the ice, as the time scale for diffusion in air is of the same order as is release of H_2O_2 from the ice. For $k_b \leq 10^{-8}$ diffusion results in a relatively uniform concentration versus depth, as changes in concentration in the ice are negligible.

Monthly latent heat flux (Figure 12) was calculated using micrometeorological data for Summit, Greenland and the energy balance model SNTHRM [*Jordan*, 1991]. Assuming that H_2O_2 leaves the snow with the H_2O in the same proportion as they occur in the snow suggests that sublimation has a relatively insignificant effect on H_2O_2 concentration in the firn. However, this calculation does suggest that it may be quite important in the seasonal water balance. The main unknown in the current calculation was surface roughness.

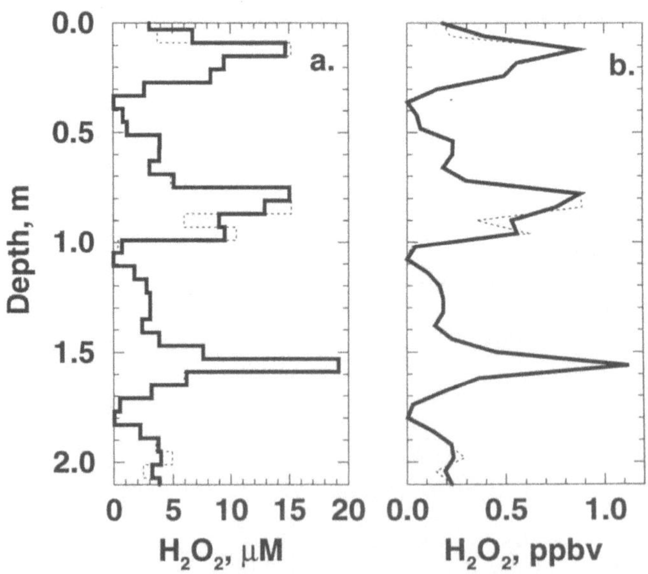

Figure 11: Concentration profile of H_2O_2 in: a) firn and b) firn gas, before and after one year of transport by diffusion with no grain-scale kinetic limitation.

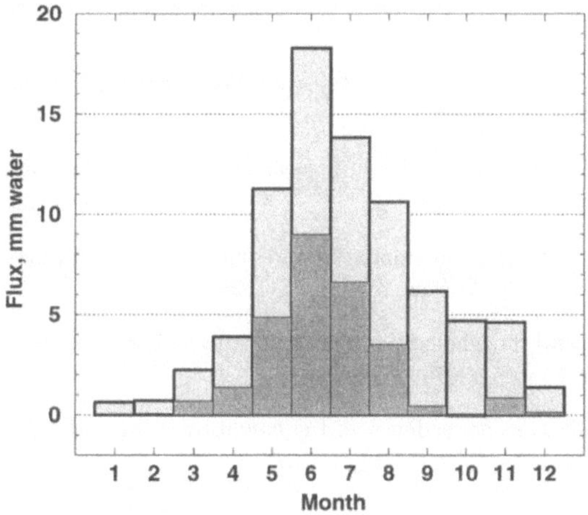

Figure 12: Sublimation of H_2O by month, based on micrometeorological data [*C. Stearns*, unpublished] and SNTHRM model [*Jordan*, 1991]. Lower, darker-shaded bars are for a surface roughness (z_0) of 0.1 mm; upper, lighter-shaded bars are for a surface roughness of 1 mm.

Acknowledgments

This work was supported in part by National Science Foundation grant DPP-9123025 to the University of Arizona, and in part by NATO collaborative research grant CRG 930929. R. Brice did the manuscript preparation. Micrometeorological data used in the analysis were from C. Stearns, University of Wisconsin.

References

Albert MR, McGilvary WR (1992) Thermal effects due to air flow and vapor transport in dry snow. J Glaciol 38(129):273–281

Alley RB (1987) Firn densification by grain-boundary sliding: a first model. J Physique 48(C1):249–254

Bales RC, Losleben MV, McConnell JR, Fuhrer K, Neftel A (1995a) H_2O_2 in snow, air and open pore space in firn at Summit, Greenland. Geophys Res Let 22(10):1261–1264

Bales RC, McConnell JR, Losleben MV, Conklin MH, Fuhrer K, Neftel A, Dibb JE, Kahl JDW, Stearns CR (1995b) The diel variations of H_2O_2 in Greenland: A discussion of the cause and effect relationship. J Geophys Res 100(D9):18661-18668

Brimblecombe P, Conklin MH (to be published) Thermodynamics of the solute layer on the surface of ice (this volume)

Bromwich DH, Robasky FM, Keen RA, Bolzan JF (1993) Modeled variations of precipitation over the Greenland Ice Sheet. J Climate:1253–1268

Chen J-P, Crutzen PJ (1994) Solute effects on the evaporation of ice particles. J Geophys Res 99(D9):18,847–18,859

Clarke GKC, Waddington ED (1991) A three-dimensional theory of wind pumping. J Glaciol 37:89–96

Clifford J (1975) Properties of water in capillaries and thin films. In: Franks F (ed) Water, A Comprehensive Treatise. vol. 5. New York, p 75–132

Colbeck SC (1983) Ice crystal morphology and growth rates at low supersaturations and high temperatures. J Appl Phys 54:2677–2682

Colbeck SC (1985) Temperature dependence of the equilibrium form of ice. J Cryst Growth 72:726–732

Colbeck SC (1989) Air movement in snow due to wind pumping. J Glaciol 35(120):209–213

Conklin MH, Bales RC (1993) SO_2 uptake on ice spheres: liquid nature of the ice-air interface. J Geophys Res 98:16851–16855

Conklin MH, Sigg A, Neftel A, Bales RC (1993) Atmosphere-snow transfer function for H_2O_2: microphysical considerations. J Geophys Res 98(D10):18367–18376

Cunningham J, Waddington ED (1993) Air flow and dry deposition of non-sea salt sulfate in polar firn: paleoclimatic implications. Atmos Environ 27A(17/18):2943–2956

Davidson CI, Bergin MH, Kuhns HD (to be published) The deposition of particles and gases to ice sheets (this volume)

Davis RE, Arons EM, Albert MR (to be published) Metamorphism on polar firn: significance of microstructure in energy, mass and chemical species transfer (this volume)

Drummey SM (1993) From snow to ice: A study of the H_2O_2 content of polar snow and ice from the Greenland and Antarctic ice sheets. M.S. thesis, University of New Hampshire, USA

Duan SX, Jayne JT, Davidovits P, Worsnop DR, Zahniser MS, Kolb CE (1993) Uptake of gas-phase acetone by water surfaces. J Phys Chem 97(10):2284–2288

Fletcher NH (1968) Surface structure of water and ice, II. a revised model. Philos Mag 18(156):1287–1300

Fuller B (1995) In preparation. M.S. thesis, University of New Hampshire, USA, personal communication from Dibb J

Golecki I, Jaccard C (1978) Intrinsic surface disorder in ice near the melting point. J Phys Chem: Solid State Physics 11:4229–4237

Goss C-U (1993) Adsorption of organic vapors on ice and quartz sand at temperatures below 0°C. Environ Sci Technol 27(13):2826–2830

Gross GW (1968) Some effects of trace inorganics on the ice/water system In: Trace Inorganics in Water, v. 73 Advances in Chemistry Series, p. 27–97 American Chemical Society, Washington, D. C.

Gross GW (1977) Concentration dependent redistribution at the ice-water phase boundary - III. Spontaneous convection, chloride solutions, J Chem Phys 67:5264–5274

Gross GW, Gutjahr A, Caylor K (1987) Recent experimental work on solute redistribution at the ice/water interface. Implications for electrical properties and interface processes. J Physique 48(C1):527–533

Gubler H (1982) Strength of bonds between ice grains after short contact times. J Glaciol 28:457–473

Harrison RM, MacKenzie AR (1990) A numerical simulation of kinetic constraints upon achievement of the ammonium nitrate dissociation equilibrium in the troposphere. Atmos Environ 24A:91–102

Hobbs P (1974) Ice Physics, Clarendon, Oxford

Hoffmann MR (to be published) Possible chemical transformations in snow and ice induced by solar (UV photons) and cosmic irradiation (muons) (this volume)

Hoffmann MR, Calvert JG (1985) in Chemical Transformation Modules for Eulerian Acid Deposition Models, Vol. II, The Aqueous Chemistry, U.S. E.P.A., N.C.A.R. Interagency Agreement DW 930237

Iribarne JV, Pyshnov T (1990) The effect of freezing on the composition of supercooled droplets — I. retention of HCl, HNO_3, NH_3 and H_2O_2. Atmos Environ 24A:383–387

Jayne JT, Duan SX, Davidovits P, Worsnop DR, Zahniser MS, Kolb CE (1991) The uptake of gas-phase alcohol and organic acid molecules by water surfaces. J Phys Chem 95:6329-6336

Jayne JT, Duan SX, Davidovits P, Worsnop DR, Zahniser MS, Kolb CE (1992) Uptake of gas-phase aldehydes by water surfaces. J Phys Chem 96(13):5452–5460

Jordan R (1991) A one-dimensional temperature model for a snow cover. Technical documentation for SNTHERM.89, Special 91-16, US Army CRREL, 49 p.

Kuroda T, Lacmann R (1982) Growth kinetics of ice from the vapour phase and its growth forms. J Cryst Growth 56:189–205

Massman J, Farrier DF (1992) Effects of atmospheric pressures on gas transport in the vadose zone. Water Resour Res 28(3):777–791

Neftel A (to be published) The record of gases and reactive species in ice cores, and problems of interpretation (this volume)

Neftel A, Bales RC, Jacob DJ (1995) H_2O_2 and HCHO in polar snow and their relation to atmospheric chemistry. In: Delmas R (ed) Ice Core Studies of Global Biogeochemical Cycles, NATO ASI Series I, vol. 30. Springer, Berlin Heidelberg New York, p 249–264

Ocampo J, Klinger J (1982) Adsorption of N_2 and CO_2 on ice. J Colloid Interface Sci 86:377–383

Ravishankara AR (to be published) Processes at ice surfaces: physical uptake and reaction (this volume)

Schuman CA, Alley RB, Anandakrishnan S, White JWC, Grootes PM, Stearns CR (1995) Temperature and accumulation at the Greenland summit: comparison of high-resolution isotope profiles and satellite passive microwave brightness temperature trends. J Geophys Res 100(D5):9165–9177

Schwander J (to be published) Gas diffusion in firn (this volume)

Sigg A (1990) Wasserstoffperoxid-Messungen an Eisbohrkernen aus Grönland und der Antarktis und ihre atmosphärenchemische Bedeutung. Ph.D. Dissertation, Physics Institute, University of Bern, Switzerland

Sigg A, Neftel A (1988) Seasonal variations in hydrogen peroxide in polar ice cores. Ann Glaciol 10:157–162

Sigg A, Neftel A (1991) Evidence for a 50% increase in H_2O_2 over the past 200 years from a Greenland ice core. Nature 351:557–559

Sigg A, Staffelbach T, Neftel A (1992) Gas phase measurements of hydrogen peroxide in Greenland and their meaning for the interpretation of H_2O_2 records in ice cores. J Atmos Chem 223–232

Thompson AM (1992) The oxidizing capacity of the earth's atmosphere: probable past and future changes. Science 256:1157–1165

Thompson AM (1995) Photochemical modeling of chemical cycles: issues related to the interpretation of ice core data. In: Delmas R (ed) Ice Core Studies of Global Biogeochemical Cycles, NATO ASI Series I, vol. 30. Springer, Berlin Heidelberg New York, p 265–297

Valdez MP, Dawson GA, Bales RC (1989) Sulfur dioxide incorporation into ice depositing from the vapor. J Geophys Res 94(D1):1095–1103

Van Doren JM, Watson LR, Davidovits P, Worsnop DR, Zahniser MS, Kolb CE (1990) Temperature dependence of the uptake coefficients of HNO_3, HCl, and N_2O_5 by water droplets. J Phys Chem 94

Waddington ED, Cunningham J, Harder S (to be published) The effects of snow ventilation on chemical concentrations (this volume)

Whillans IM, Grootes PM (1985) Isotopic diffusion in cold snow and firn. J Geophys Res 90(D2):3910–3918

Whitlow S, Mayewski PA, Dibb JE (1992) A comparison of major chemical species seasonal concentrations and accumulation at the South Pole and Summit, Greenland. Atmos Environ 26A(11):2045–2054

Worsnop DR, Zahniser MS, Kolb CE, Gardner JA, Watson LR, Van Doren JM, Jayne JT, Davidovits P (1989) Temperature dependence of mass accommodation of SO_2 and H_2O_2 on aqueous surfaces. J Phys Chem 93:1159–1172

PROCESSES AT ICE SURFACES: PHYSICAL UPTAKE AND REACTION

A. R. Ravishankara*+
National Oceanic and Atmospheric Administration
Aeronomy Laboratory
325 Broadway, Boulder CO 8030, USA

Abstract

Uptake of gases onto ice and reactions of species on the ice surface are discussed in the context of ice-cores. Because of a lack of information that is directly applicable to ice cores, occurrence of certain processes are inferred from the results of lower temperature studies carried out to understand polar stratospheric clouds (PSCs). It is suggested that strong acids will be efficiently taken up. Easily-hydrolyzable species such as N_2O_5 and $ClONO_2$ will be converted to HNO_3 on ice. The extent of absorption of weak acids and other species such as O_3 will depend on the heat of absorption and the gas phase partial pressure of the gas. The reactions of a few other species on ice at warmer temperatures of the ice sheets are proposed to be similar to those for PSCs.

Introduction

Interactions between gas phase species and condensed matter exposed to the atmosphere play important roles in various phenomena. The type of interactions depend on the phase and the composition of the condensed matter, the temperature, and the characteristics of the molecule under consideration. The major solid condensed matter for considerations are ice sheets, snow on the surface, Polar Stratospheric Cloud (PSC) particles suspended in air, and frozen aerosols in air, while the liquids include water droplets and sulfate aerosols in air and oceans. There are significant differences between the interactions of the atmospheric gases with liquids and solids.

*. Also associated with the Department of Chemistry and Biochemistry, University of Colorado, Boulder CO 80309
+. Also associated with the Cooperative Institute for Research in Environmental Sciences, University of Colorado, Boulder CO 80309.

NATO ASI Series, Vol. I 43
Chemical Exchange Between the Atmosphere and Polar Snow
Edited by Eric W. Wolff and Roger C. Bales

Reactions on solids are generally confined to the surface because of the slow rate of diffusion in a solid while reaction in the interior of liquids is easy and common. In case of the ice-cores, slow reactions in the bulk of the solid could be important because of the long times available for the reactions.

The rate and extent of uptake of atmospheric constituents onto ice depends on the phase of ice and temperature. There are very few measurements on the uptake of atmospheric molecules on ice at temperatures much above the PSC temperatures. Similarly, there are few studies of the reactive uptake of molecules on ice at T>230 K. Therefore, we will have to extrapolate the data from the PSC studies to those of the ice sheets in the polar regions.

In the case of the heterogeneous reactions on PSCs, what is of interest is the change in the composition of the gas phase abundance of chemicals due to reactions in/on the condensed matter in contact with the atmosphere. In case of the ice-core chemistry, the interest is different and the questions of interest are i) how fast and to what extent chemicals from the gas phase are taken up into the ice, (ii) are there any reactions of this species in the ice, and (iii) how fast do species diffuse in ice. Because these are not the primary interest of laboratory chemists interested in heterogeneous reactions related to the PSCs and other solids in the atmosphere, the question that are of direct relevance to ice-core studies have not been addressed. Yet, some information can be obtained from the recent studies of ice at lower temperatures.

There are a large number of laboratory studies that have been carried out on ice surfaces since the discovery of the Antarctic ozone hole. In this paper, I have mostly used examples from our laboratory simply because of my familiarity with them. This narrow choice is in no way meant to disregard the contributions of a large number of workers in this area. The picture of how the uptake occurs at the surface, what are the roles of various processes, and what determines the gas phase loss are not completely clear. There are contradictory views on how heterogeneous reactions occur. I have not attempted to present all the data and disagreements. Instead, I have made sweeping simplifications, which may not stand up when better understanding is obtained. The reason for taking such an approach is to qualitatively probe what could be happening at the surface of ice sheets and provoke the needed studies.

Ice at polar surfaces:

The stable phase of ice in the atmosphere depends on the temperature. In the stratosphere at 50 mbar and with 5 ppmv of water, for example, ice forms at temperatures less than ~187 K (see for example, Solomon [1988] and WMO [1994]). This ice could be cubic [Adamson, 1990]. Sampling from aircraft show a variety of ices [Goodman *et al.*, 1989]. At higher temperatures, pure ice cannot be formed in the stratosphere because of the limited supply of water vapor. However, co-condensation of nitric acid and water vapor at temperatures as high as 195 K leads to nitric acid trihydrate. Even though this is the thermodynamically stable phase, it is not clear if there are other phases, such as nitric acid dihydrate, that are formed first. At the much warmer temperatures characteristic of the upper troposphere, ice particles formed in the cirrus clouds are believed to be mostly hexagonal ice [Hobbs, 1974]. Since the temperature of the ice in the ice-sheet formed in the polar regions is also in the vicinity of 230 K, it is assumed that what we know about hexagonal ice is applicable to the ice cores. This is an approximation because formation of ice in the polar regions is a very complex process, which proceeds through many steps, and the temperatures vary over a wide range. It should be noted that the phase of ice used in most laboratory experiments are not characterized. Even though there are not many studies at the temperatures of the ice sheets, I believe some information can be gathered from the low temperature studies. I assume that the surface of the ice sheet rather than the interior of ice-core (which are at much higher pressures) is of interest to the gas transfer process.

One possible component that can lead to a binary solid is nitric acid. From the phase diagram of water and nitric acid, it is clear that at temperatures as high as 230 K, there are no stable two component phases [Hanson and Mauersberger,1988; Worsnop, *et al.*,1993]. Therefore, it can be assumed that the ice-sheets in the polar regions will be ice with small amounts of soluble material in it. The gases could be adsorbed either on the surface or in the grain boundaries. In any case, it is likely that species are dissolved in /adsorbed on ice rather than forming a second component in a stable phase, i.e., it does not contribute to the stability of this ice. H_2SO_4 , on the other hand, may be able to form a stable hydrate in the ice sheets.

Ice in the polar regions may be coated with some atmospheric constituents and, hence, physical adsorption and/or chemical reaction may be different from that on a pure ice surface. In the absence of information on this surface coverage, I will assume that the available surface is pure ice. Another question that has been of some concern for the PSCs and will be of concern for studies on ice sheets is the available surface area. The difference between the available surface area and the geometrical area of PSCs is not known [WMO, 1989]. There has been considerable debate regarding the surface area of even the substrates that have been studied in the laboratory [see for example, Keyser, *et al.*,1993; Hanson and Ravishankara,1993a]. Porous ice can have substantially larger surface area for uptake/reaction (See Keyser *et al.*, 1993 and references therein). It is likely that ice in the polar regions is porous and, hence, the surface area is very poorly determined. Further, as shown by Wolff *et al.* [1989], the uptake may be along the grain boundaries. Therefore, the question of where the uptake is taking place, both in the laboratory measurements and in the real world, has to be addressed if uptake of species onto ice is to be calculated based on laboratory measurements.

Another major issue that I have completely avoided is what the surface looks like; i.e., is there a quasi-liquid layer on the surface of ice and are the heterogeneous processes going to be dominated by the state of this "layer"? This issue is discussed by others at the meeting and I just note that if there is indeed a "liquid" layer on the ice, most of the concepts that are discussed here would not be applicable.

Reactive and non-reactive uptake at the surface vs. in the bulk

The diffusion coefficient of species in H_2O-ice are 10^6 to 10^{10} times smaller than in liquid water. If we assume a diffusion coefficient of 10^{-12} cm^2 s^{-1} for species in ice at the temperatures of the ice sheet, the characteristic time for diffusion into the ice would >10^3 s for a 1 μm thick surface. Even though this is usually negligibly slow for the processes of interest in the stratosphere, this diffusion could be very important for trapping species in ice and for reactions of species in the ice sheets.

A crude estimate of the bulk processing versus the surface reactions can be made using the equation [Danckwerts,1951],

$$\gamma_{bulk} = \left(\frac{4RT}{\omega}\right) H\sqrt{D_c k_1}$$

where H is the Henry's law solubility of the reactant in the solid, k_1 is the first order loss rate coefficient, D_c is the diffusion coefficient in the bulk, ω is the mean molecular speed at temperature T, and R is the gas constant. Here, I am assuming that diffusion into the bulk is taking place and the uptake is not through cracks and imperfection. Unlike the case of the liquids, H and k_1 are difficult to estimate for the solids (it may be even difficult to define it!). The solubility limit for a species such as HNO_3 or HCl in ice is ~10 ppmv. If we assume H = $1x10^5$ M atm^{-1} (i.e., [X] in the solid is 0.2 ppmv for P(X), the atmospheric partial pressure, is 10^{-10} atm) and a first order rate constant k_1 of 10^3 s^{-1}, the uptake coefficient would be in the range of 0.01 to 0.001 for D_c values of 10^{-12} to 10^{-14} cm^2 s^{-1}. (The above solubility is consistent with the results of Domine *et al.* [1994] for HCl in ice.) The calculated uptake coefficient is a very substantial number. (Note that I have used units of H as if it was a liquid and carried out this simple calculation only as a rough estimate.) If reactive uptakes as small as 10^{-6} are significant for ice sheets, reaction rate coefficients can be much smaller and still be important . Alternatively, species with very small gas phase concentrations can be taken up to a great extent if the reaction rate coefficient is higher. Thus, reactions occurring in the bulk can not be neglected in the case of ice-core chemistry.

It is also worth noting that if the reactions are fast the reacto-diffusive length, $l = \sqrt{D_c / k_1}$, the thickness over which the reaction is taking place (see for example Hanson, *et al.* [1994]) is small and the distinction between surface and bulk reaction is very ill defined. In such a case one need not consider the possibility of bulk reactions contributing to the uptake at the surface.

From the above simple arguments, it must be clear that parameters such as diffusion coefficients and solubilities are crucial quantities that need to be measured. Currently, there is very little information on diffusion coefficients in ice or, for that matter, in liquids. Recently, Domine *et al.* [1994] have measured the diffusion coefficients of HCl in ice. Such measurements for other stable species is critical. Note that the above discussion is not very applicable to the firn, but rather only to relatively stable ice.

Reactivity at ice surface: Results from PSC studies

In the following discussion, I assume that the ice surface is made up of only water molecules. Uptake of a species from the gas phase to the solid surface is controlled by its gas phase diffusion to the surface, mass accommodation at the surface, and any reaction that may take place at the surface. In general, the mass accommodation coefficients are close to unity and the over all uptake rate is controlled by the physical uptake or chemical reaction at the surface. If the uptake results solely in physi-sorption, the fraction of the substance on the surface will be dictated by the thermodynamics of physi-sorption and the concentration of the molecule in the gas phase. In such a case, the uptake can be described using the standard methods [Adamson, 1990] . I exemplify this by using our recent experiments on HOCl uptake on ice [Hanson and Ravishankara, 1992b]. If the uptake is due to chemical reaction of the molecule at the surface, the extent of uptake can be dictated by the available sites for the chemical change. Based on studies of HCl, HBr, and HF uptake on ice [Hanson and Ravishankara, 1992a], we believe that HCl dissociates upon uptake by an ice surface. It is reasonable to assume that this dissociation occurs at some special site. There are other, more recent, studies which also suggest that indeed HCl dissociates on the ice surface [Horn *et al.*, 1992]. Such a dissociation or reaction with H_2O

on ice can be reversible. In the case of HCl, we do know that HCl comes off the ice when the HCl vapor in equilibrium with this system is removed. Similarly, HNO_3 that is taken up by ice comes off the surface if the vapor above it is removed. The same can be true for other strong acids. Our findings about HCl are described below and these findings may be extended to those for uptake of strong acids. If the uptake is due to chemical reaction, as for example in the case of HOCl loss in the presence of HCl on ice where Cl_2 is formed, the reaction can lead to products that desorb.

Below, the experimental method that has been used in our laboratory to carry out the uptake of species such as HOCl and HCl are described. I will also discuss the parameters that control the uptake of species due solely to non-reactive incorporation into ice. The examples include the uptake of O_3 on to ice.

Apparatus: Coated- wall flow tube reactors, pioneered by Molina and coworkers [see for example, Molina *et al.*, 1987], have been used extensively to measure uptake coefficients (γs) onto solid surfaces by a large number of groups [see Kolb, *et al.*,1995 for references]. The value of uptake coefficients that can be measured with this method ranges from 10^{-6} to $>10^{-1}$. In addition to being sensitive to low values of γ, this method also has the advantage that the well-established mass transport theory for cylindrical flow tube reactors can be used in the data analysis [Brown,1978]. A schematic of this apparatus is shown in Figure 1.

The approach employed in flow tube studies is to measure the first-order loss rate coefficient, k_W, for the removal of the species from the gas phase to the solid or liquid surface, located on the inside wall of the reactor. The uptake coefficient, γ, can then be calculated from k_W using the procedures described in Howard [1979] and Brown [1978]. The extraction of the uptake coefficients requires knowledge of the gas phase diffusion coefficient and could restrict the accuracy of the large uptake coefficients that are measured using this method. In addition to the time dependent concentration change in the gas phase, the amount of molecules taken up by the surface can be measured by integrating the loss over exposure time. If the surface saturates, it is possible to reverse the adsorption and observe the desorption to measure the number of molecules that came off the surface.

Cylindrical flow tubes, with inner diameters of 1.5 to 3 cm and 15 to 50 cm length, are coated with a solid in these experiments [Hanson and Ravishankara,1991b; Hanson and Ravishankara,1992b]. The experiments have been carried out at temperatures that approximate stratospheric conditions, i.e., 190 to 200 K, although some have been carried out at both slightly warmer and colder temperatures. The types of surfaces under study are vapor deposited solids such as water ice and HNO_3-H_2O crystals as well as solids formed by freezing a liquid such as

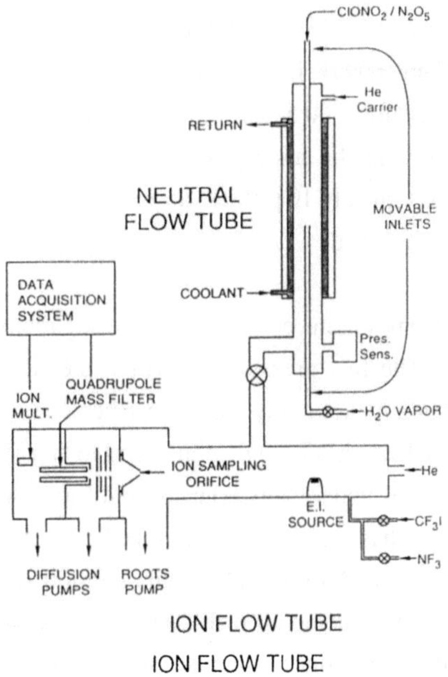

Figure 1. A schematic of the experimental apparatus used for studying physical uptake on to and chemical reactions on ice surfaces. The ion flow tube is a sensitive and selective detector for various species under study. The inside wall of the neutral flow tube is covered with ice. The ice is deposited using vapor from the movable inlet introduced from the bottom the flow tube and marked H_2O vapor. The reactant under study is introduced from the top movable inlet marked $ClONO_2$ and N_2O_5.

sulfuric acid hydrates. A movable injector, through which the reactant is introduced into the flow tube, can be manipulated from the upstream end of the tube. In some experiments, a second movable injector could be introduced at the bottom of the flow tube for adding water vapor to form the ice layers. The injectors are usually insulated and/or heated to prevent interfering heterogeneous effects on their surfaces. Often, the carrier gas flow is humidified to decrease evaporative losses of the solids.

The flow tubes are operated at pressures of 0.4 Torr; some times pressures up to 5 Torr are used. Total carrier gas mass flow is typically 5-10 (STP) $cm^3\ s^{-1}$ and the average carrier gas flow velocity is typically 1000 to 3000 cm s^{-1}. For some measurements, conducted at low pressures and high flow velocities, there is a pressure gradient along the length of the flow tube; however, we can accurately account for this gradient.

The vapor deposited ices are prepared by introducing water vapor (and also possibly HNO_3 vapor) from a room temperature reservoir into the flow tube while flowing He carrier gas through it such that the total pressure is ~0.5 Torr. Typically 50 to 100 mg of water is deposited in about one minute while moving the injector over a 10 cm length of the flow tube. A translucent ice film forms that is typically 10 μ thick; however, films as thin as 0.3 μ and as thick as 100 μ have been prepared and studied.

We have used two detection schemes in the course of our investigations: (a) optical spectroscopy and (b) mass spectrometry. The low pressures at which the flow tubes are used and the continuous nature of the detection makes many detection schemes suitable. Recently, chemical ionization mass spectrometry (CIMS) has been demonstrated to be extremely sensitive and very selective for the types of molecules under study [Hanson and Ravishankara,1991b; Hanson and Ravishankara,1991a]. Our group has greatly exploited this detection scheme in the study of the heterogeneous chemistry of solids. A drawback of this method is that it requires extensive understanding of the ion chemistry to come up with a suitable detection scheme for a particular molecule. The optical methods, laser induced fluorescence and chemi-luminescence, were used for detection of OH [Hanson, *et al.*,1992] and O_3 [Utter, *et al.*,1992; Dlugokencky and Ravishankara,1992], respectively.

Uptake/desorption of HCl onto/from ice

Our measurements of the uptake of HCl onto ice surfaces is described elsewhere [Hanson and Ravishankara,1993c]. Briefly, we exposed HCl to a ice surface inside a 80 cm long reactor. The ice was deposited over 10-20 cm of this reactor. The thickness of this ice was ~10 μm. The temperature of the flow tube was maintained between 191 and 211 K. HCl was detected using chemical ionization mass spectrometry by the reaction of SF_6^- ion with HCl to give $F^- \bullet HCl$. HCl was added to the flow tube through a movable injector. The concentrations of HCl in these experiments was ~10^{10} cm^{-3}.

The variation of the gas phase HCl concentration with exposure to ice is shown in Figure 2. When HCl was exposed to the ice surface, we observed a very rapid decrease in the gas-phase HCl reaching the detector. Then, the signal reached a minimum, which was constant for approximately 3 min. (at a gas flow rate of ~1000 cm s^{-1} and a pressure of 0.5 Torr). After this minimum, the HCl concentration in the gas phase slowly increased to the value seen when ice was not in contact with the flowing HCl gas. When the injector was moved down stream of the

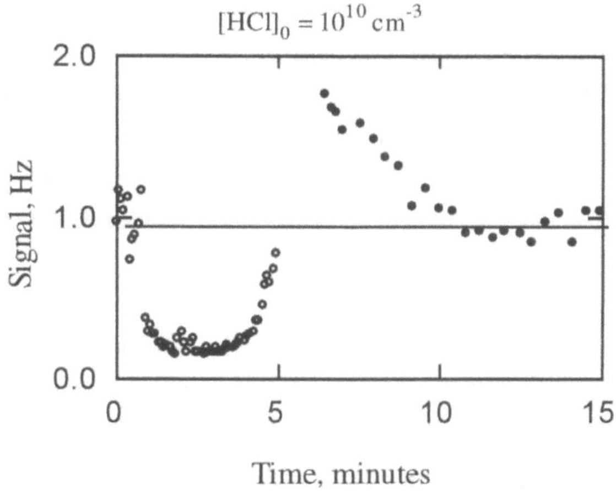

Figure 2. Plot of the HCl signal as a function of time of exposure to ice. The decrease in signal at 1 min. is due to uptake of HCl by ice. After approximately 4 min., the surface saturates and the signal returned to the pre-exposure values. After 5 minutes, the injector was moved to bypass the ice surface, while still carrying the same concentration of HCl. The desorption of HCl is evident. The amount of HCl desorbed is equal to that taken up in the initial part. This amount of HCl corresponds to approximately 5 x 10^{14} cm^{-2} of HCl on ice.

ice surface, the HCl signal increased because of desorption of HCl from the ice that was previously exposed to HCl. The amount of HCl that desorbed was roughly equal to the amount taken up during the prior exposure.

The results of these experiments show that HCl is taken up on to ice reversibly. By using the geometrical surface area of the ice that was exposed to HCl and number of molecules that were taken up (roughly same as the number that desorbed), we estimate that the number of HCl molecules that were taken up were ~5 x 10^{14} cm^{-2}. Such experiments were carried out using differing initial concentrations of HCl, 2x10^9 to 7x10^{10} cm^{-3}. In all these cases, the number of HCl molecules taken up was the same.

The results of these studies show that HCl is taken up reversibly by water ice. The number of these molecules taken up is roughly consistent with the number of H_2O molecules on the surface, or roughly speaking the number of available sites. The independence of surface coverage on pressure of HCl above the surface strongly suggests that HCl is not physi-sorbed on the ice surface. The uptake of HNO_3 on to ice surfaces also leads to approximately the same surface coverage [Hanson, 1992]. HBr is also taken up efficiently [Hanson and Ravishankara,1992a], just like HCl and HNO_3. On the other hand, the uptake of HF is much smaller and the surface saturates very quickly. The upper limit to the coverage is much smaller than those for strong acids such as HCl, HBr, and HNO_3. However, this uptake is confined to the surface as long as the ice is in equilibrium with water vapor, i.e., is being continuously replenished. These observations are consistent with HCl and HNO_3 dissociating on the ice surface at specific sites. By analogy, I suspect that all the strong acids will dissociate on the surface and the surface coverage would be ~$5x10^{14}$ cm^{-2}, as long as there are enough molecules in the gas phase to produce such a monolayer.

It is worth pointing out that experiments in other laboratories do not agree completely with our results. For example, Chu *et al.* [1993] have measured the uptake to be proportional to HCl partial pressure. Also, in our experiments, there was a small fraction of the HCl that appeared to be irreversibly taken up. I simply note there that such data exists and further work may be needed to clarify the surface coverage.

Base on the above discussions, I suggest that uptake of strong acids on ice sheets will be very efficient.

Uptake of HOCl on to ice

Our experiments which studied the uptake of HOCl on to ice surface is described elsewhere [Hanson and Ravishankara,1993b]. These experiments were very similar to that for HCl. The gas phase HOCl signal decreased when HOCl was exposed to ice, reached a minimum, and recovered to the original value. When the HOCl was not present over the ice, any HOCl that was already on ice came out. Unlike the case with HCl, we observed that the extent of HOCl uptake, i.e., the number of HOCl molecules taken up by a given ice surface, was proportional to the concentration of HOCl in the gas phase. Figure 3 shows the number of HOCl molecules taken up as a function of surface area at 191, 196, 201, and 211 K. Analysis of this data in terms of the standard Langmuir absorption isotherms yielded Δ^oH_{abs} of -14 ± 2 kcal mol^{-1} and a Δ^oS_{abs} of -40 ± 10 cal mol^{-1} K^{-1}.

These experiments showed that weak acids, like HOCl, are not taken up to make a "monolayer" unless the gas phase pressure of HOCl are very high and lead to multi-layer adsorption. Further, the heat of adsorption is small. The large change in entropy is consistent with the loss of three translation degrees of freedom in going from the gas phase to the surface. These results show that for a significant coverage on ice at warmer temperatures of the ice sheets, the heat of adsorption has to be significantly larger. The Δ^oS_{abs} for uptake on to ice will likely be in the neighborhood of -35 cal mol^{-1} K^{-1}. It is true that depending on the circumstances, the entropy change for adsorption could be different than that due simply to the loss of three degrees of

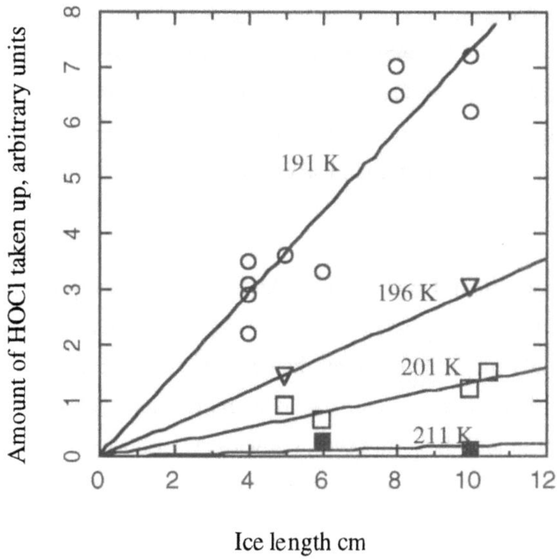

Figure 3. Plots of number of molecules of HOCl taken up as a function of surface area of ice that is exposed at a constant gas phase concentration of HOCl. Measurements at four temperatures are shown. Analysis of this data yields heat and entropy of adsorption for HOCl on ice.

translational motion [Adamson, 1990]. But, using this value for the entropy for adsorption, as a rough guide, and estimated heats of adsorption, the surface coverage of various species on to ice can be estimated.

Uptake of O_3

O_3 is an important constituent of the atmosphere. It is an oxidizing agent, which is very photochemically active. If it is photodissociated in ice, it will likely lead to H_2O_2. Since H_2O_2 is an important oxidant and a species that can be detected in ice-cores, uptake of O_3 on ice could be of interest.

We have carried out experiments to look for the uptake of O_3 on ice [Dlugokencky and Ravishankara,1992]. The uptake coefficient was extremely small. Based on the upper limits for the amount of O_3 removed from the gas phase, we deduced that O_3 is not taken up to any significant extent on the ice surface. Even at the low temperatures of these experiment, the surface coverage of O_3 had to be less than 0.01 monolayer for a gas phase O_3 concentration of $2x10^{12}$ cm^{-3}. Assuming a Δ^oS_{abs} = -40 cal mol^{-1} K^{-1} and the above surface coverage at 195 K,

I estimate the heat of adsorption of O_3 on ice to be <11.8 kcal mol^{-1}. This leads to a surface coverage of less than 5 x 10^{-6} monolayer for an O_3 partial pressure of 4x10^{-8} atm (40 ppbv) at 240 K. Therefore, the number of O_3 molecules at the surface has to be less than 2.5 x 10^9 cm^{-2}. This is an upper limit that really needs to be better quantified for ice core studies because such a coverage followed by its slow photochemical conversion may be a significant pathway for H_2O_2 formation in the ice cores.

Based on similar arguments, if hydrocarbons have heats of adsorption of ~10 kcal mol^{-1}, the surface coverage will be negligibly small. In fact, for molecules such as methane, it is clear that because they are highly insoluble in water, they will indeed have very low heats of adsorption . Thus, small hydrocarbons in the ice core has to be in the trapped air parcels (air bubbles), rather than being dissolved in ice. It is, however, known that larger hydrocarbons can adsorb on ice [Adamson, 1990]. Further studies of hydrocarbons, in particular partially oxygenated hydrocarbons, is warranted.

Chemical Reactions on ice

Some of the reactions on ice surface may be ionic in nature. For example, if HCl is dissociated on the surface, the heterogeneous reaction of $ClONO_2$ and HOCl with HCl may be the reaction of $ClONO_2$ and HOCl with Cl^-. If this is found to be true, it will be useful in extrapolating the results of the HCl studied to those of other strong acids. Recent studies by Hanson [1995] show that $ClONO_2$ can directly react with hydrocarbons.

Reactions of species such as $ClONO_2$, $BrONO_2$, and N_2O_5 with H_2O should take place on ice surfaces even at the warmer temperatures of the ice sheet just as they do on the colder surfaces. This is because what really matters for these reactions is the availability of surface water. Unlike the case of colder solids, where the surface could be contaminated and converted from ice to NAT by exposure to these species which leave behind HNO_3, it is likely that the surface will not be altered at the warmer temperatures. The extent to which the HNO_3 product will stay on the surface at the warmer temperatures is not clear.

In the troposphere, there are various other species that could be play significant roles. In particular, the uptake of sulfur containing species can be important because of the signature they leave behind in the ice. The extent to which the CH_3SCH_3, DMS, is taken up by ice is unclear. Because DMS is not very soluble in water, it is likely that it will not be taken up very efficiently. On the other hand, the oxidation products of DMS such as MSA (methane sulfonic acid) and sulfate will be taken up and incorporated into the ice.

Acknowledgment: The work discussed here would have been impossible without David Hanson. I am grateful to David Hanson and Edward Lovejoy for helpful discussions and comments on an earlier draft of this manuscript. This work was funded in part by NOAA's Climate and Global Change Research Program.

References:

Adamson, A. W. (1990). Physical chemistry of surfaces (5th edition). New York: John Wiley.
Brown, R. L., Tubular flow reactors with first-order kinetics, *J. Res. Natl. Bur. Stand.*, *83*, 1-8, 1978.
Chu, L. T., M.-T. Leu, and L. Keyser, Uptake of HCl in water ice and nitric acid ice films, *J. Phys. Chem.*, *97*, 7779-7785, 1993.
Danckwerts, P. V., Absorption by simultaneous diffusion and chemical reaction into particles of various shapes and into falling drops, *Trans. Faraday Soc.*, *47*, 1014, 1951.
Dlugokencky, E. J. and A. R. Ravishankara, Laboratory measurements of direct ozone loss on ice and doped-ice surfaces., *Geophys. Res. Lett.*, *19*(1), 41-44, 1992.
Domine, F., E. Thibert, F. V. Landeghem, E. Silvente, and P. Wagnon, Diffusion and solubility of HCl in ice: preliminary results, *Geophys. Res. Lett.*, *21*, 601-604, 1994.
Goodman, J., O. B. Toon, R. F. Pueschel, K. G. Snetsinger, and S. Verma, Antarctic stratospheric ice crystals, *J. Geophys. Res.*, *94*, 14449-16457, 1989.
Hanson, D. R., The uptake of HNO_3 onto ice, NAT, and frozen sulfuric acid, *Geophys. Res. Lett.*, *19*, 2063-2066, 1992.
Hanson, D. R., Reactivity of $ClONO_2$ on $H_2{}^{18}O$ ice and organic liquids, *J. Phys. Chem.*, *99*, 13059-13061, 1995.
Hanson, D. R., J. B. Burkholder, C. J. Howard, and A. R. Ravishankara, Measurement of OH and HO_2 radical uptake coefficients on water and sulfuric acid surfaces, *Journal of Physical Chemistry*, *96*, 4979-4985, 1992.
Hanson, D. R. and K. Mauersberger, Laboratory studies of the nitric acid trihydrate: Implications for the south polar stratosphere, *Geophysical Research Letters*, *15*, 855-858, 1988.
Hanson, D. R. and A. R. Ravishankara, The reaction probabilities of $ClONO_2$ and N_2O_5 on 40 to 75% sulfuric acid solutions, *J. Geophys. Res.*, *96*(D9), 17307-17314, 1991a.
Hanson, D. R. and A. R. Ravishankara, The reactions probabilities of $ClONO_2$ and N_2O_5 on polar stratospheric cloud materials, *J. Geophys. Res.*, *96*(D3), 5081-5090, 1991b.
Hanson, D. R. and A. R. Ravishankara, The heterogeneous chemistry of HBr and HF, *J. Phys. Chem.*, *96*, 9441-9446, 1992a.
Hanson, D. R. and A. R. Ravishankara, Investigation of the reactive and nonreactive processes involving $ClONO_2$ and HCl on water and nitric acid doped ice, *J. Phys. Chem.*, *96*, 2682-2691, 1992b.
Hanson, D. R. and A. R. Ravishankara, Response to "Comment of Porosities of ice films used to simulate stratospheric cloud surfaces", *J. Phys. Chem.*, *97*, 2802-2803, 1993a.

Hanson, D. R. and A. R. Ravishankara, The uptake of HCl and HOCl onto sulfuric acid: Solubilities, diffusivities and reaction, *J. Phys. Chem.*, *97*, 12309-12319, 1993b.

Hanson, D. R. and A. R. Ravishankara, Uptake of HCl and HOCl onto sulfuric acid: solubilities, diffusivities, and reaction, *J. Phys. Chem.*, *97*, 12309-12319, 1993c.

Hanson, D. R., A. R. Ravishankara, and S. Solomon, Heterogeneous reactions in sulfuric acid aerosols: A framework for model calculations, *J. Geophys. Res.*, *99*, 3615-3629, 1994.

Hobbs, P. V. (1974). Ice Physics. Oxford University Press.

Horn, A. B., M. A. Chesters, M. R. S. McCoustra, and J. R. Sodeau, Adsorption of atmospherically important molecules on thin D2O ice films using reflection absorption infrared spectroscopy, *J. Chem. Soc. Faraday Trans.*, *88*, 1077-1078, 1992.

Howard, C. J., Kinetic measurements using flow tubes, *Journal of Physical Chemistry*, *83*, 3-9, 1979.

Keyser, L. F., M.-T. Leu, and S. B. Moore, Comment on porosities of ice films used to simulate stratospheric cloud surfaces, *J. Phys. Chem.*, *97*, 2800-2801, 1993.

Kolb, C. E., D. R. Worsnop, M. S. Zahniser and *et al.* (1995). Laboratory studies of atmospheric heterogeneous chemistry. In J. R. Barker (Eds.), Progress and Problems in Atmospheric Chemistry. Singapore: World Scientific Publishing Co. Pte. Ltd.

Molina, M. J., T. L. Tso, L. T. Molina, F. C. Y. Wang, Antarctic stratospheric chemistry of chlorine nitrate, hydrogen chloride, and ice: Release of active chlorine, *Science*, *238*, 1253-1257, 1987.

Solomon, S., The mystery of the Antarctic ozone hole, *Rev. Geophys.*, *26*, 131-148, 1988.

Utter, R. G., J. B. Burkholder, C. J. Howard, and A. R. Ravishankara, Measurement of the mass accommodation coefficient of ozone on aqueous surfaces., *J. Phys. Chem.*, *96*(12), 4973-79, 1992.

Wolff, E. W., R. Mulvaney, and K. Oates, Diffusion and location of hydrochloric acid in ice: Implications for polar stratospheric clouds and ozone depletion, *Geophys. Res. Lett.*, *19*, 487-490, 1989.

Worsnop, D. R., L. E. Fox, M. S. Zahniser, and S. C. Wofsy, Vapor pressures of solid hydrates of nitric acid: Implications for polar stratospheric clouds, *Science*, *259*, 71-74, 1993.

WMO, Atmospheric Ozone, Scientific Assessment of Stratospheric Ozone No. WMO Global Ozone Research and Monitoring Project Report No. 20, Volume I and II, Appendix; AFEAS Report, Geneva). World Meteorological Organization. (1989)

WMO, Scientific Assessment of Ozone Depletion: 1994, WMO Global Ozone Research and Monitoring Project-Report No. 37. World Meteorological Organization (1994).

POSSIBLE CHEMICAL TRANSFORMATIONS IN SNOW AND ICE INDUCED BY SOLAR (UV PHOTONS) AND COSMIC IRRADIATION (MUONS)

Michael R. Hoffmann
Environmental Engineering Science
W.M. Keck Laboratories
California Institute of Technology
Pasadena
California 91125
USA

Introduction

Over the last decade there has been a growing interest in the chemical composition of the snow packs in the polar regions (Bales and Wolff, 1995). Delmas (Delmas, 1992; Delmas, 1994) has noted that "information recorded in polar ice cores over the last several hundred millennia is invaluable to studies aimed at understanding the pre-industrial environmental system and anticipating the future evolution of the climate and the atmosphere." For example, the isotopic composition of the ice (H_2O) matrix is a reliable paleothermometer. From the analysis of deep Antarctic and Greenland ice cores the ice age environmental conditions appeared to correspond to about 6 °C cooler temperatures and atmospheric CO_2 and CH_4 levels lower by factors of nearly 2 and 4, respectively. The biogeochemical cycles of S and N also appear to be affected by climatic changes that result in modifications in the source intensity and the transport of gaseous precursors. Even though atmospheric sulfate is derived principally from marine biogenic sources (i.e., dimethyl sulfide emission), cataclysmic volcanic eruptions can contribute sporadically to the atmospheric sulfur budget through large point source emissions of SO_2. These events are ultimately detected in polar ice as H_2SO_4 spikes. Nitrate, which is the next most abundant anion found in polar snowfall, exhibits concentration changes that are poorly understood, but which could be linked with the polar ozone hole formation. In addition to ions derived primarily from gas-to-particle conversions,

NATO ASI Series, Vol. I 43
Chemical Exchange Between the Atmosphere and Polar Snow
Edited by Eric W. Wolff and Roger C. Bales

continental dust and sea-spray aerosol components are also present in the ice at much higher concentrations during ice ages than during interglacial periods due to an intensification of their production and long range transport under glacial climatic conditions.

Wolff (Wolff, 1992) points out that "the Antarctic ice sheet is the main sink for atmospheric pollution reaching the Antarctic atmosphere from other continents." The ice preserves a historical record of the atmosphere that can be recovered in ice cores. Based on present findings, no increasing trend is observed over recent decades for nitrate and sulphate (Wolff, 1992). However, during the same period, there appears to have been an increase of perhaps eight-fold in lead concentrations in Antarctic snow, but the details of when the increase occurred have still to be defined.

An opposite conclusion with respect to nitrate and sulfate was drawn by Laj et al. (Laj, Palais et al., 1992) based on analysis of cores taken from the Greenland ice sheet. They reported that concentrations of both SO_4^{2-} and NO_3^- have increased dramatically over the last 250 years, up to three to four times the apparent 18th century levels. Large changes have also occurred in the average concentrations of several other chemical species, such as NH_4^+, excess Cl^-, and Ca^{2+}. HNO_3 appears to be deposited preferentially during summer throughout the core. This later observations emphasizes the possible role of atmospheric photochemistry in understanding nitrogen cycling in the Arctic.

Methanesulphonic acid (CH_3SO_3H; MSA) is one of the major products of the photochemical oxidation of DMS in the atmosphere . Since MSA is a stable compound, which does not undergo photochemical decomposition in the atmosphere, it is removed from the atmosphere by wet and dry deposition. Analysis of Antarctic ice cores gives evidence of its global deposition over many thousands of years. In this regard, Legrand et al. (Legrand, Fenietsaigne et al., 1992) suggested that MSA and NSS-SO_4^{2-} (non sea salt sulfate) present in antarctic ice were marine in origin and that DMS emissions have been significantly modulated by El Nino Southern Oscillation events as well as long term climatic changes in the past. Given possible biogeochemical changes coupled with possible photochemical changes in ice matrices, these conclusions may have to be tempered somewhat. For example, Baker et al. (Baker, Kelly et al., 1991) have noted that terrestrial bacteria are capable of growing on MSA. Their activities in the natural environment result in the mineralization of MSA to carbon dioxide and sulphate. These processes may, in turn, alter the apparent time history of atmospherically-derived MSA contained in polar ice cores.

Sturges et al. (Sturges, Cota et al., 1992) have reported that arctic-ice microalgae emit significant quantities of bromoform ($CHBr_3$), which may be converted photochemically into active forms of bromine. They also suggested that polar ice algae might contribute significant amounts of organic bromine compounds, on a global basis comparable with anthropogenic and macrophyte sources. In this regard, Ryan (Ryan, 1992) has noted that during the antarctic spring up to 10% of the surface UV radiation can penetrate sea ice to the algal community growing at the lower surface. Since the algal systems are photosynthetically driven, we can anticipate that other photochemical processes may be of importance in the upper layer of polar snow and ice during the polar summers.

Neftel (Neftel, 1996) addresses problems associated with the interpretation of gases and reactive species in ice cores such as hydrogen peroxide and formaldehyde in this Volume.

In spite of the wealth of apparent knowledge about our historical environment that may be contained within polar ice cores, caution must be exerted when interpreting the chemical data due to the possibility of post-depositional change. In light of these concerns, we consider two radiation-induced pathways that may result in chemical change with deposited snow and ice. These pathways for chemical change are due to penetrating cosmic radiation and absorbed incident solar radiation.

Radiation-Induced Chemical Transformations in Ice

Effects of High-Energy Cosmic Radiation on Ice

Cosmic radiation consists of high-speed sub-atomic particles that travel throughout our Galaxy and solar system (Rossi, 1964). Some of these particles originate from the sun, but most come from sources outside the solar system and are known as galactic cosmic rays. Cosmic-ray particles that arrive at the top of the Earth's atmosphere are termed primary particles; their collisions with atmospheric nuclei give rise to secondary particles such as muons. About 85 percent of the primary particles are protons (nuclei of hydrogen atoms), with approximately 12 percent consisting of alpha particles (helium nuclei). The remainder are electrons and nuclei of heavier atoms. Because cosmic-ray particles are strongly influenced by the Earth's magnetic field and the interplanetary magnetic field, most of those detected near the Earth have kinetic energies in excess of about 1 GeV (gigaelectron volts). This energy corresponds to speeds greater than about 87% the speed of light. The number of particles drops rapidly with increasing energy, but individual particles with energies as high as 10^{20} eV have been detected. Because of their deflection by magnetic fields, primary cosmic particles follow convoluted paths and arrive at the top of the Earth's atmosphere nearly uniformly from all directions.

Pions (i.e., charged mesons) are produced from collisions between primary cosmic rays and interstellar hydrogen. Pions have half-lives of about 26 ns and decay through muons to produce electrons and neutrinos. The large majority (~ 85%) of cosmic particles that impinge on the surface of the earth are muons.

The muon is a subatomic particle similar to the electron but 207 times heavier; it has two forms, the negative muon, μ^-, and its positive antiparticle, μ^+. The influx of muons in cosmic radiation at the surface are approximately equally distributed between μ^- and μ^+. A muon decays by weak interactions into an electron and two kinds of neutrinos. Before decaying, muons lose energy by knocking electrons off of atoms. At high particle velocities, close to the speed of light, ionization dissipates energy in relatively small amounts, and so muons present in cosmic radiation at the earth's surface are extremely penetrating and, thus, can travel thousands of metres below the Earth's surface (e.g., into ice cores).

In the polar regions, the flux of muons (Wolfendale, 1973) impinging on the surface snow and ice is close to 0.04 particles cm^{-2} s^{-1} or 400 particles m^{-2} s^{-1}. The energies of the muons arriving at the earth's surface range from 0.2 to 1000 GeV. The muon has a mean lifetime of 2.2 μs and decays into an electron, a neutrino and anti-neutrino. As muons travel through a condensed medium they lose kinetic energy by collisions with molecules. For example, the effective energy loss rate for 1000 GeV muons traveling through water is 2.14 MeV g^{-1} cm^{-2} (MeV/g cm^{-2}) or, in other words, each 1000 GeV muon loses approximately 2.1 MeV cm^{-1} of penetration in ice (Hayakawa, 1969). With a linear energy loss rate (i.e., $-dE/dt = \bar{a}$) , a 1000 GeV muon can travel about 4.7 km into an ice sheet before it comes to rest.

With each energy-losing collision, a highly energetic muon will have the net impact of any high-energy radiation process, which is to produce aquated electrons, H atoms, hydoxyl radicals, hydrogen, hydrogen peroxide and hydronium ions in solution (Matheson and Dorfman, 1969; Buxton, Greenstock et al., 1988). The sequence of collisionally induced chemical events is as follows:

$$H_2O + (\mu^- / \mu^+) \longrightarrow H_2O^+ + \bar{e} + (\mu^- / \mu^+)'$$

$$H_2O^+ + H_2O \longrightarrow H_3O^+ + {}^{\cdot}OH$$

$$\bar{e} + H_2O \longrightarrow \bar{e}_{aq}$$

where $\bar{e}$ is an free electron and $(\mu^- / \mu^+)'$ is the original muon pair minus the energy lost upon collision. The initial radiolysis products are generated in isolated volume elements known as "spurs." A fraction of the initial products will react within the spur while some will be able to diffuse away from the initial radiation track. The net result is to produce the following product array, which includes the aquated electron, $\bar{e}_{aq}$, the hydrogen atom, $H^{\cdot}$, the hydroxyl radical, ${}^{\cdot}OH$, molecular hydrogen, H_2, hydrogen peroxide, H_2O_2, and protons, over 0.1 μs:

$$5\,H_2O + (\mu^- / \mu^+) \rightarrow \bar{e}_{aq} + H^{\cdot} + {}^{\cdot}OH + H_2 + H_2O_2 + H_3O^+$$

A muon passing through a methane-containing gas bubble in ice may result in the following collision-induced fragmentation:

$$CH_4 + (\mu^- / \mu^+) \longrightarrow CH_4^{+} + \bar{e} + (\mu^- / \mu^+)'$$

This initial muon-induced fragmentation may be followed by

$$CH_4^{+} \longrightarrow CH_3^{+} + H^{\cdot}$$

or

$$CH_4^{+} \longrightarrow CH_2^{+} + H_2$$

The electron ejected in the initial muon-induced ionization process may further ionize and excite other molecules in its path. These secondary processes may also induce chemical changes as follows:

$$\bar{e}_{aq} + N_2O \longrightarrow N_2 + O^{-\cdot}$$

$$O^{-\cdot} + H^+ \longrightarrow {}^{\cdot}OH$$

The various sources of hydroxyl radical and hydrogen atom may lead to the following pathways for the production of hydrogen peroxide:

$$2\,{}^{\cdot}OH \longrightarrow H_2O_2$$

$$H^{\cdot} + O_2 \longrightarrow HO_2^{\cdot} \xrightarrow{HO_2^{\cdot}} H_2O_2 + O_2$$

In effect, the successive collisions of a highly energetic muon result in many ionizations and excitations. The distribution of these processes in space depends on the energy of the particle involved as well as on the pathway traversed. There is no single resultant primary process corresponding to the result of the absorption of a single photon and thus no analogue to the concept of quantum yield in photochemistry. In radiation chemistry, reaction yields are reported on an empirical basis as the number of molecules of a particular kind produced (or destroyed) per 100 eV of impinging radiation. This number is known as the "G" value. Typical G values for the pulse radiolysis of water are 4.8 for $^{\cdot}OH$ production and aquated electron production during primary collisions and 2.7 and 2.6, respectively, for secondary processes occurring during spur expansion.

We can estimate the steady-state concentration of radical and ionized intermediates produced as the result of impinging cosmic-ray muons as follows (Walker, 1983):

$$[\text{radical}] = \frac{GnE\rho 10^3}{Ar(6 \times 10^{23})}$$

where [radical] is the molar concentration of initial radical/ionization products, G = 1, n is the total number of particles of energy (e.g., 0.01 particles s^{-1}) with $E \geq 0.2$ GeV, A is the area over which the energy is deposited, r is the track radius (i.e., radius of a muon = 0.05315 nm), and ρ is the specific gravity. Using these numbers, we estimate a nominal radical production rate along the track of a penetrating muon to be 62.5 μM per second. Of course, a high percentage of the radicals generated in a solid matrix will recombine with no net chemical consequences. However, given the relatively high rates of diffusion of electrons, protons, hydroxide and hydroxyl radical through ice (Goto, Hondoh et al., 1986; Woafo, Takontchoup et al., 1995), we can given a lower estimate for net ·OH radical production on the order of 1 nM s^{-1} that will be available for subsequent chemical reactions such as hydrogen peroxide production or organic compound (e.g., formaldehyde, formate, methanesulfonate, etc.) oxidation (Buxton, Greenstock et al., 1988) as follows:

$$\cdot OH + HCHO \xrightarrow{k = 1 \times 10^9\ M^{-1}s^{-1}} H_2O + CHO\cdot$$

$$CHO\cdot + O_2 \longrightarrow \cdot O_2CHO \longrightarrow CO_2 + \cdot OH$$

$$\cdot OH + HCO_2^- \xrightarrow{k = 3.2 \times 10^9\ M^{-1}s^{-1}} CO_2^{\cdot -} + H_2O$$

$$\cdot OH + CH_3SO_3^- \xrightarrow{k = 1.3 \times 10^9\ M^{-1}s^{-1}} \cdot CH_2SO_3^- + H_2O$$

$$\cdot CH_2SO_3^- + O_2 \rightarrow \cdot O_2CH_2SO_3^- \rightarrow SO_3^{\cdot -} + HCOOH$$

Effects of Incident UV photons in Ice

Photochemical reactions of interest to environmental chemists can take place in a number of physically distinct phases. Early work in this field was focused primarily on gas-phase photochemistry relevant to photochemical smog production. After an initial focus on the gas-phase photochemistry of the polluted urban troposphere, research efforts were expanded to include reactions of importance in the remote troposphere and the upper atmosphere with a particular focus on photochemical reactions occuring in the stratosphere. An interest in aqueous-phase photochemistry of relevance to environmental systems developed at a somewhat latter date, with a primary interest in the photochemical decay of aromatic compounds by direct absorption of light. A little later in time, an interest in the indirect photochemistry of surface waters and cloud droplets evolved where the primary focus was on the production of transient reactive intermediates such as hydroxyl radical, singlet oxygen, superoxide, hydroperoxyl radical, etc. and their subsequent reactions with dissolved organic and inorganic compounds in water.

Photochemical reactions that readily occur in aqueous systems may also occur to a limited extent in solid water. For example, Pursell et al. (Pursell, Conyers et al., 1995) have reported on the photochemistry of chlorine dioxide, OClO, in amorphous ice at 200 K.

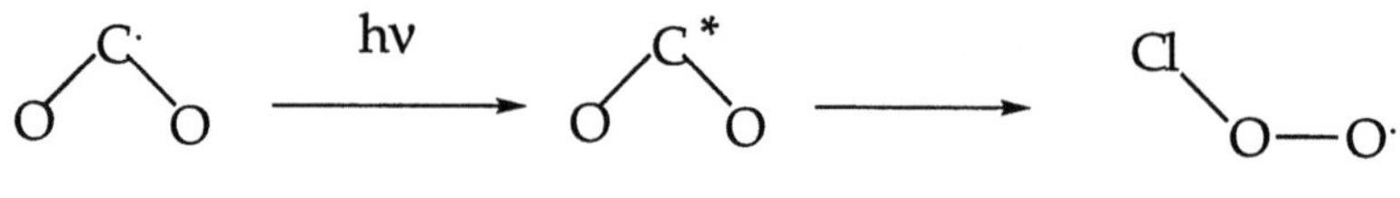

They used FTIR and UV-vis spectroscopy to follow the photoisomerization reaction. They noted that UV light (λ = 360 nm) quantitatively converted the OClO to the chlorine peroxy radical, ClOO. They used the asymmetric stretch of OClO at 1100 cm^{-1} and the O-O stretch of ClOO at 1440 cm^{-1}, along with characteristic UV absorptions at 360 and 260 nm, respectively, to monitor the photochemical process. Under conditions employed in their study, the only photoproduct appeared to be ClOO, in contrast with the gas-phase photolysis which yields predominantly ClO. Their findings suggest that the photochemical conversion of OClO to ClOO in ice particles may represent a new ozone-depleting mechanism.

Bernstein et al.(Bernstein, Sandford et al., 1994) measured the infrared specta of hexamethylenetetramine (HMT, $C_6H_{12}N_4$) isolated in an argon matrix at 12 K and frozen in ice at temperatures from 12 to 200 K. They also examined the ultraviolet photolysis of HMT frozen in ice. The half-life for HMT in ice was found to be 10 hr. Based on the observed changes in the FTIR spectra as a function of time, the products of the direct photolysis of HMT in ice at 200 K were reported to be HOCN, H_2NCN, HNCO, CH_3CN and $N(CH_3)_3$.

The first law of photochemistry states that "only light that is absorbed by a molecule can be effective in producing photochemical changes in that molecule." Thus, it is very important to know the absorption spectra of potential chromophores and their photo-action spectra in ice and snow matrices. These data are not readily available at this point in time. However, the net rate of photochemistry in the solid phase can be estimated from the intensity and spectral distribution of the actinic radiation, by the photochemical quantum yields, ϕ, and the specific absorption (extinction) coefficients of the primary chromophore.

The rate of a direct photochemical reaction can be written as

$$\frac{dC}{dt} = \phi(\lambda) I_o(\lambda)\left(1 - 10^{-\varepsilon l C}\right)$$

where C is the concentration of the reactant, $\phi(\lambda)$ is the photolysis quantum yield, I_0 is the intensity of the UV source, ε is the extinction coefficient of the primary chromophore, and l is the pathlength of irradiation. In the case of long pathlengths, high concentrations, and/or large extinction coefficients the above equation can be reduced to an apparent zero-order expression as follows:

$$\frac{dC}{dt} = \phi(\lambda) I_o(\lambda)$$

where $I_o(\lambda)$ has units of einstein $L^{-1}\ s^{-1}$

In environmental phototransformation studies, the apparent quantum efficiency is often defined as the initial measured rate of phototransformation divided by the theoretical maximum rate of photon absorption (assuming that all photons are absorbed by the chromophore and that actual light-scattering losses out of the reaction volume under consideration are negligible) as determined by chemical actinometry. This definition for species, X_i, can be expressed as follows:

$$\phi_{X_i} \equiv \frac{\pm\, (d[X_i]/dt)_o}{d[h\nu]_{inc}/dt}$$

where Φ_{X_i} = the apparent quantum efficiency for chemical species, X_i; $d[X_i]/dt$ is either the initial rate of formation or loss of chemical species, X_i; and $d[h\nu]_{inc}/dt$ is the incident photon flux per unit volume.

In the case of hydrogen peroxide production coupled to acetate oxidation, a general empirical rate expression of the following form is observed (Kormann, Bahnemann et al., 1988; Carraway, Hoffman et al., 1994; Hoffman, Carraway et al., 1994)

$$\frac{d[H_2O_2]}{dt} = \phi_o \frac{d[h\nu]_{abs}}{dt} = f(CH_3CO_2H,\ O_2)$$

Since hydrogen peroxide is also destroyed photochemically, one must also consider the photochemical rate of peroxide destruction as follows:

$$-\frac{d[H_2O_2]}{dt} = \phi_1[H_2O_2]\frac{d[h\nu]_{abs}}{dt} = f(CH_3CO_2H, O_2, H_2O_2)$$

Combining the expressions for peroxide production and destruction yields an overall empirical equation of the following form:

$$\frac{d[H_2O_2]}{dt} = (\phi_o - \phi_1[H_2O_2])\frac{d[h\nu]_{abs}}{dt}$$

where ϕ_o is the quantum yield for peroxide production and ϕ_1 is the quantum yield for peroxide destruction at the illuminated interface. During continuous photolysis a photo-stationary state is achieved, which yields the a simple steady-state relationship valid for long irradiation times:

$$[H_2O_2]_{ss} = \left(\frac{\phi_o}{\phi_1}\right)$$

This equation states that the steady-state concentration of a photochemical intermediate, which also serves as an electron donor or light absorber, in a photochemical system (i.e., an ice or snow matrix) is given by the ratios of the intrinsic quantum yields for production and destruction.

Photolytic reactions of possible importance in either the solid matrix, in aqueous microinclusions, or in the surface disordered layer on ice grains include those involving nitrous acid (Rettich, 1978), nitric acid (Graedel and Weschler, 1981), hydrogen peroxide (Taube, 1957), nitrite (Zafiriou and True, 1979), and ozone (Taube, 1957).

$$HONO \xrightarrow{h\nu} NO + {}^{\cdot}OH$$

$$HONO_2 \xrightarrow{h\nu} NO_2 + {}^{\cdot}OH$$

$$HOOH \xrightarrow{h\nu} HO^{\cdot} + {}^{\cdot}OH$$

$$NO_2^- \xrightarrow{h\nu} NO + {}^{\cdot}O^-$$

$$O_3 \xrightarrow{h\nu} O(^1D) + O_2(^1\Delta_g)$$

In the gas-phase, absorption of light by nitrous acid (pK_a = 3.2) leads to the above heterolytic cleavage (Graedel and Weschler, 1981) over the wavelength range of 300 to 400 nm with an effective quantum yield of $\phi = 0.095$. Over the same wavelength range, the extinction coefficient for HONO, ε, ranges from 9 to 74 L mole^{-1} cm^{-1}. The corresponding molar extinction coefficients for H_2O_2 from 290 to 390 range from 0.004 to 4.6 L mole^{-1} cm^{-1} with $\phi = 0.5$ over the same λ range, for NO_3^- from λ = 290 - 330 nm ε ranges from 0.9 to 7.2 L mole^{-1} cm^{-1} with $\phi = 0.04$ over the same λ range, for NO_2^- from λ = 290 - 410 nm ε ranges from 0.1 to 22.0 L mole^{-1} cm^{-1} with $\phi = 0.05$ over the same λ range, for HNO_3 from λ = 290 - 330 nm ε ranges from 0.005 to 2.22 L mole^{-1} cm^{-1} with $\phi = 0.1$ over the same λ range, and for O_3 from λ = 290 - 675 nm ε ranges from 1.6 to 905 L mole^{-1} cm^{-1} with $\phi = 0.2$ over the same λ range.

Nitrate, which is a major component of ice and snow in polar regions, serves as an excellent chromophore for the indirect photochemical production of hydroxy radical in water as follows (Schwarzenbach, Gschwend et al., 1993):

$$NO_3^- \xrightarrow{h\nu} NO_3^{-*}$$

$$NO_3^{-*} \longrightarrow NO_2 + O^{-\cdot}$$

$$O^{-\cdot} + H^+ \longleftrightarrow {}^{\cdot}OH$$

where NO_3^{-*} represents the exicted triplet-state nitrate anion. The production of hydroxyl radical indirectly by light absorption by nitrate over the wavelength range of 290 to 340 nm can lead to the free radical oxidation of other compounds present such as formaldehyde, nitrite, hydrogen peroxide, organic acids, etc. In most fresh waters and in sea water, bicarbonate and carbonate are a major competitors for hydroxyl radical. However, in ice and snow matrices the total carbonate concentrations are likely to be far less; thus hydroxyl radical reactions initiated by nitrate photolysis could be very important transformation pathway.

We can estimate the hydroxyl radical production rate from nitrate photolysis in the upper 10 cm of ice during the summer months at 70 °N latitude as follows (Schwarzenbach, Gschwend et al., 1993):

$$\sum_{\lambda = 300\text{ nm}}^{340\text{ nm}} k_a = \sum \frac{J(\lambda)\varepsilon(\lambda)(1 - 10^{-\alpha(\lambda)\ell})}{\ell\alpha(\lambda)}$$

where $J(\lambda)$ is the incident photon flux into ice at 70°N in mid-summer (e.g., 0.15 millieinstein cm^2 day^{-1}) (Leifer, 1988), $\alpha(\lambda)$ is the wavelength-dependent attentuation coefficient of UV light in ice(Perovich and Govoni, 1991) ($\alpha(\lambda)$ ranges from 0.464 m^{-1} at 300 nm to 0.263 m^{-1} at 340 nm), $\varepsilon(\lambda)$ is the molar extinction coefficent for nitrate (7 M^{-1} cm^{-1}), *ll* is the pathlength through the ice, and C is the molar concentration of nitrate.

The absorbed photon flux is given by

$$\sum_{\lambda=300}^{\lambda=340} I_a(\lambda) = \sum_{\lambda=300}^{\lambda=340} k_a(\lambda)C$$

Using an average polar nitrate concentration of 2.0 μM , we obtain a directly absorbed photon flux of 4.65 μE L^{-1} day^{-1}, which can then be used to determine the rate of hydroxyl radical production from nitrate photolysis as follows:

$$-\frac{d[NO_3^-]}{dt} = +\frac{d[\cdot OH]}{dt} = \Phi(\lambda)I_a(\lambda) = 0.017 \cdot 4.65 \frac{\mu E}{L \cdot day} = 0.08\ \frac{\mu M}{day}$$

Thus, we estimate that the hydroxyl radical production rate in the upper 10 cm of ice is 0.08 μM day^{-1}. At these production rates, significant rates of hydrogen peroxide production or organic compound destruction could take place over the summer period.

Indirect photochemical reactions (i.e., photosensitized reactions) are more complex mechanistically. For example, the rate-limiting step to form products may be a non-photochemical reaction of ground-state reactants ($^3O_2$) with intermediates generated from light absorption by the primary chromophore. Given all of the chemical and physical pathways that an excited-state chromophore such as trapped humic material (i.e., naturally-occurring organic matter or NOM) may undergo (i.e., intersystem crossing, radiationless deactivation, fluorescence, sensitization, quenching) in the presence of oxygen in water, the effective quantum yield for energy transfer can be written as:

$$\phi_{et} = \frac{k_{et}[R]}{k_{rd} + k_{so}[O_2]}$$

where k_{et} is the instrinic rate constant for energy transfer from the direct excited-state chromophore to the reactant, k_{rd} is the instrinsic first-order rate constant for

radiationless deactivation and k_{so} is instrinsic second-order rate constant for the reaction of the excited-state chromophore with molecular oxygen.
Given the specific rate of light absorption as determined above, the rate of indirect photolysis can be written as:

$$-\frac{d[R]}{dt} = \frac{k_{et}I_a[R]\phi_{tr}\phi_{ct}}{k_{rd} + k_{so}[O_2]}$$

where ϕ_{tr} is the quantum efficiency for excited-state triplet formation and ϕ_{ct} is the fraction of energy transfered from the excited-state triplet to R.

Possible Photochemical Pathways

Based on drawing analogies with photochemical reactions of importance in the gas and aqueous phases, we can consider the following photochemical processes to have some probability of ≥ 0 of occuring either within a snow or ice matrix or on the surface of either snow or ice crystals.

Photochemistry Involving H_2CO and H_2O_2 as Direct Chromophores

Fuhrer et al.(Fuhrer, Neftel et al., 1993) have reported analytical results for the continuous high-resolution measurements of H_2O_2, HCHO, NH_4^+ and Ca^{2+} concentrations performed directly on the ice cores in the field. The cores that were analyzed corresponded to a time interval between 8000 and 38,000 years B.P. They interpreted their results in terms of various climatic changes. However, one must note that both hydrogen peroxide and formaldehyde (HCHO) are sensitive to photochemical or radiochemical alteration (*vide supra*) in a variety of phases.

Hydrogen peroxide decomposes photochemically in aqueous solution (Kormann, Bahnemann et al., 1988; Hoffman, Carraway et al., 1994) and in the gas phase via a homolytic cleavage as follows:

$$H_2O_2 \xrightarrow{h\nu} 2\,{}^{\cdot}OH$$
$$290 \leq \lambda \leq 330 \text{ nm}$$

Likewise, formaldehyde in the carbonyl form (only 3% of total formaldehyde in water) photodecomposes in the atmosphere as follows (Finlayson-Pitts and Pitts, 1986):

$$H_2CO \xrightarrow{h\nu} H^{\cdot} + HCO^{\cdot}$$
$$290 \leq \lambda \leq 330 \text{ nm}$$

In an oxic gas-phase system, H atom and the HCO radical are quickly scavenged by molecular oxygen to yield the hydroperoxyl radical as follows:

$$HCO^{\cdot} + O_2 \longrightarrow HO_2^{\cdot} + CO$$

$$H^{\cdot} + O_2 \longrightarrow HO_2^{\cdot}$$

$$2\,HO_2^{\cdot} \longrightarrow H_2O_2 + O_2$$

Thus, it is conceivable that gas-phase photochemistry maybe taking place in gas bubbles of gas trapped within the ice-snow matrix that ultimately affects the actual time-dependent history of both hydrogen peroxide (Sigg and Neftel, 1991; Sigg, Staffelbach et al., 1992; Sigg, Fuhrer et al., 1994) and formaldehyde. For example, Staffelbach et al. (Staffelbach, Neftel et al., 1991) argue that the interrelationship between methane and formaldehyde in ice cores can be used to track the history of the oxidizing capacity of the atmosphere due to the conversion of methane into formaldehyde via its reaction with hydroxyl radical in the remote atmosphere.

In this regard, Bednarek and Plonak (Bednarek and Plonka, 1994) have studied the kinetics of HO_2 radicals in UV-irradiated frozen solutions of H_2O_2 by an electron spin resonance (ESR) technique. Radical disproportionations at the irradiation temperature were found to be follow a second-order equal-concentration kinetic equation with rate parameters changing non-monotonically over the temperature range 100 - 200 K. In an earlier study, Bednarek and Schlick (Bednarek and Schlick, 1991) identified with ESR several unstable radical intermediates obtained by 77 K UV irradiation of silica gel swollen by water containing a small amount of hydrogen peroxide. The identity and reactivity of the radicals have been compared with results obtained by UV irradiation of bulk H_2O/H_2O_2 solutions. In the solutions, $^{\cdot}OH$ and $HO_2^{\cdot}$ radicals were detected, with an initial intensity ratio that depended on the concentration of H_2O_2. These radicals decayed at 110 and 150 K, respectively; the decay was correlated with the corresponding relaxation regions detected in pure ice.

Photochemistry Involving Fe(III) Species as Chromophores

The deposition of heavy metals and transition metals from the atmosphere into polar regions is well documented (Boutron, Gorlach et al., 1991; Davidson, Jaffrezo et al., 1993; Davidson, Jaffrezo et al., 1993; Suttie and Wolff, 1993; Savarino, Boutron et al., 1994; Wolff and Suttie, 1994). DeAngelis (Deangelis and Legrand, 1994) has noted the role of wind blown dust in the transfer of chemical components including iron to the polar regions. Iron is one of the most abundant elements in the earth's crust, where it is present both as Fe(II) and Fe(III). Crustal iron is readily transferred into the atmosphere by wind and to a lesser extent by volcanic activity. Once in the atmosphere, iron {as Fe(II) and Fe(III)} is know to play an important role in the multiphase atmospheric chemistry of S(IV). Much of the total measured iron in atmospheric samples is predicted to be in the form of ferric oxides and oxyhydroxides based on thermodynamic considerations, and actual identification of α-Fe_2O_3, Fe_3O_4, α-FeOOH, and γ-FeOOH in airborne particles.

When iron(III)-containing aerosols come into contact with aqueous water droplets or aquated surfaces such as ice, Fe is expected to participate in electron transfer reactions within the liquid or quasi liquid phase.

Laboratory research in recent years has been focused on the photore-duction of a variety of iron phases in the presence of a wide range of electron donors such as S(IV) and organic acids (Faust and Hoffmann, 1986; Faust, Hoffmann et al., 1989; Pehkonen, Siefert et al., 1993). For example, in the case of Fe(III) species and organic acids, the following transformation has been documented by Pehkonen et al.

(Pehkonen, Siefert et al., 1993; Pehkonen, Siefert et al., 1995) and Siefert et al. (Siefert, Pehkonen et al., 1994):

$$Fe(III)(H_2O)_6^{3+} + C_2O_4^{2-} \longleftrightarrow Fe\,(III)(H_2O)_4(C_2O_4^{2-})^+ + 2H_2O$$

$$Fe(III)(H_2O)_4(C_2O_4^{2-})^+ \xrightarrow{h\upsilon} Fe\,(II)(H_2O)_4(C_2O_4^{\bullet-})^+$$

$$Fe(II)(H_2O)_4(C_2O_4^{\bullet-})^+ + O_2 \xrightarrow{2\,H_2O} Fe\,(II)(H_2O)_6 + O_2^{\bullet-} + 2CO_2$$

$$O_2^{\bullet-} + H^+ \longleftrightarrow HO_2^{\bullet}$$

$$H^+ + HO_2^{\cdot} + Fe(II) \longrightarrow H_2O_2 + Fe(III)$$

$$2\,HO_2^{\cdot} \longrightarrow H_2O_2 + O_2$$

In a similar fashion, hydroxyl radical is generated from with Fe(III) monohydroxy complexes as follows:

$$Fe(H_2O)_5OH^{2+} + H_2O \xrightarrow{h\nu} Fe(H_2O)_6^{2+} + {}^{\cdot}OH$$

The quantum yield for this latter reaction is $\phi = 0.14$ at $\lambda = 313$ nm and $\phi = 0.017$ at $\lambda = 360$ nm. Fe(II) can then act as a Fenton's reagent catalyst for the decomposition of hydrogen peroxide as follows:

$$Fe(H_2O)_6{}^{2+} + H_2O_2 \longrightarrow Fe(H_2O)_6{}^{2+} + {}^{\cdot}OH + {}^{-}OH$$

$$Fe(H_2O)_5OH^+ + H_2O_2 \longrightarrow Fe(H_2O)_5OH^{2+} + {}^{\cdot}OH + {}^{-}OH$$

$$Fe(H_2O)_6{}^{2+} + HO_2^{\cdot} + H^+ \longrightarrow Fe(H_2O)_6{}^{3+} + H_2O_2$$

The photolytic reduction of ferrihydrite (am-$Fe_2O_3{\cdot}3H_2O$), lepidocrocite (γ-FeOOH), goethite (α-FeOOH), hematite (α-Fe_2O_3), maghemite (γ-Fe_2O_3) and iron-containing aerosol particles ($Fe_{aerosol}$) in the presence of a series of substituted weak organic acids has been documented. These reactions are driven, in part, by the semiconducting properties of the iron oxide polymorphs (Faust and Hoffmann, 1986; Kormann, Bahnemann et al., 1988; Faust, Hoffmann et al., 1989; Pehkonen, Siefert et al., 1993; Hoffman, Carraway et al., 1994; Siefert, Pehkonen et al., 1994; Pehkonen, Siefert et al., 1995).

Oxalate can be photoxidized on the surface of semiconducting iron oxide polymorphs as follows (Faust, Hoffmann et al., 1989; Pehkonen, Siefert et al., 1993; Siefert, Pehkonen et al., 1994):

$$Fe_xO_y \xrightarrow{h\upsilon} h^+_{vb} + e^-_{cb}$$

where h^+_{vb} and e^-_{cb} are the solid-confined but highly mobile valence-band hole and conduction-band electron, respectively.

$$e^-_{cb} + O_2 \longrightarrow O_2^{\cdot-}$$

$$h^+_{vb} + C_2O_4^{2-} \longrightarrow C_2O_4^{\bullet-}$$

$$C_2O_4^{\cdot-} + O_2 \longrightarrow O_2^{\cdot-} + 2\,CO_2$$

Bromoacetate serves as an electron donor in a photochemical reaction with iron oxide polymorphs as follows:

$$>Fe(III)OH + BrCH_2CO_2^- + H^+ \longleftrightarrow >Fe(III)O_2CCH_2Br + H_2O$$

$$2>Fe(III)OH \xrightarrow{h\nu} >Fe(III)OH^{\cdot+} + >Fe(II)OH^-$$

$$>Fe(III)OH^{\cdot+} + >Fe(III)O_2CCH_2Br \xrightarrow{rds} >Fe(III)OH_2^+ + >Fe(III)O_2\dot{C}HBr$$

$$>Fe(III)O_2\dot{C}HBr + H_2O \longrightarrow >Fe(III)OH_2^+ + Br\dot{C}HCO_2^-$$

$$>Fe(II)OH^- + 6H_2O \longrightarrow > + Fe(H_2O)_6^{2+}$$

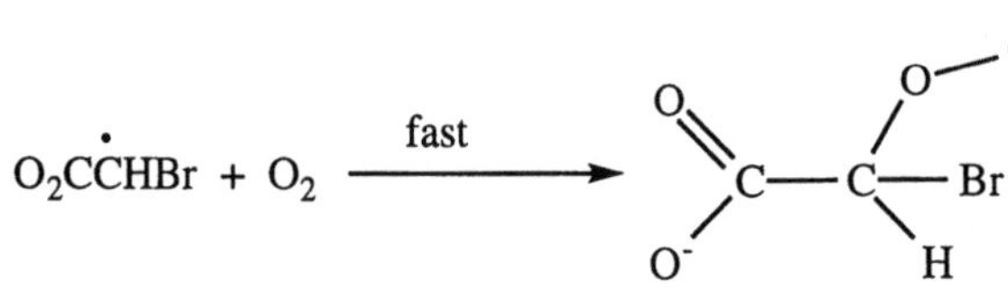

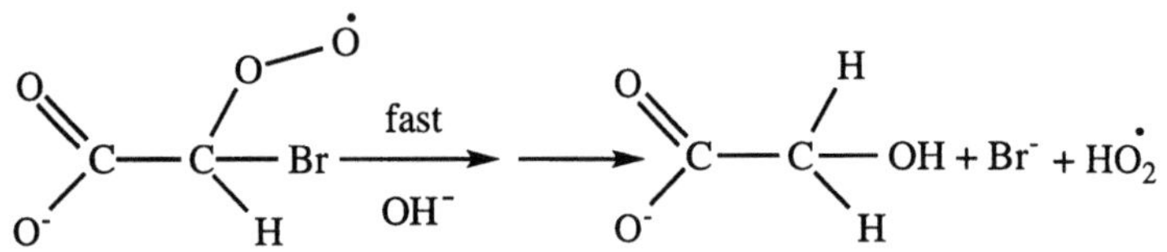

Heterogeneous photochemical reactions may also occur via a photo-Kolbe process (Carraway, Hoffman et al., 1994; Pehkonen, Siefert et al., 1995). For example, acetate can be photo-oxidized oxidized as follows (Carraway, Hoffman et al., 1994):

$$CH_3COO^- + h^+_{vb} \longrightarrow {}^{\cdot}CH_3 + CO_2$$

$$CH_3^{\cdot} + HO_2^{\cdot} \longrightarrow CH_3OOH$$

$$CH_3^{\cdot} + O_2 \longrightarrow CH_3O_2^{\cdot}$$

$$2\,CH_3O_2^{\cdot} \rightarrow\rightarrow CH_3OOH + HCO_2H$$

$$CH_3COO^- + H_2O_2 \rightarrow CH_3COOOH + OH^-$$

Carraway et al. (Carraway, Hoffman et al., 1994) reported that the photochemical oxidation of acetate on ZnO in water resulted in the following product array: CO_2, HCO_2^-, $CHOCO_2^-$, HCHO, CH_3OOH, CH_3COOOH, and H_2O_2. As can be seen by these observed products, many of the chemical species of importance in the historical record held within ice cores are susceptible to interconversion under certain environmental conditions. An additional example of these interrelationships is provided by the heterogeneous photooxidation of glyoxlate to yield formate (Carraway, Hoffman et al., 1994)

$$HCOCO_2^- + H_2O \longleftrightarrow HC(OH)_2CO_2^-$$

$$HC(OH)_2CO_2^- + h^+_{vb} \longrightarrow HC(OH)_2CO_2$$

$$HC(OH)_2CO_2^{\cdot} \longrightarrow HC(OH)_2^{\cdot} + CO_2$$

$$HC(OH)_2^{\cdot} + h^+_{vb} \longrightarrow HCO_2^- + 2H^+$$

$$O_2 + e^-_{cb} \longrightarrow O_2^{\cdot -}$$

$$O_2^{\cdot -} + H^+ \longrightarrow HO_2^{\cdot}$$

$$2\,HO_2^{\cdot} \longrightarrow H_2O_2 + O_2$$

$$HO_2^{\cdot} + e^-_{cb} \longrightarrow HO_2^{\cdot -}$$

$$HO_2^{\cdot -} + H^+ \longrightarrow H_2O_2$$

$$HCOCO_2^- + H_2O_2 \longrightarrow HCO_2^- + CO_2 + H_2O$$

This mechanism leads to an overall reaction stoichiometry of

$$2\,HCOCO_2^- + O_2 \xrightarrow{h\nu,\ Fe_xO_y} 2\,HCO_2^- + 2\,CO_2$$

Naturally Occuring Organic Matter (NOM) as a Chromophore

Since tropospheric aerosols contain a substantial fraction of humic-like material, we can assume that, in addition to the deposition of metal oxides, cations, anions, silicates, etc., a significant amount of NOM will also be deposited on polar surfaces along with wind-blown dust. NOM has a substantial aromatic character that it serves as an excellent absorber of UV and visible radiation. For example, it is well known that in solar illuminated surface waters that absorption of light by NOM will result in the indirect production of singlet oxygen according to the following mechanism (Schwarzenbach, Gschwend et al., 1993):

$$^{1}NOM \xrightarrow{h\nu} {}^{1}NOM^{*}$$

where $^{1}NOM^{*}$ represents the singlet excited state of NOM, ISC stands for "inter-system crossing from the excited singlet state to the excited triplet state, $^{3}O_2$ represents ground state molecular oxygen and $^{1}O_2$ represents excited-state singlet oxygen.

$$^{1}NOM^{*} \xrightarrow{ISC} {}^{3}NOM^{*}$$

$$^{3}NOM^{*} + {}^{3}O_2 \longrightarrow {}^{1}NOM^{*} + {}^{1}O_2$$

$$RH + {}^{1}O_2 \longrightarrow RO_2H$$

Absorption of light by NOM can lead to the indirect production of hydroperoxyl radicals and eventually organic peroxides, ROOH as follows:

$$^{1}NOM \xrightarrow{h\nu} {}^{1}NOM^{*}$$

$$^{1}NOM^{*} \xrightarrow{ISC} {}^{3}NOM^{*}$$

$$^{3}NOM^{*} + R \longrightarrow {}^{1}NOM_{1-x} + {}^{3}R^{*}$$

$$^{3}NOM^{*} + {}^{3}O_2 \longrightarrow {}^{1}NOM_{1-x} + RO_2^{\cdot}$$

$$R'H + RO_2^{\cdot} \longrightarrow R'^{\cdot} + ROOH$$

Summary

Important chemical components present in the historical ice-core and snow column records may be sensitive to chemical change due to the long-term effects of cosmic radiation in the form of high-energy (≥ 0.2 GeV) muons and with incident solar radiation at $\lambda \geq 290$ nm. The chemical effects of muons, which are similar to those of high energy electrons, penetrating into the polar ice cores may be extended as far as several kilometers in depth. The chemical effects, which are due to UV photons may occur in surface disordered layers, in aqueous microinclusions, within trapped gas bubbles, or within the solid ice matrices, will extend several meters below the air/ice surface.

Additional laboratory work is needed to explore chemical changes of important species such as H_2O_2, NO_3^-, NO_2^-, HSO_3^-, pexoxyacetylnitrate, nitrosyl chloride, Fe(III), HCHO, HCO_2H, $C_2O_4^{2-}$, and CH_3SO_3H in ice and snow through direct photolysis, in which the reactant is also the chromophore, through indirect photolysis involving chromophoric sensitizers such as naturally-occuring organic mater (NOM) or iron oxide polymorphs, and due to long-term exposure to cosmic radiation, which will induced chemical changes similar to those observed in pulse radiolysis. Of particular concern, are reactions leading to either the in situ production or destruction of hydrogen peroxide and formaldehyde. In this regard, fundamental parameters such as extinction coefficients, intrinsic quantum yields, branching ratios, wavelength dependencies, photoaction spectra, and reaction products in ice as a function of temperature need to be determined to accurately predict the stability of these important chemical components. A knowledge of the stability of these compounds in ice is critical since polar ice cores are used to probe the pre-industrial environment and to predict the evolution of climate and the atmosphere.

References

Baker, S C , D-P Kelly, et al. (1991) Microbial-degradation of methanesulfonic-acid - a missing link in the biogeochemical sulfur cycle. Nature **350**: 627-628

Bales, R C and E W Wolff (1995) Processes of chemical exchange between the atmosphere and polar snow: key to interpreting natural climate signals in ice cores.. EOS, Trans Amer Geophys Union **76**: 477-482

Bednarek, J and A Plonka (1994) ESR studies on reactivity of HO_2 radicals in polycrystalline ice - nonmonotonic changes of disproportionation rate in the temperature-range 100-200 K. Rad Phys Chem **44**: 485-489

Bednarek, J and S Schlick (1991) Stability of radical intermediates in microscopically heterogeneous media - photolysis of water adsorbed on silica-gel studied by ESR and DSC. J Phys Chem **95**: 9940-9944

Bernstein, M P , S A Sandford, et al. (1994) Infrared-spectrum of matrix-isolated hexamethylenetetramine in Ar and water at cryogenic temperatures. J Phys Chem **98**: 12206-12210

Boutron, C F , U Gorlach, et al. (1991) Decrease in anthropogenic lead, cadmium and zinc in greenland snows since the late 1960s. Nature **353**: 153-156

Buxton, G V , C L Greenstock, et al. (1988) Critical review of rate constants for reactions of hydrated electrons, hydrogen atoms and hydroxyl radicals in aqueous solution. J Phys Chem Ref Data **17**: 513-886

Carraway, E R , A J Hoffman, et al. (1994) Photocatalytic oxidation of organic-acids on quantum-sized semiconductor colloids. Environ Sci Technol **28**: 786-793

Davidson, C I , J L Jaffrezo, et al. (1993) Chemical-constituents in the air and snow at Dye-3, Greenland II. Analysis of episodes in April 1989. Atmos Environ **27**: 2723-2737

Davidson, C I , J L Jaffrezo, et al. (1993) Chemical-constituents in the air and snow at Dye-3, Greenland I seasonal-variations. Atmos Environ **27**: 2709-2722

Deangelis, M and M Legrand (1994) Origins and variations of fluoride in greenland precipitation. J Geophys Res **99**: 1157-1172

Delmas, R J (1992) Environmental information from ice cores. Rev Geophys **30**: 1-21

Delmas, R J (1994) Ice records of the past environment. Sci Total Environ **143**: 17-30

Faust, B C and M R Hoffmann (1986) Photoinduced reductive dissolution of hematite by bisulfite. Environ Sci Technol **20**: 943-948

Faust, B C , M R Hoffmann, et al. (1989) Photocatalytic oxidation of sulfur-dioxide in aqueous suspensions of α-Fe_2O_3. J Phys Chem **93**: 6371

Finlayson-Pitts, B J and J N J Pitts (1986) Atmospheric Chemistry: Fundamentals and Experimental Techniques. New York, John Wiley & Sons

Fuhrer, K , A Neftel, et al. (1993) Continuous measurements of hydrogen-peroxide, formaldehyde, calcium and ammonium concentrations along the new grip ice core from summit, central greenland. Atmos Environ **27**: 1873-1880

Goto, K , T Hondoh, et al. (1986) Determination of diffusion-coefficients of self-interstitials in ice with a new method of observing climb of dislocations by x-ray topography. Japanese J Appl Physics **25**: 351-357

Graedel, T E and C J Weschler (1981) Chemistry within aqueous atmospheric aerosols and raindrops. Rev Geophys Space Phys **19**: 505

Hayakawa, S (1969) Cosmic Ray Physics. New York, Wiley-Interscience

Hoffman, A J , E R Carraway, et al. (1994) Photocatalytic production of H_2O_2 and organic peroxides on quantum-sized semiconductor colloids. Environ Sci Technol **28**: 776-785

Kormann, C , D W Bahnemann, et al. (1988) Photocatalytic production of H_2O_2 and organic peroxides in aqueous suspensions of TiO_2, ZnO and desert sands. Environ Sci Technol **22**: 798-806

Laj, P , J M Palais, et al. (1992) Changing sources of impurities to the greenland ice-sheet over the last 250 years. Atmospheric Environment Part A General Topics **26**: 2627-2640

Legrand, M , C Fenietsaigne, et al. (1992) Spatial and temporal variations of methanesulfonic-acid and non-seasalt sulfate in antarctic ice. J Atmos Chem **14**: 245-260

Leifer, A (1988) The Kinetics of Environmental Aquatic Photochemistry. Washington, DC, American Chemical Society

Matheson, M S and L M Dorfman (1969) Pulse Radiolysis .Washington, DC, American Chemical Society

Neftel, A (1996) The record of gases and reactive species in ice cores and problems of interpretation. This Volume :

Pehkonen, S , R Siefert, et al. (1993) Photoreduction of iron oyxhydroxides in the presence of important organic compound.s Environ Sci Technol **26**: 2056

Pehkonen, S O , R L Siefert, et al. (1995) Photoreduction of iron oxyhydroxides and the photooxidation of halogenated acetic acids. Environ Sci Technol **29**: 1215-1222

Perovich, D K and J W Govoni (1991) Absorption-coefficients of ice from 250 to 400 nm. Geophysical Research Letters **18**: 1233-1235

Pursell, C J , J Conyers, et al. (1995) Photochemistry of chlorine dioxide in ice. J Phys Chem **99**: 10433-10437

Rettich, T R (1978) Some photochemical reactions of aqueous nitrous acid, Ph D Thesis, Case Western Reserve University

Rossi, B (1964) Cosmic Rays New York, Mc-Graw Hill

Ryan, K G (1992) UV-radiation and photosynthetic production in antarctic sea ice microalgae. J Photochem Photobiol Part B Biol **13**: 235-240

Savarino, J , C F Boutron, et al. (1994) Short-term variations of Pb, Cd, Zn and Cu in recent greenland snow. Atmos Environ **28**: 1731-1737

Schwarzenbach, R P , P M Gschwend, et al. (1993) Environmental Organic Chemistry New York, Wiley-Interscience

Siefert, R L , S O Pehkonen, et al. (1994) Iron photochemistry of aqueous suspensions of ambient aerosols with added organic acids. Geochem Cosmochim Acta **58**: 3271-3279

Sigg, A , K Fuhrer, et al. (1994) A continuous analysis technique for trace species in ice cores. Environ Sci Technol **28**: 204-209

Sigg, A and A Neftel (1991) Evidence for a 50-percent increase in H_2O_2 over the past 200 years from a greenland ice core. Nature **351**: 557-559

Sigg, A , T Staffelbach, et al. (1992) Gas-phase measurements of hydrogen-peroxide in greenland and their meaning for the interpretation of H_2O_2 records in ice cores. Atmos Chem **14**: 223-232

Staffelbach, T , A Neftel, et al. (1991) A record of the atmospheric methane sink from formaldehyde in polar ice cores. Nature **349**: 603-605

Sturges, W T , G F Cota, et al. (1992) Bromoform emission from arctic ice algae Nature **358**: 660-662

Suttie, E D and E W Wolff (1993) The local deposition of heavy-metal emissions from point sources in antarctica. Atmos Environ **27**: 1833-1841

Taube, H (1957) Photochemical reactions of ozone in solution. Trans Faraday Soc **53**: 656

Walker, D C (1983) Muon and Muonium Chemistry Cambridge, Cambridge University Press

Woafo, P , R Takontchoup, et al. (1995) Soliton mechanism of proton migration in hydrogen-bonded solids. J Phys Chem Solids **56**: 1277-1283

Wolfendale, A W (1973) Cosmic Rays at Ground Level London, The Institute of Physics

Wolff, E (1992) The influence of global and local atmospheric-pollution on the chemistry of antarctic snow and ice. Marine Pollution Bulletin **25**: 274-280

Wolff, E W and E D Suttie (1994) Antarctic snow record of southern-hemisphere lead pollution. Geophysical Research Letters **21**: 781-784

Zafiriou, O C and M B True (1979) Nitrite photolysis in seawater by sunlight Mar Chem . **8**: 33

METAMORPHISM OF POLAR FIRN: SIGNIFICANCE OF MICROSTRUCTURE IN ENERGY, MASS AND CHEMICAL SPECIES TRANSFER

RE Davis, EM Arons and MR Albert
Cold Regions Research and Engineering Laboratory
72 Lyme Road
Hanover, New Hampshire 03755
USA

ABSTRACT

Energy, mass transfer and grain recrystallization processes show the most dynamic variation in the upper few meters of polar firn. Our understanding of processes in dry snow and polar firn shows feedback mechanisms exist between microstructure changes and energy and mass transfer coefficients. Differences between the predominant processes in the short polar summer and the winter cause sharp textural discontinuities in the stratigraphic columns at many polar locations. This distinctive layering forms the basis of studies on accumulation rates, layer ages and other investigations pertinent to firn and ice core analysis. Processes forming the stratification of the firn are not completely understood, nor are the associated loading patterns of chemical species. Past research on energy and mass transfer in near-surface polar firn is briefly surveyed. Current research focuses on processes controlled by the geometry of the ice and pore phases in the snow and firn. An overview is given of some of this work.

INTRODUCTION

Vertical changes in the stratigraphy of polar firn reflect the history of deposition and transfer of mass and energy across the atmosphere-snow surface, which controls post-depositional changes. The general character of the strata in dry firn can be described as layers with coarse grains interspersed with layers of fine grains and slightly higher densities, which have been attributed to the processes predominating in summer and winter, respectively (Benson, 1962; Gow, 1965; Alley 1988; Alley et al, 1990). The layering and microstructure of the layers control the energy and mass exchange processes, which in turn cause metamorphic (diagenetic) changes to the layering and microstructure. These processes are more dynamic near the snow surface.

Heat is transferred by conduction through the ice-grain network, latent heat effects of water vapor diffusion across the pore spaces, and advection through the pore volume. Solar radiation penetrates the surface of the snow and the intensity below the surface depends on

NATO ASI Series, Vol. I 43
Chemical Exchange Between the Atmosphere and Polar Snow
Edited by Eric W. Wolff and Roger C. Bales

the microstructure and the presence of absorbing contaminants. The absorption of solar energy as a function of depth is spectrally dependent. Diffusion and advection of gases affect the near-surface snow mass balance and the incorporation of deposited chemical species. Advection of particulates by wind pumping into the pore space can also be important. Subsurface transmission of solar radiation can enhance metamorphic processes, which change transport coefficients in the pore and ice volume and can affect distributions of chemical species subject to photolysis. Thus processes near the polar firn surface control the atmosphere-snow transfer of chemical species, which can eventually become incorporated in the ice core record.

This paper aims to summarize key research on the metamorphism of polar firn, and to identify gaps in our understanding of the processes. We concentrate on the processes near the surface of polar firn because these are the most dynamic and least understood. This paper continues with three major parts, a background briefly summarizing what is known about dry snow processes, a brief summary of work on energy, mass transfer and metamorphism near the surface of polar firn, and an overview on ongoing and future research.

BACKGROUND

Location of Chemical Species in Dry Snow

Chemical species and other contaminants can be located inside snow crystals, be deposited on the outside or exist in the air space around the ice matrix. Rapid growth of ice crystals in the atmosphere incorporates impurities, which become defects in the crystal structure. The condensation nuclei of snow crystals are frequently solid particles, found in the central portion of the snow crystal (e.g. Kumai, 1977). Other constituents incorporated into the ice include solutes (gross defects) and substitutional species, such as deuterium and ^{18}O. Impurities in snow grains are not static with respect to their location relative to the grain boundary. The diffusion rates of defects in ice is slow, particularly at polar temperatures. However, the direction of diffusion of gross defects is generally toward the side of the crystal that is sublimating, so the combination of diffusion and migration of grain boundaries during metamorphism cause relatively rapid relocation of impurities with respect to the ice surface. Once on the grain surface, impurities are much more mobile. For example, ^{18}O may diffuse, or be advected with water vapor in the pore volume (e.g. Whillans and Grootes, 1985). Therefore it is important from a trace species point of view to understand the mechanisms and rates of ice recrystallization in snow as well as the interactions with energy and mass transfer at greater than grain scales.

Ice Grain Recrystallization

Ice particles in a snow cover undergo continual metamorphism. This depends on the geometry of the grain structure, the snow density, the pore size distribution, the temperature and the temperature gradient. The rates at polar temperatures are relatively slow compared to seasonal snow. Snow grains tend to grow, reducing their surface free energy by reducing their surface area. Growth rates increase with increasing vapor fluxes (Colbeck, 1982). At high vapor fluxes (strong temperature gradients), rapid deposition of water molecules on the ice surface gives rise to faceted crystal shapes. At the snow surface these are called surface hoar, growing during calm clear conditions when surface radiant heat loss is strong (Lang et al, 1984), and are called depth hoar within the pack. Despite the faceted and angular morphology, the surface area per unit volume of depth hoar is usually less than new snow or fine-grained wind pack. This explains the persistence of these crystals after they have formed. The crystal form depends on the configuration of the ice and pore space (Armstrong, 1985; Gubler, 1985). Depth hoar crystals grow to large sizes at the expense of smaller rounded crystals provided that pore spaces will accommodate them. Larger pore spaces cause enhanced temperature gradients across the pores, which suggests that depth hoar will fully form only in snow with lower densities.

Heat and Mass Transfer by Diffusion

The dependence of energy and mass transfer on small scale structures in heterogeneous materials such as firn has been widely studied for over a century (Maxwell, 1891; Auracher, 1978; Beasley and Torquato, 1989; Brinkman, 1947; de Quervain, 1972; Doi, 1976; Hjelström, 1890; Koplik, 1991; McCreesh, 1991; Shabtaie and Bentley, 1994; Sturm and Johnson, 1992; Arons, 1994; Dullien, 1992; Yosida and colleagues, 1955). Arons and Colbeck (1995) reviewed advances in our ability to understand this dependence on a mechanistic level for dry snow. In the absence of melt water, near surface firn is topologically similar to dry snow and so much of the knowledge that has been gained regarding dry snow can be applied to understanding heat and mass transfer in firn. The present goal is to relate the chemical composition of the atmosphere to the chemical composition of firn. The firn acts as a chemical catalyst and a sink. Its ability to function in these capacities depends on its ability to transfer heat and mass.

Mass transfer in firn primarily occurs through the diffusion in response to density gradients of water vapor (Colbeck, 1993) and in some cases by forced convection in response to pressure fields at the surface (Colbeck, 1989a; Albert, 1993). Sensible heat is transported by conduction through the ice grain lattice and by advection with air through the pore space. In the most simple approximation, each of these processes can be treated as potential flow systems in which the flux of heat or mass arises from imposed gradients. In general, flux F

$$F = -K\nabla\Phi \tag{1}$$

where K is a generalized transfer coefficient (e.g., thermal conductivity, diffusion coefficient, or permeability) and Φ is a generalized potential (e.g., thermal or mechanical). The value of K depends on the small scale structural properties of firn.

For example, there is a wide variation in values of thermal conductivity reported in the literature as reviewed by Mellor (1977). Allowing for some error in the reported measurements, thermal conductivity still varies by more than a factor of 2 at a given density, which shows that other factors must play a role. Moreover, in the density range $\rho_{snow} \leq 300\ \mathrm{kgm^{-3}}$ thermal conductivity is fairly insensitive to density and shows the widest scatter in experimental results. For example, Sturm and Johnson (1992) observed changes in thermal conductivity in response to metamorphic changes in subarctic snow. They also reported variations in published values of thermal conductivity that can be attributed to snow structure. Izumi and Huzioka (1975) also observed changes in thermal conductivity in response to snow metamorphism. Shabtaie and Bentley (1994) derived a power-law dependence of electrical conductivity on grain cluster concentration in firn; in dense firn, this analysis would apply to heat conduction as well. Yosida et al (1955) observed enhancement of water vapor diffusivity in snow due to intergranular relationships. He proposed a hand-to-hand transfer of water vapor and heat in which vapor diffuses across a pore space and deposits on an ice surface; heat is conducted across the ice and vapor is sublimated off the opposite side. These relationships have been further studied by Colbeck (1993), who showed enhancement of temperature gradients depending on geometry, Gubler (1985) who incorporated Colbeck's ideas in a numerical scheme calculate growth rates depending on the geometries of grain aggregates, and Sommerfeld (1991) who proposed a one-dimensional model. Waterhouse (1962) and Shimizu (1970) measured snow permeability and demonstrated that this property, too, depends on snow type.

Based on the observations and theoretical analyses mentioned above, it is apparent that small scale structural properties that control the value of K are both microstructural and meso-structural. Arons and Colbeck (1995) defined the former as features that occur on the scale of individual grains and the latter as features that involve more than one grain. Metamorphic processes affect both microstructural and macro-structural features, so the transport-related properties of near surface firn should vary as it is exposed to a variety of metamorphic processes that are induced by meteorological variations. Grain scale or microscale metamorphism arises from redistribution of mass on the surface of a grain due to variations in the curvature of that surface. This was analyzed in detail by Colbeck (1980). This effect is important for freshly fallen snow in the absence of large macroscopic temperature gradients and has been observed by other investigators, including Bader et al (1954). It also can be an important process in the metamorphism of depth hoar crystals in polar firn. Our observations from Summit, Greenland indicate that grains can undergo cycles during which one type of metamorphism or another dominate. This has been observed by others, including Alley et al (1990), as indicated by Colbeck et al (1990).

In the presence of macroscopic temperature gradients water vapor is transferred between grains as modeled by both Gubler (1985) and Colbeck (1983b). As reported by Colbeck (1983a), rounded forms occur at warmer temperatures and weaker vapor density gradients. Decreasing the temperature and increasing the strength of the vapor density potential in the vicinity of the growing grain causes more elaborate features to occur on the growing grain surface. Ultimately, as this effect is increased, depth hoar crystals evolve. Although it has not been thoroughly investigated, it appears that at some point of evolution, these crystals become so well-developed that the surfaces become optimal nucleation sites for new crystals. When this happens elongated chains of grains can grow. They are commonly observed in depth hoar (Akitaya, 1974, e.g.). As the chains become common, the pore space acquires elongated features, also as observed by Trabant and Benson (1972). Alley (1988) proposed this sort of microstructural anisotropy was evidence for interaction with the atmosphere in the top few meters of firn.

Heat and Mass Transfer by Advection and Convection

While diffusion of heat in the ice matrix and water vapor in the pore space are commonly presumed to be the dominant heat and mass exchange mechanisms, the potential for rapid transport of heat or gas through the snow is greatly increased if natural convection or ventilation occur. There is field evidence that air flow through snow due to natural convection and ventilation can occur in snow covers. Sturm and Johnson (1991) documented snow temperature profiles that indicate natural convection effects in a highly permeable Alaskan snow pack. Benson (1962) found that near-surface (to depths of 1.6 m) snow temperature changes in Greenland were faster than could be explained by heat conduction, and attributed the differences to heat advected by ventilation. Clarke et al (1987) found warming of the firn in windy areas of Aggasiz Ice Cap that cannot be explained by heat diffusion alone.

Numerical results give insight into the nature of the effects of ventilation processes. Albert and McGilvary (1992) presented a numerical model for advective heat and vapor transport through snow. In one-dimensional simulations, they showed that the primary shape of the temperature profile during ventilation is controlled by a balance between the heat advected by the dry air and the heat conduction due to the temperature differences at the boundaries. They also showed advection was very important to vapor transport because air moved at rates in the pore space that exceeded the capability of the pore surface to maintain saturation. They demonstrated that the major thermal effects of ventilation are manifested when the temperature gradient is low, and when the air temperature is markedly different from the snow temperature. Pressure variations on the snow surface are a forcing mechanism for ventilation of snow.

While no surface pressure measurements have been reported for ventilation, there have been relevant theoretical calculations. Colbeck (1989) showed that sastrugi on the snow surface cause pressure variations from form drag of the air flow across the surface. Clarke and Waddington (1991) showed that the depth attenuation of surface pressure fluctuations (due to turbulent air flow over flat terrain) is influenced by both the temporal and spatial frequency of the pressure signal. Colbeck (1991) showed that a fine grained layer would remove most of the higher frequency oscillations in pressure, thus dissipating the energy associated with turbulent conditions. Albert (1993) showed that resulting temperature profiles during ventilation are fairly insensitive to the frequency in time of sinusoidal surface-pressure forcing as long as the surface pressure amplitude is sufficient to sustain air movement in the snow.

Radiation Exchange Near the Snow Surface

Radiation is often the dominant component of the energy budget and in dry snow it exerts strong control on the temperature profiles near the snow-atmosphere interface. Solar radiation, mostly in the $0.35-2.5\,\mu m$ wavelength range penetrates the snow surface differentially according to the wavelength (Warren, 1982). Thermal, or longwave radiation, is only absorbed or emitted at the snow surface, and this is spectrally independent. In the visible spectral region, reflectance is controlled by the amounts and distributions of absorbing impurities while in the near-infrared wavelengths, snow reflectance and transmittance depend on the effective scatterer size of the snow (Warren and Wiscombe, 1980a,b). This is because the absorption coefficient in the visible wavelengths is low (minimum at $\lambda = 0.47\mu m$), increasing dramatically with wavelength. The effective scatterer size in snow has usually been taken to be equal to the average grain size, but in the case of irregularly shaped particles, equal to the sphere with the same surface-to-volume ratio as the bulk snow (Dozier et al 1989).

Spectral differences in the interaction of radiation with snow leads to subsurface heating under certain conditions. However, temperatures near the snow surface are exceedingly difficult to measure under sunny conditions. Published temperature profiles have led to modeling efforts aimed at predicting temperature maximums below the snow surface (Schlatter, 1972; Colbeck, 1989a; Brandt and Warren, 1993). Modeling efforts that have used a broad band extinction coefficient (Schlatter, 1972; Colbeck, 1989) show temperature maxima below the surface, which are generally too deep (Brandt and Warren, 1993), ranging from 15 to 25 cm. In broad-band calculations the bulk extinction coefficient for snow should decrease with increasing depth because at depth the overlying snow acts as a filter, leaving radiation with lower extinction.

A spectral model was used by Brandt and Warren (1993) to calculate the spectrally dependent flux extinction and couple these predictions with an energy transfer model for the snow. Subsurface heating depended on the snow having a low thermal conductivity, generally restricted to lower densities. This condition met, they show temperature maxima much closer to the surface (within 1-2 cm) than previous calculations. While they suggested that subsurface solar heating is unlikely at South Pole, fair agreement was found with temperature profiles measured by Alley et al (1990) at Summit, Greenland, which show subsurface temperature maxima. High subsurface temperatures lead to strong temperature gradients and the formation of depth hoar (Alley et al, 1990). Strong surface cooling at night due to the emission of thermal radiation can lead to the formation of surface hoar (Colbeck, 1988).

Effects of Layering

Polar snow has a characteristic layering caused by seasonal differences in deposition and post-depositional processes. The consideration of the snow as a layered system is important to understanding the changes occurring after deposition (Colbeck, 1991). When a dense layer is bounded by layers with lower density, the lower permeability and higher thermal conductivity affect the temperature gradients, and therefore the vapor density gradients near the interfaces of the layer. Colbeck (1991) showed how temperature gradients could be enhanced by the presence of a dense layer (Figure 1). Whether these effects would be enough to cause the growth of depth hoar or significant mass transfer depends on the particular situation in the snow.

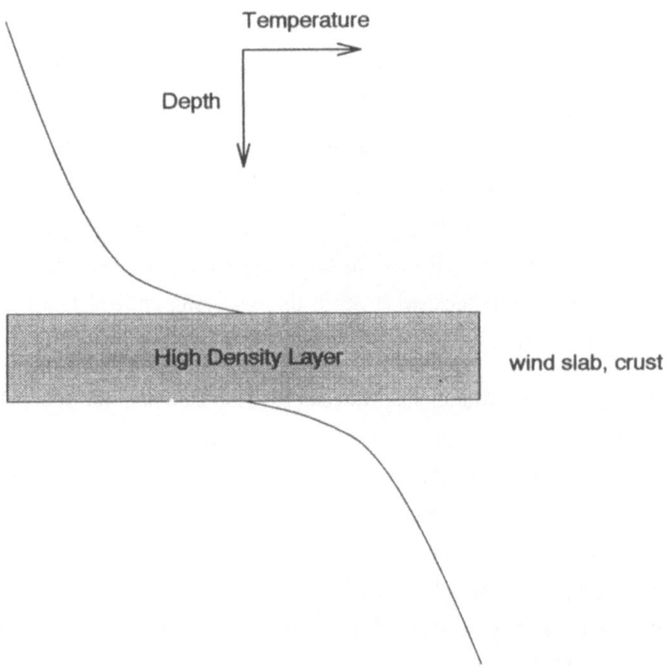

Figure 1. Simple temperature depth model showing layer with high conductivity embedded in layers with lower conductivity. (After Colbeck, 1991).

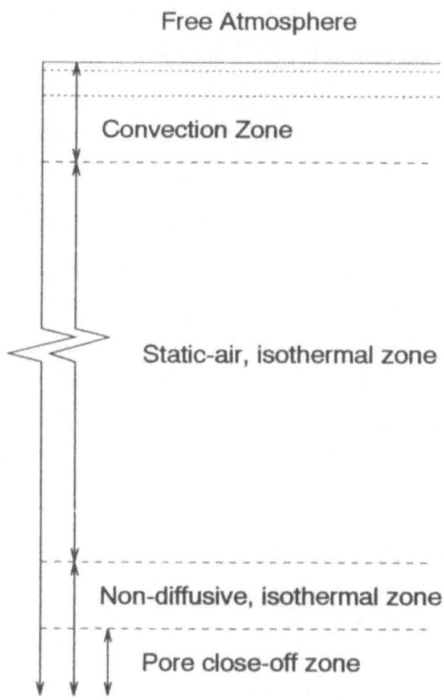

Figure 2. Schematic diagram showing zones in the column of dry polar firn. The topmost, convection zone, is subdivided into a zone where solar subsurface heating can occur (shallowest) and a zone down to the diurnal damping depth. (After Sowers et al (1993)).

NEAR-SURFACE CHARACTERISTICS OF POLAR FIRN

Processes in the Firn Column

The upper few meters in polar firn are subject to fluctuating temperature profiles, potential convection of air and relatively rapid densification. The annual damping depth in observed temperature profiles (i.e. depth at which the annual temperature signal is damped out) is usually about 10 m, while the daily damping depth is about 25 cm or less. Sowers et al (1992) divided the column of the firn from the surface down to the pore close-off region into zones depending on the character of air movement (Figure 2). At the top is the convection zone, followed by a zone where the air is relatively static, dominated by diffusive transport. Near the bottom is a region where there is a gradual close-off of the pore volume and air movement is highly restricted. While the firn remains in the upper few meters it metamorphoses at rates depending on the temperature, temperature gradient and static load (determined by accumulation rate). Relatively rapid densification takes place in this zone, where the dominant mechanisms consist of grain settling and packing (Benson, 1962; Anderson and Benson, 1963) and bond sliding (Alley et al, 1982; Alley, 1987b). Strata in the convection zone are generally regarded as the only firn subject to changes brought about by processes at the surface, while deeper firn tends to exhibit changes that vary slowly and systematically with depth. We concentrate on the convection zone where processes are driven from the firn-atmosphere surface. Most interest in this zone has been sparked by the need to specify boundary conditions (e.g. densification, diffusion, etc.) for studies of deeper layers, to the pore close off (Alley, 1982).

Near the firn-atmosphere interface some trace chemical species exhibit clear signals of a summer-winter cycle, and thus provide a means to count annual layers additional to stratigraphic methods. Some of these signals can be damped out with depth, depending on the species, and the seasonal signal can be obliterated completely depending on depth. The smoothing of the concentration gradients with depth (Langway, 1967) has been attributed to mass transfer by vapor diffusion and possibly wind pumping. Johnsen (1977) proposed a model for the mixing of ^{18}O with time, densification and temperature and attempted to reconstruct the annual signal. The deuterium and ^{18}O signals promote special interest because they appear to be correlated with the mean air temperature at the time of formation of the ice crystals in the atmosphere (Dansgaard, 1964), and thus can be used to estimate temperature records (e.g. Jouzel et al, 1983). Whillans and Grootes (1985) constructed a more sophisticated model of diffusion in both solid and vapor phases and their interactions. Vapor phase diffusion was shown to be the rate limiting process. While they were able to predict the general smoothing of profiles of concentration, they specified of a blocking term to modify water vapor diffusivity, thereby using values less than the value in free air. This is contrary to the accepted value of about 5 - 7 times the value in free air (Colbeck, 1993; Yosida, 1955). The possibility of isotopic species depositing on the cold side of an ice grain with water vapor, then sublimating into the vapor phase on the warm side was not considered. Moreover, the possibility for wind pumping was not considered. The potential effects of strong density and temperature gradients were recognized, and caused distortions in the concentration profile as predicted by the model (Whillans and Grootes, 1985). Schwander et al, (1993) conclude that molecular diffusion dominates the major part of the firn column. Their trace gas data also show the possibility of a convection zone up to 20 m below the surface and that diffusion is not apparently impeded by high density layers like wind crusts.

Depth Hoar and Crusts

The standard model of firn stratification was developed by Benson (1962), Gow (1965) and others, and consists of the annual cycle between winter deposits of wind slab and other fine-grained strata, and late-summer depth hoar layers. Figure 3 shows sections cut through fine-grained wind slab and coarse-grained depth hoar within the upper 0.5 m of firn at Summit, Greenland. Each pair of coarse-fine layers is interpreted as autumn and winter of one year in deep firn and ice core records (e.g. Meese et al, 1994), although it should be noted that Gow (1965) reported hiatuses in the stratigraphic record. The strata pairs can be seen when inspecting the core under transmitted light because more light is transmitted through the coarse-grained layers, which appear brighter.

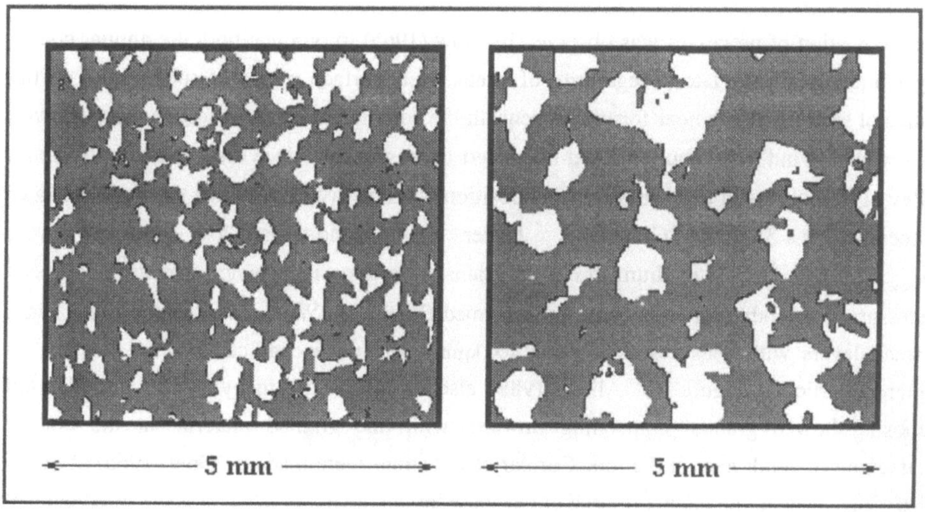

Figure 3. Section images of cuts though fine-grained wind pack (left) and coarse-grained depth hoar (right). Ice is white and the pore space is gray. The wind pack had a density of 363 kgm^{-3} and a surface area-to-volume ratio of 20 mm^{-1}, while the depth hoar had a density of 325 kgm^{-3} and a surface area-to-volume ratio of 4.5 mm^{-1}. These sections were cut from vertically adjacent samples, within 200 mm from each other.

Various conditions and mechanisms have been proposed for the formation of the annual depth hoar zone. The combination of processes or the dominant process probably depends on the location. Bader and others (1939) suggested that in polar regions wind pumping may be necessary to cause enough vapor flux for the formation of depth hoar. Gow (1965) observed depth hoar growing beneath simultaneously forming crusts at South Pole, implying some role of the crust in the hoar formation. Langway (1967) proposed that the requirements for the formation of depth hoar was the deposition of cold wind slab or winter deposit over warm snow. Koerner (1971) suggested that depth hoar originates at the surface at Plateau Station, Antarctica, possibly blurring the distinction between depth hoar and surface hoar. Palais et al, (1982) reported that the summer depth hoar zone was preserved by the formation of a thin hard crust, before autumn through spring deposition. Alley (1988) pointed out that layers of coarse faceted grains could form in a variety ways, for example by burial of surface hoar and by formation of depth hoar beneath the surface.

A combination of processes was observed by Gow (1965) to produce both the annual course layer and explain the lateral regularity of strata. Near-surface cycles of sublimation on the flanks of sastrugi, depth hoar formation beneath the common crusts, and surface hoar growth, followed by wind redistribution acted in concert to seasonally reduce the surface topography. Alley (1988) observed that the original deposition of storm layers could set the stage for later processes at Summit, Greenland. Under calm conditions, low density layers ($\rho_{snow} \leq 250\ kgm^{-3}$) with medium to coarse grains were deposited, which had the pore sizes necessary for depth hoar to become fully formed (Figure 3). Wind slabs, on the other hand, formed layers with densities as high as $400\ kgm^{-3}$, and were characterized by small, well sintered particles (Figure 3). Alley (1988) also observed that many of the buried wind deposits showed gradual coarsening upward, with fine grained material at the bottom coarsening to depth hoar at the top. Similar observations were made by Gow (1965) at South Pole. This leads to the deduction that either mass loss in the upper portion of the wind deposit took place, which opened up the pore space for depth hoar formation, or surface hoar was formed and later buried. The loss of mass from the top of a medium to fine grained layer with moderate to high densities is required for advanced formation of depth hoar, characterized by chains of grains and the fully developed morphologies. Figure 4 shows an example from Summit, Greenland of a section cut through a layer with fully developed depth hoar morphology. Given that wind deposits tend to have low permeability and high thermal conductivity, two questions can be raised. What are the factors limiting mass loss to shallow depths? And why are depth hoars of the summer so well developed?

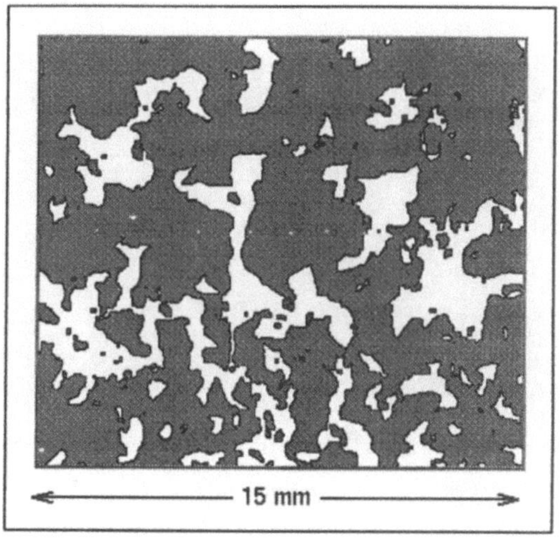

Figure 4. Section image from annual depth hoar layer near the snow surface at Summit, Greenland. Ice profiles are shown as white pore space is shown as gray.

The answers to both questions lie in the feedback between the surface or subsurface heating by solar radiation, growth of depth hoar or surface hoar and the associated dramatic decrease of thermal conductivity and increase of permeability. Alley et al (1990) measured near-surface temperature gradients responding to radiative forcing in summer at the GISP2 site, Greenland, and observed concurrent mass loss and depth hoar formation, despite the presence of a wind crust on the surface. A layer 5 - 10 mm thick beneath the surface lost almost half its mass as depth hoar developed and progressed downward. Temperature measurements showed extremely steep gradients just blow the surface, and changes in the direction of the temperature gradient, hinging on the diurnal damping depth, about 200 mm. Brandt and Warren (1993) modeled the conditions reported by Alley et al (1990): incident solar = 800 Wm^{-2}, lower boundary temperature = $-15^{\circ}C$ at 25 cm depth, and uniform snow density of 220 kgm^{-3}. They found a subsurface maximum of 3 K (i.e. above conduction alone) at 22 mm depth, and argued that the maximum would be closer to the surface if the underlying snow had higher density. They point out the model assumption of an insulating lower boundary amplifies the maximum value, but the 3 K maximum is somewhat lower than reported by Alley et al (1990). On the other hand, grain growth provides a positive feedback, increasing the subsurface temperature rise (Brandt and Warren, 1993). At South Pole, reworking of coarse crystals of surface hoar by light wind and formation of depth hoar near the surface create a layer of high permeability and low thermal conductivity, conditions ripe for mass loss by sublimation and the maintenance of steep temperature gradients.

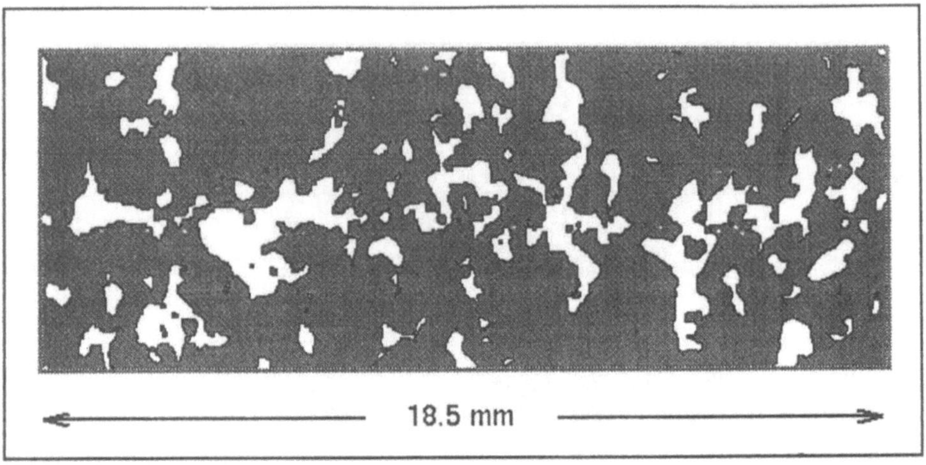

Figure 5. Section image from annual depth hoar layer with a crust (vertical center) from Summit, Greenland. Ice is white and pore space is gray. The most prominent chains of grains are just to the right of center.

The role of crusts, is unclear. They seem to do little to inhibit mass loss by sublimation or, as earlier described, restrict air movement in the upper firn column. Crusts can be described as very thin layers, often consisting of a mono-layer of grains. In the case of subsurface heating, mass sublimates away from the temperature maximum by diffusion, toward the surface and deeper into the layer. If a crust is present at the surface we might expect to observe crystals growing in chains extended into the direction of vapor migration. Crusts can be observed with chains of grains attached to one or both sides. And as an extreme case of the situation described by Colbeck (1991) and Sketched in Figure 1, we can see that strong temperature gradients might be expected near the crust. The majority of crusts we have examined on sections from Summit, Greenland show crystals growing from both sides of the crust, which shows vapor must have arrived from both sides. These crusts typically consist of almost a mono-layer of grains; a few parts of the crust consist of aggregates of a few well bonded grains. Figure 5 shows an example of a crust with elongated chains of depth hoar grain attached to both sides. This in turn suggests some type of cycling of vapor flux. Figure 6 shows an obvious case. In this example we interpret the long protrusions from the top of the crust to be buried surface hoar with chains of depth hoar extending underneath. In addition to the growth of depth hoar and surface hoar crystals on crusts, we have also observed on sections that densities of snow near the crust is significantly less than either the density of the crust or the density of the snow further away from the crust, in most cases. This clearly shows mass depletion away from the crusts for reason that

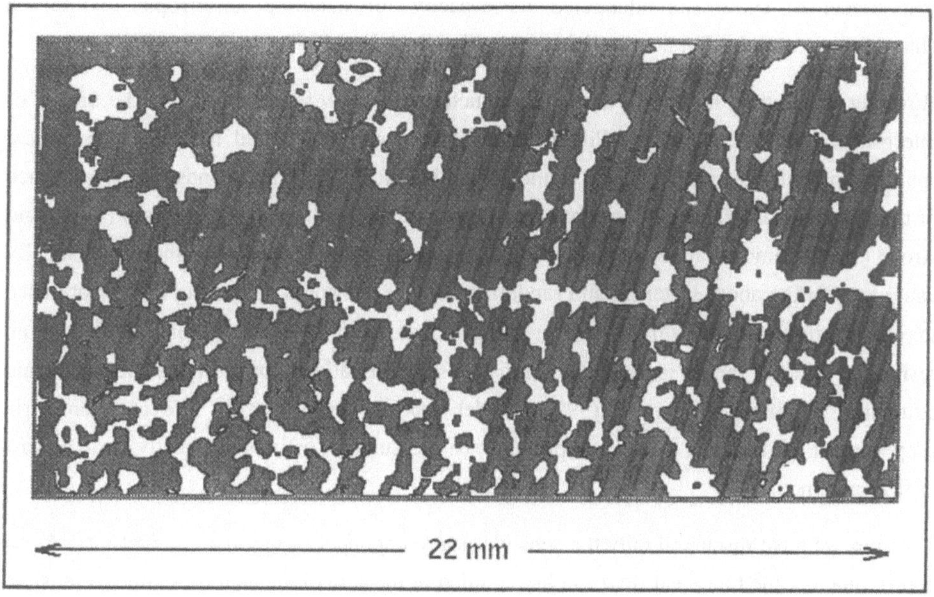

Figure 6. Section image from hoar layer with a crust (vertical center) from Summit, Greenland. Ice is white and pore space is gray. Large chains of grains extending from the top of the crust represent the profiles of buries surface hoar. are currently unclear.

RESEARCH DIRECTIONS

Summer process of energy and mass exchange over polar firn have been largely deduced from observations of snow microstructure, stratigraphy and their changes over time (or in vertical profile). Some modeling efforts have focused on particular processes, such as radiation heating, but further understanding of the magnitude and frequency of processes controlling energy and mass exchange near the firn surface are needed.

Heat and Mass Transfer by Diffusion

In current efforts to model heat and mass transfer in snow and polar firn, properties such as thermal conductivity, water vapor diffusivity, and permeability are determined for modeling purposes on the basis of empirical correlations with bulk snow density. They are not strongly tied to microstructural and meso-structural features that evolve simultaneously with density but are in general physically independent of it (Jordan, 1991; Albert and Arons, 1994, e.g.).

The evolution of our understanding of microstructural control of thermal conductivity of the ice grain lattice started with the model of Maxwell (1891). That model consisted of a dilute dispersion of conducting spheres in a non-conducting matrix. Schwerdtfeger (1963) adapted that model to snow and illustrated its capacity for handling anisotropic textures, a characteristic of polar firn that has been quantified by Alley (1987a).

However, theoretical models of thermal conductivity in snow failed to include the effects of intergranular connections until Pitman and Zuckerman (1967) used a model of identical spherical particles on a regular grid connected by identical cylindrical bonds. The importance of the degree of intergranular bonding was further illustrated by Adams and Sato (1993) and Arons (1994) demonstrated that intergranular bonding can be treated even more realistically using knowledge about sintering and random resistance networks. Sintering models are used to characterize the growth of intergranular bonds (a microscale feature) and random resistance network theory is used to quantify the spatial relationships of grains and multigrain structures. In this way, Arons' (1994) model depends on physically relevant geometric properties of the material that can in principle be measured or estimated directly from a snow or firn specimen.

Thus far, we have discussed only the sensible heat component of thermal conductivity. As an approximation, the total heat flow can be regarded as the sum of the heat flows due to sensible and latent heat transfer:

$$q_{total} = q_{lattice} + q_{vapor} \tag{2}$$

Using the equations of Fourier and Fick, this becomes

$$q_{total} = -K_{lat} \frac{dT}{dz} - LD_{eff} \frac{d\rho_v}{dz} \tag{3}$$

where K_{lat} is the lattice thermal conductivity as discussed above, L is the latent heat of water vapor, D_{eff} is the bulk coefficient of water vapor diffusion through snow, and ρ_v is water vapor density. Using the assumption that water vapor is in thermodynamic equilibrium with the ice grain lattice at a given depth, z, and the Clausius-Clapeyron equation, equation 3 becomes

$$q_{total} = -\left\{K_{lat} + K_{vap}\right\} \frac{dT}{dz} \tag{4}$$

where

$$K_{vap} = LD_{eff}\left[\frac{d\rho_v}{dT}\right]_{eq} . \tag{5}$$

The sum of K_{lat} and K_{vap} is K_{total}, which is what one typically measures in the field and laboratory. Note that as represented in equation 4 the two pathways for heat conduction occur in parallel. The question of parallelism was addressed by both Auracher (1978) and Yosida et al (1955) for snow and by Wiener (1912) for arbitrary media. We will discuss only the parallel representation in order to illustrate how to account for the latent heat effects.

The dependence of K_{vap} on micro- and meso-structural attributes is through D_{eff}. This has been treated using analogies with static electrical potential fields by Colbeck (1993). He showed that D_{eff} depends on grain spacing and decreases as mean spacing increases. Additional enhancements to D_{eff} due to meso-structural features are embodied in Colbeck's (1983b) temperature gradient enhancement factor, which takes into account grain elongation and, by extension, local spatial fluctuations in thermal conductivity that accelerate vapor diffusion in gaps between grains.

Heat and Mass Transfer by Advection and Convection

Knowledge about convective and advective processes in snow and verification of a working model are important for quantitative assessment of temperature-dependent gas flux measurements made in snow and firn. The theory of convective and diffusive heat and mass transfer through snow relies upon continuum mechanics, as is conventional for flow through porous media (e.g. Dullien, 1992; Bear, 1972). In this approach, transfer coefficients are employed to provide relationships between a scalar (e.g. temperature or vapor density) and the appropriate driving gradient. The transfer coefficients then must provide the link between phenomena at the microscale and the resulting (meso-scale) observed behavior.

Air flow through snow and firn is usually assumed to follow Darcy's law. The Darcy velocity or volumetric flux of the air, v_i, is given by

$$v_i = -\frac{k_{ij}}{\mu}\left[\frac{\partial P}{\partial x_j}\right] \tag{5}$$

where P is pressure, x_j is coordinate direction (j = 1 to 3), k_{ij} is the permeability tensor, and μ is the viscosity of the air. The permeability is the proportionality factor between the imposed pressure gradient and the resulting air flow velocity, and as such it must account for the effects that the crystal structure imposes on the blocking and drag of fluid flow through the media. Currently the best method to determine the permeability of snow or firn is to measure it directly in the field, either by measuring flow rate and pressure drop through a snow sample (and thus deducing it from equation (5)), or from acoustic measurements. While Shimizu (1970) fit an exponential curve to one type of snow to describe its permeability as a function of density and grain size, there is no reliable empirical formula today for determining permeability of an arbitrary snow sample from quantities easily measured in the field (e.g. density, grain size, and crystal classification). It is generally recognized that the microstructure controls the permeability, and attempts to model permeability by modeling pore structure are summarized in Arons and Colbeck (1995) and Dullien (1992). However, for arbitrary geometries (such as those imposed by snow and firn), there is no model that can successfully predict permeability based on realistic microstructure. A means of quantitatively evaluating preserved snow samples to predict permeability from observed microstructure is needed.

Heat and mass transport follow advection-diffusion relationships of the form, expanded from equation 5:

$$a_1 \frac{\partial C}{\partial t} + a_2 v_i \frac{\partial C}{\partial x_i} = \frac{\partial}{\partial x_i}\left[b_{ij} \frac{\partial T}{\partial x_j} \right] + S \tag{6}$$

Here C is the subject scalar (temperature, vapor density, or species concentration), t is time, a_1 is a capacitance (e.g. heat capacity), a_2 relates to porosity, b_{ij} is a diffusion tensor (e.g. thermal conductivity), and S is a source term.

In this equation the air flow velocity serves to transport temperature, vapor density, or species concentration through advection. The diffusion tensor b_{ij} is dependent upon the microstructure geometry. In practice, directionality is assumed negligible so that the tensor can be considered a scalar. Then the diffusion coefficient is usually assumed to be proportional to the diffusion coefficient in air, with the constant of proportionality being the ratio of porosity to tortuosity for non-reactive materials. There currently exists no operational model for determining diffusion coefficients from commonly made measurements on snow sections. Once again, currently the best way of determining a thermal or species diffusion coefficient for a particular snow type is to measure it, using conductivity probes or gas chambers. A means of quantitatively evaluating preserved snow samples to determine diffusion coefficients from observed microstructure would be useful.

The source term in equation (6) is especially important in mass transfer (vapor and chemical species transport), as it provides the link to metamorphic processes and chemical reaction rates. As discussed for vapor transport in Albert and McGilvary (1992), the source is dependent upon the local vapor or species gradient, the specific surface of the snow, and a transfer coefficient:

$$S = h_m a_s (C^0 - C) \tag{7}$$

where h_m is the mass transfer coefficient, a_s is the specific surface and the superscript 0 refers to a saturation value. While estimates of specific surface can be made from preserved snow samples, the mass transfer coefficients will be a function of microstructure geometry and flow rate, and must be determined experimentally.

Although snow and firn microstructure exerts a significant control on transfer phenomena, the field of quantitative microscopy is not yet sufficiently developed to provide useful input for determination of the governing parameters. We expect that progress in the field of quantitative microscopy to estimate important transfer parameters will require 3-dimensional reconstruction of serially sectioned samples.

Radiation Exchange Near the Snow Surface

Brandt and Warren (1993) coupled a spectral, two-stream, radiative transfer model of the behavior of solar radiation penetrating snow to Schlatter's (1972) model of snow conduction. They show sensitivity of light penetration to the wavelength and the grain size analogue of surface area-to-volume ratio. Generally, a large decrease in penetration depth is expected with increasing wavelength, while a slight increase in penetration depth should be expected in the visible and near-infrared with increasing grain size (lower surface area-to-volume ratio. Grenfell et al (1994) measured and modeled spectral reflectance of solar radiation by Antarctic snow. Through a series of modeling exercises they were able to show how subtle changes in the near-surface microstructure, expressed in a two-layer model representation, were required to explain measurements of spectral reflectance.

Insight to the feedback between near-surface depositional and post-depositional processes in firn and the transfer of energy by solar radiation can be gained by incorporating appropriate spectral extinction in snow with relations describing the other important processes as described above.

SUMMARY

In summary, the knowledge we have gained from analyzing transport in dry snow and arbitrary heterogeneous media is directly applicable to near surface firn. Transport properties depend on micro- and meso-scale structures. Although we have a strong understanding of the physics upon which metamorphic processes depend, we are still limited in our ability to relate those processes quantitatively to transport coefficients because of our limited ability to quantitatively characterize the geometry of snow and firn.

REFERENCES

Adams EE and Sato A (1993) Model for effective thermal conductivity of a dry snow cover composed of uniform ice spheres. Annals of Glaciology 18: 300-304.

Akitaya E (1974) Studies of depth hoar. Low Temperature Science. 26A: 1-67.

Albert MR and McGilvary WR (1992) Thermal effects due to air flow and vapor transport in dry snow. Journal of Glaciology 38(129): 273-281.

Albert MR (1993) Some numerical experiments on firn ventilation with heat transfer. Annals of Glaciology 18: 161-165.

Albert MR and Arons EM (1994) Transport processes in the Firn at Summit. Greenland (abstract). EOS. Transactions of the American Geophysical Union 75: 213.

Alley RB, Bolzan JF and Whillans IM (1982) Polar densification and grain growth. Annals of Glaciology 3:7-11.

Alley RB (1987a) Texture of polar firn for remote sensing. Annals of Glaciology 9: 1-4.

Alley RB (1987b) Firn densification by grain-boundary sliding: A first model. Journal de Physique, Supplement to 3(48) Colloque C1: 249-254.

Alley RB (1988) Concerning the deposition and diagenesis of strata in polar firn. Journal of Glaciology 34(118):283-290

Alley RB and Koci BR (1988) Ice-core analysis at Site A, Greenland: Preliminary results. Annals of Glaciology 10:1-4.

Alley RB, Saltzman ES, Cuffey KM and Fitzpatrick JJ (1990) Summertime formation of depth hoar in central Greenland. Geophysical Research Letters 17(12):2393-2396.

Anderson DL and Benson CS (1963) The densification and diagenesis of snow. in Proceedings: Ice and Snow; properties, processes and applications. Kingery W D ed. The M.I.T. Press. 391-411.

Arons EM (1994) Dependence of snow thermal conductivity on microstructure. Ph.D. thesis. Dartmouth College, Hanover, NH.

Arons EM and Colbeck SC (1995) Geometry of heat and mass transfer in dry snow: A review of theory and experiment, *in press*.

Auracher H (1978) Heat Transfer in Frost and Snow. in Proceedings of the Sixth International Heat Transfer Conference. Hemisphere Publications. Washington DC.

Bader H, Haefeli R, Bucker E, Neher J, Eckel O, and Thams C (1939) Der Schnee und seine Metamorphose. Beitrage zur Geologie der Schweiz, Geotech. Ser. Hydrol. 3 English Translation: SIPRE Trans. 14, 1954.

Bear J (1972) Dynamics of Fluids in Porous Media. American Elsevier. New York.

Benson CS (1962) Stratigraphic studies in the snow and firn of the Greenland ice sheet. SIPRE (now CRREL) Research Report 70.

Beasley JD and Torquato S (1989) New bounds on the permeability of a random array of spheres. Physics of Fluids A1: 199-207.

Brandt RE and SG Warren (1993) Solar-heating and temperature profiles in Antarctic snow and ice. Journal of Glaciology 39(131): 99-110.

Brinkman HC (1947) A calculation of the viscous force exerted by a flowing fluid on a dense swarm of particles. Applied Scientific Research A1: 27-34.

Chacho EF Jr and Johnson JB (1987) Air permeability of snow. EOS (Transactions of the AGU) 68: 1271.

Clarke GKC, Fisher DA and Waddington ED (1987) Windpumping: a potentially significant heat source in ice sheets, in: The physical basis of ice sheet modeling. IAHS Publication 170: 169-180.

Clarke GKC and Waddington ED (1991) A three-dimensional theory of wind pumping. Journal of Glaciology 37(125): 89-96.

Colbeck SC (1980) Thermodynamics of snow metamorphism due to variations in curvature. Journal of Glaciology 26(94):291-301.

Colbeck SC (1982) Growth of faceted crystals in a snow cover. CRREL Report 82-29.

Colbeck SC (1983a) Ice crystal morphology and growth rates at low supersaturations and high temperatures. Journal of Applied Physics 54: 2677-2682.

Colbeck SC (1983b) Theory of metamorphism of dry snow. Journal of Geophysical Research 88(C9):5475-5482.

Colbeck SC (1988) Ideas on the micrometeorology of surface hoar growth on snow. Boundary Layer Meteorology 44: 1-12.

Colbeck SC (1989a) Snow crystal growth with varying surface temperatures and radiation penetration. Journal of Glaciology 35:23-29.

Colbeck SC (1989b) Air movement in snow due to wind pumping. Journal of Glaciology 35:209-213.

Colbeck SC Akitaya E, Armstrong R, Gubler H, Lafeuille J, Lied K, McClung D, and Morris E (1990) The International Classification for Seasonal Snow on the Ground. The International Commission on Snow and Ice (of the International Association of Scientific Hydrology, (available from World Data Center. University of Colorado, Boulder CO).

Colbeck SC (1991) The layered character of snow covers. Reviews of Geophysics 29(1):81-96.

Colbeck SC (1993) The vapor diffusion coefficient for snow. Water Resources Research 29: 109-115.

Colbeck SC and McGilvary (1993) Dependence of vapor diffusivity, thermal conductivity, and electrical resistivity of snow on textural characteristics. US Army Cold Regions Research and Engineering Laboratory, Hanover, NH, ILIR Report.

Dansgaard W (1964) Stable isotopes in precipitation. Tellus 16: 436-468.

de Quervain MR (1972) Snow structure, heat and mass flux through snow. International Association of Scientific Hydrology Publication 107: 203-226.

Doi M (1976) A new variational approach to the diffusion and the flow problem in porous media. Journal of the Physical Society of Japan 40: 567-572.

Dozier J, Davis RE and Nolin A (1989) Reflectance and transmittance of snow at high spectral resolution. Proceedings IGARSS '89. IEEE No. 89CH2768-0: 662-664.

Dullien FAL (1992) Porous Media: Fluid Transport and Pore Structure. Academic Press. New York.

Ebinuma T and Maeno N (1985) Experimental studies on densification and pressure-sintering of ice. Annals of Glaciology 6:83-86.

Gow AJ (1965) On the accumulation and seasonal stratification of snow at the South Pole. Journal of Glaciology 5(4): 467-477.

Gow AJ (1968) Deep core studies of the accumulation and densification of snow at Byrd Station and Little America V, Antarctica. CRREL Research Report 197.

Gow AJ (1969) On the rates of growth of grains and crystals in South Polar firn. Journal of Glaciology 8(53): 241-252.

Gow AJ (1975) Time-temperature dependence of sintering in perennial isothermal snowpacks. International Association of Hyrological Sciences Publication 114. 25-41.

Grenfell T, Warren SG and Mullen A (1994) Reflection of solar radiation by the Antarctic snow surface at ultraviolet, visible and near-infrared wavelengths. Journal of Geophysical Research 99(D9): 18,669-18,684.

Gubler H (1985) Model for dry snow metamorphism by interparticle vapor flux. Journal of Geophysical Research 90(8): 8081-8092.

Hjelstroem SA (1890) Sur la conductibilite' de la neige. Ofvers. Kgl. Vetenskaps-Akad. Fvrh. 46: 669-676.

Izumi K and Huzioka T (1975) Studies of metamorphism and thermal conductivity of snow. 1, Low Temperature Sciences 33A: 91-102.

Johnsen SJ (1977) Stable isotope homogenization of polar firn and ice. International Association of Hyrological Sciences Publication 118. 210-219.

Jordan R (1991) A One-Dimensional Temperature Model for a Snow Cover: Technical Documentation for SNTHERM.89. Special Report 91-16. US Army Corps of Engineers Cold Regions Research and Engineering Laboratory, Hanover, NH.

Jouzel J, Merlivat L, Petit JR, Lorius C (1983) Climate information over the last century deduced from a detailed isotopic record in South Pole snow. Journal of Geophysical Research 88(C4): 2693-2703.

Koerner RM (1971) A stratigraphic method of determining the snow accumulation at Plateau Station, Antarctica and application to South Pole-Queen Maud Land traverse 2 1965-1966. In Cary AP ed. Antarctic Snow and Ice Studies II. American Geophysical Union. Washington DC Antarctic Research Series 16: 225-238.

Koplick J (1991) Homogenization and Effective Medium Methods for Transport in Disordered Granular Systems. in Physics of Granular Media, edited by D. Bideau and J. M. Dodds. Nova, NY.

Kumai M (1976) Identification of nuclei and concentrations of chemical species in snow crystal sampled at the south pole. Journal of Atmospheric Sciences 33: 833-841.

Lang RM, Leo BR, and Brown RL (1984) Observations on the growth process and strength characteristics of surface hoar. Proceedings of the International Snow Science Workshop, ISSW Workshop Committee, Aspen CO:188-195.

Langway CC (1967) Stratigraphic analysis of a deep ice core from Greenland. CRREL Research Report 77.

Maxwell JC (1891) A Treatise on Electricity and Magnetism. 3rd edition, Dover Publications Inc. New York. 506.

McCreesh CA, Ehrlich R and Crabtree SJ (1991) Petrography and reservoir physics. 2. Relating thin section porosity to capillary pressure and the association between pore types and throat size. American Association of Petroleum Geologists Bulletin 75: 1563-1578.

Meese DA, Gow AJ, Grootes P, Mayewski PA, Ram M, Stuiver M, Taylor KC, Waddington ED, and Zielinski GA (1994) The accumulation record from the GISP2 core as an indicator of climate change throughout the Holocene. Science 266:1680-1682.

Mellor M (1977) Engineering properties of snow. Journal of Glaciology 19(81): 15-66.

Nishimura H and Maeno N (1985) Studies on structures and physical properties of snow on Mizuho Plateau, Antarctica. Annals of Glaciology 6:105-107.

Palais JM, Whillans IM and Bull C (1982) Snow stratigraphic studies at Dome C, East Antarctica: An investigation of depositional and diagenetic processes. Annals of Glaciology 3: 239-242.

Perla R and CSL Ommanney (1985) Snow in strong or weak temperature gradients. Part I. Experiments and qualitative observations. Cold Regions Science and Technology. 11(1):23-35.

Pitman D and Zuckerman B (1967) Effective thermal conductivity of snow at -88, -27 and -5 C. Journal of Applied Physics 38: 2698-2699.

Schlatter TW (1972) The local surface energy balance and subsurface temperature regime in Antarctica. Journal of Applied Meteorology 11(10): 1048-1062.

Schwander J, Barnola JM, Andrie' C, Leuenberger M, Ludin A, Raynaud D and Stauffer B (1993) The age of the air in the firn and the ice at Summit, Greenland. Journal of Geophysical Research 98(D2):2831-2838.

Schwerdtfeger P (1963) Theoretical derivation of the thermal conductivity and diffusivity of snow. International Association of Hyrological Sciences. Publication 61: 75-81.

Shabtaie S and Bentley CR (1994) Unified theory of electrical conduction in firn and ice: Site percolation and conduction in snow and firn. Journal of Geophysical Research 99(B10): 19757-19769.

Shimizu H (1970) Air permeability of deposited snow. Contributions from the Institute of Low Temperature Science. (22): 1-32.

Sommerfeld RA, Judy C and I Friedman (1991) Isotopic changes during the formation of depth hoar in experimental snowpacks. in Stable Isotope Geochemistry: A Tribute to Samuel Epstein, edited by HP Taylor Jr., JR O'Neil and IR Kaplan. The Geochemical Society. 205-209.

Sowers T, Bender M, Raynaud D and Korotkevich YS (1992) $\delta^{15}N$ and N_2 in air trapped in polar ice: a tracer of gas transport in the firn and possible constraint on ice age-gas differences. Journal of Geophysical Research 97(D14):15,683-15,697.

Stauffer B, Schwander J and Oeschger H (1985) Enclosure of air during metamorphosis of dry firn to ice. Annals of Glaciology 6:108-112.

Sturm M and Johnson JB (1991) Natural convection in the subarctic snow cover. Journal of Geophysical Researeh 96(B7): 657-668.

Sturm M and Johnson JB (1992) Thermal conductivity measurements of depth hoar. Journal of Geophysical Research 97(B2): 2129-2139.

Wiscombe WJ and Warren SG (1980a) A model for the spectral albedo of snow. I. Pure snow. Journal of Atmospheric Science 37(12) 2712-2733.

Wiscombe WJ and Warren SG (1980b) A model for the spectral albedo of snow. II. Snow containing atmospheric aerosols. Journal of Atmospheric Science 37(12): 2734-2745.

Warren SG (1982) Optical properties of snow. Reviews of Geophysics and Space Physics. 20(1), 67-69.

Waterhouse RW (1962) Analysis of data from a snow profile. Research Report 90. US Army Corps of Engineers Cold Regions Research and Engineering Laboratory, Hanover, NH.

Whillans IM and Grootes PM (1985) Isotopic diffusion in snow and firn. Journal of Geophysical Research 90(D2): 3910-3918.

Wiener O. (1912) Die theorie des mischkörpers fü:r das feld der stationären strömung. Der Abhandlungen der Mathematisch-Physischen Klasse der Königl. Sächsischen Gesellschaft der Wissenschaften VI.

Yosida Z et al (1955) Physical studies on deposited snow. I. Thermal properties. Contributions, Institute of Low Temperature Science 7:19-74.

THE EFFECTS OF SNOW VENTILATION ON CHEMICAL CONCENTRATIONS

E.D. Waddington and J. Cunningham
Geophysics Program Box 351650

S.L. Harder
Department of Chemistry Box 351700
University of Washington
Seattle WA 98195
U.S.A.

INTRODUCTION

Ice cores contain a suite of chemical tracers which, when correctly interpreted, can reveal properties of the paleo-atmosphere and the climate system. These chemical constituents have been subjected to a series of physical and chemical processes at each step on their journey from the remote paleo-atmosphere into the ice (see Neftel, this volume; Waddington, this volume). Current research in polar snow chemistry focuses on the *transfer functions*, *i.e.* the chemical and physical processes that control the concentrations and fluxes of these atmospheric components at each step. Understanding the transfer functions is essential to the ice core - paleoclimate *Inverse Problem*, *i.e.* the derivation of the paleoclimate from the ice core data (Waddington, this volume). In this paper we focus mainly on one step in the Forward Problem; this is the transfer of dry-deposited chemical tracers from the local atmosphere into the snow by air flow through the snow (*windpumping*). Nearly two decades ago in a seminal paper entitled *The filtering effect of snow*, Gjessing (1977) summarized this topic as follows:

> "*Chemical analyses of snow and ice cores are frequently used to determine the chemical composition of the precipitation and dry deposition. Snow contains varying quantities of air, and the greater part of this air volume will constantly be replaced due to local changes of air pressure and strong winds above the surface. These ventilation processes are dependent on the snow structure and the snow surface roughness. It is reasonable to assume that the snow layers will act as a mechanical filter for aerosols so that the air leaving the snow volume is 'clean'. This air exchange will lead to an increase in the ion content of the snow.*"

In this paper we outline the current theoretical basis for calculating air flow in polar firn. (1) We first review the role of snow ventilation in the capture of aerosols from Gjessing (1977) to the present. (2) We then estimate the transport resistance for windpumping to assess whether windpumping is a rate-limiting factor in aerosol deposition, concluding that a realistic wind pumping model of aerosol deposition is necessary to reconstruct paleoclimate conditions from ice core aerosol records. (3) We show that ice core records for reversibly deposited species that have undergone post-depositional changes without achieving equilibrium with the atmosphere prior to deep burial will be difficult to interpret; ventilation is instrumental in helping reversibly deposited species reach equilibrium. (4) Finally, we speculate on the relative importance of ventilation for the concentrations of irreversibly and reversibly deposited species in Greenland and Antarctica now and at the Last Glacial Maximum (LGM), and identify areas of windpumping theory that require further research.

NATO ASI Series, Vol. I 43
Chemical Exchange Between the Atmosphere and Polar Snow
Edited by Eric W. Wolff and Roger C. Bales

EFFECTIVE SNOW SURFACE AREA

The rates of chemical or physical air-snow exchange processes depend on the rate at which the air carrying the chemical constituents can come into close proximity with a surface. Therefore the high *porosity* ϕ (*i.e.* the fractional void space) of polar snow relative to other natural surfaces can become important. If we approximate polar firn by loose-packed ice spheres with a radius of r = 0.1 mm (Figure 1), and even if only as much as the top 10 cm is accessible to the atmosphere, then the surface area A of ice available for deposition within a 1 square meter plot is given by

$$A = 5{\times}10^3 \times 5{\times}10^3 \times 5{\times}10^2 \text{ spheres} \times 4\pi r^2 = 500\ \pi\ \text{m}^2 \approx 1500\ \text{m}^2 \gg 1\ \text{m}^2\ . \qquad (1)$$

This area is more than three orders of magnitude larger than the area of a smooth impenetrable surface.

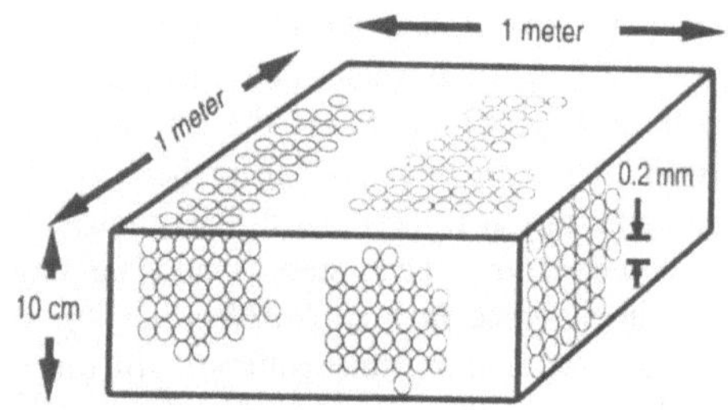

FIGURE 1: If polar snow is approximated by a collection of loose-packed ice spheres each 0.2 mm in diameter, then the surface area of ice grains in the upper 10 cm under a 1 square meter plot exceeds 1500 m^2. Much of this surface area may be accessible to airborne chemical species through wind-driven ventilation.

We must now consider whether this large ice surface area is accessible to the atmosphere for deposition or exchange of chemical species. The atmosphere carries aerosol particles that effectively undergo irreversible deposition, *i.e.* once deposition occurs, the probability of reentrainment is small. A significant fraction of this aerosol can be transported to the ice sheet by dry deposition rather than by wet deposition associated with precipitation (Ibrahim et al., 1983; Davidson, 1989) or with fog (Davidson et al., this volume). If air circulates freely in the snow, it is possible to transport a large volume of "dirty" particle-laden air into the snow where it can be cleansed by dry deposition to the large sub-surface area, as suggested by Figure 2A. This process can be important for deposition of non-seasalt (nss) sulphate aerosols (Cunningham and Waddington, 1993; Harder et al., 1991, in press) and possibly other species. Conversely, reversibly deposited components, or chemical compounds released by reactions in the snow, may be carried out of the snow by air flow, as suggested by Figure 2B. Finally, if a large sub-surface area is available to the atmosphere, it can serve as a reservoir to hold reversibly deposited chemical species whose atmospheric concentrations may have significant diurnal or seasonal swings. Averaged over longer time periods, a near-equilibrium condition could be established (*e.g.* Bales and Choi, this volume) in which comparable fluxes exist both into and out of the snow, as suggested by Figure 2C.

FIGURE 2: When snow is ventilated (arrows), (A) aerosols and solid particles in the atmosphere ("Dirty" Air) may be carried into the snow, where they irreversibly stick to the ice surfaces. Only "Clean" air leaves the snow, which has acted as a filter.
(B) Reentrained chemical species and compounds released by reactions in the snow can be carried out of the snow by initially "Clean" air.
(C) Volatile species can be carried both ways across the air-snow interface, and equilibrium may be established.

PERMEABILITY

The rate of air flow in the snow is controlled by two factors, spatial or temporal gradients in air pressure that drive flow, and the *permeability* of the snow. Intrinsic permeability K, the ability of a porous medium to allow fluid to flow through it, is defined by Darcy's Law

$$\vec{Q} = -K \frac{1}{\mu} \nabla\Psi \qquad \Psi = P_{air} - \rho_{air}\, gz \tag{2}$$

through its relationship with volumetric fluid flux $\vec{Q}$ ($m^3m^{-2}s$), fluid viscosity μ, and Ψ, the potential energy per unit fluid volume, also known as the *hydraulic potential.* $\nabla\Psi$ comprises the air pressure gradient ∇P_{air} and the gravitational potential gradient where ρ_{air} is the air density, g is gravity and z is the vertical coordinate measured positive downward. The viscosity μ of air is 1.6×10^{-5} Pa s at $-15°C$. The permeability K is a property of the porous medium alone; it does not depend on the viscosity of the fluid. A typical value of K in Antarctic snow is 10^{-9} m^2, although values can easily vary by an order of magnitude or more. Darcy's Law (2) is valid when the flow within the porous material is linear and laminar; this is generally the case for polar snow (Cunningham and Waddington, 1993).

The permeability of snow and polar firn has been studied for more than five decades (Bader, 1939; Bender, 1957; Yen et al., 1963; Shimizu, 1970; Yen, 1988; Sommerfeld and Rocchio, 1993). Significant research contributions have been made by the U.S. Army Cold Regions Research and Engineering Laboratory (CRREL), where research on snow ventilation continues (Colbeck, 1989, 1991; Albert, 1992; Albert and McGilvary, 1992; Albert and Hardy, 1995; Albert, in press; Albert et al., this issue).

Relationship between Air Flow and Dry Deposition Velocity

The dry deposition rate of irreversibly deposited aerosols at height z above the snow is often expressed (Davidson, 1989) as a *dry deposition velocity* $V_d(z)$ defined by

$$V_d(z) = \frac{F_d}{C_a(z)} \tag{3}$$

where F_d is the flux or deposition rate per unit area (g $cm^{-2}s^{-1}$) from dry deposition, and C_a is the airborne concentration of the particle species (g cm^{-3}). The rate at which air containing a uniform concentration C_a of passively advected particles must be moved through a 100% efficient filter to deposit particles on the filter at the rate F_d can be interpreted as a deposition velocity associated with ventilation. Rates of air flow into polar snow calculated below are comparable to deposition velocities reported for many chemical species reported in the literature for Greenland (Davidson, 1989).

Measurement of Snow Permeability

Since snow permeability is an extremely important parameter for ventilation transfer functions, accurate permeability measurements must be made near ice core sites. When a known pressure drop $\Delta\Psi$ is used to force air through a snow sample of thickness h at a measured flow rate Q, $\nabla\Psi$ can be approximated by $\Delta\Psi/h$, and equation (2) can be solved for K. Measurement of snow permeability presents two challenges. First, the air flow must remain one-dimensional through the sample so that the pressure gradient $\nabla\Psi$ can be approximated by $\Delta\Psi/h$, and second, air must not leak around the edges of the samples. Shimizu (1970) described an instrument that solved these problems by using a concentric double-walled sample holder. The inner chamber, through which the measured air flux Q passes, contacts the surface of the sample, but does not penetrate it. Pressure feedback control between the inner and outer walls forces the air flow through the sample into the inner chamber to be one-dimensional. Within the snow sample, "virtual walls" are imposed by the symmetry of the pressure field rather than by physical impermeable surfaces.

Factors affecting Permeability

Permeability tends to increase with porosity; however, even at constant density, data show that grain size, age, site windiness, and degree and type of post-depositional metamorphism can cause permeability to vary by a factor of ten (Shimizu, 1970). Occluded bubbles contribute to porosity, but do not enhance air flow. At the other extreme, thin ice crusts may have negligible influence on porosity, but can reduce permeability significantly by blocking air flow. Permeability generally decreases with depth when averaged over scales of several meters. However, within the upper meter or more, permeability has been observed to first increase with depth (Albert et al., this volume; Albert and Hardy, 1995; Waddington, unpublished data, Agassiz Ice Cap). Two processes contribute to this. First, the surface tends to be "self-armoring"; during high winds, small saltating snow grains tend to impact and stick in the surface pore spaces, in a process analogous to aerosol impaction (Davidson, 1989). These ice "aerosols" tend to clog the air passages in the upper few centimeters, creating a hard surface *wind slab* which reduces the near-surface permeability. Second, metamorphic processes in the snow (Colbeck, 1983, 1991; Alley at al., 1990) lead to the formation of depth hoar whose large open pore spaces provide enhanced permeability below the surface.

In natural snow, K is often neither homogeneous nor isotropic; in the most general situations, K in (1) can be replaced by a spatially-varying tensor K_{ij} (Albert et al., this issue; Albert and Hardy, 1995). Permeability approaches zero at the depth of pore closeoff. This has important implications for mixing of gases deep in the firn, and the age difference between ice and occluded gases in bubbles (Schwander, 1989). However, we treat K as a constant in this paper because many features of snow ventilation relevant to chemical deposition can be understood in the context of uniform permeability within a few meters of the snow surface. For detailed case studies, analysis of field experiments, and incorporation of layering effects such as wind slabs, a more sophisticated treatment of permeability is desirable and necessary. Finite element models of ventilation and chemical transport including thermodynamics have been developed by M. Albert at CRREL (Albert, 1992; Albert and McGilvary, 1992; Albert and Hardy, 1995; Albert et al., this volume; Albert, in press).

PHYSICAL DESCRIPTION OF FIRN VENTILATION

Air flow within permeable snow can be driven either by internal buoyancy forces (natural convection), or by externally imposed pressure fields (forced convection). Natural convection has been observed only in highly permeable seasonal snow with a large temperature gradient (Sturm and Johnson, 1991) and is not expected to occur on polar ice sheets. We restrict our discussion to forced convection.

A complete treatment of firn ventilation would include calculation of air flow and associated advective transport of chemical species in the atmosphere above the snow as well as within the snow. However, air will flow through snow only if driven by pressure gradients at the air-snow surface, and chemical species will enter the snow only if the atmosphere puts them at the snow surface. Therefore, the snow ventilation problem can be simplified immensely if we can (a) estimate the atmospheric parameters controlling air pressure and chemical concentrations at the air-snow interface, then (b) calculate air flow only within the snow. This simplification comes at a price; uncertainty associated with these parameterizations of the air-snow interface constitutes a major source of uncertainty in the calculated chemical changes caused by windpumping.

Since pressure variations in the snow are much smaller than the average air pressure P_0, the pressure distribution in the snow can be described (Clarke et al., 1987; Colbeck, 1989; Cunningham and Waddington, 1993) by a diffusion equation for p(x,y,z,t), the pressure perturbations from the average pressure P_0. This equation,

$$\frac{\partial p(x,y,z,t)}{\partial t} = \kappa_p \nabla^2 p(x,y,z,t) \qquad \kappa_p = \frac{K\,P_0}{\mu\,\phi} \tag{4}$$

is derived from Darcy's Law (2) and the mass conservation equation for air. The diffusivity κ_p for pressure perturbations is a function of snow permeability K, average atmospheric pressure P_0, air viscosity μ, and the snow porosity ϕ. A typical pressure diffusivity in snow is $\kappa_p \approx 10\ m^2\ s^{-1}$. We can now use $\nabla\Psi$ and $\nabla p(x,y,z,t)$ interchangeably, because in stagnant (unperturbed) air, $\nabla P_0 = \rho_{air}\, g$, and $\nabla\Psi = 0$.

Snow ventilation in nature can depend on complicated material properties and may be subject to complicated geometrical and temporal boundary conditions. However, for the purposes of this review, the fundamental concepts of snow ventilation can be illustrated by solutions to simple analytical problems which are free from obscuring detail associated with specific sites or situations. We investigate various features of firn ventilation by solving equation (4) with different sets of boundary conditions, *i.e.* with various combinations of temporal and spatial properties for the driving pressure at the snow surface, as shown in Figure 3. Temporally, the pressure may be steady (3A), vary sinusoidally with time t (3B), or have an arbitrary and complicated history (3C). Spatially, the pressure may be uniform (3D), vary in one horizontal dimension x (3E), or vary with both horizontal dimensions x and y (3F). Each of these simple models demonstrates important features of snow ventilation, and the results form the framework of this review paper.

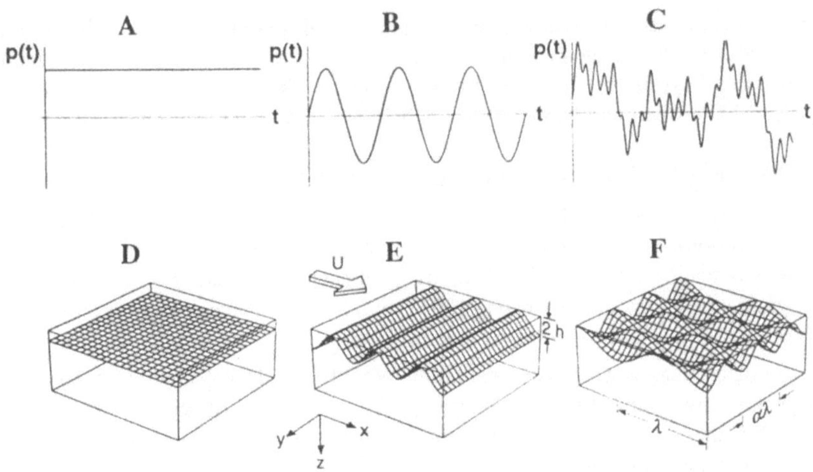

FIGURE 3: Forms of pressure variation at a snow surface in time and space used in analytical ventilation models. The pressure may (A) be steady, (B) vary sinusoidally with time t, or (C) have a more complicated time dependence. The snow microtopography may (D) be flat, (E) vary sinusoidally in x, the wind direction, or (F) undulate in both horizontal directions x and y. The surface pressure p_0 can (D) be spatially uniform, (E) vary with x only, or (F) vary with both x and y.

Pressure Oscillations in an Infinitely Thick Permeable Layer

If (a) the permeability is constant at all depths $0 < z < \infty$, and (b) the pressure in the snow is driven at the upper surface $z = 0$ by a spatially uniform pressure (Figure 3D) oscillating sinusoidally (Figure 3B) at frequency ω_0, *i.e.*

$$p(0,t) = p_0 \sin[\omega_0 t] \tag{5}$$

then the spatially simple boundary condition restricts the pressure solution to vary only with time t and depth z, and $\nabla^2 p(x,y,z,t)$ can be replaced by $\partial^2 p(z,t)/\partial z^2$. An analytical solution (Carslaw and Jaeger, 1959, p. 65) is available for this problem;

$$p(z,t) = p_0\, A(z) \sin[\omega_0 t + \theta(z)] \tag{6}$$

where the amplitude A(z) of the pressure wave decays with depth z according to

$$A(z) = A_\infty(z) = e^{-z/H} \qquad H(\omega_0) = \sqrt{\frac{2\kappa_p}{\omega_0}}\,. \tag{7}$$

$H(\omega_0)$ is a natural length scale for diffusion of pressure waves with frequency ω_0. The envelope A(z) is shown by heavy solid curves in Figure 4A. The pressure oscillates with depth within the envelope, with phase

$$\theta(z) = \theta_\infty(z) = -z/H \tag{8}$$

which is characteristic of a wave propagating downward at constant velocity $\omega_0 H$. Dashed lines in Figure 4A show the full solution (6) at time intervals of $t = \pi/4\omega_0$. Only a single oscillation is seen before the amplitude becomes small, since the wavelength $2\pi H$ exceeds the e-folding depth H. The pressure distribution throughout the firn is relatively insensitive to the permeability structure at depth z exceeding several H, because the amplitude of the pressure signal is always small there. Therefore this solution can be a useful guide when the forcing frequency ω_0 is relatively high, or the firn can be approximated by a layer that is uniformly permeable to a relatively great depth z/H.

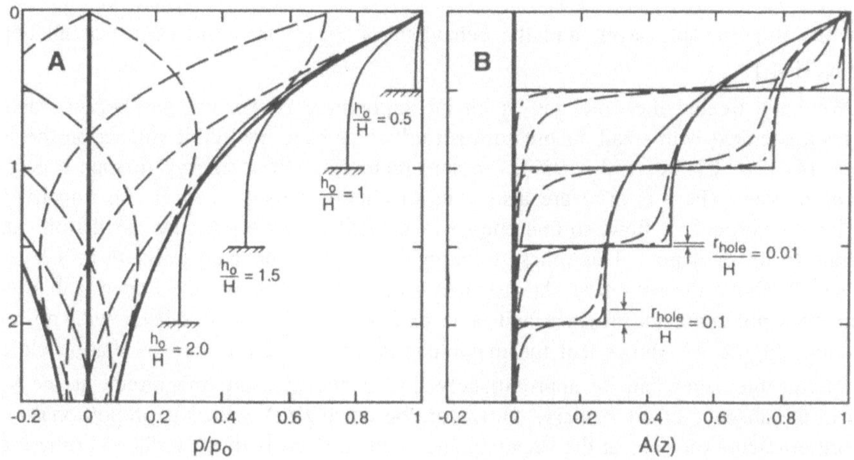

FIGURE 4: (A) Dashed lines show 1 cycle of pressure oscillations $p(z,t)/p_0$ ((6), (7) and (8)) in infinitely-thick homogeneous permeable layer responding to sinusoidal surface pressure forcing (5). Depths are scaled to the natural diffusion depth H (7). Heavy solid lines show the envelope $A_\infty(z)$. Lighter solid lines show $A_{h_0}(z)$ (9) when a single impermeable layer is located at a series of depths h_0/H.
(B) Each solid line shows the envelope (9) for pressure above a single impermeable layer. Broken lines show the pressure envelope on a vertical line when a radial pressure solution (18) for flow through a hole on the vertical line in the otherwise-impermeable layer is added. The hole has radius $r_{hole}/H = 0.1$ (dashes) or $r_{hole}/H = 0.01$ (dash-dot line).

Pressure Oscillations in a Thin Permeable Layer

A very useful, related, but more complicated ventilation problem can also be solved analytically. If (a) an impermeable barrier at depth h_0 prevents deeper pressure penetration, and (b) air pressure is driven by the same spatially uniform sinusoidal surface pressure oscillation (5), then the solution (Carslaw and Jaeger, 1959, p. 105; Colbeck, 1989) again varies only with depth and time, and again has the form (6). The amplitude of the pressure wave decays with depth z according to

$$A(z) = A_{h_0}(z) = \left[\frac{\cosh[2(h_0/H - z/H)] + \cos[2(h_0/H - z/H)]}{\cosh[2h_0/H] + \cos[2h_0/H]} \right]^{\frac{1}{2}} \tag{9}$$

where $H(\omega_0)$ is the same natural length scale given in (7). The phase of the wave changes with z according to

$$\theta(z) = \theta_{h_0}(z) = \arg\left[\frac{\cosh[(1+i)\,(h_0/H - z/H)]}{\cosh[(1+i)\,h_0/H]} \right] \tag{10}$$

where arg denotes the argument of a complex number. The amplitude $A_{h_0}(z)$ (9) and the phase $\theta_{h_0}(z)$ (10) take more complicated forms than their counterparts (7) and (8) because the air now senses the presence of the impermeable bottom. Figure 4A also shows the amplitude $A_{h_0}(z)$ (9) when a single impermeable boundary is placed at a series of depths h_0/H. When the impermeable layer is very far from the surface, *i.e.* the nondimensional number $h_0/H \rightarrow \infty$, this solution $A_{h_0}(z)$ looks much like the solution $A_\infty(z)$ for a permeable half-space. However, $A_{h_0}(z)$ (9) deviates significantly from $A_\infty(z)$ in a zone of thickness H above the impermeable layer, and the behaviour of $A_{h_0}(z)$ is unlike $A_\infty(z)$ at all depths when $h_0/H < 1$.

We have treated the pore gas as an incompressible isothermal gas because density changes associated with small air pressure perturbations have negligible impact on the solutions to (4) (*e.g.* Clarke et al., 1987). Pressure represents stored energy in such a gas; the units of pressure ($Pa \equiv N\ m^{-2}$) are also units of energy density ($J\ m^{-3}$). An impermeable layer blocks vertical air flow, so that energy associated with the pressure oscillation cannot penetrate to lower depths. This blocked energy is stored as elevated pressure in a layer of thickness H above the boundary. At distances greater than H above the impermeable layer, the air does not sense its presence and behaves like the solution in a layer with no lower boundary. Figure 4A shows that the magnitude $A_{h_0}(z)$ of the pressure oscillation close to an impermeable layer can be approximately double the pressure magnitude at the same depth in the absence of a boundary. However, the air flux (2), which is proportional to the pressure gradient, vanishes at the impermeable boundary and is much reduced in the region of thickness H above the impermeable layer relative to air flux in the infinitely-deep layer solution.

This solution is a useful guide when either (a) the frequency ω_0 is relatively low, (b) the impermeable boundary is located at a shallow depth h_0/H, or (c) a solution is needed near an impermeable lower boundary.

Hole in an Impermeable Layer

The presence of an impermeable layer amplifies the pressure variations within a depth range of H above the layer (Figure 4A), but also reduces air flow. However, air could be pumped vigorously through a small hole with radius r_{hole} in an impermeable layer, driven by the large difference between the oscillating pressure (6) above the layer and zero pressure in the zone below.

We now seek the perturbation to the pressure above and below a permeable layer due to the presence of a small hole. We assume that the hole is small, *i.e.* $r_{hole}/H \ll 1$. This assumption, combined with the boundary condition $\partial p/\partial z = 0$ at the impermeable interface, ensures that the solution (9) and (10) in the layer does not vary significantly on the scale of the hole, so that the hole can be treated as a point. Then the pressure solution due to the hole should have hemispherical symmetry both above and below the layer, so that the boundary condition $\partial p/\partial z = 0$ on the impermeable layer is still satisfied. We assume further that the hole is small relative to the layer thickness, *i.e.* $r_{hole}/h_0 \ll 1$; the pressure perturbation due to the hole should have limited spatial extent so that the boundary condition (5) at the air-snow surface can still be satisfied. With these assumptions, the pressure in the hole is given by the average of the pressure $p(h_0, t)$ (6), (9) and (10) above the layer and the pressure $p \equiv 0$ below the layer, *i.e.*

$$p_{hole}(t) = \frac{1}{2}\, p_0\, A_{h_0}(h_0)\ \sin[\omega_0 t + \theta_{h_0}(h_0)]\ . \qquad (11)$$

The perturbation solutions above and below the layer should be equal in magnitude but opposite in sign, because the hole "bleeds off" the pressure above the impermeable layer in order to generate the pressure below.

We will find the pressure field beneath the perforated layer by solving the diffusion equation (4) with an oscillating point pressure source in a permeable medium of infinite extent without an impermeable layer. The vector $\mathbf{r} = (r, \theta, \phi)$ expresses a position in spherical coordinates. We locate the origin $\mathbf{r} = 0$ in the hole, and use the lower half of the spherically-symmetric pressure solution. The boundary condition $\partial p/\partial z = 0$ along the impermeable interface is automatically satisfied by a radially-symmetric pressure solution that varies only with r. We will find the perturbation to the pressure above the impermeable interface by the same technique, with the source replaced by a sink.

When an impulsive point source $S_0(\mathbf{r}-\mathbf{r}_0, t-t_0)$ exists at position $\mathbf{r} = \mathbf{r}_0$ at time $t = t_0$, equation (4) becomes

$$\frac{\partial G[(\mathbf{r}-\mathbf{r}_0), (t-t_0)]}{\partial t} - \kappa_p \nabla^2 G[(\mathbf{r}-\mathbf{r}_0), (t-t_0)] = S_0(\mathbf{r}-\mathbf{r}_0, t-t_0) = \boldsymbol{\delta}(\mathbf{r}-\mathbf{r}_0)\, \boldsymbol{\delta}(t) \tag{12}$$

where $\boldsymbol{\delta}$ is the *Kronecker delta distribution*. The solution for this impulsive source in an unbounded continuum (Morse and Feshbach, 1953, p. 861) is

$$G[(\mathbf{r}-\mathbf{r}_0), (t-t_0)] = \frac{1}{2\sqrt{\pi}\, \kappa_p^{3/2}}\, e^{-(r-r_0)^2/[4\kappa_p (t-t_0)]}\, \frac{U(t-t_0)}{(t-t_0)^{3/2}} \,. \tag{13}$$

The *Heaviside step function* $U(t-t_0)$ prevents a response prior to the time t_0 when the impulse occurs. $G[(\mathbf{r}-\mathbf{r}_0), (t-t_0)]$ is the *Green's Function for diffusion*; it shows how a unit amount of any physical quantity introduced into the system at the point $\mathbf{r} = \mathbf{r}_0$ at time t_0 will diffuse throughout the system to other positions $\mathbf{r}$ at later times t. The solution $p(\mathbf{r}, t)$ at point $\mathbf{r}$ at time t for any more complicated spatial and temporal distribution of sources $S(\mathbf{r}_0, t_0)$ can be found by summing up the Green's Functions for all the sources at positions $\mathbf{r}_0$ and times t_0, *i.e.*

$$p(r, t) = \int_{-\infty}^{t} \iiint_{\mathbf{V}_0} G[(\mathbf{r}-\mathbf{r}_0), (t-t_0)]\, S(\mathbf{r}_0, t_0)\, d\mathbf{V}_0\, dt_0 \tag{14}$$

where $\mathbf{V}_0$ represents the entire volume, and the volume element $d\mathbf{V}_0 = r^2\sin(\theta)dr\, d\theta\, d\phi$ in spherical coordinates.

Energy flows through the hole into and out of the region below the impermeable layer; we approximate the effect below the interface by placing an oscillating power-density source distribution $S(\mathbf{r}, t)$ in the hole. $S(\mathbf{r}, t)$ is given by

$$S(\mathbf{r}, t) = \omega_0\, p_{hole}(t)\, V_{hole}\, \boldsymbol{\delta}(\mathbf{r}_0) \tag{15}$$

where V_{hole} is a characteristic volume energized by the pressure $p_{hole}(t)$. V_{hole} will be determined by the hole radius r_{hole}. The solution for this source is (see *e.g.* Morse and Feshbach, 1953, Chapter 7.4)

$$p(r, t) = 4\pi \frac{p_0\, A_{h_0}(h_0)\, V_{hole}}{H^2\, r}\, e^{-r/H} \cos[\omega_0 t - r/H + \theta_{h_0}(h_0)\,] \,. \tag{16}$$

A small sphere with radius $r_{sphere} = \sqrt{2}\, r_{hole}$ has a surface area on its lower hemisphere equal to the area of the flat hole. By specifying that the average pressure inside this sphere equals the pressure $p_1(t)$ (11) in the hole, we determine V_{hole} to be

$$V_{hole} = \frac{1}{12\pi} r_{hole} H^2 . \tag{17}$$

The solution for pressure associated with the hole is given by`

$$p(r, t) = \mp p_0 A_{h_0}(h_0) \frac{1}{3\sqrt{2}} \frac{r_{hole}}{r} e^{-r/H} \sin[\omega_0 t \pm r/H + \theta_{h_0}(h_0)] . \tag{18}$$

The solution below the impermeable layer is given when the lower signs are used. The radial solution for the perturbation above the hole is found by placing a point sink of equal and opposite magnitude to (15) in the hole, giving the solution (18) using the upper signs. The pressure solution above the impermeable layer is given by adding this solution (18) in the upper hemisphere to the solution (6), (9) and (10) for pressure in the permeable layer.

The solution (18) superficially resembles (6), (7) and (8). It consists of a travelling sine (now spherical) pressure wave which is damped by an exponential decay factor. However, a new geometrical spreading factor r_{hole}/r damps the pressure wave near the hole more severely than the exponential factor. The scale length for attenuation due to geometrical spreading near the hole is given by the hole radius $r_{hole} \ll H$, *i.e.* the magnitude of the pressure oscillation drops to very low values within a hemispherical shell with radius equal to several times the hole diameter.

Figure 4B shows the solution (18) for holes $r_{hole}/H = 0.01$ and $r_{hole}/H = 0.1$. The solutions above and below the hole are each offset vertically by the small amount $\sqrt{2}r_{hole}/3$ so that the solutions match at $z = h_0$. The wave nature of the solution (18) is not important because the pressure is attenuated in a distance r_{hole} which is very short relative to the wavelength $\lambda = 2\pi H$. Air simply blows in and out of the hole in phase with the oscillating pressure $p_{hole}(t)$ (11). The pressure gradient near the hole (Figure 4B) can be steeper than the gradient elsewhere in the permeable layer (*e.g.* compare to dashed curves in Figure 4A). The strongest air flow in the snowpack probably occurs above, through, and below small holes because air flux (2) is proportional to the gradient of the pressure field.

This solution is appropriate only when $r_{hole}/H \ll 1$. Even these small holes should be separated by at least several times the hole diameter before they can be analyzed in isolation. Air flow through a large hole disturbs the background solution for the impermeable layer in more complicated ways. The effect of the hole can no longer be decoupled from the solution for the permeable layer, and the boundary conditions at the air-snow surface (5) and at the impermeable layer ($\partial p/\partial z = 0$) may no longer be naturally satisfied. For example, Figure 4B shows that the solution (18) above an interface at $h_0/H = 0.5$ for $r_{hole}/H = 0.1$ has a significant amplitude at the surface $z = 0$, so that the boundary condition (5) is not satisfied. In this case, air flow through that hole should sense the presence of the snow surface, but our approximate solution (18) does not include this interaction.

Steady Air Flow

So far we have solved equation (4) for problems controlled by spatially simple boundary conditions (Figure 3D) that reduce the ∇^2 operator to a one-dimensional second derivative, while retaining time-dependence. However, for problems where the boundary conditions impose steady air flow (Figure 3A), we can reduce the time derivative $\partial p/\partial t$ to zero, leaving a Laplace equation *i.e.*

$$\nabla^2 p(x,y,z) = 0 . \tag{19}$$

This equation can be solved for spatially-complicated distributions of steady pressure at the snow surface (Figure 3E and 3F).

Subsurface Air Velocity and Displacement

Once the pressure distribution p(x,y,z,t) is known throughout the snowpack from any solution to (4), then the volumetric flux of air $\vec{Q}(x,y,z,t)$ ($m^3m^{-2}s^{-1}$) can be found from (2) by setting $\nabla\Psi = \nabla p(x,y,z,t)$. The air parcel velocity is then found by dividing by the porosity, *i.e.* $\vec{v}_{air}(z,t) = \vec{Q}(x,y,z,t)/\phi$. For example, when air flow in a thick permeable layer is driven by a sinusoidal surface-pressure oscillation (5) giving the pressure solution (7) and (8), then the vertical velocity of an air parcel becomes

$$v_{air}(z,t) = \frac{\sqrt{2}Kp_0}{\phi\mu H}\, e^{-z/H} \sin[\omega_0 t + \theta - \pi/4]\ . \qquad (20)$$

The air velocity decreases exponentially with depth, controlled by the length scale $H(\omega_0)$ (7). Air in the firn moves at an average speed of

$$\bar{v}_{air}(z) = \frac{\omega_0}{2\pi} \int_{(\pi/4-\theta)/\omega_0}^{(5\pi/4-\theta)/\omega_0} v_{air}(z,\tau)\, d\tau = \frac{2\sqrt{2}Kp_0}{\pi\phi\mu H}\, e^{-z/H} \qquad (21)$$

The velocity is higher for higher frequencies because $H(\omega_0)$ is proportional to $\omega_0^{-1/2}$.

Finally, air parcel trajectories can be found by integration through the velocity field $\vec{v}(x,y,z,t)$ within the snow. For example, the maximum vertical air displacement l(z) during the time $t = \pi/\omega_0$ for one half of a sinusoidal cycle (12) is given by

$$l(z) = \int_{(\pi/4-\theta)/\omega_0}^{(5\pi/4-\theta)/\omega_0} v_{air}(z,\tau)\, d\tau = \frac{2\sqrt{2}Kp_0}{\omega_0\phi\mu H} e^{-z/H} = \frac{\pi}{\omega_0}\, \bar{v}_{air}(z) \qquad (22)$$

when the total displacement l(z) is small relative to H. The result (22) shows that the air displacement l(z) decreases as frequency ω_0 rises. The reduced time available to move during each high frequency cycle more than offsets the higher average speed (21).

NATURAL SOURCES OF WINDPUMPING

Colbeck (1989) identified three sources of pressure gradients that can cause windpumping; these are barometric pressure variations, wind turbulence, and steady wind flow over topography. We now examine each of these sources using the simple analytical solutions expressed by (4) through (22) in order to assess the importance of each process to air - snow chemical transfer functions.

Slow Pressure Changes - Barometric Pumping

Synoptic weather systems moving over an ice cap can cause changes in air pressure on the order of 5% over periods of several days. Rising pressure forces air into the snow, and falling pressure draws air out. The surface pressure changes are uniform over large regions, so Figures 3B and 3D provide an appropriate model. Colbeck (1989) used the solution (6), (9) and (10) to show that for slow pressure changes, the firn column simply "breathes" as a unit. The actual value and distribution of the permeability is not important, as long as the diffusion length scale H is much greater than the firn thickness h_0, *i.e.* $h_0/H \ll 1$. Figure 4B shows that the amplitude $A_{h_0}(z)$ is nearly constant with depth when $h_0/H = 0.5$. For barometric pumping, $h_0/H \approx 0.01$, so $A_{h_0}(z) \approx 1$ and $\theta_{h_0}(z) \approx 0$. We can simply estimate the air flow through the air - snow interface from compression of the firn air as the air pressure in the firn changes. Since polar ice sheets typically contain 20 meters of air within the firn, a 5% change in air pressure will move an air column of about

1.0 meter through the atmosphere-snow interface. If the ice cap experiences one new weather system per week, barometric pumping could move about 50 meters of air into the snow each year. By integrating the pressure changes observed by Automatic Weather Stations at several polar ice cap sites, Cunningham and Waddington (1993) obtained barometric ventilation rates comparable to this crude estimate. Schwander (1989) also estimated similar flow rates due to barometric changes. Even for aerosols that are totally scavenged within the snow, this constitutes a deposition velocity about 3×10^{-5} cm s^{-1}, which is three orders of magnitude smaller than deposition velocities reported by Davidson (1989). Barometric pumping appears to be a negligible driving force for snow ventilation and aerosol deposition.

Fast Pressure Changes - Wind Turbulence

Wind transports turbulent eddies which are associated with air pressure perturbations. Eddies exist over a wide range of scales, and their pressure perturbations at the snow surface drive air in and out of the snow in a complicated way controlled by (a) the temporal and spatial spectrum of the turbulent pressure, (b) the average horizontal speed with which turbulent pressure anomalies are advected by the wind (Taylor Hypothesis, Tritton, 1988, p.308), and (c) the permeability of the snow.

When the surface pressure changes rapidly, so that h_0/H is not small, then air flow in the snow cannot keep up with the changing surface pressure, and the air exchange through the surface depends on the air pressure distribution within the snow. We can estimate the frequency above which the firn ceases to "breathe" as a unit. If we say that simultaneous "breathing" stops (h_0/H is no longer small) for $h_0/H > 0.1$, and we put $h_0 = 50$ m, then we can solve for ω_0 in

$$h_0/H = h_0 \times \sqrt{\frac{\mu\ \phi\ \omega_0}{2\ K\ P_0}} \approx 0.1\ . \tag{23}$$

We find that the firn no longer "breathes" in phase when $\omega_0 > 10^{-3}$ Hz. Most significant turbulent pressure variability exists at higher frequencies than this, *i.e.* 0.01 – 10 Hz (Elliott, 1972; Colbeck, 1989). Therefore, in order to estimate the air exchange, we must first solve (4) for the pressure distribution throughout the firn in response to the appropriate surface boundary conditions. Temporal variation (Figure 3B, 3C) is an essential aspect of the wind turbulence boundary condition. Clarke et al. (1987) and Colbeck (1989) studied aspects of the resulting ventilation assuming a uniform spatial pattern (Figure 3D), but a full treatment must also incorporate spatial variability as shown in Figure 3F (Clarke and Waddington, 1991).

In the absence of detailed pressure measurements at the ice sheet surface, how can we estimate this boundary condition? To parameterize the pressure boundary condition at the snow surface in terms of a more traditional meteorological observation, Colbeck (1989) derived a relationship between U_5, wind speed at 5 meters height (m s^{-1}), and the magnitude p_0 (Pa) of the dominant turbulent pressure fluctuations over snow. This relationship,

$$p_0 = 0.0327\ e^{0.383\ U_5} \tag{24}$$

predicted that a wind speed of 10 m s^{-1} would generate a surface turbulent pressure perturbation of only a few Pa. Based on an equation similar to (22), Colbeck concluded that the maximum vertical air displacement in the snow during each pressure cycle was on the order of 0.001 cm. Equation (21) shows that an oscillation of 5 Pa at 10 Hz would cause an average air velocity into the snow of approximately 0.05 cm s^{-1} for typical polar snow.

Deposition velocities reported for various chemical species at Dye 3 (Davidson, 1989) have comparable magnitude.

Clarke et al. (1987) solved a generalization of the problem (6) through (9) for a permeable half-space. They derived the solution for pressure perturbations driven by a spatially-uniform surface pressure field with an arbitrary spectrum over all frequencies ω, rather than the single turbulent frequency ω_0 in (5). Using turbulence data reported by Businger (1982), together with an admittedly brisk estimate (≈ 20 m s^{-1}) for the wind at Agassiz Ice Cap, Ellesmere Island, they obtained a range of 10 – 65 Pa for the average Root-Mean-Square power of the turbulent pressure spectrum. These values are not inconsistent with (24). Clarke et al. did not explicitly calculate air flux across the air - snow boundary. However, their two-fold difference in the estimated wind speed resulted in a 40-fold difference in the amplitude of the driving pressure signal at the snow surface when compared to values in Colbeck (1989). Air displacements and aerosol particle dry deposition rates due to turbulent ventilation would also increase 40-fold, compared to Colbeck's calculations. As we noted earlier, parameterization of conditions at the air - snow interface remains a source of uncertainty for windpumping calculations. In this case, the exponential dependence on wind speed in (24) makes the calculation very sensitive to atmospheric conditions that are poorly-known now, and unknown in the past at times corresponding to the ice core records.

Both the Clarke et al. (1987) and the Colbeck (1989) models were one-dimensional (Figure 3D). Colbeck (1989) noted that horizontal sub-surface flow should exist at the spatial scales of turbulent bursts, and this horizontal flow might significantly alter the character of the air flow into the snow. Clarke and Waddington (1991) measured the turbulent pressure spectrum at the snow surface and at 0.1 m depth at Agassiz Ice Cap, Ellesmere Island. Their data showed that the depth of penetration of pressure fluctuations falls off much faster with increasing frequency ω than one-dimensional theory predicts. The Taylor Hypothesis (Tritton, 1988, p.308) states that in advection-dominated flows such as strong winds, the spatial scale of turbulent eddies (wavelength Λ) can be estimated from the pressure changes recorded over time (frequency spectrum ω) at a fixed point through the mean wind speed U_0 at the surface, *i.e.*

$$\Lambda = \frac{2\pi U_0}{\omega} \tag{25}$$

because the wind moves eddies past the sensor before the eddies can change significantly. For a typical slip velocity (Bergen, 1980; Clarke and Waddington, 1991) of $U_0 = 0.5$ m s^{-1} over a permeable snow surface, turbulence recorded at 1–10 Hz by a stationary pressure sensor corresponds to horizontal wavelengths of 50 – 5 cm. This means that a two- or three-dimensional theory should be used to study flow induced in the firn by turbulence.

Clarke and Waddington (1991) addressed the more general windpumping problem in which pressure varies with both time and position on the snow surface (Figure 3C and 3F). They solved the Fourier-transformed form of the diffusion equation (4) for a surface pressure boundary condition that had an arbitrary frequency spectrum and an arbitrary spatial wavenumber spectrum in both horizontal directions x and y at the snow surface. From that result they derived properties of the time-dependent three-dimensional pressure field. They showed that the penetration depth of pressure oscillations in the firn is controlled by both the frequency ω and the wavelengths Λ of turbulence in the x and y directions, through a complicated term which, in certain limits, approximately scales with both $\omega^{-1/2}$ (as in the 1-D theory) and with Λ. The latter dependence restricts the depth of penetration to the horizontal scale of the surface pressure bursts. Subsurface horizontal pressure gradients, which are directed from regions of momentarily high surface pressure toward regions of momentarily low pressure, drive sub-surface horizontal air flow as suggested by Gjessing (1977) and Colbeck (1989).

Further understanding of time-dependent 3-D flow will probably require complicated numerical models. Although details of windpumping caused by wind turbulence are still incompletely known, Cunningham and Waddington (1993) concluded that turbulence does not contribute significantly to sulphate aerosol deposition. Since the air displacements are a fraction of 1 mm in any given pressure cycle, they argued that most of the air that moves to the snow surface comes from the concentration boundary layer adjacent to the snow surface. This layer is already depleted of aerosols. Although the air flux, and thus the air deposition velocity through the snow surface by windpumping, may be on the order of 0.01 cm s^{-1}, the low concentration C_a of aerosols means that the dry deposition flux (3) due to turbulent ventilation should be very small.

Chemical species that adhere weakly to ice surfaces, and as a result have only a weakly developed concentration boundary layer, could possibly be moved across the air - snow interface in more significant quantities in the presence of vigorous turbulent windpumping. Rapid forced oscillations of the air could enhance diffusion, impaction, or interception across the viscous sub-layer. Given the exponential nature of (24), the magnitude of ventilation-related transport would be extremely sensitive to wind speed at the site.

Using a two-dimensional numerical analysis with measured surface pressure variations, Albert and Arons (1994) calculated significant air flows in the firn at Summit, Greenland in spite of the fact that the site is relatively flat. This suggests that turbulent pumping may be an important process there.

Impermeable layers can obstruct vertical air flow, particularly at low forcing frequencies such that the layer depth is less than H (7), but these layers are seldom continuous over large areas. Impermeable layers and crusts can be found at many sites on ice caps and ice sheets, particularly where summer temperatures approach the freezing point. Figure (4B) shows that large numbers of small holes in an otherwise-impermeable crust could introduce relatively vigorous air motion at depth. This would increase the effective penetration depth of sinusoidal pressure waves in the vicinity of the holes, while decreasing the penetration elsewhere, contributing to enhanced mixing of the pore air. However, the total air flux across the air - snow interface would be relatively unaffected by the presence of the holes. Large gaps in impermeable layers would interact with each other and with the surface in more complicated ways, but would also tend to increase the air flow through the snow surface and could stimulate mixing at depth.

The solution (18) also illustrates a potential difficulty in obtaining *in situ* pressure data. Holes created by inserting sensors could generate their own vigorous localized air flow. Care is required to directly observe natural ventilation below impermeable layers.

Steady Flow - Topographic Pumping

Gjessing (1977) and Colbeck (1989) suggested that pressure patterns associated with wind flow over topographic obstacles such as sastrugi and drifts could cause steady air flow within the snow. Air pressure should be elevated over some area on the upwind face of a topographic bump (Figure 5A) due to dynamic wind pressure, and should be lowered over some area on the lee side near the crest of the bump due to the airfoil effect associated with the increased velocity. While the detailed pressure distribution would depend on the shape and orientation of the bump, these static pressure differences on the surface would induce pressure gradients and steady air flow within the bump and surrounding snow. Colbeck (1989) noted that the stagnation pressure of 0.5 ρ_{air} U^2 (Tritton, 1988) where U is background wind speed, is large enough to cause air flow through snow structures. The stagnation pressure in a wind speed of 10 m s^{-1} is ≈ 50 Pa; this is much larger than the turbulent pressure variations (24) in the same wind, and moreover can cause larger air displacements since the flow does not reverse direction frequently.

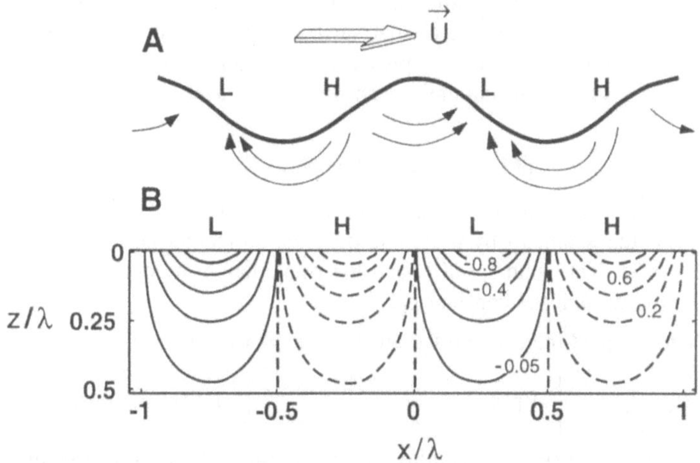

FIGURE 5: (A) In steady air flow $\vec{U}$ over a bump, high pressure zones H occur in troughs and on upwind slopes, and low pressure zones L on crests and leeward slopes. Within the snow, air flows from H to L.
(B) Colbeck's (1989) steady air pressure perturbation pattern (28) in snow under a 2-D sinusoidal surface. Colbeck applied the pressure boundary condition on a flat surface rather than on actual topography z_s. Isobars of perturbation pressure scaled to p_0 (26) are shown. Air flow is normal to the isobars.

If we can adequately parameterize the pressure distribution on the surface of topographic bumps, we can calculate the internal air flow using the Laplace equation (11). From a literature review of fluid flow over obstacles in flumes and wind tunnels, Colbeck proposed that the maximum pressure difference induced by steady flow (Figure 3A) over a wavy surface (Figure 3E) scaled with the square of the wind speed U oriented at 90° to the ripples, and with the vertical aspect ratio h/λ of the wavy surface, giving

$$p_0 = C\, \rho_{air}\, U_{10}^2\, \frac{h}{\lambda} \tag{26}$$

where h is the ripple height and λ is the ripple wavelength. The component of air flow directed along the axis of very long ripples is not expected to generate surface pressure gradients. Using data for water flow over dunes from Vanoni and Hwang (1967), he determined the constant of proportionality to be C ≈ 3; this value was close to the potential value of C = π (Colbeck, 1989). Cunningham and Waddington (1993) confirmed the validity of (26) for air flow over snow topography by measuring the pressure drop over a snow dune at Agassiz Ice Cap, and showing that the Vanoni and Hwang (1967) water flume experiments and air flow over snow ripples were dynamically similar, *i.e.* had similar Reynolds numbers (Tritton, 1988). Their data gave a value of C = 3.0. Colbeck first assumed that a typical set of ripples with height h and wavelength λ created a sinusoidal surface pressure distribution with the amplitude (26), *i.e.*

$$p(x,z_s) = p_0 \cos[2\pi x/\lambda] \tag{27}$$

on the wavy surface $z_s(x)$. He then applied the boundary condition (27) along a flat surface $z = 0$ rather than along the actual micro-topographic profile $z = z_s(x)$ that would generate such a surface pressure pattern; otherwise, an analytical solution is not in general possible. Cunningham and Waddington (1993) concluded that this approximation incorporates the essential physics of the "true" ventilation problem, in which depth z is measured relative to the wavy surface $z_s(x)$, as long as that surface appears "flat" on a horizontal scale comparable to the e-folding depth of the pressure disturbance, *i.e.* $h/\lambda < 1$. Colbeck then solved equation (11) assuming an infinitely deep permeable layer, to obtain the sub-surface pressure pattern

$$p(x,z) = p_0\, e^{-2\pi z/\lambda} \cos[2\pi x/\lambda] \tag{28}$$

which is periodic in x and decays with depth z, as shown in Figure 5B. The ripple wavelength controls the geometry by providing both the horizontal periodicity λ and the e-folding depth $\delta = \lambda/2\pi$. The wind speed U and the ripple height h control the magnitude of the pressure solution through the constant p_0.

Colbeck then obtained the sub-surface air flux $\vec{Q}(x,z)$ from (2), *i.e.*

$$\vec{Q}(x,z) = (Q_x,Q_z) = -\frac{K}{\mu}\nabla p(x,z) = \frac{2\pi}{\lambda}\,\frac{Kp_0}{\mu}\, e^{-2\pi z/\lambda} \Big(\sin[2\pi x/\lambda],\ \cos[2\pi x/\lambda] \Big) \tag{29}$$

showing that air flow is restricted to shallow depths by the same e-folding depth $\delta = \lambda/2\pi$. Colbeck noted that much of the air flow could occur within the ripples themselves. The assumption of an infinitely deep permeable layer can be justified since the air flow does not sense the permeability distribution below a few e-folding depths δ. Equation (26) shows that the magnitude of the pressure perturbation p_0 depends only on the shape h/λ of the ripples, and not on their actual size. However, the air flux (29) is proportional to the pressure *gradient*, so it scales inversely with the actual distance between high- and low-pressure areas, as manifested by λ in the denominator of (29). This means that a snow surface covered by many small ripples will pump more air than a surface of equal area covered by a few large ripples of the same shape h/λ. This trend is limited only by breakdown of the permeable medium approximation for bumps that are not large relative to snow grain size, and an aerodynamic transition to a viscous drag-dominated flow regime as bump scale λ becomes comparable to the height of the viscous sublayer (Cunningham and Waddington, 1993).

The surface of polar ice sheets is better represented by three-dimensional microtopography such as sastrugi and drifts, rather than by two-dimensional ripples. Cunningham and Waddington (1993) constructed a full 3-D ventilation model. They extended Colbeck's wavy surface model (Figure 3E) to 3-D surface bumps (Figure 3F) that generated a surface pressure distribution of the form

$$p(x,y,z_s(x)) = p_0 \cos\left[\frac{2\pi}{\lambda}\, x\right] \cos\left[\frac{2\pi}{\alpha\lambda}\, y\right] . \tag{30}$$

The x axis is aligned with the wind, and α is the horizontal aspect ratio of the topographic bumps. For $\alpha < 1$, the bumps are elongated in the direction of the wind. For $\alpha > 1$, the bumps are elongated across the wind. Like Colbeck (1989), they estimated the parameter p_0 from wind speed data, surface pressure data, and the empirical relationship (26). The pressure drop over bumps does not depend on α, which describes the microtopography transverse to the wind. This will introduce a strong dependence on wind direction into solutions for ventilation. The pressure drop p_0 arises from dynamic pressure as wind hits a surface blocking its path; however, wind blowing parallel to long ripples is not obstructed in any way, so no dynamic pressure effects are expected. When the surface boundary parameterization (30) is applied on a flat surface of a semi-infinite permeable half-space rather than on z_s, the pressure solution is

$$p(x,y,z) = p_0\, e^{-z/\delta} \cos\left[\frac{2\pi}{\lambda}x\right] \cos\left[\frac{2\pi}{\alpha\lambda}y\right], \quad \text{where} \quad \delta = \frac{\alpha}{\sqrt{\alpha^2+1}}\,\frac{\lambda}{2\pi}. \tag{31}$$

The e-folding depth δ depends on the horizontal aspect ratio α of the microtopographic bumps as well as the wavelength λ in the wind direction. Cunningham and Waddington (1993) used (2) to calculate the volumetric air flux $\vec{Q}(x,y,z)$ within the snow;

$$\vec{Q}(x,y,z) = (Q_x, Q_y, Q_z) = -\frac{K}{\mu}\nabla p(x,y,z) = \tag{32}$$

$$\frac{Kp_0}{\mu}\frac{2\pi}{\lambda}\, e^{-2\pi z/\delta}\left[\sin\left[\frac{2\pi}{\lambda}x\right]\cos\left[\frac{2\pi}{\alpha\lambda}y\right],\ \frac{1}{\alpha}\cos\left[\frac{2\pi}{\lambda}x\right]\sin\left[\frac{2\pi}{\alpha\lambda}y\right],\ \frac{\lambda}{\delta}\cos\left[\frac{2\pi}{\lambda}x\right]\cos\left[\frac{2\pi}{\alpha\lambda}y\right]\right].$$

The vertical air flux component Q_z can be integrated over the high-pressure portion of the bump surface for which the air flux is downward. After dividing by the area $\alpha\,\lambda^2$ of each bump, we find that the average volumetric air flux into the snow is

$$\bar{Q}_z = \frac{K}{\mu}\frac{p_0}{\pi^2}\frac{1}{\delta} = \frac{K}{\mu}\frac{3\rho_{air}U_{10}^2}{\pi^2}\frac{h}{\lambda}\frac{1}{\delta} = \frac{K}{\mu}\frac{6\rho_{air}U_{10}^2}{\pi}\frac{h}{\lambda}\frac{1}{\lambda}\frac{\sqrt{\alpha^2+1}}{\alpha}. \tag{33}$$

We substituted (26) for the pressure amplitude p_0 and (31) for the e-folding depth δ.

We now explore the solution (33) to understand how the surface microtopography controls ventilation. Three independent geometric parameters, the bump height h, the wavelength λ in the direction of the wind, and horizontal aspect ratio α in the cross-wind direction describe the microtopography. They in turn determine the three physical parameters on the snow surface that control ventilation, *i.e.* the magnitude of the pressure drop p_0, and the separations λ and $\alpha\lambda$ of the topographically-induced high and low pressure areas in the x and y directions. Two additional geometrical quantities, *i.e.* the e-folding depth δ, and the vertical aspect ratio h/λ in the direction of the wind, can be derived from the three geometrical parameters. In each panel of Figure 6, one of the five geometrical quantities is held constant at a reasonable value for South Pole, while two other parameters are varied. All panels show volumetric air flux in m s^{-1} calculated for parameters used by Cunningham and Waddington (1993), *i.e.* snow permeability $K = 4.0\times10^{-9}$ m^2, and wind speed $U = 4.9$ m s^{-1}. Areas of each panel in which the surface boundary condition cannot be accurately represented on a flat surface ($h/\lambda > 1$, or $h/\alpha\lambda > 1$) have been blocked out. High air flow may exist in those regimes; however, we are unable to accurately estimate the air flux with this simple analytical model.

For bumps of a constant height h = 0.1 m, Figure 6A shows that air flux increases as wavelength λ decreases while the horizontal aspect ratio α is held constant, *i.e.* as the spatial pattern of bumps (Figure 3F) is compressed in both horizontal dimensions. Increased ventilation occurs because the pressure drop p_0 increases and the high and low pressure areas get closer together, increasing the horizontal pressure gradients. For bumps elongated with the wind ($\alpha < 1$), and a constant wavelength λ in the direction of the wind, the air flux increases as α decreases, *i.e.* as the bump pattern is compressed laterally in the cross-wind direction while maintaining the same length in the wind direction. Increased ventilation arises from increased sub-surface air flow in the cross-wind direction. As dunes are eroded into fluted sastrugi, we could expect ventilation to increase. For $\alpha \gg 1$, the microtopography becomes 1-D ripples elongated across the wind direction (Figure 3E). The solution ceases to depend on α and reduces to the Colbeck (1989) 2-D solution (29).

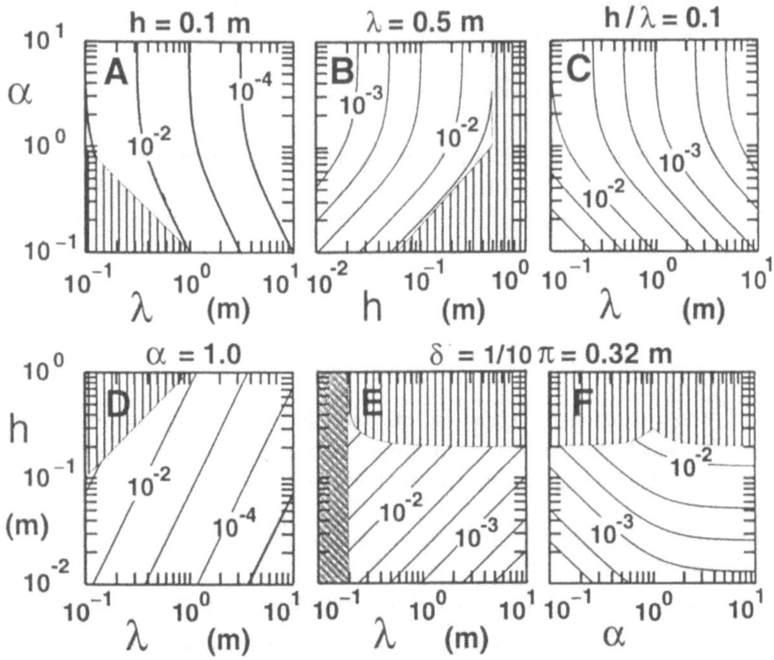

FIGURE 6: Volumetric air flow ($m^3\ m^{-2}\ s^{-1} = m\ s^{-1}$) for South Pole calculated from (33) for steady wind of $U_{10} = 4.9\ m\ s^{-1}$. Various geometric parameters of the surface microtopography are held constant in each panel. See text for details.

For microtopography with a fixed down-wind wavelength of $\lambda = 0.5$ m, Figure 6B shows that air flux increases as the bump height h increases (causing the pressure drop p_0 (26) to increase), or as α decreases, narrowing the ridges in the cross-flow direction (and bringing high and low pressure areas closer together for cross-wind subsurface flow). For microtopography with constant vertical aspect ratio h/λ in the wind direction, Figure 6C shows that the air flux increases as the wavelength λ gets shorter. Even though the bumps are also decreasing in height h, the pressure drop (26) over each bump stays constant. As α decreases and the bumps become narrower transverse to the air flow, air flux into the snow also increases. In both cases, increased ventilation is entirely due to the reduced separation of high and low pressure areas on the snow surface, in the wind direction in the first case and in the cross-wind direction in the second case. For microtopography with a fixed horizontal aspect ratio α, Figure 6D shows that air flux increases as the wavelength λ gets shorter (the bump pattern is uniformly compressed as seen from above), and as the bumps get taller. Both factors increase the magnitude p_0 of the pressure drop (26).

For a fixed ventilation e-folding depth δ as defined in (31), Figure 6E shows that air flux increases as bump height h increases for a constant wavelength λ. For constant bump height h, the air flux also increases as the wavelength λ decreases, even though the horizontal aspect ratio α is simultaneously increasing in order to maintain the same penetration depth δ. In both cases, increased pressure drop p_0 is the dominant control. Figure 6F shows an apparently counter-intuitive result; unlike the cases 6A, 6B and 6D above, air flux *decreases* as α decreases, *i.e.* as the bumps become narrower in the cross-wind direction. By holding the penetration depth δ constant, we force the bumps to develop a longer wavelength λ in the down-wind direction as α decreases, since λ, α and δ are related through (31). Since the amplitude p_0 of the pressure perturbation in (26) decreases as λ increases, the amount of air pumped through the snow surface also decreases. In particular, this shows the effect of rotating the wind direction by 90° relative to the long axis of the microtopography. Both h and δ are constant along each horizontal line in Figure 6F. Starting from any point where $\alpha = \alpha_0 < 1$, and moving to the right to the point $\alpha = 1/\alpha_0 > 1$, we go from a sastrugi-like bump with the wind blowing along its long axis to the same bump oriented as a ripple with the wind blowing across its short axis. Figure 6F shows that the air flux into the bump is greater for the ripple-like orientation.

Figures 6E and 6F remind us that the depth of air penetration is only one aspect of snow ventilation; the rate at which air is pumped into the snow to a fixed depth δ also depends on other aspects of the microtopography. Factors that enhance firn ventilation are: (1) taller bumps, (2) shorter wavelengths in either down-wind or cross-wind directions, (3) bumps oriented with their long axis across the wind.

When the air passively carries a uniform concentration of a chemical species that is 100% filtered by the snow, then (33) also gives the dry depositional velocity. All the panels of Figure 6 suggest that depositional velocities for air at South Pole could be in the range of 0.1–10 cm s^{-1} for a number of plausible surface conditions.

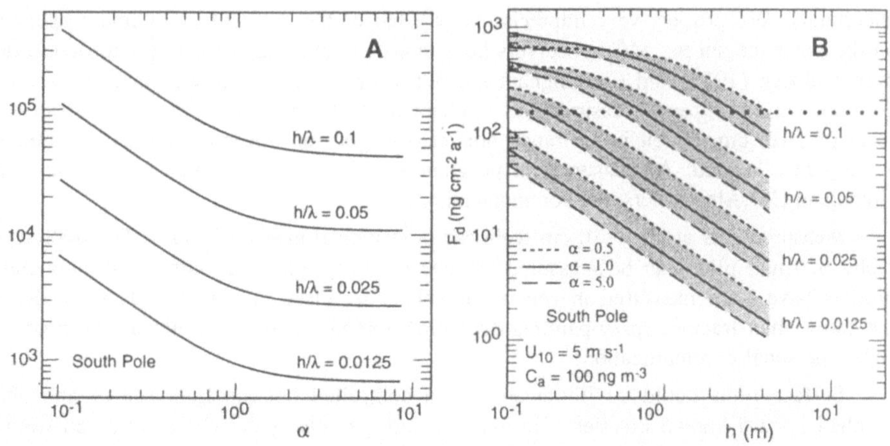

FIGURE 7: (A) Theoretical volumetric air flux $\overline{Q}_z$ at South Pole for topographic features that vary in height seasonally between 5–10 cm, as a function of aspect ratios α and h/λ.
(B) Calculated non-seasalt (nss) sulphate dry mass flux at South Pole for various microtopographies. Both panels are adapted from Cunningham and Waddington (1993).

Cunningham and Waddington (1993) also used (33) to calculate the vertical air flux through the surface at 3 polar sites for a range of shapes and sizes of surface bumps. They used measured windspeeds to determine U_{10}^2. They assumed that the height h of surface bumps varied seasonally between 0.05 and 0.10 m, with the maximum in late winter. A seasonal pattern of this form has been observed (Gow, 1965; Weller, 1969). The estimated volumetric air flux at South Pole (Figure 7A) could exceed 10^5 m a^{-1} (*i.e.* several mm s^{-1}) for sastrugi aligned with the prevailing wind ($\alpha < 1$). As we noted above, the ventilation rate would be even higher for cross-winds over the same microtopography.

All the calculations that we have described assumed that the air flow over the snow surface is rough, so that the wind speed U_{10} could be used to parameterize the pressure at the snow surface through (30), but not so rough that wake formation invalidated that same parameterization. Within the snow we assumed that the air flow is laminar and linear so that Darcy's Law (2) could be applied. Cunningham and Waddington (1993) addressed the impact of these assumptions for the calculated air flux in polar regions. They concluded that violations of the assumptions could result in uncertainties of a few tens of percent, but probably not a factor of two, if their parameterizations of the surface boundary condition are realistic. The solutions above then illustrate the concepts, the potential air flux, and the depth of snow ventilation. The largest uncertainties arise, however, from the parameterization of the snow microtopography and the pressure variations it generates, and from the permeability of the near-surface snow. The surface of a real polar ice sheet contains bumps at a wide range of wavelengths, heights, and aspect ratios, including areas that may be flat. Albert (in press) calculated that the air flux at Summit, Greenland was reduced by 80% when the observed low-permeability surface wind slab was included in a numerical model of ventilation.

Measuring Microtopography

Before we can accurately predict the influence of ventilation on concentration levels of chemical constituents in polar snow and firn, we must address a serious outstanding uncertainty. We possess very imprecise knowledge of the distribution of microtopography on the polar ice sheets. Most observers have tended to note large sastrugi and dunes; however, Colbeck (1989) and Cunningham and Waddington (1993) showed that large features are poor wind pumpers. Measurements to characterize the roughness of the ice sheet at the scale of 1–10 cm, which is the range predicted to generate the most effective ventilation, are urgently needed. A measurement program to achieve this has been started in central Greenland (M. Albert, personal communication).

Measurements at the 1–10 cm scale can be achieved in several ways. The microtopography of small plots can be measured by tape measure and string line. Microtopographic profiles have been measured in central Greenland by inserting thin boards vertically into the snow, then tracing, spray-painting or photographing the snow profile on the board. (R. Bales, personal communication).

Surface plots could be illuminated by a controlled oblique light source and photographed from a known position. The microtopography could then be determined from the shadowing function. Similar photoclinometry techniques have been used to derive the topography of planetary bodies (Davis and Soderblom 1984; Tanaka and Davis, 1988). However, light penetrates snow; scattered light re-emerging in the shadowed areas could degrade the shadowing function beyond acceptable limits. Alternatively, a light source could be used to project a grid of colored lines onto the snow surface through collimating slits. The apparent positions of the lines on the film plane could be used to determine the microtopography. This principle has been used to measure microtopography of sea ice (Briscoe et al., 1989). To make measurements during daylight periods, it is helpful to shade the site with a tarpaulin.

Detailed studies of a few small plots may be unrepresentative of large areas of the ice sheets. Airborne laser altimeter surveys from aircraft equipped with high accuracy differential GPS (Global Positioning System) receivers can now measure topography to an accuracy of about ±1 cm in a 1 m^2 footprint. A NASA-operated P-3 aircraft equipped with two laser scanners has been deployed over Icelandic glaciers (Garvin and Williams, 1993), over arctic sea ice (Wadhams et al., 1992) and over extensive areas of the Greenland ice sheet (Krabill et al., 1995). The NASA aircraft operates both a nadir tracking profiler and an Airborne Oceanographic Lidar which can rapidly scan an overlapping helical swath at 10° from the nadir, *i.e.* a 135 m wide path at typical flight altitude. Data products from both laser altimeters are available from NASA. Lasar topographic profiles could provide spectral characteristics of microtopography over large areas of the ice sheets. Since the flight tracks are repeatable with GPS navigation, temporal changes in the statistical properties of microtopography could be determined by repeat flights during different seasons.

Finally, it might be possible to obtain information about microtopography from satellite microwave scatterometer data (Ledroit et al., 1993). Scatterometers are used to estimate ocean winds by measuring the wave spectrum (*i.e.* microtopography) on the ocean surface. The polarization of microwave energy back-scattered from an ice sheet to a satellite depends on the character and orientation of surface microtopography as well as on grain structure within the near-surface snow. It is unclear whether topographic properties controlling ventilation could be uniquely determined by this technique. However, these data offer an intriguing possibility for studies of microtopography in areas where surface studies or aircraft tracks are not practical.

Shook and Gray (1994) showed that the distribution of seasonal snow depth in fields in Saskatchewan is a fractal up to a horizontal scale of 100 m. If the same result can be demonstrated for polar ice sheet topography, then the scale invariance of fractals may make it possible to characterize the microtopography at 1–10 cm by measurements of roughness at a larger scale that can be observed more easily by remote-sensing platforms. The fractal nature of snow topography has implications for measurement strategies. For example, Shook and Gray (1994) noted that random sampling of a fractal distribution is preferable to sampling at uniform intervals.

Climate and Microtopography

The wind creates and shapes dunes, ripples, and drifts from saltating snow, and erodes them into sastrugi. Sastrugi evolve throughout the year (Gow, 1965) in response to seasonal changes in wind and the frequency of saltation events. Maximum height is achieved in late winter, and the topography is smoothed during summer. Climate variables such as wind speed, temperature and accumulation rate control the microtopography. Automatic Weather Station (AWS) studies should be combined with studies of the current distribution and evolution of microtopography, with the goal of parameterizing microtopography and permeability structure in terms of climate variables.

A snow surface may be configured for more efficient ventilation at certain times than at others, as microtopography evolves. We do not need to know all the parameters describing the microtopography; we need to know only the combination of those parameters that controls the volumetric air flux. Ventilation (33) depends on the wind and on the snow surface. We will now represent $\overline{Q}_z(t)$ as a product of a wind factor W(t) and a snow factor S(t), *i.e.*

$$\overline{Q}_z(t) = W(t)\, S(t)\ . \tag{34}$$

The atmosphere affects ventilation efficiency through (a) the wind speed U_{10} which generates the pressure differences (26) over bumps, and (b) the viscosity of the air which is forced through the pore spaces by the resulting pressure gradient. These effects can be combined into a *wind index* W(t) given by

$$W(t) = C\, U_{10}^2 \frac{\rho_{air}}{\mu} = C \frac{U_{10}^2}{\nu} \tag{35}$$

where ν is the kinematic viscosity of air and the factor $C \approx 3$ comes from (26).

The second factor S(t) is a *surface index* that measures the relative receptivity of the surface snow to ventilation if the wind were to blow. When a snow surface can be adequately represented by the sinusoidal waves (Figure 3F) of the Cunningham and Waddington ventilation model, then we can use (33) to define S(t) by

$$S(t) = \frac{1}{\pi^2} K(t) \left[\frac{h(t)}{\lambda(t)} \frac{1}{\delta(t)} \right] \tag{36}$$

S(t) comprises (a) the snow permeability K, (b) a microtopography geometric factor depending on all three independent parameters h, λ and α in the Cunningham and Waddington parameterization (31), and (c) a geometrical constant π^2 that arises from the assumption of a sinusoidal pressure pattern. Permeability and microtopography depend strongly on wind, temperature and accumulation rate, and this dependence is probably nonlinear. Many surface features on the polar ice sheets are not sinusoidal (Doumani, 1967). Studies on formation and evolution of microtopography are urgently needed. Whether the climate parameters that might be found to act as proxies for microtopography can be derived for past climates is a separate question.

ROLE OF VENTILATION FOR IRREVERSIBLY-DEPOSITED SPECIES

The chemical composition of ice in a polar ice sheet depends on the deposition history and subsequent alterations. A constituent is initially deposited by wet or dry deposition. The concentration in the snow may be modified by post-depositional changes. Some species may be reentrained, chemical reactions may occur, chemical components may be concentrated by sublimation of ice, and volatile chemical constituents may move toward equilibration with the local atmosphere. Each chemical species may be affected differently by ventilation, so that a species-specific model of deposition and post-depositional changes must be coupled to the snow ventilation model.

Accumulation Mode Sulphate Aerosols

Non-seasalt (nss) sulphate is representative of irreversibly deposited atmospheric components. The transfer function for the dry-deposited fraction may depend strongly on the ventilation conditions at each site. However, ventilation should have little direct influence on the wet-deposited fraction.

Non-seasalt sulphate is formed by gas-to-particle conversion. By mass, it is distributed primarily in the accumulation mode, from 0.1 to 1 μm particle diameter (Whitby, 1978; Pruppacher and Klett, 1978; Barrie, 1985; Ito, 1995). The transport of this aerosol component to snow surfaces by both wet and dry deposition has been studied theoretically and experimentally. Ibrahim et al. (1983) developed a dry deposition model incorporating aerodynamic transport to a snow surface, and transport across the viscous sublayer by Brownian diffusion, impaction, interception, and gravitational settling. Davidson (1989), used this model to estimate the dry deposition velocity of nss sulphate aerosols at Dye 3 on the south dome of Greenland. The snow surface was represented by an array of ice needles in that model, rather than by a permeable layer. The calculated deposition rate was lower than observed, possibly because the surface area of the ice needles was much smaller than the ice surface area accessible within the firn by windpumping (Figure 1). However, the concepts developed by Davidson and others can be extended to incorporate deposition by ventilation.

Gjessing (1977) described many of the features of snow ventilation that determine the transfer function for aerosols. He concluded that (a) snow filtered particles out of the air, (b) the wind constantly replaced the air in the snow, and (c) wind flow over surface bumps caused local atmospheric pressure perturbations that were a likely source of the internal pressure gradients driving ventilation. In controlled snow ventilation experiments conducted in Norway, he pumped known volumes of air through tubes filled with snow, and observed significant increases in the concentrations of ten ionic species in the snow compared to their concentrations in his control (unventilated) tubes. Condensation nuclei counter measurements showed that some particles remained in the air exiting his 90 cm tubes after about 30 seconds residence time in the snow, suggesting that either (a) capture was not 100% effective in the snow samples studied, or (b) reentrainment occurred.

Heintzenberg and Rummukainen (1993) measured the size distribution for aerosol from 0.1 to 2 μm diameter in ambient air at a flow rate of 0.0014 cm s^{-1}, and in air filtered through snow at increasing depths in newly fallen snow in Svalbard. They found the depth profiles of number- and volume-distribution to decrease exponentially with an e-folding depth of about 1 cm. Their coefficients for exponential attenuation of aerosol varied from 0.28 to 0.4 cm^{-1}, depending on particle size. They suggested that fibrous filter theory (Fuchs, 1964) might be used to model aerosol removal by snow, but they lacked the additional physical parameters necessary to test a filter theory.

If aerosol particles are captured relatively quickly compared to the residence time of air in the snow, the volumetric air flux can be a good approximation to the dry deposition velocity. However, both a ventilation model and a pore-space deposition model are necessary to calculate the dry deposition velocity if significant numbers of particles have not yet been captured or have been reentrained when the air re-emerges from the snow.

Diffusion is an important mechanism for dry deposition of submicrometer particles with impaction and interception becoming significant for particles with diameters greater than a few tenths of micrometers (Davidson and Wu, 1990). Harder et al. (in press) measured particle removal by snow filters at South Pole. They found a greater filtering efficiency than did Heintzenberg and Rummukainen (1993), with removal coefficients of 0.3 – 0.6 cm^{-1} for particle sizes of 0.1 – 1 μm and flow rates of 0.06 – 2 cm s^{-1}, and an e-folding depth of 0.5 – 1 cm. Although near-total removal (99.86%) was observed when residence time within the filter exceeded 12 s, a very small number of particles continued to emerge downstream of the filter, indicating possible reentrainment of particles at some slight rate. This is consistent with Gjessing's findings (Gjessing, 1978).

Aerosol Particle Deposition

The capture rate for nss sulphate aerosol in air moving through snow can be modelled by a number of standard aerosol filter theories, all of which can give roughly comparable results. Harder et al. tested their results against models for fibrous filters (Fuchs, 1964). Although the models underpredicted efficiency, they confirmed that diffusion appears to be the primary mechanism for dry deposition. If diffusion is the primary mode of deposition in ventilated snow, then deposition can be modelled by air flow through a bundle of small tubes whose aerosol capture efficiency can be calculated as described by Fuchs (1964), p. 204-205) and by Reist (1993, p. 153-154). We will use this model below.

When air flux Q carrying a concentration N_0 of accumulation-mode aerosol particles with Brownian diffusivity D (Reist, 1993, p.137) enters a tube of diameter d, the fraction $(1-N/N_0)$ of aerosol removed after time t, or after travelling a distance X along the tube, can be estimated from the Gormley-Kennedy equation (Reist, 1993, p.154)

$$1-N/N_0 = 2.56\Phi^{2/3} - 1.2\Phi - 0.177\Phi^{4/3} \qquad \Phi = \frac{\pi DX}{Q} = \frac{4Dt}{d^2} \leq 0.02 \qquad (37a)$$

$$1-N/N_0 = 1.0 - 0.81905\, e^{[-3.6568\Phi]} - 0.09753\, e^{[-22.305\Phi]} - 0.0325\, e^{[-56.961\Phi]} - 0.01544\, e^{[-107.62\Phi]} \qquad \Phi > 0.02 \qquad (37b)$$

The dimensionless parameter Φ can represent either time t or distance X along the tube. Air flow at the entrance to the tube is not laminar. As air enters the tube, a laminar boundary layer forms on the tube wall, and thickens with X. Equation (37a) is applicable until the laminar boundary layers converge in the center of the tube. This point, at $\Phi = 0.02$, is known as the *entry length,* (Tritton, 1988, p.15). Beyond $\Phi = 0.02$, diffusion to the walls is no longer assisted by eddies, and the fraction $(1-N/N_0)$ of aerosol removed is then given by (37b) (Reist, 1993, p.154). Heavy lines in Figure 8 show the Gormley-Kennedy capture efficiency (37a) (solid curve) and (37b) (dashed curve) as a function of Φ (upper scale).

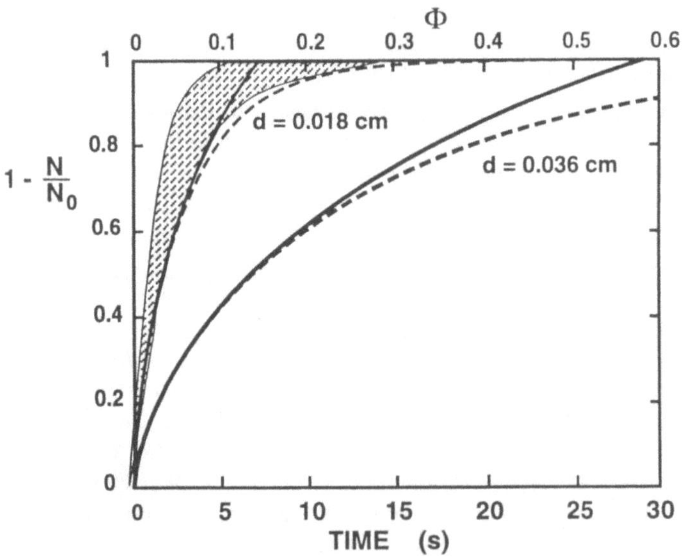

FIGURE 8: Heavy curves show capture efficiency $(1-N/N_0)$ for aerosol particles captured by Brownian diffusion in a tube as a function of the nondimensional parameter $\Phi = 4Dt/d^2$ on the upper scale using (37a) (solid curve) and (37b) (dashed curve). The lower scale shows time t (s) for 0.1 µm-diameter aerosol particles. Cunningham and Waddington (1993) used d = 0.036 cm for snow at South Pole. The permeability $K = 1.0\times10^{-9}$ m^2 measured at South Pole (Harder et al., in press) corresponds to d = 0.018 cm, giving the thinner solid and dashed curves with faster response times. The shaded area shows capture efficiency measured by Harder et al. for aerosol particles ranging from 0.1 – 1 µm in diameter.

Model for Non-seasalt Sulphate Deposition by Snow Ventilation

Cunningham and Waddington (1993) used their 3-D steady ventilation model described above to calculate air parcel trajectories through the snow for a wide range of surface microtopography and wind speeds. The length of an air parcel trajectory and the residence time of that air parcel depend on the position where the trajectory originates relative to the high and low pressure zones (see Figure 5). For each microtopography and wind speed, they numerically calculated the residence time for air parcels entering a typical bump at each position on its surface.

Many different internal configurations of pores and channels in the snow can all generate the same bulk porosity ϕ and the same permeability K. In order to estimate the aerosol particle deposition, we must choose one out of these many hydraulically-equivalent configurations that most resembles a well-understood aerosol filter model whose aerosol capture rate we can calculate. For laminar flow at low Reynolds number, Cunningham and Waddington (1993) chose to use a solid containing a series of cylindrical holes that (a) comprised a fraction ϕ of the volume, (b) were aligned parallel to the air flow, and (c) had a special diameter d called the *hydraulic diameter*, such that

$$d = 2\sqrt{\frac{8K}{\phi}} \tag{38}$$

The hydraulic diameter is a property of the porous material that measures the flow efficiency; d is selected so that the bulk permeability matches that of the snow. Based on experiments, Carman (1956) recommended a factor of ≈ 5 rather than 8 in the numerator. Cunningham and Waddington (1993), who used Carman's formula, took porosity $\phi = 0.62$ typical of snow at South Pole, and an empirical relationship between porosity and permeability of dry snow (Schwander, 1989) to estimate $K = 4.0\times10^{-9}$ m^2; this gave a hydraulic diameter (38) of $d = 0.036$ cm. Aerosol particles with a diameter of 0.1 μm have a Brownian diffusivity of $D = 6.8\times10^{-6}$ cm^{2}s^{-1} (*e.g.* Reist, 1993, p.137). According to (37b), approximately 90% of these particles should be removed after 30 s in such a tube. Impaction is negligible with 0.1 μm particles, so tortuosity of the air flow pathway in the snow or curvature of the equivalent tube should not affect the removal rate when air flow is laminar.

Cunningham and Waddington then modelled aerosol capture due to topographic ventilation by using a bundle of tubes with diameter d given by (38) to represent the permeable snow. The tubes had the same lengths as the 3-D air flux trajectories, and had air residence times calculated with the 3-D ventilation model. They used (37) to estimate the capture efficiency of each tube, then summed over all the tubes to get a bulk dry deposition rate for each surface bump pattern. They found that virtually all of the nss sulphate aerosol that entered the snow should be trapped there for most plausible combinations of permeability, wind speed and microtopography.

Cunningham and Waddington used (37a) for all residence times in their calculations; this over-estimated the amount of aerosol removed by about 10% compared to (37b) for 30 s residence times, as shown in Figure 8. The error is smaller for both shorter and longer times. Albert (in press) incorporated observed surface wind slabs into a numerical model of air flow for central Greenland, to show that these low-permeability surface layers could reduce air flux into the snow by 80%. Lower surface permeability also decreases the time for near-total capture of aerosol in the snow in the Cunningham and Waddington model.

Paleoclimate Implications

The total flux of nss sulphate has been measured at South Pole (Kirchner and Delmas, 1988), and the dry deposition component was estimated to be 160 ng $cm^{-2}a^{-1}$ (Legrand, 1987); this value is shown by a heavy dotted horizontal line in Figure 7B. Cunningham and Waddington calculated dry deposition fluxes of nss sulphate at South Pole for a range of plausible microtopographies; Figure 7B shows their results. Surface bumps in each group of 3 curves have a constant vertical aspect ratio h/λ in the direction of the prevailing wind. The profile along-wind is the same for all 3 curves within each group, but the bump width transverse to the wind increases from narrow and sastrugi-like at $\alpha = 0.5$, to wide and ripple-like at $\alpha = 5.0$. A range of realistic microtopographies with heights of $h = 0.01 - 0.1$ m can account for all of the dry nss sulphate deposition estimated by Legrand (1987). Cunningham and Waddington (1993) concluded that a change in the local windpumping efficiency would cause a change in the ice core nss sulphate record even in the absence of a change in atmospheric concentration. The dry deposition flux attributable to ventilation depends on both atmospheric concentration and ventilation efficiency at the deposition site. Therefore evidence of paleo-ventilation efficiency independent of the ice core chemical record must be supplied before atmospheric concentration can be uniquely determined.

The Vostok ice core shows an accumulation-corrected increase of 20 – 46% for nss sulphate in the Last Glacial Maximum (LGM) relative to modern conditions (Legrand et al., 1988a). Those data have been used to attribute 1 – 3 °C of cooling at the LGM to increased aerosol optical depth (Legrand et al., 1988a; Harvey, 1988), and to infer a link between CO_2 and dimethylsulphide emissions at the LGM (Legrand et al., 1988a). Legrand et al. recognized that changes in local snow accumulation rate and modifications in atmospheric transport could complicate their interpretation. However, under ice age conditions, the snow permeability, surface microtopography, and the local wind speed may also have taken different values that would have enhanced ventilation through (33). Cunningham and Waddington suggested that plausible variations in those parameters at Vostok could entirely account for the LGM - Holocene differences in the Vostok ice core nss sulphate, without requiring *any* change in the atmospheric concentration at Vostok.

Harder et al. (in press) measured the filtering efficiency of snow at South Pole by measuring particle concentrations in air before and after it was drawn through snow filters. They found an e-folding time for deposition in the snow of about 3 s (Figure 8). These data substantiate the conclusions of Cunningham and Waddington, *i.e.* (a) almost all of the nss sulphate that enters the snow remains there, and (b) capture by ventilation is capable of accounting for the estimated modern dry deposition flux on the Antarctic plateau. Clearly, we must (a) understand the impact of ventilation on the chemical transfer functions and (b) be able to estimate the relevant windpumping parameters at the ice core sites in the past before we can derive reliable atmospheric chemical concentration histories from ice core chemical records.

Toward a Transfer Function for Irreversibly Deposited Species

Ventilation of polar snow induced by steady air flow over small topographic bumps may strongly influence the deposition rates of some chemical species. If almost all airborne aerosol is captured in the snow during ventilation, then the aerosol dry deposition velocity $V_d(t)$ due to ventilation is approximately equal to the volumetric air flux $\bar{Q}_z$, (33) and (34). Then the dry mass flux F_d into the snow due to ventilation can be approximated by

$$F_d(t) = C_a(t)\,\overline{Q}_z(t) = W(t)\,S(t)\,C_a(t) \tag{39}$$

where $C_a(t)$ is the aerosol concentration above the snow surface. Chemical species are also deposited by wet-deposition or dry-deposition to the snow surface; only a fraction **f** of the total may be due to ventilation. The ventilation-driven dry flux $F_d(t)$ generates a concentration $\mathbf{f}C_{ice}(t)$ in the ice core given by

$$\mathbf{f}C_{ice}(t) = \frac{F_d(t)}{\dot{b}} = \frac{1}{\dot{b}}\,W(t)\,S(t)\,C_a(t) \tag{40}$$

where $\dot{b}$ is the snow accumulation rate (m a^{-1}) and $C_{ice}(t)$ is the measured total concentration in the ice core arising from all deposition mechanisms. The ice core concentration can also be written as

$$C_{ice}(t) = \left[\frac{1}{\mathbf{f}}\,\frac{1}{\dot{b}}\,W(t)\,S(t)\right] C_a(t) \tag{41}$$

where the quantity enclosed in square braces is the transfer function relating the concentration in ice $C_{ice}(t)$ to the concentration in the atmosphere $C_{air}(t)$. As expected, it depends on (a) a precipitation factor $\dot{b}$, (b) a snow configuration factor S, (c) a wind factor W, and (d) a deposition partition factor **f**. If ventilation dominates total deposition at a given site, then $\mathbf{f} = 1$. This is unlikely to be the case except in extremely low accumulation areas. When other wet- and dry-deposition mechanisms are also significant at the site, $\mathbf{f} < 1$. The deposition partition function **f**, which depends on all the parameters controlling wet-deposition and non-ventilation dry mechanisms, is the link through which ventilation is coupled to those other processes in the ice core inverse problem.

Application of the transfer function to the inverse problem (Waddington, this volume) may be problematic. To use the transfer function (41) to derive a record of past air concentrations $C_a(t)$ from ice core data $C_{ice}(t)$, we must first know the fraction **f** of the constituent in the ice core that was dry-deposited by ventilation. Then we must either (a) evaluate S(t), W(t) and $\dot{b}$ using climate variables from a General Circulation Model (GCM), or (b) know their functional dependence on other climate parameters that can be obtained simultaneously from other independent ice core data. We noted previously that a parameterization of S(t) in terms of climate variables such as wind speed, temperature and accumulation rates is urgently needed.

If the inverse transfer function is sensitive to unknown or poorly known paleoclimate state variables, then large errors in the inferred atmospheric concentrations can result. Paleo-accumulation may be the least troublesome; $\dot{b}$ should be derivable from ice core data through ice flow modelling (*e.g.* Cutler et al., 1995) if ice core stratigraphy can be dated. The wind index W(t) (35) depends on the square of the paleo-wind speed; this factor alone may introduce significant uncertainty into the transfer function, since the deposition rate can be more sensitive to short windy intervals than to long periods with light winds. The partition function **f** depends on other deposition processes and conditions that are beyond the scope of this paper. The form of the dependence of S(t) on paleoclimate parameters is still unknown, so we cannot assess its sensitivity to those parameters. If a wide range of climate conditions all create microtopography with comparable ventilation characteristics, then the inverse problem may be tractable, in the sense that we can estimate the role of paleo-ventilation even if large uncertainties are associated with the climate parameters in the past. However, if similar climate conditions can produce microtopographies with significantly different ventilation characteristics, the inverse problem may be unsolvable.

Seasonal cycles can also confound interpretation of an ice core record. In many areas of Antarctica, accumulation is too low to preserve annual stratigraphy, so the ice records only the average value of $C_{ice}(t)$ over intervals of a year or more. If any two of **f**, W(t), S(t), and $C_a(t)$ have strong seasonal cycles, then the nonlinear nature of (41) means that the mean annual value of C_{ice} cannot be determined from the mean annual values of the individual factors. All three factors, *i.e.* wind, a receptive surface, and airborne particles, must be present simultaneously before ventilation-related deposition can occur.

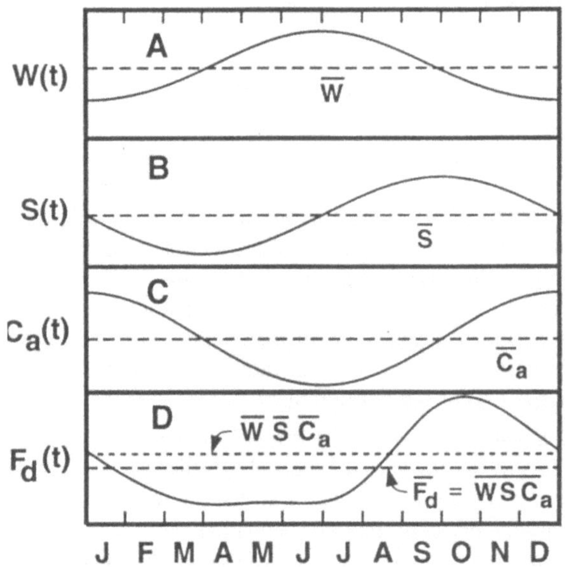

FIGURE 9: Three conditions, (A) a windy site W(t), (B) an effective snow surface S(t), and (C) significant airborne nss concentration $C_a(t)$ must be met simultaneously to give (D) significant dry deposition flux by ventilation $F_d(t)$ (39). When each factor has its own seasonal cycle, the annual average of their product $\overline{F}_d$ can differ substantially from the product of their averages $\overline{W}\times\overline{S}\times\overline{C}_a$.

Figures 9A, 9B, and 9C schematically show the seasonal cycles of W(t), S(t), and $C_a(t)$ at an Antarctic site such as South Pole. The dashed lines show the annual average values $\overline{W}$, $\overline{S}$, and $\overline{C}_a$. For this example, the highest winds U, which control W(t), are assumed to occur in midwinter. The surface configuration factor S(t) is assumed to be largest in spring, as large sastrugi decay and lighter winds build small ripples. The atmospheric concentration C_a of nss sulphate also varies seasonally over the Antarctic ice sheet. Bodhaine et al. (1986), and Shaw (1988) measured a factor of ten variation between the high levels in summer and low levels in mid-winter at South Pole, as shown schematically in Figure 9C. If all aerosol particles entering the snow are captured there, then $F_d(t)$ (39) the aerosol dry deposition flux due to ventilation in Figure 9D is the product of the other three curves. The dashed horizontal line in Figure 9D shows the annual average value of $F_d(t)$, while the dotted horizontal line shows the product of the annual average values $\overline{W}$, $\overline{S}$, and $\overline{C}_a$. Clearly, significant errors can be made if existence of strong seasonal cycles is not recognized when the transfer function is applied over time intervals longer than one year.

Depositional Resistance for Non-seasalt Sulphate Aerosol

Ventilation is only one of several processes affecting the transport of aerosol particles to ice sheets. The deposition partition factor **f** (40) depends on those other processes. We now assess the importance of ventilation for nss sulphate in relation to other dry-deposition processes. Viewing concentration $C_a(z)$ as a potential function that drives aerosol motion, and dry flux F_d as a current, then dry deposition of submicrometer aerosol particles can be described by resistances in series and parallel (Davidson and Wu, 1990; Davidson et al., this volume) by analogy with an electrical circuit. Total resistance r_t is the *depositional slowness*, given by the inverse of the deposition velocity V_d, *i.e.*

$$V_d = \frac{1}{r_t} = \frac{1}{r_a + r_b + r_c} + \frac{1}{r_g} \tag{42}$$

where r_a is the aerodynamic resistance in the first several meters above the surface, r_b is boundary layer resistance across the viscous sublayer (within 1 mm of the snow grain surface), r_c is surface resistance and r_g is the gravitational resistance as defined by Davidson et al. (this volume) and shown in Figure 10A.

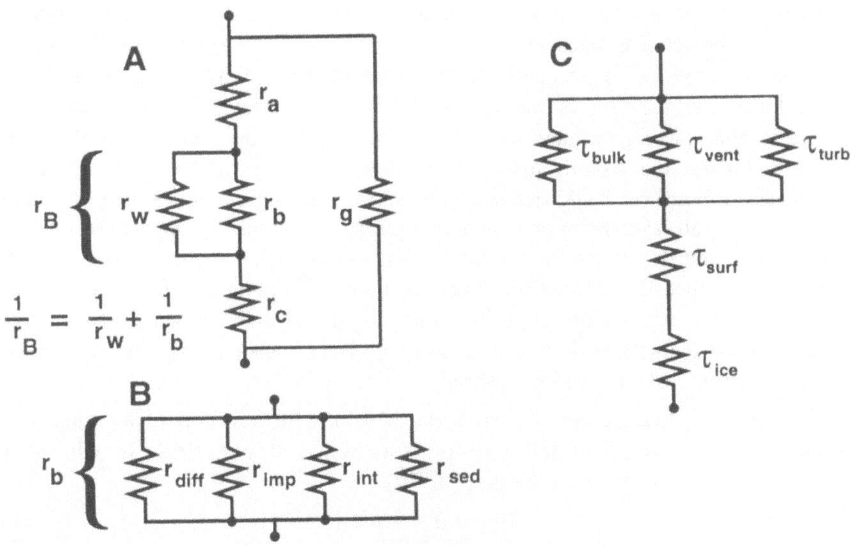

FIGURE 10: (A) Depositional resistance for aerosol particles comprises gravitational settling resistance r_g operating in parallel with a series chain of aerodynamic resistance r_a in the atmosphere, viscous sublayer resistance r_b, and resistance r_c at the ice grain surface.
(B) Processes of diffusion, impaction, interception and sedimentation act in parallel to transport particles across the viscous sublayer. Windpumping also moves particles across the viscous sublayer; it acts as an additional pathway r_w in parallel with r_b.
(C) For reversibly deposited species, resistivity to equilibration with the atmosphere for snow at depth z can be represented by a characteristic time for each process. The total resistivity to equilibration (Figure 11) is found by connecting the separate resistivities as shown here.

Boundary layer resistance r_b controls movement across the viscous sub-layer. Models developed for r_b (Davidson and Wu, 1990; Ibrahim et al., 1983) involve transport by diffusion, impaction, interception, and sedimentation as shown in Figure 10B. Windpumping promotes deposition across the viscous sub-layer by bringing aerosol particles into contact with a much larger effective area of ice. The effects of windpumping could be incorporated in r_b by modifying the individual resistances for these mechanisms. However, this approach would be complicated and we do not attempt it here. Instead we represent ventilation in a more general sense that encompasses changes to all viscous sub-layer processes, by adding a wind-ventilation resistance r_w in parallel with r_b as shown in Figure 10A. The equivalent boundary layer resistance r_B is then

$$r_B = \left[\frac{1}{r_b} + \frac{1}{r_w}\right]^{-1} \tag{43}$$

and the total resistance r_t becomes:

$$\frac{1}{r_t} = \frac{1}{r_a + fr_b + r_c} + \frac{1}{r_g} \qquad f = \frac{r_w}{r_b + r_w} . \tag{44}$$

The settling velocity is very small for submicron aerosol particles (Reist, 1993, p. 339), so r_g is very large. Surface resistance r_c expresses the likelihood that a particle will not remain bound to an ice crystal once it has been deposited to the snow. For particles that do not undergo reentrainment, $r_c \approx 0$. Therefore, neither r_c nor r_g plays a significant role for these particles in the circuit shown in Figure 10A, and the total resistance r_t depends only on the relative sizes of r_a and r_B. Davidson et al. (this volume) note that deposition (in the absence of ventilation), is frequently limited by transport across the viscous sub-layer, *i.e.* $r_a \ll r_b$. However, ventilation of permeable surface snow might decrease the viscous layer resistance so that $r_B \ll r_a \ll r_b$. In this case, r_a would be the rate-limiting factor, *i.e.* the viscous sub-layer could easily support a higher flux of aerosol particles than the atmosphere could deliver to the surface. Therefore, atmospheric processes rather than viscous sub-layer processes, would control the aerosol dry deposition rate, and research to derive transfer functions would need to focus on the atmospheric processes. However, if $r_B \geq r_a$, then the viscous sub-layer, including windpumping effects, will remain a very significant influence on dry deposition rate. If $r_B \approx r_a$, then neither process is rate-limiting; both aerodynamic transport and viscous sub-layer transport must be fully understood and incorporated into deposition models.

An estimate of r_a can be derived from the equations for transport of air momentum and boundary layer flux-gradient relationships (Davidson and Wu, 1990; Ibrahim et al., 1983; Businger, 1973, p. 67ff; Arya, 1988, p. 157ff) :

$$r_a = \frac{[\ln(z/z_0) + 5\ z/L\ ]}{k\ u_*} \tag{45}$$

where z is the height above the surface, z_0 is the surface roughness height, L is the Obukhov length, $k = 0.41$ is von Karman's constant, and u_* is the friction velocity. The term z/L is a dimensionless stability parameter representing the relative importance of buoyancy and turbulence. It can be calculated from measurements of temperature, wind speed, and pressure using the Richardson number (Arya, 1988). For conditions pertinent to the Antarctic plateau, meteorological parameters typical of South Pole are used in the calculation. A 10-meter wind speed of 5 m s^{-1}, a 10 meter temperature of -45°C, a surface temperature of -50°C, an atmospheric pressure of 680 mb, and a surface roughness of $z_0 = 0.001$ m are used to calculate deposition velocity and resistance from (45) using z = 10 m:

$$V_a \approx 0.5 \text{ cm s}^{-1} \qquad r_a \approx 2 \text{ s cm}^{-1} \tag{46}$$

The stability parameter z/L used in this calculation is based on Monin-Obukhov theory (*e.g.* Arya, 1988, Chapter 11) which is not valid for the very stable conditions that can occur over polar ice sheets with strong temperature inversions. However, when the winds are strong (the times when we are interested in windpumping), vertical mixing tends to move the atmosphere toward neutral stability. With these limitations, the method provides a rough estimate of r_a.

Cunningham and Waddington (Figure 7A) calculated that the volume flux of air entering the snow at the South Pole could be $\approx 6\times10^4$ m^3 m^{-2} a^{-1}. This corresponds to a wind-induced deposition velocity V_w and a resistance r_w of

$$V_w \approx 0.2 \text{ cm s}^{-1} \qquad r_w \approx 5 \text{ s cm}^{-1} \tag{47}$$

Cunningham and Waddington used a calculated intrinsic permeability of $K = 4\times10^{-9}$ m^2. Measured permeability at the South Pole in January 1991 (D.L. Morse and S.G. Warren, personal communication) was smaller by a factor of approximately 4, probably reflecting the existence of a surface wind slab as described by Albert (in press). This would reduce the volume flux by the same factor, giving

$$V_w \approx 0.05 \text{ cm s}^{-1} \qquad r_w \approx 20 \text{ s cm}^{-1} \tag{48}$$

Davidson and Wu (1990) reviewed several methods for determining r_b. We use the simple model of Hicks et al. (1985):

$$r_b = \frac{2}{k\,u_*}\left[\frac{Sc}{Pr}\right]^{2/3} \tag{49}$$

where $Sc = \nu/D$ is the Schmidt number, $\nu = \mu/\rho_{air}$ is the kinematic viscosity for air, D is the Brownian diffusion coefficient, $Pr = \nu/\kappa_T$ is the Prandtl number, and κ_T is the thermal diffusivity. Using conditions common to the South Pole, for a particle size of 0.1 μm, we calculate r_b = 180 s cm^{-1}. Ibrahim et al. (1983) experimentally found the resistance across the viscous sub-layer over a snow surface at −11°C to be 38 s cm^{-1} in a stable atmosphere. Using these estimates for r_b and our estimates for r_w above, we calculate a range for the total boundary layer deposition velocity and resistance as follows;

$$V_B \approx 0.06 - 0.2 \text{ cm s}^{-1} \qquad r_B \approx 4 - 18 \text{ s cm}^{-1} \tag{50}$$

Consequently $r_B \approx r_a$, and both aerodynamic transport and transport across the viscous sub-layer appear to be essential components in the air-snow transfer function. Effects of windpumping must be incorporated into the viscous boundary layer treatment before ice core records can be fully understand. However, uncertainties in parameters in the volumetric air flux model such as permeability, surface roughness distribution, and windiness suggest that further work is needed to affirm this conclusion.

VENTILATION AND REVERSIBLY-DEPOSITED SPECIES

Gases containing nitrogen are reversibly deposited in snow. They are an important source for nitrate measured in ice cores (Wolff, 1995; Silvente and Legrand, 1995). Sulphur dioxide can also be measured in snow (Bales et al., 1987; Conklin and Bales, 1993). Hydrogen peroxide H_2O_2, is also reversibly deposited (Sigg and Neftel, 1988, 1991; Conklin et al., 1993; Bales et al., (1995); Bales and Choi, this volume; Neftel, this volume).

Hydrogen peroxide is climatically important as an indicator of the oxidation state of the atmosphere. We will treat hydrogen peroxide as a representative chemical species as we describe the effect of ventilation on ice core concentrations of reversibly deposited substances.

Approach to Equilibrium

Hydrogen peroxide is incorporated within snow grains by co-deposition with water vapor (Conklin et al., 1993) and brought initially to the ice sheet by wet deposition. Although co-deposition is a non-equilibrium process, it is reasonably well understood; a relatively simple transfer function can be derived if the concentration in the snow can be preserved unaltered through the firnification process. As noted by Bales and Choi (this volume), the transfer function for reversibly deposited chemical species may also be particularly simple if equilibrium can be established between the atmosphere and the snow before the snow is buried beyond the reach of the atmosphere. This air-snow transfer function depends only on temperature-dependent chemical equilibrium partition functions. Co-deposition and equilibrium establish two end-member transfer functions.

However, bulk concentrations in surface snow have been observed to decrease or increase significantly in a complicated way in the days or months after deposition, as sublimation, surface adsorption, diffusion through ice grains and pore spaces, temperature cycles, and other processes all act to move the (initially co-deposited) chemical concentration in the snow toward equilibrium with the local atmosphere. As the snow is buried, it becomes progressively more difficult for chemical species to move to the surface in sufficient quantity to significantly alter the chemical concentrations at depth. Even though air can still move through the snow at depth, this motion becomes too slow to carry enough of the reversibly deposited species to significantly alter the bulk chemical concentration in the time available before pore close-off. In this intermediate range, burial may be slow enough that some post-depositional change takes place, but too rapid to allow full equilibration to be achieved, and the snow concentration could be "locked in" at a non-equilibrium level. In that case, the air-snow transfer function is much more uncertain than either end-member case. The transfer function must incorporate the degree to which the snow concentration has moved from co-deposition toward equilibrium. The transfer function then depends on (a) chemical properties of the species, (b) processes that attempted (incompletely) to redistribute the species in snow, and (c) the nonequilibrium concentration history of the snow and atmosphere prior to "lock-in". The inverse problem (Waddington, this volume) *i.e.* determination of the paleo-atmosphere levels from such ice core data could be difficult and might yield poorly constrained results for the atmospheric concentrations in the past. Ventilation affects reversibly-deposited species by facilitating rapid approach to equilibrium (Figure 2C), and making it possible in some cases to use the simple equilibrium end-member transfer function.

Resistivity to Equilibration

The key question for reversibly deposited species is - do they reach equilibrium with the atmosphere before they are buried beyond its reach? The role played by ventilation in deposition and approach to equilibrium with the atmosphere for reversibly deposited species can best be understood in the context of all other processes acting in the snow and firn. We cannot easily represent physical processes by a network of lumped resistors, as we did in Figure 10A for irreversibly deposited aerosols, because the firn is a continuous medium and chemical fluxes depend in a continuous way on the depth. Previously we let the dry chemical mass flux F_d play the role of current. The airborne concentration C_a played the role of the potential driving the current. The analogies in the continuous system are similar; the chemical flux F_{snow} (mol m^{-2} s^{-1}) in the snow is the current density and the bulk chemical concentration C_{snow} (mol m^{-3}) is the driving potential. However, the circuit now passes the current density F_{snow} through a continuous medium defined by its *resistivity* instead of through a series of lumped resistors. Resistivities can be defined for the physical processes that control the approach to equilibrium. Resistance for electrical current problems has units of ohm, and resistivity has units of ohm–m. Units of resistance in our dry deposition problem were depositional slowness s m^{-1}, so by analogy, we expect that the units of resistivity are slowness units multiplied by a length, *i.e.* units of time.

The total resistance **r** of a lump of resistive material with bulk resistivity ρ, length X_0 in the direction x of current flow, and cross-section area **A**(x) normal to the current flow, is then defined by

$$\mathbf{r} = \int_0^{X_0} \frac{\rho(x)}{\mathbf{A}(x)} dx \tag{51}$$

The resistivity, which can vary continuously with depth, has the form

$$\rho(x) = \mathbf{A}(x) \frac{\partial \mathbf{r}}{\partial x} = \mathbf{A}(x) \frac{\partial}{\partial x} \left[V_d^{-1}(x) \right] \tag{52}$$

which has the anticipated units of time. For each physical or chemical process affecting the approach to equilibrium with the atmosphere, we can represent the resistivity by a depth-dependent characteristic time for that process. The resistivities are then combined in a simple exploratory model of chemical equilibration in polar firn.

We will calculate characteristic times for two ice sheet regimes described in Table 1. Each regime is defined by its climatology, surface microtopography and snow properties.

TABLE 1: CLIMATE AND ICE SHEET PARAMETERS CONTROLLING RESISTIVITY TO EQUILIBRATION FOR REVERSIBLY-DEPOSITED CHEMICAL SPECIES

Site A: Dry windy environment (*e.g.* central Antarctica)
Site B: Higher precipitation and less wind (*e.g.* Greenland)

	Climatology			Microtopography		Snow and Firn	
	Accumulation Rate $\dot{b}$ (m a^{-1})	Wind Speed U (m s^{-1})	Turbulent Frequency $\omega_0/2\pi$ (Hz)	Bump Height h (m)	Bump Wavelength λ (m)	Snow Permeability K (m^2)	Closeoff Depth h_0 (m)
A	0.08	10.0	1.0	0.05	0.5	4.0×10^{-9}	80.0
B	0.75	2.5	1.0	0.05	0.5	4.0×10^{-9}	50.0

The accumulation rate $\dot{b}$ represents true snow depth rather than water equivalent. Conditions at site A could be characteristic of the dry, windy Antarctic plateau, while site B could be characteristic of areas of Greenland or small polar ice caps with higher snow accumulation and less wind.

Environmental Scaling Times

The approach to equilibrium depends on two types of time constants, (a) those that represent rates for the equilibration processes, and therefore act as resistivities, and (b) those that describe environmental time limits giving the time available for the processes to operate. We first identify the environmental time constants.

We expect that the time for sub-surface air to move through a porous snow bump will scale with the time τ_{wind} for wind to cross above the bump, given by

$$\tau_{wind} = \lambda/U = 0.05 \text{ s} \approx 2\times10^{-9} \text{ a} \tag{53}$$

for site A in Table 1. The age of snow at depth z is given by

$$\tau_{age}(z) = \frac{z}{\dot{b}} . \tag{54}$$

The ventilation e-folding depth δ (31) and the snow accumulation rate $\dot{b}$ (depth of annual snowfall) can be combined into a natural time scale $\tau_\delta(z)$ given by

$$\tau_\delta(z) = \frac{(\delta - z)}{\dot{b}} . \tag{55}$$

$\tau_\delta(z)$ estimates the remaining time during which snow will be exposed to topographically-driven ventilation. At the dry windy site A, $\tau_\delta(0) \approx 1$ a. For site B, $\tau_\delta(0) \approx 0.1$ a. After time τ_δ, further equilibration between the bulk snow and the atmosphere can be achieved only by slower, less effective transport processes such as barometric pumping or molecular diffusion through the pore spaces or ice grains.

The time remaining until pore closeoff, given by

$$\tau_{close}(z) = \frac{(h_0 - z)}{\dot{b}} , \tag{56}$$

shows how much time is left for equilibration before pore-spaces become isolated. At the surface, $\tau_{close}(0) \approx 10^3$ a for the dry windy site A in Table 1. Processes that take longer than $\tau_{close}(z)$ at depth z have insufficient time to achieve equilibrium.

Diurnal and annual temperature cycles may affect the chemical concentrations in the snow by altering the temperature-dependent vapor pressures and partition functions for chemical species. Therefore the equilibrium level is a moving target. The concentration retained in the ice reflects the last time that equilibrium was approached. Snow at depth z responds to surface temperature changes with a characteristic lag time $\tau_T(z)$ given by

$$\tau_T(z) = \frac{z^2}{\kappa_T} . \tag{57}$$

The thermal diffusivity κ_T, has the value $\kappa_T \approx 10 \text{ m}^2 \text{ a}^{-1}$ for near-surface polar snow (*e.g.* Paterson, 1994, p. 205). The thermal response at depth to a surface temperature change with period $\tau_{diel} \approx 0.003$ a is strongly attenuated at the depth z_{diel} at which $\tau_T(z_{diel}) = \tau_{diel}$. This implies that $z_{diel} \approx 0.2$ m. Similarly, $\tau_T(z_{yr}) = \tau_{yr} = 1$ a leads to $z_{yr} \approx 3$ m for the annual temperature cycle.

Characteristic Times for Equilibration Processes

Now we find the characteristic times to reach equilibrium for the processes controlling the chemical concentrations in the snow. These times are the resistivities to equilibration. The estimates that follow are extremely simple in order to set a framework for understanding the role of ventilation. Times constants can actually vary considerably from species to species, and may also be sensitive functions of temperature.

Diffusion within Ice Grains

Within an ice grain, a chemical concentration that is not in equilibrium with the concentration on the ice grain surface must equilibrate by diffusion through the solid ice. The characteristic time τ_{ice} associated with solid diffusion is controlled by the radius r of an ice grain, the bulk diffusivity κ_{ice} for that chemical species in ice, and a geometric factor G that arises because an ice grain is not a one-dimensional object. For stable isotopes of water, Whillans and Grootes (1985) used $\kappa_{ice} \approx 5\times10^{-8}$ m^2 a^{-1} at -20^oC. Values for other species are probably in the same magnitude range. They also derived the geometric factor in terms of Bessel functions, obtaining G = 0.056 for elongated ice grains. Grain size increases with depth as firn ages. Stephenson (1967) found that near-surface snow grains on the Antarctic plateau increased from $r \approx 0.025$ mm to $r \approx 0.25$ mm in the upper 1 cm. At South Pole, Gow (1969) found that the growth rate fell off steeply at greater depths. He found a linear increase in grain area πr^2 over time, with $r \approx 0.3$ mm at 2 m depth, and $r \approx 0.4$ mm at z = 10 m. We used an analytical approximation for r(z) that fits both data sets. Then,

$$\tau_{ice}(z) = G\,\frac{r^2(z)}{\kappa_{ice}} \approx 0.056\,\frac{r^2(z)}{\kappa_{ice}} \tag{58}$$

For ice grains of radius r = 0.1 mm, the equilibration resistivity is $\tau_{ice} \approx 0.01$ a. This time could be shortened by grain recrystallization.

Liquid-like Surface Layer

Significant concentrations of atmospheric chemical species can be retained in a liquid-like layer on the surface of ice grains. This reservoir attempts to equilibrate with the corresponding vapor phase in the firn pore spaces. The time constants and the concentrations in this layer may depend on chemical as well as physical processes (Conklin and Bales, 1993). Conklin et al. (1993) measured the rate of uptake of hydrogen peroxide on ice grains. Their experiment measured a combination of surface uptake time τ_{surf} and diffusion time τ_{ice} in the solid. However, we can use their results to estimate an upper limit on τ_{surf} for H_2O_2. They expressed the transfer process for H_2O_2 crossing air-ice grain boundaries by the equation

$$[1-\phi]\frac{\partial C_i}{\partial t} = \phi\, k_f \left[C_a - \frac{C_i}{K_D} \right] \tag{59}$$

where k_f is a *rate coefficient*, and K_D is a *partition coefficient*. Conklin et al. found $k_f \approx 170$ s^{-1} and $K_D \approx 5\times10^5$ for H_2O_2 and ice at -30^oC. C_a is the concentration (mol m^{-3}) in the pore-space air, and C_i is the bulk concentration (mol m^{-3}) in the ice. For our simple model we will assume that C_i is the concentration in the liquid-like surface layer. While the coefficients k_f and K_D are species-specific and temperature-dependent, we can use the values for hydrogen peroxide to estimate an order-of-magnitude upper limit on the characteristic surface resistivity τ_{surf} (diffusion in the solid ice may also have influenced the response times measured in the experiments reported by Conklin et al.). By setting the air concentration $C_a = 0$ in (59), we find an analytical solution for $C_i(t)$. The concentration on the ice $C_{ice}(t)$ relaxes exponentially toward zero with a time constant

$$\tau_{surf}(z) = \left[\frac{1-\phi}{\phi}\right]\left[\frac{K_D}{k_f}\right] \approx 3\times10^3 \text{ s} \approx 10^{-4} \text{ a} \tag{60}$$

which is the surface resistivity that we seek. Porosity ϕ decreases with depth in the firn; however, we will now neglect this influence on τ_{surf}.

Topographic Ventilation

Topographic ventilation, turbulent windpumping, and molecular diffusion in the pore spaces all act in parallel to exchange pore-space gases with the atmosphere.

A characteristic residence time τ_{vent} for air to flow through the snow under forced ventilation, from a high pressure area to a low pressure area (Figure 5) separated by λ, can be obtained from the Colbeck model (29). This time is also a characteristic flushing time for gaseous chemical species in the firn pore spaces. The speed v_{air} of an air parcel in a porous medium is defined as the magnitude $|\vec{Q}|$ of the volumetric flux (29) divided by the porosity ϕ. Therefore, τ_{vent} is given by

$$\tau_{vent}(z) = \frac{\lambda}{v_{air}} = \frac{\lambda}{|\vec{Q}|/\phi} = \frac{\lambda^2\phi\mu}{2\pi K p_0}\, e^{2\pi z/\lambda} = \frac{1}{C}\,\frac{\lambda}{h}\left[\frac{\mu/\rho_{air}}{U\lambda}\right]\frac{\phi\lambda^2}{2\pi K}\,\frac{\lambda}{U}\, e^{2\pi z/\lambda} \tag{61}$$

We used (26) to eliminate p_0. The factor in square braces is the Reynolds number $Re = U\lambda/\nu \approx 1.5\times10^5$ for free air flow over the bump, and the factor λ/U is the wind time constant τ_{wind} (53). Recognizing that $\lambda/2\pi = \delta$ is the ventilation e-folding depth, we obtain

$$\tau_{vent}(z) = \left[\frac{1}{h/\lambda}\,\frac{\phi\lambda\delta}{K}\right]\left[\frac{1}{C}\,\frac{1}{Re}\,\tau_{wind}\right] e^{z/\delta} \approx 3\, e^{z/\delta} \text{ s} \approx 10^{-7}\, e^{z/\delta} \text{ a} \tag{62}$$

for site A in Table 1. The first set of square braces characterizes the surface snow. The second factor in braces characterizes the wind generating the sub-surface air flow.

Molecular Diffusion

The characteristic time $\tau_{air}(z)$ for a single air exchange between the pore spaces and the free atmosphere by molecular diffusion depends on the depth z and the diffusion coefficient κ_{air} for gases in firn. Schwander et al. (1988) measured the diffusivity of firn samples from Siple Station, Antarctica. For low flow rates of both O_2 and CO_2, they found an approximately linear decrease of κ_{air} from $\approx 10^{-5}$ m^2 s^{-1} near the surface to vanishingly small values near the pore closeoff depth h_0. Similar numbers can be obtained from first principles by modifying the free-air value of diffusivity to account for blockage of diffusion pathways by ice grains (Whillans and Grootes, 1985). The characteristic time $\tau_{air}(z)$ is given by

$$\tau_{air}(z) = \frac{z^2}{\kappa_{air}(z)} \tag{63}$$

Using the average value of $\kappa_{air}(z)$, we find that $\tau_{air} \approx 40$ a at the pore closeoff depth. This value is within the range of firn gas ages at pore closeoff estimated by Schwander et al. (1988).

A pore gas like O_2 can equilibrate with the atmosphere in a single air exchange. However, chemical species that also reside in or on the ice require many more air exchanges because of the large chemical reservoir in and on the ice grains. If we assume that the chemical-constituent-to-water ratio in the bulk ice is approximately equal to the ratio in the pore-space air, then the number n of air exchanges required to equilibrate the firn with the atmosphere is approximately

$$n \approx \frac{(1-\phi)\, C_{ice}}{\phi\, C_a} \approx \frac{(1-\phi)\, \rho_{ice}}{\phi\, \rho_{vapor}} \approx \left[\frac{1-\phi}{\phi}\right] \frac{\rho_{ice}}{\rho_{air}\, P_{H_2O}/P_{air}} \approx 10^6 \tag{64}$$

where we used $P_{air} \approx 0.8\times10^4$ Pa and $P_{H_2O} \approx 10$ Pa, the saturation vapor pressure over ice at −25°C. This value of n is comparable to the partition coefficients K_D determined experimentally for H_2O_2 by Conklin et al. (1993).

A characteristic time τ_{bulk} for the bulk firn to equilibrate with the free atmosphere is given by

$$\tau_{bulk}(z) = n\, \tau_{air}(z) = \frac{z^2}{(\kappa_{air}/n)} \tag{65}$$

Hydrogen peroxide is known to redistribute locally within the snow (Sigg and Neftel, 1991) by the same vapor- and solid-diffusion mechanisms that control diffusion of the stable isotope ^{18}O (Whillans and Grootes, 1985). Whillans and Grootes solved the diffusion equations for stable isotope migration in firn. By this more rigorous analysis they derived a bulk isotopic diffusion coefficient $\kappa_{bulk} = 10^{-4}\ m^2\ a^{-1}$ at porosity $\phi \approx 0.3$ and temperature −30°C. Their bulk diffusion coefficient should be approximately correct for diffusion of other chemical species under similar conditions. As we might expect, $\kappa_{bulk} \approx \kappa_{air}/n$. This bulk diffusion through the gas phase can smooth out annual signals over short distance scales (Johnsen, 1977; Whillans and Grootes, 1985; Sigg and Neftel, 1991). However, the bulk concentration of H_2O_2 or other species in firn should be unable to move significantly toward equilibrium with the atmosphere in the available time $\tau_{close} \leq 10^3$ a when the depth $z > 0.3$ m in (65).

Turbulent Windpumping

The diffusivity coefficient of an ensemble of gas molecules can be estimated (Reif, 1965; p.336) from the average speed $\bar{v}$ and the mean free path length l between collisions. Turbulent windpumping might help to exchange the pore-space air by acting as a mixing process that can be represented by a diffusion coefficient κ_{turb}. We approximate $\bar{v}$ by (21) the mean speed $\bar{v}_{air}(z)$ of the air in response to surface pressure oscillations at frequency ω_0, and we use (22) the displacement $l(z)$ of the pore-space air during a single pressure cycle as an analogue to a mean free path to get

$$\kappa_{turb}(z) = \bar{v}\, l = \frac{8}{\pi}\, \frac{1}{\omega_0 H^2}\, \frac{K^2}{\phi^2}\, \frac{p_0{}^2}{\mu^2}\, \exp\left[-2z\left(\frac{1}{H} + \frac{1}{\Lambda}\right)\right] \tag{66}$$

The exponential decay factor involving Λ is added because the e-folding depth depends in an approximate way (Clarke and Waddington, 1993) on the spatial scale Λ of turbulent bursts as well as on the natural 1-D pressure diffusion scale H (7). For site A, $\bar{v}_{air}(0) \approx 1.5\times10^{-4}\ m\ s^{-1}$, $l(0) \approx 7\times10^{-5}$ m, and $\kappa_{turb}(0) \approx 10^{-8}\ m^2\ s^{-1}$.

The characteristic time that plays the role of resistivity for turbulent windpumping is given by

$$\tau_{turb}(z) = \frac{z^2}{\kappa_{turb}/n} = n\, \frac{\pi}{4} \left[\frac{\phi}{K}\, \frac{P_0\, \mu}{p_0^2}\right] \exp\left[2z\left(\frac{1}{H} + \frac{1}{\Lambda}\right)\right] z^2 \tag{67}$$

where n is the number of air exchanges (64) required to reach equilibrium. We used (4) and (7) to simplify (66). P_0 is the average atmospheric pressure P_{air}. The magnitude p_0 of the turbulent pressure variations can be expressed in terms of wind speed U through (24), and Λ can be estimated in terms of ω_0 and U using the Taylor hypothesis (25).

Total Resistivity to Equilibration

The total resistivity to equilibration $\tau_{total}(z)$ at any depth z can be formed by connecting the resistivities for all these processes using the circuit shown in Figure 10C to get

$$\tau_{total}(z) = \tau_{ice}(z) + \tau_{surf}(z) + \left[\frac{1}{\tau_{bulk}(z)} + \frac{1}{\tau_{vent}(z)} + \frac{1}{\tau_{turb}(z)}\right]^{-1} . \tag{68}$$

These total equilibration resistivities are plotted in Figure 11 for the upper 10 m for both sites A and B in Table 1.

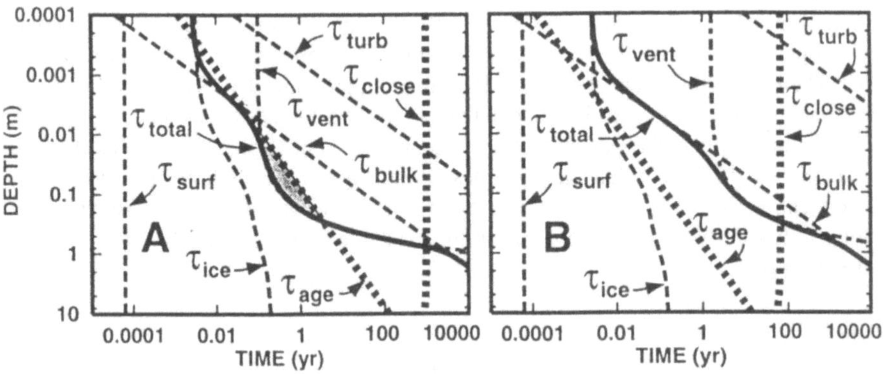

FIGURE 11: Times for transport through ice grains (τ_{ice}), across the liquid-like grain boundaries (τ_{surf}), and through the pore spaces (τ_{vent}, τ_{bulk}, and τ_{turb}), can be connected at each point as shown in Figure 9C to obtain the profile τ_{total} of total resistivity to equilibration with the atmosphere for reversibly deposited species. Heavy dotted lines show τ_{age}, the age of the snow, and τ_{close}, the time remaining until the snow reaches the firn - ice transition. Panels A and B correspond to sites A and B in Table 1. Site A reaches equilibrium because topographic ventilation τ_{vent} makes $\tau_{total} \ll \tau_{age}$ in the shaded depth range $0.01 < z < 0.30$ m; Site B probably cannot equilibrate.

First, we address conclusions applicable to both sites A and B. Transfer across the liquid-like grain surface layer τ_{surf} does not control equilibration at any depth. Diffusion in the ice grains τ_{ice} is the rate-limiting process in the upper few millimeters. Below a depth of a few millimeters, transfer through the pore spaces controls the equilibration rate. Bulk molecular diffusion in the firn τ_{bulk} controls equilibration over the depth range of 1 mm < z < 1 cm. Ventilation due to microtopography τ_{vent} controls equilibration time τ_{total} at all depths between $z \approx 1$ cm and $z \approx 1$ m. At all depths below $z \approx 1$ m, bulk diffusion τ_{bulk} again is the limiting factor.

The importance of molecular diffusion at depths near $z \approx 1$ cm could be an artifact of our microtopography which is represented with a single wavelength $\lambda = 0.5$ m. A natural surface has a spectrum of bump wavelengths and could be a fractal (Shook and Gray, 1994). The path-length that an air parcel must travel through a bump scales with λ, so shorter bumps could undergo more exchanges with fresh air in a given time interval, resulting in faster equilibration. Thus our model may underestimate the role that topographic ventilation plays in equilibration in the upper few millimeters.

Turbulent windpumping τ_{turb} is never a factor; it is always less effective than molecular diffusion τ_{bulk} by several orders of magnitude. Differences in particle energy associated with molecular diffusion and turbulent pumping cause this effect. The diffusion rate in an ensemble of particles is a function of their mean energy. For both gases and Brownian particles, this energy is given by the equipartition theorem (*e.g.* Reif, 1965, p.246) as 3/2 kT, where $k = 1.38\times10^{-23}$ J K^{-1} is the Boltzmann constant, and T is absolute temperature. At -30^oC, $3/2\ kT \approx 3.4\times10^{-21}$ J. The energy given by turbulent windpumping to a typical small gas molecule such as H_2O_2 of mass $m \approx 6.5\times10^{-26}$ kg is only $1/2\ mv_{air}^2 \approx 7\times10^{-34}$ J. In order to have energy 3/2 kT, a particle advected by turbulent windpumping must have a much larger mass M given by

$$M = \frac{3/2\ kT}{1/2\ v_{air}^2} \approx 3\times10^{-13}\ \text{kg}\ . \tag{69}$$

This mass is approximately 12 orders of magnitude larger than the mass of a hydrogen peroxide molecule. If we assume that large particles have a density of $\rho \approx 10^3$ kg m^{-3}, then the minimum radius r of particles that can be diffused as effectively by turbulent windpumping as by Brownian motion is

$$r = \left[\frac{3}{4\pi}\frac{M}{\rho}\right]^{\frac{1}{3}} \approx 3\times10^{-6}\ \text{m}\ . \tag{70}$$

Only soil or sea salt, which tend to be irreversibly deposited, take the form of particles with $r \approx 3\ \mu m$ over the ice sheets. Therefore, turbulent windpumping is unlikely to affect the rate at which any reversibly deposited species equilibrates with the atmosphere.

We now examine the differences in the equilibration resistivities for sites A and B in Figure 11. Equilibration is likely to be achieved where $\tau_{total}(z) < \tau_{age}(z)$, *i.e.* the snow can equilibrate with the atmosphere in less time than it takes to bury it. However, the snow is not stationary; it moves deeper as equilibration proceeds. Where the separation between these two curves $\Delta\tau = \tau_{age}(z) - \tau_{total}(z)$ increases with depth, equilibration becomes even more likely, the deeper the snow is buried. In Figure 11A, the total resistivity $\tau_{total}(z)$ is close to the age curve $\tau_{age}(z)$ in the upper few centimeters. Here, the concentration in the snow should tend to equilibrate with the current concentration in the air and then track the air concentration as it changes with temperature or other effects. $\Delta\tau$ is both positive and increasing with depth for $1 < z < 10$ cm where topographic ventilation τ_{vent} controls the total resistivity τ_{total}. Equilibration is almost certainly achievable in this depth range as a result of vigorous topographic ventilation. The equilibration time here is now a significant fraction of a year, so the snow concentration should reflect the average air concentration over a comparable time interval. Down to depth $z \approx 30$ cm, the equilibration time is still less than the age, so equilibration is still possible; however, this snow is carried into a region of less efficient equilibration faster than it can approach equilibrium. After the curve

$\tau_{total}(z)$ crosses the age of the snow $\tau_{age}(z)$ at $z \approx 30$ cm, equilibration is unlikely. After the equilibration time $\tau_{total}(z)$ exceeds the time $\tau_{close}(z)$ remaining until pore closeoff, further equilibration is impossible.

If a reversibly deposited species is able to achieve equilibrium with the atmosphere under conditions for which site A is typical, then (a) it must do so in the depth range of $0.01 < z < 0.3$ m, and (b) equilibration is only made possible by topographic ventilation.

Figure 11B shows the corresponding resistivities at a site with higher accumulation and less wind (site B, Table 1). The equilibration time $\tau_{total}(z)$ exceeds the age of the snow τ_{age}, and the difference $\Delta\tau$ becomes progressively more negative as depth increases. Below the depth of $z \approx 30$ cm, the snow cannot equilibrate with the atmosphere in the time τ_{close} remaining until pore closeoff (even if it could remain at $z \approx 30$ cm for that entire time). We conclude that reversibly deposited chemical species are unlikely to equilibrate with the atmosphere at sites where conditions resemble site B.

We previously estimated that daily temperature cycles could introduce a daily cycle into the equilibrium level that the snow attempts to achieve, down to the depth $z_{diel} \approx 0.2$ m. This is comparable to the deepest level at which equilibration is still possible at site A. However, the equilibration time is much greater than one day at those depths, so the snow concentration cannot change rapidly enough to track or record the daily changes in air concentrations.

Since the seasonal temperature cycle is significant down to $z_{yr} \approx 3$ m, it is still strongly felt at all levels at which equilibration with the atmosphere is possible. Therefore, ice core concentrations that reach equilibrium may reflect the temperature of the season during which the snow passes the "lock-in" depth. Only a weak annual signal would be recorded in the firn at site A, where the snow passing through the "lock-in" point remembers the atmospheric concentration averaged over a period of $\tau_{total} \approx 1$ a at the depth $z \approx 0.3$ m. A strong annual signal should be recorded in the snow at site B, because snow passes through the "lock-in" point in a period short relative to the annual cycle.

Transport by molecular diffusion or ventilation would continue to smooth these annual signals over scales of a few decimeters as they are buried deeper in the firn (Johnsen, 1977; Whillans and Grootes, 1985; Sigg and Neftel, 1991) even though the average value of the annual signal can no longer respond to changing atmospheric concentrations. This diffusion length scale can also be estimated by the depth at which the total equilibration time τ_{total} equals τ_{close} the time remaining to pore closeoff in Figure 11. This diffusion could completely eliminate the weak annual signal at site A where the annual layer thickness of only a few centimeters is less than the diffusion length. However, the strong seasonal signal could be preserved at site B where the annual layers are thicker than or comparable to the smoothing length.

At site B where burial is rapid, the concentrations established by co-condensation may be preserved, and it might be possible to infer paleo-atmospheric concentrations from ice core data using the co-deposition end-member transfer function. However, the simple model in Figure 11 assumes that snowfall is continuous and uniform. If the time between precipitation events is comparable to the equilibration time near the surface, then between snowfalls the snow concentration could move significantly toward equilibrium with the current (and variable) atmospheric level, and the co-deposition transfer function might not be an accurate way to determine atmospheric levels in the past.

Our analysis did not consider erosion and remobilization of surface snow. For snow grains that are suspended or saltating, τ_{bulk} for molecular diffusion to the free atmosphere is very short and τ_{ice} for diffusion in the solid ice controls the equilibration rate. Harder et al. (in press) estimated that the length of time that the average snow grain remains suspended by wind at South Pole is only a few hours; this should have little effect on equilibration. If the upper 1 – 2 mm of snow on average is saltating, then both the effective snow-air surface position and the total equilibration time curve τ_{total} in Figure 11 move down by the same amount. This will have a noticeable effect only at depths of less than 1 cm, where the likelihood that snow can achieve equilibration with the short term (0.01 a) atmospheric concentration is increased. However, our primary conclusion that equilibration with the annual average air concentration is achieved in the wind-pumping zone at 0.01 – 0.3 m depth is unaffected. At site B, the lower wind means that a thinner surface layer is saltating on average, and the curve τ_{total} in Figure 11B cannot be brought close to τ_{age}. Therefore, remobilization of surface snow should have a negligible influence on the equilibration state at site B.

Figure 11 represents a simple exploratory model. We do not solve the systems of coupled differential equations describing the physical, chemical and thermodynamic evolution of the snow. We have merely examined the characteristic times that appear in those coupled systems of equations. A detailed understanding of these processes and more accurate estimates of equilibration times will of course require solution of the full equations. Nevertheless, we have reached useful conclusions about the relative importance of ventilation in relation to other processes hastening equilibration with the atmosphere, and we have readily identified processes that have no significant impact on equilibration.

DISCUSSION

We now speculate on the relative roles that ventilation may have played on the chemical records of irreversibly deposited and reversibly deposited species, from Greenland and Antarctica, now and at the Last Glacial Maximum.

Ventilation and Chemical Concentrations in the Holocene

Irreversible Deposition

Davidson et al. (1985) estimated the relative contributions of wet and dry deposition to total flux for Fe, Mn, Pb and SO_4^{-2} in Greenland for particles with diameter greater than 0.5 µm. They calculated scavenging ratios and dry deposition velocities from snow and air measurements at Dye 3. They then used these values with air concentrations and precipitation rates at other sites, assuming a constant deposition velocity throughout the year, to estimate yearly wet and dry flux for different locations in Greenland. They found that dry deposition accounted for 10% of the flux in high snow accumulation sites and for as much as 35% in low accumulation sites. Bergin et al. (1994) measured wet, fog, and dry deposition of total sulphate collected on a nonporous, surrogate surface at Summit, Greenland. They found that dry deposition contributed 17% of the inventory between May and July of 1993. Deposition due to ventilation was not measured since dry deposition was collected on a nonporous surface. Therefore dry deposition may be somewhat underestimated. In many areas of Greenland, the snow accumulation rate is high and ventilation is probably relatively weak, so total dry deposition flux may not be enhanced significantly by windpumping. The seasonal distribution of snow accumulation in Greenland is poorly known (Dibb, this volume). Data on winter deposition are sparse.

In Antarctica, dry deposition may presently account for 50% of the nss sulphate flux to the ice sheet; Legrand (1987) estimated the dry contribution to be only 20% of the total flux in high accumulation areas, but as high as 70% in low accumulation areas.

Ventilation may be capable of delivering this flux if the microtopography is favorable (Cunningham and Waddington, 1993); however, this microtopography is not yet well characterized, preventing a well-constrained estimate of the modern flux attributable to ventilation.

Reversible Deposition

Ventilation may play a major role at present in equilibration with the atmosphere for reversibly deposited species. Throughout much of Antarctica and in northeast Greenland, where the accumulation rate is low and wind speed is high, wind-driven ventilation should ensure that equilibrium is established before the snow is deeply buried. Non-equilibrium chemical concentrations could be locked into the snow only in high accumulation coastal areas in Antarctica or southern Greenland,

Ventilation and Chemical Concentrations at the LGM

During the Last Glacial Maximum, the center of the Greenland ice sheet was at least 10°C colder (Dahl-Jensen and Johnsen, 1986) and significantly dryer than present (Cutler et al., 1995). Surface conditions were probably similar to those on areas of the Antarctic ice sheet today. This suggests that ventilation may have enhanced the irreversibly deposited aerosol flux to a significantly greater degree than it does today, contributing to a larger value of **f** in (40). Therefore an ice age - interglacial wind bias may exist in the Greenland ice core records. In Antarctica, colder and dryer conditions could also have resulted in an enhanced dry-deposition flux due to ventilation. According to Cunningham and Waddington (1993), the entire change in nss sulphate concentration observed across the ice age - interglacial transition in the Vostok ice core *could* be attributable to reasonable differences in windiness, permeability, and microtopography associated with the two different climate states. However, the lack of a climate-based parameterization of the ventilation process currently prevents us from assessing whether the snow surface at Vostok in the ice age had the characteristics required to enhance ventilation by 50%.

In the windier, lower precipitation conditions at the LGM, reversibly deposited species would have been even more likely to equilibrate with the atmosphere prior to burial beyond the reach of forced ventilation. Higher wind speeds on both ice sheets could also have contributed to enhanced smoothing of the annual signals in the ice core records through ventilation.

Separation of Wet- and Dry-Deposited Concentrations in Ice

We have discussed only the dry-deposition fraction of irreversibly deposited species. From an ice core record, it may be difficult to tell how much of a species such as nss sulphate is dry-deposited, or what fraction **f** (40) of the total ice-core concentration $C_{ice}(t)$ was deposited by ventilation. Legrand (1987) estimated the modern dry-deposition flux by plotting total flux against snow accumulation rate for a number of Antarctic sites. If the dry flux is spatially constant, it can be estimated by the intercept at zero accumulation rate. The same technique could be applied to paleo-climate records if enough well-dated ice cores with well-determined accumulation time-series (*e.g.* Cutler et al., 1995) were available. There are currently too few ice cores to make this approach practical.

Alley et al. (in press) present a simple model that in principle can separate relative wet- and dry-deposited fractions in a single ice core. Their method exploits the different dependence on accumulation rate for the two fractions during climate intervals (*e.g.* the Younger Dryas) over which there is a significant variance in accumulation rate at the site.

Like Legrand, they determine the dry flux by extrapolating the accumulation rate *vs* total flux curve to its intercept at zero-accumulation rate. They applied the technique to the GISP2 ice core from Greenland. If we have reason to think that most or all of the dry fraction was deposited through ventilation, then **f** is approximately equal to the dry fraction determined by the method of Legrand (1987) or Alley et al. (in press) using ice core concentration data $C_{ice}(t)$. This approach should be explored further.

Some potentially serious problems remain in separating wet and dry fractions. The main weakness of the methods of both Legrand and Alley et al. is the assumption that the dry flux is independent of the spatial or temporal changes in climate; in this paper we have argued that variations in windiness and climate-related microtopography significantly impact the dry-deposition flux. It may be possible to incorporate a physical relationship between accumulation and windiness into the Legrand and the Alley et al. techniques Then, the dry-deposition fraction of nss sulphate, the climate conditions at each site and each time, and the ventilation parameters controlling dry deposition by windpumping could be derived simultaneously. Until then, the Legrand and the Alley et al. methods can be used to estimate the dry fluxes as long as their limitations are recognized.

Wet deposition may be approximately proportional to accumulation rate; however, the wet-deposited fraction may also depend in a nonlinear way on the interaction between seasonal precipitation and aerosol concentration cycles. In addition, fog is capable of depositing chemicals to the ice sheet surface at a rate out of proportion to its contribution to the accumulation rate (Bergin et al., 1994). When annual signals are not preserved in the ice core, the method of Alley et al. will be difficult to apply. Until the dry ventilation fraction **f** can be estimated as a function of depth in a particular ice core, the transfer function (41) cannot be used to infer the paleo-atmospheric concentration C_a.

Role for Sophisticated Models

Analytical models have sufficed to describe the basic concepts of ventilation and to derive order-of-magnitude parameters. However, transfer functions will be determined from studies that relate aerosol concentration measurements in the air and in the snow at specific sites where the microtopography and climatology cannot be adequately represented by our simple analytical models. When (a) topographic bumps have asymmetric shapes, (b) permeability is variable and possibly anisotropic, or (c) temperature controls snow structure evolution or chemical reaction rates, then a more sophisticated numerical approach is needed to incorporate these important physical and chemical processes. Tracer experiments to determine *in situ* ventilation characteristics must be interpreted with models that can accurately represent the site-specific geometry and permeability. Where microtopography cannot be well-represented by the analytical functions that we have used, numerical solutions to ventilation problems with other forms of microtopography could be used to test the robustness of the conclusions that we have drawn. The finite element modelling program under development at CRREL (Albert 1992; Albert and McGilvary 1992; Albert, in press) is capable of meeting this need.

CONCLUSIONS

Our efforts to understand the effects of ventilation on chemical concentrations in polar firn are motivated by a larger problem. The ice core inverse problem, *i.e.* the determination of past atmospheric concentrations and associated climate parameters from ice core data, may depend on poorly known and potentially unknowable parameters such as microtopography and wind speed in the past. Nevertheless, ice cores clearly contain a detailed geochemical record. This record presumably reflects (a) paleo-atmospheric chemical source strengths, (b) paleoclimate conditions on the transport paths to the ice sheets, and (c) paleoclimate conditions local to the deposition sites. Therefore we must predict the appropriate influence that each of these variables exerted on deposition in the past in order to correctly interpret ice core records, and ventilation in the past falls under paleoclimate conditions at ice coring sites. Windpumping may have significantly affected the concentration of chemical species in polar snow. Irreversibly deposited species such as nss sulphates can be filtered by the large effective surface area available in the ventilated zone. Ventilation can facilitate rapid equilibration with the atmosphere for reversibly deposited species be that might otherwise be buried prior to reaching equilibrium. A complete understanding of the role played by ventilation will require improved knowledge of the relationship between the local climate and the microtopography and permeability structure on ice sheets. Effective ways to parameterize the snow surface, the snow permeability, and the pressure patterns induced on the surface of ice sheets by wind are needed.

ACKNOWLEDGEMENTS

We thank Steve Warren, Robert Charlson, Garry Clarke, Cliff Davidson, David Fisher, Roger Bales, Mary Albert, and Len Barrie for helpful discussions and suggestions during our work on windpumping and aerosols, and the many people who assisted with field measurements of snow properties and ventilation. This research was supported by grants DPP-8619265, DPP-8915924, OPP-9123660, OPP-9120380, and OPP-9421096 from the National Science Foundation. SLH has been supported by a DOE Global Change Fellowship. Arctic logistical support was provided by Polar Continental Shelf Project, Energy Mines and Resources Canada (project 167-86). Logistical support for Antarctic measurements was provided by Antarctic Support Associates.

REFERENCES

Albert M.R. (1993) Some numerical experiments on firn ventilation with heat transfer. *Annals of Glaciology* 18:161-165.

Albert M.R. (in press) A model for heat, mass, and chemical species transport in polar firn. *Annals of Glaciology.*

Albert M.R., McGilvary W.R. (1992) Thermal effects due to air flow and vapour transport in dry snow. *Journal of Glaciology* 38(129):273-281.

Albert M.R., Arons E.M. (1994) Transport processes in the firn at Summit, Greenland. *EOS* 75(44):213 (Fall Supplement).

Albert M.R., Hardy, J.P. (1995) Ventilation experiments in a seasonal snow cover. In: *Biogeochemistry of seasonally snow-covered catchments.* IAHS Publication No. 228:41-50.

Albert M.R., Arons E.M., Davis R.E. (this volume) Firn properties affecting gas exchange at Summit, Greenland: ventilation possibilities. In: Wolff E.W., Bales R.C., (eds) NATO ASI Series I. *Processes of chemical exchange between the atmosphere and polar snow.* Springer-Verlag, Berlin.

Alley R.B., Finkel R.C., Nishiizumi K., Anandakrishnan S., Shuman C.A., Mershon G.R., Zielinski G.A., Mayewski P.A. (in press) Changes in continental and sea-salt atmospheric loadings in central Greenland during the most recent deglaciation: model-based estimates. *Journal of Glaciology.*

Alley R.B., Saltzman E.S., Cuffey K.M., Fitzpatrick J.J. (1990) Summertime formation of depth hoar in central Greenland. *Geophysical Research Letters* 17(12):2393-2396.

Arya S.P. (1988) *Introduction to Micrometeorology,* Academic Press, San Diego, 307p.

Bader H., Haefeli R., Bucher E., Neher J., Eckel O. Thams Chr. (1939) *Der Schnee und siene Metamorphose (Snow and its metamorphism)* Beitreage zur Geologie der Schweiz, Geotechnische Serie, Hydrologie, Leiferung 3. U.S. Army SIPRE Corps of Engineers Translation 14, 1954. 313p.

Bales R.C., Choi J. (this volume) Conceptual framework for interpretation of exchange processes. In: Wolff E.W., Bales R.C., (eds) NATO ASI Series I. *Processes of chemical exchange between the atmosphere and polar snow.* Springer-Verlag, Berlin.

Bales R.C., Valdez M.P., Dawson G.A. (1987) Gaseous deposition to snow: II physical-chemical model for SO_2 deposition. *Journal of Geophysical Research* 92(D8):9789-9799.

Bales R.C., Losleben M.V., McConnell J.R., Fuhrer K., Neftel A. (1995) H_2O_2 in snow, air and open pore space in firn at Summit, Greenland. *Geophysical Research Letters* 22(10):1261-1264.

Barrie L.A. (1985) Atmospheric particles: their physical and chemical characteristics and deposition processes relevant to the chemical composition of glaciers. *Annals of Glaciology* 7:100-108.

Bender J.A. (1957) Air permeability of snow. *U.S. Army Snow Ice and Permafrost Research Establishment* Research Report 37.

Bergin M.H., Jaffrezo J.L., Davidson C.I., Caldow R., Dibb J. (1994) Fluxes of chemical species to the Greenland ice sheet at Summit by fog and dry deposition. *Geochimica et Cosmochimica Acta* 58(15):3207-3215.

Bergen J.D. (1980) A slip-velocity hypothesis applied to hydraulically smooth wind flow over a snow surface *Journal of Glaciology* 26(94):447-452.

Bodhaine B.A., Deluisi J.L., Harris J.M., Houmere P., Bauman S. (1986) Aerosol measurements at South Pole. *Tellus* 38(B):223-235.

Briscoe B., Brown R.J., Cihlar J., Brancker R.W.S. (1989) A field instrument for surface roughness measurement. *International geoscience and remote sensing symposium '89,* Vancouver BC July 9-13, 1989. 1177-1180.

Businger J. A. (1973) Turbulent transfer in the atmospheric surface layer. *Workshop on Micrometeorology,* 67-100, American Meteorology Society, Boston. 392p.

Businger J.A. (1982) Equations and concepts. In: Nieuwstadt F.T.M. van Dop H. (eds) *Atmospheric turbulence and air pollution modelling.* Reidel, Dordrecht. 1-36.

Carman P.C. (1956) *Flow of gases through porous media.* Academic Press, New York. 182p.

Carslaw H.S., Jaeger J.C. (1959) *Conduction of heat in solids.* 2nd edition, Clarendon Press, Oxford. 510p.

Clarke G.K.C., Fisher D.A., Waddington E.D. (1987) Wind pumping: a potentially significant heat source in ice sheets. In: *The physical basis of ice sheet modelling,* IAHS Publication No. 170:169-180.

Clarke G.K.C., Waddington E.D. (1991) A three-dimensional theory of wind pumping. *Journal of Glaciology* 37(125):89-96.

Colbeck S.C. (1983) Theory of metamorphism of dry snow. *Journal of Geophysical Research* 88(C9):5475-5482.

Colbeck S.C. (1989) Air movement in snow due to wind pumping. *Journal of Glaciology* 35(120):209-213.

Colbeck S.C. (1991) The layered character of snow covers. *Reviews of Geophysics* 29(1):81-96.

Conklin M.H., Bales R.C. (1993) SO_2 uptake on ice spheres: liquid nature of the ice-air interface. *Journal of Geophysical Research* 98(D9):16851-16855.

Conklin M.H., Sigg A., Neftel A., Bales R.C. (1993) Atmosphere-snow transfer function for H_2O_2: microphysical considerations. *Journal of Geophysical Research* 98(D10):18367-18376.

Cunningham J., Waddington E.D. (1993) Air flow and dry deposition of non-sea salt sulfate in polar firn: paleoclimate implications. *Atmospheric Environment* 27A (17/18):2943-2956.

Cutler N.N., Raymond C.F., Waddington E.D., Meese D.A., Alley R.B. (1995) The effect of ice sheet thickness change on the accumulation history inferred from GISP2 layer thicknesses. *Annals of Glaciology* 21:26-32.

Dahl-Jensen D., Johnsen S, (1986) Paleotemperatures still exist in the Greenland ice sheet. *Nature* 320(6059):250-252.

Davidson C.I. (1989) Mechanisms of wet and dry deposition of atmospheric contaminants to snow surfaces. In: Oeschger H., Langway C.C. Jr (eds) *The environmental record in glaciers and ice sheets,* Wiley, Chichester. 29-51.

Davidson C.I., Wu Y.L. (1990) Dry deposition of particles and vapors. In: Linberg S.E. (ed), *Acidic Precipitation,* Vol. 3, Springer-Verlag, New York. 103-216.

Davidson C.I., Santhanam S., Fortmann R. C., Olson M. P. (1985) Atmospheric Transport and Deposition of Trace Elements onto the Greenland Ice Sheet. *Atmospheric Environment* 19:2065-2081.

Davidson C.I., Bergin M.H., Kuhns H.D. (this volume) The deposition of particles and gases to ice sheets. In: Wolff E.W., Bales R.C., (eds) NATO ASI Series I. *Processes of chemical exchange between the atmosphere and polar snow.* Springer-Verlag, Berlin.

Davis P.A., Soderblom L.A. (1984): Modeling crater topography and albedo from monoscopic Viking Orbiter images 1. Methodology. *Journal of Geophysical Research* 89(B11):9449-9457.

Dibb J.E. (this volume) Overview of field data on the deposition of aerosol-associated species to the surface snow of polar glaciers (emphasizing recent work in Greenland). In: Wolff E.W., Bales R.C., (eds) NATO ASI Series I: *Processes of chemical exchange between the atmosphere and polar snow.* Springer-Verlag, Berlin.

Dibb J.E., Jaffrezo J.L., Legrand M. (1992) Initial findings of recent investigations of air-snow relationships in the summit region of the Greenland ice sheet. *Journal of Atmospheric Chemistry* 14(1-4):167-180.

Doumani G.A. (1967) Surface structures in snow. In: Oura H. (ed) *Physics of snow and ice* Vol. 2:1119-1136. Institute of Low Temperature Science, Hokkaido. 1414p.

Elliott J.A. (1972) Microscale pressure fluctuations measured within the lower atmospheric boundary layer. *Journal of Fluid Mechanics* 53(2):351-383.

Fuchs N.A. (1964) *The Mechanics of Aerosols,* Dover, New York. 408p.

Garvin J.B., Williams R.S. Jr. (1993) Geodetic airborne laser altimetry of Breidamerkurjokull and Skeiderarjokull, Iceland, and Jakobshavns Isbrae, West Greenland. *Annals of Glaciology* 17:379-385.

Gjessing Y.T. (1977) The filtering effect of snow. In: *Isotopes and impurities in ice and snow.* IAHS Publication No. 118:199-203.

Gow A.J. (1965) On the accumulation and seasonal stratification of snow at the South Pole. *Journal of Glaciology* 5(40):467-477.

Gow A.J. (1969) On the rates of growth of grains and crystals in South Polar firn. *Journal of Glaciology* 8(53):241-252.

Harvey L.D.D. (1988) Climatic impact of ice-age aerosols. *Nature* 334(6180):333-335.

Heintzenberg J., Rummukainen M. (1993) Airborne particles in snow. *Journal of Glaciology* 39(132):239-244.

Harder S.L., Grootes P.M., Charlson R.J. (1991) Air-snow exchange processes. *Antarctic Journal of the U.S. 26(5):277-278.*

Harder S.L., Warren S.J., Charlson R.J., Covert D.S. (in press) Filtering of air through surface snow as a mechanism for aerosol deposition to the Antarctic Ice Sheet. *Journal of Geophysical Research - Atmospheres.*

Hicks B.B. (1985) On the use of monitored air concentrations to infer dry deposition. *NOAA Technical Memorandum* ERL ARL-141, Silver Spring. MD.

Ibrahim M., Barrie L.A., Fanaki F. (1983) An experimental and theoretical investigation of the dry deposition of particles to snow, pine trees and artificial collectors. *Atmospheric Environment* 17(4):781-788.

Ito T. (1995) Nature and origin of Antarctic submicron aerosols. In: Delmas R.J. (ed). NATO ASI Series I(12) *Ice core studies of global biogeochemical cycles.* Springer-Verlag, Berlin. 23-38.

Johnsen S. (1977) Stable isotope homogenization of polar firn and snow. In: *Isotopes and impurities in snow and ice.* IAHS Publication No. 118:210-219.

Kirchner S., Delmas R.J. (1988) A 1000 year glaciochemical study at the South Pole. *Annals of Glaciology* 10:80-84.

Krabill W.B., Thomas, R.H., Martin C.F., Swift R.N., Frederick E.B. (1995) Accuracy of airborne lasar altimetry over the Greenland ice sheet. *International Journal of Remote Sensing* 16(7):1211-1222.

Ledroit M., Remy F., Minster J-F. (1993) Observations of the Antarctic ice sheet with Seasat scatterometer: relation to katabatic-wind intensity and direction. *Journal of Glaciology* 39(132):385-396.

Legrand M.R. (1987) Chemistry of Antarctic snow and ice. *Journal de Physique* 48(3):C1-77 - C1-86.

Legrand M.R., Delmas R.J, Charlson R.J. (1988a) Climate forcing implications from Vostok ice-core sulfate data. *Nature* 334(6181):418-420.

Legrand M.R., Lorius C., Barkov N.I., Petrov V.N. (1988b) Vostok (Antarctica) ice core: atmospheric chemistry changes over the last climatic cycle. *Atmospheric Environment* 22(2):317-331.

Morse P.M., Feshbach H. (1953) *Methods of theoretical physics.* McGraw-Hill, New York. 1978p.

Neftel A. (this volume) The record of gases and reactive species in ice cores, and problems of interpretation. In: Wolff E.W., Bales R.C., (eds) NATO ASI Series I: *Processes of chemical exchange between the atmosphere and polar snow.* Springer-Verlag, Berlin.

Paterson W.S.B. (1994) *The physics of glaciers.* third edition, Pergamon, Oxford. 480p.

Pruppacher H.R., Klett J.D. (1978) *Microphysics of Clouds and Precipitation.* Reidel, Dordrecht, 714p.

Reif F. (1965) *Statistical Physics.* McGraw-Hill New York. 398p.

Reist P.C. (1993) *Introduction to aerosol science,* second edition. McGraw-Hill, New York. 379p.

Shaw G.E. (1988) Antarctic aerosols: a review. *Reviews of Geophysics* 26(1):89-112.

Schwander J. (1989) The transformation of snow to ice and the occlusion of gases. In: Oeschger H., Langway C.C. Jr (eds) *The environmental record in glaciers and ice sheets,* Wiley, Chichester. 53-67.

Schwander J., Stauffer B., Sigg A. (1988) Air mixing in firn and the age of the air at pore close-off. *Annals of Glaciology* 10:141-145.

Shimizu H. (1970) Air permeability of deposited snow. *Contributions from the Institute of Low Temperature Science* Series A(22):1-32.

Shook K., Gray D.M. (1994) Determining the snow water equivalent of shallow prairie snowcovers. *Proceedings of the Eastern Snow Conference.* Dearborn Michigan, 1994. 89-95.

Sigg A., Neftel A. (1988) Seasonal variations in hydrogen peroxide in polar ice cores. *Annals of Glaciology* 10:157-162.

Sigg A., Neftel A. (1991) Evidence for a 50% increase in H_2O_2 over the past 200 years from a Greenland ice core. *Nature* 351(6327):557-559.

Silvente E., Legrand M. (1995) A preliminary study of the air-snow relationship for nitric acid in Greenland. In: Delmas R.J. (ed). NATO ASI Series I(12) *Ice core studies of global biogeochemical cycles.* Springer-Verlag, Berlin. 225-240.

Sommerfeld R.A., Rocchio J.E. (1993) Permeability measurements on new and equitemperature snow. *Water Resources Research* 29(8):2485-2490.

Stearns C.R., Keller L.M., Weidner G.A., Sievers M. (1993) Monthly mean climatic data for Antarctic Automatic Weather Stations. In: Bromwich D.H. and Stearns C.R. (eds). *Antarctic meteorology and climatology: studies based on Automatic Weather Stations.* Antarctic Research Series 61, American Geophysical Union, Washington. 1-21.

Stephenson P.J. (1967) Some considerations of snow metamorphism in the Antarctic ice sheet in the light of ice crystal studies. In: Oura H. (ed) *Physics of snow and ice* Vol. 2:725-740. Institute of Low Temperature Science, Hokkaido. 1414p.

Sturm M., Johnson J.B. (1991) Natural convection in the subarctic snow cover. *Journal of Geophysical Research* 96(B7):11657-11671.

Tanaka K.L., Davis, P.A. (1988): Tectonic history of the Syria Planum province of Mars. *Journal of Geophysical Research* 93(B12):14893-14917.

Tritton D.J. (1988) *Physical fluid dynamics* second edition. Clarendon Press, Oxford. 519p.

Vanoni V.A., Hwang L.-S. (1967) Relation between bed forms and friction in streams. *J. Hydraul. Div. Am. Soc. Civ. Eng.* 93(HY3):121-144.

Waddington E.D. (this volume) Where are we going? the ice core - paleoclimate Inverse Problem. In: Wolff E.W., Bales R.C., (eds) NATO ASI Series I: *Processes of chemical exchange between the atmosphere and polar snow.* Springer-Verlag, Berlin.

Wadhams P., Davis N.R., Comiso J.C., Kutz R., Crawford J., Jackson G., Krabill W., Sear C.B., Swift R., Tucker W.B.. III. (1991) Concurrent remote sensing of Arctic sea ice from submarine and aircraft. *International Journal of Remote Sensing* 12(9):1829-1840.

Weller G. (1969) The heat and mass balance of snow dunes on the central Antarctic plateau. *Journal of Glaciology* 8(53):277-284.

Whillans I.M., Grootes P.M. (1985) Isotopic diffusion in cold snow and firn. *Journal of Geophysical Research* 90(D2):3910-3918.

Whitby K.T.(1978) Physical characteristics of sulfur aerosols. *Atmospheric Environment* 12:135-159.

Wolff E.W. (1995) Nitrate in polar ice. In: Delmas R.J. (ed). NATO ASI Series I(12) *Ice core studies of global biogeochemical cycles.* Springer-Verlag, Berlin. 195-224.

Yen Y.-C., Tien C., Bender J.A. (1963) On the isothermal flow of air through a snowpack with variable permeability. In: *Heat budget at the snow surface.* IAHS Publication No. 61:51-61.

APPENDIX: MATHEMATICAL SYMBOLS

Symbol	Meaning
$\dot{b}$	snow accumulation rate (m a^{-1}) (40), (54)
d	hydraulic diameter of permeable medium (38)
g	acceleration due to gravity 9.8 m s^{-2}
f	weighting factor on viscous sublayer resistance (44)
f	fraction of ice core concentration C_{ice} due to ventilation (40)
h	thickness of snow sample in permeameter
h	height of surface microtopography (26)
h_0	permeable layer thickness (9)
k	von Karman constant 0.41 (45)
k	Boltzmann constant 1.38×10^{-23} J K^{-1} (69)
k_f	rate constant for exchange across vapor/solid boundary (59)
$l(z)$	displacement of pore air during a turbulent pumping cycle (22)
n	number of air exchanges to equilibrate ice with atmosphere (64)
p(x,y,z,t)	pressure perturbation (4)
p_0	magnitude of pressure perturbation p(x,y,z,t) at snow surface (5)
p(z,t)	pressure perturbation at hole in impermeable layer (14)
$p_{hole}(t)$	pressure at hole in impermeable layer (11)
r	radius of snow grain (1), (58); aerosol particle (70)
r	radial polar coordinate (13)
r_w, etc	depositional resistances (42)ff
r_{hole}	radius of hole in impermeable layer (17)
$\mathbf{r}(r,\theta,\phi)$, $\mathbf{r}_0$	position vector in spherical coordinates (12)
r	total resistance of a lump of resistive material (51)
t	time
u_*	friction velocity (49)
$v_{air}(z)$	velocity of pore air Q/ϕ (20)
$\overline{v}_{air}(z)$	v_{air} averaged over ½ turbulent pressure cycle (21)
x,y	horizontal coordinates
z	depth coordinate
z_{diel}, z_{yr}	penetration depths of daily, annual thermal wave (57)ff

z_0	surface roughness height (45)
A	accessible ice surface within a plot of 1 square meter (1)
$A(z)$, $A_\infty(z)$, $A_{h_0}(z)$	amplitude of pressure solutions (6), (7), (9)
A	area of resistive material normal to current flow (51)
C	calibration coefficient for pressure drop over bumps (26)
C_a, C_{ice}, C_{snow}	concentration of chemical species in air, ice grain, bulk snow
D	Brownian diffusion coefficient (49)
S(t)	ventilation factor of snow surface (36)
F_d	downward dry flux of aerosol (3)
F_{snow}	upward flux of gaseous species in snow
G	geometric factor for diffusion in an ice grain (58)
$G[(\mathbf{r}-\mathbf{r}_0), (t-t_0)]$	Green's Function for diffusion (13)
H	natural diffusion length (7)
K	intrinsic permeability (2)
K_D	partition coefficient for gas/solid phases (59)
L	Obukhov length (45)
M, m	masses of Brownian particle, gas molecule (70)
N, N_0	aerosol concentrations in a tube (37)
P_0	average atmospheric pressure (4)
P_{air}	total air pressure $P_0 + p(x,y,z,t)$ (4)
Pr	Prandtl number ν/κ_T (49)
$\vec{Q}(x,y,z,t)$	volumetric air flux in firn (29)
$\overline{Q}_z$	downward air flux averaged over total area of bump (33)
Q	air flux in a tube (37)
Re	Reynolds number
$S(\mathbf{r}, t)$, $S_0(\mathbf{r}, t)$	pressure source distributions (12)
Sc	Schmidt number ν/D (49)
T	temperature
U_z	wind speed at height z (5), (25)
$U(t-t_0)$	Heaviside step function (13)
$V_d(z)$	dry depositional velocity (3)
V_w etc	dry deposition velocities for various processes (42), (46)ff
V_0	volume integration variable for continuum (14)
V_{hole}	characteristic volume energized by hole in impermeable layer (15)
W(t)	atmosphere efficiency for ventilation (35)
α	horizontal aspect ratio of topographic bumps (30)
δ	e-folding depth for air flow under topographic bumps (31)
$\delta(\mathbf{r})$, $\delta(t)$	Kronecker delta distribution (12)
κ_{ice}	diffusivity coefficient for a chemical species in ice (58)
κ_{air}	diffusivity coefficient for a chemical species in air (63)
κ_{firn}	bulk diffusivity coefficient for a chemical species in firn (65)
κ_p	pressure diffusivity (4)
κ_T	thermal diffusivity (49), (57)
λ	wavelength of surface bumps (26)
μ	viscosity of air (2)
ν	kinematic viscosity of air μ/ρ_{air} (49), (61)
ϕ	porosity (4)
Φ	time parameter in Gormley-Kennedy equation (37)
τ_i	characteristic times for equilibration (53)ff
Ψ	hydraulic potential (2)
ρ_{air}	density of air (2)
ρ_{ice}	density of ice (64)
ρ	distributed resistivity (51)
$\theta(z)$, $\theta_\infty(z)$, $\theta_{h_0}(z)$	phase of pressure solution (6), (8), (10)
ω, ω_0	frequency (5)
Λ	spatial scale of turbulent bursts (25)

h_0	surface roughness height (55)
A	[illegible]
$A(z), A_0(z), A_R(z)$	amplitude of pressure solution (61), (68), (70)
[illegible]	[illegible]
[illegible]	gravitational potential (7)
ρ_{air}	density of air (2)
ρ_{ice}	density of ice (64)
ρ	distributed resistivity (51)
$\phi(z), \phi_0(z), \phi_R(z)$	phase of pressure solution (61), (68), (70)
ω, ω_0	frequency (5)
Λ	spatial scale of turbulent bursts (25)

WIND-BLOWN SNOW: SUBLIMATION, TRANSPORT AND CHANGES TO POLAR SNOW

J.W. Pomeroy and H.G. Jones[2]
National Hydrology Research Institute,
11 Innovation Blvd.,
Saskatoon, Saskatchewan,
S7N 3H5 Canada

INTRODUCTION

In their hallmark work on the Antarctic blowing snow phenomenon, Budd et al. (1966) recounted a quote of Peary (1898) that still has relevance to interpretation of glacial phenomena:

> "There is one thing of special interest to the Glacialist - the transport of snow on the ice-cap by the wind....."

Following Peary's reasoning, the blowing snow phenomenon was studied intensively in the Antarctic several decades ago (Lister, 1960; Mellor and Radok, 1960; Dingle and Radok, 1961; Budd et al., 1966) with more recent studies by Kobayashi (1978), Takahashi (1985) and Moore et al. (1994). Other work of similar nature has been conducted in the Arctic (Dyunin, 1959; Benson, 1982; Tabler et al., 1990b; Benson and Sturm, 1993). Blowing snow is quite frequent in the Antarctic, Dalrymple (1966) noted snow transport occurred from 30-55% of the time in the South Pole region, increasing to 55-65% of the time at Byrd Station. Fujii (1981) found snow accumulation to occur in only one or two out of every three years in East Antarctica because of wind erosion and sublimation. At Mizuho Station, East Antarctica, Takahashi et al. (1994) estimated that of an annual snowfall of 140-260 mm water equivalent, 100 mm/year is eroded from surface snow and transported away by blowing snow and 50 mm/year sublimates. The invariable conclusion is that snow covers in the polar regions are dramatically transformed by wind redistribution; it may therefore follow that the chemistry of

2 Institut national de la recherche scientifique - Eau
2800 rue Einstein
Sainte-Foy, Québec
G1V 4C7 Canada

NATO ASI Series, Vol. I 43
Chemical Exchange Between the Atmosphere and Polar Snow
Edited by Eric W. Wolff and Roger C. Bales

these snow covers is, as well, altered by wind redistribution. In affirmation, Delmas and Jones (1987) and Pomeroy et al. (1991a, b) measured and modelled changes to snow chemistry during drifting in temperate locations, Pomeroy et al. (1993b; 1995) presented evidence of chemical alteration of wind-blown snow in the Canadian Arctic and Osada (1994) showed strong chemical fluctuations in wind-blown snow for the Antarctic. Redistribution leads to a re-evaluation of our concepts of snowcover formation; in polar regions snow does not normally fall in a simple fashion from the atmosphere to the surface to be buried by subsequent snowfalls, but more often falls and moves horizontally, carried by the wind, to be either sublimated, resuspended or deposited. Deposition onto the snowpack occurs when and where the atmospheric surface shear stress is insufficient to continue transport. After initial deposition the snow may be eroded and transported again. In extreme blowing snow events some snow may be lifted by turbulence tens or hundreds of metres into the atmosphere where sublimation reduces its mass to insignificance. During transport the parent snow forms a turbulent mixture with the atmosphere, aerosols and snow from other sources, hence it may be mixed with older or fresher snow, is exposed to the atmosphere's chemical constituents and undergoes relatively rapid sublimation. These processes add considerable complexity to attempts to relate the chemistry of the resulting surface snow to the parent snowfall as outlined by Neftel (1996) and Bales and Choi (1996) in this volume.

Effects of wind redistribution that are important to ice core chemistry may be classified as concentration effects and deposition effects, as they affect the chemical concentration of, and thickness of, annual firn layers.

Concentration changes in snow occur during:

a) sublimation of ice where chemical species are conservative,

b) scavenging (adsorption) of particulates or vapours by the surface of blowing snow particles,

c) volatilization of chemical species unrelated to sublimation effects.

Deposition of snow may be concomitant with:

a) volatilization of chemical species,

b) snow deposition fluxes that are not equal to the snowfall flux,

c) co-deposition of atmospheric constituents.

Mass, ion and energy balances for small control volumes (i.e. over fetches of 1 m to 1 km) are used to calculate changes in concentration or deposition (Pomeroy et al., 1991a). However the application of control volumes to large areas such as the Antarctic ice sheet has not been explicitly examined, though it has been proposed for the Arctic by Pomeroy et al. (1995).

This paper reviews recent developments in our understanding of chemical changes associated with wind-blown snow. It also reassess the current conceptualisations of precipitation and accumulation in polar regions, discusses the implications for the chemistry of ice cores and attempts to link the chemistry of these ice cores with that of the atmosphere.

PHYSICS

The physical processes during wind transport and snowcover formation are extremely important for understanding the concomitant chemical transformations. This section will briefly review the physics associated with formation of snowcovers in windy, exposed locations. For recent comprehensive reviews of the phenomenon the reader is referred to Tabler et al. (1990a), Pomeroy et al. (1993a) or Pomeroy and Gray (1995).

Snowfall and Meteorological Measurements

Any discussion of the physics of snow accumulation in polar regions would be incomplete without addressing measurement problems. In extreme polar environments there are few locations where, year-round, sufficient and appropriate measurements are taken to quantify true snowfall and where wind speed, temperature, humidity and the time of occurrence of blowing snow are available for use in blowing snow transport, sublimation and accumulation calculations. The situation is particularly discouraging in central Greenland and the interior of Antarctica where even the direction of the current mass balance is not known (Houghton et al., 1990). Recent models of precipitation predict *snow accumulation* not *snowfall* for polar regions (Connolley and Cattle, 1994). Verification of climate models is conducted against snow accumulation, which is normally measured as an increase in snow depth on the ground instead of true snowfall (Marshall et al., 1994). In windy environments the difference between snowfall and snow accumulation is substantial and may vary more or less by an order of magnitude between exposed and sheltered locations. Where precipitation gauges are used, an undercatch due to wind exposure results (Goodison, 1981). This undercatch increases with wind speed; where mean wind speeds are normally in excess of 5 m/s as in the polar regions, an undercatch of over 50% is expected for unshielded gauges. Further undermeasurement, where snow collection cylinders are emptied routinely, is due to non-recording of trace snowfalls and to wetting loss of snow that is retained in the cylinder. Benson and Sturm (1993) suggest that Arctic Alaskan snowfall measurements are underestimated by a factor of three. When Goodison's techniques were used to correct winter precipitation records for Resolute, Northwest Territories, Canada, annual snowfall increased by 64% to 161% (Pomeroy and Goodison, 1995). Kobayashi et al. (1985) estimated snowfall during blowing snow by examining asymptotes of the vertical profile of blowing snow in East Antarctica at Mizuho Station. Their estimated snowfall fluxes during blowing snow ranged as high as 0.15 mm/hour. In regions of the Antarctic subject to katabatic snow-storms this flux results in annual snowfall rates during blowing snow ranging from 9.6 to 720 with an average of 140 mm/year. Because of the high wind speeds this snowfall goes largely unrecorded by standard precipitation gauges.

To calculate hourly blowing snow fluxes, wind speed, air temperature, humidity, corrected snowfall, depth of snow on the ground and the time of occurrence of blowing snow are needed (Pomeroy et al., 1993a). Monthly fluxes may be estimated using a simpler model that does not require the time of occurrence of blowing snow (Pomeroy and Gray, 1994). Unfortunately unattended anemometers in polar regions are subject to rime and snow-filled cups causing them to underestimate wind speed. Such data can result in serious errors in blowing snow calculations. The most usable data for blowing snow calculations is that collected at manned polar stations where hourly visual observations and frequent instrument inspections/corrections are made.

Wind Transport of Polar Snow

As snow is transported by wind, the crystals undergo changes to their shape and size and upon re-deposition form drifts and surfaces of higher density than the original snow. Crystals are abraded by particle impact, shattered and sublimated, to give oblong amorphous shapes with characteristic size distributions even in cold, polar environments (Budd, 1966; Schmidt, 1982).

Wind transport of snow is closely linked to synoptic winds in the Antarctic. Kodama et al. (1985) showed that the katabatic wind dropping off the ice-sheet is partially sustained by the additional weight of blowing snow in the air-flow and the cooling of the air by sublimation of blowing snow. A positive feedback develops where as more snow is entrained, the katabatic wind speed increases and the erosive force increases. Wind transports snow in two substantive modes: saltation and suspension.

Saltation - particles which move by a skipping action along the snow surface form this two-phase flow. Particle trajectories follow a near-vertical ascent with a flat descent; initial horizontal velocity is quite low but the final horizontal velocity exceeds the ambient wind speed. Saltating particles attain a maximum trajectory height of from 0.1 to 5 cm, increasing with the square of the friction velocity. The snow cover comprises a "bed" of snow from which particles are entrained and deposited and over which saltation occurs. This bed is a mixture of bonded crystals metamorphosed by particle impact and vapour transfer. Particles are normally ejected from this bed as part of a "splash" from the impact of a descending particle. In order to be ejected by splash the energy of impact must be sufficient to overcome cohesion and the work required to break bonds of surrounding and underlying particles. Over a uniform surface, erosion of snow particles occurs when the shear stress exerted by the wind, τ, exceeds the surface shear strength, τ_s where,

$$\tau_p = \tau - \tau_s \qquad (1)$$

and τ_p is the shear stress available to drive saltation. Evidence of the effect of varying shear stress on snow erosion in the Antarctic is presented by Takahashi et al. (1994); along a 850 km transect of snow surveys, snow erosion reduced the accumulation rate over large-scale convex surfaces where shear stress is at a maximum. The atmospheric shear stress, τ is equal to $\rho_a u^{*2}$, where ρ_a (kg/m^3) is atmospheric density and u^* (m/s) is the friction velocity and found over blowing snow as,

$$u^* = \frac{u_z\, k}{\ln\left(\frac{C\, z}{u^{*2}}\right)} \qquad (2)$$

k is von Kármán's constant (0.4), u_z is the wind speed (m/s) at height z (m) and C is an experimentally-derived constant found to be 163.3 m/s^2 (Pomeroy and Gray, 1990). As a result of this energy balance and measurement of saltation efficiency, Pomeroy and Gray related steady-state saltation transport to the wind speed in excess of the threshold level at the beginning or cessation of transport. Typical threshold wind speeds (10-m) range from 6 to 9 m/s for wind-hardened snow. The transport rate of saltating snow increases approximately linearly with the wind speed and at high wind speeds comprises less than 10% of the total transport rate.

Suspension - when the upwards drag on a saltating snow particle becomes greater than the gravitational force the particle becomes suspended and moves along turbulent flow lines at a mean horizontal velocity close to that of the air. Suspended snow derives from saltating snow or falling snow, never directly from the surface. Budd et al. (1966) extensively measured suspended mass flux of snow in Antarctica, describing the particle size distribution and the decrease in particle size with height. The layer of suspended snow reaches from the top of the saltation layer to up to several hundred's of metres, increasing with upwind fetch of blowing snow. The concentration of suspended snow is extremely sensitive to wind speed and declines exponentially with height, from nearly 800 g/m^3 near the top of the saltation layer to less than 0. 01 g/m^3 at several metres height. Pomeroy and Male (1992) developed a turbulent diffusion model of suspended snow, whose results agree with Budd et al. The model demonstrates that the suspended transport rate dominates the mass transport of blowing snow and increases with roughly the fourth power of wind speed.

The total transport rate of snow is a compilation of transport in its three modes and is very sensitive to the upwind fetch distance and to snow supply. A recent practical formulation for the downwind transport rate of snow, Q_T as a function of wind speed in polar regions is that by Tabler et al. (1990a). Tabler et al.'s formulation is derived from profiles of Antarctic mass flux measurements published by Mellor and Fellers (1986), integrated to a height of 10 m, and coupled with the mass concentration limitation suggested by Pomeroy and Gray (1990) for the saltation layer. The integral height of 10 m and lack of a characteristic threshold transport value are not important limitations for polar regions where upwind fetches are long and persistent snowcover may be presumed. Tabler et al.s expression has the form,

$$Q_T = 0.0000014\, u_{10}^{4.2} \quad . \tag{3}$$

As shown in Fig. 1, this transport rate increases dramatically with wind speed; given a 10-m wind speed of 8 m/s the transport rate is 0.0087 kg/s per metre width, but with a wind speed of 20 m/s the transport rate is 0.408 kg/s per metre width. Because of the non-linear increase in transport with wind speed this equation should not be used with long-term average wind speeds to estimate mean transport rates; much of the annual transport may be due to a few extreme events.

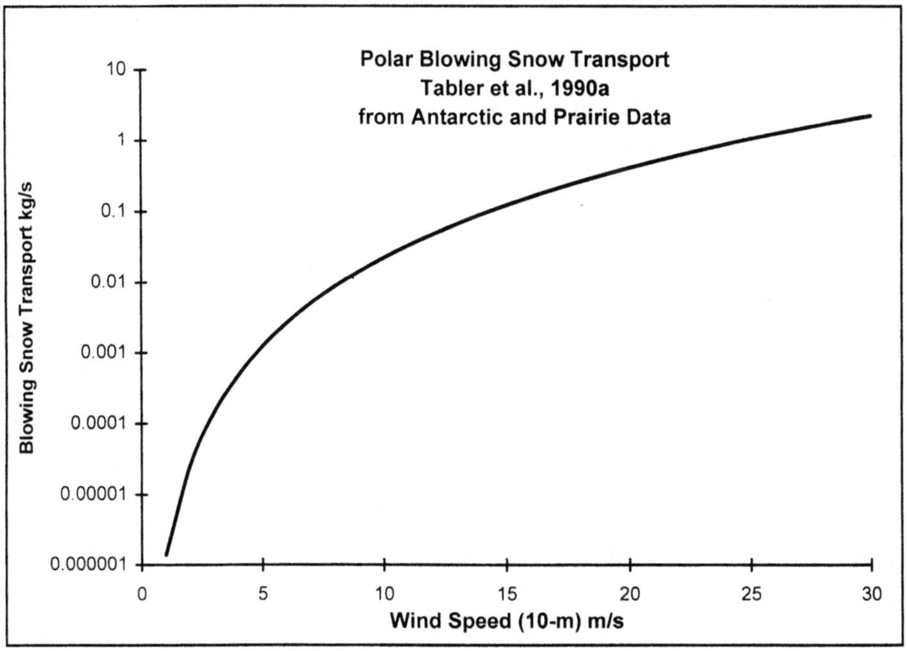

Figure 1. Blowing snow transport rate as a function of wind speed for continuous, level snow covers (after Tabler et al., 1990).

Sublimation of Polar Snow

Stationary Snow Sublimation. Sublimation of snow is driven by inputs of atmospheric sensible, Q_H, and net radiative, Q*, heat, which must be sufficient to drive the latent heat requirements of the sublimation process. For cold, level and complete snowpacks over deep snow fields the energy balance for snow may be written as,

$$Q_E = Q* + Q_H + Q_S \tag{4}$$

where Q_E is the latent heat flux. Q_S is the vertical heat flux at the snow surface and takes into account internal energy changes and any sensible heat flux from the deep polar snow. For polar snows Q_S is normally quite small and because of high albedo Q* is usually negative or slightly positive except in the warmest polar conditions. The fluxes of latent and sensible heat to the surface are controlled, in analogy to Fick's Law, by the gradients of wind speed, air temperature and humidity above the surface and the relevant turbulent diffusivities of momentum, heat and water vapour (Male, 1980). The turbulent diffusivities for heat and water vapour are normally related to the diffusivity for momentum. Granger and Male (1978) showed that for slightly-stable conditions over snow the turbulent diffusivity for heat is approximately equal to that for momentum and decreases to eventually equal half that for momentum as stability increases. In contrast they showed that the diffusivity of water vapour is roughly constant at half that of momentum for all stable conditions. Turbulent diffusion of water vapour from or to a stationary snow surface is limited by non-existent turbulence at the snow surface and the relatively small surface area of ice exposed to the wind. Wind-pumping may increase these values above previous estimates but sublimation rates are limited by the available energy (Albert and McGilvary, 1992). For polar snows, the small air temperature gradients and low rate of turbulent transfer result in the latent heat flux tending to correspond to the net radiation flux. This is analogous to cold, continuous prairie snowcovers where Male and Granger (1979) reported daily sublimation ranging from 0.02 to 0.3 mm SWE/day with a mean of 0.1 mm-SWE/day in pre-melt conditions. Most sublimation during high sun periods was counter-balanced by frost accumulation in the evening. The values reported by Male and Granger are for a higher energy environment than is normally the case for polar snow and should be considered an upper limit to sublimation from "in-situ" snowcovers. The constant exposure of polar snow can result in appreciable cumulative in-situ sublimation however. Fujii and Kusunoki (1982) measured sublimation of 50 mm/year at Mizuho Station using an ice-filled evaporation pan.

Blowing Snow Sublimation. In contrast to stationary snow, blowing snow has a high surface area to mass ratio and is subject to intense atmospheric turbulence, hence the potential for turbulent transfer of heat and water vapour is quite high. Pomeroy (1989) shows that sublimation from a column of blowing snow at air temperatures just below the freezing point, low relative humidity (70%) and high wind speed (25 m/s or severe blizzard condition) can reach 18 mm SWE/hour, well in excess of the maximum in-situ sublimation rates. Because of the small size and high albedo of blowing snow particles, their net radiative flux is small and can usually be neglected in sublimation calculations. For a single particle the energy balance is therefore one between latent and sensible heat, controlled by sensible heat transfer to the particle surface and a thermodynamic equilibrium. Pomeroy et al. (1993a) operationalised the sublimation calculation by assuming that particles remain at the ice bulb" temperature and instituting appropriate vertical profiles of particle size distribution, mass concentration (from saltation and suspension calculations), turbulent velocity, air temperature and humidity in the calculation procedure. The resulting sublimation rates for a column of blowing snow are shown in Fig. 2 for a variety of temperatures, humidities and wind speeds. The example shows overestimates sublimation in certain polar environments where temperatures are regularly lower than -30° C and relative humidities exceed 90%.

Accumulation of Polar Snow

Many conceptualizations presume that the accumulation flux is equivalent to the snowfall flux in polar regions (Connolley and Cattle, 1994; Marshal et al., 1994) however it is in fact a residual of snowfall, sublimation and blowing snow processes. The instantaneous snow accumulation flux, $Q_{surface}$, is calculated using a mass balance of snow fluxes to a control volume over the surface of interest (Pomeroy et al., 1991a) where

$$Q_{surface} = \frac{dQ_T}{dx}(x) - Q_E + Q_{snowfall} \tag{5}$$

The change in transport fluxes over horizontal distance x is an important term when fetches are short, terrain uneven or flow disturbed by buildings, etc. and can be positive or negative depending upon increasing or decreasing transport with distance. The transport change term becomes insignificant over the large homogeneous fetches typical of polar regions. Q_E is the sublimation flux expressed in kg/(m^{2}s) as are all other fluxes and $Q_{snowfall}$ is the instantaneous snowfall flux. Note that $Q_{surface}$ can be positive or negative, even when the snowfall flux is large. The quantities frequently measured in polar regions are $Q_{surface}$ and more rarely $Q_{snowfall}$ (corrected for gauge loss). A two-phase flow model such as proposed by Pomeroy et al. (1993a) is required to fully evaluate Q_E for short time periods.

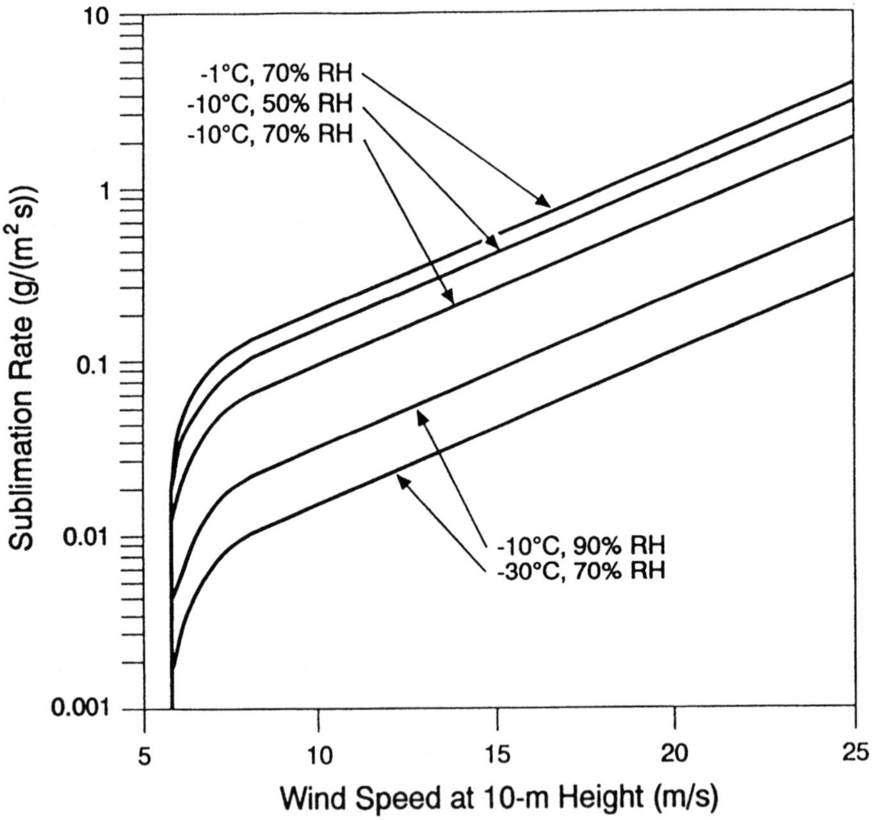

Figure 2. Blowing snow sublimation rate as a function of wind speed, air temperature and humidity for a column of blowing snow over a continuous, level snow cover (after Pomeroy et al., 1993).

Where detailed, suitable meteorological information is not available, a monthly climatological model can be used to estimate the monthly transport, sublimation and hence accumulation (Pomeroy and Gray, 1994). The model presented here was developed using data from smooth snowcovers from more temperate conditions (monthly mean of daily minimum air temperature -26.5 to -0.6 ^{o}C) than found in polar regions, though it has been verified for seasonal snowcovers in the Canadian Arctic (Pomeroy et al., 1995). For polar regions one can assume that melt is negligible and that in-situ sublimation fluxes are usually small. In this application a uniform snow covered fetch of 1 km is assumed, though this fetch length will be shorter than many fetches on the major ice sheets. Following Pomeroy and Gray (1994), the monthly transport of snow, Q_T(monthly) (mm SWE) from a 1-km fetch, ($-dQ_T/dx$ (x)) is found as,

$$Q_T(monthly) = -14.33 + 2.257\, u_{10} - 0.245\, T_{max} + 0.046\, RH_{max} + 0.0786\, P_m \tag{6}$$

where u_{10} is the means monthly wind speed (m/s), T_{max} is the monthly mean of daily maximum air temperature (^{o}C), RH_{max} is the monthly mean of daily maximum relative humidity (%) and P_m is monthly snowfall (mm SWE). The monthly sublimation of snow, Q_E(monthly) (mm SWE) from a 1-km fetch, is similarly found as,

$$Q_E(monthly) = 7.206 + 1.764\, u_{10} - 0.158\, T_{max} - 0.176\, RH_{max} + 0.191\, P_m \tag{7}$$

Equations 6 and 7 were developed to express monthly fluxes in mm SWE for a 1-km fetch; the sum of the two monthly fluxes does not vary substantially for fetches greater than 0.5 km. For fetches other than 1 km Q_T is divided by the fetch distance (km), and any difference between that and the 1-km value is added to the sublimation flux. Regardless of fetch length, the combination of these equations can provide the monthly accumulation flux in mm for polar environments where,

$$Q_{surface}(monthly) = 7.124 - 4.021\, u_{10} + 0.403\, T_{max} + 0.13\, RH_{max} + 0.73\, P_m \tag{8}$$

The third term on the right hand side of Eq. 8 indexes the frequency of wet snow for mean temperatures from -28° to 0°C and has no physical meaning at temperatures below this range; therefore a constraint is applied to T_{max} in polar regions where $T_{max} \geq$ -28°C. This model is not sensitive to fetch distance when that distance is greater than 300 m and presumes that monthly snowmelt and in-situ sublimation are negligible. For the extremely long fetches of the central Antarctic and Greenland ice sheets any difference between the accumulation predicted by Eq. 8 and precipitation may be presumed to have sublimated in transit. Figure 3 shows the monthly snow accumulation as a function of wind speed for a range of air temperatures and relative humidities and a monthly snowfall of 30 mm SWE. For typical polar "winter" conditions (maximum daily maximum temperatures less than -25°C) less than half the monthly snowfall of 30 mm accumulates when the mean monthly wind speed exceeds 4 m/s and net snow erosion occurs when mean wind speed exceeds 8 m/s. This has important implications for relating the annual snow accumulation in ice cores to the precipitation record. For instance snow might be deposited, erode months later and be redeposited in a mixture with snows of differing ages. If the chemical signature of such snows vary, then the chemistry of the final deposited layer will be more difficult to relate to specific atmospheric conditions.

CHEMISTRY

Changes in the chemical composition of wind-blown snow result from unequal fluxes, including transformations, of solute and solvent, with respect to the snow particle. Changes in the annual deposition, or load, of chemicals in snow result when the flux of chemicals, including transformations, to or from a surface snowcover is caused by blowing snow. Several potential changes to wind-blown snow chemical composition will be examined, they relate to two important aspects of the blowing snow phenomenon.

i) *Blowing snow is well-exposed to the atmosphere.* A 1-m thick layer of surface snow of average polar density (300 kg/m^{3}) has an internal surface area of 21.7 m^{2}/kg (Perla, 1985), however little of this is exposed to air. The surface area of in-situ snow exposed to the atmosphere is roughly 0.013 m^{2}/kg (assuming 1-mm grain diameters). Blowing snow has a surface area of over 60 m^2/kg exposed to the atmosphere (presuming 0.1-mm grain diameters) and is ventilated at an extremely high rate compared to surface snow.

ii) *Blowing snow undergoes sublimation.* Because of rapid energy exchange there is a phase change on the blowing snow particle surface where ice sublimates to water vapour. Intermediate liquid-like phases on the particle surface may be promoted by this transformation. Any solute or particulates in the shrinking ice lattice would be released to this liquid-like layer or possibly the atmosphere.

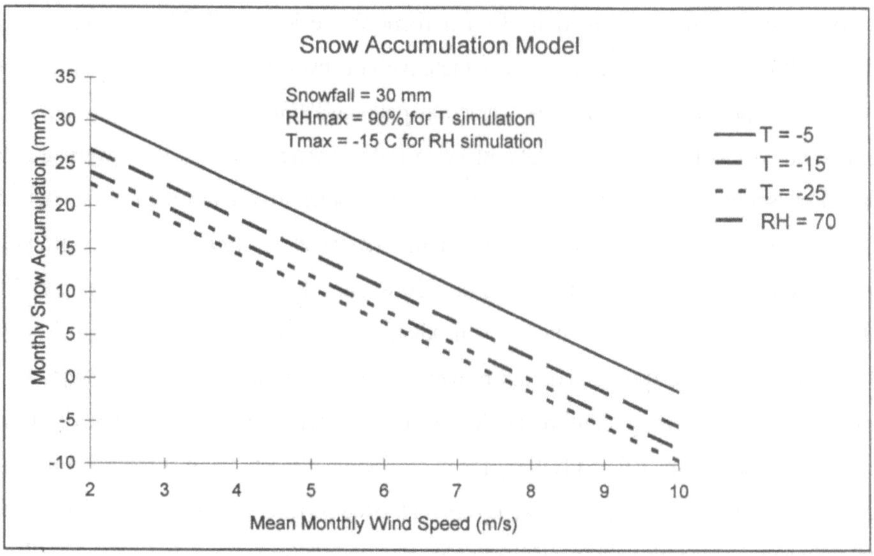

Figure 3. Monthly snow accumulation in windswept, open environments as a function of monthly snowfall, mean monthly wind speed and the monthly means of daily maximum air temperature and relative humidity.

Photochemical Reactions

Blowing snow and sublimating surface snow are particularly well-exposed to the atmosphere and to solar radiation and may therefore undergo photochemical reactions. Neubauer and Heumann (1988) suggested that photochemical reactions could cause loss of NO_3^- from surface snow in polar regions. Their hypothesis states that the loss of NO_3^- is due to the re-emission of NO_2 formed from the photodegradation of HNO_3 by solar radiation. The effective range of wavelengths that induce this reaction is approximately 300-320 nm (maximum absorption 302 nm). The reaction can actually form NO_2, the nitrite ion NO_2^-, atomic oxygen and the hydroxyl radical. Zepp et al. (1987) found that the irradiation of NO_3^- in the presence of OH-radical probe-molecules showed that the OH radical produced is a powerful oxidant. Thus the photodegradation of HNO_3 may not only result in loss of NO_3^- but also in the concomitant oxidation of other reduced species e.g. S(IV). Atomic oxygen combines with dioxygen to produce ozone which, with the hydroxyl radical, is an intermediate in the aqueous peroxide (H_2O_2 - O_2^-) reaction pathways (Gunz and Hoffman, 1990). It should be noted that these reactions take place in water; because quasi-liquid layers are also produced on snow grain surfaces from heating during solar irradiation then photochemical reactions on snow surfaces are a distinct probability (Jones et al., 1993). Conklin and Bales (1993) showed that a liquid layer thickness of 3-30 nm develops at -60° C increasing to 500-3000 nm at -1° C, so concomitant irradiation and heating are prerequisites for photochemistry on ice crystals. Though blowing snow has a high surface area that can be exposed to the sun, blowing snow particles are in transit for relatively short times (minutes or less) which may be insufficiently long for photochemical reactions to occur. Sublimating surface snow, with associated liquid-like layers on the ice surface, is a better candidate surface for photochemical transformation. The same energy source (radiation) that primarily drives sublimation of surface snow also drives the photochemical processes and warms the upper snowpack; this makes chemical changes resulting from either process difficult to distinguish in the field.

Phoresis

Beard et al. (1983) noted that evaporating drops the size of blowing snow particles (60-90 μm) can scavenge aerosols by thermophoresis, diffusiophoresis and electrostatic attraction. Sublimation creates strong gradients of temperature and water vapour density between the relatively damp, cool air near blowing snow particles and the ambient atmosphere. These gradients will cause thermophoretic and diffusiophoretic tendencies, though the magnitude of any scavenging due to these processes is not known for blowing snow. Miller and Wang (1989) indicated that for falling snow particles in undersaturated air the collection efficiency for aerosols from 0.1 to 1.0 μm in radius increases an order of magnitude as relative humidity decreased from 95% to 50% - such a decrease in humidity would cause a dramatic increase in sublimation rate and in theory thermo- and diffusiophoresis. Murakami et al (1985) and Sauter and Wang (1989) suggest that aerosol collection efficiency decreases with increasing size and compactness of a snow crystal. This is confirmed by field measurements (Lamb et al., 1986) that show small snow particles are more efficient than large particles in phoretic capture of aerosols - small particles also have higher sublimation rates per unit mass.

Electrostatic charge can increase the aerosol collection efficiency of snow crystals by two orders of magnitude, making electrophoresis an extremely important scavenging mechanism (Miller and Wang, 1989). Wishart (1970) measured the electrification of blowing snow in the Antarctic, a phenomenon which has widespread evidence in polar regions by its disruption of electronic communications during blizzards; a typical charge is -50 μC/kg. Pomeroy et al. (1991a) suggested that particle charge might contribute to scavenging of sea-salt aerosols during blowing snow and measured strong charges concurrent with rapid aerosol scavenging by blowing snow. An inverse relationship between ventilation velocity and collection efficiency due to electrophoresis has been measured, suggesting that small blowing snow particles will more effectively collect aerosols by electrophoresis (Murakami and Magono, 1983). Such a hypothesis would support electrophoretic forces being generated by sublimation as noted in the laboratory by Dong and Hallett (1990). Schmidt and Dent (1993) examined electrostatic forces on blowing snow particles and concluded that crystal fragmentation, asymmetric rubbing and transient contact of saltating particles with the surface contribute to a thermoelectric effect which is responsible for charges on blowing snow particles. The thermoelectric effect results from a temperature gradient across an ice crystal producing an ionic concentration gradient, along which H^+ ions move more rapidly than do OH^- ions (Latham and Mason, 1961). Latham and Mason showed that when two pieces of ice were put in contact and then separated, the warmer of the two developed a negative charge whilst the cooler developed a positive charge. The charge difference relates to the surface layers of the particles as affected by their temperature. A charge structure exists within the quasi-liquid layer surrounding snow particles (Baker and Dash, 1994). During transient contact between two ice particles (or a snow crystal and the snow surface) the quasi-liquid layer will flow from the thicker to the thinner layer and result in a charge transfer. The thermoelectric effect from particle impact implies that electrophoresis will occur most strongly near the surface near the saltating snow layer, however any additional electrophoresis due to sublimation would reach a maximum well above the surface. Resulting electrophoretic scavenging by blowing snow is probably most efficient in the saltation layer but still notable above this layer.

Ventilation

Ventilation of blowing snow particles contributes to scavenging of aerosols by interception and inertial impact and to turbulent diffusion of soluble gases such as HNO_3, H_2O_2, HCl and NH_3 to the snow particle surface (Dawson and Brimblecombe, 1983; Huebert et al., 1983). Concomitant with ventilation is sublimation of ice from the snow particle which may concentrate certain chemicals on the ice crystal surface and release certain chemical species from the sublimating particle (Delmas and Jones, 1987). Snow particles may be blown for distances of 3 to 5 km in temperate and low Arctic conditions (Tabler et al., 1990a) increasing to tens of kilometres in cold polar conditions (e.g. at -40 ^oC). Pomeroy et al. (1991a) provided an expression for the ventilation velocity, v_r (m/s), of blowing snow particles that includes turbulent and average components as a function of particle radius, r (m), and ambient mean wind speed, u_z (m/s), where

$$v_r = 1.1 \times 10^7 r^{1.8} + 0.01061 u_z^{1.36} \quad (9)$$

Typical mean particle radii in the saltation layer are 100 μm and at 2-m height in the suspended layer are 40 μm. Given a typical particle radius of 50 μm and ambient wind speed of 8 m/s the particle ventilation velocity would be 0.38 m/s, considerably in excess of that due to wind pumping or wind shear on surface snow.

Huebert et al. (1983) has modelled and Zinder et al. (1988) have measured gaseous scavenging of HNO_3 for falling snow - the result should be applicable to blowing snow when sublimation rates are very small (e.g. humid conditions). Zinder et al found that concentrations of NO_3^- in falling snow particles increased 50% over a 1200m fall in a humid atmosphere.

Where ventilation is accompanied by rapid sublimation then high concentrations of certain chemical species will result on the surface of a snow particle. These concentrations may exceed the solubility of these species in the liquid-like surface layer and result in volatilisation. This is particularly expected for volatile species such as HNO_3 and H_2O_2 whose rate of volatilisation may be augmented by the increase in acidity due to higher concentrations of H_2SO_4 and other non-volatile acids in the liquid-like layer. It is also likely that H^+ ions will migrate to the cool surface of a sublimating snow particle, causing further ionic imbalance and potential for volatilisation. The net effect of ventilation may be volatilisation when sublimation rates are high, but in instances of high atmospheric concentrations and low sublimation rates then gaseous scavenging by the blowing snow particle will result.

Chemical Deposition in Wind-blown Snow

The concentration change due to blowing snow may be found by a mass balance of solute and solvent, where CF_x, a dimensionless concentration factor, is the concentration of any species x, at time t, normalised to its concentration at initial time i:

$$CF_x = \frac{\left[\dfrac{m_x(i) + m_x(s) - m_x(v)}{m_m(i) - m_m(E)}\right]}{\left[\dfrac{m_x(i)}{m_m(i)}\right]} \tag{10}$$

m is the load of the chemical species (m_x, $\mu eq/m^2$) or snow (m_m, mm water equivalent) and s refers to scavenging gains, v to volatilisation losses and E to sublimation losses. Loss by chemical transformation of species is assumed to be negligible. Equations 6, 7 and 8 can be used to evaluate snow mass changes where meteorological information is incomplete. Recognising that there are potential errors which accrue in averaging snow chemistry over one month, a simple model for monthly concentration of accumulating snow, C_M (μeq/l), is

$$C_M = \frac{M_x(i) + M_x(s) - M_x(v)}{M_m(i) - M_m(E)} \qquad (11)$$

where M_x is a monthly load of chemical species in snow (μeq/(m^2 month)), M_m is a monthly snow flux (mm/month) and (i) refers to parent precipitation, (s) refers to monthly scavenging, (v) refers to monthly volatilisation and (E) to monthly sublimation.

The monthly derivation of concentration allows Eq. 10 to be recast in a monthly version,

$$CF_m = \frac{1 + \dfrac{M_x(s)}{M_x(i)} - \dfrac{M_x(v)}{M_x(i)}}{1 - \dfrac{M_m(E)}{M_m(i)}} \qquad (12)$$

where CF_M is the monthly average concentration factor. Scavenging or volatilisation must be inferred from empirical relationships, modelled results or the results of field campaigns. Equation 11 may be adapted to determine monthly chemical species accumulation or monthly load, M_x, by reference to Eq. 5 as,

$$M_x = C_M\,[M_m(i) - M_m(T) - M_m(E)]$$

$$\text{and} \qquad (13)$$

$$M_x = [M_x(i) + M_x(s) - M_x(v)]\left(1 - \frac{M_m(T)}{M_m(i) - M_m(E)}\right)$$

This mass balance calculates $dQ_T/dx = -M_m(T)$. Equation 13 embodies a mass balance where a quantity of snow is removed ($M_m(T)$) and the concentration of this snow is determined by $[M_m(i) - M_m(E)]$. Hence, sublimation plays a role in the accumulation of chemical species by controlling the concentration of ion in snow that is removed from the site by wind transport.

Presuming for the purpose of example that $M_x(s)$ and $M_x(v)$ are negligible, the effects of varying wind speed, temperature, humidity and fetch on the concentration factor, CF_m, and the net ionic load, $M_x/M_x(i)$, are shown in Fig. 4. Note that net ionic load is not equivalent to concentration as the snow removed may have a different chemical concentration than the parent or final accumulating snow. The model shows that concentration factors due to blowing snow increase sharply as sublimation increases with wind speed and less dramatically with decreasing temperature and relative humidity. CF_m can range from 1.5 to 2 for short fetches and up to 2 to 6 for longer fetches in the conditions typically found in polar regions. The net ionic load, $M_x/M_x(i)$. decreases rapidly with increasing transport and sublimation fluxes and is hence strongly related to wind speed, reaching 0 at wind speeds above 8 to 10 m/s. It decreases more rapidly with lower temperatures but is relatively insensitive to humidity. Because blowing snow sublimation increases with fetch length, the net ionic load for longer fetches remains near 1, until the wind speed is reached at which all snow is removed from the site by erosion. The transition between these states is quite dramatic and is reached at a wind speed between 8 and 10 m/s. For the longer fetches typical of polar ice caps Eq. 13 has important implications because an almost binary system develops in which relatively complete ionic load (though highly concentrated) persists for low wind speeds switching rapidly to almost no ionic load for higher wind speeds. This result is not unexpected considering Fujii's (1981) observation that the formation of an annual layer of snow accumulation occurs only once every two or three years because of sublimation and wind erosion of snow. As the transition wind speed between the two states is near to the long-term monthly average at many Antarctic stations this aspect of wind-blown snow chemical accumulation makes the process highly non-linear and sensitive to atmospheric conditions (wind speed and temperature in particular). One should note that the previous simulation presumed that all species were perfectly conserved in sublimating snow particles, subsequent evidence will show that this is not the case. For species that volatise at the same rate that ice sublimates there would be little change in concentration but a dramatic drop in load as wind speed increases.

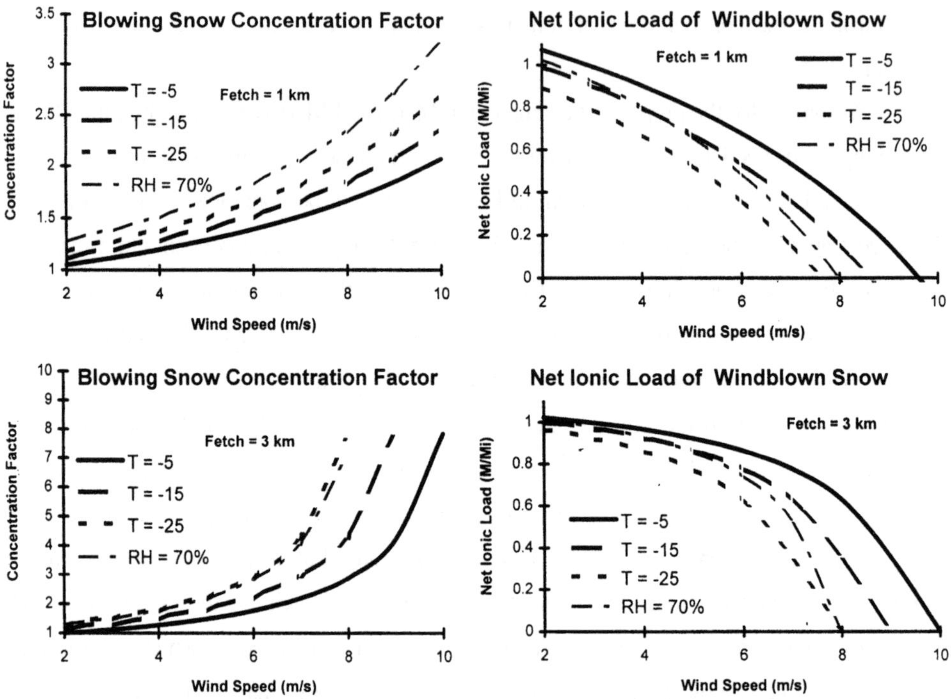

Figure 4. Effects of wind speed and atmospheric environment on Concentration Factor (CF) and ion load for wind-blown snow. All simulations are for RH_{max} = 90%, except for that at 70% for which T_{max}= -15° C.

EXPERIMENTAL EVIDENCE

Experimental evidence for changes to the chemistry of wind-blown snow exists for a few locations around the world; the small number of studies is more a reflection of the difficulty of obtaining robust field data in extreme weather conditions than lack of interest by the scientific community.

Osada (1994) measured the concentration of several major ions (Cl^-, NO_3^-, SO_4^{2-} and Na^+) in blowing snow under katabatic conditions near Mizuho Station in East Antarctica. Unfortunately the concentration of ions in falling snow was not reported for comparison. A marked seasonal change in concentration for NO_3^- and excess (to sea-salt) SO_4^{2-} was found with mean concentrations in the austral summer of 4.63 and 3.10 μeq/l respectively and in the austral winter of 0.97 and 0.37 μeq/l. A dramatic order of magnitude change in the ranges of concentrations occurs over a few weeks in March and November. Osada notes that the seasonal variation shows some correspondence with changes in the concentration of atmospheric aerosols, however the details of the variation do not match. Blowing snow concentrations show sharper rates of seasonal change than do aerosols and no evidence of the mid-winter peak of NO_3^- in aerosols. The author suggests that differences in the transport of aerosols and acidic gases or rapid scavenging of aerosols by blowing snow might account for the variation in concentration, however sublimation of blowing snow and volatilisation of chemical species were not considered.

Mulvaney and Wolff (1994) survey the spatial distribution of major ions in the Antarctic ice sheet. There is a roughly 10-fold decline in sea-salt ions as represented by Cl^- with distance from the coast; this represents a scavenging loss as the aerosols move inland from the ocean. Excess SO_4^{2-} sometimes decreases with distance from the coast (two-fold) and at other locations does not. The authors suggest that the scavenging mechanism is more efficient for sea-salts than for excess SO_4^{2-}. Of interest is a three-fold increase in SO_4^{2-} concentrations as one goes furthest inland (East Antarctica) into regions of low precipitation and high wind speeds. NO_3^- concentrations are relatively uniform and show little trend with distance from the coast but tend to increase with elevation (three-fold) and decrease with accumulation rate (three-fold). An additional seasonal variation in NO_3^- has been measured and though attributed to variable sources, post-depositional loss has not been ruled out (Mulvaney and Wolff, 1993). The spatial variation of chemical concentrations are attributed to various processes by Mulvaney and Wolff (1994), but may additionally be explained by blowing snow processes. For example, the decline of sea-salt ions from the coast is due to scavenging, part of this scavenging could be due to electrophoretic attraction to blowing snow particles. Similarly, near the coast, excess SO_4^{2-} is scavenged but not as efficiently as the sea-salt species; this is consistent with blowing snow scavenging. The increase in excess SO_4^{2-} further inland can be explained by the higher CF values due to relatively large sublimation in that low precipitation zone. The uniformity of NO_3^- concentrations is consistent with volatilisation of HNO_3 during sublimation. However the moderate increase with elevation and decrease with precipitation indicates the species does not completely volatise during sublimation.

Delmas and Jones (1987) measured the concentrations of SO_4^{2-}, NO_3^- and NH_4^+ in wind-blown and parent snow on a frozen lake in the boreal forest of Québec. Initial concentrations were 107.8, 9.5 and 1.9 μeq/l for NO_3^-, SO_4^{2-} and NH_4^+ respectively. After one day of blowing snow without precipitation the concentrations dropped 23% for NO_3^-, 20% for SO_4^{2-} and 14% for NH_4^+. The wind speeds were at the threshold for blowing snow and Québec winters are relatively humid so the sublimation rate may have been small in this case. Interestingly, atmospheric concentrations of N and S species were high during the loss from the snowpack. Delmas and Jones suggest that the decrease is due to resuspension of aerosols from the saltating snow particles. Saltating particles are abraded and in this case would undergo a small degree of sublimation which might promote detachment of aerosols from the snow crystal.

Pomeroy et al. (1991a, b) measured the chemistry of fresh, blowing and wind-blown snow on a highland plateau in the Cairngorm Mountains, Scotland. The location is in a relatively temperate environment, however extremely high wind speeds, frequent snowfalls and highly-concentrated acidic contaminants in snow, permit an unequivocal interpretation of blowing snow chemistry processes. Three types of snow were collected during a blowing snow event without concurrent precipitation and analysed for the major anions and cations: fresh undrifted snow, blowing snow at several heights above the surface and wind-blown snow that had undergone relocation by the wind. The sequence of ion concentrations from undrifted to blowing snow to wind-blown snow is shown in Fig. 5. All ion concentrations in blowing snow were higher than the parent snow concentrations, as much as three times higher in some cases (Cl^-, Na^+, K^+). Ion concentrations of Cl^-, SO_4^{2-}, Na^+, K^+ and Mg^{2+} were

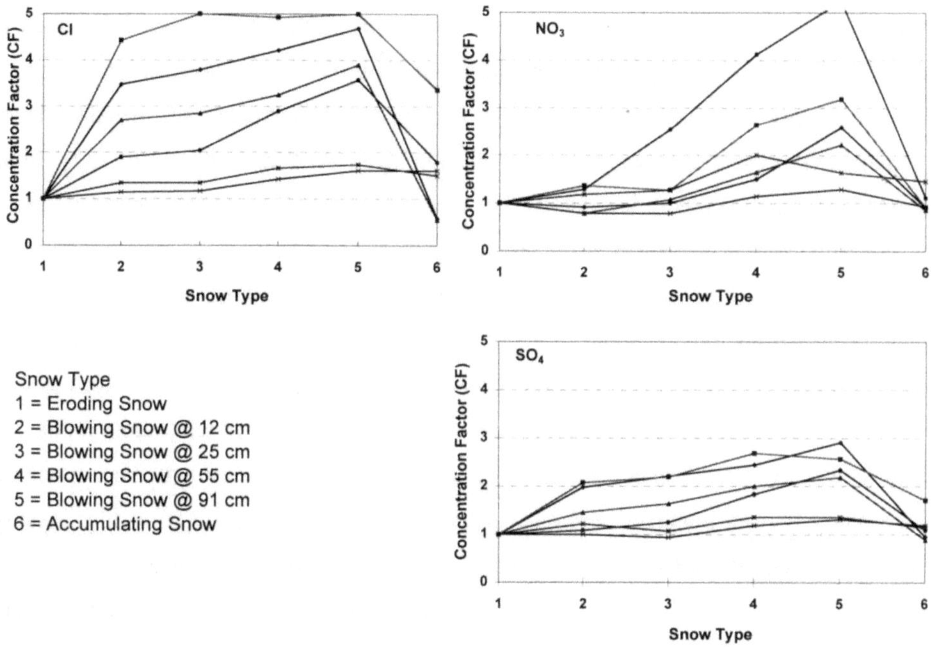

Figure 5. Sequence of ion concentration from undrifted to blowing snow to wind-blown snow as measured in the Cairngorm Mountains, Scotland (after Pomeroy et al., 1991).

significantly higher in wind-blown snow than in the parent snow (80% confidence), however NO_3^- showed little difference between the two. Hence, the elevated concentrations in blowing snow are reflected in wind-blown snow for all ions except for NO_3^-. Wind-blown snow ion concentrations are generally less than blowing snow concentrations because the wind-blown snow was a mixture of snow that had travelled at great heights (and therefore distance) with resultant high concentrations and creeping and saltating snow, which, though relocated, possessed relatively lower concentrations. A combination of ice sublimation, species volatilisation and particulate scavenging processes was suggested to account for the observed increase in concentration during wind transport. Pomeroy et al. (1991a) calculated the concentration factor due to sublimation using meteorological measurements and a range of hypothetical particle trajectories; the concentrations of SO_4^{2-}, NO_3^- and NH_4^+ in blowing snow matched that expected from the parent snow and the concentration factor (CF), however Na^+, K^+, Mg^{2+} and Cl^- concentrations were two-fold in excess of that expected from the CF due to sublimation. This excess concentration to the CF is most pronounced for blowing snow derived from the saltation layer and therefore may be due to electrophoresis; the only known scavenging process at a maximum in the saltation layer. The apparent loss of NO_3^- from wind-blown snow subsequent to deposition was dramatic and suggests a consumption or volatilisation as HNO_3. It is possible that before wind transport occurred, concentrations of NO_3^- may reflect an equilibrium between that on the ice surface and HNO_3 in the atmosphere. After rapid sublimation the NO_3^- ions are relatively concentrated on the ice particle surface and possibly in solution as HNO_3; when subject to further but much slower ventilation due to wind pumping (see Waddington this volume) the concentrations on the ice surface readjust to reach an equilibrium with the partial pressure of HNO_3 in the atmosphere via a vapour transfer from particle surface to atmosphere. Increased acidity from high H_2SO_4 concentrations on the sublimated ice crystal may further promote such a transfer. The result is a loss of NO_3^- in wind-blown snow that is roughly proportional to the loss of ice from the previously sublimating blowing snow particle.

Field evidence of the chemical transformations caused by wind-blown snow exists for the Canadian Western Arctic; a region remote from most ionic sources (except sea-salt) and where a strong gradient in annual blowing snow frequency exists over several hundred kilometres (Pomeroy et al., 1995). In the southern tundra-taiga transition (south of Inuvik, NWT) blowing snow is infrequent and concentrations of the three major ions in snow are relatively low (Cl^- = 6.2 , NO_3^- = 2.6, SO_4^{2-} = 4.5 μeq/l). In the open tundra (Tuktoyaktuk), blowing snow causes 58% of annual snowfall to sublimate before accumulation and resulting ion concentrations are much higher (Cl^- = 47.1 , NO_3^- = 4.5, SO_4^{2-} = 10.8 μeq/l). Near Inuvik and straddling the transition is Trail Valley Creek where 31% of snowfall sublimates as blowing snow. The CF due to sublimation and the measured CF for the three ions are shown in Table 1. Data derive from landscape stratified surveys with coefficients of variation for concentrations and snow water equivalent being quite consistent and increasing from 0.05 for NO_3^- south of Inuvik to 0.21 for Cl^- at Tuktoyaktuk. The higher Cl^- concentrations (7.6 times higher) are certainly related to Tuktoyaktuk's proximity to the Arctic Ocean. The highest Cl^- concentrations were found where the snow had undergone the most drifting, suggesting additional enhancement by electrophoretic scavenging and sublimation. The CF values due to blowing snow sublimation closely match those for SO_4^{2-} and overestimate those for NO_3^- in the transition from a non-transport to severe blowing snow regime. This is further illustrated by the SO_4^{2-}/NO_3^- ratios which increase with frequency of blowing snow from 1.73 south of Inuvik, to 1.94 at Trail Valley and to 2.4 at Tuktoyaktuk. These results concur with the result from Pomeroy et al. (1991a, b) that concentrations with respect to blowing snow sublimation of SO_4^{2-} are conservative, Cl^- are enhanced and NO_3^- are not fully retained in wind-blown snow.

Table 1. Concentration Factors (CF) Measured, Spring 1993 and due to Blowing Snow Sublimation, Winter 1992-93 - Canadian Western Arctic.

Site	Landscape	Sublimation	CF Subl.	CF Cl^-	CF NO_3^-	CF SO_4^{2-}
South Inuvik	Taiga	0 %	1.0	1.0	1.0	1.0
Trail Valley	Transition	31%	1.45	2.8	1.3	1.5
Tuktoyaktuk	Tundra	58%	2.38	7.6	1.7	2.4

Further field evidence from the same region shows conflicting evidence for ion conservation during sublimation. Jones et al (1993) placed weighed, homogenised snow of known chemistry in mesh bags and exposed them on open snow surfaces to wind and either sunlight or shade for 36-48 hours in late winter 1992. One set of bags was placed near Inuvik (temperature -8 to -10 ^{o}C and low humidity) and one set of bags placed in Trail Valley (temperature -5 to -15 ^{o}C and higher humidity). The proposed test of photochemical reactions indicated that changes in chemical composition after the exposure were not attributable to exposure to sunlight or shade. Of interest, however was a steady, slow sublimation from snow in the mesh bags at the Inuvik site. The bags in the warmer, low humidity environment underwent sublimation and lost on average 50 g or about 11-22% of their mass of snow during this period. The bags in the colder, high humidity environment underwent no sublimation and lost no mass. For the sublimating bags there was no statistically significant dry deposition of either NO_3^- or SO_4^{2-}, nor was there volatilisation - each species was completely conservative. For the non-sublimating bags there was no exchange of NO_3^-, but a dry deposition of SO_4^{2-} equivalent to 2.5 $\mu eq/(m^2\ day)$.

Sublimation of 11-22% over several days is slow compared to that which occurs from blowing snow particles; in this "slow sublimation" case there is complete retention of NO_3^- by the sublimating snow particles and no dry deposition.

Recently, a cold room laboratory experiment was used to test the effect of in-situ snow sublimation on chemical concentrations in snow under a controlled condition. A homogenised column of natural, cold prairie snow was placed in a climatically and atmospherically controlled cold box and maintained at -4°C, 70% relative humidity with an atmosphere of medical grade air. A bright beam of light, spectrally controlled to match the intensity and spectrum of sunlight in polar regions was shown continuously for 3 to 4 days onto part of the snow column. Ion concentrations in irradiated surface snow are similar to those in non-irradiated surface snow, hence no photochemical effects were evident in this experiment. However, sublimation of surface snow layers was caused by the radiative energy input to the snow. In the first experiment run, parent concentrations of NO_3^- and SO_4^{2-} were 10.2 and 9.4 μeq/l and CF of 1.14 and 1.36 respectively were measured after the sublimation treatment. This changed the SO_4^{2-}/NO_3^- ratio from 0.92 in parent snow to 1.10 in the sublimated snow. For the second experiment run a fan was operated to recirculate air within the chamber and increase the sublimation rate. Parent concentrations of NO_3^-, SO_4^{2-} and NH_4^+ were 9.3, 5.2 and 4.2 μeq/l and CF of 1.47, 1.88 and 1.96 were measured after the sublimation treatment. In this case the SO_4^{2-}/NO_3^- ratio increased from 0.56 in the parent snow to 0.72 in the sublimated snow. Mass losses from sublimating snow in weighed petri dishes in the chamber confirm that SO_4^{2-} and NH_4^+ were relatively conservative during sublimation but NO_3^- was lost. For the two experimental runs from 16% to 22% of NO_3^- was lost from the surface layers when sublimation removed from 1/4 to 1/2 of the snow in these layers.

IMPLICATIONS FOR POLAR REGIONS

The experimental evidence suggests that notable and complex changes occur to chemical concentrations as snow is relocated. There are five categories into which the changes can be classified, with respect to the transformation of chemical species during blowing snow:

i. *Conserved* - retained by the snow crystal during sublimation, low scavenging rates and little resuspension during transport (e.g. SO_4^{2-}),

ii *Scavenged* - retained by the snow crystal during sublimation, high scavenging rates due to primarily phoretic effects and secondarily gaseous uptake and little aerosol resuspension during transport (e.g. Cl^-, Na^+, Mg^{2+}, K^+),

iii *Volatised* - lost via vapour transfer from the snow crystal during sublimation (possibly NO_3^-),

iv *Resuspended* - lost via aerosol resuspension from the snow crystal during sublimation and abrasion (possibly NO_3^-, SO_4^{2-} and NH_4^+),

v *Consumed* - consumed by reaction within the snow crystal during blowing snow or after deposition (no confirmed examples, possible photochemical reactions, though these are usually assumed not to occur or are included in the volatilisation term).

The complex physio-chemical nature of the phenomenon means that no ion can be strictly categorised; an ion may display the tendencies of one or more categories at one time and change its behaviour under different environmental conditions. For instance NO_3^- has been observed to be conserved at low sublimation rate conditions (usually in-situ), volatised as HNO_3 during rapid snow sublimation (during blowing snow) and resuspended as an anthropogenically-derived particulate under blowing snow with low sublimation rates. The dual nature of NO_3 as part of a gas or of an aerosol, variable partial pressures, its location on the ice surface and possible particle surface reactions probably play a role in this complex behaviour. The normal conservation of SO_4^{2-} but occasional resuspension as an aerosol will relate to the position of the aerosol on the ice crystal, the depth of any quasi-liquid layer and the rate of sublimation. In most studies SO_4^{2-} is found to be completely conservative but Delmas and Jones (1987) showed that snow crystals that undergo shattering in saltation transport and little sublimation appear to resuspend anthropogenic SO_4^{2-}. When considering the accumulation in glacial snow, the behaviour of an ion may depend upon the severity of blowing snow sublimation. Following the model presented in Eq. 8 and Fig. 3 a binary system is proposed. For wind transport conditions when up to half the annual snowfall sublimates each year then many ions are either conserved or scavenged and likely undergo a concentrating effect after each blowing snow sublimation event, becoming more concentrated in accumulating snow. This corresponds to the lower wind speed condition in Fig. 3. NO_3^- is the known exception to this trend, being volatised even under relatively moderate blowing snow regimes. When blowing snow removes a large portion of the annual snowfall in several severe blizzards, then many species may be resuspended and volatised along with the remnants of the initial snow crystals. This corresponds to the higher wind speed condition in Fig. 3. These ions will then contribute little to the accumulation layer. As the transport and sublimation rates of blowing snow increase exponentially with wind speed and the wind speed undergoes seasonal and annual trends, it is quite possible that polar environments experience both types of transformation and substantial variation in the dominant mode of transformation seasonally, annually and historically.

To examine blowing snow regimes in polar environments and compare them to more intensively studied continental environments, a special run of the Hadley Centre (UK) GCM (1994 version) was used to develop a spatially-precise climatological database for the central Antarctic and central Greenland. The GCM was run for 10 years as part of a broader applications study to provide averages of *present* conditions in these remote locations. This method avoids use of wind speed data from automated weather stations which are subject to severe riming in mid-winter and snowfall measurements that are subject to undermeasurement and hence induce large errors in calculating a blowing snow climatology. Unfortunately the GCM appears to overestimate humidity in most winter conditions and underestimate winter precipitation in polar regions as it estimates accumulation not snowfall (Connolley and Cattle, 1994). Table 2 shows some average meteorological conditions in January and July as predicted by the GCM in four selected locations, central Greenland, the Canadian Prairies, coastal Antarctic and central Antarctica. The meteorological variables shown are important in estimating the monthly blowing snow fluxes using Eq. 7.

Table 2. Climatology of three blowing snow regimes (north polar, continental, south polar) as estimated from the Hadley Centre GCM.

Location	January Max. Daily Temperature	July Max. Daily Temperature	Mean Monthly Wind Speed m/s
Central Greenland $75^{o}18'N$ $40^{o}45'W$	-41 ^{o}C	-11 ^{o}C	5.9
Canadian Prairies $52^{o}38'N$ $108^{o}30'W$	-8 ^{o}C	24 ^{o}C	4.7
Central Antarctica $82^{o}05'S$ $33^{o}22'E$	-27 ^{o}C	-69 ^{o}C	5.1
Coastal Antarctica $74^{o}49'S$ $82^{o}15'W$	-7 ^{o}C	-23 ^{o}C	7.0

Annual blowing snow sublimation calculated using Eq. 7 and the GCM simulation is shown in Table 3 for the four locations, along with snowfall and the CF due to blowing snow sublimation, presuming a perfectly conservative ionic species. The CF values must be considered crude, first approximations. Sublimation is calculated for two fetch lengths; one kilometre and ten kilometres. Equation 7 provides the monthly sublimation fluxes for a 1 km fetch typical of mixed agricultural landscapes or irregular terrain. This 1-km sublimation is adjusted for the 10-km fetch more appropriate to the polar ice caps by presuming that all transported snow (1-km) except the 10% required to maintain a mass balance has sublimated over the 10-km fetch, and adding this quantity to the 1-km sublimation. Sublimation is underestimated and precipitation overestimated for the Canadian Prairies; actual values are roughly 110 mm snowfall and 40 mm sublimation (Pomeroy and Gray, 1995). The 10-km sublimation values exceed the GCM-modelled snowfall in both central Antarctica and Greenland; this is likely due to an underestimation of snowfall or an overestimation of sublimation. There has been no verification of the blowing snow sublimation routine poleward of $70^{o}N$ and it may overestimate sublimation in persistently cold conditions; however, as the GCM has been calibrated against polar *accumulation* rather than actual precipitation, the snowfall underestimation is likely to be quite large. What the modelling exercise does clearly indicate is the substantial variation possible in CF for conservative species within the present range of polar climates and the difficulty in determining the exact CF with present meteorological data sets and knowledge of polar blowing snow. For example, modelled blowing snow in central Antarctica removes most snow and ions, but in coastal Antarctica ions are deposited in a more concentrated state. In central Greenland, the modelled blowing snow regime borders between removal and concentration states, depending upon the fetch distance and probably varies in state from year to year. In all polar locations the state varies seasonally. A full-scale field investigation of the transformations of various chemical species during blowing snow in polar atmospheric conditions is required to begin to predict actual CF values in relation to that of a perfectly conserved species, to predict the onset of resuspension, and to model the transformations of volatile and scavenging species.

Table 3. Annual blowing snow sublimation for three regimes (north polar, continental, south polar) as estimated using the blowing snow model and data from the Hadley Centre GCM. All values are annual estimates. *** denotes CF impossible to calculate but estimated as quite large.

Location	Snowfall	Blowing Snow Sublimation 1-km Fetch	Blowing Snow Sublimation 10-km Fetch	CF 1 km	CF 10 km
Central Greenland	151 mm	85 mm	218 mm	2.28	***
Canadian Prairies	151 mm	12 mm	30 mm	1.09	1.25
Central Antarctica	48 mm	73 mm	242.2 mm	***	***
Coastal Antarctica	627 mm	115 mm	244 mm	1.22	1.64

SUMMARY

Blowing snow is ubiquitous in polar regions and undergoes relatively rapid sublimation whilst in transit. The erosion, transport and deposition of blowing snow dramatically transforms the physical and chemical characteristics of polar snow. Processes active during snow transport are sublimation, thermophoresis, diffusiophoresis, electrophoresis and ventilation; these processes occur at much higher rates than for "in-situ" surface snow. Conditions suitable for photochemistry exist on blowing snow particles though no strong empirical evidence exists for such reactions in snow. Annual accumulations of snow and ions on polar ice-sheets are a residual of the blowing snow phenomenon and therefore a result of the interaction of wind speed, temperature and snowfall regimes with surface topography and the meteorological history of the snowpack. Chemical transformations in blowing snow involve ion conservation, scavenging, volatilisation, resuspension and consumption. Experimental evidence from Canada, Scotland and the Antarctic shows that for moderate blowing snow regimes the first three processes dominate whilst for severe regimes resuspension dominates. In moderate blowing snow, concentration increases are greatest for scavenged ions such as those of sea-salt aerosols. Concentration increases are proportional to sublimation loss for conservative species such as SO_4^{2-} and NH_4^+ and less than proportional to the sublimation loss for volatile species such as HNO_3. Concentration increases due to sublimation increase exponentially with wind speed and vary widely with climate and terrain in polar regions. Annual ion load depends on the blowing snow regime and displays two almost binary modes of retention and resuspension. For polar atmospheres a wind speed of 8 m/s, near the mean for many ice sheets, divides the two depositional modes. Hence the process of ion deposition in snow in polar regions is inherently unstable and varies in mode seasonally and annually. The uncertainty in estimating and spatially extrapolating wind speed regimes over polar regions suggests that ion concentration factors and annual retention of ions in wind-blown snow are probably unpredictable over even recent history. This adds considerable complexity to attempts to interpret the mass balance and chemistry of recent ice cores and link this interpretation to atmospheric chemistry.

CONCLUSIONS

1. Blowing snow is ubiquitous in polar regions and dramatically transforms the physical and chemical characteristics of polar snowcovers. Associated with transport of snow particles are the processes of sublimation, thermophoresis, diffusiophoresis, electrophoresis and ventilation which occur at much higher rates than for in-situ surface snow. Conditions suitable for photochemistry may occur on sublimating snow particles.

2. Accumulation of snow and constituent ions in polar regions is a residual of blowing snow fluxes and therefore a function of the wind speed, temperature and snowfall regimes. Sublimation fluxes in polar regions exceed accumulation fluxes by several fold and approach "true" snowfall fluxes in extreme environments.

3. Transformations of blowing snow chemistry may involve species conservation, scavenging, volatilisation, resuspension and consumption. For moderate blowing snow regimes the first three transformations dominate, whilst for severe regimes resuspension dominates. Consumption is undocumented.

4. Conservative species are exemplified by SO_4^{2-} and NH_4^+, scavenging species by those of sea-salt aerosols and volatising species by HNO_3.

5. For conservative species a concentration increase due to sublimation occurs in polar regions; the increase is quite sensitive to blowing snow regime, increasing exponentially with wind speed and varying widely with climate and terrain.

6. Snow and ion loading are dependent upon the blowing snow regime and show an almost binary state of retention and resuspension. For polar atmospheres the boundary between retention and resuspension occurs at mean wind speeds of about 8 m/s above which relatively little snow or ions are retained on the ice caps. This wind speed is near the monthly averages for many interior Antarctic and central Greenland locations, suggesting that the ion deposition process in the Antarctic and Greenland is inherently unstable and varies in mode seasonally

and from year to year. The sensitivity to local climatic conditions means that even slight changes in micro-climate can cause dramatic changes in the annual deposition of chemical species in polar snow.

7. The sensitivity of blowing snow to the variable wind speed regimes over polar regions suggests that ion concentration factors and annual retention of wind-blown snow are probably unpredictable over even recent history. The variable conservation of chemical species by blowing snow is due to combinations of in-transit processes and adds another level of complexity to the system. The dependence of snow ion accumulation processes on synoptic and site specific physical factors and ice-surface chemical transformations adds considerably uncertainty to attempts to interpret the snow accumulation and snow chemistry of recent ice cores strictly in terms of atmospheric chemistry and precipitation.

ACKNOWLEDGEMENTS

The authors acknowledge the assistance of S. Adam and J. Onclin in preparing data for this report, the NHRI snow chemistry team: C. Onclin, K. Dion, R. Reid, N. Hedstrom; colleagues: Prof. T.D. Davies, Univ. of East Anglia, Dr. P. Marsh, NHRI, Dr. N.E. Peters, USGS-Water Resources Division and Dr. M. Tranter, Univ. of Bristol and chemical analysis by K. Supeene, Central Chemistry Laboratory, NHRI. The assistance of the Inuvik Research Centre, Government of the Northwest Territories and the Polar Continental Shelf Project, Department of Energy, Mines and Resources Canada in providing logistical support for field campaigns, the Hadley Centre, UK Meteorological Office in providing climate data simulations, the Natural Sciences and Engineering Research Council of Canada and NHRI for research funding and the NATO Collaborative Grants Programme for travel funds is gratefully acknowledged.

REFERENCES

Albert MR, McGilvary WR (1992) Thermal effects due to air flow and vapor transport in dry snow. Journal of Glaciology, 38(129). 273-281.

Baker MB, Dash JG (1994) Mechanism of charge transfer between colliding ice particles in thunderstorms. Journal of Geophysical Research, 99(D5). 10621-10626.

Bales RC, Choi J (1996) Conceptual framework for interpretation of exchange processes. *In this volume.*

Beard KV, Ochs HT, Leong KH (1983) Particle scavenging by evaporating cloud drops. In: Pruppacher et al. (eds) Precipitation Scavenging, Dry Deposition and Resuspension. Elsevier Science Publishing, Toronto. 517-525.

Benson CS (1982) Reassessment of winter precipitation on Alaska's Arctic slope and measurements on the flux of wind-blown snow. University of Alaska Fairbanks, Geophysical Institute Report UAG R-288.

Benson CS, Sturm M (1993) Structure and wind transport of seasonal snow on the Arctic slope of Alaska. Annals of Glaciology, 18. 261-267.

Budd WF (1966) The drifting of non-uniform snow particles. In, Rubin (ed) Studies in Antarctic Meteorology, Vol. 9. Antarctic Research Series, American Geophysical Union. Washington, DC. 59-70

Budd WF, Dingle WRJ, Radok U (1966) The Byrd snow drift project: outline and basic results. In, Rubin (ed) Studies in Antarctic Meteorology, Vol. 9. Antarctic Research Series, American Geophysical Union. Washington, DC. 71-134.

Conklin MH, Bales RC (1993) SO_2 uptake on ice spheres: liquid nature of the ice-air interface. Journal of Geophysical Research, 98(D9). 16851-16855.

Connolley WM, Cattle H (1994) The Antarctic climate of the UKMO Unified Model. Antarctic Science, 6(1). 115-122.

Dalrymple PC (1966) A physical climatology of the Antarctic Plateau. In, Rubin (ed) Studies in Antarctic Meteorology, Vol. 9. Antarctic Research Series, American Geophysical Union. Washington, DC. 13-58.

Dawson, GA, Brimblecombe P (1983) Precipitation scavenging of highly soluble gases. In: Pruppacher et al. (eds) Precipitation Scavenging, Dry Deposition and Resuspension. Elsevier Science Publishing, Toronto. 687-696.

Delmas V, Jones HG (1987) Wind as a factor in the direct measurement of the dry deposition of acid pollutants to snowcovers. In: Jones, Orville-Thomas (eds) Seasonal Snowcovers: Physics, Chemistry, Hydrology. NATO ASI Series C-211. D. Reidel, Dordrecht. 321-336.

Dingle WRJ, Radok U (1961) Antarctic snow drift and mass transport. International Association of Scientific Hydrology (IAHS) Publ. 55. 77-87.

Dong Y, Hallett J (1990) A laboratory study of ice crystal charging during growth and evaporation. In, Proceedings of the Conference on Cloud Physics. American Meteorological Society, Boston. 324-327.

Dyunin AK (1959) Fundamentals of the theory of snow drifting. Isvestia Sibirski Otdeleniya Akademii Nauk SSSR, 12. 11-24 (Trans. G. Belkov, 1961) National Research Council of Canada Technical Translation 952, 26 pp.

Fujii Y (1981) Formation of surface snow layer at Mizuho Station, Antarctica. In, Kusunoki (ed) Memoirs, 3rd Symposium on Polar Meteorology and Glaciology. National Institute of Polar Research, Special Issue 19. 280-296.

Fujii Y, Kusunoki K (1982) The role of sublimation and condensation in the formation of ice sheet surface at Mizuho Station, Antarctica. Journal of Geophysical Research, 87(C6). 4293-4300.

Goodison, BE (1981) Compatibility of Canadian snowfall and snow cover data. Water Resources Research, 17(4). 893-900.

Granger RJ, Male DH (1978) Melting of a prairie snowpack. Journal of Applied Meteorology, 17(12). 1833-1842.

Gunz DW, Hoffman MR (1990) Atmospheric chemistry of peroxides: a review. Atmospheric Environment, 24A(7). 1673-1684.

Huebert BJ, Fehsenfeld FC, Norton RB, Albritton D (1983) The scavenging of nitric acid vapour by snow. In: Pruppacher et al. (eds) Precipitation Scavenging, Dry Deposition and Resuspension. Elsevier Science Publishing, Toronto. 293-302.

Houghton JT, Jenkins GJ, Ephraums JJ (1990) Climate Change, The IPCC Scientific Assessment. Cambridge University Press, Cambridge. 365 p.

Jones HG, Pomeroy JW, Davies TD, Marsh P, Tranter M (1993) Snow-atmosphere interactions in Arctic snowpacks: net fluxes of NO_3, SO_4 and influence of solar radiation. Proceedings of the 50th Annual Eastern Snow Conference. 255-264.

Kobayashi S (1978) Snow transport by katabatic winds in Mizuho Camp area, East Antarctica. Journal of the Meteorological Society of Japan, 56. 130-139.

Kobayashi S, Ishikawa N, Ohata T (1985) Katabatic snow storms in stable atmospheric conditions at Mizuho Station, Antarctica. Annals of Glaciology, 6. 229-231.

Kodama Y, Wendler G, Gosink J (1985) Effect of blowing snow on katabatic winds in Antarctica. Annals of Glaciology, 6. 59-62.

Lamb D, Mitchell D, Blumenstein R (1986) Snow chemistry in relation to precipitation growth forms. In: Preprint Volumes, 23rd Conference on Radar Meteorology and the Conference on Cloud Physics, Snowmass, Colorado. American Meteorological Society, Boston. 77-80.

Latham J, Mason BJ (1961) Electric charge transfer associated with temperature gradients in ice. Proceedings of the Royal Society of London, Series A, 260(1303). 523-536.

Lister H (1960) Solid precipitation and drift snow. Trans-Antarctic Expedition 1955-1958, Science Report 5.

Male DH (1980) The seasonal snowcover. In, Colbeck (ed), Dynamics of Snow and Ice Masses. Academic Press, New York. 305-395.

Male DH, Granger RJ (1979) Energy and mass fluxes at the snow surface in a prairie environment. In, Colbeck and Ray (eds), Proceedings, Modelling of Snow Cover Runoff. US Army Cold Regions Research and Engineering Laboratory, Hanover, New Hampshire. 101-124.

Marshall S, Roads JO, Glatzmaier G (1994) Snow hydrology in a general circulation model. Journal of Climate, 7. 1251-1269.

Mellor M, Radok U (1960) Some properties of drifting snow. In, Antarctic Meteorology. Pergamon Press, Oxford. 333-346.

Mellor M, Fellers G (1986) Concentration and flux of wind-blown snow. US Army Cold Regions Research and Engineering Laboratory, Hanover, New Hampshire. Special Report 86-11. 16 p.

Miller NL, Wang PK (1989) A theoretical determination of the efficiency with which aerosol particles are collected by falling columnar ice crystals. Journal of Atmospheric Sciences, 46(12). 1656-1663.

Moore I, Mobbs SD, Ingham DB, King JC (1994) A numerical model of blowing snow around an Antarctic building. Annals of Glaciology, 20. 341-346.

Mulvaney R, Wolff EW (1993) Evidence for winter/spring denitrification of the stratosphere in the nitrate record of Antarctic firn cores. Journal of Geophysical Research, 98(D3). 5213-5220.

Mulvaney R, Wolff EW (1994) Spatial variability of the major chemistry of the Antarctic ice sheet. Annals of Glaciology, 20. 440-447.

Murakami M, Magono C (1983) An experimental study of electrostatic effect on aerosol scavenging by snow crystals. In: Pruppacher et al. (eds) Precipitation Scavenging, Dry Deposition and Resuspension. Elsevier Science Publishing, Toronto. 541-549.

Murakami M, Magono C, Kikuchi K (1985) Experiments on aerosol scavenging by natural snow crystals part iii, the effect of snow crystal charge on collection efficiency. Journal of the Meteorological Society of Japan, 63(6). 1127-1137.

Neftel A (1996) The record of gases and reactive species in ice cores, and problems of interpretation. *In this volume.*

Neubauer J, Heumann KG (1988) Nitrate trace determinations in snow and firn core samples of ice shelves at the Weddel Sea, Antarctica. Atmospheric Environment, 22. 537-545.

Osada K (1994) Seasonal variation of major ionic concentration levels in drifting-snow samples obtained from east Dronning Maud Land, East Antarctica. Annals of Glaciology, 20. 226-230.

Peary EA (1898) A journey in north Greenland. Geographical Journal, 11. 213-240.

Perla RI (1985) Snow in strong or weak temperature gradients, part ii: section-plane analysis. Cold Regions Science and Technology, 11. 181-186.

Pomeroy JW (1989) A process-based model of snow drifting. Annals of Glaciology, 13. 237-240.

Pomeroy JW, Gray DM (1990) Saltation of snow. Water Resources Research. 26(7). 1583-1594.

Pomeroy JW, Male DH (1992) Steady-state suspension of snow. Journal of Hydrology, 136. 275-301.

Pomeroy JW, Gray DM (1994) Sensitivity of snow relocation and sublimation to climate and surface vegetation. In, Jones, Davies, Ohmura and Morris (eds) Snow and Ice Covers: Interactions with the Atmosphere and Ecosystems. IAHS Publ. No. 223, IAHS Press, Wallingford, UK. 213-225.

Pomeroy JW, Goodison BE (1995) Winter and snow. In Surface Climates of Canada. McGill-Queen's University Press, Montreal. In press.

Pomeroy JW, Gray DM (1995) Snowcover Accumulation, Relocation and Management. NHRI Science Report No. 7. Supply and Services Canada, Saskatoon. 144 p.

Pomeroy JW, Davies TD, Tranter M (1991a) The impact of blowing snow on snow chemistry. In: Davies, Tranter, Jones (eds), Seasonal Snowpacks: Processes of Compositional Change. NATO ASI Series G, Vol. 28. Springer-Verlag, Berlin. 71-114.

Pomeroy JW, Davies TD, Tranter M (1991b) Relationships between snow chemistry and blowing snow: initial findings. In, Prowse and Ommanney (eds), Northern Hydrology: Selected Perspectives. NHRI Symposium No. 6. Minister of Supply and Services Canada, Saskatoon. 277-292.

Pomeroy JW, Gray DM, Landine PG (1993a) The Prairie Blowing Snow Model: characteristics, validation, operation. Journal of Hydrology, 144. 165-192.

Pomeroy JW, Marsh P, Lesack L (1993b) Relocation of major ions in snow along the tundra-taiga ecotone. Nordic Hydrology, 24. 151-168.

Pomeroy JW, Marsh P, Jones HG, Davies TD (1995) Spatial distribution of snow chemical load at the tundra-taiga transition. In, Tonnessen, Williams, Tranter (eds), Biogeochemistry of Seasonally Snow-covered Catchments. IAHS Publ. No. 228. IAHS Press, Wallingford, UK. 191-206.

Sauter DP, Wang DK (1989) An experimental study of the scavenging of atmospheric aerosols by natural snow crystals. Journal of the Atmospheric Sciences, 46(12). 1650-1655.

Schmidt S, Dent JD (1993) A theoretical prediction of the effects of electrostatic forces on saltating snow particles. Annals of Glaciology, 18. 234-238.

Schmidt RA (1982) Vertical profiles of wind speed, snow concentration and humidity in blowing snow. Boundary-layer Meteorology, 23. 223-246.

Tabler RD, Pomeroy JW, Santana BW (1990a) Drifting snow. In, Ryan and Crissman (eds), Cold Regions Hydrology and Hydraulics. American Society of Civil Engineers, New York. 95-146.

Tabler RD, Benson CS, Santana BW, Ganguly P (1990b) Estimating snow transport from wind speed records: estimates versus measurements at Prudhoe Bay, Alaska. In, Proceedings of the 58th Annual Meeting of the Western Snow Conference. 61-78.

Takahashi S (1985) Characteristics of drifting snow at Mizuho Station, Antarctica. Annals of Glaciology, 6. 71-75.

Takahashi S, Ageta Y, Fujii Y, Watanabe O (1994) Surface mass balance in east Dronning Maud Land, observed by Japanese Antarctic Research Expeditions. Annals of Glaciology, 20. 242-248.

Wishart ER (1970) Electrification of Antarctic drifting snow. In, International Symposium on Antarctic Glaciological Exploration. IAHS Publ. No. 86. IAHS Press, Wallingford, UK. 316-324.

Zepp RG, Hoigui J, Bader H (1987) Nitrate-induced photo-oxidation of trace organic chemicals in water. Environmental Science and Technology, 21. 443-450.

Zinder B, Schumann T, Waldvogel A (1988) Aerosol and hydrometeor concentrations and their chemical composition during winter precipitation along a mountain slope ii, enhancement of below-cloud scavenging in a stably stratified atmosphere. Atmospheric Environment, 22(12). 2741-2750.

TURBULENT EXCHANGE OF MOMENTUM AND SCALARS IN THE SURFACE LAYER OVER ANTARCTIC SNOW AND ICE*

Richard Bintanja and Michiel R. van den Broeke
Institute for Marine and Atmospheric Research Utrecht
Utrecht University
Princetonplein 5
3584 CC Utrecht
The Netherlands

Abstract. Vertical profiles of wind speed, temperature and humidity were used to estimate the roughness lengths for momentum (z_0), heat (z_H) and moisture (z_Q) over Antarctic surfaces. The vertical profile-measurements were performed on and near a blue ice field in Queen Maud Land, East Antarctica. The roughness lengths are generally used to compute the fluxes of momentum, heat and moisture from standard meteorological observations using the flux-profile relationships. The value of z_0 is easily evaluated from the wind speed profile. The scalar roughness lengths are evaluated using a new method, which circumvents the difficult measurement of the surface temperature. It is found that the vertical exchange processes are strongly influenced by the saltating snow particles close to the surface during snow-drift events. The scalar roughness lengths seem to be approximately equal to z_0 for a large range of roughness Reynolds numbers, despite the frequent occurrence of drifting snow. It is suggested that snow-drift processes are important for the turbulent transport of scalar quantities such as heat and water vapour, and presumably also for the transport of other atmospheric constituents.

1. Introduction

Knowledge of the transfer coefficients of momentum, heat and moisture is necessary to evaluate the turbulent fluxes of momentum, heat and moisture from standard meteorological

* This contribution is based on the paper '*Momentum and scalar transfer coefficients over aerodynamically smooth Antarctic surfaces*' by R. Bintanja and M.R. van den Broeke, which will be published in Boundary-Layer Meteorology. The material used here is reprinted by permission of Kluwer Academic Publishers.

NATO ASI Series, Vol. I 43
Chemical Exchange Between the Atmosphere and Polar Snow
Edited by Eric W. Wolff and Roger C. Bales

observations. Over Antarctica, the fluxes of sensible and latent heat play an important role in the surface energy and mass balance. The sensible heat flux is interrelated with the katabatic wind system: negative buoyancy at the surface forces the katabatic winds while the strength of the inversion depends, among others, on the wind speed. The importance of sublimation on the surface mass balance of Antarctica is a matter of great interest: in a modelling study of the surface mass balance, Fortuin and Oerlemans (1992) found that the sublimation rate was an important component of the surface mass balance, especially near the coast. Relatively high sublimation rates were also found in local experimental studies in the near-coastal region (e.g. Wendler et al., 1988; Bintanja and Van den Broeke, 1995). On a local scale the sublimation rate can be extremely large, which can result in a negative surface mass balance. These regions are generally known as blue ice areas.

Recently, Bintanja and Van den Broeke (1995) showed that the momentum roughness at neutral stability of bare ice is more than two orders of magnitude smaller than that of snow. Obviously, this can be attributed to the differences in the mechanisms of dissipation of turbulent kinetic energy in rough and smooth flow conditions. From theoretical considerations (e.g. Hinze, 1975), it is known that momentum decreases due to the turbulent form drag at the surface elements in rough flow and to viscous shear stress in smooth flow. Very close to the surface, the transport of scalars (heat, moisture, chemical species) is governed by viscous diffusion, independent of the type of flow. According to Andreas (1987), the values of the momentum and scalar roughness lengths should be within one order of magnitude under smooth flow conditions.

Once the turbulent transport mechanisms of scalars such as water vapour are known, one can use this knowledge to estimate the transfer of chemical species. Chemically inert species can be treated as scalars and will be subject to the standard flux-profile relationships. On the other hand, the turbulent transfer of chemically reactive species follows an adapted version of this theory, in which the difference in time scales between chemical reaction and turbulence is taken into account (Vilà-Guerau de Arellano et al., 1995). In either case, this study provides insight into the important mechanisms of the turbulent transport of any scalar quantity in the surface layer. This is useful if one is interested in atmosphere-to-snow deposition rates of certain gases in connection with ice-core or firn-core analyses.

Very few observational studies on the transfer coefficients over snow and ice surfaces exist to date, e.g. Liljequist (1957), Hicks and Martin (1972), Joffre (1982), Kondo and Yamazawa (1986), King and Anderson (1994). In this paper we present the transfer coefficients for momentum, heat and moisture over smooth ice and snow surfaces at low roughness Reynolds number. Also, we will show how the turbulent transfer is affected by drifting snow processes.

2. Instrumentation and field location

The measurements were performed in the western part of Queen Maud Land, Antarctica (Fig. 1) in the austral summer 1992/93. The Scharffenbergbotnen valley, which is part of the Heimefrontfjella, contains a 2x5 km^2 large blue ice area. The base camp was Swedish research station Svea (74° 35' S, 11°, 13' W), and the expedition was hosted by SWEDARP (Swedish Antarctic Research Programme). The main meteorological goals of the field experiments were: (1) evaluate the surface energy and mass balance differences between the blue ice area and the adjacent snow fields and (2) study the summertime variability of the boundary layer flow.

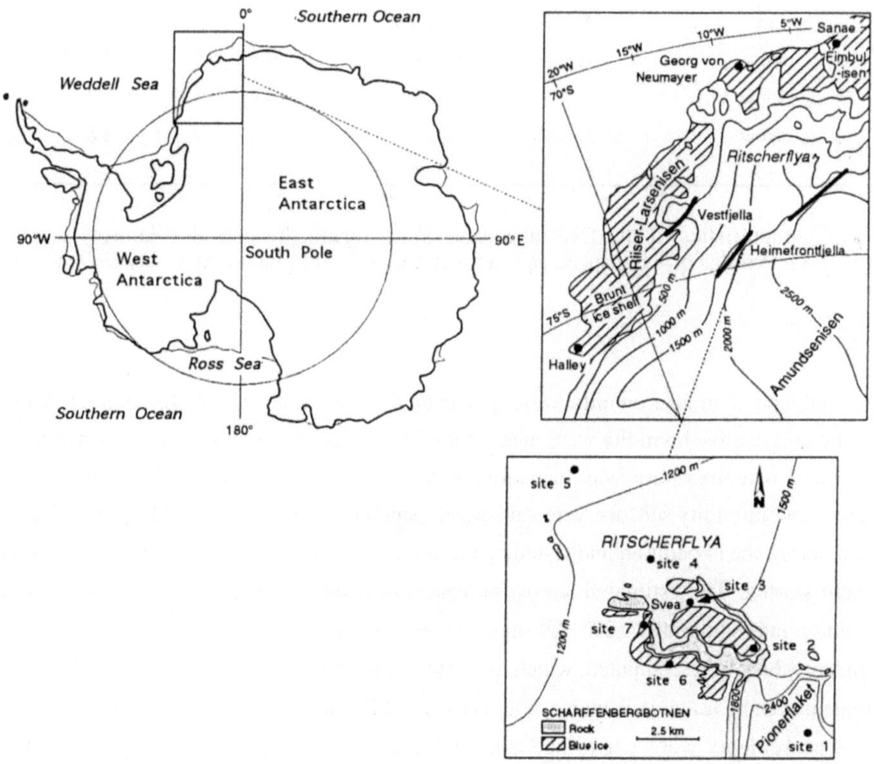

Figure 1. Map of the Scharffenbergbotnen valley in the Heimefrontfjella, Queen Maud Land, Antarctica. The locations of the measuring sites are given in the lower panel.

Seven meteorological stations were placed in a small area around Svea. In this study, we will focus on measurements made at sites 2, 3, 4, 5 and 6 (Fig. 1, lower panel). An overview of these sites is given in Table 1. It was chosen to place a dense net of measuring sites to capture spatial variations in meteorological conditions and surface characteristics of a few kilometer. Because of the limited amount of sensors available, meteorological quantities could then only be measured at two heights (2 and 6 m) at each site. As a result, the curvature of the vertical profiles cannot be obtained and standard log-linear profiles have necessarily to be assumed. This also implies that high sensor accuracy is required for an accurate determination of the vertical gradients.

Site	Surface type	Number of measuring days	Profile	2 m temperature (°C)	6 m wind speed (m s^{-1})
2	ice	37	*U, T, RH*	-8.3	4.9
3	snow	41	*U, T, RH*	-9.8	4.2
4	snow	41	*U, T*	-10.8	5.2
5	snow	27	*U, T*	-11.4	5.8
6	ice	41	*U, T*	-10.1	3.9

Table 1. Characteristics of the five measuring sites during the period 1 January to 10 February 1993. Profiles of wind speed (*U*), temperature (*T*) and relative humidity (*RH*) were measured.

The wind speed measurements were performed with Aanderaa 2740 sensors while temperature and relative humidity were measured with ventilated Rotronic YA-100 sensors. At sites 4, 5 and 6 temperature was measured with ventilated Aanderaa 2775 sensors. All temperature and humidity sensors were sufficiently shielded from radiation. The sensors were calibrated before the expedition and tested again in the field during two days before the actual experiment started. The estimated sensor precision is wind speed ±0.2 m s^{-1}, temperature ±0.2 °C and relative humidity ±2%. All sensors were sampled every two minutes from which hourly mean values were computed, which are used in the present analyses. More details of the measurements can be found in Bintanja et al. (1993) and Bintanja and Van den Broeke (1995).

The measurements were performed during January-February 1993, during which the weather was rather variable. Midlatitude cyclonic systems occasionally entered the region, causing fierce easterly winds, drifting and blowing snow and overcast conditions. In between, calm, clear-sky conditions prevailed. Table 1 gives mean temperatures and wind speeds at the various sites. The wind-exposed sites 4 and 5 experience relatively high winds and low temperatures, whereas the wind-shielded sites 2, 3 and 6 are warmer and less windy. Nearly all observations used in this analysis are associated with strong, easterly winds. A thorough description of the various mesoscale and local flow types occurring in this region is presented by Van den Broeke and Bintanja (1995).

3. Theory

Various observations have revealed that, within the surface layer, the turbulent fluxes of sensible and latent heat can be described by the Monin-Obukhov similarity theory. Under a wide range of meteorological conditions this theory successfully explains the variation of the vertical profiles by changes in the typical scales, provided that the requirements of stationarity and horizontal homogeneity are met. The characteristic turbulent scales of velocity (u_*), potential temperature (θ_*) and specific humidity (q_*) are generally taken as:

$$u_* = \sqrt{\frac{\tau}{\rho}},\ \theta_* = -\frac{H}{\rho c_p u_*},\ q_* = -\frac{LE}{\rho L_v u_*} \qquad (1)$$

with τ is the surface stress, ρ is the air density, c_p is the specific heat at constant pressure, L_v is the latent heat for sublimation and H and LE are the fluxes of sensible and latent heat at the surface, respectively. Above the interfacial or viscous sublayer (the layer closest to the surface where diffusive and viscous processes are dominant), which depth is approximately equal to $30\nu/u_*$ (with ν the molecular viscosity) over smooth surfaces (Hinze, 1975), the mean vertical profiles of wind speed (u), potential temperature (θ) and specific humidity (q) are related to the appropriate scales by:

$$\frac{\kappa z}{u_*}\frac{\partial u}{\partial z} = \phi_m(\zeta),\ \frac{\kappa z}{\theta_*}\frac{\partial \theta}{\partial z} = \phi_h(\zeta),\ \frac{\kappa z}{q_*}\frac{\partial q}{\partial z} = \phi_h(\zeta) \qquad (2)$$

in which κ (= 0.4) is the Von Kármán constant, z is the height above the surface, ϕ_m and ϕ_h are the stability correction functions for momentum (m) and heat and moisture (h). $\zeta = z/L$ is the non-dimensional stability parameter with L the Obukhov length scale:

$$L = \frac{u_*^2}{\kappa\left(g/\theta_0\right)(\theta_* + 0.62\,\theta_0\, q_*)} \qquad (3)$$

where θ_0 is a reference temperature. The quantity g/θ_0 is often referred to as the buoyancy parameter. Since the characteristic scales u_*, θ_* and q_* are independent of z, Eqs. (2) can be integrated between two heights z_1 and z_2 ($z_1 < z_2$). This yields expressions for the vertical profiles in the surface layer:

$$\kappa\frac{\delta u}{u_*} = \ln\left(\frac{z_2}{z_1}\right) - \psi_m(\zeta_2) + \psi_m(\zeta_1) \tag{4a}$$

$$\kappa\frac{\delta \theta}{\theta_*} = \ln\left(\frac{z_2}{z_1}\right) - \psi_h(\zeta_2) + \psi_h(\zeta_1) \tag{4b}$$

$$\kappa\frac{\delta q}{q_*} = \ln\left(\frac{z_2}{z_1}\right) - \psi_h(\zeta_2) + \psi_h(\zeta_1) \tag{4c}$$

where δu, $\delta\theta$ and δq are the differences in wind speed, potential temperature and specific humidity between z_2 and z_1. $\psi_m(\zeta)$ and $\psi_h(\zeta)$ are the integrated diabatic correction functions. If these equations are applied between a specific height in the surface layer and the surface, z_1 in Eqs. (4a) - (4c) has to be replaced by the surface roughness lengths z_0, z_H and z_Q, respectively.

For the stability functions ϕ_m and ϕ_h under unstable conditions ($\zeta < 0$) we use the expressions given by Dyer (1974) and Högström (1988):

$$\begin{aligned} \phi_m(\zeta) &= (1 - \gamma_m\zeta)^{-1/4} \\ \phi_h(\zeta) &= (1 - \gamma_h\zeta)^{-1/2} \end{aligned} \tag{5}$$

with $\gamma_m = 20$ and $\gamma_h = 15$. For stable conditions ($\zeta > 0$) we use the expressions given by Duynkerke (1991) which are very similar to those proposed by Högström (1988) for moderately stable cases:

$$\begin{aligned} \phi_m(\zeta) &= 1 + \beta_m\zeta\,(1 + \beta_m\zeta/a)^{a-1} \\ \phi_h(\zeta) &= 1 + \beta_h\zeta\,(1 + \beta_h\zeta/a)^{a-1} \end{aligned} \tag{6}$$

with $\beta_m = 5$, $\beta_h = 7.5$ and $a = 0.8$. The expressions of Duynkerke are preferred here because they seem to better represent observations in very stable conditions. The specific form of the expressions (5) and (6) and the empirical determination of the constants has received considerable attention in the past. Since we will mainly use measurements made during moderately stable conditions, the constants β_m and β_h are important parameters in our analysis. From turbulence measurements made in the lowest parts of the stable surface layer over the Antarctic ice shelf near Halley, King (1990) concluded that values of β_m (= 8.0) and β_h (= 5.4) did not significantly deviate from those found in other studies. From turbulence measurements over glacier ice, Munro (1989) obtained $\beta_m = 4.5$.

From the measured wind speeds, the surface roughness length (z_0) can be calculated as being the height at which the downward extrapolated wind speed profile (4a) reaches its surface value, i.e. $u = 0$. Note that since z_0 in smooth flow is smaller than the depth of the viscous sublayer, the extrapolation proceeds into a region where the logarithmic profile is not valid. The surface drag in wall-bounded shear flows is generally dependent on the roughness Reynolds number (Re), which is defined as:

$$Re = \frac{z_0 u_*}{\nu} \tag{7}$$

In accordance with Andreas (1987), the aerodynamic roughness is used in this formulation, although strictly speaking, the 'real' surface roughness, which is dependent only on the geometry and distribution of the roughness elements, should be used. However, for practical purposes, the definition (7) with z_0 suffices to characterise the flow. Three different flow regimes can be distinguished, dependent on the value of Re: smooth flow ($Re < 0.135$), transitional flow ($0.135 < Re < 2.5$) and rough flow ($Re > 2.5$). According to Hinze (1975), in smooth flow the roughness elements are entirely submerged in the viscous sublayer, which means that the surface drag is due entirely to viscous shear stress. In fully rough flow, the surface elements protrude through the viscous sublayer into the inertial sublayer. Then, the decrease in momentum is caused mainly by turbulent friction at the roughness elements (form drag). In the transitional regime, neither the dissipation of turbulent kinetic energy by viscous stresses nor the dissipation by form drag can be neglected entirely.

4. Data selection

Strictly, the Monin-Obukhov similarity theory can only be applied to 'ideal' measurements, i.e. measurements performed at locations with long unobstructed upstream fetches during stationary conditions. The 'fetch'-requirement concerns the fact that the surface layer profiles are influenced by only one surface type, i.e. no internal boundary layers. At sites 4 and 5, where winds blow mainly from easterly directions, we expect this requirement to be met. However, since sites 2, 3 and 6 are surrounded by mountain ridges it is not clear whether or not they were located sufficiently far from upstream obstacles. According to Garratt (1992), a rough estimate of the depth of the internal boundary layer (h_b) is $h_b \sim 0.1x$, where x is the distance to the upstream roughness change. However, the depth of the layer that is only influenced by the local surface stress (h_l) is $h_l \sim 0.1h_b$. Thus, regarding sites 2, 3 and 6, where the distance to the upstream rock outcrops is approximately 400 - 800 m, h_l is of the same order as z_2 (the heighest level of measurement). Therefore, it may be questionable whether the surface layer below 6 m is fully adjusted to the local surface type at these sites.

The requirement of stationarity is discussed by Joffre (1982). From considering characteristic time scales of turbulence, he concluded that deviations from equilibrium are due mainly to temperature variations rather than to variations in wind speed. In other words, the time scale at which buoyant production/destruction processes (T_b) vary is generally smaller than the time scale of variation of turbulent shear production (T_s). This means that particularly periods with large temperature variations have to be excluded when only (quasi-) stationary conditions are to be considered. Therefore, in (quasi-) stationary conditions, turbulence is in equilibrium with its production/destruction processes when $T_s > T_b >> T_t$, where T_t is a rough measure of the typical time scale of turbulence: $T_t \sim z/u_* = 20$ s for $z = 6$ m and $u_* = 0.3$ m s^{-1}.

According to Joffre (1982), an effective way to eliminate non-equilibrium periods is to look for strongly departing values of the roughness lengths from their mean trend. Here, we have eliminated the profiles that yielded roughness lengths (z_0, z_H, z_Q) smaller than 10^{-8} m. This seemingly ad-hoc requirement is based on the fact that roughness lengths smaller than 10^{-8} m are considered to have no physical meaning, since they probably result from non-steady periods. This requirement also ensures that erroneous profiles due to measuring uncertainties, especially at low winds, will be excluded from the analysis. As a result, most of the data in daytime in unstable conditions (light winds) are removed. The remaining data mostly comprises periods with high wind speeds and large amount of clouds. These periods were characterised by a small diurnal temperature cycle, which suggests that the requirement of stationarity may be fulfilled.

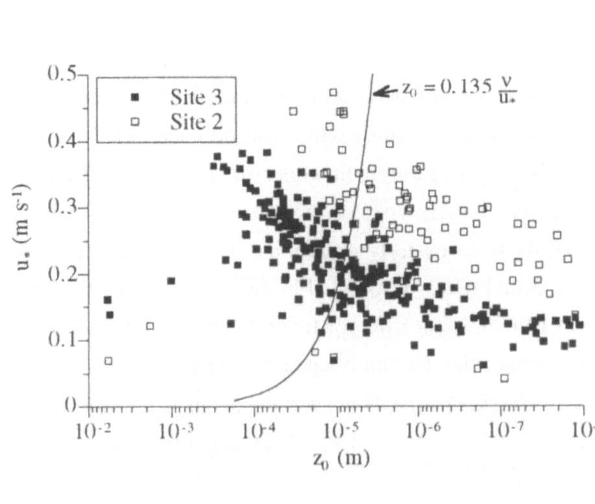

Figure 2. Dependence of the momentum roughness length z_0 on friction velocity u_* at sites 2 and 3. The curve represents the theoretical z_0 - u_* relation for non-saltating, aerodynamically smooth flows (Eq. (8)).

As can be inferred from Figs. 2 and 3, the erroneous profiles which produces values of z_0 smaller than 10^{-8} m are represented by a long, horizontal tail at small u_* (not shown) in the z_0 - u_* plot. Obviously, this is due to the fact that errors in u_* (by errors in wind speed) at small u_* produce large errors in z_0, which can be easily understood from the profile equation (4a). Wind speed errors at large u_* produce only small errors in z_0, which makes an upper limit to the roughness lengths unnecessary. The results of Joffre (1982) (his figure 5) seem to confirm this reasoning.

During periods with weak large scale forcing, shallow, nocturnal katabatic flows developed, especially along the local slopes inside the valley. This resulted in wind speed maxima at heights lower than 6 m. These periods are neglected in the present analyses for practical purposes, although it may be possible that the shear stress is maintained by substantial changes in wind direction at the height of the wind speed maximum.

The above described procedure of data selection significantly decreases the amount of profiles, i.e. a reduction of 91.5%, 77.2%, 47.1%, 59.2% and 65.1% at sites 2, 3, 4, 5 and 6, respectively. The large reduction at sites 2 and 3 is due mainly to the frequent occurrence of wind speed maxima at low elevation. The number of hourly mean profiles that remains for analysis is 81, 246, 559, 274 and 352 at sites 2, 3, 4, 5 and 6, respectively.

5. Momentum transfer

In rough, non-saltating flows, the value of z_0 depends only on the geometry and abundance of the roughness elements. Wind tunnel measurements have shown that, in smooth flow, the logarithmic wind profile (4a) is valid with (Hinze, 1975):

$$z_0 \sim 0.135 \frac{\nu}{u_*} \tag{8}$$

in which the proportionally factor 0.135 has been determined empirically. Then, z_0 is determined entirely by the flow characteristics. Fig. 2 shows the dependence of z_0 on u_* for sites 2 and 3. Comparing the bulk of the points of both sites, it appears that low values of z_0 at site 2 compared to site 3 are associated with larger values of u_* at site 2, a tendency which is in qualitative agreement with Eq. (8). This is in sharp contrast with the form of the z_0 - u_* relation at one specific site, which obviously disagrees with Eq. (8). The small values of z_0 indicate that the flow at both sites is in the smooth regime ($Re < 0.135$) for most of the time, especially at site 2. There, the extremely low values of z_0 can be attributed to the presence of a smooth ice surface. At site 3, most measurements fall within the transitional regime

(0.135 < *Re* < 2.5). From visual observation at site 3 it appeared that the snow surface is rather flat, with only very small surface elements, which results in the relatively low values of z_0 for a snow surface. From Fig. 3, in which the z_0 - u_* relation is plotted for site 4 (snow surface), it can be concluded that outside the valley, z_0 is much larger than at site 3 and that z_0 is virtually independent of u_* (for $u_* > 0.2$ m s^{-1}), especially in near-neutral conditions. At the wind-exposed sites 4 and 5, fierce, persistent winds cause the formation sastrugi (some sort of snow dune) of approximately 10 cm height at which turbulent kinetic energy easily dissipates (form drag), although these surface elements are often oriented in the direction of the strongest winds. Consequently, outside the valley rough flow (*Re* > 2.5) prevails.

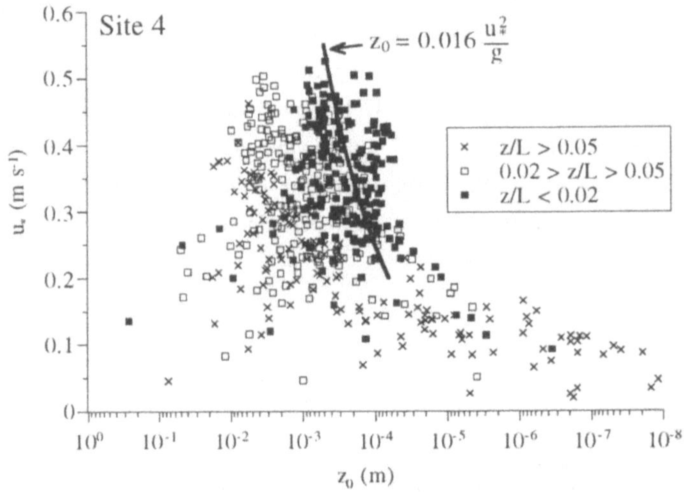

Figure 3. Dependence of the roughness length z_0 on friction velocity u_* at site 4. The different symbols are for stable (crosses), moderately stable (open squares) and near-neutral (black squares) cases. The curve represents Charnock's (1955) relation for saltating flow (Eq. (9)).

As stated earlier, a positive correlation between z_0 and u_* is apparent at sites 2 and 3, in contrast to what might be expected from Eq. (8) under smooth flow conditions (Fig. 2). It is suggested that this deviation from the standard theory is due to processes associated with drifting snow. Above a certain threshold value of u_*, snow particles are lifted from the surface and start to drift. Owen (1964) demonstrated that in flows with uniform saltating particles, z_0 is proportional to u_*^2. Chamberlain (1983) showed that Charnock's (1955) relation

$$z_0 = \alpha \frac{u_*^2}{g} \tag{9}$$

is applicable for saltating flows over water, sand and snow surfaces with α approximately equal to 0.016. Snow particles are accelerated and extract kinetic energy from the air flow, especially close to the surface. The particle momentum is lost at the surface when it is deposited again, increasing the shear stress close to the surface. Also, friction between the saltating particles exerts an influence on the dissipation. Equation (9) is plotted in Fig. 3 as the solid line. Data at site 4 do not seem to disagree with Eq. (9), especially at near-neutral stability ($z/L < 0.02$), although the scatter is too great to confirm it (the same holds for data from site 5). Thus, it cannot be entirely ruled out that Eq. (9) is applicable for saltating flows over rough Antarctic snow surfaces. Joffre (1982) also found only qualitative agreement between his data and Eq. (9) at large wind speeds. Inoue (1989a, 1989b) found a clear relationship between z_0 and wind direction on the sastrugi-covered snow fields in the escarpment region near Mizuho Station. However, his data did not reveal a significant dependence of z_0 on u_*, in accordance with our data from sites 4 and 5. On the other hand, wind profile measurements over a snow surface at Maudheim in the Antarctic coastal region (Liljequist, 1957) indicate that z_0 is a very strongly dependent on wind speed (a 100-fold increase of z_0 when wind speed increased from 13 to 24 m s^{-1}). He attributed this behaviour to the increased frictional surface drag due to drifting snow.

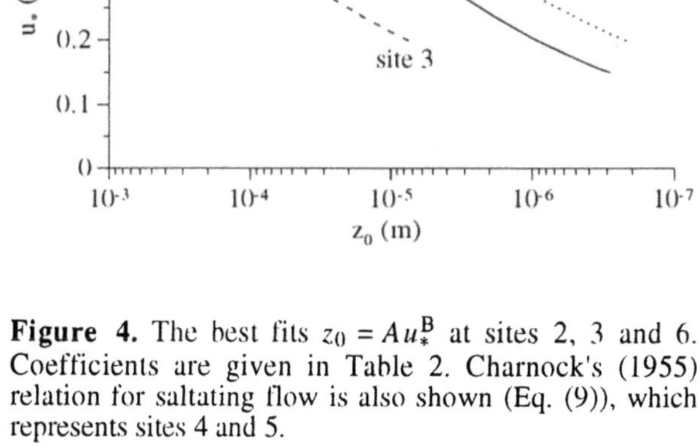

Figure 4. The best fits $z_0 = Au_*^B$ at sites 2, 3 and 6. Coefficients are given in Table 2. Charnock's (1955) relation for saltating flow is also shown (Eq. (9)), which represents sites 4 and 5.

Probably the most important effect of erosion and deposition of snow is that it effectively alters the surface characteristics, i.e. snow surface elements can be formed or eroded and bare ice can become (partially) snow-covered. To get a better view on the different z_0 - u_* relationships at the various sites, the best fits of the equation $z_0 = Au_*^B$ for data from sites 2, 3 and 6 are presented in Fig. 4 together with Charnock's (1955) relation, which is assumed to represent sites 4 and 5. Only values of u_* larger than a threshold value were used to ascertain that drifting snow occurred. Obviously, the z_0 - u_* relationships at sites 2 and 6, both on blue ice, are very similar. Sites 2, 3 and 6 all show a strong increase of z_0 with increasing u_*. The increase seems to be far too large to be explained entirely by pure snow-drift processes (momentum decrease associated with saltating particles), since the loss of momentum over bare ice and smooth snow is probably much smaller than over snow (smaller particle-surface friction over ice). The mean snow-drift transport rates and particle speeds over ice are thus probably larger than over snow (Schmidt, 1982). A more obvious reason for the strong dependency of z_0 on u_* at sites 2 and 6 is that bare ice becomes partially covered by snow during snow-drift events. Occasional lulls allow temporary snow deposition on the ice, thus forming an aerodynamically rougher surface. It is observed that a percentage snow cover of 50% is not uncommon during snow drift events. This can be inferred also from Fig. 5a, in which the variation of the daily mean albedo (used here as a rough measure of the amount of snow on bare ice) during the measuring period is shown. Evidently, the surface albedo increases significantly due to snow deposition on bare ice during snow drift events, indicating that considerable portions of the blue ice surface can occasionally become covered by snow. These are the periods with strong winds, during which most of the data points at the top-left corner of Fig. 2 were obtained. This can be inferred from Fig. 5b, in which the albedo is given as a function of $\ln(z_0)$ for daytime measurements at site 2. Although the scatter is large, there seems to be a tendency for an increase in z_0 with increasing albedo (and hence, with snow cover on ice).

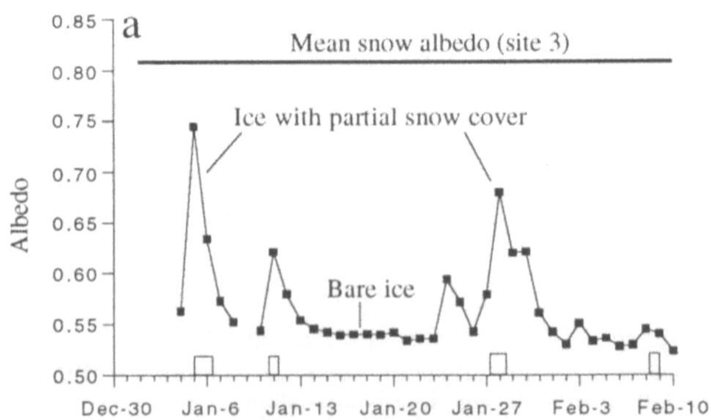

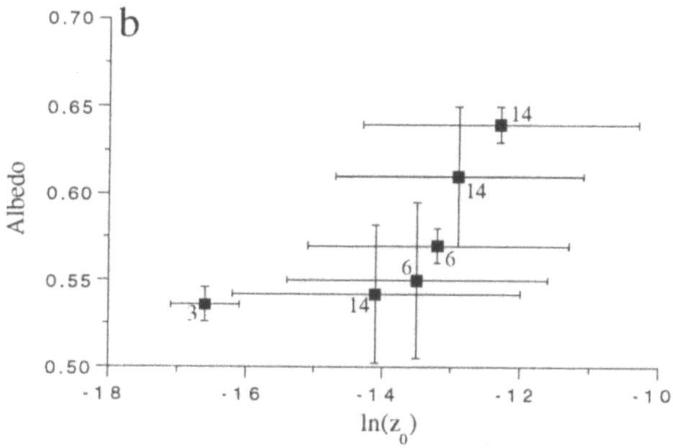

Figure 5. (a) Daily mean albedo of blue ice at site 2 for the entire measuring period (squares). The thick, solid line represents the mean snow albedo at site 3. The rectangles at the bottom indicate periods during which most of the profiles yielding high u_* values were obtained (i.e. the data points at the top-left corner of Fig. 2). (b) Albedo versus momentum roughness lengths for six periods with more than three consecutive hours of measurements for site 2. The error bars represent the standard deviation. The numbers near the data points represent the number of profiles used for each period.

By calculating roughness lengths of a partially snow-covered frozen lake, Tabler (1980) showed that z_0 increases by two orders of magnitude when snow coverage increases from 0% to 75%. From Fig. 5b we can crudely estimate that for a snow cover increase from 0% to 30% z_0 increases by one order of magnitude (it is assumed that snow coverage is linearly dependent on albedo with albedo values of 0.54 and 0.81 representing 0% and 100% snow coverage, respectively). If snow is present on ice, Tabler (1980) showed that Charnock's relation (Eq. (9)) is valid, which further contributes to the strong z_0 - u_* relationship over ice.

At site 3, the increase of z_0 with u_* is even larger than on ice (Fig. 4). Probably, this may be caused also by changing surface conditions, i.e. a flat surface during light winds and higher roughness elements during high winds. The coefficients of the best fits for sites 2, 3 and 6 are given in Table 2. Note that the values of B are much larger than 2. A thorough description of the influence of snow-drift processes on the z_0 - u_* relationship is given by Schmidt (1986). He clearly demonstrated the positive correlation between roughness length and snowdrift transport rate.

In conclusion of this section, we state that the momentum transfer over smooth surfaces (sites 2, 3 and 6) differs significantly from that over rough surfaces (sites 4 and 5). This is caused not only by the well-known differences in the dissipation of turbulent kinetic energy in non-saltating flows, but is also associated with snow-drift processes, which cause z_0 to be a strong function of u_*, especially over smooth surfaces.

Site	u_* (m s^{-1})	log(A)	B
2	> 0.20	-3.2	4.9
3	> 0.20	-2.0	4.4
6	> 0.15	-3.2	4.1

Table 2. Coefficients of the best fits of $z_0 = Au_*^B$ for sites 2, 3 and 6 as displayed in Fig. 4. Threshold values of u_* have been chosen as being the approximate starting velocity for drifting snow.

6. Scalar transfer

Unlike momentum, the scalar transports are governed solely by molecular diffusion processes close to the surface. There is no 'scalar' analogy for the dissipation of turbulent kinetic energy at the roughness elements in rough flow. Therefore, the Reynolds analogy is only valid under smooth flow conditions (small Re). The important parameters for heat and water vapour transport are the Prandtl ($Pr = \nu/K_h$) and Schmidt ($Sc = \nu/K_q$) numbers, respectively, in which K_h and K_q are the molecular diffusion coefficients for heat and water vapour.

With one level in the atmosphere and the other at the surface, the relationships (4b) and (4c) relate θ_* to z_H and q_* to z_Q, respectively. Subtracting (4c) from (4b) yields an expression for the ratio z_H/z_Q:

$$B_{HQ}^{-1} \equiv \frac{1}{\kappa} \ln\left({}^{z_H}/_{z_Q}\right) = \frac{\delta q}{q_*} - \frac{\delta\theta}{\theta_*} \tag{10}$$

in which we have used the fact that the stability corrections for heat and water vapour are commonly taken equal. The differences in potential temperature ($\delta\theta$) and specific humidity (δq) are taken between 2 m and the surface. Moreover, the surface air over ice and snow is assumed to be saturated (with respect to ice). As a result, the specific humidity at the surface depends only on the surface temperature (θ_s). Since θ_* and q_* are determined by the measured vertical profiles between 6 and 2 m, Eq. (10) is a relation between θ_s and B_{HQ}^{-1}. Specifying one determines the other. Brutsaert (1975) and Andreas (1987) showed that z_H and z_Q do not differ more than one order of magnitude for a wide range of Reynolds numbers. This is confirmed by experimental studies of Garratt and Hicks (1973) and King and Anderson (1994). In general, the difference between the scalar roughness lengths will be considerably smaller than the difference between a certain scalar roughness length and the roughness length for momentum. This means that *Pr* and *Sc* are nearly equal under normal atmospheric conditions (according to Andreas (1987), $Pr = 0.71$ and $Sc = 0.63$ for temperatures below 0 °C). Thus, by lack of further knowledge, a logical step seems to be to take z_H equal to z_Q, i.e. $B_{HQ}^{-1} = 0$. Then, for each profile, θ_s can be easily solved using the Newton-Raphson iterative method from the relation:

$$\frac{\theta_2 - \theta_s}{\theta_*} = \frac{q_2 - q_{sat}(\theta_s)}{q_*} \tag{11}$$

where $q_{sat}(\theta_s)$ is the saturated specific humidity at the surface temperature and θ_2 and q_2 are the potential temperature and specific humidity at 2 m. The value of z_H is then evaluated with Eq. (4b) (or (4c)).

From a practical point of view, this method provides a way to estimate the scalar roughness lengths without measuring θ_s. This is an important advantage, since surface temperature measurements are usually associated with large errors and uncertainties, which negatively influence the accuracy of the calculated scalar roughness lengths.

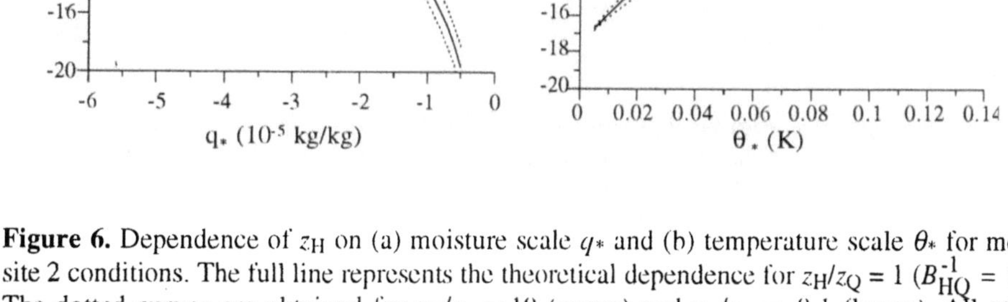

Figure 6. Dependence of z_H on (a) moisture scale q_* and (b) temperature scale θ_* for mean site 2 conditions. The full line represents the theoretical dependence for $z_H/z_Q = 1$ ($B_{HQ}^{-1} = 0$). The dotted curves are obtained for $z_H/z_Q = 10$ (upper) and $z_H/z_Q = 0.1$ (lower). All other quantities (e.g. θ_2, q_2 and u_*) are kept fixed at their mean site 2 values.

However, the method described here requires precise profile measurements of temperature and humidity, since they determine to a large extent the value of z_H, as can be inferred from Fig. 6. This figure shows the relationship between z_H and the turbulent scales θ_* and q_* as calculated with Eqs. (10) and (4b). All other quantities needed to calculate θ_s and z_H (e.g. θ_2, q_2, u_*) are kept fixed at their mean site 2 values except q_* and θ_* (Fig. 6a and 6b, respectively), which have been varied over the range representative for site 2 conditions. Notice that when θ_* is varied q_* is kept fixed at its mean site 2 value and vice versa. Thus, θ_s and z_H are evaluated for given values of θ_* (or q_*) (Eq. (10) yields the surface temperature which is inserted in Eq. (4b) to give z_H).

Obviously, the dependence of z_H on θ_* or q_* becomes stronger as θ_* or q_* (and hence the gradients) approaches zero. Therefore, estimates of z_H are most accurate for large (absolute) values of θ_* and q_*. Furthermore, the exact value of B_{HQ}^{-1} is far less important than the values of θ_* or q_*, as indicated in Fig. 6 by the small deviation of the two dotted curves that represent $z_H/z_Q = 10$ (upper) and $z_H/z_Q = 0.1$ (lower) from the line representing $z_H/z_Q = 1$. Since both profiles of temperature and of humidity are needed, the above described method for evaluating the scalar roughness lengths can be applied only to measurements performed at sites 2 and 3.

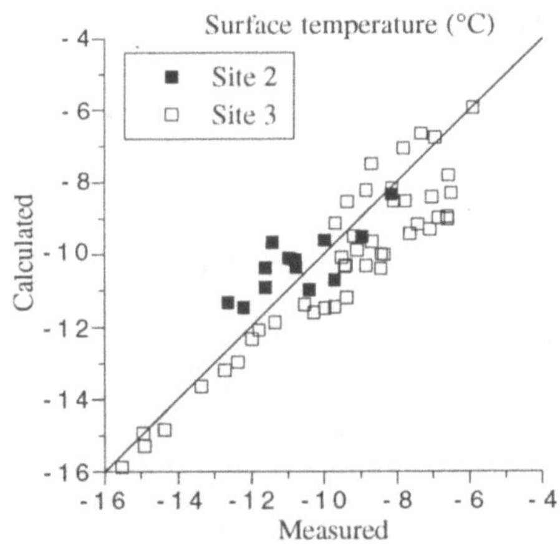

Figure 7. Comparison of calculated and measured surface temperatures at sites 2 and 3. Also shown is the 1:1 line.

In order to obtain insight in the accuracy of the method, calculated surface temperatures are compared with measurements (Fig. 7). It must be noted that the surface temperature measurements, performed with a downward-facing infrared-sensor and a resistance thermometer placed at the surface, were rather experimental. Therefore, surface temperatures were measured only during some selected periods, which enables direct comparison for only a limited amount of data. The overall agreement between measured and calculated surface temperatures seems to be reasonably good at both sites, although individual differences can amount to 2.5 °C. Obviously, the scatter can be attributed to uncertainties in the (surface) temperature and humidity measurements. From Fig. 7 we conclude that the approach used here seems to be able to produce correct surface temperatures, and, therefore, correct scalar roughness lengths.

The dependence of B_H^{-1} (= $(1/\kappa)\ln(z_0/z_H)$) on the roughness Reynolds number is shown in Fig. 8 for sites 2 and 3, together with the theoretical relationship of Andreas (1987). His calculations for smooth flow were based on a surface-renewal model: small eddies continuously enter the interfacial sublayer, transferring heat and moisture by an advection-diffusion process. By matching the interfacial sublayer and surface layer profiles, he arrived at B_H^{-1} = -3.9 and B_Q^{-1} = -4.0. The same approach was used by Brutsaert (1982), although he assumed the scalar transfer to be governed by molecular diffusion only, which led to B_H^{-1} = -2.7 and B_Q^{-1} = -3.5.

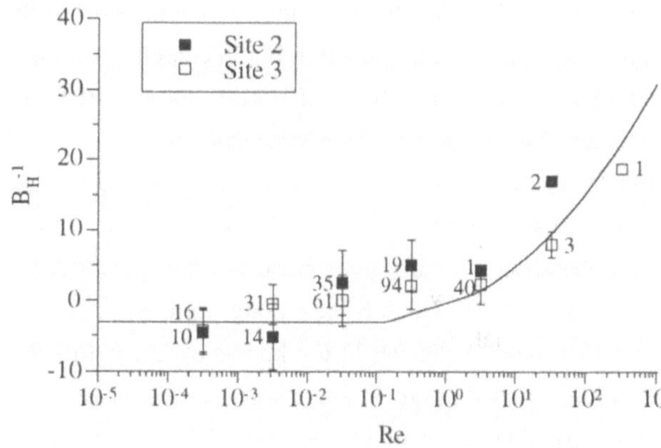

Figure 8. Dependence of B_H^{-1} on the roughness Reynolds number for site 2 and 3. Vertical bars represent the standard deviation around the means of values of B_H^{-1} for $\log(Re)$ classes of 1. The numbers near the data points represent the number of profiles within each class. The curve represents the theoretical relationship of Andreas (1987).

Thus, the scalar transfer coefficients will be somewhat larger than the transfer coefficient of momentum in smooth, non-saltating flows. Most measurements fall within the smooth/transitional regime, where B_H^{-1} is approximately equal to zero. In this region, the data agree roughly with the theoretical values of B_H^{-1}, although there seems to be a slight overestimation at the intermediate Reynolds numbers. This may be partly explained by the assumption of $Pr = Sc = 1$ in Eqs. (5) and (6), which leads to $B_H^{-1} = 0$. For larger Reynolds numbers, the higher values of B_H^{-1} of Andreas (1987) are roughly followed by our data, although the number of measurements is too small to draw firm conclusions. At least, the results indicate that over smooth ice and snow surfaces the interfacial transfer coefficients for momentum and heat and moisture do not differ more than one order of magnitude.

The reasonable agreement of our values of B_H^{-1} with those predicted by the theory of non-saltating flows is remarkable, given the strong correlation between z_0 and u_* due to snow-drift processes. However, it is argued that z_H and z_Q will also increase due to drifting snow. Since the temperature of drifting snow is lower than that of the air (in stable conditions), the sensible heat flux will cool the air surrounding a snow particle. Since snow-drift rates are maximum close to the surface, the temperature gradient in the surface layer is increased, and, consequently, z_H becomes larger. Similar arguments can be given with regard to the moisture gradient. Schmidt (1986) stated that significant sublimation rates occur within the drifting-snow layer. As a result, the moisture gradient increases, and so does z_Q. These relationships can be inferred also from Fig. 6. Although changes in the roughness lengths z_0, z_H and z_Q due to snow-drift processes are thus in the same direction, they do not necessarily have to be of equal magnitude. In this light, the agreement displayed in Fig. 8 can also be judged as rather fortuitous, especially at larger Reynolds number. Otherwise, snow-drift processes apparently change all roughness lengths by approximately the same amount.

Up to date, not much experimental estimates of B_H^{-1} have been given under atmospheric smooth flow conditions. This results from the fact that very few permanent natural smooth surfaces comparable to Antarctic blue ice areas exist. Further, potentially smooth areas are located mainly in the polar regions, which brings along high expedition costs and difficult measuring conditions. Experimental estimates of B_H^{-1} known to date at low Re over ice or snow surfaces are schematically illustrated in Fig. 9. There seems to be a large scatter in the available data, indicating that the difficulties involved in estimating B_H^{-1} are large, which are believed to result largely from uncertainties in surface temperature measurements. Munro (1989) suggested that differences between theory and his experimental results are due to an uncertainty in measuring height, as his data do confirm the shape of Andreas' (1987) theoretical curve. King and Anderson (1994) imputed their departing B_H^{-1} values to a near-surface divergence of longwave radiation which could influence the temperature profile.

We conclude that more measurements over natural smooth surfaces are needed to investigate the momentum, heat and moisture transfer processes at low Reynolds numbers in the atmosphere. Particularly, their relationship with snow-drift processes deserves more attention.

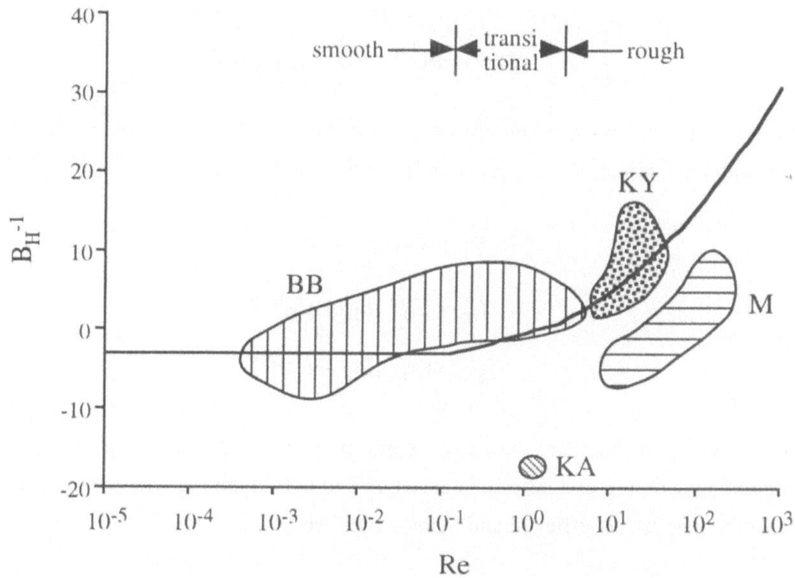

Figure 9. The dependence of B_H^{-1} on the roughness Reynolds number as determined in this and other studies. Values plotted are KY (Kondo and Yamazawa, 1986), M (Munro, 1989), KA (King and Anderson, 1993) and BB (this study). In case of the values of KY and M, the indicated areas represent the bulk of their data points. KA do not give values of B_H^{-1} as a function of Re, and, therefore, the circle represents a mean value. The area BB represents the values of B_H^{-1} within the standard deviation from the mean for sites 2 and 3.

7. Near-neutral bulk transfer coefficients

In this section we will ignore the dependence of the roughness lengths on whatever variable and treat them as if they are constant in order to compare our values with those from other studies. Mean bulk transfer coefficients are presented here, since, generally, atmospheric models use bulk transfer formulations to represent the turbulent exchange processes at the surface. The bulk transfer coefficients for momentum, heat and moisture (C_D, C_H and C_Q, respectively) are given by:

$$C_D = \left(\frac{u_*}{u_r}\right)^2, \quad C_H = \frac{u_*\theta_*}{u_r\left(\theta_r - \theta_s\right)}, \quad C_Q = \frac{u_* q_*}{u_r\left(q_r - q_s\right)} \tag{12}$$

where subscript r refers to a reference (measuring) height. C_H and C_Q are also referred to as the Stanton and Dalton number, respectively. In neutral conditions, they can be written as:

$$C_D = \kappa^2 \left(\ln \frac{z_r}{z_0}\right)^{-2} \tag{13}$$

$$C_H = C_D^{1/2}\left(B_H^{-1} + C_D^{-1/2}\right)^{-1} \tag{14}$$

$$C_Q = C_D^{1/2}\left(B_Q^{-1} + C_D^{-1/2}\right)^{-1} \tag{15}$$

For each site, we have selected near-neutral profiles using the criterion $|z/L| < 0.02$. From the remaining data, regressions of u_*^2 against u_6^2 and $u_*\theta_*$ against $u_6(\theta_6-\theta_s)$ have been made to evaluate the 6 m drag coefficient and Stanton number.

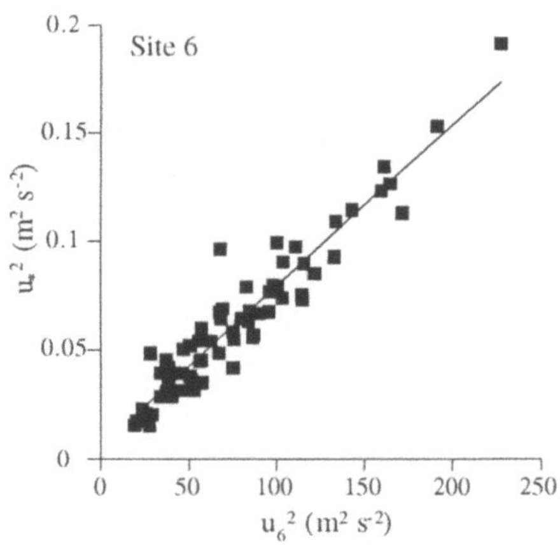

Figure 10. Dependence of u_*^2 on u_6^2 for site 6 under near-neutral conditions ($|z/L| < 0.02$). The line is the best linear fit to the data, which slope is equal to the 6 m drag coefficient.

As can be seen from Fig. 10, where the u_*^2 - u_6^2 regression is shown for site 6, relatively low-scattered data remain. As shown in section 4.5, the apparent relation between z_0 and u_* caused by snow-drift processes induces a dependence of C_D on u_* (or u_6), an issue addressed by several authors (e.g. Joffre, 1982). From Eqs. (13) and (14), z_0, z_H (= z_Q) and the bulk exchange coefficients (10 m) are calculated for each site (Table 3).

Site	Number of profiles	z_0 (m)	$z_H \equiv z_Q$ (m)	C_D (10^{-3})	$C_H \equiv C_Q$ (10^{-3})
2	33	$2.8 \cdot 10^{-6}$	$1.2 \cdot 10^{-6}$	0.70	0.66
3	105	$3.0 \cdot 10^{-5}$	$2.9 \cdot 10^{-5}$	1.0	1.0
4	252	$2.5 \cdot 10^{-4}$		1.4	
5	191	$7.5 \cdot 10^{-4}$		1.8	
6	80	$3.7 \cdot 10^{-6}$		0.73	

Table 3. Near-neutral values of the roughness lengths for momentum (z_0), heat (z_H) and moisture (z_Q) and the associated 10 m bulk transfer coefficients.

Study	Surface type	C_D	C_H	C_Q
Liljequist (1957)	snow	1.2		
Hicks and Martin (1972)	snow	1.2	0.9	2.5
Tabler (1980)	frozen lake without snow	0.72		
Tabler (1980)	frozen lake with >75% snow cover	1.9		
Joffre (1982)	sea-ice with snow	1.5	1.0	
Kondo and Yamazawa (1986)	snow	1.4	1.2	1.3
Inoue (1989b)	snow	1.3		
King and Anderson (1994)	snow	1.2	2.8	2.8

Table 4. 10 meter drag coefficient (C_D), Stanton number (C_H) and Dalton number (C_Q) (10^{-3}) over ice and snow surfaces as determined in other studies. Note that the value given by Liljequist (1957) represents the drag coefficient for non-saltating flows. He observed an increase in momentum roughness length of maximum two orders of magnitude during drifting snow conditions (equivalent to an increase in drag coefficient of $2.3 \cdot 10^{-3}$).

Obviously, the results confirm the findings of the previous sections: large drag coefficients at the wind-exposed sites 4 and 5, where sastrugis dominate the shape of the surface and smaller drag coefficient at the wind-shielded sites 2, 3 and 6. Inside the valley the Stanton number and drag coefficient are of equal magnitude, in accordance with results described in the previous section. In order to compare these values, Table 4 lists near-neutral 10 m values of C_D, C_H and C_Q from other studies. Our values over snow roughly agree with the other estimates over snow surfaces. The small values of C_D over ice are in perfect agreement with those of Tabler (1980), although our 'ice'-value is presumably representative for an ice surface covered by patches of snow, as discussed in section 5. No other estimates of the transfer coefficients over smooth ice surfaces have been found in literature. There are large discrepancies between the (very few) estimates of C_H and C_Q known to date. However, one must keep in mind that comparison of near-neutral transfer coefficients from different locations has meaning only if differences in flow conditions (Reynolds numbers) and in local snow-drift processes are taken into account.

8. Discussion and conclusions

In this paper we have evaluated the transfer coefficients of momentum, heat and moisture over Antarctic blue ice and snow surfaces using vertical profile data. The measurements were performed at five sites, two on snow with large roughness elements (sastrugi), one on a relatively flat snow surface and two on blue ice. Under near-neutral conditions, the aerodynamic roughness length (z_0) varied from $2.8 \cdot 10^{-6}$ m over smooth blue ice to $7.5 \cdot 10^{-4}$ m over rough snow. The extremely low roughness of ice cannot be understood from local momentum decrease alone, since at near-neutral conditions the ice was partially covered with snow.

Measurements from only two levels have been used in this analysis. It cannot entirely be ruled out that the highest level of measurement (6 m) was influenced by the upwind located rock outcrops, since, according to Garratt (1992), the internal boundary layer (which is a layer that is influenced by only one surface type) deepens only slowly after a rough-to-smooth transition. As a result, the 6 m wind speed may be underestimated, which reduces the vertical velocity gradient. Therefore, it is possible that z_0 over smooth surfaces (especially sites 2 and 6) is somewhat underestimated in this study. Results from these sites presented in this paper should therefore be interpreted as rough estimates rather than accurate values.

The dissipation of turbulent kinetic energy close to the surface is significantly influenced by snow-drift processes, i.e. the variability of the surface elements by snow deposition and erosion and the loss of momentum at saltating snow particles. This causes a positive correlation between the momentum roughness length and the turbulent friction velocity, especially over smooth surfaces. This is totally opposite of what might be expected from wind tunnel experiments, which indicate that z_0 is proportional to u_*^{-1} under smooth flow conditions. Apparently, the dissipation of turbulent kinetic energy at drifting snow particles is much larger than the dissipation due to the viscous shear stress.

The roughness lengths for heat and moisture were estimated using vertical profiles of temperature and humidity. By specifying the ratio z_H/z_Q, the surface temperature and z_H

(and z_Q) can be evaluated from the vertical profiles. This method circumvents surface temperature measurements, which are usually very difficult to perform accurately. It appears that, under smooth flow conditions (i.e. small Reynolds number), z_0 and z_H are of equal magnitude, in agreement with theory and wind tunnel measurements that predict interfacial transfer coefficients of momentum and scalars to be of the same order of magnitude. However, processes associated with drifting snow not only influence the wind speed profile, but also the vertical gradients of temperature and humidity, and, therefore, the scalar roughness lengths. Apparently, changes in z_0, z_H and z_Q due to snow-drift processes are of approximately equal magnitude so that the dependence of ratio z_0/z_H on roughness Reynolds number roughly follows non-saltating interfacial transport theory.

The strong influence of snow-drift processes on not only the momentum transfer, but also on the scalar transport in the surface layer suggests that also the transport of other atmospheric constituents may be affected. For instance, in case of a downward transport of such a species, the occurrence of snow-drift will enhance the uptake by the snow (the snow particle has a greater chance to get in contact with the gas molecule). Consequently, the vertical gradient of such a species will increase. This means that the surface layer transport increases in case of snow-drift, equivalent to the effect of snow-drift on the transfer of heat and moisture. As a result, the equivalent roughness length is likely to increase significantly during snow-drift conditions. Therefore, one may speculate that the strengthening of the vertical transport due to snow-drift can be extended more generally to other scalar quantities, such as chemical species. This phenomenon might be of importance for the interpretation of firn and ice cores. This is an important consideration since snow-drift is a very regular phenomenon over the windy dry snow surfaces of Antarctica.

Acknowledgements. M. Portanger is thanked for electronic and technical support during the field campaign. We are grateful to our Swedish companions in the field who have contributed in some way or another to the success of the expedition. Logistic support was given by L. Conrads, the members of NARP/SWEDARP 1992-93, the Norsk Polarinstitutt and the Swedish Polarforsknungssekretariatet. We thank J. Oerlemans, P. Duynkerke and W. Greuell for discussions, which enabled us to improve the paper. Financial support was provided by the Netherlands Antarctic Research Programme (GOA) and the Dutch National Research Programme on Global Air Pollution and Climate Change (NOP).

References

Andreas EL (1987) A theory for the scalar roughness and the scalar transfer coefficient over snow and sea ice. Boundary-Layer Meteorol. 38: 159-184

Bintanja R, Van den Broeke MR, Portanger MJ (1993) A meteorological and glaciological experiment on a blue ice area in the Heimefront Range, Queen Maud Land, Antarctica, Svea Field Report, I.M.A.U. report, Institute for Marine and Atmospheric Research Utrecht, Utrecht University, 29 pp.

Bintanja R, Van den Broeke MR (1995) The surface energy balance of Antarctic snow and blue ice. J. Appl. Meteorol. 34: 902-934

Bintanja R, Van den Broeke MR (1995) Momentum and scalar transfer coefficients over aerodynamically smooth Antarctic surfaces. Boundary-Layer Meteorol. (in press)

Brutsaert WH (1975) The roughness length for water vapor, sensible heat and other scalars. J. Atmos. Sci. 32: 2028-2031

Brutsaert WH (1982) Evaporation into the atmosphere. Reidel, Dordrecht, 299 pp.

Chamberlain AC (1983) Roughness length of sea, sand and snow. Boundary-Layer Meteorol. 25: 405-409

Charnock H (1955) Wind stress on a water surface. Quart. J. Roy. Meteorol. Soc. 81: 639-640

Duynkerke PG (1991) Radiation fog: a comparison of model simulation with detailed observations. Mon. Wea. Rev. 119: 324-341

Dyer AJ (1974) A review of flux-profile relationships. Boundary-Layer Meteorol. 7: 363-372

Fortuin JPF, Oerlemans J (1992) An atmospheric model for simulating the mass balance and temperature on the Antarctic ice sheet. Zeitschr. Gletscherk. Glazialgeol. 26: 31-56

Garratt JR (1992) The atmospheric boundary layer. Cambridge University Press, 316 pp.

Garratt JR, Hicks BB (1973) Momentum, heat and water vapour transfer to and from natural and artificial surfaces. Quart. J. Roy. Meteorol. Soc. 99: 680-687

Hicks BB, Martin HC (1972) Atmospheric turbulent fluxes over snow. Boundary-Layer Meteorol. 2: 496-502

Hinze JO (1975) Turbulence, 2nd ed., McGraw-Hill, New York, 790 pp.

Högström U (1988) Non-dimensional wind and temperature profiles in the atmospheric surface layer: A re-evaluation. Boundary-Layer Meteorol. 42: 55-78

Inoue J (1989a) Surface drag over the snow surface of the Antarctic plateau. 1: Factors controlling surface drag over the katabatic wind region. J. Geophys. Res. 94: 2207-2217

Inoue J (1989b) Surface drag over the snow surface of the Antarctic plateau. 2: Seasonal change of the surface drag in the katabatic wind region. J. Geophys. Res. 94: 2219-2224

Joffre SM (1982) Momentum and heat transfers in the surface layer over a frozen sea. Boundary-Layer Meteorol. 24: 211-229

King JC (1990) Some measurements of turbulence over an Antarctic ice shelf. Quart. J. Roy. Meteorol. Soc. 116: 379-400

King JC, Anderson PS (1994) Heat and water vapour fluxes and scalar roughness lengths over an Antarctic ice shelf. Boundary-Layer Meteorol. 69: 101-121

Kondo J, Yamazawa H (1986) Bulk transfer coefficient over a snow surface. Boundary-Layer Meteorol. 34: 123-135

Liljequist GH (1957) Energy exchange over an Antarctic snow field. Norwegian-British-Swedish Antarctic Expedition 1949-52, Scientific Results, Vol. 1-4

Munro DS (1989) Surface roughness and bulk heat transfer on a glacier: comparison with eddy correlation. J. Glaciol. 35: 343-348

Owen PR (1964) Saltation of uniform grains in air. J. Fluid Mech. 20: 225-242

Schmidt RA (1982) Vertical profiles of wind speed, snow concentration and humidity in blowing snow. Boundary-Layer Meteorol. 23: 223-246.

Schmidt RA (1986) Transport rate of drifting snow and the mean wind speed profile. Boundary-Layer Meteorol. 34: 213-241

Tabler RD (1980) Self-similarity of wind profiles in blowing snow allows outdoor modeling. J. Glaciol. 26: 421-434

Van den Broeke MR, Bintanja R (1995) Summer time katabatic flow and local circulation in the vicinity of a blue ice area in Dronning Maud Land, Antarctica. Boundary-Layer Meteorol. (in press)

Vilà-Guerau de Arellano J, Duynkerke PG, Zeller KF (1995) Atmospheric surface layer similarity theory applied to chemically reactive species, J. Geophys. Res. 100: 1397-1408

Wendler G, Ishikawa N, Kodama Y (1988) The heat balance of the icy slope of Adelie Land, Eastern Antarctica. J. Appl. Meteorol. 27: 52-65

THERMODYNAMICS OF THE SOLUTE LAYER ON THE SURFACE OF ICE

Peter Brimblecombe
School of Environmental Sciences
University of East Anglia
Norwich NR4 7TJ
U. K.

Martha H. Conklin
Department of Hydrology and Water Resources
University of Arizona
Tucson, Arizona 85721
U. S. A.

The question of gas adsorption or absorption at the ice-air interface as function of temperature is central to understanding how and to what extent ice cores preserve a representative record of atmospheric chemistry. Gases can either adsorb at the two-dimensional interface between air and the "surface region" on ice, partition to the three-dimensional surface region, partition into the bulk ice, or some combination of these. Ice has a "rough" surface near 0°C, and as it approaches 0°K should become perfectly faceted. At some intermediate temperature much of the rough surface, which has some liquid-like properties, becomes small. Acidic gases such as SO_2 [*Conklin et al.*, 1993] and CO_2 [*Ocampo and Klinger*, 1982] show greater uptake with increasing temperature, whereas non-reactive species such as NO [*Sommerfeld et al.*, 1992] show the opposite trend; H_2O_2 shows both trends depending on the temperature range. There is currently no method of predicting this behavior based on known properties of gases and our limited knowledge of the nature of the air-ice interfacial region. It is thought that the increase observed with acidic gases is due to trapping of solute molecules at the surface by mobile water molecules (i.e. a surface liquid-like layer) as the temperature approaches the melting point. If the ice surface is curved, as in the case of snow, the freezing point is also a function of grain curvature. The addition of impurities can allow liquid layers to be present in a snowpack at temperatures below 0°C.

1 Liquid layers on the ice surface

The addition of impurities can allow liquid layers to be present in a snowpack at temperatures below 0°C.

The existence of a liquid layer on the surface of ice was proposed by Michael Faraday in the mid 19th century [*Faraday*, 1860]. The approximate thermodynamics of this layer has been addressed for planar and spherical ice grains by *Lacman and Stranski* [1972] and *Takagi* [1990].

NATO ASI Series, Vol. I 43
Chemical Exchange Between the Atmosphere and Polar Snow
Edited by Eric W. Wolff and Roger C. Bales

They show how thin water layers result from the fact that the surface tension of the vapour-ice interface is greater than the sum of the water-ice and water-vapour interfacial tensions. The equilibrium thickness of such water layers is only 10^{-8}m about 0.001°C below the freezing point and declines as temperature falls further. By -1°C the layer is close to molecular thickness.

Water can be held within capillaries of snow formed as grains clump together. *Colbeck* [1973, 1980] has shown how curvature and capillary pressure can give equilibrium pressures below zero, as expressed in the simple formula:

$$T_m = -\frac{T_o}{L}(v_l p_c + \frac{s_{sg}}{r_{sg}}) \tag{1}$$

where T_m is the triple point temperature for the geometry under consideration, T_o the triple point temperature for the planar case, v_l the liquid volume, p_c the capillary pressure, s_{sg} the interfacial energy between the solid and gas, r_{sg} the radius of the gas ice surface and L the latent heat of fusion. This reduces to:

$$T_m = -0.814 \times 10^{-6} p_c - \frac{194 \times 10^{-6}}{r_{sg}} \tag{2}$$

where p_c is in pascals and r_{sg} is in mm.

Liquids can exist in snow packs by virtue of the freezing point depression that arises from the presence of solutes. Sodium chloride and sulphuric acid are common solutes in snow, and are addressed in this paper. The thickest liquid layers would result when all the solute within the ice is transferred to the surface film. An estimate of the volume of these solutions can be made from freezing point depression.

There have been many attempts to measure the thickness of liquid surface layers on ice. The recent measurements of *Elbaum et al.* [1993], using interference microscopy show the liquid film to be thinner than previously thought and extant over a narrow temperature range. Their experiments suggest that air is an important contaminant that predisposes ice to form liquid layers and that surface layers tend to be incomplete. They argue against thick liquid films at low temperature (i.e. -15°C).

Proton channeling experiments [*Golecki and Jaccard*, 1978] confirm rather thicker layers at the ice surface for temperatures as low as 30°C below the melting point. However they are insistent that this layer is not liquid like, as it retains residual structure to at least -1.8°C, the highest temperature of their experiments.

Conklin and Bales [1993] have attempted to probe the liquid film using gas solubility on 200-μm ice spheres. They are careful to point out that this probe may examine a phase that has a very different nature to those examined by previous techniques. This may be true as the equivalent liquid phase defined by their solubility measurements give a substantially thicker liquid film than those estimated by other, physical techniques.

Figure 1 shows the fraction of liquid water (F_w) present in the layers defined by the theoretical and experimental estimates discussed above. The quantity F_w has been adopted in the current work because the volume of liquid rather than surface area will determine the concentration of chemical species. Film thicknesses have been converted to volumes on the assumption that the snow has a surface area of 18 m^2 kg^{-1} which follows the surface area found in the system studied by *Conklin and Bales* [1993]. Putting these many observations and calculations together on one figure is really comparing apples and oranges. Each is measuring a different property of the ice, or the ice-air interface. It has not been shown that any of the physical measurements or thermodynamic calculations of ice surface properties are relevant to explaining or estimating gas uptake on ice. It may be that we lack even a sufficiently complete conceptual picture of the system to model temperature-dependent changes for even simple systems.

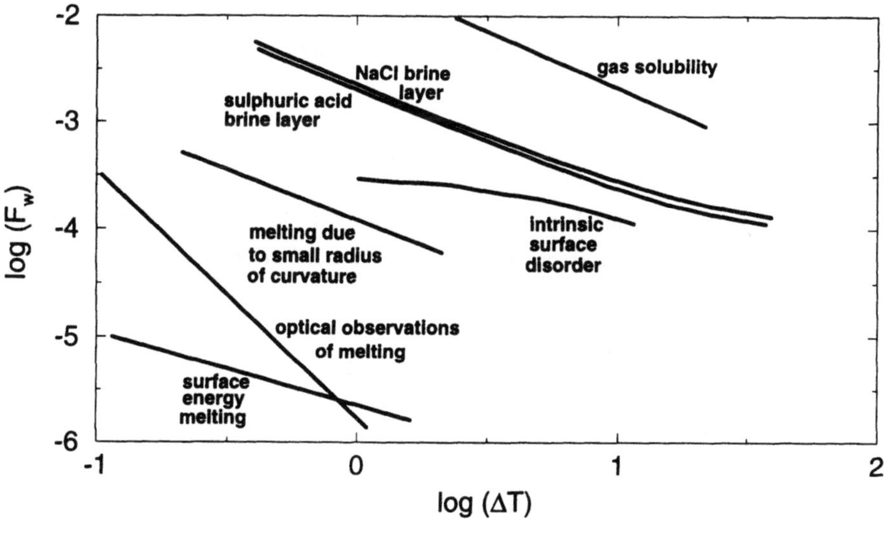

Figure 1: Fraction of snow present in surface or quasi-liquid layers from a range of experiments, compared with that of thermodynamic calculations.

In this paper, we use thermodynamic examples to examine the possible contribution of gas adsorption to developing liquid layers on ice. That is, we use thermodynamic calculations to determine if observations of gas uptake on ice can be explained by partitioning to a surface brine, or liquid-like layer, versus adsorption or partitioning into bulk ice.

2 Surface brines

The observations in Figure 1 suggest that surface brines offer the possibility of contributing the largest amount of liquid to a snow pack below the freezing point. These solutions are also relatively easy to describe using thermodynamics, which simplifies this type of analysis.

Calculations of the volume of these surface brines have often been made the assumption that the solutes are completely rejected by the ice. Although it is true that ions tend to distribute into the liquid rather than the ice phase, the distribution coefficients are of the same order of magnitude as the water-to-ice volume. This means that simple calculations potentially over-predict the amount of solute available for brine formation.

Let us define the water fraction (F_w) in terms of the mass of the aqueous layer (M_a) divided by the total weight ($M_T = 1\text{kg} = M_a + M_i$):

$$F_w = \frac{M_a}{M_T} \tag{3}$$

The number of moles of solute N will depend on the amounts present in the ice and aqueous layers and can be written:

$$N = M_a m_a + M_i m_i \tag{4}$$

where m_i and m_a are the molal solute concentrations in the ice and aqueous layers, respectively. Allowing for distribution according to the distribution coefficient ($D = m_i/m_a$) and mass conservation ($M_T = M_a + M_i$) yields:

$$N = M_a m_a + (1 - M_a) D m_a \tag{5}$$

Combining with [3] and rearranging gives:

$$F_w = \frac{m_a D - N}{m_a(D-1)} \tag{6}$$

This requires that N (effectively the molal concentration of the water-ice system) must be greater than $m_a D$ for realistic systems. As substantial freezing point depression requires m_a to be unity or greater, then N needs to be greater than D.

This turns out to be a restrictive requirement for real systems. If we adopt a value of 10^{-3} for the distribution coefficient then the sodium chloride concentration of ice would have to exceed 10^{-3} mol kg^{-1} to ensure substantial brine formation under equilibrium conditions. It would imply that any brine layers formed in the *Conklin and Bales* [1993] could not have been at equilibrium with the bulk ice. The range of solutions possible for sodium chloride is shown in Figure 2. Here we see that solving the equation in a realistic way (ie F_w positive) suggests that the formation of brines at low temperature require substantial concentrations of solute.

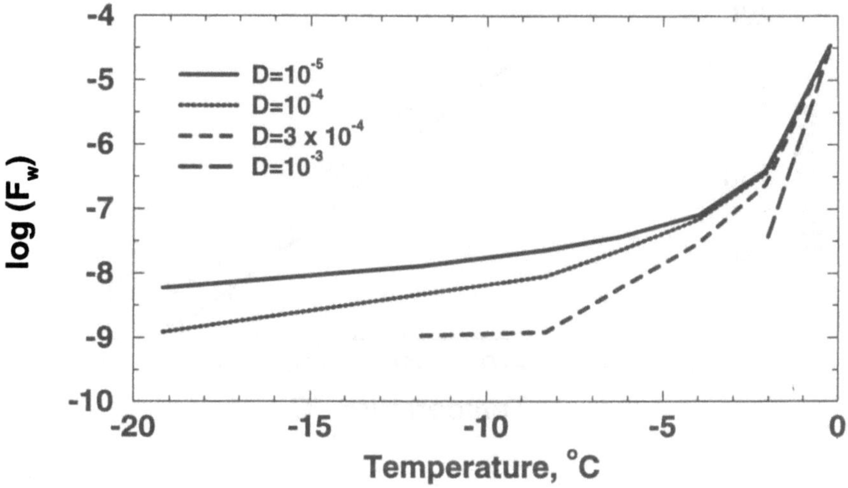

Figure 2: The fraction of water present as a solution layer with sodium chloride present at 0.001 mol kg^{-1} in the bulk snow assuming equilibrium at various distribution coefficients.

The amount of liquid that would be present under equilibrium conditions depends on the way in which these solutes distribute between the ice and aqueous phases. Assumption of partition between the phases places a lower limit on the solute concentration in bulk snow capable of producing brine layers.

3 Volatile solutes

Brines derived from volatile solutes represent a special case of liquid-layer formation. If we had a pure ice then it would be reasonable to imagine a gas (hydrogen chloride for example) forming a brine at the surface. Let us assume no transfer to the ice phase. Gibbs phase rule ($P+F=C+2$) reminds us that a two-component three-phase system will have only one degree of freedom. This means that if we choose temperature as a variable then all other extensive variables are defined. Thus, at any specified temperature, the concentration and pressure of HCl (P_{HCl}) are fixed as shown in Figure 3. Presumably if a higher P_{HCl} is chosen, then it will melt the ice until the pressure is reduced or the ice phase removed entirely.

The figure shows that high P_{HCl} are required to form brine layers on ice. These concentrations in the ppb range are much higher than the ppt concentration range typical of the atmosphere. This implies that hydrochloric acid solution layers are unlikely to form as a result of exposure of snow to ambient air. As nitric acid has a similar Henry's Law constant the same argument would follow for this gas.

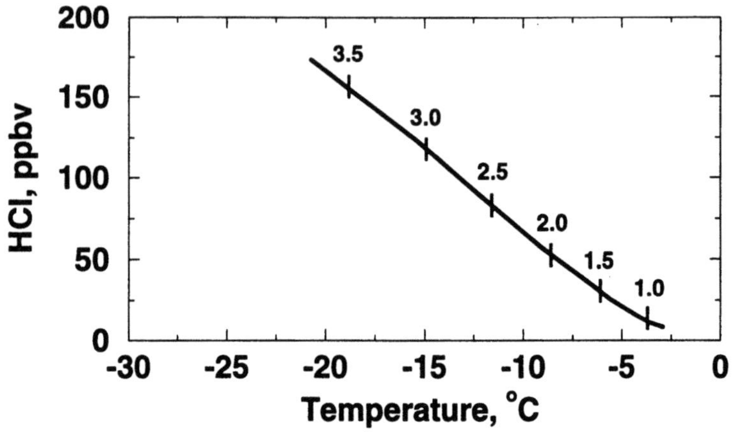

Figure 3: Hydrogen chloride partial pressure above ice required to induce a brine layer at various temperatures, assuming Henry's law equilibrium and freezing-point depression. The tick marks on the line show the equilibrium hydrochloric acid concentration in the brine ($\mathrm{mol\,kg^{-1}}$).

A further observation, concerning the deposition of gases to snow packs, arises from the Gibbs Law restriction emphasized above. This is, that in any idealized isothermal snow pack there should be no pressure gradient despite very different bulk concentrations of dissolved HCl in the solute layer. In practical terms this means that if we impose an HCl gradient in the interstitial air of a snow pack there are two ways this gradient can be reduced: (i) the gas can diffuse through the intergranular spaces (ii) it can form a solution on the grains. This second process would effectively impose resistance to interstitial gas phase diffusion that would be needed to be added to that induced by the tortuosity of the system.

4 Dissolution in a hypothetical solution layer

A more general scenario for gas dissolution would be the transfer of a gas to a pre-existing surface brine. Such a system would have three phases (gas, ice, brine) and three components (water, the dissolving gas, the brine solute) which would give two degrees of freedom. Thus, we would be free to choose both temperature and gas pressure.

Having established the somewhat less restrictive nature of this system over the one discussed in section 3, let us consider the general process of absorption of a gas by a snow pack. In an ideal surface liquid layer the concentration of dissolved gas would be given by the Henry's Law relation:

$$c = K_H p \tag{7}$$

where p is gas partial pressure. This can be expressed in terms of the concentration relative to the mass of snow (c') as a whole:

$$c' = K_H p F_w \tag{8}$$

We can imagine a pseudo-Henry's Law constant $K*$ that relates the partial pressure of the gas to the concentration found in the bulk snow:

$$K* = \frac{c'}{p} = K_H F_w \tag{9}$$

We can now explore the variation of $K*$ with temperature that is important in considerations of the role liquid layers playing absorbing gases. We can do this in a simple way by considering that K_H will vary with temperature. This can be done using an Ahrrenius-type relationship and F_w related to the total mass of brine solute (i.e. $F_w = 3.7N/(T_o - T)$, where 3.7 is a typical freezing point depression for a 1:1 electrolyte at 1 mol kg^{-1}). Thus we can imagine that the pseudo-Henry's Law constant, at a given temperature, will vary according to:

$$K* = exp(A + \frac{B}{T})\, 3.7 \frac{N}{(T_o - T)} \tag{10}$$

where T_o is the freezing-point temperature and A and B Ahrrenius-type constants. Under conditions where B is positive (ie solubility decreases with increasing temperature) we can postulate systems where $K*$ falls to some minimum value as the volume of the solute layer declines with falling temperature, but then increases again due rising gas solubility. This predicted result turns out to be found when we do thermodynamic calculations for more complex real systems (as seen in below in section 5 for hydrogen fluoride).

5 Dissolution of gases the solution layer

The discussion above has neglected non-ideal effects imposed by the presence of high concentrations of non-volatile electrolytes in the surface brine. Once these effects are considered, it is possible to undertake more-realistic calculations of the solubility of gases in the brine layer on ice. Activity coefficients of ionic and neutral species were determined adopting the formalism of *Pitzer* [see *Pitzer*, 1991; *Clegg and Whitfield*, 1991] in a computer program PITZ93. The variation of K_H and other equilibrium constants with temperature was also calculated within the program. Figure 4a shows the estimated value of $K*$ for oxygen dissolving in a sodium chloride brine on the surface of ice as a function of temperature. The total amount of sodium chloride in the ice was set to 0.001 mol kg^{-1}. Oxygen is not very soluble and additionally the decreasing volume of brine with declining temperature outweighs the increasing solubility.

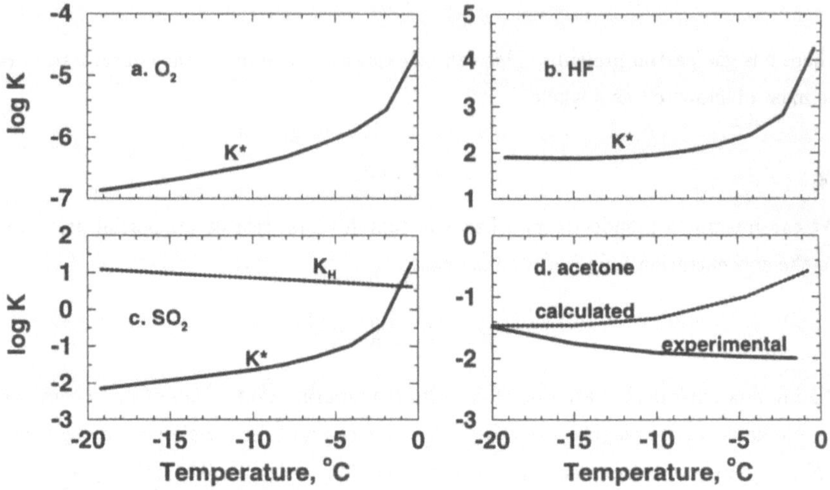

Figure 4: Predicted pseudo-Henry's law constants for snow ($K*$) with a bulk sodium chloride concentration of 0.001 mol kg^{-1} under the assumption of complete rejection of sodium chloride by ice, for: a) oxygen, b) hydrogen fluoride, c) sulphur dioxide (Henry's constant for SO_2 also shown), and d) acetone. For HF and SO_2, the counter ions sodium and hydrogen are present in the ratio 999:1. The acetone value is compared with observed values derived from the equation of *Goss*, [*1993*] as a dotted line.

Dissolution of weakly dissociating gases such as hydrogen fluoride are more complicated because of the different forms of dissolved species. Here the term c' in equation [10] is taken to represent the total concentration of species in solution. Thus, with hydrogen fluoride this would be both F^- and the undissociated HF. Calculations require the concentration of hydrogen ion to be specified and here that has been taken as 0.001 of the concentration of chloride in the brine. The pseudo-Henry's Law constant for hydrogen fluoride declines with temperature and then reaches a minimum as mentioned above (Figure 4b).

The dissolution of sulphur dioxide in surface brines on ice is important because of the interest in sulphur dioxide uptake on snow surfaces. Although this system is more complicated than that for hydrogen fluoride we can write an expression for the pseudo-Henry's Law constant:

$$K* = \frac{c'_{SO_2}}{p_{SO_2}} = c_{H_2SO_3} + c_{HSO_3^-} + c_{SO_3^{2-}} = F_w K_H\left(\frac{1}{f_{H_2SO_3}} + \frac{K'}{a_{H^+}}\left(\frac{1}{f_{HSO_3^-}} + \frac{K''}{a_{H^+} f_{SO_3^{2-}}}\right)\right) \quad (11)$$

where K' and K'' are the first and second dissociation constants of sulphurous acid and f the activity coefficients of the various sulphur species. Figure 4c shows the calculated values of $K*$ with hydrogen ion concentration at 0.001 that of chloride in the surface brine. The effective solubility decreases throughout with temperature, in spite of the increasing in solubility of the gas (as shown by the Henry's Law constant).

A measured pseudo-Henry's Law constant for sulphur dioxide can be calculated from the data of *Conklin and Bales* [1993] using measured sulphur accumulation in snow from their experiments that used 36.5 cm^3 of ice spheres [*Conklin et al.*, 1993]. At temperatures around -4°C the interpolated experimental $K*$ is 18.5 mol kg^{-1} atm^{-1}. This can be compared to the value calculated from equation [12], which comes to 0.66 mol kg^{-1} atm^{-1}. Lack of agreement is consistent with the high fraction of liquid water estimated by the *Conklin and Bales* [1993] experiments (see Figure 1). It seems that their experiments are likely to measure a more complex process than simply dissolution in surface brines.

In the case of soluble organic substances such as acetone there is even less agreement between a model based on the dissolution of the gas by a surface brine and experimental observations. Experiments [*Goss*, 1993] suggest that the pseudo-Henry's Law constant for acetone increases markedly with temperature, while predictions based on its absorption into a liquid layer decline with temperature (Figure 4d).

6 Conclusions

Liquid layers can be maintained at sub-zero temperatures on the surfaces of ice crystals in snow packs. Although they seem an attractive explanation for the uptake of water-soluble gases by snow, the layers appear to have insufficient volume to explain the observed absorption of gases such as SO_2 at temperatures significantly below zero (i.e. snow packs that aren't melting. However, the model correctly predicts a decline in uptake with decreasing temperature. However other gases, such as acetone, show increasing uptake on ice with decreasing temperature. This suggests that absorption onto the ice-air interface may be a more relevant process with these gases.

For modeling gas uptake on polar snow, the question that remains is: what are the liquid volumes and interfacial areas in snow and firn [*Bales and Choi*, this volume]. That is, what are the "equivalent" liquid volumes for absorption and chemical reactions? Physical measurements or thermodynamic calculations alone fail to offer a consistent picture of the temperature and ionic strength dependence. We suggest that the operational or equivalent surface properties defined by gas adsorption studies offer a promising approach to be explored by the ice-core community.

Acknowledgments

This work was supported in part by National Science Foundation grant OPP-92-24192 to the University of Arizona, and in part by NATO cooperative research grant CRG 930929. R. Brice did the manuscript preparation.

References

Bales RC, Choi J (to be published) Conceptual framework for interpretation of exchange processes (this volume)

Clegg SL, Whitfield M (1991) Activity coefficients in natural waters in Pitzer KS (ed) Activity Coefficients in Electrolyte Solutions, CRC Press, Boca Raton, FL, 279–4341

Colbeck SC (1973) Theory of metamorphism of wet snow. US Cold Regions Research Laboratory, Research Report No. 313

Colbeck SC (1980) Thermodynamics of snow metamorphism due to variations in curvature. J Glaciol 26:291–301

Conklin MH, Bales RC (1993) SO_2 uptake on ice spheres: liquid nature of the ice-air interface. J Geophys Res 98:16851–16855

Conklin MH, Sommerfeld RA, Laird K, Villinski JE (1993) SO_2 reactions on ice surfaces: implications for dry deposition to snow. Atmos Environ 27A:159–167

Elbaum M, Lipson SG, Dash JG (1993) Optical study of surface melting on ice. J Cryst Growth 129:491-505

Faraday M (1860) Note on regelation. Proc R Soc London 10:440-450

Golecki I, Jaccard C (1978) Intrinsic surface disorder in ice near the melting point. J Phys Chem: Solid State Physics 11:4229–4237

Goss C-U (1993) Adsorption of organic vapors on ice and quartz sand at temperatures below 0°C. Environ Sci Technol 27(13):2826–2830

Lacman R, Stranski IN (1972) The growth of snow crystals. J Cryst Growth 13/14:236-240

Ocampo J and Klinger J (1982) Adsorption of N_2 and CO_2 on ice. J Colloid Interface Sci 86:377–383

Pitzer KS (ed) (1991) Activity Coefficients in Electrolyte Solutions, CRC Press, Boca Raton, FL

Sommerfeld RA, Conklin MH and Laird K (1992) NO adsorption on ice at low concentrations. J Colloid Interface Sci 149:569–574

Takagi S (1990) Approximate thermodynamics of the liquid-like layer on an ice sphere based on an interpretation of the wetting parameter. J Colloid Interface Sci 137:446–455

GAS DIFFUSION IN FIRN

Jakob Schwander
Department of Climate and Environmental Physics
Physical Institute, University of Bern
Sidlerstrasse 5
CH-3012 Bern
Switzerland

Introduction

Different components and properties of the ice deposits on the earth (glaciers, ice sheets) store information on the climate and environment. We can mainly distinguish between three storage types: a) properties and isotopic composition of the ice itself, b) solid (and liquid) trace substances in the ice, and c) gaseous and volatile components. This article reviews some aspects of the transfer of gaseous components from the atmosphere to the ice.

The porosity of the firn decreases gradually from the surface to the depth of firn-ice transition where the pores are pinched off to form isolated air bubbles (Fig. 1 and 2). Near the snow surface the porosity (defined as the void volume fraction) is about 0.6 to 0.7, corresponding to a density of about 350 kg/m^3. Under cold, polar conditions there are only few bubbles closed-off at porosities above 0.15 (density 780 kg/m^3) which is reached about 10 m above the final close-off depth.

More than 90% of the gases are closed-off into bubbles in a relatively short depth range at the so called firn-ice transition (Schwander and Stauffer, 1984). The depth of the transition depends mainly on temperature and snow accumulation rate (Herron and Langway, 1980). For the polar ice sheets the depth lies between about 60 meters and 110 meters. The interval where about 90% of the bubbles are formed spans about the lowest 10% of the firn column.

The firn above the transition is open-porous and the gases in it can exchange with the atmosphere. For most 'permanent' gases the mixing processes can be looked at in a purely geometrical way, i.e. due to the low solubility of the gases in the ice the interaction with the solid ice can be neglected. The physical processes determining the local concentrations are restricted to diffusion, convection/advection, gravitation, and in case of radioactive substances, the rate of decay. In case of reactive and/or soluble gases, like NO_x, SO_2, H_2O_2, HCHO etc., sorption and chemical reactions must also be considered. The behaviour of those gases is still relatively poorly understood and few laboratory experiments have been carried out to assess the involved processes quantitatively (Conklin and Bales, 1993; Conklin et al., 1993).

NATO ASI Series, Vol. I 43
Chemical Exchange Between the Atmosphere
and Polar Snow
Edited by Eric W. Wolff and Roger C. Bales

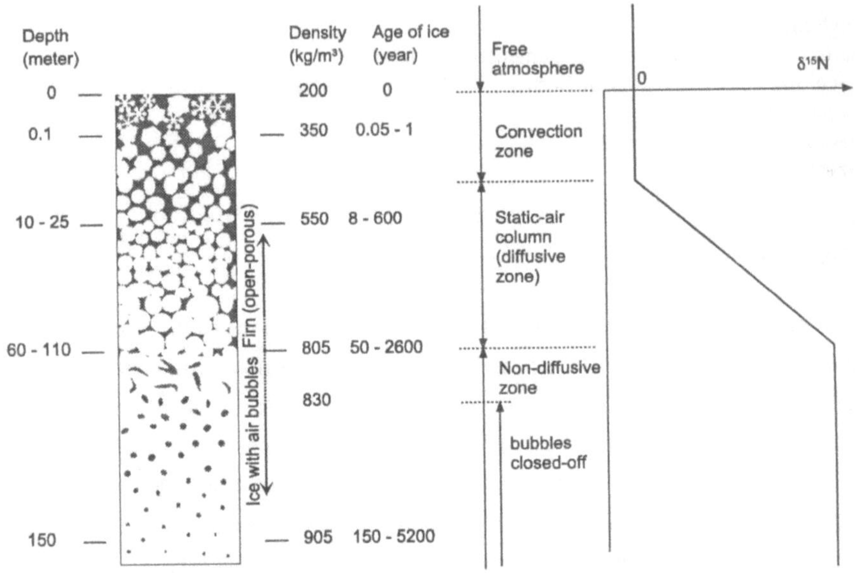

Fig. 1. Schematics of the firn column. The indicated depth and age ranges are typical for polar ice sheets and do not necessarily apply for mountain sites. Subdivision into zones suggested by Sowers et al., 1992 and the corresponding $\delta^{15}N$ profile are given on the right side. Convection in the uppermost layer results in a $\delta^{15}N$ plateau. In a molecular-diffusive regime the $\delta^{15}N$ increases linearly with depth due to gravitation. Diffusion is expected to stop somewhat above the final close-off of air bubbles. Under stationary conditions $\delta^{15}N$ remains constant in this non-diffusive zone.

Various processes are involved in the mixing of the air in the open-porous part of the firn:

Air mixing processes:

- Molecular diffusion
- Diffusion pumping
- Barometric pressure variations
- Expelling air from densification
- Wind pumping
- Thermal convection

The relative importance of these processes varies from site to site and depends on:

- Surface topography
- Wind speed

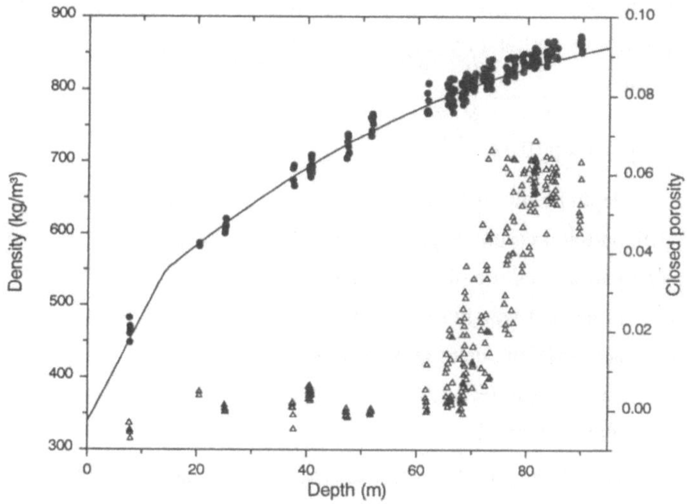

Fig. 2. Density (dots) and closed porosity (triangles) versus depth at Summit, Central Greenland (Schwander et al., 1993). The line through the density data is calculated with the Herron-Langway model (Herron and Langway, 1980)

- Wind direction distribution
- Permeability (icy layers)
- temperature
- pressure
- accumulation rate

Molecular diffusion plays a dominant role at many polar sites, as will be discussed later. The rate at which molecular diffusion mixes the air components depends on the individual diffusivities of the different air components. Generally the diffusivity decreases with depth due to lower porosity and higher tortuosity. Probably there occurs also diffusion pumping in case of a strong gradient in one air component, e.g. water vapour near the surface. Diffusion pumping has for example been observed in cold traps of our gas extraction systems, where the water vapour flux preferentially carries along CO_2-molecules leading to an enrichment of CO_2 inside the trap. Also, if persisting temperature gradients exist we must expect thermo-diffusion effects.

A direct consequence of a dominance of molecular diffusion is that each gas component tends to reach its individual equilibrium concentration profile, that is, the concentration of heavier gases is increased deeper in the firn due to gravity (Craig et al., 1988; Schwander, 1989). The equilibrium partial pressure of each gas component adjusts according to the barometric formula

$$p_i(z) = p_i(0) \cdot e^{\left(\frac{M_i g z}{RT}\right)},$$

where: z: depth below surface (m)

$p_i(0)$: partial pressure of gas i at the surface

$p_i(z)$: partial pressure of gas i at depth z

M_i: molecular weight of gas i (kg/mol)

g: acceleration due to gravity ($m \cdot s^{-2}$)

R: gas constant (= 8.134 J/(K mol))

T: temperature (K)

Generally the age of the ice at the firn-ice transition is large compared to the diffusive mixing time of the air in the firn. Atmospheric components with constant concentrations, like for example the isotopes of nitrogen or argon, will thus practically reach the equilibrium state in the diffusive zone of the firn. Only at very high accumulation sites we must correct for an incomplete equilibrium.

As the exponent in the barometric formula is very small, the fractionation δc_{ij} of two gas components with a molecular mass difference of ΔM_{ij} is approximately

$$\delta c_{ij}(z) = \Delta M_{ij} g z/(RT) = 1.18 \cdot \Delta M_{ij} z/T.$$

Any air flow through the firn enhances the dispersion of gases. Several processes can cause air fluxes in the firn. There are processes, for example barometric pressure variations, that move the air up and down without net displacement of any gas species. Other processes, such as wind pumping, will locally displace the air in a distinct direction.

Atmospheric pressure variations make the firn 'breathe'. Low frequency pressure oscillations (e.g. synoptic variations) propagate undamped into the firn, causing a one-dimensional, vertical air flow which is in phase over the whole firn column. The amplitude of the vertical air displacement at a certain depth is therefore proportional to the total air volume between this depth and the firn-ice transition, and inversely proportional to the open porosity. At higher frequencies there is additional damping due to frictional losses and in general the flow is no longer one-dimensional because of horizontal pressure differences.

Air is also forced to flow through the firn pores by the densification process itself. As the pore volume decreases with depth air is squeezed out and moves upwards relative to the ice matrix. The corresponding air flow speeds are largest near the surface where they are of the order of the snow accumulation rate. Local high air flows can however also occur at the firn-ice transition or in melt layers when the permeable fraction of a dense, partly impermeable layer becomes very small. Air from less dense layers underneath is then squeezed through the few permeable 'holes' in the dense layer.

Wind blowing over an uneven surface leads to local pressure differences inducing air flow along pressure gradients. Depending on the size of the surface undulations the induced flow penetrates more or less deeply into the firn (Colbeck, 1989).

During summer the stratification of the air in the firn pores is stable, whereas in winter cold air at the snow surface tends to sink into the firn. At the beginning of the winter season the horizontal temperature gradients are probably too small to initiate any substantial convection flux. Wind pumping could however induce such temperature gradients and thereby density gradients that might eventually cause small convection flows which could maintain themselves even after the wind has calmed down. Free convection has so far only been observed in shallow snow covers (Sturm and Johnson, 1991) and it is not known whether such convection cells also occur in glacier firn. There, the conditions will probably be rather close to the stability limit and, based on firn permeability and maximum possible temperature differences we can estimate that convective air fluxes will probably not exceed 1 m/day.

All mixing processes are strongly influenced by the presence or absence of icy layers resulting from surface melting during summer. Surface melting is rare if the mean annual temperature is below -30°C. If icy layers exist the geometry of air flow paths is highly complex, and, to my knowledge, no quantitative investigations exist for this case.

Based on the dominant mixing process and the dispersion rate we can discern three different zones between the snow surface and the firn-ice transition (Sowers et al., 1992): Near the surface the pore space may be well mixed by wind induced air fluxes. Below there is a zone where molecular diffusion dominates. In the lowest part of the firn diffusion may be impeded by nearly impermeable winter layers to such a degree that practically no gas exchange between layers occurs any longer in the vertical direction, although the pores may still not be completely pressure tight. The mean depth where bubbles become pressure tight has been investigated by total gas measurements in the ice. The resulting close-off densities as a function of the mean annual temperature is shown in Fig. 3.

A consequence of the gas exchange between the atmosphere and the firn is that the age of the air in the pore space and in the bubbles of the ice below the firn-ice transition is younger than the surrounding ice. The age distribution of the air components in the bubbles of the ice is controlled by two processes: 1) the air mixing in the firn leads to a characteristic distribution of the age of gas molecules as a function of depth (the age of a gas molecule is defined as the time elapsed since it has for the last time crossed the atmosphere-snow boundary). 2) There is another process, which is often as important for the width of the total age distribution: The formation of isolated air bubbles or clusters occurs gradually over a depth range of typically 10 m, i.e. at a certain depth in the ice we have a bubble population of

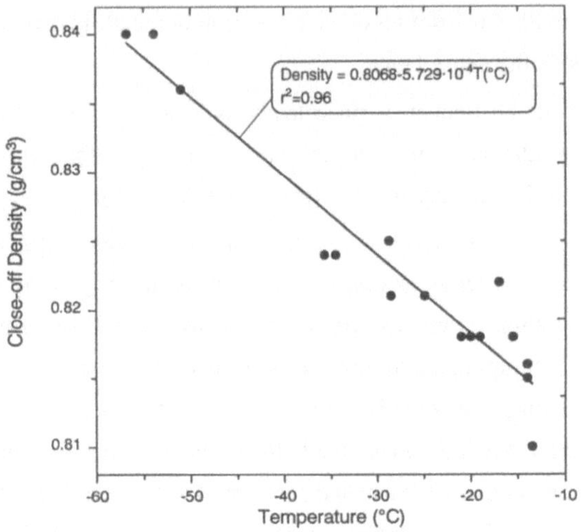

Fig. 3. The dependence of the close-off density on temperature of 14 Antarctic sites, Camp Century (Greenland) and Mt. Logan (Canada) (Martinerie et al., 1992).

which the individuals had been formed at different times. This means that there is a second convolution of the age distribution. How much this increases the width of the age distribution depends mainly on the time the ice needs to cross the bubble close-off zone, which is about inversely proportional to the accumulation rate.

Methods and measurements

In order to assess the properties of firn quantitatively various methods have been used:

Density: The easiest way to characterise the firn column is by its density profile. It provides a first estimate of the depth of the firn-ice transition zone. Since the more precise immersion method is not applicable for porous material the density of firn is usually determined by weighing and measuring the size of a sample.

Porosity: By measuring the open and closed porosity of firn samples we can determine the bubble formation zone. A method to determine the open and closed porosity of firn is based on volume measurements using the law of Boyle-Mariotte. Such measurements can be easily carried out at the drilling site (Schwander and Stauffer, 1984; Stauffer et al., 1985).

Porosity data have been published for Siple Station, West Antarctica and Summit, Central Greenland (Schwander and Stauffer, 1984; Schwander et al., 1993). Most of the bubbles are formed over a depth range of about 10 m within a density band between about 790 and 830 kg/m^3. Rather strong seasonal variations of the porosity are present throughout the whole firn profile. The bubbles in the denser winter layer are formed at shallower depths than in the summer layers (Schwander, 1989; Martinerie et al., 1992).

Air permeability: The major use of the air permeability is to assess air fluxes in the firn (Albert et al., this volume; Waddington et al., this volume). It also provides information on the depth where first impermeable (winter) layers occur. Various methods have been proposed to measure the air permeability of firn (Bender, 1958; Maeno et al., 1978; Conway and Abrahamson, 1984).

Diffusion coefficient: The diffusivity profile of the firn column is the basis for a quantitative assessment of the air mixing using numerical models. A method to measure diffusion coefficients in firn is by measuring the broadening of a gas packet in a carrier gas flowing though a firn sample (Schwander et al., 1988).

Diffusion coefficients along a whole firn column have so far only been determined at Siple Station. As expected the diffusivity in the firn column decreases as we move down the firn column and reaches zero at a density of about 805 kg/m^3 (Fig. 4), which is interestingly above the end of the bubble formation range. This means that the air is actually stagnant in the firn before all bubbles have been formed. As discussed earlier, there is also an upward flow in the small permeable fraction of the denser layers, isolating the layers underneath. This implies that only the upper part of the bubble formation range (above about density 805 kg/m^3) leads to a broadening of the age distribution.

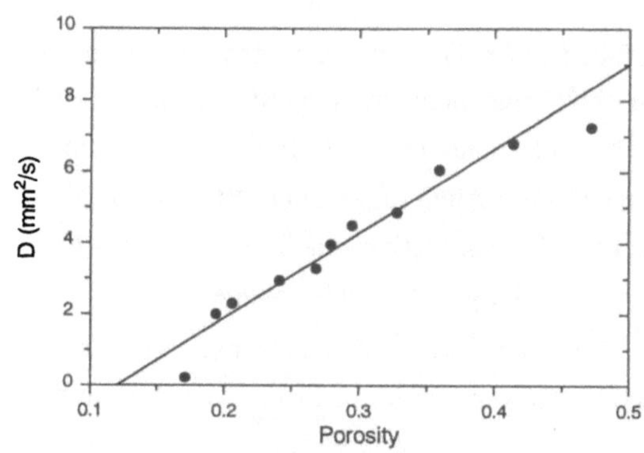

Fig. 4. Diffusion coefficients of CO_2 in N_2 measured on firn samples from Siple Station, West Antarctica.

Firn air analyses: A more direct way to investigate the air exchange in the firn is by measuring trace gas concentrations and the isotopic composition in air extracted from the pore volume. Uncontaminated samples of firn air can be taken by pumping air from the bottom of a freshly drilled hole. For that purpose the lower end of the bore-hole must be sealed immediately after the drill has been lifted.

Any trace gas of which the past concentration is known and has increased (or decreased) monotonically can be used to determine the age of the gas as a function of depth. e.g. ^{85}Kr, CO_2, CH_4, CFCs, SF_6.

Another important use of firn air analyses is the demarcation of the diffusive zone by means of the gravity effect. Good tracers for this purpose are the ratios of pairs of components with different molecular/atomic mass and with a constant concentration in the past. A prominent candidate is the $^{15}N/^{14}N$ ratio, usually expressed as

$$\delta^{15}N = \left(\frac{\left({}^{15}N^{14}N / {}^{14}N_2 \right)_{sample}}{\left({}^{15}N^{14}N / {}^{14}N_2 \right)_{ref}} - 1 \right) * 1000 ‰$$

where $\left({}^{15}N^{14}N / {}^{14}N_2 \right)_{sample}$ and $\left({}^{15}N^{14}N / {}^{14}N_2 \right)_{ref}$ are the isotopic ratios in the sample and in the mean atmosphere, respectively. $\delta^{15}N$ measured on the air extracted from ice samples below the close-off depth can also be used to assess the height of the diffusive column in the past (Sowers et al, 1989; Sowers et al., 1992).

Firn air analyses have been carried out during the EUROCORE drilling project at Summit (Schwander et al., 1993) and during projects from French, US, and Australian colleagues. (Bender et al., 1994; Sturges et al., this volume).

The trace gas profiles from polar sites are in good agreement with profiles calculated using diffusion models. $\delta^{15}N$ measurements suggest a thickness of the convective zone of roughly 10 meters at flat polar sites (Fig. 5). Trace gas measurements made on firn air sampled on Col du Dome (French Alps) suggest that at this alpine site convection acts deeper down than at flat polar sites. The convective zone extends roughly to 30-m depth or half way down to the firn-ice transition (Sturges et al., this volume).

Pressure and air flow distribution: Attempts have been made to investigate wind pumping directly. Since the flow velocities in firn are expected to be very small, especially below depths comparable to the typical surface roughness (sastrugi), their direct measurement would be rather ambitious. An easier method is to measure local pressure differences and air

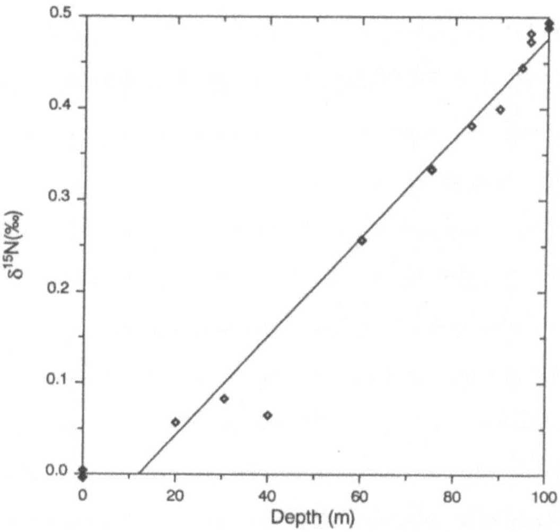

Fig. 5. Gravitational settling of ^{15}N in the firn observed at Vostok Station (Bender et al., 1994). The solid line has a slope of 0.542 ‰/100 m, according to the predicted gravitational settling at -55.5 °C and intercepts the x-axis at 12 m. It is an excellent trend-line of the measured data points if we disregard the outlier at 40 m depth, which also showed an anomalous O_2/N_2 ratio. The measurements thus suggest a convection zone of somewhat over 10 m and a diffusive zone down to 100 m, which is near the base of the bubble close-off zone.

permeabilities and to model the flow based on this information. (Colbeck, 1989; Albert et al., this volume).

Modelling the diffusion : Based on a densification model (Herron and Langway, 1980; Barnola et al., 1991) and measured diffusion coefficients we were able to model the gas diffusion in the firn at Summit, Central Greenland (Schwander et al., 1993).

It turned out that the model which is based entirely on molecular diffusion describes the measured trace gas profiles fairly well. The mixing over the whole firn column is clearly dominated by molecular diffusion. This is in agreement with a relatively small convective zone as inferred by the $\delta^{15}N$ results. The air is virtually stagnant below a density of 805 kg/m^3. This is in agreement with the diffusion-coefficient versus porosity relationship determined from Siple Station firn.

At Summit, the CO_2 in the air near the firn-ice transition has an age distribution with a mean of about 12 years and a standard deviation of about 7 years, as depicted in Fig. 6 (Schwander et al., 1993). For other trace gases these values are inversely proportional to their

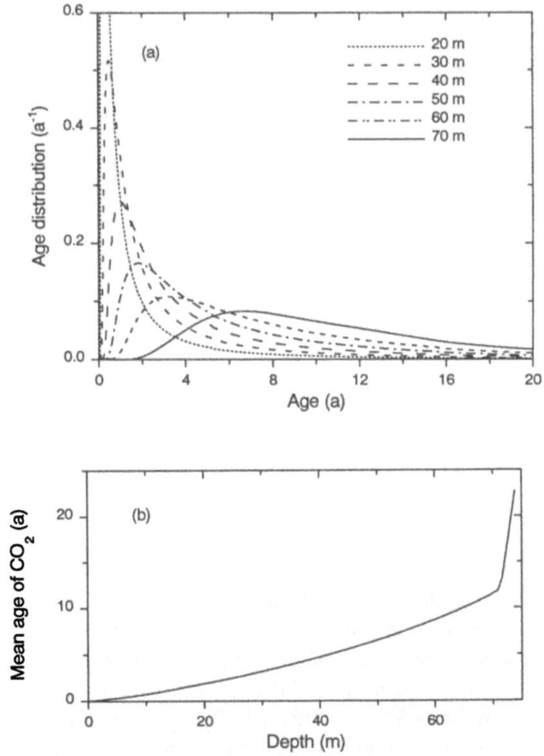

Fig. 6. (a) Age distribution of CO_2 in the firn of Summit, Central Greenland calculated with a diffusion model. (b) Mean age of CO_2 versus depth. The mean age at the close-off depth (70 m) is 12 years.

diffusivities. Such width of the age distribution also implies that seasonal variations in trace gas concentrations are practically lost after close-off.

Discussion of importance of processes

Air flows through the firn have two effects on the distribution of gases. First, like diffusion, air flows equalize concentration differences by dispersion arising from the different paths that individual gas molecules travel. Second, unlike diffusion, a persistent air flow in one direction shifts the position of whole packets of air. A crucial difference between the two mechanisms is that in the first case the mean distance covered by a molecule increases with the square root of time, while in the second case it increases linearly with time. That is, even flow speeds as low as 10 m/a can lead to air displacements of the magnitude of the firn thickness after the typical diffusive mixing time of the firn of 10 years, and thus compete with molecular diffusion. At smaller scales accordingly higher flow speeds are required to be of similar importance as molecular diffusion. In the special case, when the time averaged flow vector is zero, like for example for flows induced by large scale atmospheric pressure variations, there is no net displacement of the air, and the only effect is the increased dispersion.

The influence of the flow velocity on the dispersion coefficient has been determined with firn samples from Siple Station. The results clearly demonstrate that air fluxes induced by atmospheric pressure variations do not significantly increase the dispersion of permanent gases in the firn. (Schwander, 1989). Atmospheric pressure variations do however have an effect on dry deposition of aerosol below the snow surface (Cunningham and Waddington, 1993; Waddington et al. this vol.). Air flow induced by the densification of the firn is likewise very small, and it cannot compete with molecular diffusion.

In the case of wind pumping we may experience a net transport of air over considerable distances. Whether there is an net flux depends partly on the variability of the wind direction. In case of a dominant wind direction dunes induce air fluxes that point in the same direction for a prolonged period, whereas with variable wind direction the air moves back and forth, i.e. we have a similar situation as with barometric pressure variations. The closer to the surface the shorter can these periods of steady wind direction be in order to effectively ventilate that layer. The ventilated zone is important for the dry deposition of aerosols and the exchange of reversibly deposited species, whereas the influence on the concentration of gases that do not interact with the ice crystals can generally be neglected except when looking at small signals, e.g. from the gravity effect.

At locations with sufficiently flat surfaces, wind induced air flow will penetrate to between 10 and 20 percent of the total firn thickness. The firn air analysis carried out at Summit has shown that the extent of the non-diffusive zone is also of the order of only a few meters. At these sites the diffusive column height extends therefore over a large part of the firn thickness.

At polar sites the thickness of the firn is typically 70 m. If molecular diffusion is the dominant process then the mixing time of such a firn column is of the order of 10 years. This mixing time is inversely proportional to the diffusion coefficient and thus depends mainly on the temperature and the gas species. The mixing time is also about proportional to the square of the firn-ice transition depth. The age difference between the ice and the gases enclosed in the ice is therefore about the age of the ice at bubble close-off minus the diffusive mixing time. The standard deviation of the age distribution is typically of the same order of magnitude as the mean age.

Conclusions

In general, gas transport in firn is achieved by a complicated three-dimensional flux pattern, especially if low-permeability layers, like melt layers, exist. Our present knowledge does not yet allow us to mathematically model the mixing of the air for any site based on the local climatic and topographic conditions alone. For the simpler conditions we can however make statements about the importance of the processes involved. Phenomenologically we can discern three firn zones: A convective zone, a diffusive zone and a non-diffusive zone (from top to bottom). The non-diffusive zone is relatively small and is located within the firn-ice transition. The mass transport in the convective zone is most likely dominated by fluxes due to pressure gradients generated by the wind (wind pumping), although other processes, like free convection, cannot be ruled out. The extent of the convective zone depends therefore strongly on the firn surface topography, the wind speed and the wind direction. At flat polar sites with low to moderate winds the convective zone is estimated to be about 10 m deep. On alpine-type glaciers with a strongly shaped surface the convective zone can be considerably thicker.

In terms of the mixing of individual air components we must distinguish between species with little interaction with the ice crystals and species that are reversibly deposited on the ice. Species of the first category are located primarily in the pore space and therefore they disperse more rapidly in the firn than species of the second category. Convection in the uppermost layers certainly increases dispersion but molecular diffusion alone is rather efficient to remove concentration differences in gases of the first type. At flat polar sites molecular diffusion is therefore the most important process that controls the mixing of species of the first category. This is e.g. the case at Summit, Central Greenland (average wind speed below 5 m/s; unpublished meteo station data), as shown by the close fit between measured and model calculated trace gas concentrations using a pure molecular diffusion approach. If, in addition, the accumulation rate is high then the time span to cross the bubble formation zone is short, resulting in little extra width in the age distribution of the air. In this case we can estimate the age distribution of air trapped in the bubbles of ice using the age distribution at the firn base obtained with a molecular diffusion model. If, however, the accumulation rate is low like in the central parts of Antarctica so that the time to cross the bubble formation zone is long compared to the diffusion time then the gradual close-off becomes the dominant term in the age distribution. If summer melting is important, horizontal ice layers impede diffusion and air flow in the vertical direction, leading to an earlier isolation of the air.

Convection in the upper firn layers does not contribute significantly to the mixing of 'permanent' gases, however prevents or reduces gravitational settling. The convective zone may also be very important for the dispersion of reversibly-deposited species, especially if the fraction of the absorbed phase predominates, requiring a multiple air exchange for the transport of such substances. In this case wind pumping could strongly enhance the dispersion as compared to molecular diffusion alone.

Based on paleo-records of climatic conditions we may extrapolate to the past: During Holocene the climate was relatively constant. In particular, temperature and accumulation rates did not change significantly (Meese et al., 1994) and therefore the gas mixing was not subject to large variations. However, during the ice age, temperature and accumulation rate varied rapidly, at least in Greenland (Alley et al., 1993; Kaspner et al., 1995). During the coldest stages the temperatures were by 10 to 20 K colder than today and precipitation rates were roughly only half as high. Winds are believed to be generally stronger. The firn ice transition is expected to be up to 50% deeper than under present conditions and the ice age - gas age difference could have been up to five times as big. The stronger winds probably enhanced the air mixing in the upper firn strata (Waddington et al., this volume) and thus favoured the dispersion of reversibly-deposited species.

References

Albert MR, Arons EM, Davis RE (this volume) Firn properties affecting gas exchange at Summit, Greenland: Ventilation possibilities. In: E.W. Wolff, R.C. Bales (eds.) NATO ASI Series I: Chemical exchange between the atmosphere and polar snow. Springer Verlag, Berlin

Alley RB, Meese DA, Shuman CA, Gow AJ, Taylor KC, Grootes PM, White JWC, Ram M, Waddington ED, Mayewski PA, Zielinski GA (1993) Abrupt increase in Greenland snow accumulation at the end of the Younger Dryas event. Nature 362: 527-529

Barnola J-M, Pimienta P., Raynaud D, Korotkevich YS (1991) CO_2-climate relationship as deduced from the Vostok ice core: a re-examination based on new measurements and on a re-evaluation of the air dating. Tellus 43B: 83-90

Bender JA (1958) Air permeability of snow. IUGG-IASH General Assembly, Toronto, 3-14 Sept. 1957, IUGG-IASH (Gentbrugge) Publ 46: 46-62

Bender ML, Sowers T, Barnola J-M, Chappellaz J (1994) Changes in the O_2/N_2 ratio of the atmosphere during recent decades reflected in the composition of air in the firn at Vostok Station, Antarctica. Geophys Res Lett 21: 189-192

Colbeck SC (1989) Air movement in snow due to wind pumping. J Glaciol 35: 209-213

Conklin MH, Bales RC (1993) SO_2 Uptake on Ice Spheres: Liquid Nature of the Ice-Air Interface. J Geophys Res 98(D9): 16851-16855

Conklin MH, Sigg A, Neftel A, Bales RC (1993) Atmosphere-Snow Transfer Function for H_2O_2: Microphysical Considerations. J Geophys Res 98(D10): 18367-18376

Conway H, Abrahamson J (1984) Air permeability as a textural indicator of snow. J Glaciol 30: 328-333

Craig H, Horibe Y, Sowers T (1988) Gravitational Separation of Gases and Isotopes in Polar Ice Caps. Science 242: 1675-1678

Cunningham J, Waddington ED (1993) Air flow and dry deposition of non-sea salt sulfate in polar firn: paleoclimatic implications. Atmos Environment 27A: 2943-2956

Herron MM, Langway CC, Jr (1980) Firn densification: an empirical model. J Glaciol 25: 373-385

Kaspner WR, Alley RB, Shuman CA, Anandakrishnan S, Grootes PM (1995) Dominant influence of atmospheric circulation on snow accumulation in Greenland over the past 18,000 years. Nature 373: 52-54

Maeno N, Narita H, Arakoa K (1978) Ice-coring project at Mizuho Station, East Antarctica, 1970-1975. Memoirs of National Institute of Polar Research, Spec Issue 10: 62-76

Martinerie P, Raynaud D, Etheridge DM, Barnola J-M, Mazaudier D (1992) Physical and climatic parameters which influence the air content in polar ice. Earth Planet Sci Lett 112: 1-13

Meese DA, Gow AJ, Grootes P, Mayewski PA, Ram M, Stuiver M, Taylor KC, Waddington ED, Zielinski GA (1994). The Accumulation Record from the GISP2 Core as an Indicator of Climate Change Throughout the Holocene. Science 266: 1680-1682

Schwander J (1989) The transformation of snow to ice and the occlusion of gases, in The Environmental Record in Glaciers and Ice Sheets, eds H. Oeschger and C.C. Langway, Jr., pp. 53-67 John Wiley, New York

Schwander J, Stauffer B (1984) Age difference between polar ice and the air trapped in its bubbles. Nature 311: 45-47

Schwander J, Stauffer B, Sigg A (1988) Air mixing in firn and the age of the air at pore close-off. Ann Glaciol 10: 141-145

Schwander J, Barnola J-M, Andrié C, Leuenberger M, Ludin A, Raynaud D, Stauffer B (1993) The Age of the Air in the Firn and the Ice at Summit, Greenland. J Geophys Res 98(D2): 2831-2838

Sowers T, Bender M, Raynaud D (1989) Elemental and isotopic composition of occluded O_2 and N_2 in polar ice. J Geophys Res 94(D4): 5137-5150

Sowers T, Bender M, Raynaud D, Korotkevich YS (1992) $\delta^{15}N$ of N_2 in Air Trapped in Polar Ice: a Tracer of Gas Transport in the Firn and a Possible Constraint on Ice Age-Gas Age Differences. J Geophys Res 97(D14): 15683-15697

Stauffer B, Schwander J, Oeschger H (1985) Enclosure of air during metamorphosis of dry firn to ice. Ann Glaciol 6:108-112

Sturges WT, Penkett SA, Barnola J-M, Chappellaz JA (this volume) Fluorocarbon tracers of the age of air in Alpine firn. In: E.W. Wolff, R.C. Bales (eds.) NATO ASI Series I: Processes of chemical exchange between the atmosphere and polar snow. Springer Verlag, Berlin

Sturm M, Johnson JB (1991) Natural convection in subarctic snow cover. J Geophys Res 96(B7): 11657-11671

Waddington ED, Cunningham J, Harder SL (this volume) The effects of snow ventilation on chemical concentrations. In: E.W. Wolff, R.C. Bales (eds.) NATO ASI Series I: Chemical exchange between the atmosphere and polar snow. Springer Verlag, Berlin

LOCATION, MOVEMENT AND REACTIONS OF IMPURITIES IN SOLID ICE

Eric W. Wolff
British Antarctic Survey
Natural Environment Research Council
High Cross
Madingley Road
Cambridge CB3 0ET
UK

1. Introduction

The aim of drilling ice cores is to obtain paleoatmospheric information of wide significance. However, the concentrations of chemicals found in the ice are determined both by the atmospheric concentrations, and by depositional and post-depositional processes. Other papers in this volume discuss the depositional processes, and the processes that subsequently alter the concentrations in near-surface (up to a metre or so depth) snow. However, there are a number of documented cases where further changes occur below the surface snow. Additionally, there is the possibility that chemical changes can occur, at least to more complex species, over the long timescales that are relevant for ice. Diffusion is bound to take place to some extent, affecting the apparent rate of temporal change inferred from ice core profiles. Discussion of all these factors requires an understanding of the way in which impurities are held in the ice. This paper discusses all these items, concentrating on processes in solid ice, and in firn below the top metre or two.

2. Known examples of changes within the firn and ice

Most ice core researchers assume that the profiles they look at are locked firmly in place by a snow depth of about 2 metres (I exclude here the gases trapped in air bubbles, which form only at the pore close-off, and are discussed by Schwander (1996)). However, a number of examples have been presented in the literature where this is clearly not the case. These are summarised below without any deep exploration of the causes, which will be discussed in later sections. Naturally, the list includes only examples that have been observed and reported: it is conceivable that other important cases await discovery.

NATO ASI Series, Vol. I 43
Chemical Exchange Between the Atmosphere and Polar Snow
Edited by Eric W. Wolff and Roger C. Bales

2.1. H_2O_2 losses

Hydrogen peroxide is the clearest example of a species that suffers diffusive smoothing throughout the firn layer (Neftel 1991; Sigg and Neftel 1991). The original strong seasonal signals are almost completely lost (at Summit, Greenland, an original max/min ratio of about 10 is reduced to <2 by 70 m (Neftel 1991)) (Fig. 1A). This is caused by movement of H_2O_2 in the vapour phase through the pore space, and implies that most of the H_2O_2 is at some time available at the grain surfaces. Oxygen isotope ratio seasonal signals are attenuated in an apparently similar way to this (Johnsen 1977), and one can assume that other molecules with a relatively high volatility would behave similarly.

While diffusion will destroy the seasonality of signals, long-term average concentrations will not be affected. However, other processes seem to destroy H_2O_2 completely. In alpine glaciers, H_2O_2 is lost rapidly, probably by a light-enhanced catalytic destruction on dust particles (Neftel 1991). In ice from the polar regions, H_2O_2 is also lost, but much more slowly. Nonetheless, H_2O_2 is completely absent from the very dusty, alkaline ice of the last ice age at Summit (Fuhrer et al. 1993) (Fig. 1B). H_2O_2 "holes" are also seen in shallower ice, and cannot have been due to a decrease in atmospheric deposition, because diffusion would have smoothed out the sharp edges of the holes. Again, the holes seem to be associated with the presence of alkaline dust. Other holes are associated with sulphate from volcanic eruptions, and it is unclear whether they are of atmospheric or post-depositional origin (Laj et al. 1990; Sigg and Neftel 1991).

In summary, it is clear that H_2O_2 suffers disintegration, probably on dust particles, even in the dark, and this makes it currently impossible to deduce the atmospheric concentrations of H_2O_2, at least in Greenland, from any period earlier than the Holocene.

2.2. MSA movement

Methanesulphonic acid (MSA) is an oxidation product of dimethylsulphide, which is produced by marine organisms. It is generally considered to be deposited as part of the primary aerosol, and its concentrations in Antarctic air and snow always peak in summer (eg (Prospero et al. 1991; Wagenbach 1996)), more or less in phase with excess sulphate (also largely a DMS oxidation product).

However, at a site in the Antarctic Peninsula (Dolleman Island), it has been observed that, although most MSA peaks in the top 10 m occur in summer, in phase with excess sulphate peaks, by 50 m the peaks are clearly out of phase, and by 120 m the MSA peaks are centred in the winter trough of excess sulphate, while the excess sulphate peaks remain at the isotopic summers (Mulvaney et al. 1992) (Fig. 2). Similar, but less extreme, behaviour was seen in another Antarctic Peninsula core. It is extremely unlikely that MSA deposited in

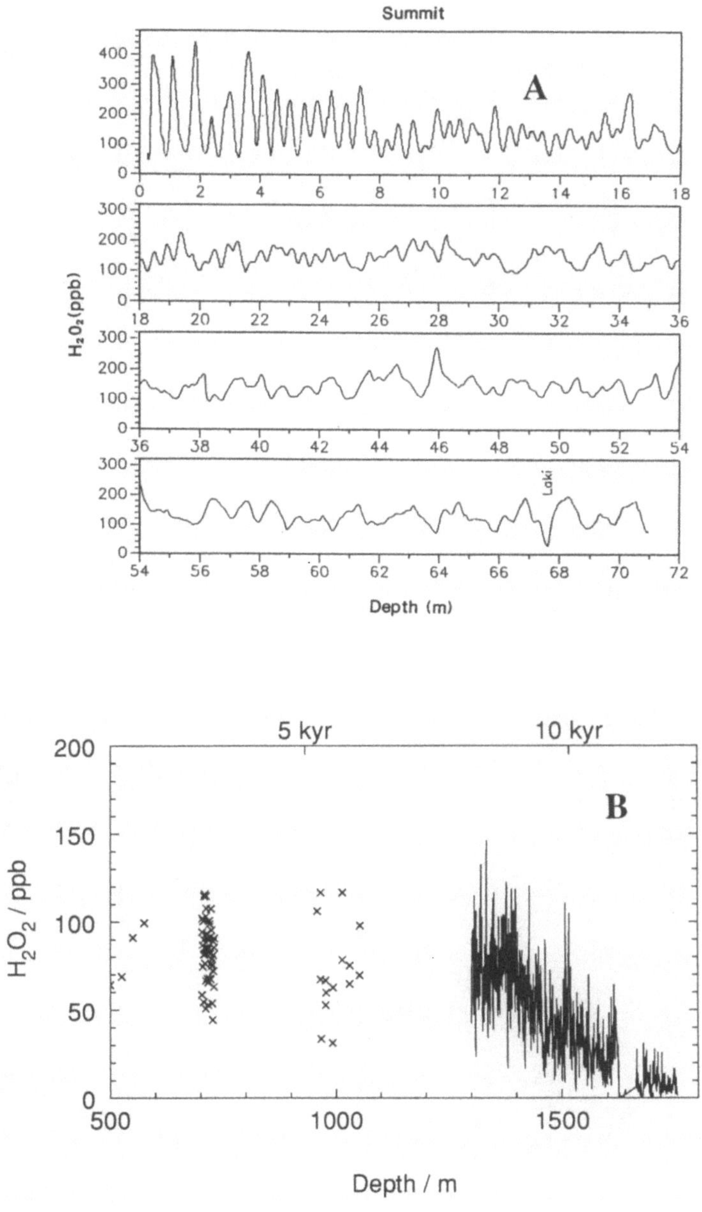

Fig. 1. Changes in H_2O_2 concentrations with depth at Summit, Greenland. (A) Diffusive smoothing in the firn (Sigg and Neftel 1991) (Reprinted with permission from Nature, copyright (1991) Macmillan Magazines Limited), (B) Loss of H_2O_2 in older, dusty ice from the GRIP core. Particularly low values are seen in ice from the Younger Dryas period (1625-1661 m), and the last glacial (below 1752 m) (Fuhrer et al. 1993).

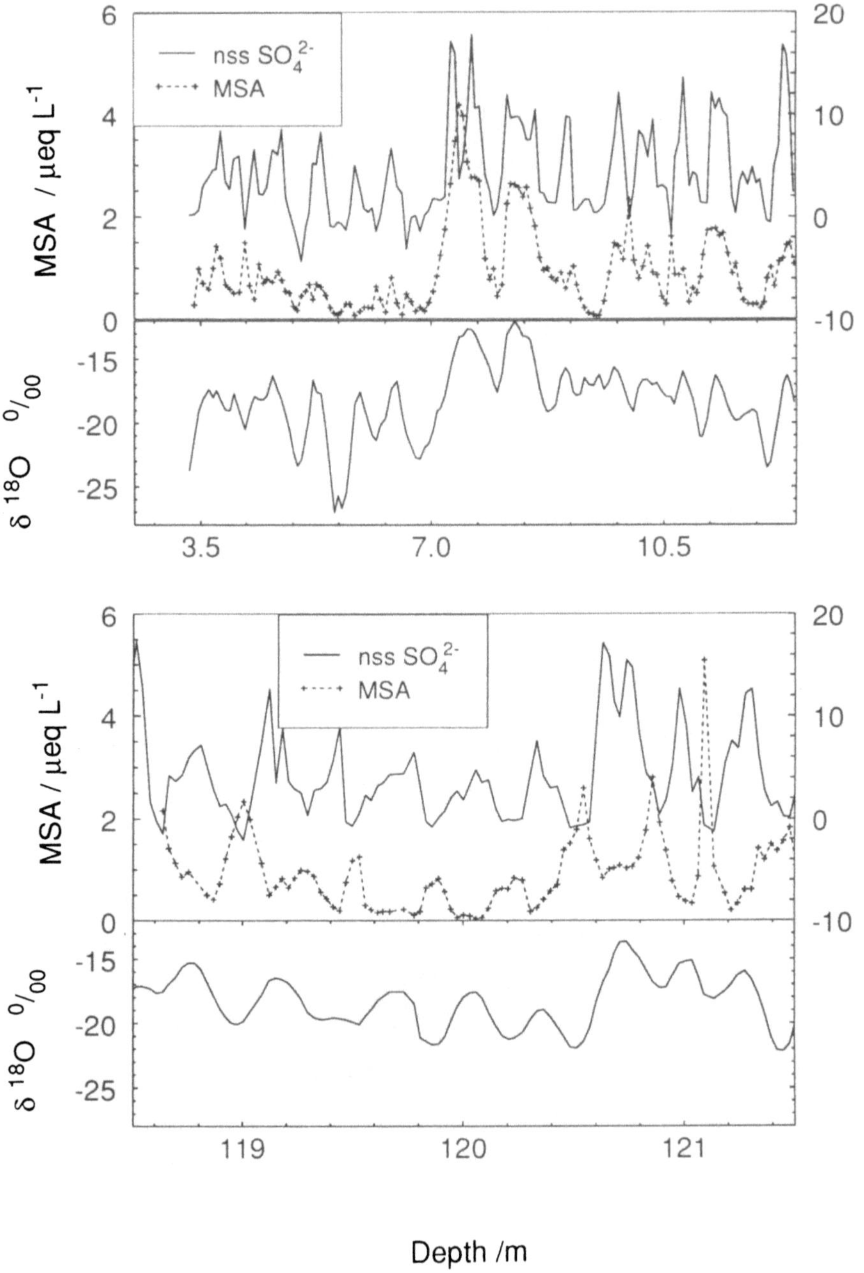

Fig. 2. MSA and excess (non-sea-salt) sulphate at Dolleman Island, Antarctic Peninsula. (A) 2.5-12.5 m, with MSA and excess sulphate still partly in phase; (B) around 120 m, with MSA peaking at the troughs in excess sulphate (Mulvaney et al. 1992).

winter in the past, so it appears that post-depositional movement has occurred. The same effect has been observed at Berkner Island, a site where summer melting is completely absent, in the Ronne-Filchner Ice Shelf (Wagenbach et al. 1994). Although it was not commented on in the original paper, similar behaviour, with out of phase peaks occurring by 10 m depth, can be seen in the MSA record from Byrd Station, Antarctica (Langway et al. 1994). It can thus be seen that this phenomenon is quite widespread, and appears to occur in the firn.

This phenomenon appears not to affect the multi-year average of MSA, and is therefore not too detrimental to the MSA record in ice. However, it does mean that extreme caution is needed in the use of MSA/excess sulphate ratios, which can be highly biased if short sampling steps are used. Additionally, this may be related to the observation that, where rapid changes from clean, acidic ice to dusty, impurity-laden, alkaline ice occur in the very deep "Eemian" section of the Summit (Greenland) ice core, MSA is entirely missing from both shoulders of the peak (Clausen and Steffensen 1993) with elevated values in the peak (also J-P Steffensen, pers. comm.). This is either an extremely unusual change in the chemistry of the atmosphere, or a sign of post-depositional movement, with the MSA apparently sucked into the dustier ice. Clearly, the difference between an atmospheric and a post-depositional interpretation of these events is critical.

2.3. Possible displacement of nitrate

Nitrate is a potentially very interesting species, whose interpretation in ice cores is still very difficult (Wolff 1995). Nitrate is another species that shows some signs of diffusive smoothing of seasonal patterns in the firn. However, this must be rather limited, as clear seasonal cycles of nitrate have been shown at 820 m (approx. 2000 years old) depth in ice from Dye 3 (Herron 1982), and at far greater depths at Summit, Greenland (C. Hammer, pers. comm.). There is also good evidence that nitric acid is lost from near-surface snow after deposition (Legrand 1996). At low accumulation rate sites such as Vostok, nitrate concentrations both smooth and decrease with depth in the top metre or so of firn (Mayewski and Legrand 1990), suggesting that losses occur in the long timescales when the snow is close to the surface. Fresh snow shows much higher concentrations than those seen in old snow or in ice cores (Neubauer and Heumann 1988a, b). Overall, this suggests that nitrate losses do occur, although movement in both directions (air to snow and snow to air) may take place, and that the nitrate concentrations seen in ice are the result of a complex series of processes and equilibria. The changes discussed in this paragraph probably occur in nitric acid, and we have no specific evidence that they occur in the alkaline nitrate present in Greenland ice of the last ice age.

Most of the nitrate changes discussed here occur in the near-surface snow. However, there are two possible cases of nitrate movement in deeper ice. Firstly, nitrate "holes" are seen across the depths of some volcanic eruption peaks with large sulphate peaks present (Laj et al. 1993). These may be a genuine atmospheric depletion of nitrate caused by reactions or changes in deposition rate from the stratosphere (Bekki 1994). However, it is also possible that they represent movement of nitrate from the heavily acidic ice into the shoulders of the eruption peak (Wolff 1995). Additionally, there is a slight reduction in nitrate concentrations on the shoulders of some large alkaline ice peaks within the "Eemian" section of the Summit ice core (Clausen and Steffensen 1993)(also J-P Steffensen, pers. comm.). These reductions occur at the same place as MSA losses mentioned earlier, and could represent a post-depositional movement or loss of nitrate.

2.4. Movement of Mg relative to Na.

Sea salt cations such as Na and Mg are considered very conservative species - reaction is impossible, and diffusion is considered slow. However, in the Dolleman Island, Antarctica, ice core, there is clear evidence of movement of one species relative to the other. Near the surface, Mg is in phase with Na, as expected from two sea salt-derived elements, and the ratio is close to that of sea water. At greater depths (below about 10 m), Mg forms two peaks on either side of a drop that coincides with the Na peak (Fig. 3). K shows similar, but less marked behaviour. Mg has clearly moved in the firn layer at some depth. This behaviour has not until now been reported at other sites, but Dolleman Island has unusually high sea salt concentrations compared to other ice core sites. It also has slight but significant melt in summer, but as the movement takes place at least a few metres below the surface, this cannot be the cause. It might be thought that this behaviour is anomalous and confined to a few special sites. However, it does indicate the possibility of movement of even this simple species. Since ice age ice from large parts of both Greenland and Antarctica contains high levels of sea salts, we should try to understand this phenomenon before we discard it as being of limited significance.

2.5. Diffusion of HF

HF in strongly acidic volcanic layers seems to be displaced to either side of the main acidic input (de Angelis and Legrand 1994). This process seems to occur very near the surface in the upper layers of firn. The same authors also noted a broadening of volcanic HF

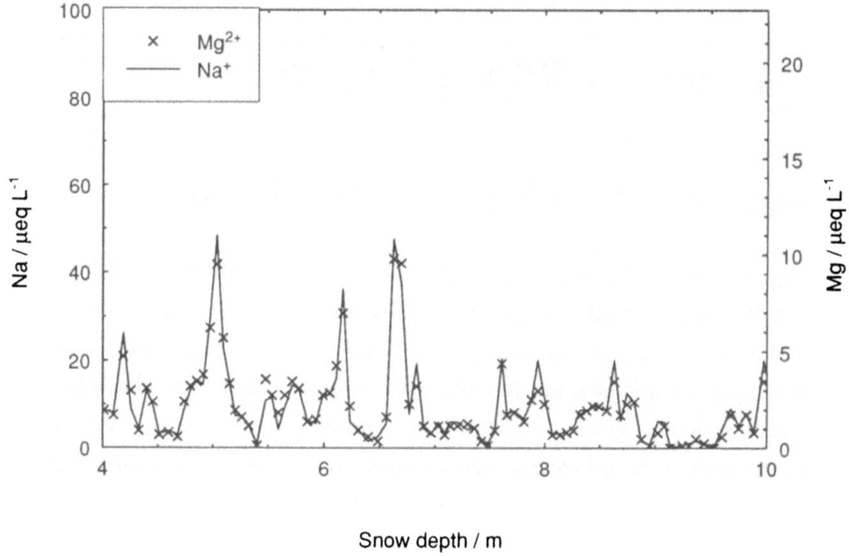

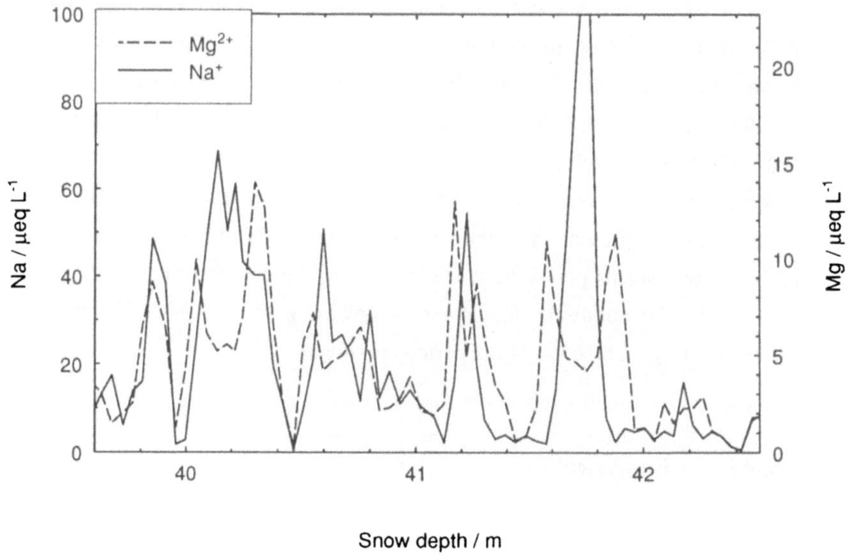

Fig. 3. Mg and Na concentrations at Dolleman Island, Antarctica. The Mg have been plotted at a scale so that they would coincide with the Na measurements if the two elements were in the ratio found in sea water. At 4-10 m, Mg is in phase with Na, but by 40 m, many of the Mg peaks have split to form a doublet around the Na peaks.

peaks with depth, so that an HF peak at 1210 m depth at Summit, Greenland, was approximately 11 years "wide". This was interpreted as showing diffusion of HF, at a much faster rate than the HCl also present (de Angelis and Legrand 1994).

2.6. Rapid variation in CO_2 concentration during the last glaciation

Large and rapid changes in CO_2 concentration were observed in ice from the last glaciation at Dye 3, Greenland (Stauffer et al. 1984). However, these changes occur on very fast timescales, are parallel to oxygen isotope changes in the ice (despite the expected age difference between ice and air), and are not seen at Antarctic sites even though CO_2 should give a globally coherent signal. One possibility, which now seems rather probable, is that CO_2 concentrations in the ice have actually altered since enclosure (Raynaud et al. 1993), with increased concentrations in Greenland ice from the acidic, warm, phases of the glaciation. The mechanism is unknown, but one suggestion (Delmas 1993; Anklin et al. 1995) is that the reaction of (warm) acidic ice with "cold" alkaline, carbonate-laden ice results in production of CO_2 ($2H^+ + CaCO_3 \rightarrow Ca^{2+} + H_2O + CO_2$). The interstadial ice in Greenland is often quite close to neutral, so that acidic and alkaline layers are in close proximity, and the components may gradually interact as time passes. Since some interstadials are alkaline throughout, but others alternate between acidic and alkaline ice (Wolff et al. 1995), high-resolution sampling for CO_2 across various interstadial sections might be used to test the theory. This reaction would be unimportant in Antarctica, where much less carbonate is present, and the ice stays acidic throughout the glacial period. Although this reaction seems plausible, it is unproven as the cause of the variations, and, we should also consider the possibility that other C-containing species (HCHO, organic acids) may react to form CO_2 over the long time periods involved.

3. Location of impurities in ice

Impurity molecules and ions can be present within the ice itself, either aggregated as a microparticle (which may be insoluble, or soluble but not yet dissolved), or dispersed through the lattice. The latter case includes the situation where molecules may be substitutionally inserted into the lattice in place of a water molecule, with hydrogen bonds to the surrounding molecules. Additionally, impurities may exist at grain boundaries between crystals. The location of impurities will have a major effect on their ability to move along the ice core, to diffuse, and to "meet" other species so that reactions can occur. For instance, the likelihood of a chemical reaction is far greater for two species that exist only at the limited number of sites in a grain boundary, and even greater if the grain boundary is liquid.

The location in ice depends on the thermodynamically-preferred location, and also on the ability of the impurity to reach that location.

In the original snowflakes, impurities may have been trapped as solid aerosol particles within the ice (e.g. terrestrial dust particles and sea salt), they can have been dispersed through the lattice, or they can be attached (as an adsorbed monolayer or a liquid film) to the outside of the ice. Dry-deposited material, including gases adsorbed after snow deposition, is certain to start on the outside of the snowflakes. By a depth of 2 metres, the snow will be between about 0.5 and 20 years old, depending on the accumulation rate at the site, and considerable metamorphosis will have taken place.

In order to be bonded within the ice lattice, a species must be of a size and type to fit into the lattice without causing excessive strain. The most obvious species that can certainly fit within the lattice are H_3O^+ and OH^-. These can take the place of a water molecule, but with one extra (or one less) H atom. They produce ionisation defects, and normally also Bjerrum D or L defects (two or zero H atoms on a bond) (Hobbs 1974). The only other species that are generally believed to be incorporated into the lattice are F^- (or HF), Cl^- (or HCl), and NH_3. The first two of these may replace an O atom, leaving one or more L defects in the ice, while NH_3 will leave a D defect. In each case the atom replacing O is a similar size to O. More complex anions such as nitrate and sulphate cannot reasonably be incorporated into lattice sites, and nor can cations such as Na^+, which are far larger than the H atom they would need to replace.

Many experimenters have assumed or inferred that species such as HF, HCl and NH_3 are incorporated into the lattice, at least up to a given solubility limit (Hobbs 1974; Gross et al. 1975, 1987; Camplin et al. 1978). Few experiments have shown directly that the species are indeed incorporated into the lattice, or explored differences between laboratory-grown and natural ices.

There is direct evidence that some S in natural ice from the polar regions is localised at grain boundaries, and particularly at triple junctions. Using ice with relatively high concentrations from Dolleman Island, Antarctica, Mulvaney and others (1988) carried out electron microscopy with X-ray microanalysis and a cold stage at liquid nitrogen temperatures. They found S at high concentrations at triple junctions in the ice (Fig. 4). They assumed that this S was sulphate, and calculated that between 40 and 100% of the S in the ice was at these locations, and that the concentrations were high enough that the sulphate, if present as sulphuric acid, would have been on the freezing point line at normal ice temperatures. In other words, it would have been present as liquid sulphuric acid solutions.

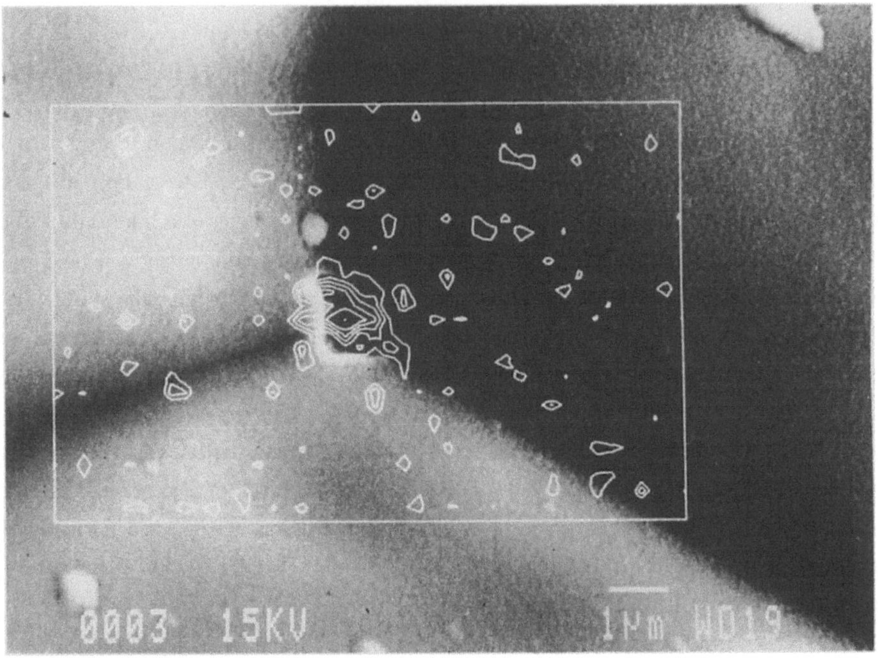

Fig. 4. A contour map (from X-ray microanalysis) showing the distribution of sulphur around a triple junction in ice from Dolleman Island, Antarctica (Mulvaney et al. 1988) (Reprinted with permission from Nature, copyright (1988) Macmillan Magazines Limited). The scale bar is 1 μm.

If liquid is present, there are thermodynamic arguments as to why it should be at triple junctions in preference to two-grain boundaries (Wolff and Paren 1984).

In subsequent experiments, we have found S at triple junctions in a volcanic layer corresponding to the Laki (1783) eruption in a core from near Summit, Greenland (Wolff and Mulvaney 1990). However, S is not consistently found at junctions in all samples containing sulphate, and it is not clear whether this is due to sample preparation inconsistencies, or because S is sometimes absent from the triple junctions. We have occasionally seen evidence for marine cations at triple junctions in ice with very high marine ion concentrations (A.P. Reid, pers. comm.). We have never seen any Cl at triple junctions in polar glacier ice, although it is seen in triple junctions and inclusions in more saline ice formed from the freezing of seawater to the underside of an ice shelf (Moore et al. 1994a).

Electrical evidence can also be used to infer the location of impurities in ice. Only strong acids contribute to the DC conductivity of polar ice (Hammer 1980; Moore et al. 1992b), while acids, marine chloride, and ammonium all contribute to the conductivity at higher frequencies (Moore et al. 1992b, 1994b; Wolff et al. 1995). One interpretation of this is that ammonium and chloride are incorporated into the lattice, giving D and L defects respectively, and that these give rise to an AC conductivity only; acids give a DC conductivity either because they are in the lattice and cause ionisation defects, or because they are at grain boundaries and conduction is through a liquid network of veins at triple junctions (Wolff and Paren 1984). Above a certain limit (possibly about 400 μM, or 400 $\mu mol\ L^{-1}$) in marine ice under ice shelves, conductivity no longer rises as Cl^- increases (Moore et al. 1992a), and this is interpreted as a solubility limit for Cl^- in the lattice.

Combining the above evidence, we can form an idea of the likely preferred location of impurities in polar ice, though there is still considerable uncertainty. At the concentrations found in glacier ice, both ammonium and chloride are probably mainly incorporated into the lattice. It is to be expected that fluoride will also be incorporated, although there is no direct evidence on this. Large molecules such as sulphate, nitrate, and organic anions are likely not to be in the lattice in any appreciable concentrations, and could be at grain boundaries, sometimes as liquid. We have no direct evidence to decide whether H^+ is at the boundaries or in the lattice at present, but metal cations would be expected to be excluded from the lattice. The question of the extent of liquid at boundaries is unresolved at present.

We need also to consider how the impurities reach their preferred locations. From theoretical considerations of grain growth processes, Alley and others (Alley et al. 1986a, b) suggested that bubbles and microparticles would not be held at grain boundaries, but that dissolved impurities would be carried with grain boundaries. Impurities that start on the outside of snowflakes will stay at grain boundaries unless they can be incorporated into the lattice over a period of time. Soluble impurities that start as particles within a snowflake will slowly become dissolved into the lattice and segregated to the grain boundaries above the solubility limit. Insoluble microparticles should stay randomly distributed within ice grains.

We now have a very imperfect model for impurity locations that allows us to discuss possible processes within the ice.

4. Reactions

Reactions within the ice make up the most dangerous process for our interpretation of ice core data, because they have the potential to affect long-term average concentrations.

In the top metre or so of ice, photochemistry (Hoffmann 1996) may be able to occur (e.g. possibly involving HCHO or HNO_3). However, insufficient light penetrates below this level to be likely to have any significant effect (Neftel 1991). For greater depths, the only other documented reaction is the disintegration of H_2O_2 with age/depth. The possibility also exists that CO_2 is produced by some reaction in the ice.

Reaction clearly depends on the likelihood of intimate contact between the reactant molecules. For two reactants that are within the lattice, this is quite unlikely. If there is liquid at grain boundaries, then reactants that are both in the liquid should react very rapidly. Even without impurity stimulation, there is likely to be a liquid or liquid-like film at grain boundaries well below the melting point (Dash et al. 1995). The intermediate case is where one reactant is at a grain boundary, while the other is in the ice (in the lattice or e.g. a microparticle or bubble). Where there is grain growth, grain boundaries will occasionally sweep across bubbles, particles, and material in the lattice, providing the opportunity for reaction.

We have no direct information on the location of H_2O_2 in ice. The original co-condensation model (Sigg and Neftel 1988) would suggest that this species would be uniformly dispersed through the lattice (Sigg et al. 1992). Experiments also show that uptake of H_2O_2 on ice is reversible (Conklin et al. 1993), implying that H_2O_2 can sit on the surface of the ice crystals. The extent to which it will partition into the bulk is not yet clear. The loss mechanism in solid ice is not known at all. H_2O_2 in aqueous solution is a strong oxidising agent, but there are probably rather few species present that can be oxidised. H_2O_2 also spontaneously decomposes (to water and molecular oxygen) in aqueous solution, particularly if the solution is basic, and in the presence of rough surfaces and metals. This is perhaps its most likely loss mechanism in dusty, alkaline ice.

Reactions that could create or destroy CO_2 are also not well-studied. The possible reaction of acidity with carbonate to produce CO_2 (Delmas 1993) would depend on acidic ice coming into very close contact with alkaline, carbonate-laden ice. It is true that interstadial ice is, on average, only mildly acidic, so that there are both alkaline and acidic layers within it. The extent to which this is so varies from interstadial to interstadial, and detailed CO_2 analysis within a number of interstadials might show whether the anomalies were indeed stronger in ice where the acid-alkali budget was most finely balanced. Reactions to transform CO_2 to or from other organic molecules need further investigation. For instance, formaldehyde concentrations are lower in interstadial ice at Summit, Greenland, than in the remaining ice age ice (Fuhrer et al. 1993). The reduction is approximately 0.2 μM. The increases in CO_2 in interstadial ice are about 50 ppmv, which is also of order 0.2 μM. Although there is no mechanism advanced for converting formaldehyde into CO_2, even the highly improbable must be considered over the very long timescales involved here. Organic acids are the other C-containing species present in sufficient quantities that it has the potential to influence CO_2 values, if a reasonable mechanism were proposed.

One other type of reaction that can take place is the reaction of two soluble ions to precipitate an insoluble component. This will be discussed later as one explanation for the movement of species in the ice.

5. Diffusion

If diffusion of impurities occurs, then it would smooth out seasonal variations, and reduce the apparent speed of rapid changes between climate states. We have already seen that vapour phase diffusion occurs to smooth the annual cycles of H_2O_2, oxygen isotopes and perhaps HNO_3. This section discusses possible diffusion in solid ice (including grain boundaries).

Here again, the location of impurities is very important. For species in the lattice, the diffusion rate through the ice matrix itself will be the governing factor. If impurities are localised at grain boundaries, then the diffusion rate through the grain boundary material (which is more disordered so that rates may be faster) is important. If they are in a liquid matrix, then very fast diffusion through the liquid is important.

However, in this last case, the analysis is rather different. For a given ice temperature, the concentration of impurities in the liquid vein is fixed according to the freezing point depression line. If there are two ice layers with the same temperature next to each other, and one has more impurity in the grain boundaries, the one with more impurity will have larger vein widths or a greater density of veins (meaning smaller grain size), but the concentration of impurity in the two sets of veins will be the same. Thus there is no strictly chemical driving force for diffusion in such a system. It is possible to have relatively slow apparent diffusion rates even if liquid is present at grain boundaries. However, to accommodate different impurity loadings in adjacent sections of ice, either veins must have different diameters, or their number density must alter (through changing grain size). In the former case, there will ultimately be a flow of heat to equalise the vein diameters (Nye 1991), and this second-order effect will cause small chemical gradients and slow diffusion. The potential for diffusion would be limited if there are any necks where veins are closed off completely.

Diffusion coefficients have been measured for oxygen-18, deuterium and molecular oxygen in ice, and are of the order 2×10^{-15} m^2s^{-1} at -10°C (Hobbs 1974). However, much faster rates have been reported for the diffusion of HF (Haltenorth and Klinger 1977) and other molecules in ice (Barnaal and Slotfeldt-Ellingsen 1983). It is possible that these latter results were partly due to diffusion in small-angle grain boundaries rather than through the lattice. By direct SEM observation of ice doped with HCl, an upper limit for the diffusion rate of HCl through ice of 10^{-17} m^2 s^{-1} at -20°C was suggested (Wolff et al. 1989), but again this experiment can be interpreted in terms of solubility limits rather than diffusion rates, so the conclusions are equivocal.

Ice core data in fact provide their own information on apparent diffusion rates through polar ice. For instance, in the GRIP ice core, annual cycles are still clearly seen in ammonium, Ca and H_2O_2 throughout the Holocene (giving a maximum diffusion length over this period of probably less than 5 cm) (Fuhrer et al. 1993). The resolution of the continuous flow instrument used prevents a definitive statement about cycles in these species in older ice. We have additionally seen clear annual cycles in nitrate, chloride and sulphate in a section of ice 18000 years old in the GRIP core (Fig. 5). Here, the annual cycles are only of the order 2 cm thick, and, remembering to allow for layer thinning, diffusion must be less than 1 cm over 10 000 years for these species. ECM (electrical conductivity method) signals from Byrd Station, Antarctica (Hammer et al. 1994), show seasonal signals in acidity to at least 30 kyr BP, again implying diffusion lengths of less than 1 cm in 10 kyr for H^+.

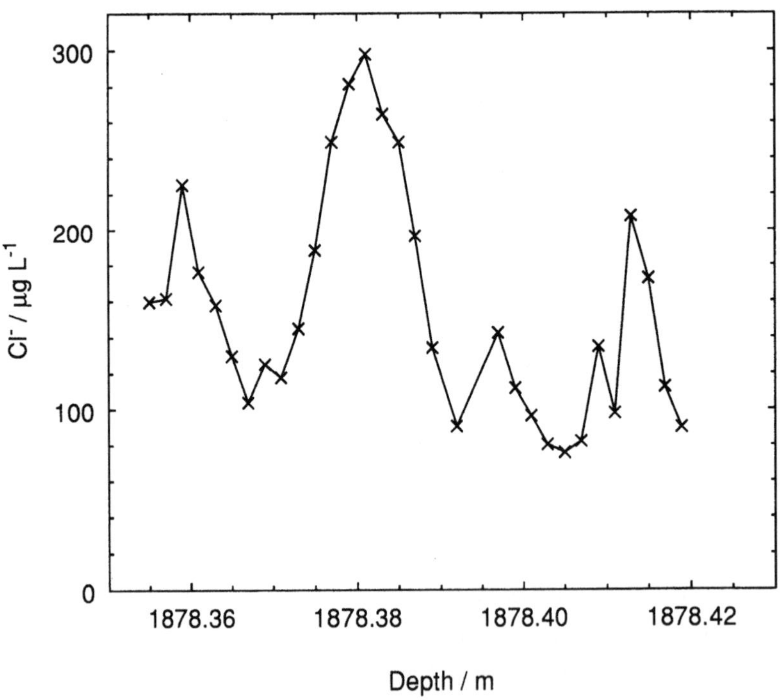

Fig. 5. Seasonal variation of Cl^- at 18000 years BP in the Summit, Greenland, ice core. Three or possibly four years are represented here.

Such diffusion lengths imply maximum apparent diffusion coefficients of order 10^{-16} $m^2\ s^{-1}$ at temperatures of approximately -30°C for all these species. Depending on the location of impurities, this may or may not represent the diffusion coefficient through ice. In contrast, Dominé et al. (1996) report the much higher experimental diffusion coefficient of $10^{-14}\ m^2\ s^{-1}$ for HNO_3 at -15°C. The apparent diffusion rate of HF from volcanic peaks at Summit, Greenland (de Angelis and Legrand 1994) is $2 \times 10^{-12}\ m^2\ s^{-1}$, which compares well with laboratory measurements. Thus, HF appears to diffuse far faster than other species, faster indeed than water molecules themselves. Further detailed study of the evolution of seasonal signals may allow differences between the diffusion rates of different species to be elucidated. It is safe to say that annual cycles, if they exist initially, are preserved in all species mentioned above except fluoride for at least 20 kyr at temperatures typical of the centre of the ice sheets, and that large scale loss of signals representing century-scale climatic change is therefore unlikely even at the lowest accumulation sites and the oldest ice.

6. Other movement

We have still not treated two cases where post-depositional change has been observed - the movement of MSA relative to sulphate (Fig. 2), and of Mg relative to Na (Fig. 3).

In both cases, the cause of the movement cannot be simple diffusion of a species, because new peaks are created where they were not present before; diffusion would be expected merely to broaden the existing peaks.

For MSA, movement is into the winter layers, where sea salt concentrations are high. One possibility is that MSA may form a salt (with one of the sea salt cations) that is insoluble at the temperatures in the ice sheet. In that case, the MSA ion (or acidic MSA) can diffuse outwards from the original peak, and then forms the insoluble salt, which is a different species. Diffusion can then continue as the concentration of MSA itself (as opposed to the salt) in the winter layer is still low. In this way, a new peak can build up. However, I should emphasise that we know of no data on the low-temperature solubility of MSA salts, and have no indication that this is the mechanism of movement.

For Mg, the same possibility that an insoluble salt may precipitate out, allowing continued diffusion from a peak, must exist, but in that case we have to find an anion with sufficient concentration to neutralise the Mg. Gravitational movement of a grain-boundary liquid is not a possible explanation, as the Mg peaks are on both sides of the Na peak. A second possible hypothesis is that there are a limited number of sites for cations, either within the grain (unlikely location), or at the boundaries. As metamorphosis occurs, we expect that sea salt is initially present as discrete particles, but will slowly dissolve to give Cl in the lattice, and Na and Mg on the outside (the grains could be negatively charged with cations on the outside, or H^+ could be incorporated into the lattice with the Cl^-). As grains grow, the surface area decreases, and this could limit the number of sites available for cations. This opens up the possibility that Na could preferentially take up the external sites (either attracted by the negative charge or by the OH^- left behind by the H^+ incorporation). The Mg could then be imagined to be driven away from the high Cl regions towards the edges, where sites could exist for it. Again, there is no evidence for this imaginative mechanism, and the results described will be assessed more deeply soon.

These processes may be confined to sites with particularly high concentrations of sea salt, such as Dolleman Island, but have wider significance because high concentrations are seen in ice age ice throughout the polar regions. In addition, the apparent loss of MSA and perhaps nitrate seen on the shoulders of major changes in very old ice at Summit may be due to some related process, with ions possibly moving from or to preferred sites along the ice core (the apparent movement would be for nitrate and MSA to be sucked from acidic into alkaline ice regions). It is obviously crucial that any such movements are not misinterpreted in terms of major climate change.

7. Conclusion

Although it is tempting to assume that the ice core record is locked in place after a few metres depth, there are a few clear examples where this is not so. It is important to understand these cases, which may provide clues to what may happen to a lesser extent in other cases, confusing the desired record of atmospheric changes.

It seems likely that some species are generally found within the lattice structure, while others are found normally in the grain-boundary, which may in some cases be liquid. This should strongly affect the ability of species to take part in chemical reactions. The loss of H_2O_2 and changes in CO_2 concentration are two very important cases, and careful laboratory work is likely to be needed to assess what the reactions are and how fast they are. Reactivity within the ice may become more important as ice core researchers move on to measuring less common and less stable species that are important for atmospheric chemistry. Any work done on this will need to pay special attention to deriving rates that are sensitive to temperature, alkalinity/acidity, and any other factors that may vary from warm to cold periods; without such corrections, past chemistry of the atmosphere cannot be deduced.

Compared to the small number of laboratory measurements, diffusion rates actually seen in ice cores are surprisingly slow. Where annual cycles occur in the original record, they appear (with the exception of fluoride) to survive at least tens of thousand of years in the ice.

This is an encouraging result, which leaves the possibility that ice cores can tell us something about the season when a chemical change occurred, not just the gross long-term trends. This would be crucial for testing models of atmospheric chemical change.

There clearly are some cases where movement of species does occur, by some mechanism that is not true diffusion. These need to be understood so that we can be sure they are not important in some kinds of ice found in deeper cores. There are indications that some confusing movements might occur where different types of ice are in close proximity.

We have not considered the case (which appears to occur at Summit (Taylor et al. 1993; Alley et al. 1995)) where the chemical record has been altered by purely mechanical movements of ice (folding, boudinage, etc.). We have also not considered what may happen near the base of the ice sheet, where temperatures become much warmer (at some deep sites in Antarctica, such as Vostok, the melting point is reached). The warm temperatures and long timescales will of course increase diffusion rates, and any other temperature dependent processes. Additionally, if there is liquid at triple junctions between crystals, these veins will increase in size, and it may be possible for them eventually to drain from the ice sheet. We would expect the symptoms of such drainage to be preferential loss of sulphate and other species expected to be at boundaries. This should be looked for near the bed of ice sheets, and caution exercised in interpreting records near the bed.

Problems still to be understood clearly do affect some very interesting short-term records that have been obtained in coastal regions of Antarctica. Despite this, neither diffusion nor movement will have any effect on the long-term record (interpreted over years and metres of ice, and well above the bed), so that we can safely ignore them when looking at the broad picture of climate change over the last climatic cycle. The same is not true of the possible reactions mentioned earlier, so their study is the more crucial. We now have sufficient evidence to know that post-depositional changes do occur, even deep in the ice, but also to know that they seriously affect only a limited number of species. With a few relatively simple field and laboratory studies, this particular set of processes need not damage the magnificent record that ice cores offer us for interpreting past changes in climate.

Acknowledgements

I thank Richard Alley for his helpful and encouraging comments on this paper, and Robert Mulvaney for the use of unpublished data.

References

Alley RB, Perepezko JH, Bentley CR (1986a) Grain growth in polar ice: I. Theory. J Glaciol 32:415-424

Alley RB, Perepezko JH, Bentley CR (1986b) Grain growth in polar ice: II. Application. J Glaciol 32:425-433

Alley RB, Gow AJ, Johnsen SJ, Kipfstuhl J, Meese DA, Thorsteinsson T (1995) Comparison of deep ice cores. Nature 373:393-394

Anklin M, Barnola J-M, Schwander J, Stauffer B, Raynaud D (1995) Processes affecting the CO_2 concentrations measured in Greenland ice. Tellus 47B:461-470

Barnaal D, Slotfeldt-Ellingsen D (1983) Pulsed nuclear magnetic resonance studies of doped ice Ih. J phys Chem 87:4321-4325

Bekki S (1994) Denitrification mechanism of the polar winter stratosphere by major volcanic eruptions. J geophys Res 99:18871-18878

Camplin GC, Glen JW, Paren JG (1978) Theoretical models for interpreting the dielectric behaviour of HF-doped ice. J Glaciol 21:123-142

Clausen HB, Steffensen JP (1993) The chemical signature of two cold events in the Eemian. EOS Trans 74:84

Conklin MH, Sigg A, Neftel A, Bales RC (1993) Atmosphere-snow transfer function for H_2O_2: Microphysical considerations. J geophys Res 98:18367-18376

Dash JG, Fu H, Wettlaufer JS (1995) The premelting of ice and its environmental consequences. Reports on Progress in Physics 58:115-167

de Angelis M, Legrand M (1994) Origins and variations of fluoride in Greenland precipitation. J geophys Res 99:1157-1172

Delmas RJ (1993) A natural artefact in Greenland ice-core CO_2 measurements. Tellus 45B:391-396

Dominé F, Thibert E, Chaix L (1996) Interactions of gas phase HCl and HNO3 with ice. In: Wolff EW, Bales RC (eds) NATO ASI Series I: Chemical exchange between the atmosphere and polar snow. Springer-Verlag, Berlin (this volume)

Fuhrer K, Neftel A, Anklin M, Maggi V (1993) Continuous measurements of hydrogen peroxide, formaldehyde, calcium and ammonium concentrations along the new GRIP core from Summit, central Greenland. Atmos Environ 27A:1873-1880

Gross GW, Wu C, Bryant L, McKee C (1975) Concentration dependent solute redistribution at the ice/water phase boundary. II. Experimental investigation. J chem Phys 62:3085-3091

Gross GW, Gutjahr A, Caylor K (1987) Recent experimental work on solute redistribution at the ice/water interface. Implications for electrical properties and interface processes. J Physique 48(Mar Suppl. 3):C(1-533)

Haltenorth H, Klinger J (1977) Solubility of hydrofluoric acid in ice Ih single crystals. Solid State Commun 21:533-535

Hammer CU (1980) Acidity of polar ice cores in relation to absolute dating, past volcanism, and radio echoes. J Glaciol 25:359-372

Hammer CU, Clausen HB, Langway CC Jr (1994) Electrical conductivity method (ECM) stratigraphic dating of the Byrd Station ice core, Antarctica. Ann Glaciol 20:115-120

Herron MM (1982) Impurity sources of F^-, Cl^-, NO_3^- and SO_4^{2-} in Greenland and Antarctic precipitation. J geophys Res 87:3052-3060

Hobbs PV (1974) Ice Physics. Clarendon Press, Oxford

Hoffmann MR (1996) Possible chemical transformations in snow and ice induced by solar (UV photons) and cosmic irradiation (muons). In: Wolff EW, Bales RC (eds) NATO ASI Series I: Chemical exchange between the atmosphere and polar snow. Springer-Verlag, Berlin (this volume)

Johnsen SJ (1977) Stable isotope homogenization of polar firn and ice. In: IAHS Publication, 118: Isotopes and Impurities in Snow and Ice. IAHS, Grenoble

Laj P, Drummey SM, Spencer MJ, Palais JM, Sigurdsson H (1990) Depletion of H_2O_2 in a Greenland ice core: implications for oxidation of volcanic SO_2. Nature 346:45-48

Laj P, Palais JM, Gardner JE, Sigurdsson H (1993) Modified HNO_3 seasonality in volcanic layers of a polar ice core: Snow-pack effect or photochemical perturbation. J atmos Chem 16:219-230

Langway CC Jr, Osada K, Clausen HB, Hammer CU, Shoji H, Mitani A (1994) New chemical stratigraphy over the last millenium for Byrd Station, Antarctica. Tellus 46B:40-51

Legrand M (1996) Acidic gases (HCl, HF, HNO3, HCOOH, CH3COOH): a review of ice core data and some preliminary discussions on their air-snow relationships. In: Wolff EW, Bales RC (eds) NATO ASI Series I: Chemical exchange between the atmosphere and polar snow. Springer-Verlag, Berlin (this volume)

Mayewski PA, Legrand MR (1990) Recent increase in nitrate concentration of Antarctic snow. Nature 346:258-260

Moore JC, Paren JG, Oerter H (1992a) Sea salt dependent electrical conduction in polar ice. J geophys Res 97:19803-19812

Moore JC, Wolff EW, Clausen HB, Hammer CU (1992b) The chemical basis for the electrical stratigraphy of ice. J geophys Res 97:1887-1896

Moore JC, Reid AP, Kipfstuhl J (1994a) Microstructural and electrical properties of marine ice and its relationship to meteoric ice and sea ice. J geophys Res 99:5171-5180

Moore JC, Wolff EW, Clausen HB, Hammer CU, Legrand MR, Fuhrer K (1994b) Electrical response of the Summit-Greenland ice core to ammonium, sulphuric acid, and hydrochloric acid. Geophys Res Lett 21:565-568

Mulvaney R, Wolff EW, Oates K (1988) Sulphuric acid at grain boundaries in Antarctic ice. Nature 331:247-249

Mulvaney R, Pasteur EC, Peel DA, Saltzman ES, Whung P-Y (1992) The ratio of MSA to non-sea-salt sulphate in Antarctic Peninsula ice cores. Tellus 44B:295-303

Neftel A (1991) Use of snow and firn analysis to reconstruct past atmospheric composition. In: Davies TD, Tranter M, Jones HG (eds) NATO ASI Series G, 28: Seasonal Snowpacks. Compositional Change. Springer-Verlag, Berlin and Heidelberg

Neubauer J, Heumann KG (1988a) Determination of nitrate at the ng/g level in Antarctic snow samples with ion chromatography and isotope dilution mass spectrometry. Fres Z Anal Chem 331:170-173

Neubauer J, Heumann KG (1988b) Nitrate trace determinations in snow and firn core samples of ice shelves at the Weddell Sea, Antarctica. Atmos Environ 22:537-545

Nye JF (1991) Thermal behaviour of glacier and laboratory ice. J Glaciol 37:401-413

Prospero JM, Savoie DL, Saltzman ES, Larsen R (1991) Impact of oceanic sources of biogenic sulphur on sulphate aerosol concentrations at Mawson, Antarctica. Nature 350:221-223

Raynaud D, Jouzel J, Barnola JM, Chappellaz J, Delmas RJ, Lorius C (1993) The ice record of greenhouse gases. Science 259:926-934

Schwander J (1996) Gas diffusion in firn. In: Wolff EW, Bales RC (eds) NATO ASI Series I: Chemical exchange between the atmosphere and polar snow. Springer-Verlag, Berlin (this volume)

Sigg A, Neftel A (1988) Seasonal variations in hydrogen peroxide in polar ice cores. Ann Glaciol 10:157-162

Sigg A, Neftel A (1991) Evidence for a 50% increase in H_2O_2 over the past 200 years from a Greenland ice core. Nature 351:557-559

Sigg A, Staffelbach T, Neftel A (1992) Gas phase measurements of hydrogen peroxide in Greenland and their meaning for the interpretation of H_2O_2 records in ice cores. J atmos Chem 14:223-232

Stauffer B, Hofer H, Oeschger H, Schwander J, Siegenthaler U (1984) Atmospheric CO_2 concentration during the last glaciation. Ann Glaciol 5:160-164

Taylor KC, Hammer CU, Alley RB, Clausen HB, Dahl-Jensen D, Gow AJ, Gundestrup NS, Kipfstuhl J, Moore JC, Waddington ED (1993) Electrical conductivity measurements from the GISP2 and GRIP Greenland ice cores. Nature 366:549-552

Wagenbach D (1996) Coastal Antarctica: Atmospheric chemical composition and atmospheric transport. In: Wolff EW, Bales RC (eds) NATO ASI Series I: Chemical exchange between the atmosphere and polar snow. Springer-Verlag, Berlin (this volume)

Wagenbach D, Graf W, Minikin A, Trefzer U, Kipfstuhl J, Oerter H, Blindow N (1994) Reconnaissance of chemical and isotopic firn properties on top of Berkner Island, Antarctica. Ann Glaciol 20:307-312

Wolff EW (1995) Nitrate in Polar Ice. In: Delmas RJ (ed) NATO ASI Series I, 30: Ice core studies of global biogeochemical cycles. Springer-Verlag, Berlin

Wolff E, Mulvaney R (1990) Impurity distributions in ice under different environmental conditions (Abstract). Ann Glaciol 14:362

Wolff EW, Paren JG (1984) A two-phase model of electrical conduction in polar ice sheets. J geophys Res 89:9433-9438

Wolff EW, Mulvaney R, Oates K (1989) Diffusion and location of hydrochloric acid in ice: implications for polar stratospheric clouds and ozone depletion. Geophys Res Lett 16:487-490

Wolff EW, Moore JC, Clausen HB, Hammer CU, Kipfstuhl J, Fuhrer K (1995) Long-term changes in the acid and salt concentrations of the GRIP Greenland ice core from electrical stratigraphy. J geophys Res 100:16249-16264

FIRN PROPERTIES AFFECTING GAS EXCHANGE AT SUMMIT, GREENLAND: VENTILATION POSSIBILITIES

Mary R. Albert, Edward M. Arons, Robert E. Davis
Cold Regions Research and Engineering Laboratory
Hanover, N.H. 03755
USA

Introduction

The processes by which chemical species in the atmosphere become incorporated in the firn depend both upon the nature of the forcing from the atmosphere and upon the properties of the firn itself. These processes include both diffusion and advection (the transport of heat, vapor, and chemical species by air flow within the snow and firn). In this paper we present recent field measurements of firn properties relevant to the transport processes, and use simplified model calculations to investigate the possibility of advection at Summit.

Advective processes include the result of wind-driven air flow through the snow and firn, also called ventilation or wind-pumping. If present in the firn, the air flow would advect heat, mass, and chemical species through the firn at rates far greater than diffusive processes permit. In a theoretical analysis, Colbeck (1989) concluded that surface topography would be the main driver of ventilation, and suggested that the majority of the flow may be within the surface features themselves. It has been commonly assumed that the surface roughness present at Summit is insufficient to induce ventilation. In another theoretical analysis, Clarke and Waddington (1991) showed that pressure disturbances due to wind turbulence over flat terrain attenuate with depth, depending on both the temporal and spatial frequency content of the pressure signal. Albert and McGilvary (1992) showed that the temperature profile resulting from ventilation is controlled by a balance between heat carried by the dry air flow and heat conduction due to boundary temperatures.

Field Measurements

Field measurements of near-surface firn permeability, microstructure, temperature, tracer gas diffusivity, surface pressure variations due to the wind, and local meteorology were conducted at Summit in the summer of 1994. For this paper we focus on the measurements of the physical characteristics of the top meter of snow and firn in July that are relevant to transport processes.

NATO ASI Series, Vol. I 43
Chemical Exchange Between the Atmosphere and Polar Snow
Edited by Eric W. Wolff and Roger C. Bales

The snow pit was located in an area of undisturbed snow. Upon excavation, the stratigraphic layers were identified, density and permeability profiles were measured, grain size was estimated, photographs were taken of crystals, selected firn samples were preserved for later section analysis, and a portion of the pit was back-lighted to confirm stratigraphic horizons. Figure 1 depicts measured firn stratigraphy, density, and permeability. The average annual snow depth accumulation rate at Summit is approximately 60 cm; depth hoar layers from the current summer (22 to 32 cm depth) and the previous summer (65 to 81 cm) manifest themselves as lower-density regions.

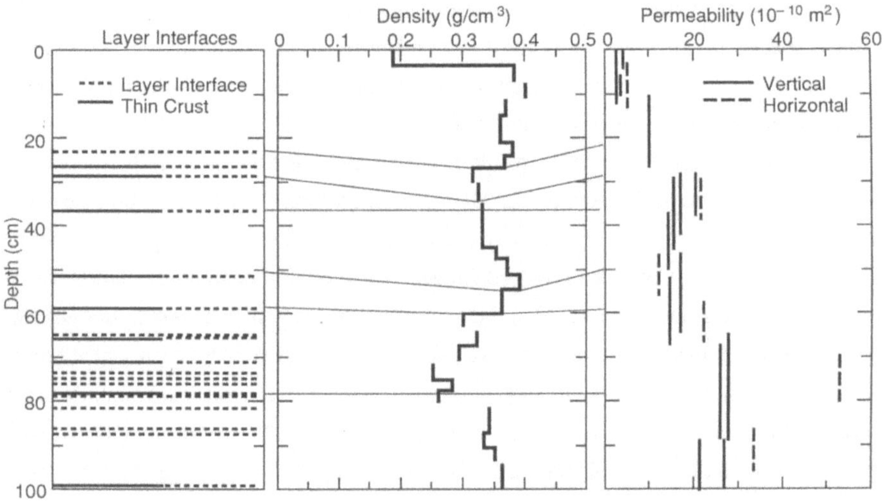

Figure 1. Layering, density, and permeability of the top meter of firn.

The permeability is the property of the firn that controls the amount of air that can flow through the firn for any given driving pressure difference. The permeability was deduced from Darcy's law using measured air flow rates and pressure drops through cylindrical snow samples. The permeameter was developed at CRREL (Chacho and Johnson 1987, Hardy and Albert 1993) and employs a concentric cylindrical head (Shimizu 1970) in order to minimize edge effects of the sample. From Figure 1, it is evident that permeability generally increases with depth over the top meter of firn. In this illustration, the length of the line on the depth scale represents the region of firn that was included in the experimental sample. The thick lines show vertical permeabilities, that is, the air flow was forced through the sample in a vertical direction relative to the original sample orientation in the firn. The dashed lines represent horizontal permeabilities. These permeabilities represent the first time that the directional nature of snow or firn permeability has been reported. The most striking directional difference in permeability is evident in the sequence of depth hoar layers below 70 cm depth. While the vertical permeabilities in this region are the highest of that type in the top meter of firn, the horizontal permeabilities in that region are as much as double the vertical. This provides opportunity for more lateral flow and transport in those layers. Future model simulations of firn transport processes will need to treat the permeability as a tensor.

Examples of the firn microstructure are depicted in Figure 2. Samples were collected in the field for preparation for sectioning in the cold laboratory, using the basic procedure described by Perla (1982). Measurements on the images consist of line counting statistics accumulated for each image and subimages, as described by DeHoff (1987).

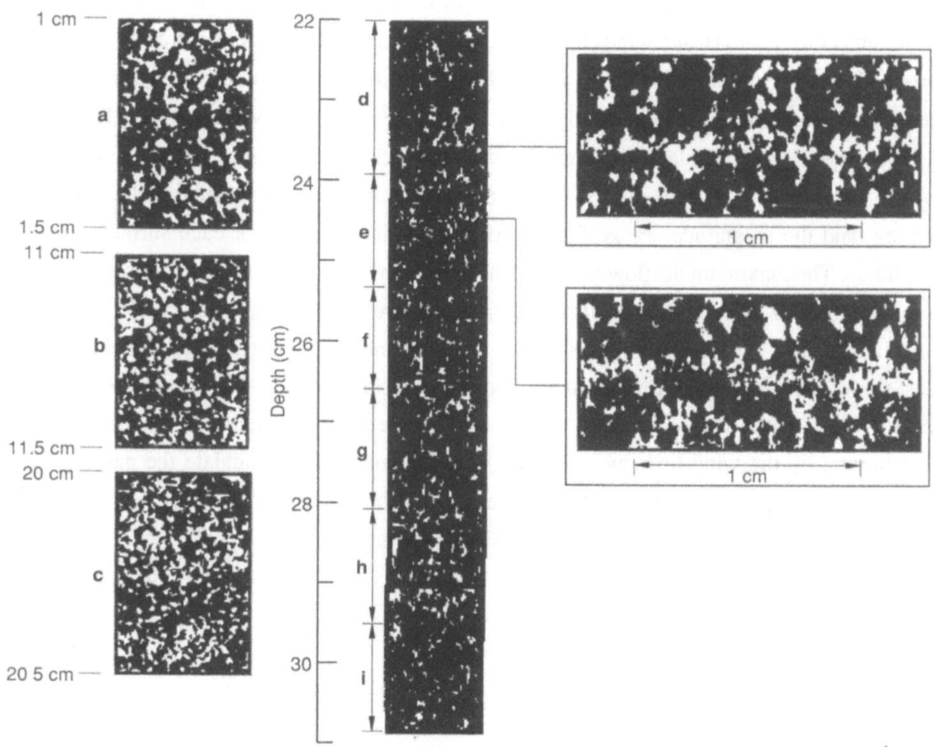

Figure 2. Microstructure of selected samples from the top 30 cm of firn. Mean pore intercepts (mm) are a: 0.27, b: 0.22, c: 0.21, d: 0.92, e: 1.0, f: 1.2, g: 1.1, h: 0.98, I: 1.3.

Crusts are identified on the sections as lines lighter than the surrounding material, and on close inspection are shown to comprise mostly single-grain layers. Most of the crusts that we identified occurred in the vicinity of the summertime depth hoar layers. The mean pore intercept distances were measured on test lines scanned horizontally across the subimages. In all cases the mean pore intercept of a crust is significantly smaller (by 36 %) than one or both of the layers above and below the crust, and the crust intercepts are similar to those of the surface wind pack. These stereological data strongly suggest that the crusts affect the vertical continuity of the firn pore space, which in turn will affect mass and vapor transport.

The permeability measurements discussed above raise the question of whether these permeabilities are sufficient to allow for the possibility of ventilation at Summit. While a detailed discussion of pressure measurements and complete modeling efforts is beyond the scope of this paper, we offer a simple model calculation to show that the measured permeabilities will allow air flow under reasonable assumptions of pressure forcing. The two-dimensional finite element model discussed in Albert (1993) is used to solve for two-dimensional Darcy air flow in firn. For this simplified analysis, we assume a steady state, sinusoidal surface pressure forcing with wavelength 1 m (estimated average wavelength) and peak amplitudes of 10 Pa (measured by Farrell and others (1966) with strong winds over a grassy flat area) and 0.15 Pa (measured by Albert and Arons (1994) at Summit with light (4 m/s) winds). In these calculations, the permeability is treated as a scalar that increases linearly from $5 \times 10^{-10} m^2$ at the snow surface to $30 \times 10^{-10} m^2$ at 1 meter depth. A rectangular finite element mesh is employed. The calculated air flow vectors follow two-dimensional paths through the snow in response to the sinusoidal surface pressure forcing, and the magnitudes range from zero to a maximum value for each surface pressure amplitude. The maximum air flow velocities at 1 m depth range from 30 mm/s for surface pressure amplitude of 10 Pa to 0.04 mm/s for surface pressure amplitude of 0.15 Pa. These calculations indicate that the firn at Summit is sufficiently permeable to permit air flow and advection at least through the top meter. Further analysis of 1994 surface pressure forcing data and refined measurements in the summer of 1995 will be used to determine whether there is sufficient pressure forcing by the wind to induce ventilation at Summit, and to elucidate the nature of the convective zone. Modeling efforts are being refined for the simulation of heat, mass, vapor, and chemical species transport through the firn (Albert 1995).

Conclusions

Recent field measurements of firn properties relevant to the transport processes at Summit, Greenland, have been presented. Firn permeability generally increases with depth over the top meter of firn. The directional nature of snow or firn permeability has been reported for the first time. Vertical permeabilities in the region of the previous year's depth hoar layers are the highest of that type in the top meter of the firn, and horizontal permeabilities in that region are as much as double the vertical, providing opportunity for more lateral flow and transport in those layers. The stereological data strongly suggest that the crusts affect the vertical continuity of the firn pore space, which will affect mass and vapor transport. Simplified two-dimensional finite element calculations indicate that the firn at Summit is sufficiently permeable to permit air flow and advection at least through the top meter. Further analysis of this year's data and next summer's experiments will further elucidate the nature of transport processes and microstructure evolution at Summit.

References

Albert MR, McGilvary WR (1992) Thermal effects due to air flow and vapor transport in dry snow. J Glaciol 38(129): 273–281

Albert MR (1993) Some numerical experiments on firn ventilation. International Symposium on Snow and Snow-Related Problems, Annals Glaciol 18: 161–165

Albert MR, Arons EM (1994) Transport processes in the firn at Summit, Greenland.

EOS Trans AGU 75(44): 213

Albert MR (in prep) Modeling heat, mass, and chemical species transfer in firn. To be presented at the International Symposium on Ice Sheet Modeling, Strasbourg, France, Sept 1995

Chächo EF Jr, Johnson JB (1987) Air permeability of snow. EOS, Trans AGU 68: 1271

Colbeck SC (1989 Air movement in snow due to windpumping. J Glaciol 35(120): 209–213

Clarke GKC, Waddington ED (1991) A three-dimensional theory of wind pumping. J Glaciol 37(125): 89–96

DeHoff RT (1987) The estimation of particle size distribution from simple counting measurements made on random plane sections. Trans AIME 233: 975

Farrell DA, Greacen EG, Gurr CG (1966) Vapor transfer in soil due to air turbulence. Soil Sci 102(5): 305–315

Hardy JP, Albert DG (1993) The permeability of temperate snow: Preliminary links to microstructure. Proc 50th East Snow Conf, Quebec City, Canada, 149–156

Perla R (1982) Preparation of section planes in snow specimens. J Glaciol 28(98): 199–204

Shimizu H (1970) Air permeability of deposited snow. Contributions from the Inst of Low Temp Sci, Series A 22: 1–32

INTERACTIONS OF GAS PHASE HCL AND HNO_3 WITH ICE

Florent Dominé, Emmanuel Thibert and Laurent Chaix
CNRS, Laboratoire de Glaciologie et Géophysique de l'Environnement
BP 96, 54 rue Molière
38402 St Martin d'Hères Cedex
France

INTRODUCTION

The analysis of ice cores from polar ice caps has the potential to yield information on the evolution of atmospheric composition during the past 400,000 years (Legrand, 1995). Before such information is obtained, however, we need to understand the relationship linking atmospheric composition to snow composition, i.e. the transfer function. For gases that do not interact with ice, such as CH_4, the transfer function is simple, and the CH_4 mixing ratio in air bubbles trapped in ice is the same as in the atmosphere (Chappellaz et al., 1990). Gases such as HCl and HNO_3, however, interact strongly with ice, and occluded gases contribute to the chloride or nitrate content of ice in a negligible manner. These compounds are then mainly included in the ice volume, but their mechanism of incorporation is not understood. HCl is an indicator of sea salt fractionation. Its mixing ratio yields information on atmospheric transport and acidity. HNO_3 is the final stage of oxidation of NO_y species. Its determination is useful to understand the NO_y budget and the oxidizing capacity of the atmosphere.

To contribute to our understanding of the HCl and HNO_3 transfer functions, we have performed laboratory experiments to study the adsorption, diffusion, and solubility of these acids in ice. The data obtained are then used to attempt to understand the HCl and HNO_3 content of summer Greenland snow.

EXPERIMENTAL METHODS AND RESULTS

HCl adsorption on ice was measured at 195 K using the apparatus described in Figure 1. Equal pressures of HCl are admitted in volumes V_1 and V'_1. The pressure in V'_1 is measured with a mass spectrometer and/or a high precision capacitance manometer. Volume

NATO ASI Series, Vol. I 43
Chemical Exchange Between the Atmosphere and Polar Snow
Edited by Eric W. Wolff and Roger C. Bales

V_2 is immersed in a cooled ethanol bath. Its walls have been coated with ice formed by stirring pure degassed water under vaccuum in it while in a cold room at -34°C . The valve between V_1 and V_2 is opened, exposing the ice to HCl and resulting in adsorption. The valve is closed after allowing 40 minutes for equilibration and the pressure in V_1 is measured with the mass spectrometer. A molar budget yields the amount of adsorbed gas, and adsorption isotherms are deduced. A fresh ice sample was used for each data point in order to limit diffusion of HCl into the ice, which would result in an apparent increased adsorption. The surface area of the ice was about 340 cm^2, as determined by the dimensions of V_2. Only preliminary data have been obtained at the moment, and a typical adsorption isotherm is shown in Figure 2.

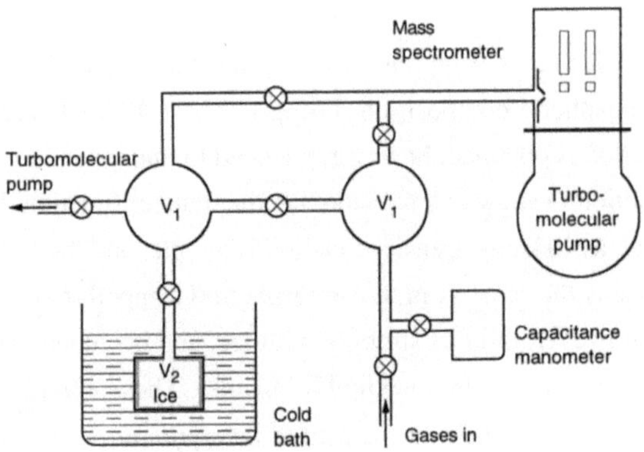

Figure 1. Apparatus used to study HCl adsorption on ice.

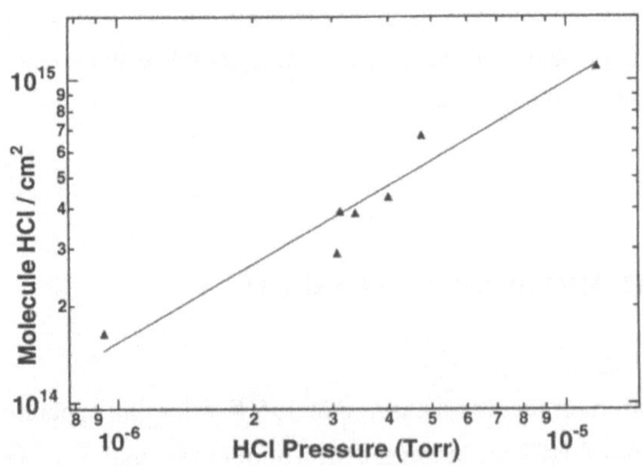

Figure 2. Adsorption isotherm of HCl on ice at 195 K (-78°C).

The diffusion and solubility of HCl and HNO_3 in ice were measured as described in Dominé et al. (1994). Briefly, laboratory-grown cylindrical ice single crystals were exposed to a flow of HCl or HNO_3 diluted in water-saturated nitrogen. The ice crystals were then sliced in sections 25 μm thick which were melted and analyzed by ion chromatography.

Analysis of successive slices gave diffusion profiles, which were fitted to equation (1) :

$$X(x,t) = X_0 \left[1 - erf\left(\frac{x}{2\sqrt{Dt}} \right) \right] \tag{1}$$

where $X(x,t)$ is the trace gas mole fraction at a distance x from the surface after a diffusion time t, X_0 is the equilibrium solubility, D is the diffusion coefficient, and *erf* is the error function. The fits yielded D and X_0 values. Experiments were done under various conditions of temperature (T) and partial pressure (P_{HCl} or P_{HNO3}). For HNO_3, only a few data points were obtained, at -15°C: two with $P_{HNO3}=4\times10^{-5}$ Torr, yielding $X_{HNO3}=2.0\times10^{-7}$ (700ng/g) and $X_{HNO3}= 2.25\times10^{-7}$, and one with $P_{HNO3}=8\times10^{-5}$ Torr, yielding $X_{HNO3}=2.86\times10^{-7}$ (1000 ng/g). D_{HNO3} is about 10^{-10} cm^2s^{-1} at -15°C. The HCl data show that D_{HCl} is in the range of 10^{-12} cm^2s^{-1}, increasing slightly with T over the range studied.

The HCl solubility data allowed the determination of the pressure and temperature dependance of HCl solubility in ice. Equation (2)

$$X_{HCl} = 3.8\times10^{-9}\, e^{\frac{2845}{T}}\, P_{HCl}^{1/2.65} \tag{2}$$

links X_{HCl} to P_{HCl} (in Torr) and T. The resulting solubilities are shown in Figure 3.

DISCUSSION AND APPLICATION TO THE ATMOSPHERE

The adsorption data presented in Figure 2 are similar to those that Chu et al. (1993) obtained at 193 K using a dynamic method with a flow tube reactor. Hanson and Ravishankara (1992) and Abbatt et al. (1992) also studied HCl adsorption on ice and found similar surface coverages, but they observed no surface coverage dependence on P_{HCl}. Our data are too preliminary to dicuss the differences between these various studies. An important point, however, is that our work confirms that around 195 K, for P_{HCl} of about 10^{-6} Torr, the HCl surface coverage on ice is of the order of one monolayer (about 6×10^{14} molecule cm^2).

Application of these data to the polar troposphere is difficult because we have not yet obtained HCl adsorption data at relevant temperatures. Dominé et al. (1995), however, studied the composition of fresh and aged summer Greenland snow and discussed the possibility that HCl was located on the snow surface. From the observation that snow aging resulted in an increased HCl concentration and in a decreased surface area, they concluded that HCl in summer Greenland snow, at a temperature around -15°C, was not located on the snow surface. We plan to measure HCl adsorption on ice at -15°C to test this suggestion, but in the absence of such measurements, we assume that the snow surface does not contribute significantly to the HCl content of Greenland snow.

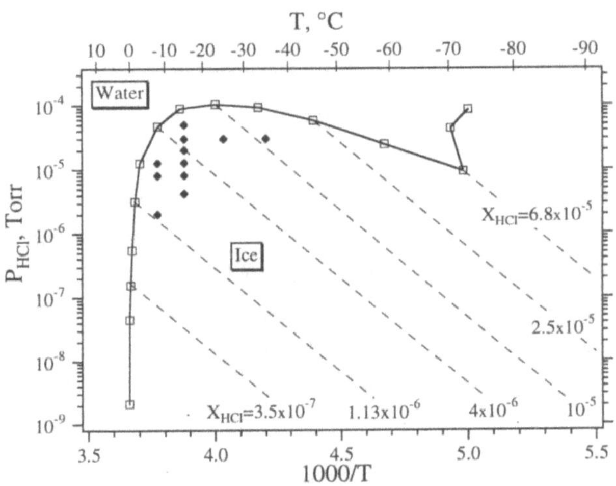

Figure 3. Iso-solubility curves for HCl in ice, as a function of P $_{HCl}$ and inverse temperature. (◆) : experimental conditions. The solid line is the ice-water boundary.

The HCl content of Greenland snow then has to be explained in terms of volume content. HCl is then mostly located in the ice crystal lattice in fresh unrimed snow and possibly at grain boundaries in the firn. Typical Greenland values measured for fresh unrimed snow were $X_{HCl}=5x10^{-9}$ (≈10 ng/g) while P_{HCl} was $1.3x10^{-8}$ Torr at ground level and T was about -15°C (Dominé et al., 1995). Figure 3 and equation (2) indicate that X_{HCl} in fresh Greenland snow is much lower than the equilibrium value of $2.5x10^{-7}$. HCl incorporation in Greenland snow may then have been ruled by kinetics and could be explained by considering the number of sticking collisions at the surface of the forming crystal. If this is the case, X_{HCl} could be predicted by:

$$X_{HCl} = \frac{P_{HCl}\ \alpha_{HCl}}{P_{H2O}\ \alpha_{H2O}} \sqrt{\frac{M_{H2O}}{M_{HCl}}} \qquad (3)$$

where α is the sticking coefficient (0.2 for HCl and 0.5 for H_20, Dominé et al., 1995, and references therein) and M the molecular mass. P_{H2O} is the saturating vapor pressure of ice at -15°C and is 1.24 Torr. P_{HCl} in the free troposphere where the snow formed was probably higher than at ground level and we use $P_{HCl} = 2.6 \times 10^{-8}$ Torr. Equation (3) then predicts $X_{HCl}=5.9 \times 10^{-9}$ (≈12 ng/g) which is similar to the measured value of $X_{HCl}=5 \times 10^{-9}$.

Our laboratory measurements show that HNO_3 equilibrium solubility in ice is much lower than that of HCl. Typical Greenland measurements on fresh snow (Silvente and Legrand, 1995) are $X_{HNO3}=3 \times 10^{-8}$ (≈105 ng/g) for $P_{HNO3}= 3.9 \times 10^{-9}$ Torr at ground level, for T=-15°C. Despite its lower solubility and lower vapor pressure, HNO_3 is much more concentrated in Greenland snow than HCl. Assuming, as for HCl, that HNO_3 is not located on the surface of Greenland snow (which still has to be tested), a calculation following equation (3), using $\alpha_{HNO3}=0.2$ shows that HNO_3 incorporation is not determined by collision kinetics. We propose than HNO_3 in Greenland fresh snow is in equilibrium with the atmosphere. A confirmation of this speculative suggestion of course requires more laboratory measurements. However, it is consistent with our observation that the diffusion coefficient of HNO_3 in ice is 100 times greater than that of HCl, which may have allowed equilibration during the growth and fall of snow.

REFERENCES

Abbatt JPD, Beyer KD, Fucalora AF, McMahon JR, Woolridge P.J, Zhang R, Molina MJ (1992) Interactions of HCl with water-ice: implications for the stratosphere. J. Geophys. Res. 97: 15819-15826.

Chappellaz J, Barnola JM, Raynaud D, Korotkevich YS, Lorius C (1990) Ice-core record of atmospheric methane over the past 160,000 years, Nature 345: 127-131.

Chu LT, Leu MT, Keyser LF (1993) Uptake of HCl in water ice and nitric acid films, J. Phys. Chem. 97: 7779-7785.

Dominé F, Thibert E, Van Landeghem F, Silvente E, Wagnon P (1994) Diffusion and solubility of HCl in ice: preliminary results. Geophys. Res. Lett. 21: 601-604.

Dominé F, Thibert E, Silvente E, Legrand M, Jaffrezo JL (1995) Determining past atmospheric HCl mixing ratios from ice core analyses. J. Atmos. Chem. (In press).

Hanson DR, Ravishankara AR (1992) Investigation of the reactive and nonreactive processes involving $ClONO_2$ and HCl on water and nitric acid doped ice. J. Phys. Chem. 96: 2682-2691.

Legrand M (1995) Ice core chemistry: implications for our past atmosphere, In: NATO ASI Series: Low-temperature chemistry of the atmosphere. (ed. by G. K. Moorgaat et al.) 421-445, Springer-Verlag, Berlin.

Silvente E, Legrand M (1995) A preliminary study of the air-snow relationship for nitric acid in Greenland. In: NATO ASI, Ice core chemistry of global biogeochemical cycles, (ed by R.J. Delmas) 233-248, Springer-Verlag, Berlin.

CLIMATE AND ATMOSPHERIC TRACERS MODELLING WITH GCM, POLAR APPLICATIONS

Christophe Genthon, Alexandre Armengaud, and Gerhard Krinner
Laboratoire de Glaciologie et Géophysique de l'Environnement
(CNRS and Université Joseph-Fourier)
Domaine Universitaire BP96
F-38402 Saint Martin d'Hères Cedex
France

Introduction.

The global distribution of atmospheric constituents (wether major or minor species, gas or aerosol, or even energy and momentum) is determined by the distribution of the sources, sinks, and conversion within the atmosphere, and by transport and mixing by the atmospheric circulation. General circulation models (GCMs) simulate, on a global scale, the transport and mixing of air, heat, moisture and momentum. All numerical formulations for these processes should be universal, but of course they are not, or otherwise one would never have to use the terms "parameterization" and "tuning". However, if a process parameterization is physically-based, it should be usable for all species equally affected by this process. Implementing transport and mixing of atmospheric species within a GCM should thus be fairly straightforward.

GCMs are useful for the statistics of climate (means, diurnal, seasonal and stochastic variability...) they produce. Therefore, unlike an off-line Chemical Tracer Model (CTM), a tracer model developed in-line with a GCM is also limited to providing statistics of tracer distributions. Yet, the quantity and time density of climate data produced by a GCM is higher than the content of any data-base used to drive CTMs. The information might be questionable, yet it can all be used to drive tracers, and formulations for tracer processes can therefore be very detailed. As an example, the simulation of rain by GCMs is often moderately successful, yet no data-base or climatology provides the frequencies and vertical distribution of precipitation formation and re-evaporation as GCMs do. Implementing tracers within GCMs also helps study, understand and improve the GCMs. Shortcomings in tracers simulation can be due to deficiencies in either tracers modelling or climate modelling. With simple enough tracers, e.g. tracers only significantly affected by transport and mixing, and provided that the formulations set

NATO ASI Series, Vol. I 43
Chemical Exchange Between the Atmosphere and Polar Snow
Edited by Eric W. Wolff and Roger C. Bales

up for tracers resemble the formulations used for climate variables, then tracer results might help identify and characterize GCM deficiencies (e.g., Genthon and Armengaud [1995a]).

The longer the transport range is, the more intimate the relations between tracers and the atmospheric circulation are, and to some extent the more sensitive a model result is to modelling deficiencies. Most species which reach the interior of polar ice sheets, or even the coasts for Antarctica, have to travel long distances. Polar ice sheets are therefore adequate observatories for large scale air monitoring, not only in the real world but also in the models. In addition, because the snow surface acts as an automatic sampler and preserver of atmospheric species, long term records of atmospheric composition can be retrieved by firn and ice analyses. This provides the statistics which can be compared with model results, provided that the ice-atmosphere transfer function, which is certainly not trivial, is adequately simulated.

The development and results of two purely physical (no chemistry involved) tracers/climate models based on existing GCMs are briefly described here, with emphasis on the polar ice sheets, some aspects of the ice-atmosphere transfer function, and model shortcomings. A third model is currently being set-up on the basis of a new GCM, the characteristics of which are also presented.

Dust, sea-salt, ^{222}Rn and ^{210}Pb in the GISS and/or LMD GCMs.

The tracers/climate models. Numerical formulations of boundary layer turbulent diffusion, convective mixing and large scale transport of atmospheric species have been implemented in the Goddard Institute for Space Studies (GISS, NASA, New York) and the Laboratoire de Météorologie Dynamique (LMD, CNRS, Paris) GCMs. The formulations, which are fully described in Genthon and Armengaud [1995a], are largely inspired by the formulations used for the climate variables in the base GCMs, except for transport for which a scheme with better performances (in particular, with little spurious numerical diffusion) is implemented. Turbulent diffusion, convective mixing and large scale transport are considered universal and therefore apply to all the tracers we tentatively simulate. On the other hand, sources and sinks, precipitation scavenging, and some aspect of dry deposition are tracer dependent. For these, specific parameterizations are set up, or at least parameters are adjusted separately for each tracer.

Radon 222. ^{222}Rn is an inert natural gas, and its only significant sink is radioactive decay with a period ($\tau_{1/2}$=3.8 days) typical of the life time of tropospheric aerosol. Figure 1 shows the simulated and observed mean seasonal cycle in the surface atmosphere at Bombay (India) in the tropics and at Dumont d'Urville, the French station at the coast of Terre Adelie, Antarctica. These are not locations where the models are more successful, but these are cases particularly representative of large divergence between the two models and difference with the observation, which are most probably due to deficiencies in climate modelling rather than tracers modelling. GISS does very poorly at Bombay, but the model horizontal resolution is much coarser than that of LMD in the tropics. At Dumont d'Urville, where resolution is similarly coarse, the two models are about equally wrong. However, not only both climate models apparently fail to simulate some aspects of the atmospheric circulation which strongly impact on ^{222}Rn (i.e., highest concentration in local summer), but they do not even fail in the same way. ^{222}Rn is a good example of a tracer appropriate for testing the atmospheric circulation in climate models.

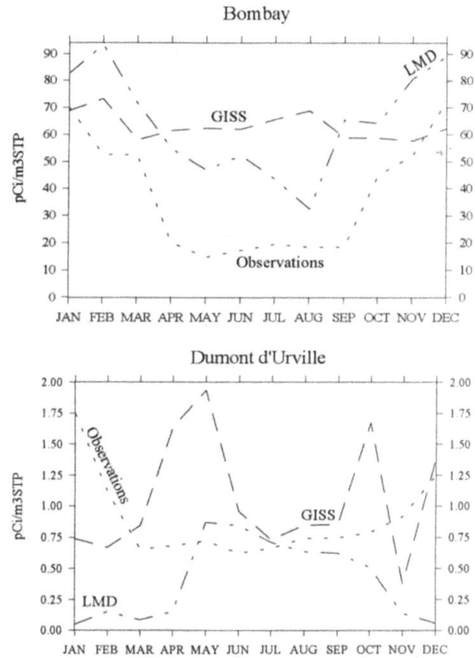

Figure 1. The seasonal cycle of ^{222}Rn at Bombay (India) and Dumont d'Urville (Antarctica), as observed and simulated by the GISS and LMD tracer models.

Dust and sea salt. For mineral dust and marine salt, parameterizations of sources (arid regions for dust, ice-free ocean for sea-salt), wet scavenging, and dry deposition had to be developed and implemented in the GISS GCM, in addition to transport and mixing. The formulation for precipitation scavenging is straightforward, yet it intimately interacts with the modeled precipitation. For any GCM simulated condensed water, formed (condensation scavenging) or sweeping through (impaction scavenging) a given grid box, the quantity of tracer removed Q_r is related to the quantity Q_i initially in the box by a decreasing exponential function of the precipitation rate P:

$$Q_r = Q_i\ [1. - \exp(-\alpha.P)]$$

α is an efficiency coefficient, set empirically according to precipitation type (solid or liquid), scavenging mode (nucleation or impaction) and particle size and type [Genthon, 1992]. The grounds and advantages of this formulation, and the loose setting of efficiency parameters on the basis of measurements during temperate rain events are discussed in Genthon [1992]. By lack of appropriate observation, α is not specifically adjusted for the particular characteristics of precipitation over polar ice sheets, and the simulated scavenging ratios turn out to be larger than observed.

Dry deposition is by turbulent vertical mixing and sedimentation. A bulk aerodynamic formulation with superimposed gravitational settling is used to estimate the atmosphere-to-surface flux ϕ:

$$\phi = \rho_a C_d V_s (Q_s - Q_g) + \phi_g$$

where the atmospheric parameters ρ_a (density of surface air), C_d (drag coefficient) and V_s (surface wind speed) are provided by the GCM. Q_s and Q_g are the tracer atmospheric mixing ratio a few meters above the surface and at ground level, respectively. In the GISS model, Q_g is set to 0, which means that any particle which reach ground level is instantly and irremediably captured by the surface, probably inducing dry deposition to be to high for very fine particles for which sedimentation is very slow (this point is discussed for ^{210}Pb, next). In fact, near the ground, a large fraction of dust and sea-salt reach the surface by gravitational sedimentation, the corresponding flux ϕ_g, being calculated with Stokes law for spherical particles.

The model reproduces the present day seasonal cycle of atmospheric dust and sea-salt at South Pole, but fails to simulate a significant increase of concentration in the snow during an ice age (Genthon [1992]; Andersen [1994] obtains a similar negative result for Greenland). Because the modern source regions were used in the ice age simulation, it is tempting to conclude that the observed increase of dust deposition is mostly due to a migration and/or an extension of the sources. However, although some aspects of the sources of sea-spray could have been overlooked (e.g., within the ice pack), the fact that the model does not simulate an increased salt deposition raises questions about the modelling of either tracers or climate. There are good reasons to have limited confidence in some tracer formulations (particularly deposition), yet ^{222}Rn now suggests that the simulated atmospheric dynamics is not to be trusted either, even for modern climate.

Lead 210. ^{222}Rn decays in ^{210}Pb which attaches on available atmospheric particles (typically sub-micron sulphates) immediately after formation, so that ^{210}Pb behaves like an inert aerosol. The parameterizations of wet removal and dry deposition developed for dust and sea-salt are also used for ^{210}Pb in the GISS model, and similar parameterizations are set up in the LMD model [Genthon and Armengaud, 1995b]. The gravitational component of dry deposition is neglected because the particles traced by ^{210}Pb are small. The other component, turbulent diffusion as simulated by the GCMs, results in deposition velocities on the ice sheets which are an order of magnitude too high and surface concentrations conversely too low. Very near the surface where turbulence is weak, if sedimentation is negligible, then the transfer is by particle (or Brownian) diffusion (unless another process like impaction, snow ventilation, fog deposition... takes over). Particle diffusion is now implemented in the LMD model, and dry deposition occurs in two steps:

$\phi_{si} = \rho_a C_d V_s (Q_s - Q_i)$

(Bulk aerodynamic)

and

$\phi_{ig} = \rho_a K_B (Q_i - Q_g) / \Delta z$

(Brownian diffusion)

Level i is the interface between the turbulent and Brownian layers. K_B is the Brownian diffusion coefficient and Δz the Brownian layer width. In steady state, the atmosphere-to-interface (ϕ_{si}) and interface-to-ground (ϕ_{ig}) fluxes are equal, and the system can be solved for Q_i (mixing ratio at interface). The second step generally strongly limits deposition.

Figure 2 shows the monthly mean surface atmospheric concentration at South Pole (Antarctica) and Summit (Greenland). Because particle dry deposition is controlled by turbulence in the GISS model, deposition is much faster and concentration much too low.

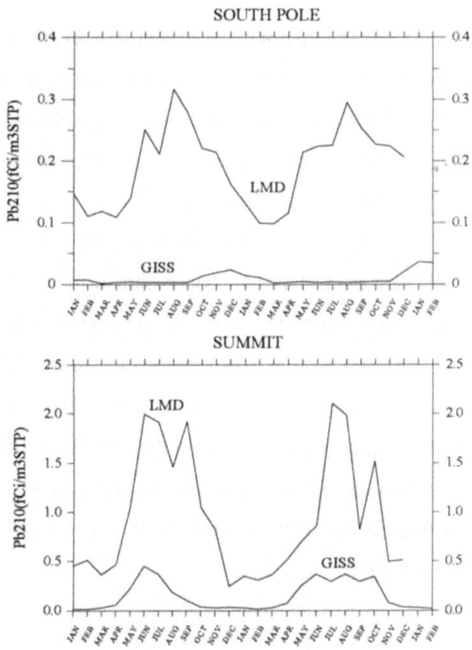

Figure 2. LMD and GISS simulated surface concentration of ^{210}Pb at South Pole (Antarctica) and Summit (Greenland).

A new tracers/climate model.

For tracer studies over the ice sheets, one must first gain confidence in the GCM performances in the polar regions. If necessary, the climate model must be improved to an acceptable level before or while tracers are simulated. Spatial resolution is an important aspect of climate modelling over ice sheets [Genthon et al., 1994], and resolution limitation is one reason why, rather than attempt to improve further the versions of the GISS and LMD climate models described so far, we have decided to develop a new tracers/climate model on the basis of a new GCM.

In fact, the base GCM is a new version of the LMD climate model which allows grid stretching (or "zooming") over a given region of the world. For pure climate studies over the Antarctic region, we aim at reaching a meridional grid step of 0.5° over the ice sheet with a global number of grid points small enough to make multi-year simulations possible at an affordable computing cost. The mean resolution should be about 4° and the maximum grid step (in the northern hemisphere for a "zoom" over Antarctica) should not reach beyond 10°. Work is in progress to set up the new grid, and analyze and compare the circulation over and around the ice sheet both in regular and stretched resolution [Krinner, 1994]. The model performance is very good over the "zoomed" region. However, because the GCM had never been used at such high resolution before, some of its components need to be revised (e.g., flux computations over sea ice).

Meanwhile, but so far in regular grid only, the tracer couple $^{222}Rn/^{210}Pb$ is being used to implement and test the new tracers/climate model. The same tracer formulations as in the former LMD model are set up in the new one, to the extent that the corresponding formulations for climate processes (e.g., turbulence, convection) are the same in the old and new GCMs. The model will also be used with locally refined grid, but not necessarily with the finest resolution over the ice sheets. In some cases, it might be more appropriate to set the finest grid somewhere along the pathway of transport toward the polar regions, or even in the source regions if resolution is particularly necessary there (e.g., for dust uptake). In fact, with the new model, it is even theoretically possible to zoom over several regions in one same run.

Conclusions and perspectives.

Modelling atmospheric tracers within GCMs is a double challenge since results can be equally sensitive to deficiencies of the tracer formulations and to shortcomings of the simulated climate. In fact, simple tracers, when implemented within GCMs, provide diagnostics useful to identify and understand deficiencies of the simulated climate. For other more complex species, for instance aerosols, one should have confidence in the GCM performances, or at least one should be aware of its intrinsic flaws in order to adequately interpret tracers results. To take full advantage of the approach, one should be equally interested in climate modelling and in tracers modelling.

Studying tracers over the polar ice sheets adds to the difficulty. Because tracers must be transported over long distances, they are more likely to be affected by climate or tracer modelling deficiencies. On the other hand, snow and ice analyses provide long term information on past atmospheric environment, and GCMs are the most appropriate tools to study the components of the atmospheric circulation in different climate, and relations with the distribution of atmospheric impurities. In addition, modelling tracers within GCMs leaves wide open the possibility of studying the impact of atmospheric minor constituents and their global distribution on climate. However, this adds one more layer of uncertainty which we are not yet ready to address.

Indeed, particularly in the polar regions and even with simple tracers, our results so far are moderately successful. On-going work aims at improving both the climate and tracer aspects of the problem, and this is being done through the development of a new and promising stretchable grid model. Some of the tracer formulations used so far are very empirical and not validated for the polar regions (wet deposition). Others may be incomplete, or a dominant aspect may not be adequately taken into account (dry deposition). In principle, new results from experimental programs can be tested within a GCM-based tracer model. For instance, provided that a climate model can reproduce surface fog events (it theoretically should), then fog deposition of particles can be tentatively formulated and evaluated. Given the scales involved, and the fact that only parameterized microphysics can be handled in GCMs, it is unlikely that formulations of particle deposition can be developed from theory alone. Rather, on-going and future field investigations are expected to provide more hints on the processes at work in the real world, which can eventually be implemented in models for large scale testing.

REFERENCES

Andersen K. (1994) Atmospheric dust in relation to climate and modelling. Master Thesis, Dept. of Geophysics, University of Copenhagen, Denmark, 100 p.

Genthon C. (1992) Simulations of desert dust and sea-salt aerosols in Antarctica with a general circulation model. Tellus 44: 371-389.

Genthon C., J. Jouzel, and M. Déqué (1994) Accumulation at the surface of polar ice sheets: Observation and modelling for global climate change. In "Global Precipitations and Climate Change", edited by M. Desbois and F. Desalmand, NATO ASI Series I, Springer Berlin Heidelberg New York, Vol. 26: 53-76.

Genthon C., and A. Armengaud (1995a) Radon 222 as a comparative tracer of transport and mixing in two general circulation models of the atmosphere. J. Geophys. Res. 100: 2849-2867.

Genthon C., and A. Armengaud (1995b) GCM simulations of atmospheric tracers in the polar latitudes: South Pole (Antarctica) and Summit (Greenland) cases. Sci. Total Environ. 160/161: 101-116.

Krinner G. (1994) Contribution à la mise au point d'une grille localement raffinée dans un modèle de circulation générale de l'atmosphère. Internal LGGE report (in French), Laboratoire de Glaciologie et Géophysique de l'Environnement, Universtite Joseph-Fourier, Grenoble, France, 35p.

ATMOSPHERIC RESIDENCE TIMES INFLUENCE ON TRACER CONCENTRATIONS IN REMOTE POLAR AREAS

Margareta Hansson
Department of Meteorology
Stockholm University
S-106 91 Stockholm
Sweden

INTRODUCTION: Records from polar ice sheets contain invaluable information about past variations in the climate and the composition of the atmosphere. Past temperatures and precipitation rates can be deduced from stable water isotope records. Soluble (ion) and insoluble (dust) compounds forming the atmospheric aerosol are scavenged by the precipitation from the air mass above the ice sheet, or dry-deposited on the ice sheet surface, and become incorporated in the ice. Concentration records of ions and dust from ice cores can reveal the past aerosol composition and concentration in the atmosphere provided that the air-snow transfer functions are known. These are, at present, poorly known and the topic of this workshop. However, the atmospheric aerosol has a high spatial variability due to short residence times in the atmosphere of most compounds forming the aerosol. This complicates any attempt to extrapolate information from polar ice cores to airborne concentrations on a global scale.

The airborne concentration of any tracer in the air mass above the ice sheet is dependent on the strength and location of the source, the atmospheric residence time of the tracer and the transport time of the air parcel from the source area to the deposition site. The pattern of general circulation of the atmosphere determines which source areas influence the deposition site. Tracer concentration in the ice is dependent on the airborne concentration and the local total scavenging (by wet and dry deposition). If ice-core tracer concentrations representing such different climatic conditions as glacials and interglacials are compared, possible changes in all these parameters have to be taken into account.

Increased ice-core tracer concentrations in glacial stages have often been attributed to expanded source areas and increased production rates however this hypothesis has not

NATO ASI Series, Vol. I 43
Chemical Exchange Between the Atmosphere and Polar Snow
Edited by Eric W. Wolff and Roger C. Bales

been verified. General circulation models fail to reproduce expanded source areas and increased production rates for dust that could account for the observed increases (Joussaume, 1993). The residence time in the atmosphere of a tracer is determined by wet and dry deposition and chemical transformation processes along the transport path to the deposition site. The residence time is probably longer in a glacial atmosphere as an effect of a lower water-vapour capacity in a colder atmosphere and, hence, a lower wet scavenging rate of the tracer. The transport time of an air parcel is probably shorter in a glacial atmosphere due to higher mean wind speeds, provided that the distance to the influencing source area is identical. A change in the pattern of the general circulation (Anderson et al., 1988; Joussaume, 1993) may alter the effective distance to the source area or make other source areas influence a specific receptor area. Both increased residence times and decreased transport times will enhance the airborne concentrations in the remote polar areas.

Any change in source strength and location will affect the tracer concentration individually for each tracer, while changes in the physical conditions of the atmosphere can be expected to influence all tracers systematically. Taking advantage of recent multi-ion ice-core analyses, where many different sources and source types are represented, may reveal systematic changes of ice-core tracer concentrations with changes in the climate. A simple model (Hansson, 1994; 1995) is formulated to test the effects of changes in residence times and transport times (two parameters which can respond quickly to climatic changes) while keeping source areas and production rates constant. The continuous concentration profiles of dust and nine ions in the period 10-120 ka B.P. and in the 19th century, from the Renland ice core, East Greenland (Hansson, 1994), is used in this test.

MODEL: The relative airborne concentration (the reference state is denoted by an asterisk) at Renland in different climatic stages is calculated from:

$$C_{air}/C^*_{air} = C(0)_{air}/C(0)^*_{air} \exp[(1-n/k)\, t^*/\tau^*] \qquad (1)$$

where $C(0)_{air}$ is the airborne concentration at the source and thus reflects the source strength, n is the relative transport time ($n=t/t^*$) and k is the relative residence time ($k=\tau/\tau^*$). Relative concentrations make it easy to compare tracers of very different concentration ranges. It also makes real numbers of the air-snow transfer functions unnecessary, although knowledge about how they may vary with the climate is desirable. If dry deposition is negligible, changes in the concentration in the ice can be approximately related to changes in the airborne concentration at the deposition site according to:

$$C_{ice}/C^*_{ice} = (C_{air}/C^*_{air})\,(W/W^*) \qquad (2)$$

where W is the local scavenging ratio, and the total deposition flux can be approximated with the wet deposition flux Φ_{wet} which changes can be calculated from:

$$\Phi_{wet}/\Phi^*_{wet} = (C_{air}/C^*_{air})\,(W/W^*)\,(P/P^*) \qquad (3)$$

where P is the local precipitation rate. If wet deposition is negligible, the concentration in the ice is inversely dependent on P, while the total deposition flux (approximated with the dry deposition flux) is independent of P. Wet deposition is the dominant process on Greenland under present climatic conditions (Davidson et al., 1993) but dry deposition may be important under colder climatic conditions.

The average residence time for wet scavenged tracers is inversely proportional to the precipitation rate which may vary by different degrees along the transport path. The precipitation rates will vary most at high latitudes and more over the continents than over the oceans. P on Renland was lower by roughly a factor-of-five (Johnsen et al., 1992) during the last glacial maximum (LGM). An overall decrease of the precipitation rates by a factor-of-five along the transport path corresponds to a factor-of-five increase of the residence time (τ) which gives a relative residence time (k) of 5. If the mean wind speeds were doubled (S. Harrison, 1994, personal communication) during the same time, the transport times (t) may have been reduced by a factor-of-two which gives a relative transport time (n) of 0.5. This gives an estimated variation of the ratio n/k from 1 in the present climate to a minimum of 0.1 during LGM. The gradual change of k with climate is, in this model simulation, set inversely proportional to the accumulation rate at Renland (which is exponential related to the $\delta^{18}O$ record). The gradual change of n is set proportional to the temperature at Renland (which is linear related to the $\delta^{18}O$ record).

The actual ratio between transport time and residence time (t^*/τ^*) is individual for each tracer, depending on the location of the source and the individual average residence time. However, the gradual change with climate and thus the ratio between relative residence time and relative transport time (n/k) can be expected to be similar for all tracers. The shorter the actual residence time (larger particles) and/or the further away the sources, the more important are these changes in residence time and transport time for the concentration in the ice.

OBSERVATIONS: The variations in the Renland ice core of concentrations in the ice and total deposition fluxes for the different tracers can be divided into three groups; 1) increasing concentrations and fluxes with colder climate (dust, nss Ca^{2+} and nss Mg^{2+}), 2) increasing concentrations but decreasing fluxes (Na^+, Cl^-, nss SO_4^{2-} and nss K^+), and 3) decreasing concentrations and fluxes (NO_3^-, NH_4^+ and $CH_3SO_3^-$). The patterns of the first two groups can be explained entirely by a general change in residence times and transport times due to decreased precipitation rates along the transport paths and a stronger general circulation in colder climatic stages. No increase in source emissions is necessary to explain also the ten-fold increased tracer concentrations in the Renland ice core in cold glacial stages. With the same general change in residence times and transport times affecting the third group, specific changes (e.g. decreased source emissions with colder climate) are necessary to explain the patterns of this group. The concentrations and fluxes of the biogenic tracers in the third group are found to increase in glacial stages in ice cores from Antarctica (Legrand et al., 1988; 1991). This difference may be attributed to shorter distances (which gives shorter transport times and hence lower t^*/τ^* ratios) from the sources in the Northern Hemisphere to Greenland than from sources in the Southern Hemisphere to Antarctica. Thus, the Antarctic deposition sites are more sensitive to changes in residence times and transport times and increased concentration and fluxes may be seen even if the source emissions decreased. This shows that neither concentrations nor fluxes directly reflect emission changes. Only changes in the local precipitation rate are accounted for when fluxes are calculated. The local precipitation rate is important for the concentration of dry deposited tracers and for the flux of wet deposited tracers.

IMPLICATIONS FOR THE GLOBAL ATMOSPHERE: The sulphate concentration is here taken as an example. The sulphate concentration in the Renland ice core is increased by a factor-of-two during LGM. The pattern of the sulphate data can be reproduced with the model using a decrease of n/k to 0.1 during LGM and a t^*/τ^* ratio between 0.5 and 1.0. The relative atmospheric loading can be described by:

$$M/M^* \propto C(0)_{air}/C(0)^*_{air}\, k/n \qquad (4)$$

As a consequence, the change of n/k to 0.1 (k/n=10) leads to a factor-of-ten increase of the atmospheric loading of sulphate although it is seen as a factor-of-two increase of the airborne concentration at Renland. A factor-of-two higher source emission leads to a factor-of-two increase of the atmospheric loading. This will also be seen as a factor-of-two increase of the airborne concentration at Renland. Thus, different interpretations of the reason behind the increased ice core sulphate concentration lead to a factor-of-five difference in the estimated atmospheric loading.

CONCLUSIONS AND PERSPECTIVES: This simple model simulation shows that for many of the soluble and insoluble tracers measured in polar ice cores, the changes in the physical conditions of the atmosphere are sufficient in themselves to account for all the observed variations in the concentrations. This means that interpretations of changes in source emissions have to be made with caution from records of deposition. Variations in the atmospheric cleansing can have a large impact on the total content of tracers in the atmosphere and hence on the radiative properties of the atmosphere. This implicates that a climate forcing mechanism may be found in the dynamics of the atmosphere by the course of events: lower temperatures → lower precipitation rates → less atmospheric cleansing of tracers → longer atmospheric residence times → higher tracer concentrations in the atmosphere → more reflected solar flux → lower temperatures.

The residence time in the atmosphere is a key parameter in this discussion. Changes in residence time directly imply changed atmospheric loading. When residence times are changed, changes in airborne concentrations at one single deposition site cannot be taken as representative for changes of the total atmospheric loading. This is essential for the discussion about climate forcing of aerosols in the atmosphere and the use of ice core data for global estimates. One way of bringing the knowledge further is to compare tracer data from different ice cores at varying distances from the main sources. It is also necessary to attain a better knowledge of the atmospheric conditions during glacial stages to take the full advantage of the ice core records.

REFERENCES:

Anderson, P.M. and 32 others. (1988) Climatic changes of the last 18,000 years: Observations and model simulations. Science 241, 1043-1052.

Davidson, C.I. and 14 others. (1993) Chemical constituents in the air and snow at Dye 3, Greenland: 1. Seasonal variations. Atmos. Environ. 27A, 2709-2722.

Hansson, M.E. (1994) The Renland ice core. A Northern Hemisphere record of aerosol composition over 120,000 years. Tellus 46B, 390-418.

Hansson, M.E. (1995) Are changes in atmospheric cleansing responsible for observed variations of impurity concentrations in ice cores? Ann. Glac. 21 (in press).

Johnsen, S.J., Clausen, H.B., Dansgaard, W., Gundestrup, N., Hansson, M., Jonsson, P., Steffensen, J.P. and Sveinbjörnsdottir, A.E. (1992) A "deep" ice core from East Greenland, Meddr. Grönland Geosci. 30, 22pp.

Joussaume, S. (1993) Paleoclimatic tracers: An investigation using an atmospheric general circulation model under ice age conditions. 1. Desert dust. J. Geophys. Res. 98 (D2), 2767-2805.

Legrand, M.R., Lorius, C., Barkov, N.I. and Petrov, V.N. (1988) Vostok (Antarctica) ice core: Atmospheric chemistry changes over the last climatic cycle (160,000 years). Atmos. Environ. 22, 317-331.

Legrand, M., Feniet-Saigne, C., Saltzman, E.S., Germain, C., Barkov, N.I. and Petrov, V.N. (1991) Ice-core record of oceanic emissions of dimethylsulphide during the last climatic cycle. Nature 350, 144-146.

THE BEHAVIOUR OF ORGANIC CHEMICALS IN SNOW

Don Mackay[(1)], Charles Jia[(1)], John Hoff[(2)], Dennis Gregor[(2)] and Frank Wania[(3)]
(1): Department of Chemical Engineering and Applied Chemistry
University of Toronto,
Toronto,
Ontario,
Canada

Introduction

There are several incentives for improving our understanding of the role of snow and ice as determinants of the behaviour of organic chemicals in the environment. Snow scavenges organic chemicals from the atmosphere both by adsorbing gaseous chemical to the ice surface and by scavenging aerosol particles with their associated chemical. Snow undoubtedly influences soil-air transfer of chemicals, and ice probably prevents water-air transfer in lakes, rivers and oceans. The period of snow melt may cause a pulse in organic chemical loading to receiving waters and to groundwater. This is obviously of particular importance in arctic and sub-arctic regions. Finally, it is possible that the glacial record of concentration may contain valuable information about chemical concentrations in past atmospheres, similarly to that of lead. Unlike lead, organic chemicals have an appreciable volatility. It is thus likely that when they are present in snowpack they are subject to evaporation back to the atmosphere. This complicates their behaviour and the interpretation of glacial records.

Past attempts to model the global distribution of chemicals virtually ignore the effects of snow, simply because the partitioning, transport and transformation processes to which chemicals are subject in snow are too poorly understood to permit reliable equations to be written for inclusion in models. Accordingly in this paper we describe an on-going project which seeks to improve our understanding of the behaviour of organic chemicals in snow, and the role which snow plays in influencing chemical behaviour on a local, regional and even global scale, and we discuss research needs.

---

(2). Groundwater Research Institute, University of Waterloo, Waterloo, Ontario, Canada;
(3). NILU, Tromso, Norway.

NATO ASI Series, Vol. I 43
Chemical Exchange Between the Atmosphere and Polar Snow
Edited by Eric W. Wolff and Roger C. Bales

Physical Chemistry of Air-Ice Surface Partitioning

Fundamental to the understanding of the depositional and post depositional behaviour of contaminants in the presence of snow is information about gas phase to ice surface partitioning. Accordingly we have undertaken studies to determine the magnitude of this partitioning, it being well established that adsorption at the air-water interface is important (Hoff et al., 1993).

Various investigators have pointed out that at temperatures close to the melting point, the surface of ice is covered by a quasi-liquid transition layer (Fletcher, 1973; Jellinek, 1967a and 1967b). It thus seems likely that air-water interface partitioning information will be valuable for assessing air-ice partitioning. In the air-water system if there are concentrations of chemical C_A in air and C_W in water (mol/m^3), then at equilibrium they will be related by the air-water partition coefficient (a dimensionless Henrys law constant) K_{AW}, which is C_A / C_W. There is an excess "concentration" at the interface C_I (mol/m^2) which can be related to C_A and C_W by an adsorption partition coefficient k_{IA} which equals C_I/C_A, or k_{IW} which equals C_I/C_W. These coefficients have dimensions of length (m) and represent the depth of the bulk phase (e.g. air) which will contain the same quantity of chemical as is present at the interface. This is readily demonstrated since for an area of A (m^2), the quantity at the interface is AC_I (mol) or AC_Ak_{IA}(mol) and is thus the amount in a layer of air k_{IA} metres thick. Obviously k_{IW}/k_{IA} equals K_{AW}. Measurements of partitioning to the air-ice interface suggest that the air-ice and air-water partition coefficients are similar in magnitude (Hoff et al., 1995; Goss, 1992 and 1993); thus it is useful to examine some of the data for the air-water interface in more detail.

Hoff et al. (1993) have measured air-water interface partitioning for 44 polar and non-polar chemicals. For non-polar chemicals a good correlation was obtained between k_{IW} (m) and the solubility of the liquid chemical in water C^S_W (mol/m^3), namely,

$$logk_{IW} = -5.97 - 0.769logC_W^S$$

over the range $-4 < \log C_W^S < 2.5$.

Thus

$$\log k_{IA} = -0.769\log C_W^S - 5.97 - logK_{AW}$$

An alternative expression is to calculate the "pseudo solubility" at the interface C_I^S such that C_I^S/C_W^S is k_{IW} or C_I^S/C_A^S is k_{IA}, where C_A^S is the saturation concentration in air (mol/m^3). Substituting gives

$$C_I^S = k_{IW} C_W^S = 10^{-5.97} C_W^{S\,0.231}$$

Apparently, as the solubility of a substance in water increases, so too does its "solubility" at the interface, but the increase is only to a power of 0.231. The result is that C_I^S is fairly constant since a 1000 fold change in solubility C_W^S causes only 5 fold change in C_I^S. The coefficient k_{IA} is given by,

$$k_{IA} = C_I^S/C_A^S = 10^{-5.97} C_W^{S\,0.231} C_A^{S\,-1}$$

This correlation is valid only for non-polar chemicals and it has not been verified for chemicals less volatile and less soluble than n-decane, but it is believed to give an order of magnitude estimate.

The implication is that partitioning to the ice interface is most significant when k_{IA} is large, which occurs when C_A^S or vapour pressure P^S is small since C_A^S is P^S/RT. A substance such as hexachlorobenzene is slightly less "soluble" at the interface than octane but its very low vapour pressure causes k_{IA} to be much larger. Solution of nonpolar organics in water is usually endothermic, but the enthalpy of vaporization is larger, causing k_{IA} to increase as the temperature decreases. Accordingly the quantity of chemical adsorbed to snow will be large relative to the quantity present in an adjacent volume of air. In our view, it is very important to measure and correlate this partitioning phenomenon for a variety of chemicals over a range of temperatures.

Scavenging Process by Snow in the Atmosphere

Organic chemicals in the atmosphere exist in gaseous form or associated with aerosols. Falling snow flakes can scavenge chemicals from the atmosphere by adsorption and/or absorption of gaseous organic chemicals, and by removal of aerosols.

The level of understanding of the removal of aerosols by snow is better than that of scavenging of organic vapours. Gaseous scavenging by snow can be quantified by a dimensionless ratio W_g,

$$W_g = C_{S,g}/C_{A,g}$$

where C_{Sg} is the concentration of chemical sorbed by snow, in which the quantity of snow is expressed on the basis of 1 m^3 of meltwater; C_{Ag} is the concentration of chemical in air in gaseous form during the snow event. Few data exist on this phenomenon. We believe that prediction of the effectiveness of gaseous scavenging is best approached by estimating the interfacial area of the snow, and calculating the concentration adsorbed to the interface from a knowledge of the gaseous concentration in air, the interfacial area and the air-ice interface partition coefficient as discussed earlier.

The specific surface area of freshly fallen snow can be measured by adsorption of nitrogen gas at temperatures close to the boiling point of liquid nitrogen. The few published data extant suggest a range of 0.1 to 10 m^2/g. We have obtained several measurements for local samples of snow using a commercial surface area analyzer (Micromeritics Gemini 2375), and our results fall between 0.1 to 0.3 m^2/g. We thus presently estimate that the specific surface area of freshly fallen snow is in the range of 0.1 to 1.0 m^2/g.

According to Franz (1994), scavenging of aerosols can be envisioned as occurring in successive stages. Initially, nucleation on aerosols forms either cloud droplets or ice crystals. These droplets or crystals may then grow by water vapour diffusion and condensation, and aggregate into precipitation elements, usually near the cloud top. They then settle out of the cloud. The process of aerosol scavenging from the atmosphere by snow events has been shown to be complex. Its efficiency is a function of variety of factors, including the intensity of the event, the concentration of the contaminants, the size distribution of both hydrometeors and aerosol, the hygroscopic nature of the particulate matter and the ambient temperature (Schümann et al., 1988; Tsai et al.,1991). To quantify aerosol scavenging, a particle (aerosol) scavenging coefficients can be defined as

$$W_P = C_{Sp}/C_{Ap}$$

where C_{Sp} and C_{Ap} are the particle concentration in snow and the atmosphere prior to or during the snow event. Coalescence of cloud droplets into rain drops has been estimated to produce a W_p about 10^6, while a W_p not more than 10^5 is indicative of unrimed snow or scavenging of either insoluble particles or particles with diameters of 0.1 to 1.0 μm (Scott, 1981). Field measurements of W_p range from 10^3 to 10^6 (Slinn, 1983; Bidleman, 1988; Eitzer and Hites, 1989; Koester and Hites, 1992). Franz's recent study (1994) indicates that aerosol scavenging by snow is a major contributor to the removal of both PCBs and PAHs from the atmosphere.

Preliminary modelling results suggest that W_P will often be larger than W_g, but the relative importance of the two processes is very sensitive to the fraction of the chemical which is associated with aerosols. This is determined by the aerosol concentration and area and the subcooled liquid vapour pressure of the chemical. The relative contributions of gaseous and aerosol scavenging are likely to be time-dependent or event-dependent.

Post-deposition Fate of Organic Chemicals

Snow is a finely dispersed aggregate of ice, water vapour, air and possibly water depending on temperature (Colbeck, 1987). After deposition, snow undergoes many physical changes, such as compacting, sintering, freezing and melting. There is a continual reduction in specific surface area with the corresponding increase in grain size and the decrease in grain population by vapour transfer. When temperatures exceed the freezing point, meltwater can percolate through the snowpack by capillary action and refreeze among the snow grains (LaChapelle, 1969). These changes are likely to have a significant effect on the behaviour of organic chemicals contained in the snow. In a field study, it has been established that semi-volatile organic compounds volatilize from snow pack with the loss apparently proportional to the vapour pressure (Gregor, 1991). In a study of the accumulation of organochlorine pesticides in Canadian Arctic snow, Gregor and Gummer (1989) reported that release of organic chemicals to the Arctic aquatic environment occurs at the time of snowmelt.

Models are under development in our group to describe this post depositional behaviour. The first and simplest model deduces the partitioning of a specified quantity of chemical between air, water, and ice surface in a defined snowpack. Sintering is then allowed to occur reducing the surface area. The effect is to drive chemical from the snow surface into the air phase where it is more susceptible to evaporation. The second model incorporates the first model but also includes melting. The aim is to deduce what fractions of the chemical initially present in the snow are lost by evaporation, in meltwater and are retained in the pack at various times during the entire post depositional process from snowfall to complete melting. A complementary laboratory program to determine the post depositional fate of chemicals is also underway. Although it is too early to give results of the modelling and laboratory work, it is clear that a very complex set of processes prevail and that different chemicals are likely to experience quite different fates in snowpacks, i.e some will primarily leave in meltwater, whereas others will primarily evaporate.

Conclusions

This work has, we believe, important implications for (i) assessing the usefulness of the glacial record as an indicator of past atmospheric levels, (ii) developing reliable sampling protocols and (iii) for compiling more reliable estimates of net atmospheric deposition to soils and lakes. To a first approximation we believe that the air-ice partition coefficient of non-polar chemicals can be estimated by extrapolating adsorption coefficients for the air-water interface to below freezing temperatures, but more data are needed, especially for less volatile chemicals. Our experimental and modelling results to date provide strong support for the suggestion that snow is very important for determining the fate of relatively involatile organic chemicals in cold climates. In a snowpack, the air-ice interfacial area serves as a significant reservoir for the chemicals scavenged by snow. The fraction of chemical adsorbed at the air-ice interface can be over 90%. As a result of sintering, which leads to a rapid decrease of air-ice interfacial area, organic chemicals initially removed by fresh snow could evaporate back to the atmosphere. This process is likely to be a strong function of temperature which controls the rate of sintering, and the rate of diffusion. Modelling the deposition and post depositional process to which chemicals are subject in snow is a challenging task and is, we believe, best addressed by a combination of controlled experiments in the laboratory, field measurements, and by modelling.

References

Bidleman, T.F. *Environ. Sci. Technol.* (1988) 22, p361-367.

Colbeck, S.C. in Seasonal Snowcovers:Physics, Chemistry, Hydrology (1987) Jones, H.G. and Orville-Thomas, W.J. ed., D.Reidel: Dordrecht; Boston; Lancaster; Tokyo, p 1-35.

Eitzer, B.D. and Hites, R.A. *Environ. Sci. Technol.* (1989) 23, p1396-1401.

Fletcher, N. in Physics and Chemistry of Ice, Whalley, E.; Jones, S.; Gold, L., Ed,; Royal Society of Canada: Ottawa, 1973.

Franz, T.P. Jr. Ph.D. Thesis (1994) University of Minnesota.

Goss, K.-U. *Environ. Sci. Technol.* (1992) 26, p2287-2294.

Goss, K.-U. *Environ. Sci. Technol.* (1993) 27, p2826-2830.

Gregor, D. J.; Gummer, W. D. *Environ. Sci. Technol.* (1989) 23, p561-565.

Gregor, D. J. in Pollution of the Arctic Atmosphere, Sturges, W.T. (ed.), Elsevier Science Publishers Ltd., Essex, England (1991) p217-254.

Hoff, J.T.; Gillham, R.; Mackay, D. and Shiu, W.Y. *Environ. Sci. Technol.* (1993) 27, p2789-2794.

Hoff, J.T.; Wania, F. and Mackay, D and Gillham, R., accepted by *Environ, Sci. Technol.* (1995).

Jellinek, H.J. *Colloid Interface Sci.* (1967a) 25, p192-205.

Jellinek, H.; Ibrahim, S.J. *Colloid Interface Sci.* (1967b) 25, p245-254.

Koester, C.J. and Hites, R.A. *Environ. Sci. Technol.* (1992) 26, p1375-1382.

LaChapelle, E.R. Field Guide to Snow Crystals (1969) University of Washington Press, Seattle, p101.

Schumann, T.; Zinder, B. and Waldvogel, A. *Atm. Environ.* (1988) 22, p1443-1459.

Scott, B.C. *J. Appl. Met.* (1981) 20, 619-625.

Slinn, W.G.N. in Air-Sea Exchange of Gases and Particles (1983) Liss, P.S. and Slinn, W.G.N. ed., D.Reidel Publ. Co., Boston, MA. p299-405.

Tsai, W.; Cohen, Y.; Sakugawa, H. and Kaplan, I.R. *Environ. Sci. Technol.* (1991) 25, p2012-2023.

MASS SIZE DISTRIBUTIONS FOR ATMOSPHERIC PARTICULATE ELEMENTS AT THE ZEPPELIN BACKGROUND STATION IN NY ÅLESUND, SPITSBERGEN

Willy Maenhaut[1], Vladimír Havránek[1], Geert Ducastel[1] and Jan Erik Hanssen[2]
[1]University of Gent, Institute for Nuclear Sciences, Proeftuinstraat 86, B-9000 Gent, Belgium
[2]Norwegian Institute for Air Research, P.O. Box 100, N-2007 Kjeller, Norway

Introduction

Measurements of the multielemental composition of size-fractionated aerosols provide information on the source processes of the various elements and they allow one to examine the physico-chemical transformation processes that take place during the atmospheric transport or as a result of local meteorology and atmospheric conditions. Furthermore, such measurements provide the necessary input data for estimating the scavenging probabilities and dry deposition velocities of the elements. Several studies on elemental mass size distributions have already been performed in the Arctic, e.g., by Pacyna et al. [1984], Li and Winchester [1990], Hillamo et al. [1993] and Barrie et al. [1994], but virtually all former studies were done in the winter and were limited to time periods of a few months. To investigate changes in sources, source processes and/or particle size modification processes over the course of the year, in late 1990 we started long-term (and still ongoing) samplings with a cascade impactor at the Zeppelin mountain station in Spitsbergen. Selected preliminary results of these samplings are presented and briefly discussed.

Experimental

The device used for the size-fractionated aerosol samplings is a 1-l/min, 8-stage, Battelle-type PIXE International cascade impactor (PIXE Int. Corporation, P.O. Box 2235, Tallahassee, FL, USA) [Mitchell and Pilcher, 1959; Bauman et al., 1981]. It contains 7 impaction stages with nominal cut-points (d_{50}-values) of 16, 8, 4, 2, 1, 0.5 and 0.25 μm equivalent aerodynamic diameter (EAD), respectively. A filter stage with a 0.4 μm pore size Nuclepore filter collects the particles <0.25 μm EAD. The PIXE cascade impactor has been calibrated by Hillamo et al. [1993], and the actual cut-points for the 7 impaction stages were found to be 14.9, 9.4, 4.7, 2.3, 1.3, 0.62 and 0.19 μm EAD (for operation at 1013 hPa and 293 K). The samplings at the Zeppelin station are done on a continuous basis according to a 2-2-3-day schedule, so that about 150 samples are obtained per year. All samples collected up to the end of 1993 were analyzed by particle-induced X-ray emission analysis (PIXE), using procedures which were described before [Maenhaut et al., 1981; Maenhaut and Raemdonck, 1984; Maenhaut and Vandenhaute, 1986]. Raw mass size distributions (with mass per stage and per m^3) were obtained for the elements Na, Mg, Al, Si, S, Cl, K, Ca, Ti, Mn, Fe, Zn, Br and Pb. From these raw size distribution data mass median aerodynamic diameters (MMADs) were calcu-

NATO ASI Series, Vol. I 43
Chemical Exchange Between the Atmosphere and Polar Snow
Edited by Eric W. Wolff and Roger C. Bales

lated for each element in every sample. To allow a comparison with MMADs from previous studies with the same type of PIXE cascade impactor [Ducastel et al., 1991], the nominal cut-points were adopted. The MMADs thus derived are about 15% lower than would have been obtained with the actual (calibrated) cut-points. The raw size distribution data were also converted into smooth size distributions using the inversion code MICRON [Wolfenbarger and Seinfeld, 1990]. These calculations were done with kernel functions for the actual cut-points. Finally, lognormal curves were fitted to the inverted size distributions, so that the elemental mass concentration, geometric mean diameter and standard deviation of the different contributing modes could be obtained [Winklmayr et al., 1990].

Results and discussion

Spitsbergen is a receptor site for long-range transport of continental and anthropogenic aerosols (and aerosol precursors) from source regions which are at a distance of 1000 km or more. Alterations in elemental mass size distributions as a result of such long-range transport can be examined by comparisons with data from receptor sites (or areas) which are much closer to the source regions and which at the same time are situated along possible transport pathways. One such receptor area is southern Norway. Detailed size distribution measurements with PIXE and Berner [Berner, 1984] cascade impactors were conducted at two southern Norwegian sites during the spring (April-May) of 1987, 1988 and 1989 [Ducastel et al., 1991]. The median MMADs derived from these measurements are presented in Table 1, together with median MMADs for the Zeppelin station. For the latter site, separate medians are given for the second quarter of the year (months April-May-June) and for the third quarter (July through September). In both cases, the median MMADs are based on the data from the three years 1991 through 1993. In southern Norway (and during spring), MMADs in the range 3-4 µm are noted for the typical coarse mode sea-salt, mineral dust and mixed sea-salt/mineral elements (Na, Cl, Al, Si, Ti, Fe, Mg, K, Ca); the anthropogenic metals Zn and Pb have MMADs of 0.6-0.8 µm, and the secondary aerosol constituent S exhibits the lowest MMAD (i.e., 0.50 µm). Mn has both crustal and local anthropogenic sources in southern Norway, and, as a consequence, its MMAD is somewhat lower than that of the more pure crustal elements. The Zeppelin MMADs from the second quarter of the year (spring) are systematically lower than the Norwegian spring data for the coarse mode elements (the difference is about 1 µm). For the fine mode anthropogenic elements, on the other hand, the MMADs are slightly, but clearly higher at the Arctic site. These observations are consistent with our current knowledge of removal and growth processes of aerosol particles during long-range transport. Dry deposition of very coarse particles and coagulation processes of very fine particles will lead to narrower particle size distributions and to a decrease in the supermicrometer MMADs and an increase of the submicrometer MMADs. During the third quarter of the year (summer), the MMADs of the crustal and mixed sea-salt/crustal elements at Zeppelin are all systematically higher than the corresponding second quarter values, and they have become similar to and even slightly larger than the Norwegian spring MMADs. This is an unexpected result, and one can speculate about the causes for this. It is possible that the

Table 1. Median values of the mass median aerodynamic diameter (in μm), as obtained from cascade impactor samplings in southern Norway and at Zeppelin. The two columns for the latter site apply to the second and third quarter of the year, resp.

Elem.	S. Norway[a]	Zeppelin 2nd q.	Zeppelin 3rd q.
Na	3.1	2.4	2.9
Mg	3.2	2.4	3.2
Al	3.7	2.3	5.1
Si	3.5	2.4	4.0
S	0.50	0.71	0.67
Cl	4.0	3.2	3.3
K	3.0	2.2	3.2
Ca	3.9	2.7	4.0
Ti	4.2	2.4	4.1
Mn	2.0	2.0	3.7
Fe	3.8	2.2	4.0
Zn	0.8	0.9	(1.5)[b]
Br	0.7	1.0	2.5
Pb	0.6	0.75	(1.0)

[a] Based on results reported by Ducastel et al. [1991]
[b] The values in parentheses are questionable, because the concentrations were too close to the detection limit

contribution from local (and coarse) mineral dust at Zeppelin is more important during summer than during spring, but changes with season in local meteorological conditions and/or in atmospheric processes during long-range transport may also have played a role. It is interesting to note in this respect that similar unexpectedly large MMADs for the crustal elements during summer were also found in the very high Arctic [Maenhaut et al., 1994b]. Sulfur at Zeppelin will be discussed in more detail below, but Table 1 already shows that its summer MMAD is slightly lower than the spring value, but still clearly larger than the MMAD in southern Norway. For the metallic elements Zn and Pb, the Zeppelin summer data were of very poor quality because of the low concentrations, and it is therefore unclear whether there is a real change in their MMAD from spring to summer. Finally, for Br at Zeppelin, the MMAD is much larger in summer than in spring. The low MMAD in spring is due to the presence of large amounts of fine non-sea-salt particulate Br during that season.

The MICRON data inversions of the raw size distributions and the subsequent log-normal fits generally gave at most two really significant modes. The mass present in the third mode was usually less than 5% of that in the first mode (The first mode is defined here as the mode with most of the mass). Furthermore, the second mode contained typically only around one third of the mass of the first mode. The major exception here was K, which showed an average (mode 2/mode 1) mass ratio of 0.45. In order to examine the changes in the size distributions over the course of the year, the geometric mean diameters (GMDs) and associated standard deviations of the different modes for each sample were plotted as a

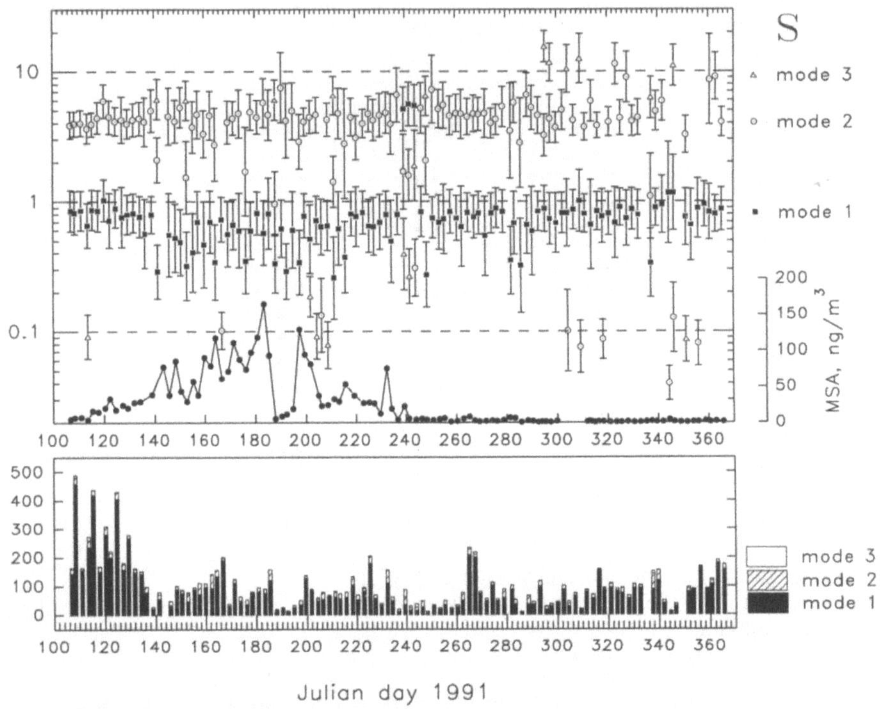

Fig. 1. Geometric mean diameters and associated standard deviations of the different modes for S as a function of sampling date for days 100-365 of 1991. The bottom part (bar graph) gives the S concentration in each of the modes. The line graph represents the MSA concentration in <2.5 μm EAD particles. See text for further details.

function of sampling date for each of the 3 years and for each element. Because of space limitations, only one example of such plots is shown here, i.e., for S in part of 1991 (Fig. 1). In order not to overload the graph with questionable data, the GMD of the third mode is only shown when the (mode 3/mode 1) ratio was larger than 0.05. The bottom part of the Figure (i.e., the bar graph) presents the atmospheric concentration in each of the three modes. Also included is a line graph with the atmospheric levels of methanesulfonate (MSA), as obtained from samplings with a modified Sierra-Andersen Hi-Vol cascade impactor [Maenhaut et al., 1994a]. MSA is a particulate conversion product of the marine biogenic precursor gas dimethylsulfide (DMS). The MSA data shown apply to the size fraction <2.5 μm EAD, but this size fraction contained 90% or more of all particulate MSA collected (at least during summer). As can be seen from Fig. 1, the GMD of the first S mode is generally around 0.8-0.9 μm, but during summer, the values tend to be lower, occasionally even down to about 0.3 μm. A very substantial fraction of the fine sulfate at Ny Ålesund is of marine biogenic origin during summer, and Maenhaut et al. [1994a] found that, on the average, almost 50% of the fine sulfate was attributable to that marine source during the summer of

1991. As the Zeppelin site is much closer to the DMS source regions than to the large continental source areas of anthropogenic SO_2, it is tempting to attribute the lower S GMDs during summer to the shorter "average" transport time from source to receptor. However, Fig. 1 also shows that the low S GMDs are not really correlated with the high MSA levels, so that the true cause of the lower GMDs is probably more complex.

Acknowledgements

We gratefully acknowledge the financial support from the Belgian NFWO and a research grant from the Impulse Programme "Global Change" supported by the Belgian State - Prime Minister's Service - Federal Office for Scientific, Technical and Cultural Affairs.

References

Barrie LA, Staebler R, Toom D, Georgi B, den Hartog G, Landsberger S, Wu D (1994) Arctic aerosol size-segregated chemical observations in relation to ozone depletion during Polar Sunrise Experiment 1992. *J Geophys Res* 99:25,439-25,451

Bauman S, Houmere PD, Nelson JW (1981) Cascade impactor aerosol samples for PIXE and PESA analysis. *Nucl Instr and Meth* 181:499-502

Berner A (1984) Design principles of the AERAS low pressure impactor. in: *Aerosols. Science, Technology, and Industrial Applications of Airborne Particles* (Liu BYH, Pui DYH, Fissan HJ, eds), Elsevier New York, pp 139-142

Ducastel G, Maenhaut W, Hillamo R, Pakkanen T, Pacyna JM (1991) Detailed size distributions of atmospheric trace elements in Southern Norway during spring 1988 and 1989. in: *Heavy Metals in the Environment, International Conference, Edinburgh, September 1991, Vol 1* (Farmer JG, ed), CEP Consultants Ltd Edinburgh, pp 270-273

Hillamo RE, Kerminen V-M, Maenhaut W, Jaffrezo J-L, Balachandran S, Davidson CI (1993) Size distributions of atmospheric trace elements at Dye 3, Greenland - I. Distribution characteristics and dry deposition velocities. *Atmos Environ* 27A:2787-2802

Li SM, Winchester JW (1990) Particle size distribution and chemistry of late winter arctic aerosols. *J Geophys Res* 95:13,897-13,908

Maenhaut W, Raemdonck H (1984) Accurate calibration of a Si(Li) detector for PIXE analysis. *Nucl Instr and Meth* B1:123-136

Maenhaut W, Vandenhaute J (1986) Accurate analytic fitting of PIXE spectra. *Bull Soc Chim Belg* 95:407-418

Maenhaut W, Selen A, Van Espen P, Van Grieken R, Winchester JW (1981) PIXE analysis of aerosol samples collected over the Atlantic Ocean from a sailboat. *Nucl Instr and Meth* 181:399-405

Maenhaut W, Ducastel G, Beyaert K, Hanssen JE (1994a) Chemical composition of the summer aerosol at Ny Ålesund, Spitsbergen, and relative contribution of natural and anthropogenic sources to the non-sea-salt sulfate. in: *The Proceedings of EUROTRAC Symposium '94* (Borrell PM, Borrell P, Cvitas T, Seiler W, eds) SPB Academic Publishing bv The Hague The Netherlands, pp 467-471

Maenhaut W, Ducastel G, Leck C, Nilsson ED, Heintzenberg J (1994b) Multielemental composition and sources of the high Arctic atmospheric aerosol during summer and autumn 1991. *Tellus B*, submitted

Mitchell RI, Pilcher JM (1959) Improved cascade impactor for measuring aerosol particle sizes in air pollutants, commercial aerosols and cigarette smoke. *Ind Eng Chem* 51:1039-1042

Pacyna JM, Vitols V, Hanssen JE (1984) Size differentiated composition of the arctic aerosol at Ny Ålesund, Spitsbergen. *Atmos Environ* 11:2447-2459

Winklmayr W, Wang H-C, John W (1990) Adaptation of the Twomey algorithm to the inversion of cascade impactor data. *Aerosol Sci Technol* 13:322-331

Wolfenbarger JK, Seinfeld JH (1990) Inversion of aerosol size distribution data. *J Aerosol Sci* 21:227-247

[illegible]

[illegible] (eds) [illegible] Kluwer Academic Publ., Dordrecht, pp 467-477

[illegible] Millennium [illegible] Weather [illegible]

AIR/SNOW TRANSFER STUDIES AT THE HIGH-ALPINE SITE JUNGFRAUJOCH, SWITZERLAND

Margit Schwikowski
Urs Baltensperger
and Heinz Gäggeler
Paul Scherrer Institute
CH-5232 Villigen-PSI
Switzerland

1. Introduction

Ice cores are commonly used to reconstruct the chemical composition of historical deposition, and there is great interest to link the concentrations of various trace components in the ice with their concentrations in the atmosphere at the time of deposition. Therefore it is important to understand the atmosphere/snow transfer mechanisms for the different species. In mixed phase clouds, i.e. clouds containing ice crystals, it is assumed that supercooled cloud droplets are first formed. After ice nucleation, ice crystals grow by water vapour deposition at the expense of the supercooled cloud droplets (Wegener-Bergeron-Findeisen mechanism) and/or by direct accretion of cloud droplets (called riming) onto the crystal's surface. The first mechanism is often the dominant physical process but insignificant with respect to the removal or transfer of pollutants resulting in very clean ice crystals. The latter mechanism determines to a great extent the chemical composition of the ice crystals in polluting them with activated aerosol particles. However, the degree of riming depends on a variety of parameters, such as cloud droplet size, snow crystal size and snow crystal settling velocity.

It was a major goal of the European project ALPTRAC, a subproject of EUROTRAC to provide information on the interrelations between air, cloud and snow chemistry in mixed phase clouds. Therefore, field experiments were conducted at Jungfraujoch with special emphasis on the investigation of the occurrence and the impact of riming on spring and early-winter precipitation at a high-alpine site. Results from six snowfall events are presented and compared to results from another high-alpine site (Sonnblick, Kalina and Puxbaum, 1994) and from a prealpine site (Mount Rigi, Dixon *et al.*, 1994).

NATO ASI Series, Vol. I 43
Chemical Exchange Between the Atmosphere
and Polar Snow
Edited by Eric W. Wolff and Roger C. Bales

2. Experimental

In-cloud scavenging experiments were carried out at the high-alpine research station Jungfraujoch, Switzerland (3450 m a.s.l.) in April and May 1992 as well as in October and November 1993. Emphasis was laid on the investigation of the interactions between aerosol particles, cloud droplets and ice crystals. The aerosol number concentration for particle diameters d > 0.01 μm and the size distribution for particles with 0.1 < d < 7.5 μm were measured by a condensation particle counter (CPC) and an optical particle counter (OPC). Aerosol samples for chemical analysis were collected by a filter pack. Cloud liquid water (LWC) was measured by a particulate volume monitor (PVM), and droplet size distributions were determined by a forward scattering spectrometer probe (FSSP). Cloudwater samples for chemical analysis were collected with a Berner cloudwater impactor. Snow crystals were sampled by a collector designed after Borys *et al.* (1988), which allowed the determination of the mass concentration of ice crystals in air (ice water content). Chemical analyses were performed by ion chromatography. Formvar replicas of the ice crystals were collected on slides and visually classified into six classes of riming (from unrimed to graupel) after Mosimann *et al.* (1994). The degree of riming was determined for each slide as the surface weighted average of the collected crystals. Altogether, 12 snowfall events were investigated with a time resolution of 0.5 to 2 h for the chemical measurements. The results from 6 selected events are presented here.

3. Results and Discussion

In order to corroborate the role of riming on the overall transfer of trace species from cloudwater into snow, the rimed mass fraction was estimated for each snow sample from the degree of riming using a method recently developed by Mosimann *et al.* (1994). The rimed mass fraction (RMF) is defined as the ratio of the mass of accreted cloud droplets and the mass of the whole ice crystal and serves therefore as a direct measure of the amount of water transferred to the crystal by the riming process. For the transfer of chemical impurities not only the RMF but also the respective concentration in cloudwater must be known. To prove that riming is the dominant process for the incorporation of impurities into the ice crystal, the concentrations of the impurities in snow were calculated by multiplying the concentration of the respective component in cloudwater by the rimed mass fraction. The calculated concentrations versus the actually measured concentrations are shown in Figure 1 for the trace species sulphate. The figure includes data from four snowfalls in November 1993 as well as from one snowfall in April and one in May 1992. In addition, winter data from the prealpine site Mount Rigi (1620 m a.s.l., Switzerland; Dixon *et al.*, 1994) as well as summer and winter data from Sonnblick (3106 m a.s.l., Austria, Kalina and Puxbaum, 1994) are shown. Although there is a certain degree of scatter, a reasonable agreement between measured and calculated concentrations of sulphate in snow can be seen.

The generally higher concentrations at Mount Rigi are on the one hand due to higher degrees of riming and on the other hand due to higher concentrations of sulphate in supercooled cloudwater. Both facts seem to be caused by the lower elevation of the Mount Rigi site. A possible reason for the different degrees of riming is the greater vertical extent of the precipitation layer above the sampling site (or the lower position of the site relative to the cloud top). The larger distance to the cloud top results in larger settling times of the ice crystals favouring the collection of a large number of supercooled droplets. For the cases where cloud top height data were available from radar measurements (at Mount Rigi and Jungfraujoch) these cloud top heights were comparable and it is assumed that this was the case also for the other events. Another reason for enhanced degrees of riming could be the different precipitation systems predominating at the sites. Stratiform precipitation is the more frequent type at Mount Rigi, whereas convective precipitation (due to orography) prevails at Jungfraujoch. Cloudwater with high concentrations of impurities results when during activation a large number of aerosol particles compete for the existing water vapour and, therefore, the formed cloud droplets remain small. Particularly in winter time with increased atmospheric stability, the lower site Mount Rigi can more likely be reached by planetary boundary layer air parcels with elevated concentrations of aerosol particles than the high-alpine sites. In addition, cloud liquid water content (LWC) at Mount Rigi was comparable or even lower than at the two high-alpine sites. This also favours the formation of small droplets with high concentrations of impurities which where, however, still large enough to be collected by the ice crystals.

Generally, at Jungfraujoch and Sonnblick only light to moderate degrees of riming were observed and so far there is no indication of a systematic variation of this parameter with season. The lower concentrations of sulphate in snow at Jungfraujoch compared to Sonnblick reflect the lower concentrations of this component in supercooled cloudwater which might be due to the higher elevation of the Jungfraujoch (by 400 m). The concentration in cloudwater seems to be governed by the atmospheric stability. In early winter enhanced stability decouples the site from the planetary boundary layer leading to very low aerosol concentrations (Baltensperger *et al.*, 1991) and thus, very low concentrations in the cloud droplets. In spring and summer, convective processes transport more polluted air up to the site, leading to cloudwater with higher concentrations of pollutants.

To summarise, the concentration of sulphate in snow is determined by the riming process. However, the difference of a factor of ten in the concentrations in snow at Jungfraujoch and Mount Rigi, for instance, is to two thirds due to the difference in cloudwater concentration and only to one third due to differing rimed mass fractions. Thus, the variability of the concentration in snow is to a large extent caused by the variability of the concentration in accreted cloud droplets and to a lesser extent by the degree of riming.

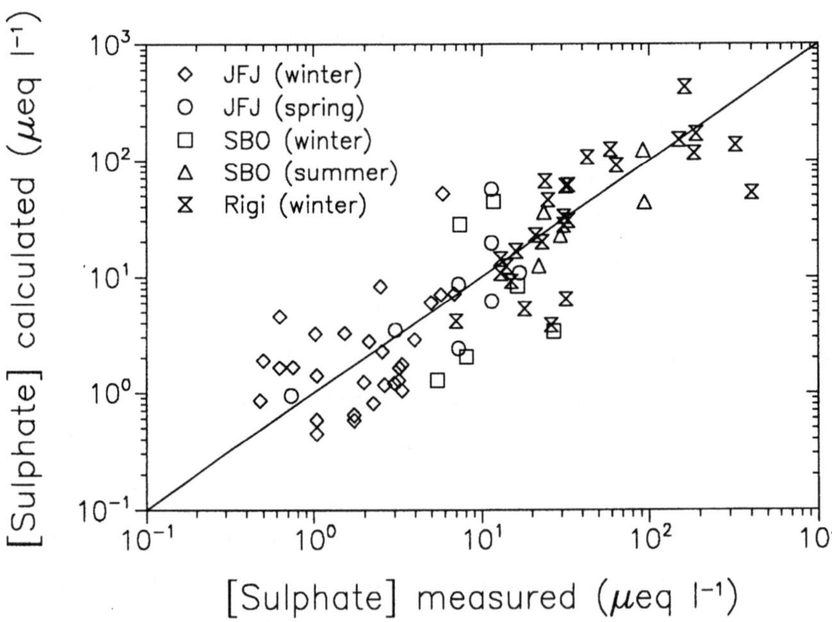

Figure 1. Measured and calculated concentrations of sulphate in snow. Data are from snowfalls at Jungfraujoch in November 1993 (JFJ winter), in April and May 1992 (JFJ spring), from the prealpine site Mount Rigi (Rigi winter; Dixon *et al.*, 1994) and from Sonnblick (SBO winter and SBO summer; Kalina and Puxbaum, 1994). The solid line represents the 1:1 relationship.

References

Baltensperger U, Gäggeler HW, Jost DT, Emmenegger M, Nägeli W (1991) Continuous background aerosol monitoring with the Epiphaniometer. Atmos. Environ. 25A: 629-634.

Borys RD, Hindman EE, DeMott PJ (1988) The chemical fractionation of atmospheric aerosols as a result of snow crystal formation and growth. J. Atmos. Chem. 7: 213-239.

Dixon RW, Mosimann L, Oberholzer B, Staehelin J, Waldvogel L (1994) The effect of riming on winter precipitation ion concentrations Part I. A quantitative investigation of field measurements. submitted to J. Geophys. Res.

Kalina MF, Puxbaum H (1994) A study of the influence of riming of ice crystals on snow chemistry during different seasons in precipitating continental clouds. Atmos. Environ. 28: 3311-3328.

Mosimann L, Weingartner E, Waldvogel A (1994) An analysis of accreted drop sizes and mass on rimed snow crystals. J. Atmos. Sci. 51: 1548-1558.

ON THE SPATIAL VARIABILITY OF IMPURITY CONTENT AND STABLE ISOTOPIC COMPOSITION IN RECENT SUMMIT SNOW

J.P.Steffensen, H.B.Clausen and J.M.Christensen
Department of Geophysics
University of Copenhagen
Haraldsgade 6
DK-2200 Copenhagen N
Denmark

In recent years the study of how atmospheric aerosols and gases are deposited in the snow pack of the interiors of the Greenland and Antarctic ice sheets has become a growing field, exemplified by the proceedings of this volume and the proceedings of the NATO ARW in Annecy in 1993 (Delmas, 1995). If the functions of transfer of various aerosols and gases from the atmosphere to snow/ice strata were known, then ice core data could be used to calculate past atmospheric compositions.

So far, most ice core data on precipitation chemistry have been interpreted mainly by assuming that the transfer functions have varied only little over time, so that observed changes in the chemical composition of ice core strata reflect changes in the composition of the atmosphere or changes in rates of snow precipitation. This way past atmospheric composition can be inferred from ice core data without detailed knowledge of the transfer functions. With this approach however, calculated past atmospheric compositions contain a lot of uncertainties, and the scientific interpretation has to be done carefully. Therefore, ice core analysts rarely publish any quantitative estimates on past atmospheric composition. Instead they discuss the evolution of the atmospheric composition qualitatively, with the focus on changes (Dansgaard et al., 1989; Mayewski et al., 1993; Alley et al, 1993; GRIP members, 1993).

There is no doubt, however, that if the transfer functions of atmospheric components to ice core strata were known it would provide a tremendous improvement of the interpretation of ice core data in terms of past atmospheric/climatic changes. But how far can we go in our interpretation of an ice core record? How well does one ice core represent the composition of the local snow strata? These questions are relevant since even if the transfer functions were known, it still leaves the problem that snow is deposited unevenly on the surface of the ice sheet causing depositional noise, for instance due to the formation of sastrugi on scales from a few cm to meters. These questions are also addressed by J.Dibb, 1996.

NATO ASI Series, Vol. I 43
Chemical Exchange Between the Atmosphere and Polar Snow
Edited by Eric W. Wolff and Roger C. Bales

One of the few exceptions of the above are the calculations of the total release of sulfuric acid into the stratosphere from large volcanic eruptions based on acid layers in Greenland ice cores. In this case, the pattern of deposition of stratospherically transported radioactive isotopes (e.g. ^{90}Sr and ^{137}Cs) on the Greenland ice sheet from nuclear bomb tests in the 50's and 60's could be used as reference (Clausen and Hammer, 1988).

In order to investigate the problem of depositional noise, we have analysed several series of samples taken from a wall of a 2.30 m snow pit 4.4 km N of the Summit drilling site in Central Greenland. The pit was 1.1 km S of the automatic aerosol sampling site (Wåhlin, 1996). After cleaning of the pit wall, a horizontal and vertical grid was laid out, and 7 sequences of samples were taken with a vertical resolution of 2.5 cm and with a horizontal spacing of 30 cm between the sequences (Fig.1a). The samples were taken at the grid points without any consideration of the visible crust and hoar layers that could be seen as slightly wavy lines along the pit wall. The D-sequence (fig.1a) is 2.20 m and covers the entire depth of the pit. The other sequences were sampled from 1.10 m to 2.20 m. This gave a total of 352 samples for analysis.

The samples were analysed for stable isotopic composition in the mass spectrometer, for ionic composition using ion chromatography and for dust concentration using the Coulter Counter technique. All analysis were performed on aliquots of the same sample. The δ^{18}O profile of the long D-sequence was used to date the profiles, and the other profiles were plotted on top for comparison (fig. 1b). The profiles are slightly shifted relative to one another due to the uneven thickness of certain layers. The depth of some isochrones, as defined by characteristic features of the δ^{18}O profiles, is shown in fig. 1a. The bottom isochrone, autumn 1989, was deepest in the D profile.

Snow layers of uneven thickness were only detected at three depth intervals, all in the vicinity of winter snow with low δ^{18}O values, and all in connection with observed layers of loose hoar-type snow. In order to compare the seven profiles on a sample by sample basis, it was necessary to remove the effects of uneven thicknesses. Therefore, the depth scales of the other profiles were adjusted so that their δ^{18}O profiles matched the δ^{18}O profile of the D profile. First, the A,G,J,M,P and S profiles had to be shifted downwards by 2.5 cm at 112.5 cm depth, then the A,G,J and S profiles were shifted downwards 2.5 cm at 155 cm depth, the A,G and S profiles were shifted downwards 2.5 cm at 207.5 cm and finally the J,M and P profiles were shifted downwards by 5.0 cm at 207.5 cm depth. This was done by inserting gaps without values in the sample sequences. The adjusted δ^{18}O profiles are shown in fig. 1c, where the dating is also indicated.

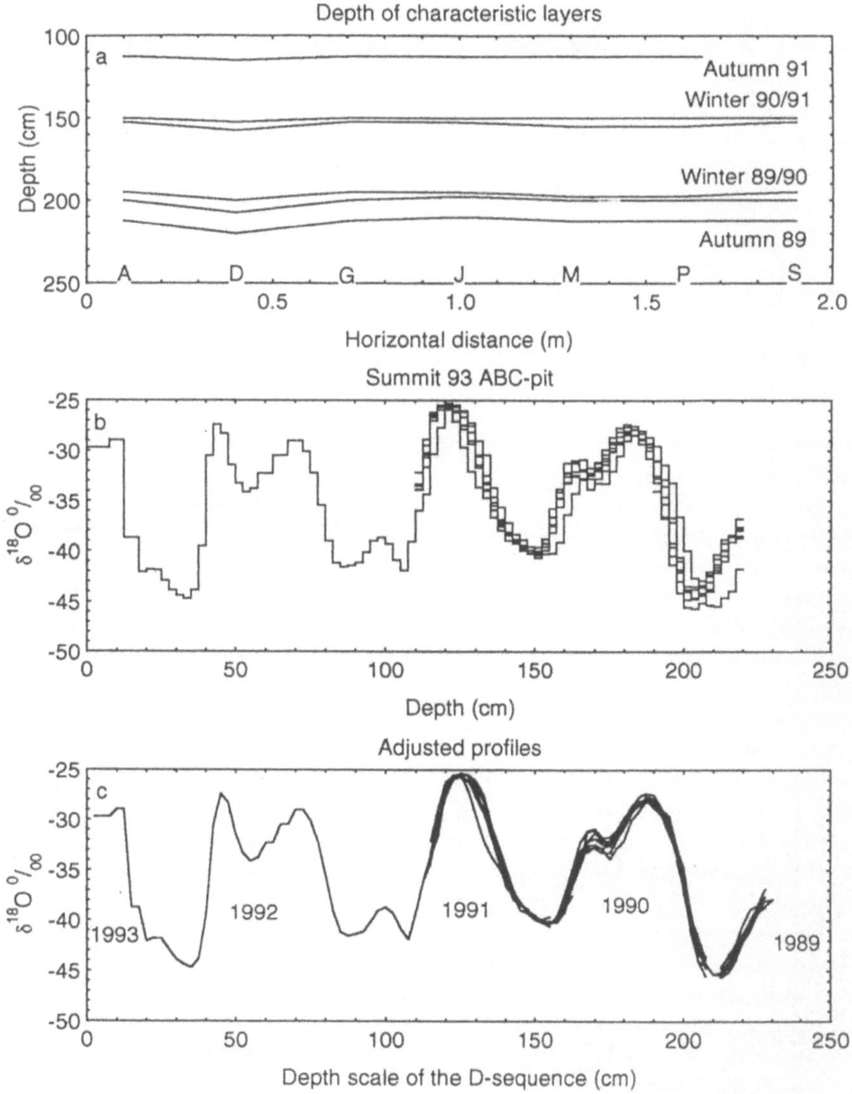

Figure 1: (a)Vertical cross section of the pit wall showing the depth of isochrones as defined by $\delta^{18}O$. The letters along the horizontal axis indicate the location of the 7 vertical profiles. (b) The $\delta^{18}O$ profiles of the 7 sequences plotted along the depth scale of the pit. (c) The $\delta^{18}O$ profiles after adjustment of 6 profiles to the depth scale of the D-sequence. This was done to remove the effects of the uneven isocrones.

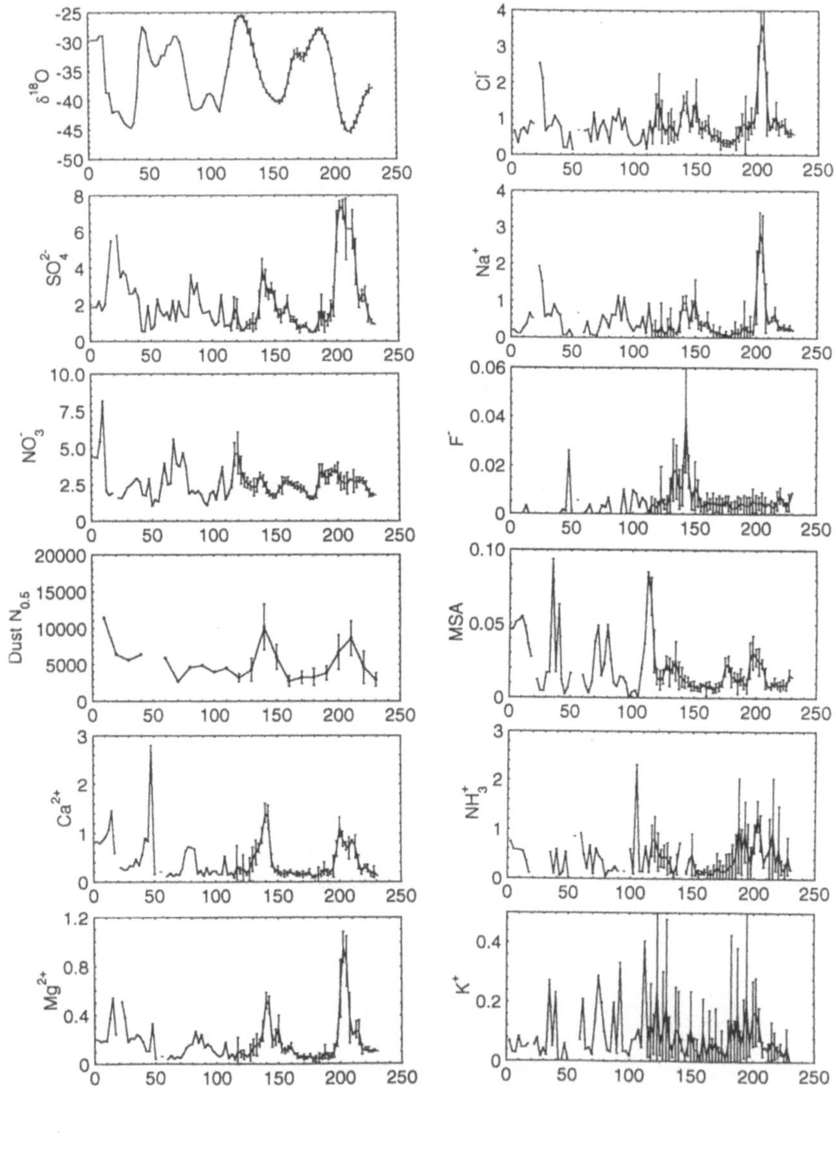

Figure 2: Vertical profiles of 12 components. The horizontal axes show the depth in cm on the depth scale of the D-sequence. $\delta^{18}O$ values are in per mille. Dust concentrations are in number of particles larger than 0.5 μm radius per ml. All other concentrations are in μequiv./kg. Between 110 cm and 230 cm the curves show the mean of the 7 profiles. The error bars show the single sample standard deviation at each depth interval. Between 0 cm and 110 cm only the D-sequence

For each component in our experiment, the seven profiles were now treated as seven independent snow records of the same time interval. The profiles were stacked, and the resulting mean curves are shown in fig. 2. The standard deviations of the individual samples contributing to the mean curves is indicated in fig 2 as error bars. The mean curves and the error bars thus show the signal to noise ratio of the individual components. In other words, if a new snow profile were to be analysed in close vicinity of the seven there is a 70 % likelihood that this new profile will fall within the region defined by the error bars.

Two factors contributing to the noise in the records are known: 1) measurement uncertainty, which can be quantified by calibration and repetitive measurements, and 2) depositional noise from the uneven distribution of the components in the snow pack. In Table 1 we give a comparison of the measurement uncertainty and the mean of the observed noise of the individual samples. The depositional noise is generally 2-6 times higher than the measurement uncertainty, but it varies between the components.

	Na^+	NH_4^+	K^+	Mg^{2+}	Ca^{2+}	$\delta^{18}O$
Mean values of all profiles	0.45	0.4	0.08	0.19	0.39	-34.53
Average single sample std. dev.	0.23	0.31	0.09	0.06	0.12	0.6
Single sample std.dev. (in % of mean)	51	78	110	32	31	-
Measurement uncertainty at mean values	0.05	0.06	0.04	0.02	0.04	0.1
Measurement uncertainty (in %)	10	15	50	10	10	-

	F^-	MSA	Cl^-	NO_3^-	SO_4^{2-}	Dust
Mean values of all profiles	0.007	0.016	0.95	2.12	2.62	5000
Average single sample std. dev.	0.005	0.005	0.3	0.44	0.39	1500
Single sample std.dev. (in % of mean)	71	31	32	21	15	30
Measurement uncertainty at mean values	0.002	0.002	0.05	0.11	0.13	500
Measurement uncertainty (in %)	25	10	5	5	5	10

Table 1: The average standard deviations between individual samples of the 7 profiles compared to the mean concentrations and the measurement uncertainty at mean concentration levels. Concentrations of the ionic species are in μequiv./kg, of dust in number of particles larger than 0.5μm radius per ml. $\delta^{18}O$ values are in per mille.

Although the standard deviation of $\delta^{18}O$ is 0.6 per mille, i.e. six times the measurement error of 0.1 per mille, this does not seriously affect the determination of the $\delta^{18}O$ profile. The

$\delta^{18}O$ profile is the best determined in our experiment. This is due to the fact that the seasonal signal of $\delta^{18}O$ has a large amplitude, and that the signal at this depth already has a smooth seasonal behaviour.

The profiles of nitrate, sulphate, calcium, magnesium and MSA (methane sulphonic acid) all show a well defined signal. The noise does not increase when concentrations go up. We suggest two reasons for this: 1) These components are all determined with a high accuracy in the ion chromatograph, and 2) The components are evenly distributed within the snow pack relative to the sampling resolution of 2.5 cm. The sodium, chloride and fluoride profiles show increased noise levels when concentrations are high. From ice cores we have the experience, that although the sea salt concentration has a winter high and a summer low, we frequently encounter narrow layers with high sea salt concentrations like the one in fig.2 at 202.5 cm. Examination of the individual profiles reveals, that occasionally the concentration of these three components is highly variable between samples, and when profiles are compared, concentration peaks only show up in the vicinity of each other, but not always at the same depth. This indicates that these components are not always evenly distributed in the snow pack relatively to the 2.5 cm resolution.

The dust concentrations are given as $N_{0.5}$, i.e. number of particles with radii larger than 0.5 μm per ml (As a rule of thumb, $N_{0.5}/150$ is approximately the dust concentration in μg/kg). Due to limited amounts of sample, dust analysis was performed with a depth resolution of 10 cm. The standard deviation of the dust profile is about 20-30% of the particle count. The measurement uncertainty of $N_{0.5}$ has been experimentally determined to be about 10% of $N_{0.5}$. The dust results also appear to reflect uneven distribution in the snow pack. The high standard deviations of the two spring concentration peaks are caused by the almost absence of the 1990 peak in the S-profile, and a large variation between the profiles of the peak concentration in the narrow spring 1991 peak. Still, the seasonal signal in the dust concentration record is resolved, due to the character of the signal: autumn/winter low concentrations and spring/summer high concentrations with relatively long temporal duration.

The ammonium and potassium records are the most noisy, and the seasonal signals of ammonium and potassium are poorly defined. Examination of the individual profiles reveals erratic variations in concentration up to a factor of 10 between 2.5 cm samples. Although there is a risk of sample contamination by gaseous ammonia in laboratory air, and often potassium concentrations are close to the detection limit of the ion chromatograph, the variations are much too large to be ascribed to contamination and experimental uncertainty. They must reflect that ammonium and potassium is very unevenly distributed in the snow pack with alternating snow layers of high and low concentrations with thicknesses less than 2.5 cm. Only around 202.5 cm depth did all seven profiles show elevated potassium concentrations in three consecutive samples, this concurs with the sodium and chloride concentration peak, and is probably due to the presence of sea salt potassium (in sea salt, the molar ratio of potassium and sodium is 0.02).

Two full seasonal cycles were covered by the seven profiles: 1989/1990 and 1990/1991. For each profile we calculated the two annual mean values by taking the mean of the results. In Table 2 we show the mean of the 7 annual means from both years and the calculated standard deviations on the mean values of the individual profiles. If we assume that measurement uncertainty is unbiased, then this contribution to the standard deviation becomes negligible as we now consider the mean values of approx. 25 measurements. The standard deviation on the annual mean values therefore reflect the noise from the inhomogeneous snow pack.

		1989/1990				**1990/1991**	
Component:	**Mean**	**Std.dev.**	**in %**		**Mean**	**Std.dev.**	**in %**
Na^+	0.49	0.03	6		0.40	0.09	23
NH_4^+	0.44	0.16	36		0.35	0.1	29
K^+	0.08	0.04	50		0.08	0.05	63
Mg^{2+}	0.20	0.02	10		0.18	0.01	6
Ca^{2+}	0.35	0.04	11		0.44	0.04	10
F^-	0.004	0.002	50		0.010	0.004	40
MSA	0.015	0.001	7		0.017	0.002	12
Cl^-	0.97	0.07	7		0.91	0.12	13
NO_3^-	2.45	0.06	2		1.71	0.07	4
SO_4^{2-}	2.56	0.09	4		2.70	0.12	4
dust ($N_{0.5}$)	4800	1500	31		5300	1000	19
$\delta^{18}O$	-35.33	0.4	-		-33.54	0.2	-

Table 2: The mean of the annual means of the 7 profiles, and the standard deviation of the individual annual means. Concentrations of the ionic species are in µequiv/kg, of dust in number of particles larger than 0.5 µm radius per ml. $\delta^{18}O$ values are in per mille.

Our results have relevance when ice core profiles are considered. In analyses with sub annual sample resolution to resolve the seasonal variations of the various components, the seasonal signals of $\delta^{18}O$, sulphate, nitrate, calcium, magnesium, sodium, chloride, MSA and dust can be considered to be representative of the local snow strata. This is less the case for fluoride and ammonium , and not the case for potassium. If sampling of annual layers has been performed to analyse the inter-annual variation of the components in an ice core, then the local depositional noise has to be considered before the results are interpreted. Different parameter values of two annual layers only reflect significant changes of the composition of the local snow pack if the values differ by more than 1-2 times the standard deviations given in Table 2. The percentages of the annual signal to noise ratio in Table 2 indicate how well mean annual values represent local snow of individual years. If mean values of several years are considered the effects of the local depositional noise will be reduced.

In this work we have only dealt with local depositional noise, i.e. how well a sample profile represents the snow composition in an area of approx. 5 m^2 around the sampling site in the region around Summit, Greenland. Comparison of our coefficients of variation (standard deviation/mean) of the annual mean values in Table 2 with the 1994 survey of J.Dibb (1996) shows that local depositional noise is comparable to the variability found over a 1 km^2 area. However, our coefficients of variation of the sea salt and dust derived components (sodium, chloride, magnesium and calcium are lower than the values of J.Dibb. We believe this is due to fact that our values are based on full years of precipitation, including winter and spring peak concentrations of sea salt and dust, whereas the values of J.Dibb are based on summer snow.

The results of this investigation of local depositional noise only reflect conditions in the Summit area in Greenland, where the rate of precipitation is 0.23m ice equiv./year. At locations with lower rates of precipitation the effects of depositional noise on the annual mean values will be higher especially for impurity components with uneven distribution within the snowpack.

References

Alley, R.B., D.A.Meese, C.A.Shuman, A.J.Gow, K.C.Taylor, P.M.Grootes, J.W.C.White, M.Ram, E.D.Waddington, P.A.Mayewski and G.A.Zielinski, 1993, Abrupt increase in Greenland snow accumulation at the end of the Younger Dryas event. Nature 362, p 527 - 529.

Clausen, H.B. and C.U. Hammer, 1988, The Laki and Tambora eruptions as revealed in Greenland ice cores from 11 locations. Annals of Glaciology 10, 1988, p 16 - 22.

Dansgaard, W., J.W.C.White and S.J.Johnsen, 1989, The abrupt termination of the Younger Dryas climate event. Nature 339, p 532 - 534.

Delmas, R.J. (ed), 1995, Ice core studies of global biogeochemical cycles. NATO ASI Series I, 30, Springer Verlag, Berlin, Heidelberg, 1995.

Dibb, J.E., 1996, Overview of field data on the deposition of aerosol-associated species to the surface snow of polar glaciers. In Wolff E.W., Bales, R.C. (eds) Chemical exchange between the Atmosphere and Polar snow. Berlin Heidelberg, Springer-Verlag, NATO ASI Series I (this volume).

GRIP project members, 1993, Climate instabillity during the last interglacial period recorded in the GRIP core. Nature 364, p 203- 220.

Mayewski, P.A., L.D.Meeker, S.Whitlow, M.S.Twickler, M.C.Morrison, R.B.Alley, P.Bloom field and K.Taylor, 1993, The atmosphere during the Younger Dryas. Science 261, p 195 - 197.

Wåhlin, P., 1996, One year's continuous aerosol sampling at Summit in Central Greenland. In Wolff E.W., Bales, R.C. (eds) Chemical exchange between the Atmosphere and Polar snow. Berlin Heidelberg, Springer-Verlag, NATO ASI Series I (this volume).

FLUOROCARBON TRACERS OF THE AGE OF AIR IN ALPINE FIRN

William T Sturges and Stuart A Penkett
School of Environmental Sciences
University of East Anglia
Norwich NR4 7TJ
England

Jean-Marc Barnola and Jerome A Chappellaz
CNRS Laboratoire de Glaciologie
BP 96, 38402 St Martin d'Hères Cedex
France

Introduction

As snow accumulates on ice sheets it is slowly compressed and sintered, ultimately forming solid ice, typically at depths of between 50 and 100 m. Above this depth, in the unsolidified firn, the interstitial air is still in connection with the surface atmosphere. Molecular diffusion, however, retards the attainment of equilibrium at the base of the firn with any changes in surface concentration (Raynaud *et al.* (1993); Schwander *et al.* (1993)). As a result, concentrations at depth in the firn may reflect atmospheric conditions years, or even decades, earlier.

There have been few reported studies to date on the composition of air in firn (Raynaud *et al.* (1993); Schwander *et al.* (1993); Bender *et al.* (1994)). We report here the first measurements of a number of trace halocarbons in firn air. Some of these compounds are very long-lived in the atmosphere (e.g. >1000 y for CFC-115 and C_2F_6) and have exhibited marked growth rates over the past twenty years, making them ideal candidates for determining the age of firn air. The work described here was a pilot study, aimed primarily at validating collection and analytical procedures, and assessing the potential for such measurements of trace species in firn and ice. Work is now in hand to study older firn air in Antarctica.

Methods

Firn air samples were collected with a core hole pumping system (Schwander *et al.* (1993)) from Col du Dome, below the summit of Mont Blanc, in June of 1994. Electropolished stainless steel flasks were pressurized, without drying, to c. 2 bars above ambient. Samples were collected at two depths: 31 and 45 m; and also at the surface, through

NATO ASI Series, Vol. I 43
Chemical Exchange Between the Atmosphere and Polar Snow
Edited by Eric W. Wolff and Roger C. Bales

the entire sampling train, as a control. The flasks had been shipped to Col du Dome with c. 2 bars of air, collected the previous month at the UEA Weybourne Atmospheric Observatory (north Norfolk coast) during humid, clean northerly air flow, to condition the flasks. The fourth bottle was returned unopened, and analysed as a comparison to background marine atmospheric composition.

On return to UEA, the flasks were analysed within a few weeks with a VG 'Autospec' gas chromatograph-triple-sector mass spectrometer combination (GC-MS). For each measurement, 200 ml of air was dried through magnesium perchlorate, trapped on glass beads at -185°C, and injected at ~90°C into the GC-MS. The compounds reported here were all separated with an Al-PLOT capillary column (except HCFC-141b, from a separate analysis with a CP-Sil 13 CB column). Calibrations were performed by sequential static dilution of pure gases (Penkett *et al.* 1979). Absolute concentrations reported here are preliminary and subject to final calibration.

Sub-surface concentrations were corrected for gravitational enrichment according to Craig *et al.* (1988). The transition from turbulent to diffusion-dominated transport was assumed to occur at 13 m in common with Bender *et al.* (1994). A mean temperature of -15°C in the firn was assumed.

Results

Figure 1 shows concentration profiles of selected fluorinated gases in the firn. Separate measurements of CO_2 and CH_4 (Barnola and Chappellaz, unpublished data) showed that pore close-off occurred at around 65 m. The results here, therefore, represent relatively recent air in the upper part of the firn. Results for the "background" air sample are plotted at -5 m for comparison. Open symbols are the mean concentrations at each depth, whereas individual analyses are marked with crosses (two or three measurements per depth); although in many cases these are hidden by the open symbols. A notable exception to the otherwise high precision was the HCFC-141b measurement in the background air sample, possibly linked to excessive humidity. A vertical dashed line has been drawn through the surface firn sample as a visual aid. An important observation is that, for the species reported here, good agreement was observed between the background air sample and the surface firn sample. This indicates an absence of artifact effects (contamination or absorptive losses) from the firn air pumping system.

We will use CFC-115 here as the primary dating tool, since it has been reported to exhibit an almost linear growth rate in the Northern Hemispheric (NH) since 1987 (Daniel *et al.* (1995)), increasing by about 0.44 pptv y^{-1} between 1992 and 1994. CFC-115 decreased

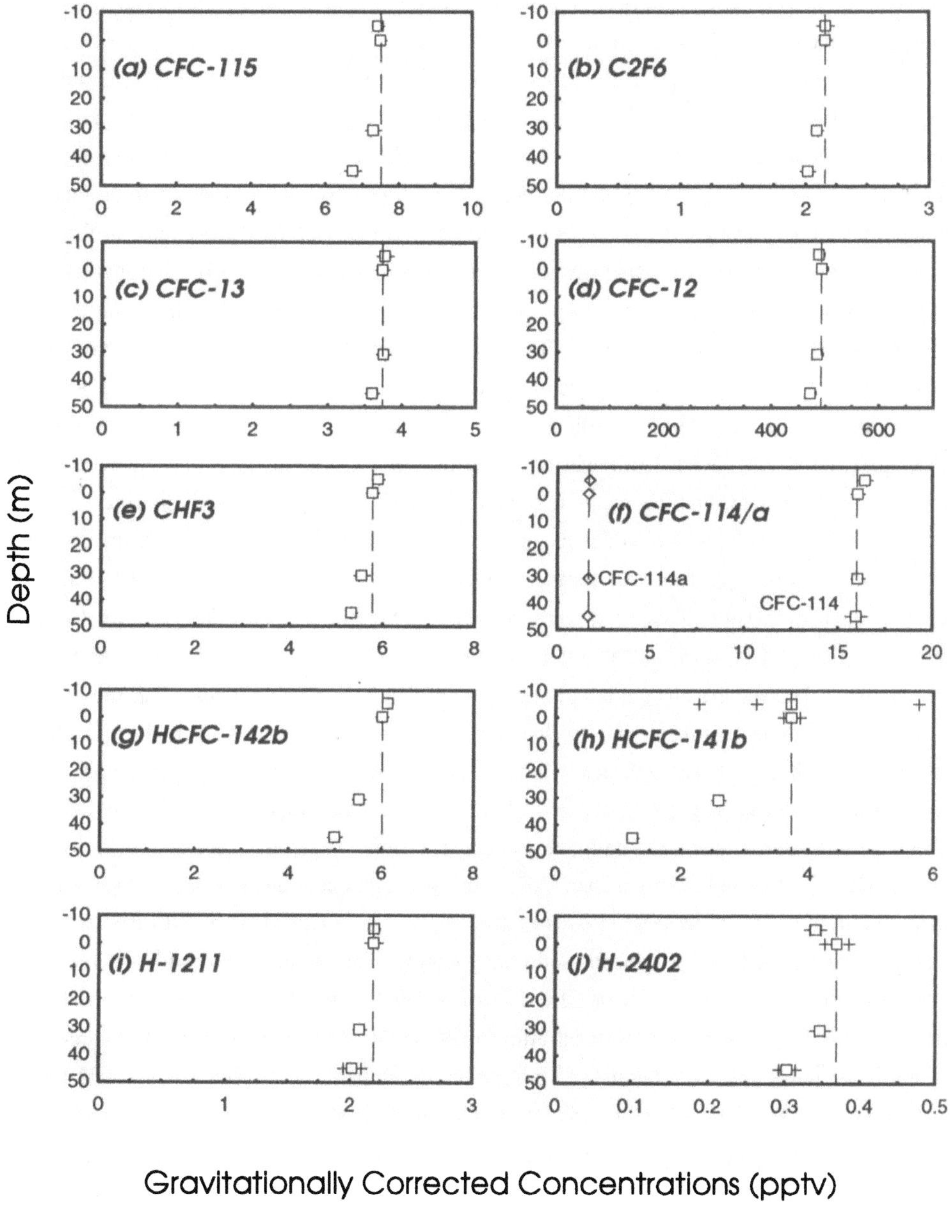

Figure 1. Concentration profiles of selected fluorocarbons in firn air from Col du Dome. The sample plotted at -5 m represents clean maritime air. Open symbols are mean concentrations at each depth. Crosses represent individual analyses.

from 7.50 pptv in the surface firn to 6.74 pptv at 45 m (Figure 1a). Scaling our observed concentrations to those of Daniel *et al.* (1995) (5.6 pptv in mid-1994) equates to an age of 15 months for the CFC-115 concentration at 45 m.

The age of other gases in the firn air will not be the same as that of CFC-115, and will depend on such parameters as the relative diffusion rates in firn, the concentration gradients, and variations in the tropospheric trend over time. To properly assess the age profiles of trace gases in the firn requires the application of a detailed transport model in which these factors, along with the physical characteristics of the firn, are known. The diffusion rates in firn are, however, unknown for the majority of trace gases, and ideally should be determined experimentally for each particular firn profile. Some estimate may be made by using the gas phase diffusion coefficient (D) instead (Table 1). Bender *et al.* (1994) suggested that a first approximation of the age of a trace gas in firn can be made by correcting the age of a 'known' gas by the ratio of the diffusion coefficients of the two gases, e.g. if D for a gas is half that of CO_2, it's age at any given depth will be twice that of CO_2. This is applicable where the concentration gradients and tropospheric trends are similar (and approximately linear).

Table 1 shows the percentage changes in composition between the surface firn and the firn at 45 m ($-\Delta C_{45m}$) from our measurements. Also shown are the estimated diffusion coefficients in air where values of η were found in the literature. The value of D for C_2F_6 is similar to that of CFC-115, and therefore should have a similar age, all other factors being equal. The smaller change in concentration at depth therefore suggests that the growth of C_2F_6 was roughly half the c. 8% y^{-1} change observed for CFC-115. An actual tropospheric trend for C_2F_6 has not been published. Our own preliminary measurements of air samples from Cape Grim, Tasmania suggest an increase of about a factor of 2.5 between 1978 and 1992 (D Oram and P Fraser, unpublished data), equivalent to a trend of roughly 4% y^{-1} in 1992.

CFC-12 declined even less than C_2F_6 in the firn, again consistent with the observed low growth rate of c. 2% y^{-1} in 1993 (Elkins *et al.* (1993)). Both CFC-114, and it's isomer CFC-114a, showed no change in concentration with depth. This is consistent with the reported large reduction in emission rates of CFC-114 since 1987 (Kaye *et al.* (1994)), which should have stabilized the atmospheric concentrations of these long-lived compounds. The values of η and D are not known for Halon-1211 but, given the similarity in molecular mass with CFC-115, we might expect similar values for D. This would infer a growth rate of about 6% y^{-1} for Halon-1211, in agreement with the reported tropospheric trend of 3-7% y^{-1} in 1993 (Thompson *et al.* (1994)).

The two HCFCs showed very high apparent growth rates from the firn profiles. Recent large increases in tropospheric concentrations have been reported by Montzka *et al.* (1994). They noted that mean global HCFC-142b concentrations were increasing by about 26% y^{-1} in mid-1993. They also reported a three-fold increase in HCFC-141b during 1993. The relative sizes of these increases are consistent with those observed in the firn, but we must be cautious in our interpretation of these data due to the very different concentration gradients. Tropospheric trends of the other trace gases in Table 1 are unknown, although we are beginning to determine these from the Cape Grim archive.

Table 1. Summary of diffusion coefficients (D) and the measured decrease in gravitationally-corrected concentrations from the firn surface to 45 m ($-\Delta C_{45m}$). Diffusion coefficients were estimated from gas phase viscosities (η) at STP.

Compound	Molar mass (g)	D (10^{-6} m^2 s^{-1})	$-\Delta C_{45m}$ (%)
CFC-115	155	2.04[a]	10.1
C_2F_6	138	2.57[b]	6.5
CFC-13	105	3.42[b]	3.7
CFC-12	121	2.56[b]	4.3
CHF_3	70	5.07[b]	7.6
CFC-114	171	1.73[b]	0.5
CFC-114a	171	nk	0.6
HCFC-142b	122	2.61[c]	17.0
HCFC-141b	108	2.09[c]	66.9
Halon-1211	165	nk	7.8
Halon-2402	260	nk	18.9

Viscosity data references: (a) Touloukian *et al.* (1975), (b) Gallant (1968), (c) ICI Chemicals and Polymers, unpublished data (A McCulloch, personal communication); nk = not known.

Conclusions

We have shown the potential for making high-precision measurements of a number of fluorinated gases in firn air. There was little evidence of sampling artifacts. Although only shallow firn, containing quite recent air, was examined, concentration trends could still be detected, and reasonable convergence was obtained between these and atmospheric trends, where these are known.

Acknowledgments. We gratefully acknowledge the cooperation of D Oram, P Fraser, J Daniel, S Schauffler, J Butler, S Montzka and A McCulloch. This work was supported in part by the Natural Environment Research Council. We thank the Région Rhône-Alpes and the Programme Phase Atmosphérique des Cycles Biogéochimiques du CNRS for financial support of the drilling operations at Col du Dôme.

References

Bender ML, Sowers T, Barnola J, Chappellaz J (1994) Changes in the O_2/N_2 ratio of the atmosphere during recent decades reflected in the composition of air in the firn at Vostok Station, Antarctica. Geophys Res Lett 21:189-192

Butler JH, Elkins JW, Hall BD, Cummings SO, Montzka SA (1992) A decline in the growth rates of atmospheric Halons. Nature 359:403-405

Craig H, Horibe Y, Sowers T (1988) Gravitational separation of gases and isotopes in polar ice caps. Science 242:1675-1678

Daniel JS, Schauffler SM, Pollock WH, Solomon S, Weaver A, Atlas E, Heidt LE, Garcia RR, Vedder JF (1995) On the age of stratospheric air and inorganic chlorine release. J Geophys Res, submitted

Elkins JW, Thompson TM, Swanson TH, Butler JH, Hall BD, Cummings SO, Fisher DA, Raffo AG (1993) Decrease in the growth rates of atmopsheric chlorofluorocarbons 11 and 12. Nature 364:780-783

Gallant RW (1968) Physical Properties of Hydrocarbons. Gulf Publishing Houston

Kaye JA, Penkett SA, Ormond FM (1994) Report on Concentrations, Lifetimes, and Trends of CFCs, Halons, and Related Species. NASA Reference Publication 1339. NASA Washington DC

Montzka SA, Myers RC, Butler JH, Elkins JW (1994) Early trends in the global tropospheric abundance of hydrochlorofluorocarbon-141b and -142b. Geophys Res Lett 21:2483-2486

Penkett SA, Brice KA, Derwent RG, Eggleton AEJ (1979) Measurements of CCl_3F and CCl_4 at Harwell over the period January 1975 - November 1977. Atmos Environ 13:1011-1019

Raynaud D, Chappellaz J (1993) The record of atmospheric methane, *In* Atmospheric Methane: Sources, Sinks, and Role in Global Change. Springer-Verlag Berlin Heidelberg, 38-61.

Schwander J, Barnola J, Andrie C, Leuenberger M, Ludin A, Raynaud D, Stauffer B (1993) The age of air in the firn and the ice at Summit, Greenland. J Geophys Res 98(D2):2831-2838

Thompson TM, Elkins JW, Butler JH, Montzka SA, Myers RC, Baring TJ, Cummings SO, Dutton GS, Gilligan JM, Hayden AH, Lobert JM, Swanson TH, Hurst DF, Volk CM (1994) Chapter 5. Nitrous Oxide and Halocarbon Division, *In* Climate Monitoring and Diagnostics Laboratory Summary Report 1993 (No. 22), Peterson JT, Rosson RM (eds). US Department of Commerce, Boulder Colorado, 72-91

Touloukian YS, Saxena SC, Hestermans P (1975) Viscosity. IFI/Plenum Press New York

THE PEROXIDE RECORD FROM THE DSS ICE CORE, LAW DOME, ANTARCTICA: PRELIMINARY RESULTS

Tas van Ommen and Vin Morgan
Antarctic CRC and Australian Antarctic Division
GPO Box 252C
Hobart Tasmania 7001
Australia

Introduction

Hydrogen peroxide is a major trace constituent of polar ice cores, and is the only archived species so far detected that reflects atmospheric oxidation capacity (Neftel & Fuhrer 1993). As has been noted by Neftel (this volume) and Bales & Choi (this volume) our ability to interpret this archived record is limited by our lack of understanding of the atmosphere to snow transfer function. The Law Dome site, and in particular the Dome Summit South (DSS) ice core, offers distinct features which may well shed light on the difficult issue of the atmosphere/ice relationship. Firstly, the high accumulation rate (0.7 m/yr ice equivalent) leads to well preserved seasonal cycles which, in conjunction with atmospheric measurements or models, should provide useful constraints on exchange processes. Secondly, the climatology of the site (mean temperature -20.8°C) leads to a near total absence of summer melt, which is known to interfere with both measured peroxide levels and diffusion processes in the firn. Thirdly, and related to the first point, there is a significant gradient in accumulation across Law Dome. This means that the timescale of firnification differs significantly between the DSS core and other sites which sample the same air mass (e.g. DE08, 16 km from DSS, with approximately twice the accumulation rate). Consequently if post-industrial changes in H_2O_2 levels are detected (Sigg & Neftel 1991), these can be distinguished from firnification effects.

The DSS site (lat. -66°46′, long. 112°48′) is located 4.6 km SSW of the summit of Law Dome in Wilkes Land, Antarctica. The core was drilled over successive summers from 1988/9 to 1992/3 when drilling reached silty ice near the base of the ice sheet, 1200 m below surface (mbs) and provides a record which extends back in time beyond the last glacial maximum (LGM). The preservation of seasonal cycles in $\delta^{18}O$ and H_2O_2 in the DSS core permits accurate dating by layer counting through most of the Holocene period. Further site and drilling details may be found elsewhere (Morgan et al. 1994; Morgan et al. 1995).

Analysis

Measurements of H_2O_2 concentration have been made using the enzymatic fluorometric method (Sigg & Neftel, 1988; Lazarus et al. 1985) at intervals along the full 1200 m length of the ice core. Discrete samples have been cut at intervals to provide typically 10-12 samples per annual layer (fine sampling), as determined by initial $\delta^{18}O$ measurements, electrical conductivity

NATO ASI Series, Vol. I 43
Chemical Exchange Between the Atmosphere and Polar Snow
Edited by Eric W. Wolff and Roger C. Bales

measurements (ECM) and ice sheet modelling. Tests of the measurement system show that the absolute accuracy of individual measurements is approximately 1.3 ng/g, with a relative accuracy over several successive samples of approximately 0.5 ng/g.

Long-term record

Figure 1 shows the long-term peroxide and oxygen isotope record for the DSS core and Figure 2 shows detail of the peroxide signal in the region ~220–240 mbs. The oxygen isotope record clearly shows that the majority of this ice corresponds to the Holocene period, with the transition to the LGM at around 1120-1130 mbs. There is no clear indication in the data so far of any in-ice chemical transformation of H_2O_2 over the past 3-4 kyr, with mean H_2O_2 concentrations near 40 ng/g. This contrasts with the data from Greenland ice cores discussed by Neftel (this volume). In particular, there is no sign of the narrow depletions seen in the Greenland ice and it seems likely that this is the result of the differing accumulation and impurity levels at the two sites.

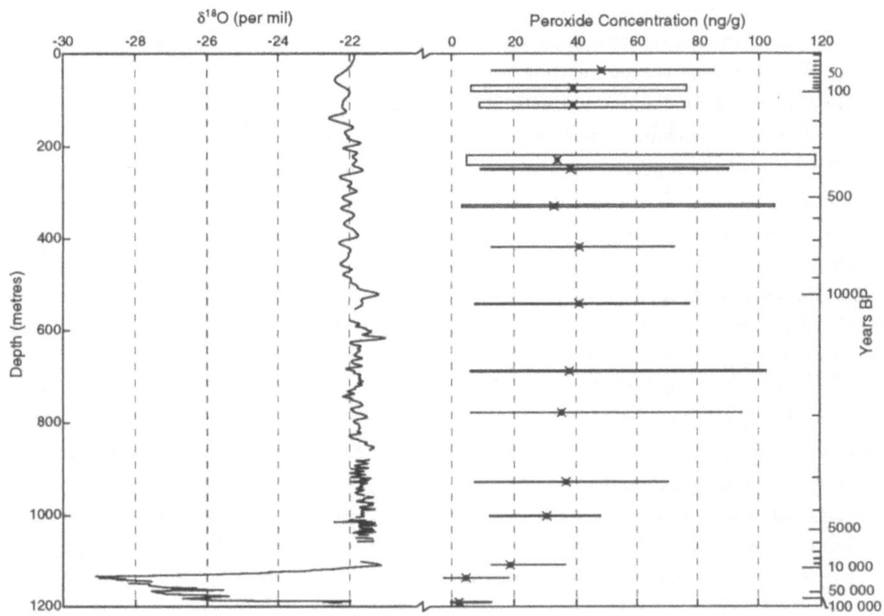

Figure 1: Average $\delta^{18}O$ and H_2O_2 records from the DSS core. The raw average $\delta^{18}O$ measurements have been smoothed with a gaussian filter having an rms of ~7–10 years to reveal the underlying trend. The markers on the peroxide plot denote mean peroxide levels at each depth. The superimposed boxes indicate the depth range measured at each point and the total spread of measurements in that interval. Above 1000 mbs this spread defines the envelope of the seasonal variation. Below 1000 mbs, the sampling is not sufficiently fine to resolve annual cycles. At depths above 400 mbs, the age scale is accurately determined by counting annual layers. Below 400 mbs, the age-scale is determined by fitting an ice flow model to observed layer thicknesses (see text).

There is some indication of a reduction in mean H_2O_2 levels beyond 3-4 kyr BP, and values just above the Wisconsin/Holocene transition (estimated age ~9 kyr) are unambiguously reduced, at approximately 19 ng/g.

As with other ice cores (see for example Neftel et al. 1986), Wisconsin ice shows very low levels of peroxide, typically less than 5 ng/g. There is some uncertainty as to the absolute values measured in this range as they may be significantly affected by handling-induced impurities. The magnitude of such contamination is a subject of on-going investigation. Detailed analysis of recent changes in H_2O_2 levels is planned, using the upper sections of the DSS core and proposed firn cores to be collected in the austral summer of 1995/6. The seasonal variations, as indicated by the peak to peak amplitudes of the boxes in Figure 1, show maximum summer levels occasionally reaching almost 120 ng/g, and winter minima below 10 ng/g. It should be noted that the data have not been deconvolved to correct for diffusion.

Dating

Currently, fine $\delta^{18}O$ measurements provide accurate dating by counting of annual cycles at depths down to approximately 400 mbs. These have been used in conjunction with fine H_2O_2 measurements, sulphate (SO_4^{2-}) measurements and electrical conductivity measurements (ECM) to count individual annual cycles. The presence of irregularities due to surface transport of the snow and accumulation variability results in ambiguities which contribute an estimated uncertainty of ±5 years at 400 mbs (1304 AD), and proportionally less at shallower depths.

The ice flow has been modelled (Dansgaard & Johnsen 1969) to extend the dating to the lower sections of the core. The model has been used to fit measured annual layer thicknesses from spot $\delta^{18}O$ measurements down to 1000 mbs. Consequently, dating down to this level is well constrained (estimated uncertainty of ~5% at 4000 years BP). Dating below this level is by extrapolation of this fit. Reasonable parameter adjustments give ages at the Wisconsin/Holocene transition in the range 10000-11500 years BP. The adopted values place the transition near the middle of this range at 10900 years BP and the LGM at ~14600 years BP (±20%).

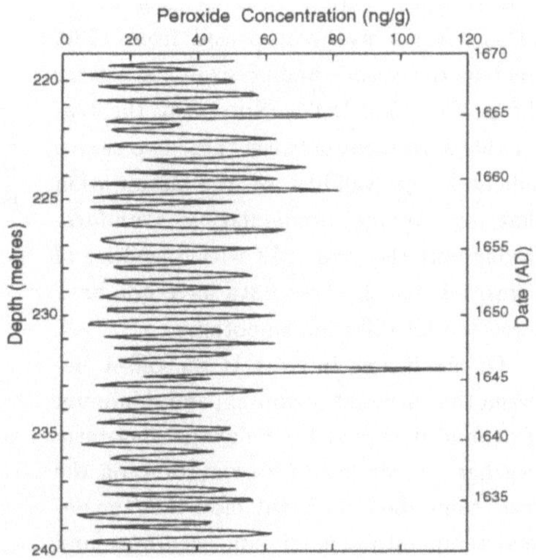

Figure 2: Detail showing fine-sampled (10-12 per year) H_2O_2 measurements over a ~20 m section of core.

Seasonal variation

Figure 3 shows mean seasonal variations in H_2O_2 and $\delta^{18}O$, computed from fine-sampled data over a span of 262 years. The original data, sampled at typically 10-12 samples per year have been dated using the $\delta^{18}O$ maxima as the defining points for the start of each year. The precise depth assigned to the start of each year was computed from a parabolic fit to the $\delta^{18}O$ curve in the region of the seasonal maximum. Sub-annual dating was achieved by assuming uniform annual accumulation. Finally, to facilitate time-series analysis and remove biases induced by the way in which dates were ascribed, the data have been interpolated onto an equally spaced grid using spline-curve interpolants. These data were then 'folded' to compute mean annual accumulation-weighted curves for both parameters. This comparatively long span of continuous H_2O_2 data come from preliminary field measurements which are subject to more poorly quantified errors and have somewhat lower accuracy than other data presented here. The $\delta^{18}O$ curve shows clearly the asymmetrical seasonal signature of polar temperature variations, with a broad shallow ('core-less') winter and a shorter peaked summer. This is confirmed by Figure 4, which shows this same curve superimposed on temperature data from an automatic weather station ~5 km from DSS. The weather station data (Allison et al. 1993) cover a five year period from 1986, whereas the isotope and peroxide data cover the period 1580-1842. Note that the reasonable agreement between these two curves indicates the validity of the assumption that, on average, accumulation is uniform throughout the year. As with the data of Figures 1 and 2, these data have not been corrected for diffusion smoothing.

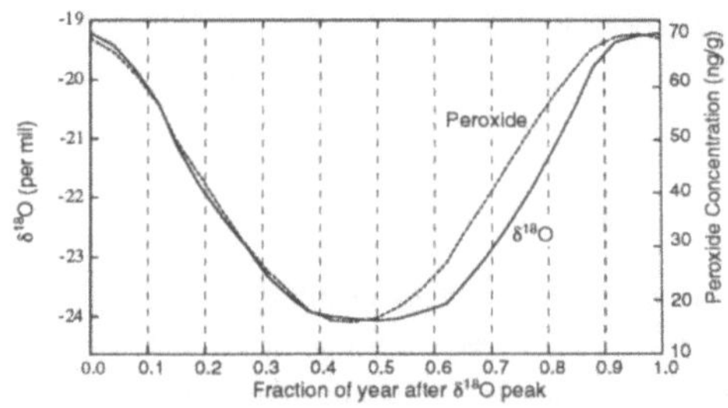

Figure 3: Average seasonal curves (262 seasons) for H_2O_2 and $\delta^{18}O$. The 'start' of the year is defined as the peak in the mean $\delta^{18}O$ curve.

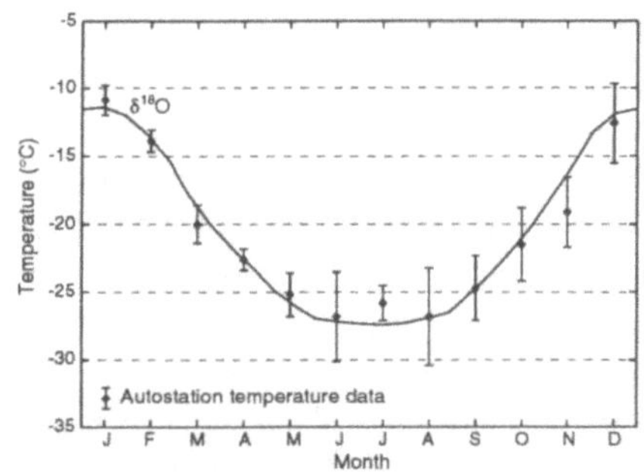

Figure 4: Automatic weather station temperature data at Law Dome over a 5 year period, with the mean $\delta^{18}O$ curve of Figure 3 overlaid for comparison.

Of significant interest is the offset between the curves at maximum and minimum of around 0.05 year (~18 days). Long-term weather records from Casey station, on the coast some 100 km distant show that the annual temperature maximum occurs around January 10, and so this offset fits well with the difference between the radiation maximum (December 22) and the temperature maximum. Published data from Greenland (Beer et al., 1991; Sigg and Neftel, 1988) show H_2O_2 lagging $\delta^{18}O$, while data from Siple in Antarctica (Sigg & Neftel 1988) show more or less coincident maxima and minima.

Ayers et al. (1995) have observations of atmospheric peroxides at Cape Grim, Tasmania, and photochemical modelling which show seasonal maxima in December and minima in June. Modelling for Macquarie Island (Ayers, private communication) shows the same timing, but suggests a more prolonged winter drop in peroxides with increasing latitude. Comparisons between the shape of the modelled seasonal atmospheric signal and the signal in Figure 3 are difficult to make since diffusion effects have smoothed out most sub-seasonal variability in the ice core record.

Bales & Choi (this volume) model snow H_2O_2 signals at Summit, Greenland, and predict a spring peak, produced as a consequence of rising atmospheric levels with increasing radiation while temperature remains low (for which the air-snow equilibrium favours increased snow uptake). The timing of the H_2O_2 peak in Figure 3 tells a different story, suggesting that for DSS, radiation dominates despite the equilibrium temperature dependence.

REFERENCES

Allison, I., Wendler, G. & Radok, U. (1993) Climatology of the East Antarctic Ice Sheet (100°E to 140°E) Derived From Automatic Weather Stations. J. Geophys. Res. 98:8815–8823

Ayers, G.P., Penkett, S.A., Gillett, R.W., Bandy, B., Galbally, I.E., Meyer, C.P., Elsworth, C.M., Bentley, S.T. & Forgan, B.W. (1995) The annual cycle of peroxides and ozone in marine air at Cape Grim, Tasmania. J. Atmos. Chem. (submitted).

Beer, J., Finkel, R.C., Bonani, G., Gäggeler, H., Görlach, U., Jacob, P., Klockow, D., Langway, C.C.Jr., Neftel, A., Oeschger, H., Schotterer, U., Schwander, J., Siegenthaler, U., Suter, M., Wagenbach, D., & Wölfi, W. (1991) Seasonal variations in the concentration of ^{10}Be, Cl^-, NO_3^-, SO_4^{2-}, H_2O_2, ^{210}Pb, ^3H, mineral dust and $\delta^{18}O$ in Greenland snow. Atm. Env. 25A:899–904

Dansgaard, W. & Johnsen, S.J. (1969) A flow model and a time scale for the ice core from Camp Century, Greenland. J. Glaciol. 8:215–223

Lazarus, A.L., Kok, G.L., Gitlin, S.N., Lind, J.A. & McLaren, S.E. (1985) Automated Fluorometric Method for Hydrogen Peroxide in Atmospheric Precipitation. Analyt. Chem. 57:917–922

Morgan, V., Wehrle, E., Fleming, A., Richardson, M., Elcheikh, A. & Brand, R. (1994) Technical aspects of deep ice drilling on Law Dome. Mem. Natl Inst. Polar Res., Spec. Issue 49:78–86

Morgan, V., Wookey, C.W., Li, J., Skinner, W. & Fitzpatrick, M.F. (1995) Site information and some initial results from deep ice drilling on Law Dome. J. Glaciol. (submitted).

Neftel, A. & Fuhrer, K. (1993) A record of atmospheric oxidant from polar ice cores over the past 100,000 years: Dream or real possibility? in The Tropospheric Chemistry of Ozone in the Polar Regions, Niki, H. & Becker, K.H. (eds.) NATO ASI Series, Vol. 17, Springer-Verlag

Neftel, A., Jacob, P. & Klockow, D. (1986) Long-term record of H_2O_2 in polar ice cores. Tellus 38B:262–270

Sigg, A. & Neftel, A. (1988) Seasonal variations in hydrogen peroxide in polar ice cores. Ann. Glaciol. 10:157–162

Sigg, A. and Neftel, A. (1991) Evidence for a 50% increase in H_2O_2 over the past 200 years from a Greenland ice core. Nature 351:557–559

WHERE ARE WE GOING?
THE ICE CORE - PALEOCLIMATE INVERSE PROBLEM

E.D. Waddington
University of Washington
Geophysics Program
Box 351650
Seattle WA 98195-1650
U.S.A.

INTRODUCTION

Significant progress has been made recently (*e.g.* Bales and Choi, this volume; Neftel this volume) toward understanding the *transfer functions* that describe how stable isotopes, chemical species, and other impurities get from the polar atmosphere into deep polar ice. This research is motivated by the reverse process, *i.e.*, a need to derive quantitative paleo-climate information from ice core data. Even when the transfer functions from atmosphere to ice are known, the *Inverse Problem*, *i.e.* inference of concentrations in the paleo-atmosphere from concentrations in ice cores, may be neither straightforward nor simple. The need to solve Inverse Problems in many different fields has lead to the creation of *Inverse Theory*. Inverse Methods (a) provide a conceptual framework in which to understand the limitations common to all efforts that use data to infer properties of physical systems, (b) determine how much information can be obtained from the data, and (c) assess the reliability of that information. The ability of Inverse Methods to assess the *Resolving Power* of the data (*i.e.* to determine how well the Inverse Problem can be solved) is a key feature of great value to the ice core Inverse Problem. Inverse Theory takes many forms; this paper outlines a few concepts to show how Inverse Methods can provide a useful focus for ice core - atmosphere paleoclimate studies.

WHAT IS AN INVERSE PROBLEM?

Most research in the earth sciences can be broadly assigned to one of two categories. In a very general sense, Forward Problems are theoretical, while Inverse Problems interpret experimental data. In a Forward Problem, we know a situation and we predict the outcome, *i.e.* we solve the evolution of a system for which the mechanical workings (*e.g.* the physics and chemistry) are understood. Starting the system from a known initial configuration, we try to predict the evolution and final state, as the system is pushed and prodded by various driving influences. When the system is defined mathematically, the mechanical workings are expressed by the controlling differential equations, the known initial configuration is the initial condition and the pushes and prods are the boundary conditions. For example, predicting the pattern of tree rings over the past 300 years by running a tree ring growth model with a specified history of rainfall and temperature is a Forward Problem.

NATO ASI Series, Vol. I 43
Chemical Exchange Between the Atmosphere and Polar Snow
Edited by Eric W. Wolff and Roger C. Bales

An Inverse Problem goes in the other direction. We know an outcome, and we want to learn about the situation from which it arose. Observations of a system in some final state constitute the *data*. A scientist, with some knowledge of how the system works, tries to use those data to determine what is called a *model* which can comprise (a) the initial state of the system, (b) the history of pushing and prodding that the system has undergone (boundary conditions), or (c) the numerical values of parameters that control the system (coefficients in the governing equations). For example, a scientist with a tree ring growth model might observe the densities of rings in a 300 year-old tree, and want to infer (a) the temperature during the Little Ice Age when the tree germinated (an initial condition), or (b) the temperature and precipitation histories over the past 300 years (boundary conditions), or (c) the (potentially changing) sensitivity of ring growth to summer temperature (a parameter in the governing equations). The Inverse Problem is conceptually distinct from the corresponding Forward Problem, and it presents its own special conceptual and computational challenges. The Inverse Problem can be solved only after the physics and chemistry of the Forward Problem are well-understood.

THE INVERSE PROBLEM FOR ICE CORES

The atmosphere - ice transfer functions describe the physical and chemical processes by which airborne tracers get from the atmosphere into the snow (and possibly out again) and how they ultimately become trapped in the ice. This is the heart of the ice core Forward Problem, as formulated by Neftel (this volume). Figure 1, following Neftel, shows a framework for both the Forward and the Inverse Problems. From a chemist's viewpoint, the concentrations of each species ${}_iC_j$ are the *state variables*. The concentrations ${}_iC_{atm}$ of the N constituents represent the initial condition of the atmospheric chemical system, and the transfer functions ${}_iF_j$ for each constituent i at step j express the evolution of the state variables, incorporating (a) their dependence on concentrations of other species ${}_iC_j$ through chemical reactions or physical processes, and (b) their dependence on external climate conditions such as temperature, wind, and accumulation rate. The coefficients or *functional parameters* in the transfer functions are determined experimentally, and have uncertainties associated with them. Some of these functional parameters may also be time-dependent or climate-dependent. The upper balloons in Figure 1 remind us that external climatic boundary conditions and physical and chemical functional parameters can affect the output of the transfer function. Dynamic meteorologists studying past climate might prefer to treat climate variables such as pressure, temperature, precipitation and insolation as the primary variables, with atmospheric chemistry included only insofar as it influences the dynamics. However, the chemist's and the meteorologist's viewpoints are just different perspectives on the ice core - paleoclimate Inverse Problem. Following Neftel (this volume) and others, I will take the chemists' viewpoint.

When we (1) know the history of concentrations ${}_iC_{atm}$ of N chemical species, isotopes, or other tracers in the free atmosphere, (2) fully understand the physical and chemical processes encapsulated in the transfer functions ${}_iF_1$, and (3) know the values of all the relevant climate parameters and other functional parameters that control ${}_iF_1$, we can then use those transfer functions to predict the concentrations ${}_iC_{atm\ local}$ of the same N chemical species in the local atmosphere above the ice.

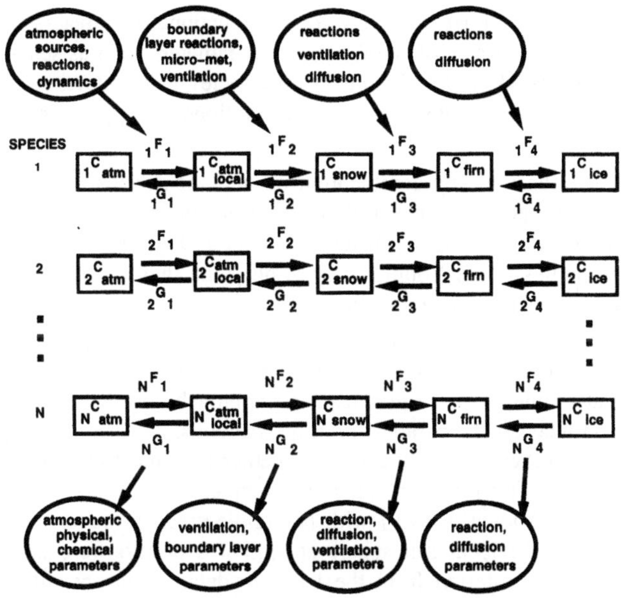

FIGURE 1: The Forward and Inverse Problems relating atmospheric chemical composition to ice core data. Boxes show the concentrations ${}_iC_j$ of N chemical or isotopic tracers as they move by steps from the atmosphere into the ice. The transfer functions ${}_iF_j$ represent the physical and chemical processes that control the concentrations at each step in their evolution. The balloons represent external and potentially uncertain functional parameters (including the climate state). The inverse functions ${}_iG_j$ use data to estimate the concentrations at previous steps. Since atmospheric composition and climate are coupled, the ${}_iG_j$ may also provide estimates of the functional parameters (including the climate) as indicated by the balloons. Where the resolving power of the data is poor, the ${}_iG_j$ may be unable to recover useful information about concentrations ${}_iC_j$ at the previous level.

Similarly, when we understand the functional form of all the physical and chemical processes encapsulated in the transfer functions ${}_iF_2$ acting in the local meteorological boundary layer and at the snow interface, and we know all the functional coefficients and relevant climate parameters controlling those processes (as indicated in the second balloon), we can predict the concentrations ${}_iC_{snow}$ in the near-surface snow. We can calculate the concentrations ${}_iC_j$ at each level j in turn, until we know the concentrations ${}_iC_{ice}$ in the ice cores. This is the ice core Forward Problem.

However, motivation for research on transfer functions is provided by the potential to use ice core data to derive (a) past atmospheric concentrations ${}_iC_{atm}$, (b) the climate state in the past, and possibly (c) the values of the functional parameters in the transfer functions. This is the corresponding ice core - paleoclimate Inverse Problem, as illustrated by the inverse functions ${}_iG_j$ in Figure 1. We can think of the set of all the ${}_iG_j$ in Figure 1 as one gigantic Inverse Problem for the simultaneous derivation of all the paleo-atmospheric concentrations ${}_iC_{atm}$, all the controlling climate conditions at all times in the past, and the values of all the functional parameters in the transfer functions. This complete

Inverse Problem has not been attempted for any ice core; however it offers a framework to develop a consistent quantitative determination of those unknowns. Any subset of the ${}_iG_j$ in Figure 1 also constitutes an Inverse Problem, right down to the jth individual step ${}_iG_j$ for a particular species i. In such an inversion we can solve for any or all of the initial concentrations ${}_iC_j$, the climate variables, or the functional parameters in the transfer function. For example, estimation of the functional parameters such as sensitivity to accumulation rate and wind speed in ${}_1F_2$ (Figure 1) for Species 1 constitutes an Inverse Problem, if the climate variables at the site and the concentrations ${}_1C_{atm\ local}$ and ${}_1C_{snow}$ are all provided by data.

We can also consider any combination of steps or species to be an Inverse Problem, particularly if we expect the concentrations of different species to be coupled through the transfer functions. For example, Mayewski et al. (1993) used multiple species to infer atmospheric properties in the Younger Dryas, although they did not attempt to follow concentrations back through each step in Figure 1.

Working backward from the data, *i.e.* the concentrations ${}_iC_{ice}$ in the ice core, through a series of smaller Inverse Problems, allows the concentrations at each previous level to be estimated one at a time; this is a practical first approach. At each step in the inversion we must also decide whether the parameters in the balloons, *i.e.* the climate parameters and the transfer function coefficients are known, or whether they must also be determined from the data as a part of the Inverse Problem. For example, reaction and diffusion parameters in the ice may be well-known *a priori* with negligible error; however, ventilation parameters and the dynamics of the local atmosphere at time of deposition (*e.g.* see Waddington et al., this volume), must be inferred from the ice core data. It may be difficult to separate the inversion for chemical concentrations from the inversion for paleoclimate conditions such as wind strength. Therefore, the Inverse Problems for paleoclimate conditions and for atmospheric concentrations may be inextricably coupled. Even worse, the atmospheric concentrations ${}_iC_{atm}$ may be unresolvable for some chemical species in time periods in which uncertainties in the climate variables are large.

While we hope that it is possible to determine the values of the functional parameters in the transfer functions *a priori*, they may also be coupled into the Inverse Problem. For example, the stable isotope $\delta^{18}O$ in ice is known to be a function of paleotemperature. The sensitivity of $\delta^{18}O$ to temperature (the functional parameter) is often assumed to be given by the modern spatial correlation of temperature and isotopes along an ice sheet surface. This relationship has often been used both to derive paleotemperatures and to decouple the paleoclimate inversion from the atmospheric concentrations Inverse Problem. However, Cuffey et al. (1995) used borehole temperature measurements to show that the sensitivity of $\delta^{18}O$ to temperature is significantly smaller than the modern spatial data suggest for climate variations that occur over time scales of thousands of years; this implies that temperatures in the last ice age in Greenland were possibly $10^o C$ colder than the temperatures derived from the stable isotope record calibrated with modern spatial gradients. Furthermore, the sensitivity may have varied in the past in ways that can no longer be tested by borehole thermometry. Using incorrect or unverified functional parameters can introduce large errors into the inverse solution, both for paleoclimate and for atmospheric concentrations.

Even when we fully understand the Forward Problem (*i.e.* the transfer functions ${}_iF_j$ are known), it is not always clear that we can solve the Inverse Problem. If the resolving power of the data is inherently poor at a particular step, the ${}_iG_j$ may be unable to recover the concentrations ${}_iC_j$ at the previous step in Figure 1.

First, the Inverse Problem may have a nonunique solution. A *Model Space* (Figure 2) is a way to display all possible combinations of the unknowns (chemical concentrations ${}_iC_j$ and/or climate parameters); each axis of the multi-dimensional Model Space corresponds to one of those unknowns. Any combination of the unknown parameters that produces the observed data (represented by the solid line time series in Figure 2) is a solution to the Inverse Problem and it can be represented by a point in the Model Space.

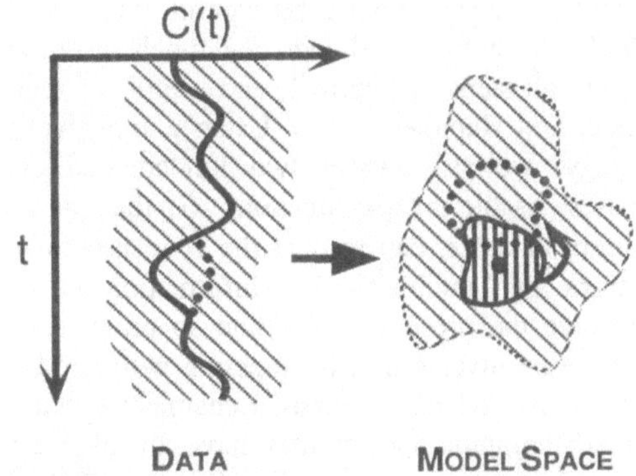

FIGURE 2: An infinite number of perfectly accurate ice core data (solid curve) could possibly resolve a unique set of atmospheric concentrations and climate parameters represented by a single point in a Model Space of inverse solutions. However, a typical finite number of data can always be matched by an infinite number of possible solutions covering a region in the Model Space. (vertically hatched region). If the data are also not perfectly accurate, the region of possible solutions is even larger (slant hatching). The size of the region indicates the Resolving Power of the Inverse Method. Furthermore, if the Inverse Problem is unstable, small changes in the data (dotted line) can cause large shifts in the region of the inferred solutions in the Model Space.

Inverse Problems would be simpler if only a single set of model parameters (the single point in the Model Space in Figure 2) could produce the observed data in the ice core. However, in practice an infinite number of points in the Model Space can all exactly reproduce the observed data set. These points occupy a region of the Model Space. The size of that region (vertical hatch marks in Figure 2) is an indication of the uncertainty associated with a solution, *i.e.* indicates the Resolving Power of the Inverse Problem. As an example of a situation with poor resolving power, consider the concentration of a particular chemical that is readily altered by remobilization and loss to the atmosphere from near-surface snow before it can reach equilibrium. There may be little hope of recovering the concentration in the air from its concentration in the snow or ice; a wide range of concentrations in the air may all be capable of producing the observed concentration in the ice.

The size of the inferred solution region increases further as the level of uncertainty associated with the data also increases, as indicated by the shaded regions with slanted hatch marks in Figure 2.

Second, the Inverse Problem may be unstable. Many physical and chemical processes in nature are diffusive, so that information is continually lost at each step, and insufficient information might survive to accurately determine the initial atmospheric concentrations. The presence of small errors in the ice core measurements ${}_iC_{ice}$ could then generate extremely large changes in the inferred atmospheric concentrations ${}_iC_{atm}$, *i.e.* the inferred region of possible solutions in the Model Space (Figure 2) may move to a new location (enclosed by dots) in response to a small change (dotted line) in the ice core data.

Third, finite sampling intervals also contribute to further information loss. Fourth, data may be inconsistent with physical and chemical approximations made in the transfer functions (poor model). For example, if an unanticipated chemical reaction has taken place, we cannot hope to derive the correct initial concentrations by running the wrong chemical reactions backwards, and the solution we obtain may occupy a totally wrong region in the Model Space.

Some of these difficulties can be avoided by careful solution of the Inverse Problem by formal Inverse Methods. History has shown that much information about the paleo-atmosphere survives through all steps in Figure 1, and useful inferences can be made from ice core data (*e.g.* Paterson and Hammer, 1987). Using a framework such as Figure 1 to guide our research strategy, we can answer two fundamental questions. First, which transfer functions ${}_iF_j$ possess usable inverse functions ${}_iG_j$ that can ultimately contribute to the inversion of the ice core data, *i.e.* can resolve the concentrations ${}_iC_j$ at earlier levels, and which transfer functions are "annihilators" of information, *i.e.* have inverse functions with limited ability to resolve the preceding ${}_iC_j$, and so produce non-unique or unstable inversions? Dust should have an inverse transfer function with good resolving power, since it is relatively inert in the snow and ice, whereas a chemical species that undergoes rapid photo-dissociation into other common species may have an inverse transfer function with limited ability to resolve earlier concentrations ${}_iC_j$. Second, of the inverse transfer functions that have some resolving power, which ones most restrict our ability to resolve the paleo-atmosphere concentrations? These must be understood in the most detail. For example, diffusion in firn may attenuate an annual signal but not eliminate it (Johnsen, 1977), and so must be understood if the annual signal is to be recovered.

In addition, a framework such as Figure 1 also serves the ice core community. It can be used to determine which ice core data should be collected to answer particular questions about the paleoclimate, and to learn how accurately these data must be measured to make a useful inversion possible.

Fortunately, we do not need to invent techniques to cope with this Inverse Problem. A number of formal inversion techniques have been described in the applied mathematics and geophysical literature. Backus Gilbert Inversion, Control Methods, Monte Carlo Methods, and Simulated Annealing are just a few. Parker (1977; 1994), Tarantola (1987) and Menke (1989) review Geophysical Inverse Methods that are capable of deriving inverse solutions and quantifying the non-uniqueness and stability problems discussed above. A linear Inverse Problem can be expressed as a matrix equation

$$\underline{\mathbf{F}}\,\vec{\mathbf{m}} = \vec{\mathbf{d}} \tag{1}$$

where the vector $\vec{\mathbf{d}}$ contains the N data (*e.g.* the ice core observations), and the vector $\vec{\mathbf{m}}$ contains the M unknown "model" parameters (the chemical concentrations at a previous level in Figure 1, and possibly unknown boundary conditions and physical and chemical parameters in the transfer functions). The $N \times M$ matrix $\underline{\mathbf{F}}$ represents the physical and chemical processes encapsulated in the transfer functions ${}_iF_j$. One obvious way to find an estimate $\vec{\mathbf{m}}_{est}$ of the desired solution $\vec{\mathbf{m}}$ is to multiply the data vector $\vec{\mathbf{d}}$ by the inverse of the matrix $\underline{\mathbf{F}}$, *i.e.*

$$\underline{\mathbf{F}}^{-1}\,\vec{\mathbf{d}} = \underline{\mathbf{F}}^{-1}\,\underline{\mathbf{F}}\,\vec{\mathbf{m}} = \vec{\mathbf{m}}_{\mathbf{est}} \tag{2}$$

However, because of the ubiquitous non-uniqueness and unstable character of many Inverse Problems, the matrix $\underline{\mathbf{F}}$ is often singular, and it does not have an inverse in the usual sense.

Any inversion scheme that does not recognize this fact is likely to produce nonsensical answers. A recommended approach is to use Singular Value Decomposition (SVD) (*e.g.* Press et al., 1986) to find a Generalized Inverse denoted by $\underline{\mathbf{F}}^{-g}$. This matrix does not have the identity property of a true inverse matrix, *i.e.*

$$\underline{\mathbf{F}}^{-g}\,\underline{\mathbf{F}} \neq \underline{\mathbf{I}} \tag{3}$$

where $\underline{\mathbf{I}}$ is the identity matrix (ones on the diagonal and zeroes elsewhere). However, $\underline{\mathbf{F}}^{-g}\,\underline{\mathbf{F}}$ has the property that it comes close to the identity matrix $\underline{\mathbf{I}}$.

What is going on here? If we try to drive the transfer function backward from the data $\vec{\mathbf{d}}$ by brute force, we quickly run afoul of the non-uniqueness and instability, such that we may calculate an answer $\vec{\mathbf{m}}_{est}$ that is total nonsense. This danger is exacerbated by the possibility that the transfer functions ${}_iF_j$ may incorporate incorrect or incomplete physics and chemistry, such that they are incompatible with the data $\vec{\mathbf{d}}$. However, in a Formal Inverse Method, we back off slightly and don't squeeze so hard on the data; we settle for inverse solutions that do not exactly reproduce the data. The tradeoff is a stable answer to the Inverse Problem that should be "close to" the correct unknown solution, while still incorporating the essential physical and chemical processes. The meaning of "close to the correct solution" can also be formalized in Inverse Methods.

Typically, we want to use ice core data to try to recover as much information as possible about a "true" atmospheric concentration history indicated by the solid curve in Figure 3. Often, however, the data inherently contain insufficient information to resolve both the detailed sequence of changes in concentration, and the timing of those changes. Some Inverse Methods incorporate a formal tradeoff curve that can be used to balance temporal resolution of the parameters sought against the uncertainty in the values of those parameters. If we demand small error bars in the timing of events, an Inverse Method will tell us that the recovered concentration levels at those highly resolved times are highly uncertain. On the other hand, if we are willing to accept less time resolution, we can recover a temporally averaged time series (dashed line) where the uncertainty in the timing is large but the error bars on concentration are small.

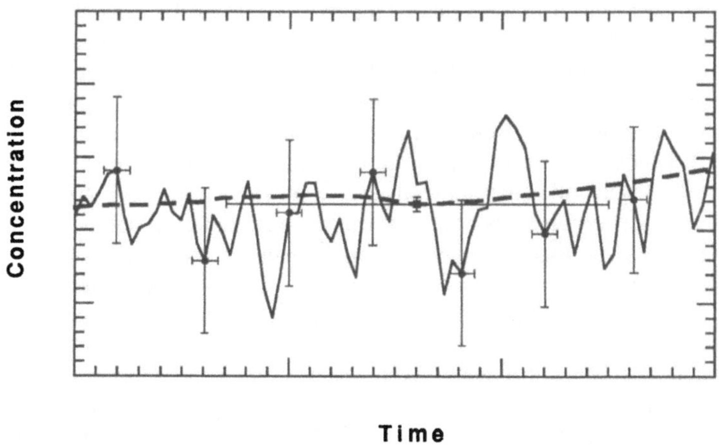

FIGURE 3: Tradeoff of temporal resolution and accuracy. If we seek a solution for a detailed concentration history at well-resolved times (solid curve), then the uncertainty in the concentrations is large, whereas if we seek only the temporally averaged concentration value (dashed line), then that averaged concentration can be found with very small error.

The Forward Model containing the essential physics and chemistry is a central component of every Inverse Problem. Most Formal Inverse Methods attempt to choose a set of values $\vec{m}_{est}$ for the unknown parameters $\vec{m}$ that will minimize the mismatch between the observed data $\vec{d}$ and the estimated data values $\vec{d}_{est}$ that would be predicted by the Forward Model operating on those chosen parameters $\vec{m}_{est}$.

The matrix $\underline{\mathbf{F}}^{-g}\,\underline{\mathbf{F}}$ can thus provide extremely valuable information about the Inverse Problem. An identity matrix $\underline{\mathbf{I}}$ has unity values on the diagonal and zeroes everywhere off the diagonal. The matrix $\underline{\mathbf{F}}^{-g}\,\underline{\mathbf{F}}$ is a square matrix with dimension $M \times M$, where M is the number of parameters we are trying to determine. It looks similar (we hope) to the identity matrix $\underline{\mathbf{I}}$, but the diagonal elements are smeared out into adjacent rows and columns, as shown schematically in Figure 4.

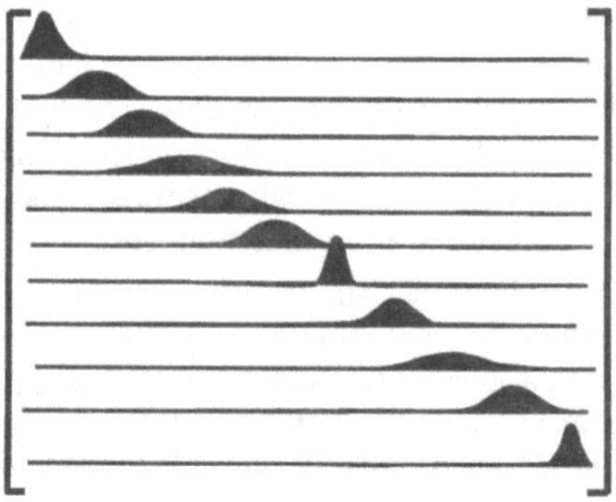

FIGURE 4: A near-identity matrix in which the diagonal elements are not fully resolved. The Model Resolution Matrix $\underline{\mathbf{F}}^{-g}\,\underline{\mathbf{F}}$ and the Data Resolution Matrix $\underline{\mathbf{F}}\,\underline{\mathbf{F}}^{-g}$ both have this form. Highly smeared diagonal elements indicate respectively paleoclimate parameters that cannot be well-resolved, or data that are not as useful for estimating those parameters.

The degree of smearing around each diagonal element i indicates the degree of uncertainty that will still remain in the value of the ith element of the desired "model" solution vector $\vec{m}$ after the inversion, even with the best possible data. The smearing is caused by destruction of information in the Forward Problem by the transfer functions ${}_iF_j$. In Geophysical Inverse Theory terminology, the matrix $\underline{\mathbf{F}}^{-g}\,\underline{\mathbf{F}}$ is called the *Model Resolution Matrix* because it shows how well each model parameter in the solution can be resolved. The Model Resolution Matrix expresses the degree to which the inverse transfer functions ${}_iG_j$ can actually resolve the concentrations ${}_iC_j$ at previous levels in Figure 1. It is a function only of the transfer functions ${}_iF_j$, so it can be evaluated even before any ice core data are available. It can be used to understand which parts of the overall Inverse Problem in Figure 1 are feasible to solve, and which are not. Clow (1992) used this approach to investigate the extraction of paleotemperature data from borehole temperature logs. For example, the fourth and the ninth parameters, represented in rows 4 and 9 of Figure 4, are highly smeared, indicating that under even the best of circumstances, these atmospheric concentrations can be only poorly resolved. On the other hand, the first, the seventh and eleventh parameters or concentrations are well resolved since the corresponding diagonal elements of the Model Resolution Matrix are only slightly smeared out.

A Model Resolution Matrix can also serve another useful function. The transfer functions ${}_iF_j$ can be complicated. A successful inversion can be numerically intensive, and any simplifications we can make without degrading the solution should be attempted. By formulating the Model Resolution Matrix for successively simpler transfer functions until the smearing becomes unacceptable, it is possible to determine which processes are the limiting ones in the Forward Problem, and which processes can be neglected because they have restricted influence on the inferred atmospheric history.

We can also use Formal Inverse Methods to ask which data sets from the ice core are most useful for determining the atmospheric concentrations of chemical species or the atmospheric characteristics, and which data sets are redundant or useless. To do this, we can test the consistency of our inversion by closing the loop, *i.e.* by running our estimated paleo-concentrations $\vec{\mathbf{m}}_{est}$ through the forward transfer functions $_iF_j$ represented by the matrix $\underline{\mathbf{F}}$ to "predict" values $\vec{\mathbf{d}}_{est}$ for the data set $\vec{\mathbf{d}}$ that has already been measured in the ice core. We hope that $\vec{\mathbf{d}}_{est}$ and $\vec{\mathbf{d}}$ are similar. Doing this, we find

$$\vec{\mathbf{d}}_{\mathbf{est}} = \underline{\mathbf{F}}\,\vec{\mathbf{m}}_{\mathbf{est}} = \underline{\mathbf{F}}\,\underline{\mathbf{F}}^{-g}\,\vec{\mathbf{d}} \tag{4}$$

From this we see that $\underline{\mathbf{F}}\,\underline{\mathbf{F}}^{-g}$ should also be a near-identity square matrix, but this time with dimensions $N \times N$, where N is the number of data values. It is called the *Data Resolution Matrix* because the amount of smearing around diagonal element i shows us how well data from observation i can be predicted from our knowledge of the physical and chemical processes encapsulated in the matrix $\underline{\mathbf{F}}$. If we cannot predict a datum value accurately, we can anticipate that measuring that particular property at that particular depth in the ice core may contribute little to the inverse solution, *i.e.* to our knowledge of the paleo-atmosphere and its loading. For example, if Figure 4 represents a Data Resolution Matrix, then a data value represented by the fourth diagonal element contributes little to the inverse procedure, whereas the first and seventh data points are valuable ones to measure.

A Simple Example of Resolving Power

Consider a chemical species "X" whose concentration in the ice core C_{ice} depends on the concentration in the air C_{atm} and on the temperature T. A simple linear transfer function for X might be

$$C_{ice}(t) = a\ C_{atm}(t)\ +\ b\ T(t) \tag{5}$$

where a and b are constants. If we thought that the temperature was rising, but the airborne concentration $C_{atm}(t)$ was relatively constant during the period of interest, then we could approximate $T(t)$ by a linear function

$$T(t)\ =\ T_0 + \dot{T}\ t \tag{6}$$

where T_0 and $\dot{T}$ are the unknown initial value and the rate of change of temperature. The transfer function then would become

$$C_{ice}(t) = a\ C_{atm}\ +\ b\ T_0\ +\ b\ \dot{T}\ t \tag{7}$$

If we have 5 ice core data with dates t_1 through t_5, then the left side of (1) takes the form

$$\underline{\mathbf{F}}\,\vec{\mathbf{m}} = \begin{bmatrix} a & b & bt_1 \\ a & b & bt_2 \\ a & b & bt_3 \\ a & b & bt_4 \\ a & b & bt_5 \end{bmatrix} \begin{bmatrix} C_{atm} \\ T_0 \\ \dot{T} \end{bmatrix} \tag{8}$$

If the functional parameters took the values $a = 0.4$ and $b = 0.6$, and the ice core samples are spaced at intervals of one time unit, then we could form $\underline{\mathbf{F}}$ and find $\underline{\mathbf{F}}^{-g}$ by SVD (Press et al., 1986) to get the Model Resolution Matrix:

$$\underline{\mathbf{F}}^{-g}\,\underline{\mathbf{F}} = \begin{bmatrix} 0.31 & 0.46 & 0.0 \\ 0.46 & 0.69 & 0.0 \\ 0.0 & 0.0 & 1.0 \end{bmatrix} \tag{9}$$

The right column and bottom row look like elements of an identity matrix; therefore, the third parameter $\dot{T}$ is well-resolved. However, the upper left 2×2 submatrix is smeared out, telling us that C_{atm} and T_0 cannot be resolved well.

Although the number of ice core data exceeds the number of parameters sought, the problem is still inherently under-determined, because none of the data is capable of distinguishing a temperature-offset effect from an air concentration effect. We would be unable to resolve the air concentration no matter how many ice core data points $C_{ice}(t)$ we were to use.

The Data Resolution Matrix for this problem is

$$\underline{\mathbf{F}}\,\underline{\mathbf{F}}^{-g} = \begin{bmatrix} 0.6 & 0.4 & 0.2 & 0.0 & -0.2 \\ 0.4 & 0.3 & 0.2 & 0.1 & 0.0 \\ 0.2 & 0.2 & 0.2 & 0.2 & 0.2 \\ 0.0 & 0.1 & 0.2 & 0.3 & 0.4 \\ -0.2 & 0.0 & 0.2 & 0.4 & 0.6 \end{bmatrix} \tag{10}$$

This matrix is highly smeared out compared to an identity matrix. This means that the best temperature history and airborne concentration $C_{atm}(t)$ that we can derive from these data would still do a poor job of predicting the ice core concentrations even when the transfer function parameters a and b were known exactly. The first and last data are resolved best; they are the most valuable data to collect because together they provide the longest possible time baseline for estimating the rate of warming $\dot{T}$.

The ice core data from species "X" in the previous example were incapable of separating the influence of initial temperature from air concentration, regardless of how many ice core data we used. However, the design of the experiment could be changed based on what we have learned. We could add just a single observation at time t_6 of a different parameter "Y" that responds only to the temperature T_0. This new parameter Y could be $\delta^{18}O$ calibrated by borehole temperatures (Cuffey et al., 1995).

$$\delta^{18}O = \alpha T + \beta \tag{11}$$

where α and β are constants. If we treat $\delta^{18}O(t_6) - \beta$ as the datum, then the left side of (1) becomes

$$\underline{\mathbf{F}}\,\vec{\mathbf{m}} = \begin{bmatrix} a & b & bt_1 \\ a & b & bt_2 \\ a & b & bt_3 \\ a & b & at_4 \\ a & b & at_5 \\ 0 & \alpha & \alpha t_6 \end{bmatrix} \begin{bmatrix} C_{atm} \\ T_0 \\ \dot{T} \end{bmatrix} \tag{12}$$

and the Model Resolution Matrix becomes

$$\underline{\mathbf{F}}^{-g}\,\underline{\mathbf{F}} = \begin{bmatrix} 1 & 0 & 0 \\ 0 & 1 & 0 \\ 0 & 0 & 1 \end{bmatrix} \tag{13}$$

All three parameters C_{atm}, T_0 and $\dot{T}$ are fully resolvable in this modified Inverse Problem.

The Data Resolution Matrix for this modified problem is

$$\underline{\mathbf{F}}\,\underline{\mathbf{F}}^{-g} = \begin{bmatrix} 0.6 & 0.4 & 0.2 & 0.0 & -0.2 & 0.0 \\ 0.4 & 0.3 & 0.2 & 0.1 & 0.0 & 0.0 \\ 0.2 & 0.2 & 0.2 & 0.2 & 0.2 & 0.0 \\ 0.0 & 0.1 & 0.2 & 0.3 & 0.4 & 0.0 \\ -0.2 & 0.0 & 0.2 & 0.4 & 0.6 & 0.0 \\ 0.0 & 0.0 & 0.0 & 0.0 & 0.0 & 1.0 \end{bmatrix} \qquad (14)$$

showing that even this improved solution for C_{atm}, T_0, and $\dot{T}$ will do no better at predicting the observed values of the concentration of species X. However, it will be able to reliably estimate the observed value of $\delta^{18}O$. We conclude that $\delta^{18}O(t_6)$ is a valuable datum to collect.

ADDITIONAL FEATURES OF INVERSE METHODS

A number of additional characteristics of Formal Inverse Methods can be exploited to design experiments and to interpret results.

Most problems encountered in nature turn out to be nonlinear. Nonlinear Inverse Problems can be solved by an iterative approach in which the problem is linearized about an estimated approximate solution.

A key element of Inverse Theory is its ability to carry error estimates on the data through the physical operations of the inverse transfer functions ${}_iG_j$. Thus reliable uncertainty estimates on the inferred paleoclimate are derived naturally. For example, if all the M data points are independent and each point j has standard deviation given by σ_j, then we can form a variance matrix $\underline{\mathbf{S}}_d$ that has $\sigma_j{}^2$ on the diagonal in row j and zeroes off the diagonal. Then the corresponding error estimates for the model $\vec{\mathbf{m}}$ are given by (Menke, 1989, p. 65)

$$\underline{\mathbf{S}}_m = \underline{\mathbf{F}}^{-g}\,\underline{\mathbf{S}}_d\,\underline{\mathbf{F}}^{-gT} \qquad (15)$$

where the T superscript indicates the transpose of the matrix.

Inverse Methods often work best when the number M of parameters sought is much smaller than the number N of data values used. This keeps the problem computationally manageable; the Forward Model must generally be run many times to solve an Inverse Problem. It also allows the data redundancy to be exploited through least squares or some other measure of mismatch to enhance the reliability of the inverse solution. This means, for example, that Inverse Methods are better suited to deriving the average climate over long periods rather than its high frequency variations. This is good news for the many climatologists who feel that as a first step in understanding past climates, it is most important to understand how the earth breathes, rather than how it coughs.

SUMMARY

Formal Inverse Methods can provide a framework for designing the air-snow transfer function research underway to investigate the atmosphere-ice core connection. This framework can show (a) which transfer functions are most important in controlling the ice core concentrations, (b) which atmospheric concentrations can be resolved by the ice core data, and (c) which data are important to collect to achieve this. This atmospheric/snow chemistry community together with the ice core community can benefit from formulating such a formal framework in order to most effectively determine the paleoclimate history which is their joint research goal.

ACKNOWLEDGEMENTS

I am indebted to Gary Clow for many discussions clarifying concepts of Inverse Theory while writing this paper. Suggestions and comments from Roger Bales and Joe McConnell have also been valuable. Thanks to Susan Harder and Susan Douglass for help with figure design and manuscript layout. This work was supported by grants OPP-9123660 and OPP-9221261 from the National Science Foundation.

REFERENCES

Bales R.C., Choi J. (this volume) Conceptual framework for interpretation of exchange processes. In: Wolff E.W., Bales R.C., (eds) NATO ASI Series I. *Processes of chemical exchange between the atmosphere and polar snow.* Springer Verlag, Berlin.

Clow G.D. (1992) The extent of temporal smearing in surface-temperature histories inferred from borehole temperature measurements. *Palaeogeography, Palaeoclimatology, Palaeoecology* 98:81-86.

Cuffey K.M., Clow G.D., Alley R.B., Stuiver M., Waddington E.D., Saltus R.W. (1995) Large arctic temperature change at the Wisconsin-Holocene glacial transition. *Science* 270:455-458.

Johnsen S.J. (1977) Stable isotope homogenization of polar firn and ice. In: *Isotopes and impurities in ice and snow.* IAHS Publication No. 118:210-219.

Mayewski P.A., Meeker L.D., Whitlow S., Twickler M.S., Morrison M.C., Alley R.B., Bloomfield P., Taylor K. (1993) The atmosphere during the Younger Dryas. *Science* 261:195-197.

Menke W. (1989) *Geophysical Data Analysis: Discrete Inverse Theory.* Academic Press, San Diego. 289p.

Neftel A. (this volume) The record of gases and reactive species in ice cores, and problems of interpretation. In: Wolff E.W., Bales R.C., (eds) NATO ASI Series I. *Processes of chemical exchange between the atmosphere and polar snow.* Springer Verlag, Berlin.

Parker R.L. (1977) Understanding inverse theory. *Ann. Rev. Earth Planet. Sci.* 5:35-64.

Parker R.L. (1994) *Geophysical Inverse Theory.* Princeton University Press. Princeton. 386p.

Press W.H., Teukolsky S.A., Vetterling W.T., Flannery B.P. (1986) *Numerical Recipes.* Cambridge University Press, Cambridge. 818p.

Paterson W.S.B., C.U. Hammer (1987) Ice core and other glaciological data. In: Ruddiman W.F., Wright H.E. Jr. (eds) *North America and adjacent oceans during the last deglaciation.* Geological Society of America, Boulder. 91-110.

Tarantola A. (1987) *Inverse Problem Theory.* Elsevier, Amsterdam. 613p.

Waddington E.D., Cunningham C., Harder S.L. (this volume) The effects of snow ventilation on chemical concentrations. In: Wolff E.W., Bales R.C., (eds) NATO ASI Series I. *Processes of chemical exchange between the atmosphere and polar snow.* Springer Verlag, Berlin.

WORKING GROUP REPORT - IRREVERSIBLE SPECIES

Cliff I. Davidson and Robert J. Delmas

This group focused on aerosol species that deposit irreversibly on an ice sheet. Irreversibly depositing species include both primary aerosols emitted directly from sources and also secondary aerosols formed in the atmosphere from precursor gases. These species can be conveniently divided into several categories:

1. Soil-derived inorganic aerosols, including those containing Al, Ti, Si, Ca, Fe and several other elements. Al, Ti and Si are considered to be the best tracers of soil dust, although some of the Si in the polar regions may originate from coal combustion.
2. Seasalt inorganic aerosols, including those containing Na, Cl and other elements. Na is considered to be one of the better tracers for this category. Cl is prone to chemical reactions and subsequent losses, and a fraction of the Cl in the polar regions may be from anthropogenic sources.
3. Anthropogenic inorganic aerosols, including those containing S, Pb, V, In, Zn, As, Cd and several other elements. The seven elements listed are all suitable tracers under many conditions.
4. Sulphate in older snow, and in Antarctica, derives mainly from marine biogenic emissions of DMS (producing also MSA), and from volcanic emissions.
5. Organic aerosols, including natural organics such as biogenically emitted aerosols as well as anthropogenic organics such as combustion products. However, not all organics can be considered irreversible.

The group considered three steps in the overall transfer of chemical species from source regions to their final resting place in the deep ice. The first step involves long-range atmospheric transport to the air over the ice sheet. This is followed by the second step in which chemical species reach the ice sheet surface. Finally, the third step involves physical and chemical changes that occur as the chemical species become buried by successive snowfall, and the snow changes to firn and ultimately to ice.

NATO ASI Series, Vol. I 43
Chemical Exchange Between the Atmosphere
and Polar Snow
Edited by Eric W. Wolff and Roger C. Bales

Long-Range Atmospheric Transport

The group generally agreed that airmass back-trajectories, although they have limitations, can provide useful information for identifying source regions of chemical species reaching the polar regions. Back-trajectories can provide indications of transport at low elevations as well as in the free troposphere. Although the method can at times identify the continent of origin for constituents reaching the polar regions, it is tenuous to use the results to claim the relative importance of regions on a sub-continental scale. Trajectories are more reliable for ice sheets in the Northern Hemisphere than they are in Antarctica, where meteorological stations are very scarce and transport times from source regions are considerably longer. A limitation of trajectories is that they have reasonably accuracy for 5-10 days at most. Any constituent that has an atmospheric residence time longer than this cannot be traced reliably to its source. Note that winter polar aerosols can have lifetimes of several weeks, making source identification difficult.

Applying these results to firn core samples requires one to account for trajectories occurring over the time periods represented by each core sample. For example, core samples often involve snow which originally fell over a period of several months, representing a full season of the year or longer. It is necessary to compute average trajectories corresponding to the same season, preferably derived from data for several years, in order to identify possible source regions that may have influenced the core sample. Of course, attributing modern-day source regions to older ice requires the assumption that global cirçulation patterns have not changed significantly over the years. This may or may not be a reasonable assumption, depending on circumstances. Global Climate Models may be useful for investigating this issue.

The group also considered physical and chemical changes affecting the chemical species of interest during transport to the polar regions. Firstly, the efficiency of losses en route to the ice core site could determine the concentrations eventually seen. This will depend on the ratio of the residence time to the transit time; both of these would have been substantially different during the last ice age. Changes in size distribution of the aerosols can be significant over travel times of several days. Very small particles may coagulate while the largest particles can settle out, resulting in a more narrow distribution of sizes. Gas-to-particle conversion can occur, e.g., SO_2 forming SO_4^{2-}, while some aerosols such as soil dust or seasalt may act as sites for the sorption of SO_2, HNO_3, or other gases. Some aerosol species such as certain polynuclear aromatic hydrocarbons (PAH) can react to change chemical form. Since the time for changes in chemistry are known, identification of certain PAH compounds in the polar atmosphere can lead to estimates of residence time of the species en route to the sampling site.

Air-Snow Exchange

The frequency and amount of precipitation are key parameters affecting processes which control air-snow transfer. Particle characteristics such as size and chemical composition can also have a major effect.

The consensus of the group was that we are now beginning to understand the relative importance of different deposition processes (e.g. precipitation scavenging, dry deposition, fog deposition) in determining overall rate of transfer of chemical species to ice sheets. For example, in central Greenland it appears that wet deposition may be the dominant mechanism, while dry deposition is more likely to dominate in the central Antarctic plateau due to very low snow accumulation rates. However, the available information is generally limited to summer field campaigns and may apply only to selected sites where measurements have been made. Furthermore, only certain chemical species have been measured; of particular concern is the paucity of data on organics. Even with the best information, the group acknowledged that air-snow exchange rates will provide merely ranges of local airborne concentrations on the basis of chemical species data for the ice sheet.

Some specific deficiencies in our current knowledge were identified. For example, we do not have detailed knowledge of frequency and amount of precipitation during different times of the year at most ice coring sites, despite the importance of this information to interpreting the ice core record. There is also a lack of information on airborne concentrations of chemical species throughout the year. We do not yet have reliable data on coarse particles due to the difficulties in sampling them. These particles will have the greatest dry deposition velocities and may be most efficiently scavenged by precipitation and fog, leading to major impacts on the chemical content of ice cores. There are also some uncertainties in estimating dry deposition velocities of submicron particles over snow surfaces.

Aside from dry deposition to the surface, it is recognised that deposition may occur within the porous upper layers of snow. Recent work on "wind pumping" suggests that ventilation of snow layers can deliver chemical species that may deposit efficiently below the surface. The amount of ventilation depends on the presence of surface irregularities such as sastrugi, as well as the porosity of the snow. The effect can be highly variable; at times when relatively impermeable crusts from on the surface, the amount of ventilation can be negligible. Linked with the wind pumping process is the rate of delivery of chemical species to the surface. If the rate of delivery is slow, e.g. during stable atmospheric conditions, then the effect of wind pumping may be small even if filtering by subsurface snow is efficient.

Also important to the deposition process is the effect of blowing snow and sublimation of snow crystals. Evidence suggests that blowing snow can scavenge aerosols as it moves over the surface, enhancing the total deposition in a region. On the other hand, blowing snow in dry conditions may sublime, releasing previously scavenged aerosol back to the atmosphere. The net effect depends on specific characteristics of a sampling site and the region surrounding it. However, since a large fraction of the annual snowfall is believed to sublime at some sites, the effect cannot be ignored.

Physical and Chemical Changes in the Snow and Ice

Careful consideration of snowpit and ice core data has identified several anomalies that are not understood. Comparing chemical species concentrations in fresh snowfall with concentrations in snowpits over the same time period shows that overall features are generally but not always preserved. For example, characteristic springtime maxima in crustal and anthropogenic species observed in aerosols and in fresh Greenland snow are usually identified in snowpit strata, although winter peaks in seasalt seen in airborne concentrations are often not found in snowpits. It is unknown whether there is a lack of precipitation in winter or conversely whether snow that falls in winter does not accumulate due to high winds and dry conditions. As another example, aerosol chemical species as well as gases may be affected by a variety of processes that alter their profiles in the snowpack. We have indications that alterations occur for $\delta^{18}O$, where profiles in upper snowpit layers are sometimes highly irregular; the irregularities become much less noticeable as the snow ages, leading to a very regular pattern at depth that enables dating of annual layers. The extent to which such processes affect irreversibly deposited aerosols is unknown.

Data from deep ice cores also show that changes are occurring. MSA and SO_4^{2-} concentrations are in phase in both air and recent core layers in coastal Antarctica, with peaks of both species prevalent in summer. However, deeper ice shows that the MSA profile has shifted its peak into winter; long-term changes in the firn must have occurred. Similarly, at one coastal Antarctic site, it has been shown that Mg and Na peaks are in phase in upper layers of ice cores but become out of phase in deeper ice. Such changes indicate that it is tenuous to interpret profiles of new chemical species until the physical and chemical processes that may have influenced them are investigated. However, while shifts in the timing of certain chemical species in ice cores may be problematic, they are less damaging to the ice core record than chemical reactions that can deplete concentrations of some species in the core, leading to faulty interpretations of time trends.

Both concentration and flux profiles can be valuable in interpreting ice core data. Depending on the snow accumulation rate (and hence the importance of wet versus dry deposition), either might be more representative of airborne concentrations.

Recommendations

On the basis of discussions at the working group meeting and overall discussions at the NATO workshop, the group would like to make the following recommendations:

1. Of highest importance are year-round measurements at ice-coring sites that provide information on air-snow chemical exchange. These should include concentration measurements of aerosol (and gaseous) species of interest in air, snow, and fog as well as acquisition of meteorological data. Frequency and amount of precipitation and fogs, overall net snow accumulation as a function of time, and presence of blowing snow are extremely important to measure. If feasible, soundings should be conducted to get both concentration and meteorological data aloft. Experiments to quantify ventilation through porous snow layers and snow filtration of aerosols as well as effects of blowing snow should be conducted.
2. Global climate models should be used in conjunction with airmass back-trajectories to identify possible continents of origin for measured chemical species in ice cores. Modelling of the movement of tracers to follow airmass pathways would be valuable for testing GCM's against present conditions.
3. Use of satellite photos to identify the movement of clouds and hence airmass motions should be investigated. These can be used to confirm airmass back-trajectory computations.
4. Snowpit studies represent a valuable way of identifying the extent to which chemical signals in the atmosphere are preserved in the uppermost snow layers. Detailed stratigraphic snowpit studies should be continued and expanded for this purpose. The snowpit studies should be conducted in coordination with collection of surface snow and fresh snow, if possible. Comparing chemical species concentrations in fresh snow with concentration profiles in snowpits covering the same time period can help establish the

extent to which individual events are preserved in the snowpack. Such studies may also quantify spatial inhomogeneities in the snow, and may identify when and where blowing snow has resulted in loss of layers in the snowpack. These issues are important in assessing representativity of an ice core.

5. To identify and quantify processes occurring during ice age conditions, field campaigns are needed in winter time periods. Although extremely difficult to conduct, such measurements would be valuable even if they are very limited and conducted over short time periods to provide a basis for estimating air-snow exchange in very cold and dry conditions.

6. Field experiments at mid-latitude sites that are relatively easy to access can provide information on fundamental processes relevant in the polar regions. Examples include studies of precipitation scavenging in and below clouds, studies of dry deposition using ambient particles and tracer releases, and measurement of deposition by fog. Such experiments may be performed with less expense and easier logistical arrangements compared with those at polar ice coring sites. Similarly, information already obtained on mid-latitude, high altitude glaciers or ice caps should be explored to its fullest extent to help understand processes occurring in the Arctic and Antarctic.

7. The extent to which ice coring sties are representative of broader areas of the polar regions should be explored. Differences between high altitude polar sites and locations at sea level should be identified, as well as differences among sites with varying precipitation/accumulation rates. For example, sites of major glaciers in the Northern Hemisphere other than Greenland need to be included in this type of analysis. These might include Agassiz at 80°N and 1.6 km altitude and Logan at 65°N and 5 km altitude, which would complement Summit, Greenland at 72°N and 3 km altitude.

8. Finally, laboratory experiments may be valuable for identifying certain fundamental processes. For example, quantification of snow filtration of aerosols under controlled ventilation conditions and well-defined snow properties would be valuable. As another example, experiments can focus on examination of snow, ice and firn samples to determine where in the microstructure chemical species are located e.g. with respect to crystal grain boundaries. Limited laboratory work has already been conducted on these and similar topic areas and such work should be continued.

WORKING GROUP REPORT - ACIDIC GASES

Michel Legrand and Len A. Barrie

1. Relevance of considered species

Nitric Acid (HNO_3)

Nitric acid studies may help to evaluate natural (eg lighting, stratospheric N_2O oxidation, terrestrial biogenic) and anthropogenic contributions to the NO_x budget in remote regions. For instance, the role of the stratosphere in contributing via the denitrification process to the nitric acid burden of the polar boundary layer, if confirmed, could be useful to investigate minimum temperatures of the lower stratosphere in the past. In polar ecosystems nitrate is the limiting nutrient. In addition, nitrate paleodata are useful in the study of the carbon cycle because of the interactions between the C and N cycles.

Hydrochloric acid (HCl)

HCl in ice cores is generally calculated as the excess of Cl^- over that expected from ratios in sea water. Natural sources include acidification of sea salt (see below), volcanoes and to a lesser extent, the photochemistry of methyl chloride (CH_3Cl emitted by marine biogenic activity, biomass burning and waste incineration). Although the role of anthropogenic emissions is not clearly established, direct combustion emissions and an enhanced loss of chloride from sea salt in polluted regions (enhanced activity) are suspected. HCl in ice may be another indicator of anthropogenic perturbations of the Earth's atmosphere. The hypothesis to be tested is as follows: If $H_2SO_4 + 2NaCl \rightarrow 2HCl + Na_2SO_4$ is a reaction that occurs in the marine boundary layer and produces HCl gas, then since H_2SO_4 has increased during the industrial era from anthropogenic sources in the northern hemisphere but not in the southern hemisphere, we should see an increase in excess HCl in northern hemispheric glaciers provided post depositional effects are predictable. In addition, snowpack heterogeneous and photochemical reactions that effect HCl need to be investigated.

Hydrofluoric acid (HF)

Natural emissions of HF within the troposphere are mainly from volcanic activity (eruptions or passive outgassing) and are likely small relative to anthropogenic inputs to the atmosphere dominated by coal combustion, aluminium production and stratospheric degradation of freons.

NATO ASI Series, Vol. I 43
Chemical Exchange Between the Atmosphere and Polar Snow
Edited by Eric W. Wolff and Roger C. Bales

Formic (HCOOH) and acetic (CH_3COOH) acids

Formic acid can be emitted directly into the atmosphere by vegetation as well as by biomass burning. Furthermore, the gas phase atmospheric oxidation of various hydrocarbons (e.g. terpenes) emitted by biogenic activity as well as in cloud photochemistry of formaldehyde (HCHO) may contribute to atmospheric formic acid. Both primary emissions and gas phase oxidation of hydrocarbons (no aqueous chemistry producing acetic acid has been identified) contribute to the acetic acid budget. A link to the global nitrogen cycle as reaction products of atmospheric organic nitrogen compounds such as PAN and alkyl nitrates is potentially important. Because of human activities, increased emissions of hydrocarbons likely influence the carboxylic acid budget.

2. What we know about air/snow relationships

Nitric acid

In general nitrate and sulphate are the major anions in polar precipitation. On an equivalents basis, particularly at high elevation sites in summer, nitrate is often more abundant than sulphate.

Atmospheric measurements have been performed during summer at Summit in the boundary layer. Using different trapping devices (denuder tubes, nylon filters, mist chambers), observed mixing ratios ranged between 10 and 50 pptv. In general more than 90% of the nitrate is in the gas phase. However, nitric acid comprises only a small fraction of total NO_y which is as high as 1 ppbv. The rest is presumably peroxyacetyl nitrate (PAN), alkyl nitrates, particulate NO_3^- and NO_x.

From pit studies, a post depositional loss of HNO_3 is evident. The fraction of post depositional loss seems to depend inversely on the accumulation rate. It is greatest, for example, in East Antarctica where accumulation rates are relatively low (<10 g/cm^2/year). At Summit where accumulation rates are considerably higher, post depositional losses in summer take place on time scales greater than a week. These losses likely involve processes such as evaporation during blowing snow, snowpack production and destruction of HNO_3 and loss by snowpack ventilation during ice metamorphosis.

Based on preliminary laboratory determinations of the solubility, and diffusion rate of HNO_3 in pure ice, it has been suggested that the content of nitrate in summer fresh snows fit rather well with a thermodynamic equilibrium between atmospheric and ice phase HNO_3.

One major question related to the study of the air-snow relationship is the possible role of PAN and other alkyl nitrates in contributing to snowpack nitrate and nitrite.

HCl

Ice and snow data suggest that excess chloride (ie as HCl) is present in summer snow layers deposited both in central Antarctica and Greenland. Marked seasonal variations in surface snow that exhibits a maximum in summer, generally survive well with time as the snow is buried deeper in the firn layer. Until now, the existence of rapid post depositional loss of HCl is not clearly observed although it seems that a loss of HCl occurs slowly from firn in East Antarctica.

Rare measurements of HCl achieved in the Summit summer atmospheric boundary layers indicate an average mixing ratio close to 50 pptv.

In contrast to HNO_3, preliminary laboratory determinations suggest that HCl in snow is far from thermodynamic equilibrium with the atmosphere, likely because of a lower diffusion rate in ice.

Carboxylic acids

The lack of a global budget for carboxylic acids is, in part, limited by the study of carboxylic acids present in precipitation of remote regions. While concentrations of formate and acetate in snow are one order of magnitude lower than those of nitrate, atmospheric mixing rations of carboxylic acids at Summit are much higher than those of HNO_3 with values ranging from 200 to a maximum of 6000 pptv (biomass burning plumes reaching Greenland).

Concentrations in snow generally depend on acidity, decreasing as acidity increases. Several observations suggest that these acids are lost rapidly (< 1 day) from the surface after a fresh snowfall.

Some important issues emerge both from ice core data and studies of the relationship between the content of air and snow. First, ice core studies suggest that the main source of formic acid apart from biomass burning is biogenic emissions from vegetation. Such studies are still unable to identify the main source of acetate which, in contrast to formic acid, has a source that persisted during the last glacial maximum at high northern latitudes. Second, a very good stoichiometry is observed between ammonium and formate in snow layers highly impacted by biomass burning plumes. Finally, the role of PAN as a possible reservoir of acetic acid needs to be investigated.

Concerning the relationship between air and snow content, laboratory determinations of the solubility and the diffusion speed of these two carboxylic acids in ice needs to be made. The rapid losses of carboxylic acids after deposition needs to be demonstrated by studies of net vertical fluxes. Finally, as for nitrate, the role of organic nitrate at Summit needs to be elucidated.

3. Measurement and modelling studies

It appears that for some of the acidic gases, post-depositional processes are more important than depositional processes in determining ice core concentrations, at least for low accumulation rate sites, so studies need to focus on these processes. Recognising the gaps in our understanding of the air-snow transfer processes the following studies are recommended:

- laboratory and field measurements of the distribution of impurities in ice. This is to better understand the exchange rate and absorption capacity of acidic gases by glacial ice. Reactions at ice crystal surfaces and at grain boundaries are of interest.
- Laboratory studies using artificially prepared and real ice need to be undertaken that elucidate:
  a) the process of transfer from air to ice of gases such as HNO_3, HCl, HCOOH, and CH_3COOH.
  b) mechanisms of transfer of ice impurities during snow and ice metamorphosis.
  c) heterogenous chemical transformations that take place on snow and ice

Specifically with regard to each class of acid gas considered here are the following priorities.

HNO_3

A comprehensive field measurement programme is needed during the warmer sunlit half of the year that characterises simultaneously air composition immediately above a glacier surface and in snowpack interstitial air and ice while measuring the composition of fresh snowfall and gas/particle exchange. Parameters that should be measured include: NO_y species (PAN, alkyl nitrates, HNO_3, HONO, particulate NO_3^- and NO_2^-), the flux of NO_y and NO_x (eddy correlation), volatile organic compounds (VOCs) that are reactive in ozone chemistry, O_3, the chemical composition of snow (major ions, organics, halogen ions), physical properties of the snowpack such as snow surface area, permeability, crystal structure relevant to atmosphere-snowpack exchange and changes in snowpack composition before and after blowing snow events in which little precipitation has fallen.

Post-depositional effects during the non-summer season have not been studied for HNO_3. It should be emphasised that an integral part of such a comprehensive study of air/snow exchange processes is a model of the chemical and physical processes that affect the atmospheric boundary layer and snowpack. $\delta^{15}N$ studies could be very important for elucidating losses of HNO_3.

HCl

Measurements of gas/particle partitioning of Cl^- in air above ice caps is needed throughout the year immediately above most glacial sites. Only one measurement has been made. It showed most of the Cl^- in the gas phase.

There is a need for model and laboratory studies of the chemical reactions of HCl and NaCl with acidic sulphate aerosols ranging in composition from NH_4HSO_4 and H_2SO_4 that are in equilibrium with atmospheric water vapour between temperatures of 0 and -90°C (273 and 183°K).

Organic Acids

Organic acid products such as formate, acetate and oxalate ion as well as the dicarboxylic acids such as succinic and maleic are beginning to be measured in snow. There is a need for care to be taken in preventing contamination of these substances during sample collection and handling prior to analysis. Plastic containers are particularly prone to contaminate a snow sample. Stainless steel or glass are most appropriate. Now liquid ion chromatography is often used to analyse snow. However, there are sometimes problems with interferences with this technique. More specific techniques, such as liquid chromatography coupled with mass spectrometry and, for less volatile organics such as dicarboxylic acids, organic extraction, and analysis by gas chromatography should be used when interferences are unknown.

Based on limited evidence, the post-depositional changes of organics acids after fresh snow fall are fast (several hours) and substantial. More observations of this phenomenon are needed at several glacial locations. Preferably, these studies should be done in parallel with measurements of snowpack physical properties (such as snow surface area) and using an atmosphere/snowpack chemical exchange model.

Other important species that need to be measured in snow and ice

Many new species have been suggested for measurement in ice cores. Some of these can add important new information to atmospheric chemistry studies, but many of them are expected to be reversible, and therefore extensive study of their uptake by snow and ice will be required before their concentrations can be interpreted.

Measurements of PAN and alkyl nitrates in air and possibly in snow would help with the interpretation of nitrate profiles, as would further measurements of nitrate N^{15}/N^{14} ratios, as mentioned above. As well as being important as an environmental parameter in its own right, ammonia will affect the uptake of acidic gases, and therefore requires study at glacial sites. Acetone is an indicator of chemical reactions in the atmospheric boundary layer and snowpack, and should be measured at polar sites.

Finally, the other halogen acids could be measured in polar ice. HF may be important as an indicator of anthropogenic emissions. HBr is of interest because of links with tropospheric ozone destruction at polar sunrise. HI should act as a marine biogenic marker.

WORKING GROUP REPORT - OXIDANTS

Martha Conklin and Albrecht Neftel

The goal of this working group was to determine what data and analyses are needed to develop a transfer function between atmospheric concentrations and snow/ice concentrations of H_2O_2 and HCHO. The proposed suggestions are directed towards the interpretation of ice cores collected in Greenland and Antarctica.

The relevance of trying to determine how the concentrations preserved in the ice core relate to atmospheric concentrations (i.e. determine the transfer function) is that atmospheric H_2O_2 and HCHO are linked with the oxidative capacity of the atmosphere. If a transfer function was developed, historical records of these species contained in ice cores would give valuable information about past atmospheric composition. In addition, the H_2O_2 record, with its clear seasonal signal, is a useful tool for glaciology.

To put everything in perspective, we developed a working hypothesis: before industrialization, there was no change in the oxidative capacity of the atmosphere for approximately 8,000 years (i.e. the Holocene). The annual mean, free tropospheric concentration of H_2O_2 during this period is therefore assumed to have been constant so that the variability of H_2O_2 in the ice core record would be a direct indicator of variability in the atmosphere to snow to ice transfer function. Hence, the variability in the H_2O_2 record is an image of different stochastic influences (e.g. meteorological factors, loss processes, etc.) on the transfer function. Time series analysis in the frequency domain should yield information on the stability of the transfer function. Additional insight can be obtained by conducting the same analysis on species that experience different loss or transformation processes. Species to consider include: $\delta^{18}O$, δD, δD excess, NH_3 and Ca^{2+}.

In addition to using past records to obtain information about the transfer function, measurements of H_2O_2 in the free troposphere concurrent with ground-based measurements are necessary. As snow concentrations may also be influenced by local atmospheric concentrations (defined as the concentration 2 m above the surface), both free troposphere and local concentrations need to be measured. We propose that O_3, NO_x, CO, PAN, ethane, alkyl nitrates and acetone should also be measured concurrently with H_2O_2 and HCHO both locally and in the free troposphere since these other gases can provide information about free tropospheric

NATO ASI Series, Vol. I 43
Chemical Exchange Between the Atmosphere and Polar Snow
Edited by Eric W. Wolff and Roger C. Bales

processes. Both local and free troposphere concentration measurements should be made at Summit and Law Dome.

The purpose of local atmospheric measurements is to help interpret snow concentrations, snow and ice being in direct contact with the atmosphere. Measurements of the reactive trace gases (O_3, NO_x, CO, ethane, etc.) are used in an atmospheric photochemical model to compute gas-phase HCHO and H_2O_2. The degree to which these computed values agree with measurements will help to constrain the mathematical statement of the transfer function. For example, scavenging by particles and deposition processes will modify simple gas-phase equilibrium among constituents O_3-NO_x-CO-hydrocarbons and intermediates such as HCHO and H_2O_2.

Similarly, local atmospheric concentrations and free tropospheric composition are not identical due to chemical and physical processes in the surface layer and in haze and clouds that are typically present in the free troposphere. The relationship between the local atmosphere and free troposphere must also be determined in order to define quantitatively the relationship among ice/snow-local atmosphere-free troposphere. For gases with multi-year lifetime, CO_2, CH_4, N_2O, the ice-core abundance is a straightforward reflection of changes in global composition. Gases with lifetimes of days (e.g. HCHO, H_2O_2) to weeks (e.g. PAN, alkyl nitrates) may have significant differences between free tropospheric abundance and the local atmosphere, as well as great geographical variability. Because the principal motivation for trace gas ice-core research is to understand global chemical and climate change, it is critical to have a quantitative relationship between what is measured in ice cores and the free troposphere by filling in all steps in the transfer process.

The development of an overall transfer function will involve determining transfer functions for individual processes. To date, a considerable amount of effort has been aimed at characterizing the transfer function between the firn and the local atmosphere. This existing model should be expanded to include ventilation and sublimation processes. There are large differences of H_2O_2 concentrations at different locations (e.g. Summit and South Pole). We hypothesize that losses of H_2O_2 in the interior of Antarctica are due to ventilation and sublimation processes.

An additional way to test this hypothesis is to compare H_2O_2 and HCHO records at neighbouring sites which are expected to have been exposed to the same air mass, but which have different snow accumulation rates. Three pairs of sites that could be compared are Law Dome (DSS and DE08), Summit and North Central, and Byrd and Siple stations.

The H_2O_2 record in ice cores (e.g. Dye 3 and Summit) has shown an increase of H_2O_2 concentrations over the last 150 years. Our hypothesis is that this increase in H_2O_2 reflects a true atmospheric increase. Proposed experiments to test this hypothesis include: 1) returning to Summit to remeasure H_2O_2 firn concentrations for the last 50 years (six years have passed since those layers were analyzed so more current measurements would help confirm if there is a persistent trend) and 2) make similar measurements at Law Dome for both H_2O_2 and HCHO.

One criticism about the validity of the H_2O_2 and HCHO records as indicators of past atmospheric concentrations, is that these two species can be destroyed in the snow by photochemical processes. We believe that these processes are not significant destruction mechanisms since the H_2O_2 record at Summit matches trends predicted independently by atmospheric modelers. Two ways to estimate losses due to photochemical processes are 1) to try to correlate H_2O_2 and HCHO levels in freshly fallen snow with cloud cover and solar intensity and 2) to analyze HCHO and H_2O_2 levels in surface blue ice.

Although we think all areas of inquiry that we have presented are important, our ordered list of research priorities is: 1) recheck H_2O_2 and HCHO concentrations in the firn at both Summit and Law Dome; 2) conduct a time series analysis of the H_2O_2 record; and 3) expand the current model of snowpack processes to include sublimation and ventilation. The current field program to directly measure the transfer functions in polar regions should continue. In addition, more gaseous species that are tracers for free tropospheric processes should be measured. These include: O_3, NO_x, CO, ethane, acetone, PAN and alkyl nitrates. These species should be measured at sites to include Summit, South Pole and Law Dome.

SUMMARY OF MEETING AND CONCLUSIONS FOR ICE CORE COMMUNITY

The current state of knowledge on the topic of the meeting can be found in the papers and Working Group reports that precede this conclusion. These provide a summary of progress in the last few years, and an agenda for the next few years. In the final overview session of the meeting, and in the period since the meeting, it has been possible also to stand back and view some of the larger issues that were highlighted at the meeting. These have been discussed in greater detail by Bales and Wolff (1995).

One major goal of the workshop was to present some conclusions to the ice core community, who are currently working on very exciting data from recent Greenland deep ice core drillings, and on plans for Antarctic deep drillings. The meeting offered these comments to that community:

1. Ice sheets certainly contain information on parameters such as volcanic activity, terrestrial dust sources, sea ice location, terrestrial and marine biological activity, pollution, atmospheric oxidation capacity, etc., that can multiply the value of ice core climatic information many times. These parameters would define many of the most important feedbacks to the climate system.

2. However we cannot yet have confidence that we can separate effects of changing source strength, transport, and air-snow exchange.

3. The separation of source strength and transport is in the realm of atmospheric models and meteorological study. The current meeting considered mainly the relationship between local atmospheric concentrations and resultant ice core concentrations.

For aerosol, irreversibly-deposited species, the relationship between air and snow concentrations has been measured at a few sites under a limited range of conditions, and is known quantitatively to within a small factor. We do not yet have the knowledge to estimate the relationship for different climatic conditions that prevailed in the past, although we understand which factors affect the relationship.

For reversibly-deposited (partly gas-phase) species such as H2O2, HCHO, and acidic gases, we cannot yet state the form of the transfer function with any confidence. We do have a conceptual framework, and understand that different processes and timescales control the concentrations seen at different sites.

4. The aim of future work must be to identify which processes CONTROL transfer for

each species. This is essential if the more difficult inverse problem (deducing atmospheric concentrations from ice core concentrations) is to be solved.

5. Certain key information that could help solve these uncertainties is missing:

* Field measurements of the air/snow relationship in cold/dark conditions (winter)
* Field measurements of the air/snow relationship at low accumulation rates (central Antarctic)
* Field measurements of typical atmospheric concentrations of many species (especially for the Antarctic)
* Laboratory measurements of uptake of chemicals by ice at relevant temperatures
* Knowledge of the location of impurities in ice
* Knowledge of rates of slow reactions over long time periods
* Better understanding of the influence of wind pumping and sublimation
* Models to identify which processes are key

6. For future ice core drillings, it is recommended that the large investment in obtaining the cores should be accompanied by an air/snow exchange programme to fill some of the information deficits defined above. Such programmes have the capability to greatly increase the value of the data derived from the cores themselves. Field, laboratory and modelling efforts are required.

7. Despite the uncertainties, we recommend that all species with a signal should continue to be measured in major new cores, so that the records are available when the transfer function is understood.

Reference

Bales RC, Wolff EW (1995) Interpreting natural climate signals in ice cores. Eos 76:477-483

ARW PARTICIPANTS

Mary Albert	Geophysical Sciences Branch, US Army CRREL, 72 Lyme Road, Hanover NH 03755, USA
Martin Anklin	Department of Hydrology and Water Resources, University of Arizona, Tucson, AZ 85721, USA
Roger C Bales	Department of Hydrology and Water Resources, University of Arizona, Building 11, Tucson, Arizona 85721, USA
Len A Barrie	Atmospheric Environment Service, Environment Canada, 4905 Dufferin St, Downsview, Ontario M3H 5T4, Canada
Mike Bergin	Department of Civil Engineering, Carnegie Mellon University, Schenley Park, Pittsburgh, PA 15213, USA
Richard Bintanja	Institute for Marine and Atmospheric Research, Utrecht University, PO Box 80005, 3508 TA Utrecht, Netherlands
Jan Bottenheim	Atmospheric Environment Service, Environment Canada, 4905 Dufferin St, Downsview, Ontario M3H 5T4, Canada
Jungyill Choi	Department of Hydrology and Water Resources, University of Arizona, Tucson, AZ 85721, USA
Henrik Clausen	Department of Geophysics, University of Copenhagen, Juliane Maries Vej 30, DK 2100 Copenhagen, Denmark
Martha Conklin	Department of Hydrology and Water Resources, University of Arizona, Tucson, AZ 85721, USA
Cliff Davidson	Department of Civil Engineering, Carnegie Mellon University, Pittsburgh, PA 16213, USA

Robert Delmas	Laboratoire de Glaciologie et, Géophysique de l'Environnement, BP 96, 38402 Saint Martin d'Hères Cedex, France
Jack Dibb	Glacier Research Group, EOS/SERB, Morse Hall, University of New Hampshire, Durham, NH 03824-3525, USA
Florent Dominé	Laboratoire de Glaciologie et, Géophysique de l'Environnement, BP 96, 38402 Saint Martin d'Hères Cedex, France
Katrin Fuhrer	Physikalisches Institut, Universität Bern, Sidlerstrasse 5, CH-3012 Bern, Switzerland
Christophe Genthon	Laboratoire de Glaciologie et, Géophysique de l'Environnement, BP 96, 38402 Saint Martin d'Hères Cedex, France
Dennis Gregor	Waterloo Center for Groundwater Research, University of Waterloo, Waterloo, Ontario, N2L 3G1, Canada
Gerardo W Gross	Department of Geoscience, New Mexico Tech, Socorro, New Mexico 87801, USA
Julie Hall	British Antarctic Survey, High Cross, Madingley Road, Cambridge CB3 0ET, UK
Margareta Hansson	Department of Meteorology, Stockholm University, Arrhenius Laboratory, S-10691 Stockholm, Sweden
Susan Harder	Department of Chemistry Box 351700, University of Washington, Seattle, WA 98195, USA
Klaus Heumann	Institut für Anorganische Chemie, Universität Regensburg, Universitätstrasse 31, D 8400 Regensburg, Germany
C Nick Hewitt	Institute of Environmental Sciences, University of Lancaster, Bailrigg, Lancaster, LA1 4YQ, UK
Manuel Hutterli	Physikalisches Institut, Universität Bern, Sidlerstrasse 5, CH-3012 Bern, Switzerland

Trond Iversen	University of Oslo, Institute of Geophysics, PO Box 1022 Blindern, 0314 Oslo, Norway
Peter Jacob	Institut für Spektrochemie und angewandte Spektroskopie, Bunsen-Kirchhoff-Str 11, Postfach 10 13 52, D 44013 Dortmund, Germany
Jean-Luc Jaffrezo	Laboratoire de Glaciologie et, Géophysique de l'Environnement, BP 96, 38402 Saint Martin d'Hères Cedex, France
Charles Jia	Department of Chemical Engineering, 200 College Street, University of Toronto, Toronto, Ontario M5S 1A4, Canada
Jon Kahl	Department of Geosciences, University of Wisconsin-Milwaukee, PO Box 413, Milwaukee, WI 53201, USA
Kimi Kawamura	Department of Chemistry, Faculty of Science, Tokyo Metropolitan University, 1-1 Minami-Ohsawa, Hachioji, Tokyo 192-03, Japan
Michael Kuhn	Institut für Meteorologie und Geophysik, Universität Innsbruck, Innrain 52, A-6020 Innsbruck, Austria
Paolo Laj	Istituto FISBAT-CNR, via Gobetti 101, I-40129 Bologna, Italy
Michel Legrand	Laboratoire de Glaciologie et, Géophysique de l'Environnement, BP 96, 38402 Saint Martin d'Hères Cedex, France
Willy Maenhaut	Institute for Nuclear Sciences, University of Gent, Proeftuinstraat 86, B-9000 Gent, Belgium
Joe McConnell	Department of Hydrology and Water Resources, University of Arizona, Tucson, AZ 85721, USA
Andreas Minikin	Alfred Wegener Institut für Polar u Meeresforschung, Columbusstrasse, D27515 Bremerhaven, Germany
Mike Mozurkewich	Department of Chemistry, York University, 4700 Keele Street, North York, Ontario M3J 1P3, Canada

Robert Mulvaney — British Antarctic Survey, High Cross, Madingley Road, Cambridge CB3 0ET, UK

Bill Munger — Engineering Sciences Laboratory, Harvard University, 40 Oxford Street, Cambridge, MA 02138, USA

Albrecht Neftel — Eidgenössische Forschungsanstalt für Agrikulturchem und Umwelthygiene, CH 3097 Liebefeld-Bern, Switzerland

John Pomeroy — Hydrological Sciences Division, National Hydrology Research Institute, 11 Innovation Boulevard, Saskatoon, Saskatchewan S7N 3H5, Canada

Hans R Pruppacher — Institut für Physik der Atmosphäre, Becherweg 21, Johannes Gutenberg-Universität, 55099 Mainz, Germany

A R (Ravi) Ravishankara — ERL R/E/AL2, NOAA, 325 Broadway, CO 80303, USA

Jakob Schwander — Physikalisches Institut, Universität Bern, Sidlerstrasse 5, CH-3012 Bern, Switzerland

Margit Schwikowski — Paul Scherrer Institut, Wurenlingen and Villigen, CH5232 Villigen PSI, Switzerland

Jorgen-Peder Steffensen — Department of Geophysics, University of Copenhagen, Juliane Maries Vej 30, DK 2100 Copenhagen, Denmark

Bill Sturges — School of Environmental Sciences, University of East Anglia, Norwich NR4 7TJ, UK

Anne Thompson — NASA Goddard Space Flight Center, Code 916, Greenbelt, MD 20771, USA

Thorsteinn Thorsteinsson — Alfred Wegener Institut für Polar und Meeresforschung, Columbusstrasse, D27578 Bremerhaven, Germany

Tas van Ommen — Glaciology Section, Australian Antarctic Division and CRC, University of Tasmania, GPO Box 252C, Hobart, Tasmania 7001, Australia

Job van Roijen	Department of Subatomic Physics, Utrecht University, PO Box 80000, 3508 TA Utrecht, Netherlands
Ed Waddington	Geophysics Program AK-50, University of Washington, Seattle, WA 98195, USA
Dietmar Wagenbach	Institut für Umweltphysik, Universität Heidelberg, Im Neuenheimer Feld 366, D69120 Heidelberg, Germany
Peter Wåhlin	National Environment Research Institute, PIXE Laboratory, c/o Niels Bohr Institute, Blegdamsvej 17, DK 2100 Copenhagen O, Denmark
Frank Wania	Norsk Institutt for Luftforskning, PO Box 1245, N-9001 Tromsoe, Norway,
Eric Wolff	British Antarctic Survey, High Cross, Madingley Road, Cambridge CB3 0ET, UK

AUTHOR INDEX

SUBJECT INDEX

The ASI Series Books Published as a Result of Activities of the Special Programme on Global Environmental Change

This book contains the proceedings of a NATO Advanced Research Workshop held within the activities of the NATO Special Programme on Global Environmental Change, which started in 1991 under the auspices of the NATO Science Committee.

The volumes published as a result of the activities of the Special Programme are:

Vol. 1: **Global Environmental Change.** Edited by R. W. Corell and P. A. Anderson. 1991.
Vol. 2: **The Last Deglaciation: Absolute and Radiocarbon Chronologies.** Edited by E. Bard and W. S. Broecker. 1992.
Vol. 3: **Start of a Glacial.** Edited by G. J. Kukla and E. Went. 1992.
Vol. 4: **Interactions of C, N, P and S Biogeochemical Cycles and Global Change.** Edited by R. Wollast, F. T. Mackenzie and L. Chou. 1993.
Vol. 5: **Energy and Water Cycles in the Climate System.** Edited by E. Raschke and D. Jacob. 1993.
Vol. 6: **Prediction of Interannual Climate Variations.** Edited by J. Shukla. 1993.
Vol. 7: **The Tropospheric Chemistry of Ozone in the Polar Regions.** Edited by H. Niki and K. H. Becker. 1993.
Vol. 8: **The Role of the Stratosphere in Global Change.** Edited by M.-L. Chanin. 1993.
Vol. 9: **High Spectral Resolution Infrared Remote Sensing for Earth's Weather and Climate Studies.** Edited by A. Chedin, M.T. Chahine and N.A. Scott. 1993.
Vol. 10: **Towards a Model of Ocean Biogeochemical Processes.** Edited by G. T. Evans and M.J.R. Fasham. 1993.
Vol. 11: **Modelling Oceanic Climate Interactions.** Edited by J. Willebrand and D.L.T. Anderson. 1993.
Vol. 12: **Ice in the Climate System.** Edited by W. Richard Peltier. 1993.
Vol. 13: **Atmospheric Methane: Sources, Sinks, and Role in Global Change.** Edited by M. A. K. Khalil. 1993.
Vol. 14: **The Role of Regional Organizations in the Context of Climate Change.** Edited by M. H. Glantz. 1993.
Vol. 15: **The Global Carbon Cycle.** Edited by M. Heimann. 1993.
Vol. 16: **Interacting Stresses on Plants in a Changing Climate.** Edited by M. B. Jackson and C. R. Black. 1993.
Vol. 17: **Carbon Cycling in the Glacial Ocean: Constraints on the Ocean's Role in Global Change.** Edited by R. Zahn, T. F. Pedersen, M. A. Kaminski and L. Labeyrie. 1994.
Vol. 18: **Stratospheric Ozone Depletion/UV-B Radiation in the Biosphere.** Edited by R. H. Biggs and M. E. B. Joyner. 1994.
Vol. 19: **Data Assimilation: Tools for Modelling the Ocean in a Global Change Perspective.** Edited by P. O. Brasseur and J. Nihoul. 1994.

Vol. 20: **Biodiversity, Temperate Ecosystems, and Global Change.**
Edited by T. J. B. Boyle and C. E. B. Boyle. 1994.

Vol. 21: **Low-Temperature Chemistry of the Atmosphere.**
Edited by G. K. Moortgat, A. J. Barnes, G. Le Bras and J. R. Sodeau. 1994.

Vol. 22: **Long-Term Climatic Variations – Data and Modelling.**
Edited by J.-C. Duplessy and M.-T. Spyridakis. 1994.

Vol. 23: **Soil Responses to Climate Change.**
Edited by M. D. A. Rounsevell and P. J. Loveland. 1994.

Vol. 24: **Remote Sensing and Global Climate Change.**
Edited by R. A. Vaughan and A. P. Cracknell. 1994.

Vol. 25: **The Solar Engine and Its Influence on Terrestrial Atmosphere and Climate.**
Edited by E. Nesme-Ribes. 1994.

Vol. 26: **Global Precipitations and Climate Change.**
Edited by M. Desbois and F. Désalmand. 1994.

Vol. 27: **Cenozoic Plants and Climates of the Arctic.**
Edited by M. C. Boulter and H. C. Fisher. 1994.

Vol. 28: **Evaluating and Monitoring the Health of Large-Scale Ecosystems.**
Edited by D. J. Rapport, C. L. Gaudet and P. Calow. 1995.

Vol. 29: **Global Environmental Change Science: Education and Training.**
Edited by D. J. Waddington. 1995.

Vol. 30: **Ice Core Studies of Global Biogeochemical Cycles.**
Edited by R. J. Delmas. 1995.

Vol. 31: **The Role of Water and the Hydrological Cycle in Global Change.**
Edited by H. R. Oliver and S. A. Oliver. 1995.

Vol. 32: **Atmospheric Ozone as a Climate Gas.**
Edited by W.-C. Wang and I. S. A. Isaksen. 1995.

Vol. 33: **Carbon Sequestration in the Biosphere.**
Edited by M. A. Beran. 1995.

Vol. 34: **Climate Sensitivity to Radiative Perturbations: Physical Mechanisms and Their Validation.**
Edited by H. Le Treut. 1996

Vol. 35: **Clouds, Chemistry and Climate.**
Edited by P. J. Crutzen and V. Ramanathan. 1996.

Vol. 36: **Diachronic Climatic Impacts on Water Resources.**
Edited by A. N. Angelakis and A. S. Issar. 1996.

Vol. 37: **Climate Change and World Food Security.**
Edited by T. E. Downing. 1996.

Vol. 38: **Evaluation of Soil Organic Matter Models.**
Edited by D. S. Powlson, P. Smith and J. U. Smith. 1996.

Vol. 39: **Microbiology of Atmospheric Trace Gases.**
Edited by J. C. Murrell and D. P. Kelly. 1996.

Vol. 40: **Forest Ecosystems, Forest Management and the Global Carbon Cycle.**
Edited by M. J. Apps and D. T. Price. 1996.

Vol. 41: **Climate Variations and Forcing Mechanisms of the Last 2000 Years.**
Edited by P. D. Jones, R. S. Bradley and J. Jouzel. 1996.

Vol. 42: **The Mount Pinatubo Eruption: Effects on the Atmosphere and Climate.**
Edited by G. Fiocco, D. Fuà and G. Visconti. 1996.

Vol. 43: **Chemical Exchange Between the Atmosphere and Polar Snow.**
Edited by E. W. Wolff and R. C. Bales. 1996.

Vol. 44: **Decadal Climate Variability: Dynamics and Predictability**
Edited by D. L. T. Anderson and J. Willebrand. 1996.

Vol. 45: **Radiation and Water in the Climate System: Remote Measurements**
Edited by E. Raschke. 1996.

Springer-Verlag and the Environment

We at Springer-Verlag firmly believe that an international science publisher has a special obligation to the environment, and our corporate policies consistently reflect this conviction.

We also expect our business partners – paper mills, printers, packaging manufacturers, etc. – to commit themselves to using environmentally friendly materials and production processes.

The paper in this book is made from low- or no-chlorine pulp and is acid free, in conformance with international standards for paper permanency.

MIX
Papier aus verantwortungsvollen Quellen
Paper from responsible sources
FSC® C105338
FSC
www.fsc.org

If you have any concerns about our products,
you can contact us on
ProductSafety@springernature.com

In case Publisher is established outside the EU,
the EU authorized representative is:
Springer Nature Customer Service Center GmbH
Europaplatz 3, 69115 Heidelberg, Germany

Printed by Libri Plureos GmbH
in Hamburg, Germany